Protection and Switchgear

Second Edition

Bhavesh Bhalja
Associate Professor
Department of Electrical Engineering
IIT Roorkee

R.P. Maheshwari
Professor
Department of Electrical Engineering
IIT Roorkee

Nilesh G. Chothani
Associate Professor
Department of Electrical Engineering
A. D. Patel Institute of Technology, Gujarat

OXFORD
UNIVERSITY PRESS

Oxford University Press is a department of the University of Oxford.
It furthers the University's objective of excellence in research, scholarship,
and education by publishing worldwide. Oxford is a registered trade mark of
Oxford University Press in the UK and in certain other countries.

Published in India by
Oxford University Press
22 Workspace, 2nd Floor, 1/22 Asaf Ali Road, New Delhi 110002, India

First Edition published in 2011
Second Edition published in 2017
Digitally Printed in 2023

ISBN-13: 978-0-19-947067-9
ISBN-10: 0-19-947067-7

Typeset in Times New Roman
by Ideal Publishing Solutions, Delhi
Printed at Manipal Technologies Limited, Manipal

Cover image: Actor / Shutterstock

To my wife Jital, son Ansh, and other family members
for being a part of my life

Bhavesh Bhalja

To my beloved teacher, mentor, and guide Dr H.K. Verma
who taught me power system protection

R.P. Maheshwari

To my wife Anjana and son Jainil
for their patience and encouragement

Nilesh G. Chothani

Features of

Learning Objectives   Each chapter of the book has a section 'Learning Objectives', which briefs about all the topics discussed in the chapter.

Figures and Tables   Numerous well-illustrated figures and tables are given for better understanding of concepts.

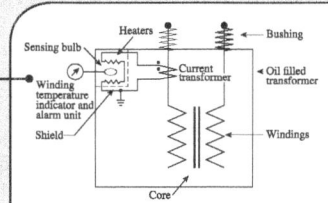

Table 6.1   Transformer category (ANSI/IEEE Standard C57.109-1985 curves)

Category	Minimum nameplate (kVA)		Reference protective curve
	Single-phase	Three-phase	
I	5–500	15–500	Fig. 6.8
II	501–1667	501–5000	Fig. 6.9
III	1668–10,000	5001–30,000	Fig. 6.10
IV	Above 10,000	Above 30,000	Fig. 6.11

Fig. 6.10   Connection of WTI and alarm unit with transformer

Example 2.15   The relays shown in Fig. 2.36 are directional and non-directional relays. Identify the directional relays. The breaking capacities of the breakers are given in Table 2.10. The PS of relays are given in Fig. 2.36; the TDS value of the relays R_2 and R_5 is set at 0.1. Determine the TDS of other relays.

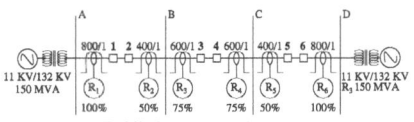

Fig. 2.36   Single line diagram of power system

Solution:
The relays R_1 and R_4 are non-directional, whereas all others relays are directional because the fault current direction can change on bus B and bus C.
   We will split the network into two radial networks.
   The TDS value of relays R_2 and R_5 is given as 0.1. The TDS of the other relay can be calculated on the basis of its operation in the

Table 2.10   Breaking capacities of breakers

Breaker no.	Capacities (MVA)
1, 6	1000
3, 4	800
2, 5	500

Examples   Each chapter supports numerous well-illustrated numerical examples.

Summary   Recapitulation at the end of each chapter enables quick revision of important concepts discussed in the chapter.

the Book

ADDITIONAL MULTIPLE CHOICE QUESTIONS

Additional Multiple Choice Questions Besides the multiple choice questions provided at the end of each chapter, the book also supports Additional Multiple Choice Questions that has as many as 130 questions along with answers.

ADDITIONAL MULTIPLE CHOICE QUESTIONS

1. Selectivity, which is one of the requirements of protection system, is also known as
   (a) dependability
   (b) relay coordination
   (c) security
   (d) none of the above

2. Economics criteria of the protective scheme indicates to combine features of
   (a) maximum protection with minimum cost
   (b) maximum protection with maximum cost
   (c) minimum protection with maximum cost
   (d) none of the above

   (c) cables
   (d) all of the above

8. Electromechanical relays are still used by the utilities due to their
   (a) ruggedness and withstanding capacity of voltage spikes
   (b) lower cost
   (c) simple construction
   (d) all of the above

9. Operating torque is provided in single input relay with the

Review Questions

1. Explain the different ratings and functions of CBs.
2. Explain the isolating and load-break switches.
3. Enumerate the different types of fuses for low voltage applications and write a detailed note on HRC fuse.
4. Explain the construction and operating principle of MCB.
5. Why is the ELCB used in domestic supply?
6. Write a note on ACBs.
7. Discuss the advantages of oil as an insulating medium in a CB.
8. Draw a well-labelled diagram of an MOCB and explain each part and how it works.
9.
10. State the merits of SF_6 gas compared to other arc-quenching mediums.
11. Explain the working of the following SF_6 CB:
    (a) Non-puffer type SF_6 breaker
    (b) Puffer type SF_6 breaker
12. Explain the maintenance procedures for medium voltage CBs.
13. Give the classification of tests to be carried out on CB.
14. Explain the short circuit test plant and procedures of testing for high voltage CBs.

Review Questions Each chapter supports a wealth of short and long answer questions to help students during exam preparation.

Numerical Exercises

1. A 100 MVA, 11 kV, generator with $x_d'' = 25$ % is connected to a transformer rated 125 MVA, 13.8/220 kV with leakage reactance of 10%. If the base of 150 MVA and 230 kV is used on HV side of transformer, determine the per unit value to be used for the generator and transformer.
2. The single line diagram of an unloaded power system is shown in Fig. 19.40. The rating of each component is given as below.

Generator1 (G1): 30 MVA, 18 kV, $x_d'' = 0.2$pu, Generator2 (G2): 30 MVA, 15 kV, $x_d'' = 0.15$pu, Transformer (T1) is composed of three single phase unit, each rated 10 MVA, 127/18 kV, X = 10%, Transformer (T2): 35 MVA, 230/15 kV, X = 10%, Transmission line-1 has total reactance of 50Ω and line-2 has total reactance of 70Ω. Compute the per unit reactance of all components and draw reactance diagram marking all reactances in per unit.

Numerical Problems Numerical problems are also given in the book in relevant chapters.

Books
Anderson, P.M., *Power System Protection*, IEEE Press, New York, 1999.
Blackburn, J.L., *Applied Protective Relaying*, Westinghouse Electric Corporation, New York, 1982.
Burrus, C.S., R.A. Gopinath, and H. Guo, *Introduction to Wavelets and Wavelet Transform: A Primer*, New Jersey, Prentice Hall, 1998.
Chakrabarti, A., M.L. Soni, P.V. Gupta, et al., *A Text Book on Power System Engineering*, Dhanpat Rai & Co. Pvt. Ltd, Delhi, 2010.
Chui, C.K., *An Introduction to Wavelets*, Academic Press Inc., San Diego, 1992.
Elmore, W.A. *Protective Relaying*, New York, Marcel Dekker Inc., 1994.
Garzon, R.D., *High Voltage Circuit Breakers Design and Applications*, Marcel Dekker Inc., New York, 1996.
GEC Measurement, *Network Protection and Automation Guide*, Morrison & Gibb Ltd; Edinburgh, Scotland, 1987.
Grigsby, L.L., *Electric
Gupta, B.R., *Power

Bibliography For those who wish to gather some additional information on certain topics, Bibliography at the end of the book provides a list of books and journals references.

Others
'Controlled Switching of HVAC Circuit-Breakers: Benefits & Economic Aspects', *Cigré Working Group A3.07*, January 2004.
'Controlled Switching of HVAC Circuit-Breakers: Guidance for further applications including unloaded transformer switching, load and fault interruption and circuit-breaker uprating', *Cigré Working Group A3.07*, December 2004.
'IEEE Guide for AC Generator Protection', ANSI/IEEE C37.1021995.
'IEEE Recommended Practice for Protection and Coordination of Industrial and Commercial Power Systems', IEEE Industry Applications Society, IEEE Std 242-1986.
'IEEE Standard Common Format for Transient Data Exchange (COMTRADE) for Power Systems', Sponsored by the Power System Relaying Committee of the Power Engineering Society, IEEE C37.111-1991.
'Power Swing and Out-of-step Considerations on Transmission Lines', *IEEE Power System Relaying Committee*, 2005 Report, [online]. Available: http://www.pes-psrc.org.

Foreword

In spite of all the care and precautions taken in the design, installation, and operation of power systems and power equipment, abnormal conditions and faults do occur in the system. Some faults such as short circuits can prove highly damaging, not only to the component that develops the fault but also to adjacent components and, sometimes, to the entire power system. Fault occurrence and component damage can be minimized through careful design of the protection system, which primarily includes protective relays, current and voltage transformers feeding these relays, and the switchgear responsible for disconnecting the faulted element(s). As the size of generating stations and complexity of power systems (in terms of interconnections) go up, the demand on the protection system in terms of sensitivity, selectivity, speed, and reliability increases.

This book, *Protection and Switchgear*, authored by Bhavesh Bhalja, R.P. Maheshwari, and Nilesh G. Chothani, deals with this complex subject holistically. It is a fine combination of theory and practice. In my assessment, it will serve as a good reference book to undergraduate and postgraduate students, as also to subject teachers. With a pro-student style and a number of exercises, multiple choice questions, and solved and unsolved questions in every chapter, the book can also become a textbook for colleges, institutes, and universities. I believe that because of an emphasis on practical aspects and the coverage of modern protective equipment, practising engineers will also find it a good handbook/guide.

I congratulate the authors for writing this complete reference book on power system protection, including switchgear, and sincerely hope that it will benefit numerous students, teachers, and practising engineers.

Dr H. K. Verma
Professor Department of Electrical Engineering
India Institute of Technology Roorkee
Roorkee, India

Preface to the Second Edition

Developments in advanced signal processing, information and communication, and Intelligent Electronic Devices (IEDs) have a tremendous potential to refine or even redefine the formation and implementation of switchgear and protection technology. Increased computing power at low cost has provided opportunities to implement more computation-intensive methods/algorithms in real time. At the same time, Phasor Measurement Units (PMUs) providing faster and diverse synchronized measurements over a wide area, and new communication options have also emerged. It is necessary to design a reliable protection system which can detect and locate events, analyse system integrity, and take corrective action. Subsequently, processing and communication delays, erroneous data, and cyber-attacks pose challenges to security as well as dependability of protection.

Due to major reforms in protection technology since the publication of the first edition in 2012, a sizable portion of the existing content had to be expanded and rewritten in the second edition. In the second edition, we have tried to include more topics which have emerged as important due to the developments that took place during the last five years. We hope students, practising engineers, and faculty members will welcome this second edition.

Salient Features of Second Edition

- Three new chapters — Chapter 18 Smart Grid Technologies and Applications, Chapter 19 — Symmetrical and Unsymmetrical Faults in Power Systems, and Chapter 20 Basic Concept and Application of Controlled Switching.
- Static differential and distance relay with various types of amplitude and phase comparator for static relays.
- Different types of digital filters along with their comparison.
- Large number of multiple choice questions especially for students/faculty and utility engineers.
- Large number of solved and unsolved examples in each chapter for practice and self-evaluation.
- Case study on overcurrent relay coordination along with its source code.
- Comparison among various distance relay characteristics, different types of circuit breakers and their selection is also incorporated.
- Description of earthing transformer and frequency protection is included which provides better understanding.
- In-depth explanation of phasor measurement unit which works as the heart of the wide area protection, monitoring, and control.
- Reclosing scheme used in practice for transmission lines is included which improves reliability of the power system network.
- Various technical challenges due to integration of renewable energy sources with the existing grid are discussed. In addition, the issue of islanding, its hazards and risk of islanding are also added.
- PSCAD examples on inverters and converters which are extensively being used in the field are included in this edition.

New to this Edition

As gradually protection system is moving from electromechanical relays to static and numeric relays, the following are included in chapter 1:

1. Types of amplitude and phase comparator for static relays
2. Static differential relay for unit protection
3. Static distance relay

In this chapter, a new section on digital filters is also introduced along with the description of tools for distorted relaying signals.

In chapter 2, a case study for overcurrent relay coordination along with its source code is included to help the reader to develop his own code.

For better understanding from selection point of view, comparison among various distance relay characteristics is included in chapter 3.

In chapter 4, we have included permissive inter-tripping scheme and carrier-aided distance scheme for acceleration and pre-acceleration of zone II.

New rate of change of frequency protection as well as additional solved and unsolved examples have been included in chapter 5.

As we considered description of earthing transformer an important issue, it is included in chapter 6.

In order to minimize outage time and also to improve reliability of power and distribution network, the current practice is to use auto-reclosing feature with the existing circuit breaker. In chapter 11, auto-reclosing scheme used in practice for transmission lines is included.

Today, more emphasis is being given to tap renewable energy sources. However, integration of renewable energy sources with the existing grid imposes many technical challenges. The same are presented in chapter 12. In addition, the issue of islanding, its hazards, and risk of islanding are also incorporated in this chapter.

A comparison of different types of circuit breakers and its selection is included in chapter 14.

In chapter 15, stability, overshoot, and voltage withstand test are included to make the presentation complete as per present requirements.

The industry is moving towards more robust protection systems and for that PMUs are deployed. The description of PMUs is included in chapter 16.

As power system cannot be experimented, simulations of the system with various types of equipment are the only option for its behavioural study. For the simulation study of power system, PSCAD has emerged as an important tool. This was included in the first edition itself. But over the period, as inverters and converters are extensively being used, their case study is also included in chapter 17 of the revised edition.

To keep pace with development in power system protection, three new chapters have been added in the revised edition.

Chapter 18 Smart Grid Technologies and Applications discusses the various components and benefits of smart grid—an improved electricity supply chain that runs from a major power plant all the way inside to your home, and also highlights the challenges that may arise while designing a smart grid.

Chapter 19 Symmetrical and Asymmetrical Fault Analysis deals with per unit system, symmetrical components, transformation of unbalanced phasors into balanced symmetrical components and vice versa, transient phenomenon that occurs in transmission line and formation of sequence networks.

Chapter 20 Basic Concept and Application of Controlled Switching discusses controlled switching which is a recent practice followed by most of the utilities for reduction in inrush current and voltage across circuit breaker assembly.

Online Resources

To aid the faculty and students using this book, additional resources are available at www.india.oup.com/orcs/9780199470679

For Faculty

Solutions Manual and Lecture PPTs, Chapter-wise MCQs

For Students

MCQ test generators, PSCAD simulations

Acknowledgements

During revision of this book, we received valuable suggestions and positive feedbacks from many students, faculty members, and practising engineers working in the utility/industry. Influence of all these readers has had a major impact on the revision of this book. We hope that their support will continue in future as well.

The publishers and authors would like to extend their special thanks to the following reviewers who spared their valuable time to review this book.

Dr. D. P. Kothari	Former Director i/c IIT Delhi
Dr. B. K. Panigrahi	Indian Institute of Technology Delhi, Delhi (India)
Dr. S. R. Samantaray	Indian Institute of Technology Bhubaneshwar, Bhubaneshwar (India)
Dr. K. S. Swarup	Indian Institute of Technology Madras, Chennai (India)
Dr. U. B. Parikh	ABB India Limited, Vadodara, Gujarat (India)
Dr. Manohar Singh	Central Power Research Institute, Bangalore (India)
Dr. P. Jena	Indian Institute of Technology Roorkee, Roorkee (India)
Dr. Sanjay R. Joshi	Principal, Government Engineering College, Valsad
Dr. Saurav Pandya	Professor, Government Engineering College, Bhavnagar
Dr. Vivek Pandya	Professor, Pandit Deendayal Petroleum University, Gandhinagar, India
Dr. Pragnesh Bhatt	Professor, Charutar University, Changa, Gujarat, India
Prof. Kunal Bhatt	Assistant Professor, Government Engineering College, Dahod
Mr. Vishal Gaur	Research Scholar, Indian Institute of Technology Roorkee

Feedback

We are happy to welcome any encouraging criticism of the book and will be thankful for an evaluation by the readers. The suggestions can be sent to bhaveshbhalja@gmail.com

Bhavesh Bhalja
R. P. Maheshwari
Nilesh G. Chothani

Preface to the First Edition

A modern power system is a complex arrangement of machinery, with countless interfacing control loops and distribution and transmission channels, along with automated protection and safety support systems. Although engineers may take due care, faults in the protection scheme are inevitable.

Statistical data reveals that a large number of relay trippings occur due to improper or inadequate relay settings rather than due to actual faults. Hence, it is the duty of a protection engineer to design a protection scheme, either for the apparatus or for the lines, which provides maximum protection features at minimum cost. Therefore, it is extremely important for power system engineers and students of electrical engineering to study the various protection schemes in detail. These include power and control circuits of the equipment to be protected, various relay characteristics, relay design, and construction of the relays. Moreover, protective devices cannot perform their task without the support of instrument transformers such as current and potential transformers and switchgears such as circuit breakers, isolators, fuses, earthing switches, etc. Thus students of electrical engineering and engineers working in the industry should have adequate theoretical and practical knowledge of protective devices and switchgears. This knowledge is also helpful during design, erraction, procurement, and maintenance of various power system components. Study of relays and switchgears is also important to understand the procedure of actual relay setting in the practical scenario.

Today, in order to economize, each component of a power system is operated with relatively small margins from stipulations. This can cause rapid damage, with safety implications and huge penalties resulting in the loss of revenue due to system downtime and repair costs. Component failure either at the micro or macro levels is a vital factor that affects the reliability of power supply to the end users. Keeping this in mind, this book discusses in detail the implementation of sophisticated protection systems at each hierarchy of the power system and also the utility of switchgear.

About the Book

This book aims to give a comprehensive, up-to-date presentation of the role of protection safety systems, switchgears, and their advances in modern power systems. It begins with a survey of the theories and methods of protection and switchgear. Additionally, it provides a theoretical summary along with examples of real life engineering applications to a variety of technical problems. It bridges the gap among the theoretical advances, experimental validations, and engineering in real life.

This book is designed as a textbook for undergraduate students of engineering for a course on protection and switchgear. It will also be immensely useful for power system engineers seeking information about the principles and working of protection and switchgear systems.

Salient Features

- Provides in-depth coverage of apparatus protection, circuit breaking fundamentals, and selection and testing of circuit breakers using actual field data.
  This book covers analytical techniques, selection, and testing of switchgears in an easily comprehensible manner. It also covers transformer, generator, induction motor, and busbar protection in detail. For each apparatus, digital protection is also discussed. It also discusses various digital relaying schemes for line and equipment protection.

- Contains a chapter on recent developments in protection relays.
  This chapter discusses topics such as wide-area protection, synchronized sampling, wide-area phasor measurement technology, application of artificial intelligence in protective relays, and application of wavelet transform in protective relaying.
- Contains a chapter on power systems computer aided design (PSCAD)
  PSCAD is a powerful and flexible graphical user interface of the world renowned EMTDC solution engine. It enables the user to schematically construct a circuit, run a simulation, analyse the results, and manage the data in a completely integrated, graphical environment. This chapter provides a detailed discussion on this interface with screen shots to help students understand this software.
- Includes solved examples, numerical exercises, review exercises, and multiple choice questions at the end of each chapter
  The problems have been included with the intention of helping students realize that many problems that will be faced in practice will require careful analysis, consideration, and some approximations.
- Appendices at the end of the book
  A number of appendices have been provided at the end of the book such as the international code list for protective relaying schemes with a description of each device, data sheets of different types of relays, system line parameters for overcurrent relay coordination, and simulation of transmission line systems.

Content and Coverage

The book is divided into 17 chapters. A brief description of these chapters is given here.

Chapter 1 starts with the fundamentals of protective relaying, which include the history and incremental developments, followed by the classification of protective relays. It also includes construction of various protective relays. The concept of digital/numerical relay is also discussed along with a block diagram and the function of each block. At the end, various algorithms used in digital relays are discussed.

Chapter 2 focusses on overcurrent protection of the transmission line. It covers various characteristics of bidirectional and directional overcurrent relays. Guidelines for phase and ground relay settings along with the relay coordination procedure are explained with suitable examples.

Chapter 3 gives special emphasis to problems and remedies of distance protection. Various distance relay characteristics are discussed along with the derivation of quantities fed to the phase and ground distance unit.

Chapter 4 discusses the importance of the pilot relaying scheme used for transmission lines. It covers various pilot relaying, carrier blocking, and transfer tripping schemes along with the control circuits and R–X diagrams.

Chapters 5, 6, and *7* deal with apparatus protection, which includes the generator, transformer, and induction motor. Various types of protection such as overcurrent, earth fault, and differential are discussed in detail along with relevant circuit diagrams and examples. For each apparatus, digital protection is also discussed.

Chapter 8 discusses different arrangements of the busbar and the concept of busbar protection. Further, recent trends in double bus arrangement are explained. Special schemes such as centralized and decentralized busbar protection are also elaborated.

Chapter 9 explains the principle, construction, and performance of the current transformer (CT) and the potential transformer (PT). Further, specifications of CT and PT are given, which will be helpful during their procurement.

Chapter 10 presents various neutral grounding schemes and their effect on the power system. This chapter focusses on the sources of transient surges and the protective measures against them. It also covers various devices used in the field for protection against overvoltage owing to switching and lightning.

Chapter 11 presents various types of reclosing relays and the procedure for automatic reclosing and synchronizing.

Chapter 12 covers the behaviour of the power system during severe upsets such as islanding and under frequencies. Different load shedding techniques and islanding schemes are also discussed.

Chapters 13 and *14* discuss the fundamentals of circuit breaking, arc phenomenon, and the factors affecting the arc interruption process. Moreover, construction and working of various types of switches, fuses, and circuit breakers are explained in detail with their relative merits and demerits. The chapter concludes with different testing methods of circuit breakers.

Chapter 15 discusses testing, commissioning, and maintenance of relays used in the field. Different relay testing methods and relay test setups are also covered.

Chapter 16 presents applications of soft computing techniques in protective relays. Recent trends in the development of relay algorithms are also discussed in detail.

Chapter 17 provides an introduction to a new computational tool in power system engineering, PSCAD. It covers different library components and procedures for constructing a sample case. Various case studies related to the power system are discussed as tutorials at the end of this chapter.

Codes of protective devices used in control circuits as per IEC standards, fundamentals of symmetrical component theory, data sheets of various relays, and line and system parameters of simulated systems are given in the appendices.

Bhavesh Bhalja
R.P. Maheshwari
Nilesh G. Chothani

Brief Contents

Detailed Contents

Protective Relaying Fundamentals

1

Learning Objectives

After going through this chapter, the students will be able to:

- List the key requirements of protective devices against overload
- Differentiate between unit protection and non-unit protection
- Explain primary and backup protection of power systems
- Explain the use of thermal relay in the protection of equipment against overload
- Discuss the advantages and disadvantages of static relays and compare with electromechanical relays
- Explain the concept of adaptive relaying
- Discuss half-cycle and full-cycle discrete Fourier transform algorithm

1.1 General Background

Socio-economic growth and rapid industrialization have resulted in the fast increase in per capita consumption of electricity the world over. Modern electric power systems catering to huge energy demands are spread over wide areas and contain several major components such as generators, transformers, and transmission and distribution lines. They are designed to provide uninterrupted electrical power supply. The increase in demand has necessitated the use of large-capacity power equipment and complex interconnections among them. This has increased the pressure on the protection systems multifold.

The advent of large generating stations and highly interconnected power systems has made early fault identification and rapid equipment isolation imperative to maintain system integrity and stability. It is evident that in spite of all the precautions taken in the design and installation of such systems, there are possibilities that abnormal conditions or faults may arise. Some faults such as short circuits may prove extremely damaging not only to the faulty components, but also to the neighbouring components and to the power system as a whole. So it is of vital importance to limit the damage to a minimum by speedy isolation of the faulty section, without disturbing the working of the rest of the system.

A *fault* is a condition that causes abnormal stoppage of current in the desired path or makes the current to flow towards an undesired path. Faults include, but are not limited to, short or low-impedance circuits, open circuits, power swings, overvoltages, elevated temperature, and off-nominal frequency operation. They are generally caused by the failure of insulation, breaking of conductors, or shorting of two supply wires by birds, kite string, tree limbs, etc. Occurrence of a fault can cause the following problems:

1. Interruption in the power supply to the consumers
2. Substantial loss of revenue due to interruption of service

3. Loss of synchronism
4. Extensive damage to equipment
5. Serious hazard to personnel

All power system equipment must, therefore, be protected to avoid system collapse and the associated consequences. The protective relays stand watch and in the event of a failure such as short circuit or abnormal operating conditions, de-energize the unhealthy section of the power system and restrain interference with the remainder of it. They are also used to indicate the type and location of failure so as to access the effectiveness of the protective schemes.

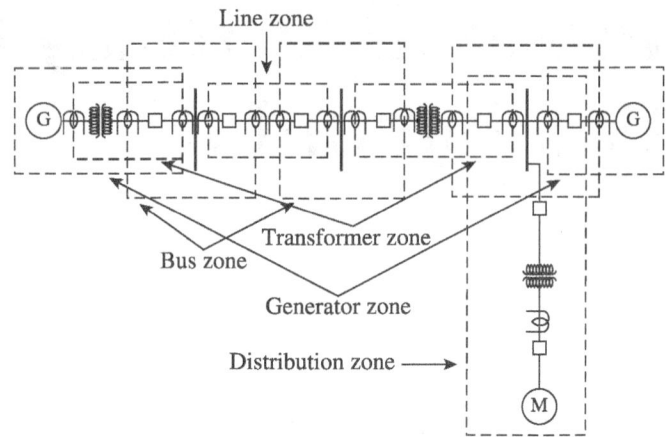

Fig. 1.1 Typical primary relay protection zones in a power system

1.2 Zones of Protection

A power system is normally segmented into a number of protective zones as shown in Fig. 1.1. A zone is protected by a system of relays, circuit breakers, and associated equipment. The circuit breakers are arranged in a manner that makes it possible to isolate the protected zone while the remaining system continues to supply energy to the customers. Each zone covers one or more components of the system. To provide complete protection for the entire system, that is, to avoid having an unprotected region, neighbouring zones are arranged to overlap each other. When a system is experiencing an abnormal condition, the relays first identify this condition and then send trip signals to appropriate circuit breakers that open to isolate the affected zone.

1.3 Requirements of Protection Schemes

Protection schemes are required to possess the following properties to perform their functions.

Selectivity This is the ability of protection devices to isolate only the faulty network of the power system from the healthy part to minimize the outage area and also to maintain normal power supply for the rest of the power system. The possibility of failure to operate and failure of protective relays and circuit breakers should be considered in determining the selectivity of protective relays. Hence, selectivity is also known as *relay coordination*. The coordination of primary relays and backup relays can be achieved by different operation zones and operation time delays.

Reliability Reliability is the ability of protection devices to operate properly during the period they are in service. It is also defined as the ability of protective devices to operate properly during their operational life. It can be categorized as follows:

Dependability It is the certainty of correct operation in response to system trouble.

Security It is the ability of the protection schemes to avoid maloperation between faults.

Speed It is apparent that quick disconnection of the faulted area or the elements can significantly improve the stability of the power system, reduce outage duration, and minimize the damage of faulted elements. Therefore, when a fault occurs, the protective relays should identify the fault and operate as fast as possible. The total time to remove the fault is determined as the sum of operation time of relays and circuit breakers. Typically, a high-speed relay can operate in the range of 10 to 30 ms. However, high operation time is not always required, especially in low-voltage systems for economic reasons.

Discrimination A protection system should be able to discriminate between fault and loading conditions even when the minimum fault current is less than the maximum full load current.

Simplicity The term *simplicity* is often used to refer to the design quality of a protective relay system. It is obvious that the simplest relay design is not always the most economical. Hence, the protective system should be as simple and straightforward as possible without disturbing its basic tasks. This improves system reliability as there are fewer elements that can malfunction and require less maintenance.

Sensitivity It is the ability of the protective device to operate correctly to the faults or abnormal conditions inside the zone of protection. It refers to the minimum level of fault current at which the protective device operates. Protective devices with good sensitivity can sense any faults within the zone of protection with respect to different fault locations, different fault types, and even different fault resistance. The sensitivity factor usually determines the sensitivity of protective relays, which depends on the parameters of protected elements and operating condition of the power system.

Economics Besides the six factors mentioned, economics of protective relays is another important factor that should be considered. A good protective relay system should combine features of both maximum protection and minimum cost. Moreover, some of these properties are contradictory to one another, and it is the duty of the protection engineer to maintain a balance among them, when choosing a protection scheme for a particular application.

1.4 Unit and Non-unit Protection

Unit protection scheme is a scheme that operates for a fault within its zone. Here, zone of protection is decided on the basis of current transformers (CTs), and includes every fault point inside the CTs where measurement of currents is carried out. This type of protection scheme is widely used in generators, transformers, and large induction motors. Differential protection scheme is the best example of this type of protection scheme.

It is universally accepted that the current-based relaying scheme is not a good choice for transmission line protection as it does not give instantaneous operation throughout the entire line. Distance relaying scheme is a good replacement for current-based relaying scheme for transmission line protection. This scheme is not affected by the ratio of source impedance to the impedance from the relaying point to the fault point. Moreover, it is less sensitive to system conditions and does not require a communication channel. A scheme that achieves protection using grading of successive relays is known as *non-unit protection scheme*. Overcurrent and distance relays are the best examples of non-unit protection schemes. However, the reach of distance relays is highly affected by fault resistance, mutual zero sequence coupling, shunt capacitances, and remote in feed. Moreover, the first zone reach of distance relays is restricted up to 80%–90% of the line because of transient overreach. More details regarding transient overreach can be found in Chapter 3. Therefore, it is not possible to achieve instantaneous operation throughout the entire line using non-unit protection schemes. This can be achieved by unit protection scheme. This concept is known as *differential protection of transmission line*.

1.5 Primary and Backup Protection

Two sets of relays, primary and backup, are usually provided for each zone of protection. Main or primary protection schemes are always there as the first line of defence. Equally important and essential is a second line of defence provided by backup schemes, which will clear the fault if the primary protection schemes fail to operate for some reason. In order to give ample time to the primary relays to make a decision, backup relays are time delayed. The measures taken to provide backup protection vary widely, depending on the value and importance of the power system equipment and the consequence of its failure. Normally, primary relays have a small operation zone but operate instantaneously, whereas backup relays have a large operation zone, namely, overreached area, and operate with a particular time delay. There are two kinds of backup relaying.

Local backup In this relaying scheme, a separate duplicate set of primary relays is used. Recently, it has been observed that local backup is required at the local station to open all the breakers around the bus, rather than at the remote terminals.

Remote backup Remote backup is provided by a relay on the next station towards the source. This remote relay will trip in a delayed time if the breaker in the faulty section has not tripped because of some reason. This is the most widely used form of backup protection.

1.6 Classification of Protective Relays

Various types of protective relays are used in practice depending on the function, actuating quantities, or component that is used. The following is the classification of protective relays.

According to the quantities by which the relay operates These are thermal relays, overcurrent relays, overvoltage/under voltage relays, under/over frequency relays, over fluxing relay, and power relays.

According to their construction These are attracted armature type relay, induction disc or induction cup type relays, and balanced beam type relays.

According to the number of sensing quantities Protective relays can be classified as single input and multiple input relays, based on this parameter. A single input relay measures (senses) only one quantity, and it responds when input quantities exceed the predetermined threshold. A multiple input relay measures two or more than two quantities and responds when the output of mixing device exceeds the predetermined threshold.

According to its function in protective scheme Relay may be divided into main relays, auxiliary relays, and signal relays.

According to components and devices used These are electromagnetic relays (mechanical devices), static relays (electronic devices), microprocessor relays (sophisticated algorithm), and digital/numerical relays (fast processor with communication facilities).

According to the characteristic they adopt Instantaneous relay, time delayed relay, and inverse time delayed relay are the best examples of this type.

1.7 Electromechanical (Electromagnetic) Relay

The earliest protective devices were fuses that were, and are, used in many situations to isolate the faulted equipment. This development was followed by the evolution of circuit breakers equipped with series trip coils. Later, first generation electromechanical relays came in the industrial market in 1901. These relays operate on the regulation of a mechanical force generated through the flow of current in windings wounded on a magnetic core and hence the name *electromechanical relay*.

Advantages
The following are the merits of electromechanical relays:

1. They are reliable in nature and still used by the utilities.
2. This relay provides isolation between the input's and output's quantities.
3. They are rugged in nature as they can withstand voltage spike due to surges and can carry substantial currents.

Disadvantages
The demerits of electromechanical relays are as follows:

1. They consist of moving parts and suffer from the problem of friction.
2. They produce low torque.
3. They suffer from the problems of high burden and high power consumption for auxiliary mechanisms.

1.7.1 Thermal Relay

Overload situation occurs many times during the operation of electrical equipment. Any electrical equipment has the ability to withstand the overload condition for a definite period of time depending on the severity of overload. Thermal relays are required to protect the equipment against the overload condition. The name *thermal relay* itself suggests that the device operates on the principle of heating effect of electrical current. The characteristic of thermal relay should match with the thermal withstanding characteristic of an equipment to be protected. Thermal relay requires a longer time (in seconds) to operate compared to overcurrent relay used for overcurrent detection, which requires a very small time (in millisecond). Figure 1.2 shows the time–current characteristic of thermal relay, overcurrent relay, and thermal withstand capability of the equipment to be protected.

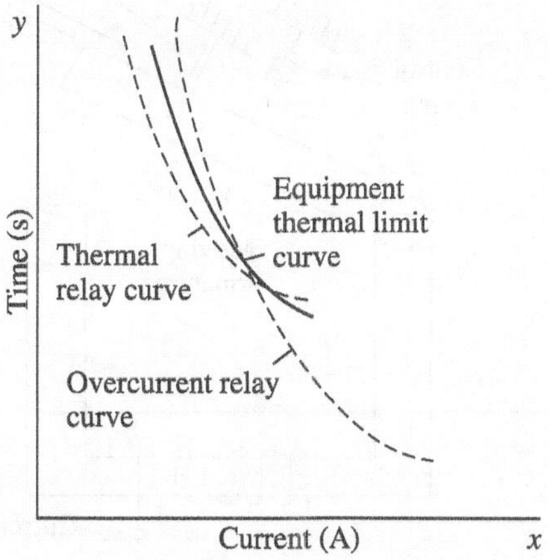

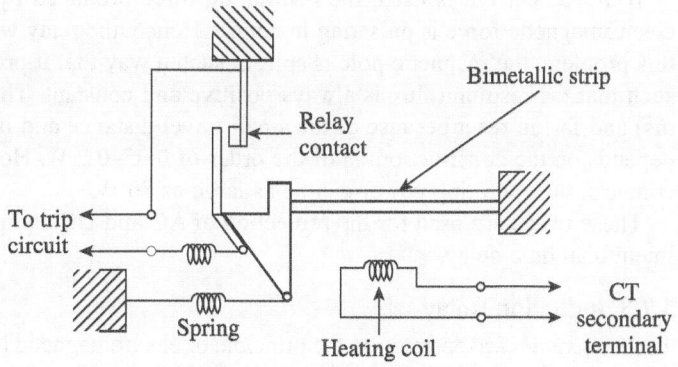

Fig. 1.2 Thermal relay characteristic

It has been observed from Fig. 1.2 that overcurrent relays cannot be used for overload protection of equipment. This is because overcurrent relays cannot fully exploit the thermal withstand capability of the equipment as it operates in the range of milliseconds. Such fast operation of overcurrent relay is not desirable for an overload condition of the equipment.

Figure 1.3 shows the replica-type thermal relay. It consists of bimetallic strips made up of nickel alloyed steel. These are heated by a heater element that absorbs the output of a current transformer in a power circuit. At one end of a bimetal strip, an insulated arm with trip contacts is provided. The arm is connected to a spring, which provides a tension against the closing of trip contacts. The characteristic of the heater element and bimetallic strips is in approximation to the heating curve of the equipment to be protected. Under normal operating condition, the

Fig. 1.3 Replica-type thermal relay

bimetal strips remain in straight position against the action of spring tension. When the overload condition is detected (120% to 140% of the rated current), the bimetal strips bend and allow the trip contact to energize the trip circuit. Thermal relay is normally used for low-voltage and low-power-rating induction motor and DC motor where resistance temperature detectors (RTDs) are not generally built-in in the protected motor.

1.7.2 Attracted Armature Relay

Attracted armature relay is a simple type of protective relay, which generally consists of an electromagnet and a hinged armature or plunger/solenoid. It can be energized either by AC or DC supply. The attracted armature relay operates on the principle of electromagnetic force produced, which attracts the plunger or hinged armature. A restraining force is provided by means of a spring so that the armature returns to its original position when the electromagnet is de-energized. Whenever the force developed by the electromagnet exceeds the restraining force, the moving contact closes due to movement of the armature. Sometimes,

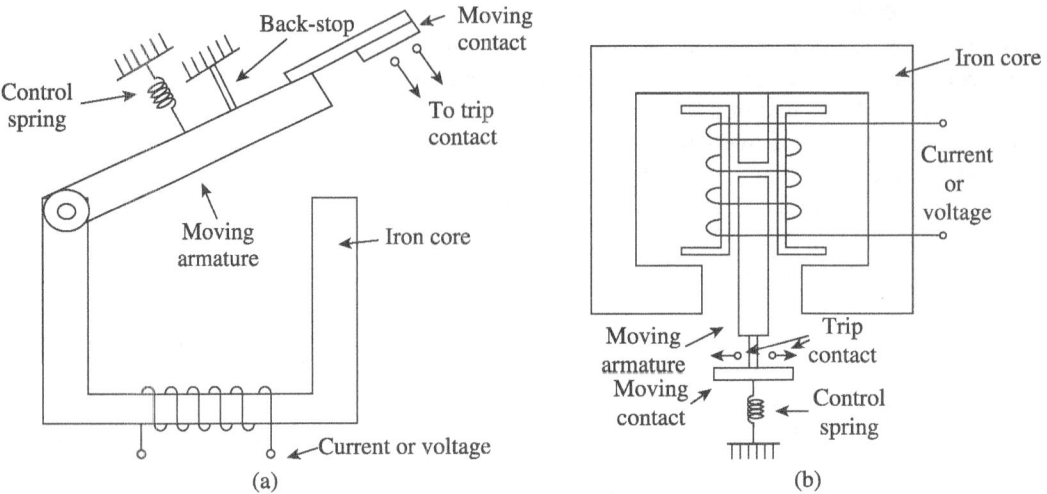

Fig. 1.4 Attracted armature relay (a) Hinged armature type relay (b) Plunger type relay

multiple contacts are mounted in parallel, which cause a single input to actuate the number of outputs. Figure 1.4 shows a hinged type (Fig. 1.4(a)) and a plunger type (Fig. 1.4(b)) attracted armature relay.

If an AC current is used, the restraining force produced by the spring is constant and the developed electromagnetic force is pulsating in nature. Hence, the relay will chatter and produce noise. To overcome this problem, the magnetic pole is split in such a way that it produces two phase-shifted fluxes in the pole such that the resultant flux is always positive and constant. These relays are fast in operation (10 and 50 ms) and fast in reset because of the small travel distance and light moving parts. Operating power, which depends on the construction, is of the order of 0.05–0.2 W. However, for a relay with several heavy duty contacts, the operating power can be as large as 80 W.

These relays are used for the protection of AC and DC equipment as an instantaneous relay that has no intentional time delay.

1.7.3 Induction Relay

The *induction relay* operates on the principle of electromagnetic induction. Hence, it is a split-phase induction motor with contacts. They are the most widely used relays for protection of lines or apparatus. Operating force is developed due to the interaction of two AC flux displaced in time and space in a movable element (rotor). Depending on the type of rotor, whether a disc or a cup, the relay is known as *induction disc relay* or *induction cup relay*.

Induction Disc Relay

Figure 1.5 shows the most commonly used shaded pole type induction disc relay. This relay is generally activated by current flowing in a single coil placed on a magnetic core having an air gap. The main air-gap flux caused because of the

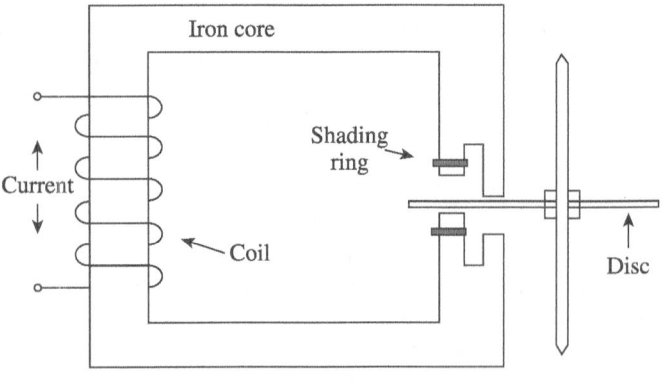

Fig. 1.5 Induction disc relay

flow of current is split into two out-of-phase components by a *shading ring*, which is made up of copper that encircles the portion of the pole face in each pole. The air-gap flux of shaded pole lags behind the flux of non-shaded pole. The rotor (made up of copper or aluminium disc) is pivoted in such a way that it rotates in the air gap between the poles. The phase angle between the two fluxes, piercing the disc, is decided at the design stage.

Induction Cup Relay

Figure 1.6 shows the constructional view of induction cup relay. In this relay, the rotating magnetic field is produced by the pair of relay coils. A rotor is a hollow metallic cylindrical cup that is arranged between two/four/eight electromagnets and a stationary iron core. The cup (looks like an induction rotor) is free to move in the gap between the electromagnet and the stationary iron core. The rotating field induces current into the cup, which then causes the cup to rotate in the same direction. The rotation depends on the magnitude of the applied AC quantities and phase displacement between them. Induction cup relay is more efficient than the induction disc relay as far as torque is concerned. Moreover, induction cup relay is faster than induction disc relay. Further, it is also used in systems where directional control is required.

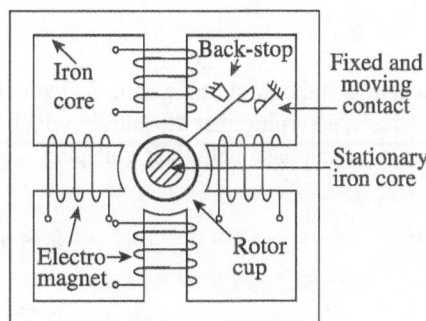

Fig. 1.6   Induction cup relay

1.7.4 Balance Beam Relay

Balance beam relay is one type of attracted armature device. As shown in Fig. 1.7, the relay with two coils surrounding the iron core is used to compare two quantities, P and Q. Operating coil produces operating torque, whereas the restraining coil produces restraining torque. These two coils are connected in such a way that their electromagnetic forces are in opposition. The electromagnetic force produced is proportional to the square of the supplied quantity (Ampere-turns). When the operating torque exceeds the restraining torque, the movement of armature closes contacts. This relay has the tendency to overreach because of a low ratio of reset to the operating current. Balance beam relay is widely used as a differential relay to compare two AC quantities.

1.7.5 Universal Torque Equation

Electromagnetic relay operates on the principle of mechanical force produced in a current conducting material because of the interaction of magnetic fluxes with their eddy currents. Figure 1.8 shows how force is produced in a part of the rotor (aluminium disc) that is penetrated by two adjacent AC fluxes. Various quantities are shown at an instant when both fluxes are directed downward and are increasing in

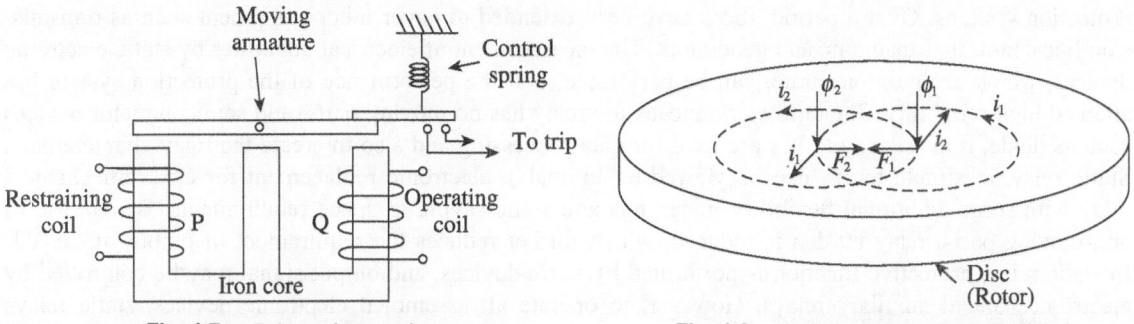

Fig. 1.7   Balance beam relay   **Fig. 1.8**   Force production in a rotor (disc)

magnitude. Individual voltages are produced because of each flux around itself in the rotor, and currents flow in the rotor under the influence of the two voltages. The mechanical forces produced by the reaction of two fluxes act on the rotor.

With reference to Fig. 1.8, the two fluxes are given by

$$\Phi_1 = \Phi_{1max} \sin \omega t$$

$$\Phi_2 = \Phi_{2max} \sin(\omega t + \theta)$$

where θ is the angle by which Φ_2 leads Φ_1.

Now, assuming that the path in which the rotor currents flow has negligible self-inductance, the rotor currents are in phase with their voltages.

$$i_1 \propto e_1 \propto \frac{d\Phi_1}{dt} \propto \Phi_{1max} \cos \omega t$$

$$i_2 \propto e_2 \propto \frac{d\Phi_2}{dt} \propto \Phi_{2max} \cos(\omega t + \theta)$$

As the two forces F_1 and F_2 are in opposition, the resultant force (F) acting on the rotor is given by

$$F = (F_2 - F_1) \propto \Phi_2 i_1 - \Phi_1 i_2$$
$$F \propto \Phi_{1max} \Phi_{2max} \{\cos \omega t \sin(\omega t + \theta) - \cos(\omega t + \theta) \sin \omega t\} \quad (1.1)$$
$$F \propto \Phi_{1max} \Phi_{2max} \sin \theta$$

The resultant force is the same at every instant because the ωt component is not involved in Eq. (1.1). It is clear from Eq. (1.1) that the magnitude of force developed on the rotor depends on the phase angle θ between two fluxes. Greater the phase angle between the two fluxes, greater the magnitude of force on the rotor. With $\theta = 90°$, the net force is maximum. The direction of force and hence the direction of rotor depends on the flux that leads the other.

1.8 Solid State Relay

With the advent of electronic devices such as diode, transistor, ICs, chips, and many more static circuits, second generation of relays, that is, the static relays, came into operations in 1950s. The development of advanced protection schemes have been started on extensive experience in the use of electronics in simple protection systems. Over a period, these have been extended to cover other equipment such as transmission lines, motors, capacitors, and generators. The measurement of electrical quantities by static electronic devices, which are more accurate, can be performed, and the performance of the protection system has attained high reliability. The term *static* means the relay has no moving parts, and semiconductor devices such as diode, transistors, and ICs are used for data processing and also to create the relay characteristic. Static relay, in simple terms, can be viewed as an analog electronic replacement for electromechanical relay with some additional flexibility in settings and some saving in space requirements. By the use of non-moving parts, relay burden is reduced, which further reduces the requirement of output of CT/VT. In static relay, protective function is performed by static devices, and output signal may be controlled by electromechanical auxiliary relays. However, to operate all assembled electronic devices, static relays require separate DC power supply.

1.8.1 Types of Amplitude and Phase Comparator for Static Relays

The static relay senses the magnitude of voltage, current, and their phase angle to detect the fault. In static relays, amplitude or phase angle of any electrical quantity is compared with the set value of threshold to issue trip signal. Hence, the static relay has either amplitude comparator or phase comparator or both in its deriving circuit. Here, various methods of static comparators are discussed. The conversion between amplitude and phase comparator is described in Section 3.7, Chapter 3.

Amplitude Comparators

An amplitude comparator compares the magnitude of two or more quantities. It does not utilize phase angle value of input quantities. Amplitude comparator is classified into (i) bridge rectifier based comparators, (ii) averaging type comparators, (iii) phase splitting based comparator, and (iv) sampling comparators.

Bridge rectifier based comparators The overcurrent and differential relay operation is carried out by the use of bridge rectifier type amplitude comparators. In this comparator, rectified operating and restraining signals are given to the polarised relay or static integrator. The relay operates when the operating quantity exceeds the restraining quantity. Figures 1.9 (a) and (b) show the circulating current type and opposing voltage type bridge rectifier based amplitude comparators.

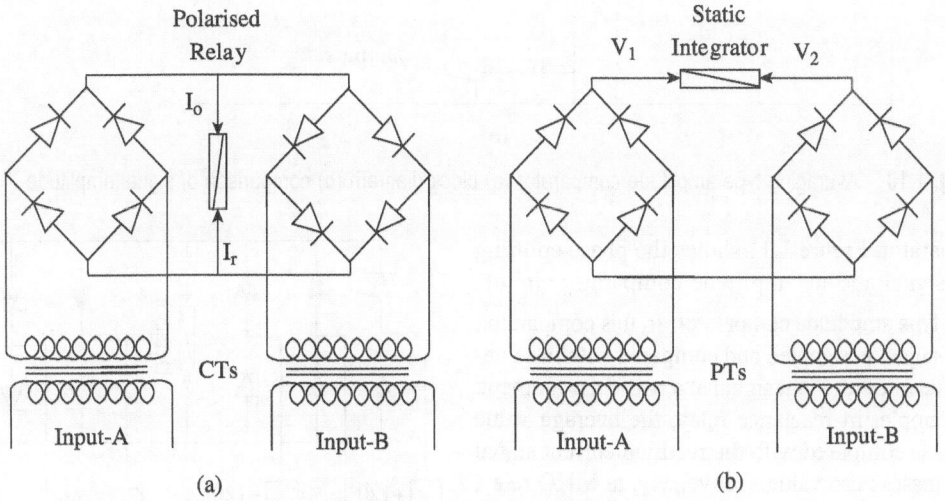

Fig. 1.9   (a) Circulating current type (b) opposing voltage type bridge rectifier based amplitude comparator

Averaging type amplitude comparator In averaging type amplitude comparator, to provide enough restrain level, the restricting quantity is rectified and smoothened to near DC value. The peak value of operating signal is compared with rectified (DC) restraining quantity. If the amplitude of operating signal exceeds the level of restrain, tripping signal is generated. Figures 1.10 (a) and (b), respectively, show the block diagram and comparison of signal for averaging type amplitude comparator.

Phase splitting type amplitude comparator In phase splitting comparator, the input quantities are divided into six components. The phase angle difference of 60°, among the six components, results in smooth rectified output. The amplitude of rectified outputs of operating and restraining signals are compared in terms of polarity detector. The phase splitting circuit and its time constant decides the operating time of

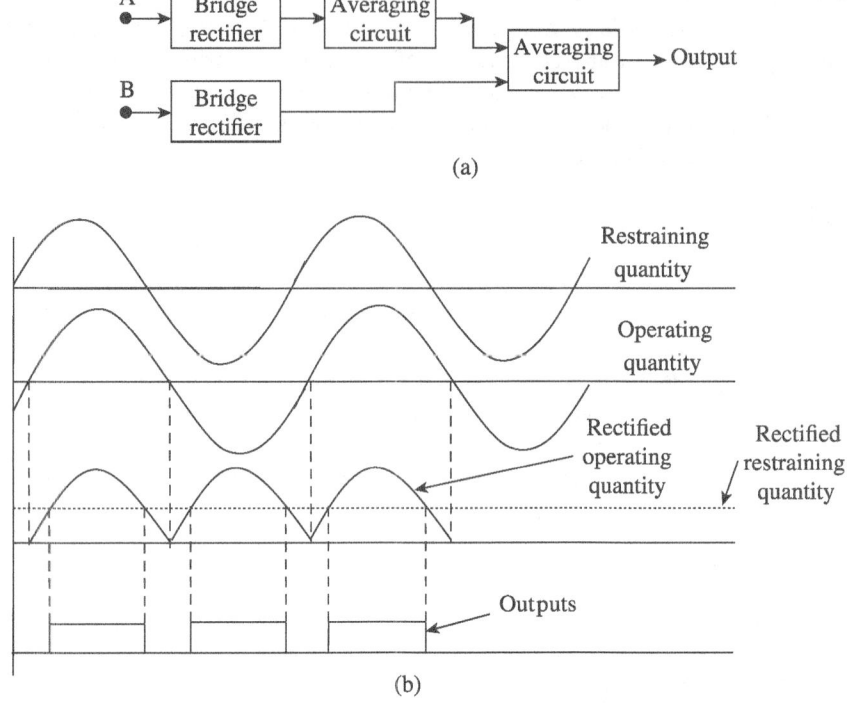

Fig. 1.10   Averaging type amplitude comparator (a) block diagram (b) comparison of signal amplitude

the comparator. Figure 1.11 shows the phase splitting of input signal and its amplitude comparator circuit.

Sampling type amplitude comparator In this comparator, one input signal is sampled and compared with instantaneous value of other input signal at a particular moment. As an example, in reactance relay, the average value of voltage is compared with the rectified current signal when it crosses zero value. Conversely, in MHO relay, the instantaneous value of current is compared with the rectified voltage signal when it crosses zero value.

Phase Comparators

A phase comparator equates the phase angles of input signals. The output of this comparator is based on the period of coincidence of given input quantities. The

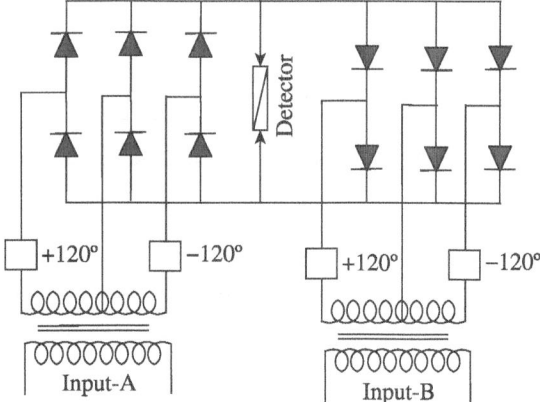

Fig. 1.11   Phase splitting type amplitude comparator

phase comparator used to measure the phase difference in static relays is classified into (a) block spike type (b) phase splitting type, and (c) integrating type.

Block spike method of phase comparison Figures 1.12 (a) and (b) show the block diagram and corresponding waveform for block spike method of phase comparison. With reference to Fig. 1.12, 'α' is the period of coincidence for the given input signals A and B. These two input signals have phase angle difference of 'θ'. Hence, the period of coincidence is given by $\alpha = 180 - \theta$. The decision of trip signal is based on whether α is

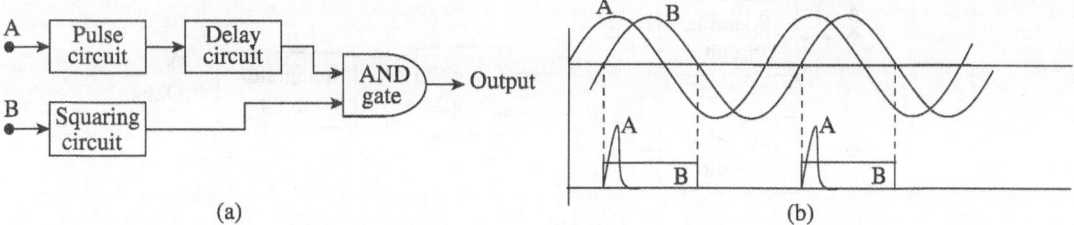

Fig. 1.12 Block spike method of phase comparison (a) block diagram (b) comparison of phase

greater or less than 90°. The output depends on the time of spike generated on coincidence of the input signals A and B. The positive rising edge of the given input signals are compared for their coincidence.

Phase splitting method In this method, as shown in Fig. 1.13 (a), the input signals A and B are divided into two components each shifted by phase displacement of $\pm 45°$ from the original. These four components are fed into the AND logic gate. As shown in Fig. 1.13 (b), the output is available only when all four inputs become simultaneously positive at any time (coincidence) in a cycle. The trip signal is obtained for $-90° < \alpha < 90°$, where α is the angle by which one signal (B) lags behind the other signal (A).

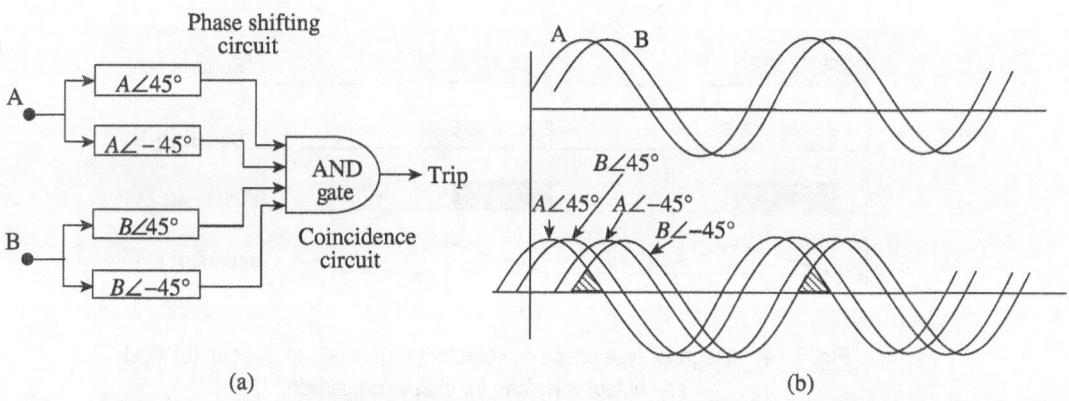

Fig. 1.13 Phase splitting comparator (a) block diagram (b) waveform of phase split input

Integrating type phase comparator In this comparator, the time of coincidence of two input signals is measured using AND logic and integrator circuit. As shown in Figs 1.14 (a) and (b), the output of AND logic is a square pulse when the two input signals (A and B) overlap in their respective positive half cycle. The integrator is R-C charging circuit whose output is given to the level detector. If the output of integrator is more than 900, the relay issues trip signal.

1.8.2 Comparison between Static and Electromagnetic Relays

The following subsections discuss the many advantages and limitations that static relays have in comparison with electromagnetic relays.

Advantages of static relays Static relays have many advantages in comparison to the corresponding electromagnetic relays. These are as follows:

1. Static relays do not contain moving parts. Therefore, they are free from problems such as contact bouncing, arcing, erosion, friction, and maintenance.
2. They have high operating torque with respect to electromechanical relays.

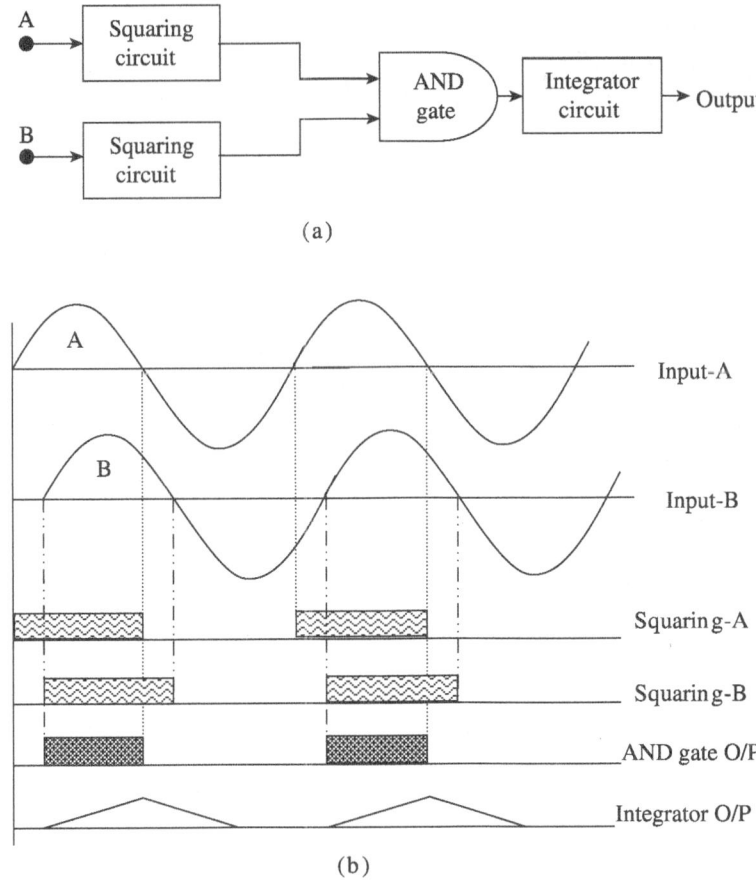

Fig. 1.14 Integrator type phase comparator (a) generalised diagram (b) input and output waveform for phase comparison

3. They place significantly less burden on instrument transformers than that placed by electromechanical relays.
4. They are compact in size.
5. They can incorporate variety of functions in a single unit.
6. Absence of moving parts in static relays leads to quick response and quick reset action. Further, they are free from the problem of overshoot owing to the absence of mechanical inertia.
7. Greater sensitivity can be obtained in static relays owing to the provision of amplification block.
8. The use of electronic devices enables achieving a greater degree of superiority in determining the operating characteristic closer to ideal characteristic as per requirement.

Limitations of static relays However, static relays also have certain shortcomings as listed here.

1. Electronic components are more sensitive to voltage spike and other transients that cause malfunctioning of static relays.
2. Auxiliary DC supply is required to operate the static relay.
3. The characteristics of electronic devices are affected by variation in temperature and ageing of semiconductor devices.
4. Static relays have low short time overload capacity compared to electromagnetic relays.

5. The reliability of static relays depends on the quality and number of small components and their electrical connection.
6. Static relays, for a particular function, are costlier compared to the corresponding electromagnetic relays.
7. Complex protective functions require highly trained persons for the servicing of static relays.

Therefore, the prime problem with electromechanical and static relays is that there is no continuous check on their operational integrity.

1.8.3 Classification of Static Relays

Various static relays have been designed by various manufacturers. They are classified according to the type of measuring unit and comparator they possess.

Electronic relays These relays use electronic valves for measuring unit and electronic tubes for comparator.

Magnetic amplifier relays (Transductor) These relays possess operating winding and controlling winding. Both are wounded on a common magnetic core. Restraining quantities are applied to the control winding, whereas relay quantities are applied to the operating winding. When operating value exceeds the magnitude of restraining value, a voltage is induced in the output winding wounded on the same core.

Rectifier bridge relays These relays make use of semiconductor devices. They consist of two rectifier bridges and a moving coil element. These are arranged in such a way that they work either as amplitude comparators or as phase comparators.

Transistor relays As transistors are able to perform amplification, summation, switching, and comparison tasks, they overcome many problems. Hence, it is possible to develop sensitive, high-speed, and precise static relays. Transistor circuit provides the necessary flexibility to outfit various relay requirements and to design various relay characteristics. These static relays are most widely used for the protection of electrical equipment and distribution feeders.

Static relays are also classified according to the protection requirement.

1. Static overcurrent relays
   (a) Static instantaneous overcurrent relays
   (b) Static definite minimum time overcurrent relays (DMT)
   (c) Static inverse definite minimum time overcurrent relays (IDMT)
2. Static directional relays
3. Static differential relays
4. Static distance relays

1.8.4 Generalized Static Time Overcurrent Relays

Different manufacturers have designed many static relays. However, due to space limitation, it is not possible to cover all the static relays here. Hence, the most widely used generalized static time overcurrent relay is discussed here.

Figure 1.15 shows the block diagram of a generalized static time overcurrent relay. This relay can be designed to achieve any characteristic such as instantaneous time overcurrent relay, DMT or IDMT.

Initially, the secondary current of CT is rectified and filtered. The filtered output of rectifier is supplied to the timing and curve shaping circuit,

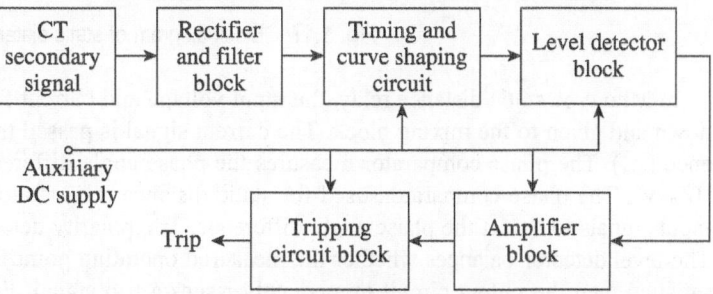

Fig. 1.15   Generalized block diagram of static time overcurrent relay

which contains non-linear resistors and RC networks to shape the time–current characteristic. The output of timing circuit is given to the level detector, which compares the relay quantities with reference quantities. If the magnitude of the relaying quantities exceeds the magnitude of the reference quantities (threshold value), it generates a voltage signal. The generated voltage signal is amplified by an amplifier block and fed to the tripping circuit. The tripping circuit may be an electromagnetic one or a static one. At last, the tripling circuit generates a tripping command, which will be given to the trip coil of circuit breaker. Suitable DC auxiliary power supply is provided to static relay from separate rectifier or from station battery.

1.8.5 Static Differential Relay

Figure 1.16 shows the basic block diagram of a simple static differential relay. The input signals (A and B) are initially scaled down to nominal value of current by current transformer. The rectifier bridge type amplitude comparator, as discussed previously, is used as a static comparator to check the difference of input signals. The output of the comparator is given to the integrator and the level detector followed by the driving circuit. The driving circuit issues a trip signal if the desired condition is fulfilled in terms of magnitude comparison.

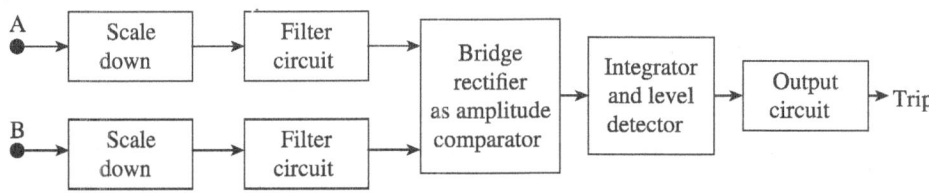

Fig. 1.16 Block diagram of static differential relay

1.8.6 Static Distance Relay

In static distance relaying scheme, both amplitude and phase comparator are used as per the requirement of distance characteristic. Figure 1.17 shows the schematic block diagram of phase comparator based static distance relay.

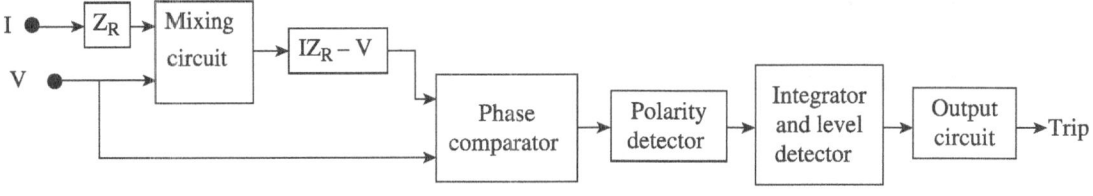

Fig. 1.17 Block diagram of static distance relay

In Mho type static distance relay, the input voltage and current from CT and PT, respectively, are scaled down and given to the mixing block. The current signal is passed through a predefined replica type imped-ance (Z_R). The phase comparator measures the phase angle difference (θ) between voltage signal (V) and 'IZ_R-V'. The phase comparator, used for static distance relay, follows the principle of coincidence of two input signals to detect the phase angle difference. The polarity detector detects whether θ lies within ±90°. The level detector balances whether the measured operating point falls below Z_R. If the above condition is satisfied then the output circuit immediately issues a trip signal. For discrimination of zone fault, a timer circuit is integrated into the output circuit so as to provide intentional time delay.

1.9 Digital Relaying

By 1970, advances in the very large scale integrated (VLSI) technology and software techniques led to the development of third generation microprocessor-based relays. Early designs used the fundamental approaches that were previously used in the electromechanical and solid-state relays. Though complex algorithms for implementing protection functions were developed, the microprocessor-based relays marketed till 1980s did not incorporate them. However, these relays, which performed basic functions, took advantage of the hybrid analog and digital techniques, and offered good economical solution. Though the performance of these relays was just adequate their introduction was appreciated by the highly conservative world of power system protection. Continuous advances in electronics, combined with extensive research conducted in the microprocessor-based systems, led to a few applications where multiple functions were performed by a microprocessor relay. By late 1980s, multifunction relays were introduced in the market. These devices reduced the production and installation costs drastically. Utilities are using microprocessor-based, dedicated, and economic protective modules in the protective relaying schemes.

With the advent of digital computing technology, digital relaying is a very promising area of thought. In 1969, Rockefeller came out with his landmark paper on the use of digital computer for protection purposes. Since then, digital relays also have gone through revolutionary changes, thanks to the advent of low-cost, high-performance, high-density, large-scale integrated digital circuits, particularly microprocessor and related devices. With the change in technology, it became evident that a single computer for the protection of all the equipment in a substation was not an efficient approach in view of the presently available computer hardware. A probable solution to this problem is to use a number of microprocessors dedicated to individual equipment relaying tasks with an inter-computer data exchange facility. The concept of digital computer relaying has grown rapidly as digital computers have become more powerful, cheaper, and sturdier. It is to be noted that digital relays can realize some very useful functions, which are not possible with electromechanical or analog circuits such as mathematical functions, long-term storage of pre-fault data, and, they also inherit all the features of microprocessor-based relays. However, these computer relays do not have successful solutions to cumbersome problems such as high fault resistance, mutual coupling, remote infeed, time delay, and so on, which have been bothering relay engineers for many years. The fourth generation of relays, that is, digital relays, came into the market in the 1990s. Let us now discuss some features of these relays.

1.9.1 Merits and Demerits of Digital Relay

Digital/numerical relays have many advantages over electromechanical/static relays:

1. They provide many functions such as multiple setting groups, programmable logic, adaptive logic, sequence-of-events recording, and oscillography.
2. Digital relays have the ability of self-monitoring and self-testing, which were not available in electromechanical/static relays.
3. Digital relays have the ability to communicate with other relays and control computers.
4. The cost per function of digital/numerical relays is lower as compared to the cost of their electromechanical and solid-state counterparts. The digital relays include all the relay characteristics in one group. For example, in IDMT relay, digital relay includes normal inverse, very inverse, extremely inverse, and many more characteristics in one group. On the other hand, in case of electromechanical/static relays, one has to purchase a separate unit for each characteristic.
5. A major feature of digital/numerical relays, which was not available in previous technologies, is the ability to allow users to develop their own logic schemes, including dynamic changes in that logic.
6. Digital/numerical relays place significantly less burden on instrument transformers than the burden placed

by the relays of the previous technologies.

7. Digital protection systems require significantly less panel space than the space required by electromechanical and solid-state systems that provide similar functions.

8. Reporting features, including sequence-of-events recording and oscillography, are another feature of digital relays.

However, digital/numerical relays have certain shortcomings:

1. Digital/numerical devices, including the protection systems, have short life cycles. While each generation of microprocessor-based systems increases the functionality compared to the previous generation, the pace of advancements makes the equipment obsolete in shorter times. This makes it difficult for the users to maintain expertise with the latest designs of the equipment.

2. Another variation of this shortcoming is in the form of changes in the software used on the existing hardware platforms. Sometimes, these changes effectively generate newer relay designs. This requires that a software tracking system be used for each device owned by a utility.

3. Electromechanical relays are inherently immune to electrical transients such as electromagnetic interference (EMI) and radio frequency interference (RFI). Early designs of solid-state relays were susceptible to incorrect operations owing to transients, but later designs included adequate countermeasures. Because of a better understanding of the problems, digital/numerical relays were designed in a manner that provided excellent reliability under the said conditions as long as they conform to the IEEE Standard C37.90 or IEC 61000 series of standards. However, digital/numerical relays will always remain more susceptible to such problems because of the nature of the technology compared to the systems built with the electromechanical technology.

4. Many digital/numerical relays, which are designed to replace the functions of several electromechanical and static relays, offer programmable functions that increase the application flexibility compared to the fixed function relays. The multifunction digital/numerical relays, therefore, have a significant number of settings. The increased number of settings may pose problems in managing the settings and in conducting functional tests. Setting-management software is generally available to create, transfer, and track the relay settings. Special testing techniques, specifically the ability to enable and disable selected functions, are generally used when digital/numerical relays are tested. This increases the possibility that the desired settings may not be invoked after testing is completed. Proper procedures must be followed to ensure that correct settings and logic are activated after the tests are completed.

1.9.2 Generalized Block Diagram of Digital Relay

Figure 1.18 shows the basic block diagram of a digital relay. Analog signals, such as currents and voltages acquired from the power system network, are processed by a signal conditioning device, which consists of isolation transformer, surge protection circuit, and anti-aliasing filter (AAF). Isolation transformer provides the electric isolation, whereas surge protection circuit gives protection to the digital component against transients and spikes. AAF is a low pass filter that blocks the unwanted frequency component. Further, it also avoids aliasing error. According to Nyquist criterion, the sampling frequency must not be less than two times the maximum frequency contained in original signal.

$$f_s \geq 2 \times f_m \tag{1.2}$$

where,

f_s = sampling frequency and f_m = maximum significant frequency within the signal sample.

It is to be noted that this processing is true if conventional transducers are used. On the other hand, these input signals can be given directly to the central processing unit (CPU) if electronic CTs and CVTs

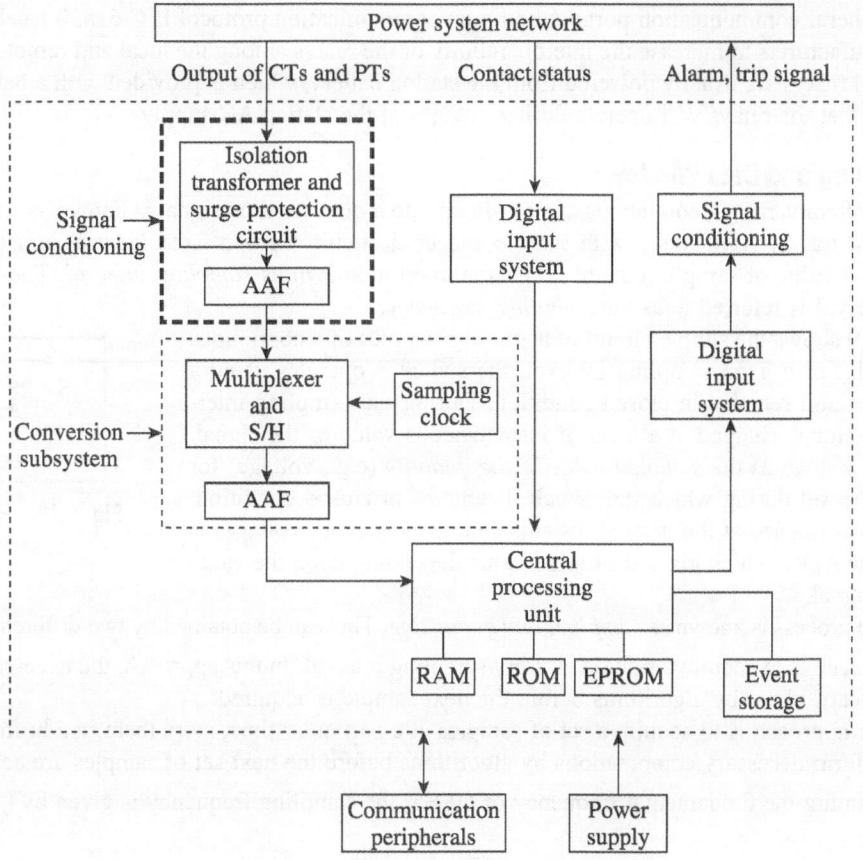

Fig. 1.18   Basic block diagram of digital relay

are used. These signals are given to the CPU through multiplexer and analog to digital converter (ADC), which samples, combines, and converts the analog signal into digital form. The input signals are frozen by sample and hold circuit to achieve synchronized sampling between all the acquired signals. The digital input, such as status of circuit breaker contacts, status of local and remote end relays, and reset signals are acquired by the digital input system and transferred to the CPU. CPU is the core component of digital relay, where all processes regarding different logics/algorithm have been carried out. CPU executes the relay programme with a different characteristic, maintains different timing function, and communicates with external devices.

Several memory units are allocated for data storage and data processing purposes. The random access memory (RAM) stores the input sample data temporarily and buffer data permanently. Further, the stored data in RAM is processed during the execution of relay algorithm. The read only memory (ROM) is used to store the relay algorithm permanently. EPROM is used to store certain parameters such as relay setting. These parameters may change in case of change in external system conditions. The event storage block is used for storing historical data such as fault related data, transient data, and event time data.

The digital output system provides the tripping, alarm, and other control signals to activate the external devices in the power system. A self-diagnosis software available in the digital relay checks integrity of the relay at regular intervals. This feature allows the relay to remove itself from service when a malfunction occurs and to alert the control centre. Relay setting, data uploading, and event data recording are done through the various peripheral communication ports. A common communication protocol IEC 61850 has been adopted by relay manufacturers to increase the interoperability of the relays among the local and remote substations.

The digital relays are usually powered from the station battery, which is provided with a battery charger. This ensures that the relays will operate during outages of the station AC supply.

1.9.3 Sampling and Data Window

The process of converting a continuous analog signal into a discrete-time signal is known as *sampling*. This task is carried out by ADC along with sample and hold circuit. Certain fixed interval is used to acquire the next (new) value of sample (quantity). This interval is known as *sampling interval*. The reciprocal of sampling interval is referred to as the *sampling frequency*.

Figure 1.19 shows the simple circuit of acquiring samples of a continuous analog signal. For a fixed sampling interval, the switch S operates using a periodic pulse and remains in closed condition. During the sampling interval, the capacitor is charged at a level of instantaneous value of the signal. This value is known as the *sampled value of the quantity* (e.g., voltage) for a particular period during which the switch S remains in closed condition. The switch S is opened at the desired instant. The quantity (e.g., voltage) is then fed to the ADC, which gives the digital value depending upon the value of sampled signal.

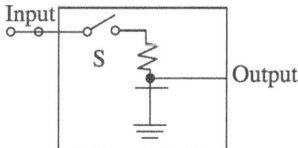

Fig. 1.19 Basic circuit of sampling

This entire process is known as *sampling and quantizing*. This can be obtained by two different approaches.

1. One approach is to acquire a sample at every sampling interval. In this approach, the necessary computations are carried out by algorithms before the next sample is acquired.
2. The other approach is to acquire a set of samples at a particular time, store them in a buffer, and thereafter, perform necessary computations by algorithms before the next set of samples are acquired.

Now, assuming the fundamental frequency of 50 Hz, the sampling frequency is given by

$$f_s = f \times n \tag{1.3}$$

where,

f_s = sampling frequency (Hz)
f = fundamental frequency (Hz)
n = number of samples/cycle

Data window is the window having a set of acquired samples that are used to obtain an estimate of the acquired signal/quantity. Figure 1.20 shows the concept of data window, which uses three samples at a time in a window. It is to be noted that in each data window, the number of samples remain constant (three samples in the case). Therefore, when the next sample is acquired, the previous sample is discarded. Whenever a new sample is taken, the data window advances, that is, slides ahead. Hence, this concept is also known as *sliding window* concept.

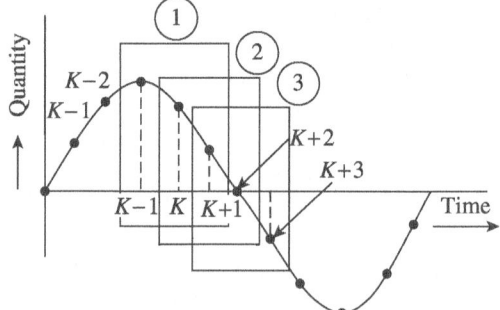

Fig. 1.20 Concept of data window

1.10 Adaptive Relaying

Conventionally, the relays are provided with setting switches and other means that can be selected/adjusted by the operator, depending on the operating condition of the system. Since changes and events occur quite rapidly in a power system, human intervention to change setting switches to cope with every system change is not possible. Therefore, the settings are usually selected on the basis of the worst case and changed only when a major change in the system configuration is made. This requires high degree of professionalism on the part of the user to decide as to when and what changes to make in the settings. Furthermore, the relay settings that are selected for the worst case would generally give slow speed, low sensitivity, or poor selectivity on other conditions in the protected system. Last, but not the least, a fixed operating characteristic of a given relay may not be able to give the requisite speed, selectivity, and sensitivity on all the operating conditions of the protected system. Relay engineers have dreamed that relay could adapt to the system changes. With the development of high-speed microprocessors, new tools for signal processing and digital communication techniques, this dream is fast turning true. With the use of programmable devices in digital relays, it is possible to design a relay such that it changes its settings, parameters, or even the characteristic automatically and appropriately in accordance with the changed condition of the system protected by it. A relay having such a feature is called an *adaptive relay*. The idea of modifying relay settings to correspond to changing system conditions, as a preventive action to improve system stability, was first proposed by DyLiacco in 1967. Thereafter, different researchers have given different definitions of adaptive protection. All these definitions narrate the same facts in different forms. Therefore, adaptive relaying is defined as 'changing relaying parameters or functions automatically depending upon the prevailing system condition or requirements'. The adaptive relaying philosophy can be made fully effective only with computer-based relays.

1.11 Tripping Mechanism of Relay

The relay is always connected in the secondary circuit of CT and potential transformer (PT) irrespective of the type of relay. The main function of any type of relay is to detect or sense the inception of fault, whereas the tripping task is carried out by the auxiliary relay and circuit breaker. Since the relay only does the function of sensing, the speed of the relay is increased, and hence, it operates at a very high speed. *Auxiliary relay* is a relay that carries high value of trip coil current during a fault. Moreover, it also gives signals to perform certain other functions associated with relays such as alarms and interlocking. Figure 1.21 shows the basic tripping circuit of any type of relay.

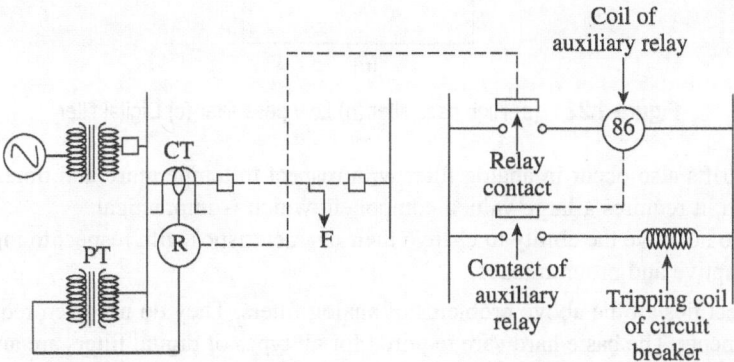

Fig. 1.21 Power circuit and control circuit of operating mechanism of any relay

If single input relay is used (current-based relay or voltage-based relay), then the relay receives a signal from the secondary of CT or PT only. Conversely, for two input relays, it receives signals from the secondary of both CT and PT. As shown in Fig. 1.20, the relay R senses the fault F within a fraction of second (in millisecond) and gives signal to the auxiliary relay through its contact. The contact of auxiliary relay closes owing to energization of the coil of auxiliary relay. This will further energize the trip coil of the circuit breaker.

1.12 Digital Filters

A digital filter is a device which executes mathematical operations on a discrete-time signal to decrease/increase definite aspects of that signal. Filtration of signals is important in protective relaying schemes for the evolution of signal whether it has such information or not like CT saturation, Magnetising inrush, Power swing, Harmonic and noise analysis etc. For filtering a signal in digital relaying schemes, various types of digital filters are utilized. Also, there is a wide difference between analog and active filters.

Normally, filters are formed by utilizing combinations of resistance, inductance, and capacitance. Figures 1.22 (a) and (b) show the basic circuit of low pass and high pass filter, respectively. These filters are known as analog filters. If the output of analog filter is added with amplifier then this filtering circuit is known as active or digital filters. The circuit of digital filter is shown in Fig. 1.22 (c) in which X_n is the input signal and Y_n is the output signal.

Some of the limitations of analog filter are as follows.

1. Due to the large space requirement of inductor, analog filters are bulky in size. Hence, they become expensive due to the requirements of highly precise components.

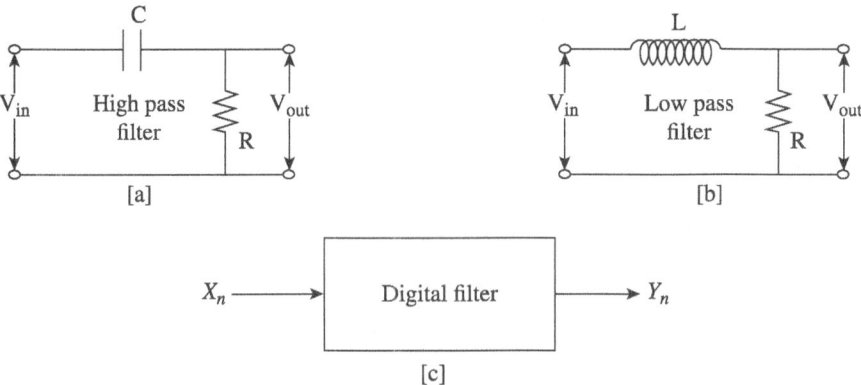

Figure 1.22   (a) High pass filter (b) Low pass filter (c) Digital filter

2. Characteristic drifts also occur in analog filter with respect to temperature and time. Even for low frequency filtration, it requires a large valued component which is impractical.
3. Analog filters do not have the ability to change their characteristics with respect to input signals. Hence, they are not adaptive and programmable.

A digital filter rectifies all the above problems of analog filters. They do not even require highly precise R, L, and C components. The basic hardware required for all types of digital filters are an anti-aliasing filter with sample and hold (S/H) circuit, analog to digital (ADC) and digital to analog (DAC) converter with digital processor, and a reconstruction filter for smoothing the signals. The simplified representation of a digital filter is shown in Fig. 1.23. As shown in Fig.1.23, anti-aliasing filter is always placed before the sample and hold (S/H) circuit to prevent the detrimental effect widely known as aliasing. Then, the signal is processed through S/H circuits in which analog to digital conversion is carried out by the processor. Finally, it converts the digital value to analog value. Since the output of DAC is like a staircase the smoothing filter is required.

As shown in Fig. 1.23, $x[n]$ is the continuous time signal whereas $y[n]$ is the processed discrete time signal. Moreover, $x_a(t)$ and $y_a(t)$ are the continuous and discrete time equivalent of the signal, respectively.

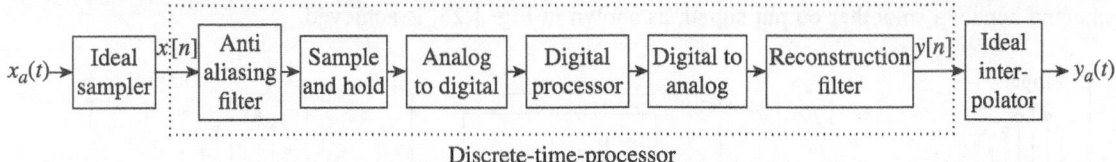

Discrete-time-processor

Fig. 1.23 Simplified representation of digital filter

The several benefits of digital filters include easy to change filter characteristics through programming, immune to ageing and drift against time and temperature variations, and no maintenance and tuning.

Digital filters are classified as follows

1. Low pass filters
2. High pass filters
3. Finite Impulse Response (FIR) filters
4. Infinite Impulse Response (IIR) filters

1.12.1 Simple Low Pass Filter

A low pass filter is a device which allows signal with a frequency lower than a definite cutoff frequency and attenuates signals with frequencies higher than the cutoff frequency. The operation of a low pass filter depends on the running average of the last two samples. The internal structure of a digital low pass filter is shown in Fig. 1.24 (a). In this filter, the R-C circuit behaves as a filtering circuit and its output is given to the amplifier. 'A_n' is the sample for filtration and 'B_n' is the filtered output.

The filtered output of low pass filter is given by Eq. (1.4),

$$B_n = \frac{A_n + A_{n-1}}{2} \tag{1.4}$$

The frequency against amplitude characteristic of a low pass filter is shown in Fig. 1.24 (b). In this figure, 'F_H' indicates pass band which filters higher frequency signal. The region between pass band and stop band is known as transition region.

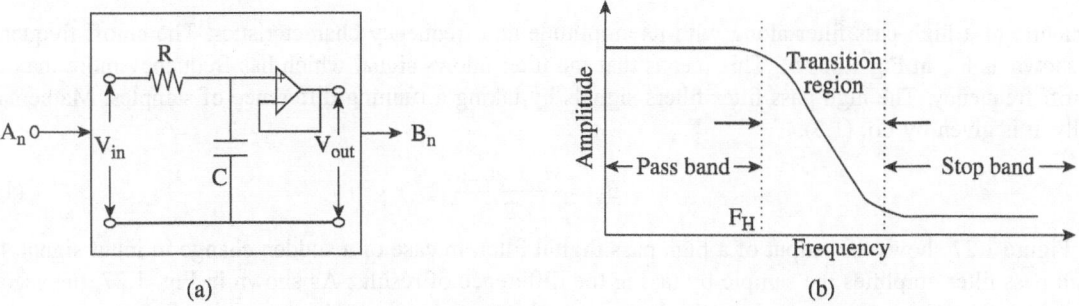

(a)   (b)

Fig. 1.24 (a) Internal structure of low pass filter (b) Frequency vs amplitude characteristic

The output of a low pass digital filter is shown in Fig. 1.25. As shown in Fig. 1.25, the second sample at the input of the digital low pass filter has a large magnitude and has positive polarity, whereas the third sample has an equal magnitude but opposite polarity. Here, as noise signal is considered as a high frequency signal it rides over the low frequency signal. However, as the output of digital low pass filter is formed by taking the running average of last two samples, the effect of positive spike cancels the effect of negative spike and hence, a smoother output signal, as shown in Fig. 1.25, is achieved.

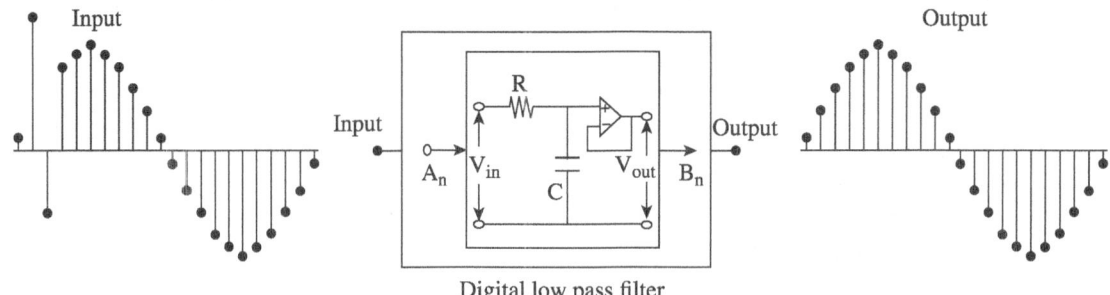

Fig. 1.25   Output of low pass filter

1.12.2 Simple High Pass Filter

A high pass filter is an electronic filter that allows signals with a frequency higher than a certain cutoff frequency. At the same time, it also attenuates signals with frequencies lower than the cutoff frequency. The amount of attenuation for each frequency depends on the design of filter. Figure 1.26 shows the internal

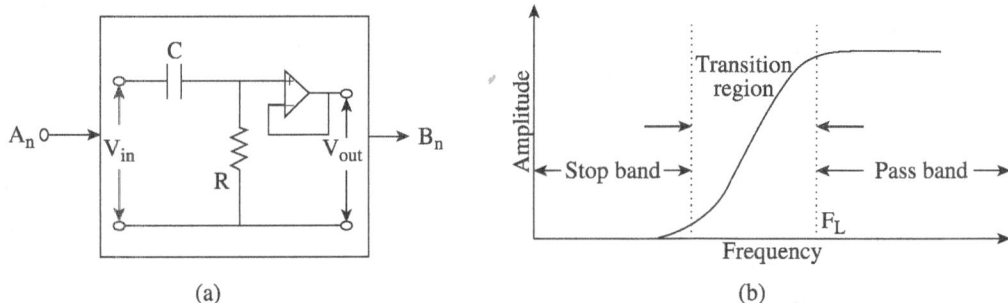

Fig. 1.26   (a) Internal structure of high pass filter (b) Frequency v/s Amplitude characteristic

structure of a high pass filter along with its amplitude and frequency characteristics. The cutoff frequency is shown as F_L, in Fig 1.26 (b). This means that the filter allows signal which has frequency more than the cutoff frequency. The high pass filter filters signals by taking a running difference of samples. Mathematically, it is given by Eq. (1.5).

$$B_n = \frac{A_n - A_{n-1}}{2} \tag{1.5}$$

Figure 1.27 shows the output of a high pass digital filter. In case of a sudden change in input signal, the high pass filter amplifies the sample by taking the difference of results. As shown in Fig. 1.27, the second sample with positive polarity and the third sample with negative polarity having higher frequencies are visible in input signals. After filtering, only high frequency signals are highlighted in the output.

The combination of low pass filter and high pass filter behaves as a band pass filter. The characteristic of band pass filter is shown in Fig. 1.28. The frequency spectrum is bounded by the limit $F_L > F > F_H$. This indicates that the filtering must be carried out between F_L (lower frequency limit) and F_H (higher frequency limit).

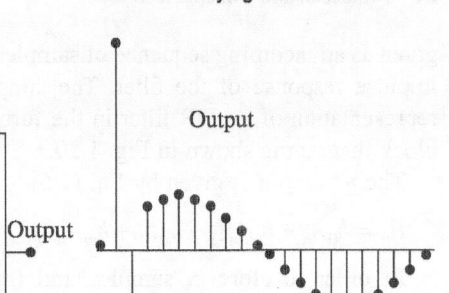

Fig. 1.27 Output of high pass digital filter

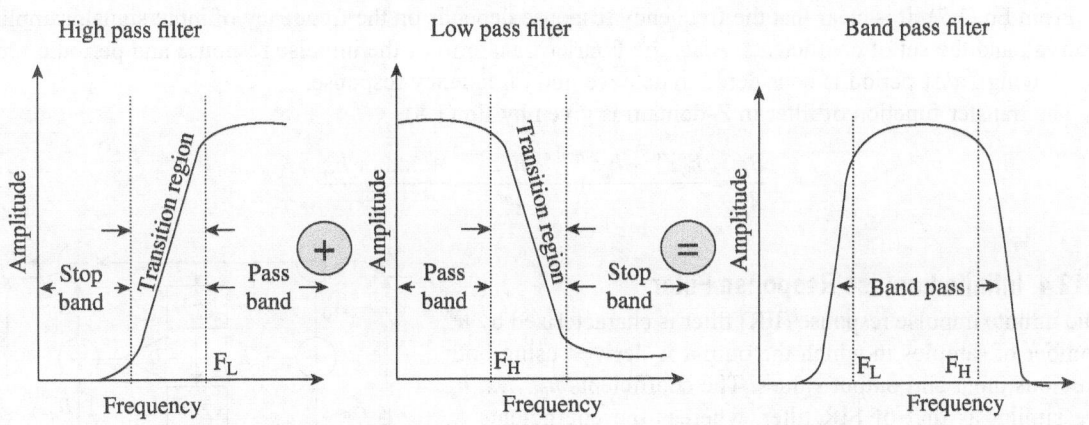

Fig. 1.28 Band pass filter as combination of low and high pass filter characteristics

1.12.3 Finite Impulse Response Filter

When the impulse response is followed by a finite number of terms it is known as finite impulse response (FIR) filter. The standard FIR filter is also known as *transversal filter*. Figure 1.29 shows the input–output response of a digital FIR filter. The output samples are formulated by filtering the weighted sum of the input samples and a limited number of previous input samples. Hence, samples of the impulse response are important for analysing and decision-making of the system. Thus, the output of FIR filter with length 'm' is

Fig. 1.29 FIR filter input–output results

given as an incoming sequence of samples with impulse response of the filter. The simplified representation of an FIR filter in the form of a block diagram is shown in Fig. 1.30.

The nth output is given by Eq. (1.6)

$$B_n = h_0 A_n + h_1 A_{n-1} + \dots\dots + h_m A_{n-m} \quad (1.6)$$

In order to store m samples and $(m + 1)$ numbers of coefficients, the memory requirement should be sufficient.

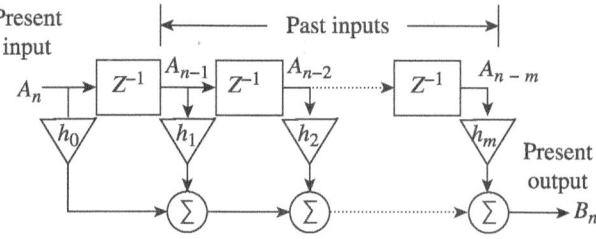

Fig. 1.30 Block diagram and simplified representation of FIR filter

The frequency response of FIR filter is given by Eq. (1.7)

$$f(j\omega) = \sum_{n=0}^{m} e^{(-j\omega n \Delta t)} . h_m \quad (1.7)$$

where Δt = sampling inverter, h_m = set of co-efficients, ω = frequency.

From Eq.(1.7), it is clear that the frequency response depends on the frequency of input signal, sampling interval, and the set of coefficients. Also, the Fourier transform of the impulse response and periodic function having $2\pi/\Delta t$ period is considered to achieve good frequency response.

The transfer function of filter in Z-domain is given by Eq (1.8),

$$\frac{B_n}{A_n} = \frac{h_0 z^m + h_1 z^{m-1} + h_2 z^{m-2} + \dots\dots + h_m}{z^m} \quad (1.8)$$

1.12.4 Infinite Impulse Response Filter

The infinite impulse response (IIR) filter is characterized by n^{th} number of samples in which the output is derived using both previous input and output values. The coefficients $h_0, \dots, h_m$ are similar as that of FIR filter, whereas the coefficients $i_1, \dots, i_k$ form the recursive part of the filter. Figure 1.31 shows the block diagram of input–output signals of an IIR filter. Many alternative ways are available for the implementation of IIR filter. Figure 1.32 shows the basic circuit diagram of an IIR filter.

The output with n samples is given by,

$$B_n = h_0 A_n + h_1 A_{n-1} + \dots\dots + h_m A_{n-m}$$
$$+ i_1 B_{n-1} + i_2 B_{n-2} + \dots\dots + i_k B_{n-k} \quad (1.9)$$

The output at any instant is a function of m number of past inputs and k number of past outputs. Due to the presence of closed loop feedback, this filter is known as *recursive filter*.

The transfer function of IIR filter in Z-domain is given by,

$$\frac{B_{(Z)}}{A_{(Z)}} = \frac{h_0 + h_1 Z^{-1} + h_2 Z^{-2} + \dots\dots + h_m Z^{-m}}{1 - i_1 Z^{-1} - \dots\dots - i_k Z^{-k}} \quad (1.10)$$

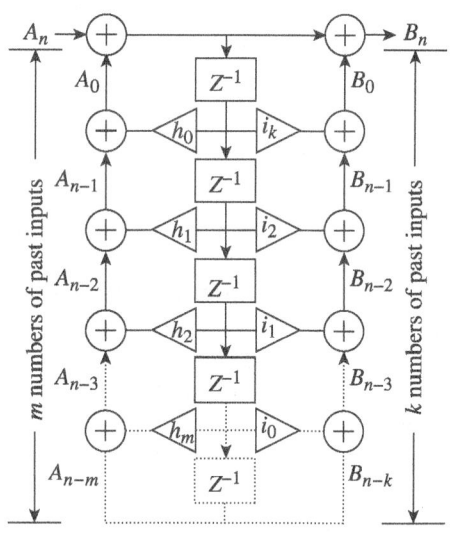

Fig. 1.31 Block diagram of IIR filter

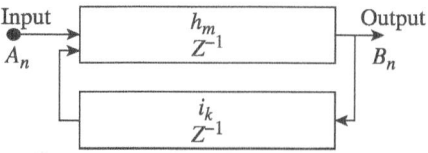

Fig. 1.32 Basic circuit diagram of IIR filter

1.12.5 Comparison Between FIR and IIR Filters

Base of comparison	FIR filter	IIR filter
Output	Non-recursive	Recursive
Response	Finite impulse response	Infinite impulse response
Stability	Always stable since there is no feedback	Because of feedback, possibility of instability exists
Coefficients	Less	More than FIR
Transfer function	Has only numerator terms	Both numerator and denominator terms
Order of filter required for a given frequency response	Higher order	Lower order
Response characteristics	Linear	Non-linear
Simplicity for implementation	Very simple	Complex compared to FIR filter.
Basic circuit diagram	Input $\xrightarrow{}$ $\boxed{\dfrac{h_m}{Z^{-1}}}$ $\xrightarrow{}$ Output $\quad A_n \qquad\qquad\qquad B_n$	Input $\xrightarrow{}$ $\boxed{\dfrac{h_m}{Z^{-1}}}$ $\xrightarrow{}$ Output $\quad A_n \qquad \boxed{\dfrac{i_k}{Z^{-1}}} \qquad B_n$
Equation	$B_n = h_0 A_n + h_1 A_{n-1} + \ldots\ldots + h_m A_{n-m}$	$B_n = h_0 A_n + h_1 A_{n-1} + \ldots\ldots + h_m A_{n-m}$ $+ i_1 B_{n-1} + i_2 B_{n-2} + \ldots\ldots + i_k B_{n-k}$
Transfer function	$\dfrac{B_n}{A_n} = \dfrac{h_0 z^m + h_1 z^{m-1} + h_2 z^{m-2} + \ldots\ldots + h_m}{z^m}$	$\dfrac{B_{(Z)}}{A_{(Z)}} = \dfrac{h_0 + h_1 Z^{-1} + h_2 Z^{-2} + \ldots\ldots + h_m Z^{-m}}{1 - i_1 Z^{-1} - \ldots\ldots - i_k Z^{-k}}$
Conversion of FIR filter from IIR filter	Infinite impulse response	Delayed and scaled IIR $\ominus$ $\quad$ Transduced IIR = FIR $\equiv$

1.13 Different Relay Algorithms

Protection of transmission lines is a very important area. Owing to the increase in demand, the lines are heavily loaded, because of which the margin between load and fault currents is often small. Sometimes, the magnitude of fault current may be less than the maximum full load current in the line.

Long EHV (extra high voltage) and UHV (ultra high voltage) lines are protected by the modern digital/numerical relay that contains hardware and software as two main parts. The software part includes a digital algorithm that is based on a set of mathematical equations. It basically involves the estimation of line parameters in frequency domain, time domain, or both domains by monitoring the voltage and current at the relaying point. These estimated parameters are compared with their preselected thresholds, and a trip decision is taken. A number of algorithms for the digital protection of transmission lines have been proposed in the literature. On the basis of special applicability and domain of analysis used, these algorithms can be classified into four groups:

1. Algorithms assuming pure sinusoidal relaying signal
2. Algorithms based on the solution of system differential equations
3. Algorithms applicable to distorted relaying signals
4. Algorithms based on travelling wave approach

In the following sections, these algorithms are discussed in detail. In the said four categories, each algorithm contains many algorithms. However, we have discussed only one or two main algorithms. Besides the development of algorithms, efforts have been made by different researchers to study the applicability of the said algorithms on the basis of speed of convergence of the estimated values of parameters to their post fault values and accuracy. It has been found that high-speed algorithms have poor accuracy in the presence of distortion in signals due to transients. On the other hand, the algorithms giving more accurate results tend to be slower because of the larger data window and complex computations involved.

1.13.1 Algorithms Assuming Pure Sinusoidal Relaying Signal

This was the first approach used by the researchers to compute apparent real and imaginary parts of the impedance from current and voltage samples. As the inputs to the relay are assumed to be pure sinusoidal, this approach has certain advantages.

1. The sampling process is not required to be synchronized with the phase position of the sine wave being measured.
2. Shorter data window is sufficient.
3. Owing to low computational requirements, the decision process is fast.

On the other hand, as the relaying signals contain DC offset, harmonics, and noise, especially during a few cycles following the fault inception, this approach lacks accuracy.

This algorithm is derived as follows:

As the relaying signals are assumed to be pure sinusoidal, at any sampling instant k, the current, i_k, is given by

$$i_k = I_P \times \sin \omega t \tag{1.11}$$

The rate of change of the current with time, i'_k, is given by

$$i'_k = w \times I_P \times \cos \omega t \tag{1.12}$$

Therefore, the peak I_P, and the phase angle, ϕ_i, of current can be expressed as

$$I_P^2 = i_k^2 + \left(\frac{i'_k}{w}\right)^2 \text{ and } \phi_i^2 = \tan^{-1}\left(\frac{\omega i_k}{i'_k}\right)^2 \tag{1.13}$$

The derivative at any sampling instant can be calculated from

$$i'_k = \frac{i_{k+1} - i_{k-1}}{2h} \tag{1.14}$$

where, h is the sampling interval.

Here, in this algorithm, digital filtering technique, namely, averaging of samples over a short span, is used to attenuate harmonics. For the suppression of decaying DC offset, series R–L circuit across CT is used.

1.13.2 Algorithms Based on Solution of System Differential Equations

Transmission line is modelled by a set of first order linear differential equations. It is given by

$$v(t) = R \times i(t) + L\frac{di(t)}{dt} + E(t) \tag{1.15}$$

where, $v(t)$ and $i(t)$ are instantaneous value of voltage and current measured by the relay, and $E(t)$ is the error.

This equation is solved for R and L using numerical techniques, and relaying decisions are taken accordingly. For a solution of this model, different algorithms have been proposed by different researchers. However, we have given only one algorithm.

In order to determine the parameters of line, that is, R and L, we have to take the derivation of the current signal. This can be achieved by two methods. The first method is to use the derivative approximation using the input signal samples. It is given by

$$v(t_k) = R \times i(t_k) + L \times i'(t_k) + \Delta E(t_k) \tag{1.16}$$

where, $v(t_k)$ and $i(t_k)$ are the input voltage and current samples and $i'(t_k)$ is the derivative approximation in the t_k time. The derivative of current ($i'(t_k)$) is expressed using backward or forward or central difference approaches.

$$\left.\begin{array}{l} i'(t_k) = \dfrac{i(t_k) - i(t_{k-1})}{h} \text{ using backward approach} \\[3mm] i'(t_k) = \dfrac{i(t_{k+1}) - i(t_k)}{h} \text{ using forward approach} \\[3mm] i'(t_k) = \dfrac{i(t_{k+1}) - i(t_{k-1})}{2h} \text{ using central approach} \end{array}\right\} \tag{1.17}$$

The second method uses integration to eliminate the derivative approximation. It is given by

$$\int_{t_1}^{t_2} v(t)\, dt = R \int_{t_1}^{t_2} i(t)\, dt + L \times \left[i(t_2) - i(t_1)\right] + \int_{t_1}^{t_2} \Delta E(t)\, dt \tag{1.18}$$

In order to evaluate the said equation, which involves digital integration, trapezoidal approach is used. This is given by

$$\int_{t_1}^{t_2} X(t)\, dt = \sum_{K=n-N}^{n-1} \frac{(X_k + X_{k+1})}{2} h \tag{1.19}$$

where,

N = number of samples per cycle, n and $n - N$ are samples corresponding to the times t_2 and t_1, and are given by,

$t_1 = (n - N) \times h$ and $t_2 = n \times h$

where, h is the sampling interval.

Now, neglecting the error term (ΔE) in Eq. (1.18) and using two equations over two successive time periods, we get the following two equations.

$$\int_{t_0}^{t_1} v(t)\, dt = R \int_{t_0}^{t_1} i(t)\, dt + L \times \left[i(t_1) - i(t_0)\right] \tag{1.20}$$

$$\int\limits_{t_1}^{t_2} v(t)\, dt = R \int\limits_{t_1}^{t_2} i(t)\, dt + L \times \left[i(t_2) - i(t_1) \right] \tag{1.21}$$

Using Eq. (1.21), we obtain the following two equations.

$$\frac{h}{2}\left(v_k + v_{k-1}\right) = R\,\frac{h}{2}\left(i_k + i_{k-1}\right) + L\left(i_k + i_{k-1}\right)$$

$$\frac{h}{2}\left(v_k + v_{k-2}\right) = R\,\frac{h}{2}\left(i_{k-1} + i_{k-2}\right) + L\left(i_{k-1} + i_{k-2}\right)$$

Therefore, the line parameters R and L are estimated by

$$R = \frac{\left(i_{k-1} - i_{k-2}\right)\left(v_k + v_{k-1}\right) - \left(i_k - i_{k-1}\right)\left(v_{k-1} + v_{k-2}\right)}{\left(i_k + i_{k-1}\right)\left(i_{k-1} - i_{k-2}\right) - \left(i_{k-1} + i_{k-2}\right)\left(i_k - i_{k-1}\right)} \tag{1.22}$$

$$L = \frac{h}{2} \times \frac{\left(i_k + i_{k-1}\right)\left(v_{k-1} + v_{k-2}\right) - \left(i_{k-1} + i_{k-2}\right)\left(v_k + v_{k-1}\right)}{\left(i_k + i_{k-1}\right)\left(i_{k-1} - i_{k-2}\right) - \left(i_{k-1} + i_{k-2}\right)\left(i_k - i_{k-1}\right)} \tag{1.23}$$

1.13.3 Algorithms Applicable to Distorted Relaying Signals

Owing to switching and faults, the voltage and current signals to the relay get distorted. The algorithms discussed in this section assume that the relaying signals can be modelled by an expression containing the fundamental frequency, high frequency, and DC components. All the algorithms discussed in this section use data from one half/full cycle of the fundamental frequency. Hence, these techniques are also known as *long window techniques*. Most of the modern digital relays use different phasor estimation algorithms depending upon the requirements and applications such as discrete Fourier transform (DFT), least square error, and Walsh function.

Fourier Analysis-based Algorithm

In this algorithm, the acquired quantities (voltage and current) are transformed into the frequency domains, which are then used to obtain the apparent value of impedance from the relaying point to the fault point.

Assumption As this algorithm does not reject DC and even harmonics completely, digital filtering is required to pre-process the signals for removal of DC offset and harmonics before the extraction of fundamental frequency components.

Any periodic function (say voltage or current) $f(t)$ can be represented by the Fourier series as

$$f(t) = \frac{a_0}{2} + \sum_{n=1}^{\infty} a_n \cos n\omega_0 t + \sum_{n=1}^{\infty} b_n \sin n\omega_0 t \tag{1.24}$$

where,

$\omega_0 = 2\pi f_0 = $ angular fundamental frequency

$n\omega_0 = n^{\text{th}}$ harmonic angular frequency

$T = \dfrac{1}{f_0} = $ time interval of fundamental component

If the periodic function $f(t)$ is assumed as a current quantity, then using Eq. (1.24), it is given by

$$i(t) = \frac{a_0}{2} + \sum_{n=1}^{\infty} a_n \cos n\omega_0 t + \sum_{n=1}^{\infty} b_n \sin n\omega_0 t \tag{1.25}$$

The coefficients of the current wave are given by

$$a_n = \frac{2}{T} \int_{t_1}^{t_1+T} i(t) \cos n\omega_0 t \, dt \tag{1.26}$$

where n starts from 0, 1, ….

$$b_n = \frac{2}{T} \int_{t_1}^{t_1+T} i(t) \sin n\omega_0 t \, dt \tag{1.27}$$

where $n = 1, 2, \dots.$

Full-cycle algorithm Extraction of fundamental component of current and voltage quantity during fault is the main theme of this algorithm. This is achieved by correlating one cycle of faulted waveform of current or voltage with stored reference sine and cosine waves.

Let us assume that I_x and I_y are the real and imaginary parts of the fundamental component of the faulted current waveform $i(t)$.

Using Eqs (1.26) and (1.27), I_x and I_y are given by

$$I_x = a_1 = \frac{2}{T} \int_{t_0}^{t_0+T} i(t) \cos \omega_0 t \, dt \tag{1.28}$$

$$I_y = b_1 = \frac{2}{T} \int_{t_0}^{t_0+T} i(t) \sin \omega_0 t \, dt \tag{1.29}$$

where,

t_0 = time under consideration

If M is the number of samples per cycle of fundamental component and h is the sampling time interval, then

$t_k = k \times h$ = the time of the kth sample, and

$T = M \times h$ = the period of fundamental component.

From these expressions, Eqs (1.28) and (1.29) can be evaluated as follows:

$$I_x = \frac{2}{Mh} \left[\begin{array}{l} i(t_0) \cos \omega_0 t_0 + i(t_1) \cos \omega_0 t_1 + \cdots + i(t_k) \cos \omega_0 t_k + \\ \cdots + i(t_{M-1}) \cos \omega_0 t_{M-1} + i(t_M) \cos \omega_0 t_M \end{array} \right] h$$

$$I_x \cong \frac{2}{M} \sum_{k=0}^{M} i_k \cos \left(\frac{2\pi k}{M} \right) = \frac{2}{M} \sum_{k=0}^{M} W_{xk} i_k \tag{1.30}$$

where,

$i_k = i(t_k)$ is the kth sample of current waveform

W_{xk} and W_{yk} are weighting factors of the kth sample.

This expression is given by

$$W_{xk} = \cos \omega_0 t_k = \cos \frac{2\pi}{T} kh = \cos(2\pi k/M) \tag{1.31}$$

where, $k = 0, 1, \ldots, M$

$$W_{yk} = \sin \omega_0 t_k = \sin \frac{2\pi}{T} kh = \sin(2\pi k/M) \text{ , where, } k = 0, 1, \ldots, 1$$

$$I_y = \frac{2}{Mh}\left[\begin{array}{l} i(t_0) \sin \omega_0 t_0 + i(t_1) \sin \omega_0 t_1 + \cdots + i(t_k) \sin \omega_0 t_k \cdot \\ \cdots + i(t_{M-1}) \sin \omega_0 t_{M-1} + i(t_M) \sin \omega_0 t_M \end{array}\right]$$

$$I_y \cong \frac{2}{M} \sum_{k=0}^{M} i_k \sin\left(\frac{2\pi k}{M}\right) = \frac{2}{M} \sum_{k=0}^{M} W_{yk} i_k$$

Figure 1.33 shows the frequency response of a full-cycle algorithm. It is to be noted that if the full-cycle window algorithm uses k samples, then only $\left(\frac{k}{2}-1\right)$ harmonics can be estimated. To reduce higher frequency harmonics, an initializing filter is required.

Half-cycle algorithm This algorithm uses information corresponding to one half cycles in contrast to the one cycle information used by the full-cycle algorithm. If *Ix, hc* and *Iy, hc* are the real and imaginary parts of the fundamental component of the phasor derived from a half-cycle window, then their expressions are given by

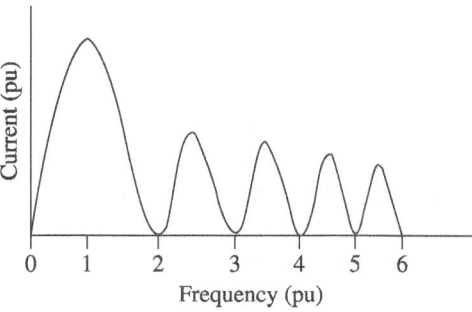

Fig. 1.33   Frequency response of a full-cycle algorithm

$$I_{x,hc} = \frac{2}{(T/2)} \int_{t_0}^{t_0+T/2} i(t) \cos \omega_0 t \, dt \qquad (1.32)$$

$$I_{y,hc} = \frac{2}{(T/2)} \int_{t_0}^{t_0+T/2} i(t) \sin \omega_0 t \, dt \qquad (1.33)$$

Following the same procedure as done in Eq. (1.30) of full-cycle algorithm, we get

$$I_{x,hc} = \frac{4}{M} \sum_{k=1}^{M/2} W_{xk} i_k \qquad (1.34)$$

$$I_{y,hc} = \frac{4}{M} \sum_{k=1}^{M/2} W_{yk} i_k \qquad (1.35)$$

Figure 1.34 shows the frequency response of a half-cycle Fourier algorithm.

The main advantages of this algorithm over the full-cycle algorithm are as follows:

1. It is faster than full-cycle algorithm.
2. It can easily remove odd harmonics.

However, the prime limitation of the half-cycle algorithm is the increase in error due to even harmonics.

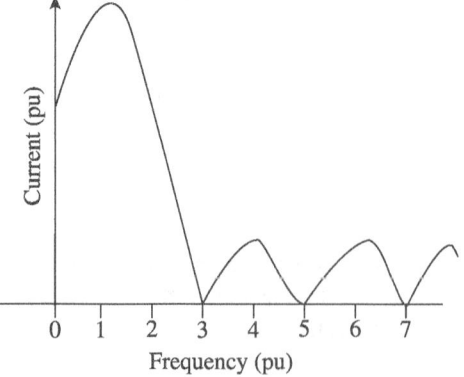

Fig. 1.34   Frequency response of half-cycle algorithm

Walsh Function Technique

The fundamental part of odd and even square waves is added to harmonically related square waves to obtain the *Walsh function*. This is correlated with the fault signal to extract the fundamental frequency components.

Any periodic function, say current between an interval of t_1 to $t_1 + T$, can be expanded using Fourier series.

$$i(t) = F_0 + \sum_{n=1}^{\infty} \left(\sqrt{2} F_{2n-1} \sin n\omega_0 t + \sqrt{2} F_{2n} \cos n\omega_0 t \right) \tag{1.36}$$

where,

$$F_0 = \frac{1}{T} \int_0^T i(t)\, dt, \ F_{2n-1} = \frac{\sqrt{2}}{T} \int_0^T i(t)\sin n\omega_0\, dt, \text{ and}$$

$$F_{2n} = \frac{\sqrt{2}}{T} \int_0^T i(t)\cos n\omega_0 t\, dt \tag{1.37}$$

Here, F_0 is the DC component, whereas F_{2n} and F_{2n-1} are the real and imaginary components of nth harmonic, respectively. Equation (1.36) can be further extended by using Walsh series as

$$i(t) = \sum_{k=0}^{\infty} W_k\, W_{al}\left(k, \frac{t}{T} \right) \tag{1.38}$$

where, W_k is the kth Walsh coefficient and is given by,

$$W_k = \frac{1}{T} \int_0^T i(t)\, W_{al}\left(k, \frac{t}{T} \right) dt, k = 0, 1, 2, \dots \tag{1.39}$$

With M as the number of samples per cycle, Walsh coefficient $W_k(\text{s})$ is given by

$$W_k(\text{s}) = \frac{1}{M} \left[\frac{1}{2} f_\text{s} + f_{1+\text{s}} + \dots + f_{j+\text{s}} + \dots + f_{M-1+\text{s}} + \frac{1}{2} f_{M+\text{s}} \right] \tag{1.40}$$

The fundamental frequencies of sine and cosine components in terms of Walsh coefficients for a sampling rate of $M = 16$ are given by

$$F_\text{s} = 0.9 W_1 - 0.373 W_5 - 0.074 W_9$$
$$F_\text{c} = 0.9 W_2 + 0.373 W_6 - 0.074 W_{10}$$

where, W_s represents the Walsh coefficients.

Least Error Square Technique

Least square fitting technique estimates the impedance from the relaying point to the fault point, which includes fundamental component, decaying DC offset component, and certain harmonics of fault current and voltage wave.

Any signal (either voltage or current) that includes the decaying DC offset component, fundamental component, and harmonics can be given by

$$f(t) = K_1 e^{-t/\tau} + \sum_{m-1}^{M} \left(k_{2m} \cos m\omega_0 t + k_{2m+1} \sin m\omega_0 t \right) \tag{1.41}$$

where,

$K_1, K_2, \dots, K_{2M+1}$ are the unknown constants

M = the highest harmonic considered

τ = time constant of the decaying DC component

ω_0 = angular frequency of the fundamental component

Now, to determine the constants $K_1, K_2, ..., K_{2M+1}$, the integral of error is given by

$$S = \int [v(t) - f(t)]^2 \, dt \tag{1.42}$$

$$S = \int \left[v(t) - K_1 e^{-t/\tau} - \sum_{m-1}^{M} k_{2m}\left(\cos m\omega_0 t + k_{2m+1}\sin m\omega_0 t\right) \right]^2 dt$$

In order to achieve the necessary condition for S to be a minimum, take the partial derivation of S. Therefore,

$$\frac{ds}{dk_i} = 0 \, (i = 1, \, 2 \text{ and } 2M + 1)$$

$$\frac{ds}{dk_i} = 0 = -2 \int_{t_1}^{t_1+T} \left[v(t) - K_1 e^{-t/\tau} - \sum_{m-1}^{M}\left(k_{2m}\cos m\omega_0 t + k_{2m+1}\sin m\omega_0 t\right) \right] \times e^{-t/\tau} \, dt$$

$$\frac{ds}{dk_{2m}} = 0 = -2 \int_{t_1}^{t_1+T} \left[v(t) - K_1 e^{-t/\tau} - \sum_{m-1}^{M}\left(k_{2m}\cos m\omega_0 t + k_{2m+1}\sin m\omega_0 t\right) \right] \cos m\omega_0 t \, dt$$

$$\frac{ds}{dk_{2m+1}} = 0 = -2 \int_{t_1}^{t_1+T} \left[v(t) - K_1 e^{-t/\tau} - \sum_{m-1}^{M}\left(k_{2m}\cos m\omega_0 t + k_{2m+1}\sin m\omega_0 t\right) \right] \sin m\omega_0 t \, dt$$

These three equations can be further simplified and are given as

$$K_1 = \int_{t_1}^{t_1+T} v(t) \, e^{-t/\tau} \, dt$$

$$K_{2m} = \frac{2}{T} \int_{t_1}^{t_1+T} v(t) \cos m\omega_0 t \, dt \tag{1.43}$$

$$K_{2m+1} = \frac{2}{T} \int_{t_1}^{t_1+T} v(t) \sin m\omega_0 t \, dt$$

Using trapezoidal method, Eq. (1.43) can be written as

$$K_1 = W_{11}s_1 + W_{1N}s_N + \sum_{n=2}^{N-1} 2W_{1N}s_N \tag{1.44}$$

where,

W_{1N} = weighting factor of the ith sample

N = number of samples per cycle

$$K_{2m} = \frac{1}{N}\left[W_{2m,1}\,s_1 + W_{2m,N}\,s_N + \sum_{n=2}^{N-1} 2W_{2m,n}s_n \right] \tag{1.45}$$

$$K_{2m+1} = \frac{1}{N}\left[W_{2m+1,1}s_1 + W_{2m+1,N}s_N + \sum_{n=2}^{N-1} 2W_{2m+1,n}\,s_n\right] \qquad (1.46)$$

where, $W_{2m,n}$ and $W_{2m+1,n}$ are the weighting factors of the nth sample.

Finally, the impedance is estimated by

$$Z = \frac{K_{2V} + jK_{3V}}{K_{2I} + jK_{3I}} \qquad (1.47)$$

where, K_{2V}, K_{2I} and K_{3V}, K_{3I} are the real and imaginary parts of the fundamental component of voltage and current.

Recapitulation

- Speed, selectivity, discrimination, and time of operation are the prime requirements of protective devices.
- Unit protection is based on absolute selectivity, whereas non-unit protection is based on relative selectivity.
- In primary protection relays operate in the first line of defence, whereas in backup protection relays work as the second line of defence.
- Relays have progressed from electromechanical, static, and microprocessor, to digital/numerical.
- It is very important to examine the technology used in modern digital/numerical relays and analyse their application in the protection of power systems.
- Sampling frequency in digital relays is decided using Nyquist criterion, $f_s \geq 2 \times f_m$.
- Sampling frequency is given by $f_s = f \times n$.
- The analytical approach of different relay algorithms such as Fourier analysis-based algorithm, Walsh function technique, and least error square technique used in practice have immense importance in designing digital relaying schemes for particular protection functions.

Multiple Choice Questions

1. Unit protection is based on the concept of
   - (a) absolute selectivity
   - (b) relative selectivity
   - (c) both (a) and (b)
   - (d) none of the above

2. Blind spot is a point in zones of protection where
   - (a) partial protection is available
   - (b) complete protection is available
   - (c) no protection is available
   - (d) none of the above

3. The function of anti-aliasing filter is
   - (a) to remove high frequency components
   - (b) to remove both low and high frequency components
   - (c) to allow low frequency components
   - (d) to remove low frequency components

4. Which relay is more susceptible to electromagnetic interference?
   - (a) Digital relay
   - (b) Electromechanical relay
   - (c) Static relay
   - (d) All of the above

5. The operating time of modern digital relay is of the order of
   - (a) 20–30 ms
   - (b) 10–20 ms
   - (c) 400–600 ms
   - (d) 1–10 s

6. Which type of backup protection scheme is widely used in the field?
   - (a) Relay backup
   - (b) Breaker backup
   - (c) Remote backup
   - (d) None of the above

7. As the sampling frequency increases, the computational requirements
   - (a) increase
   - (b) remain constant
   - (c) decrease
   - (d) none of the above

8. Distance relay is the best example of
   - (a) unit protection scheme
   - (b) non-unit protection scheme
   - (c) independent protection scheme
   - (d) none of the above

9. The function of trip isolation circuit is
   - (a) to avoid maloperation of relay during periodic testing of relay
   - (b) to trip the circuit breaker
   - (c) to trip the main relay
   - (d) none of the above

10. The function of auxiliary relay is
    - (a) to carry high fault current
    - (b) to sense the inception of fault
    - (c) to provide backup
    - (d) none of the above

Review Questions

1. Explain how the protection zone of various types of equipment is decided.

2. Enlist the various requirements of protection systems.

3. Explain the concept of unit and non-unit protection.

4. What do you mean by primary and backup protection of power system?

5. What is the function of a bimetallic strip in a thermal relay?

6. Explain how thermal relay is used for the protection of equipment against overloading condition.

7. Why can an overcurrent relay not be used in place of a thermal relay for the protection of equipment against overloading condition?

8. Why is induction cup relay superior to induction disc relay?

9. Discuss the advantages and disadvantages of static relays compared to electro-mechanical relay.

10. Explain the working of a generalized static relay.

11. Discuss the various components of digital relays used in power systems.

12. Explain the function of the following with reference to digital relay.
    - (a) Anti-aliasing filter
    - (b) Analog-to-digital converter
    - (c) Isolation transformer
    - (d) Surge protection circuit
    - (e) Signal condition device
    - (f) Digital output system

13. Discuss various merits and demerits of digital relays, with reference to electro-mechanical and static relays.

14. Explain the concept of adaptive relaying.

15. What are the different types of structures/equipment required to implement the concept of adaptive relaying?

16. Discuss half-cycle and full-cycle discrete Fourier transform algorithms.

Answers to Multiple Choice Questions

1. (a) 2. (c) 3. (d) 4. (a) 5. (a) 6. (c) 7. (a) 8. (b) 9. (a) 10. (a)

Current-based Relaying Scheme

2

2.1 Introduction

Transmission and distribution lines connect various parts of a power system network. The main function of transmission and distribution lines is to transmit and distribute power from power stations to the high-tension (HT) and low-tension (LT) consumers. They are subjected to various types of faults due to harsh environmental conditions. Whenever conductors of transmission line are accidentally shorted together because of wind, ice, and falling of trees, there is a possibility of faults. Sometimes, transmission line faults also occur because of the flashover of insulators caused by contamination of insulators. The impedance of system changes from high value (load condition) to low value during fault resulting in large amount of current to flow in the transmission line. If these faults are not cleared promptly, then they may result in serious hazards due to fire, and also damage the transmission line and the substation equipment.

Overcurrent protection is generally used to isolate the fault section of transmission and distribution line. An overcurrent relay which is connected on secondary side of CTs is used to detect the fault condition. This unit picks up when the magnitude of current goes beyond the threshold value. Overcurrent protection includes overload and short-circuit condition having abnormal behaviour than normal operating condition of equipment. Quick isolation of fault is desirable and is achieved by adjusting the operating time of overcurrent relay. In power systems depending of the geographical situation and line configures, various non-directional and directional protection schemes are used. This chapter deals with the types of overcurrent relays, their operating characteristic, parameter setting guidelines and relay coordination in interconnected power system.

2.2 Overcurrent Protection of Transmission Line

Protection of transmission line using overcurrent relays is one of the cheapest and simplest forms of protection. The overcurrent indicates that the relay operates and closes its contacts when the current exceeds the predetermined threshold, pickup value, or plug setting. This *pickup* or *plug setting* (PS) is defined as the threshold beyond which a relay operates. However, the characteristic of overcurrent relays can be plotted between multiple of pickup current (MP) and time instead of current versus time.

Overcurrent relays can be classified on the basis of the type of characteristic used:

1. Instantaneous overcurrent relay
2. Definite minimum time (DMT) relay
3. Inverse time overcurrent relay
4. Inverse definite minimum time (IDMT) overcurrent relay

2.2.1 Instantaneous Overcurrent Relay

As the name suggests, this relay operates instantaneously, that is, without an intentional time delay. However, in practice, no relay can operate instantaneously, that is, in zero time. Therefore, a relay that operates within 20 to 40 ms is known as *instantaneous relay*. This type of relay has only current setting and no time setting. This relay can be made using different construction philosophies such as attracted armature type, induction disc type, and induction cup type. Figure 2.1 shows the typical characteristic of instantaneous overcurrent relay.

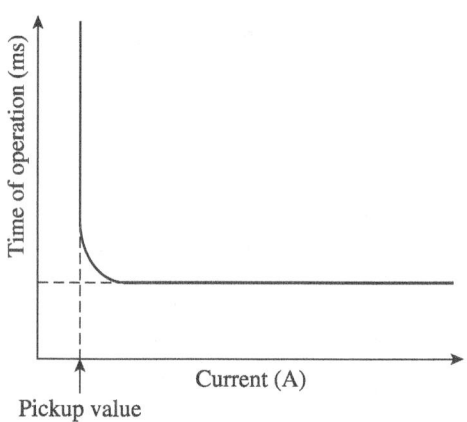

Fig. 2.1 Characteristic of instantaneous overcurrent relay

2.2.2 Definite Minimum Time Relay

Definite minimum time relay is a relay that operates after a definite period of time once the current exceeds the pickup value. Hence, this relay has current setting range as well as time setting range. The current setting range is of the order of 50–200% of I_n, where I_n is the relay-rated current. Time setting ranges are of the order of 0.1–1 s, 1–10 s, or 6–60 s. In an electromechanical relay, the relay characteristic can be achieved by the following equation.

$$T_{op} = \frac{A}{(MP)^B - 1} \times TDS + C \tag{2.1}$$

where,

T_{op} = time of operation of the relay in seconds

MP is the multiple of pickup current and is also known as plug setting multiplier (PSM). It is given by the following equation.

$$MP = \frac{I_{f(CTS)}}{I_{pickup}}$$

where,

$I_{f(CTS)}$ = fault current referred to as current transformer (CT) secondary current,
I_{pickup} = relay pickup current, and TDS is the time dial setting.

A, B, and C are the circuit constants that decide the relay characteristics. For definite minimum time relay $C = 0$, and A and B are very small (close to zero).

For static relays, the characteristic of the relay is given by the following equation.

$$T_{op} = \frac{a}{(MP)^n - C} \times TDS + b \times TDS + K \qquad (2.2)$$

where,

T_{op} = time of operation of a relay in seconds,
MP = the multiple of pickup current,
n = an exponent,
a, b, C, and K (preferably 0.01) are constants, and TDS is the time dial setting.

Figure 2.2 shows the typical characteristic of definite time overcurrent relay.

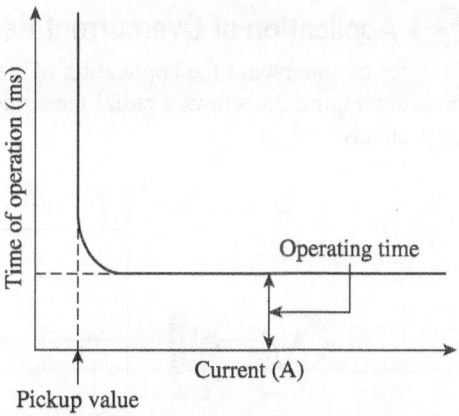

Fig. 2.2 Characteristic of definite minimum time relay

2.2.3 Inverse Time Overcurrent Relay

As the name suggests, the time of operation of this relay is inversely proportional to fault current. This is the most widely used relay as it operates very quickly for a fault near the source. This is very important as the more severe faults are cleared quickly. With the advent of digital relays, it is possible to generate any type of inverse time overcurrent relay characteristics. However, as electromechanical relays are still widely used in substations, one has to use a characteristic that can easily be matched with the available electromechanical relay characteristic. Figure 2.3 shows normal inverse, very inverse, and extremely inverse characteristics of the inverse time overcurrent relay. These characteristics are obtained in electromechanical and static relays using Eqs (2.1) and (2.2), respectively, by selecting different values of constants. Table 2.1 shows the values of constants for different characteristics.

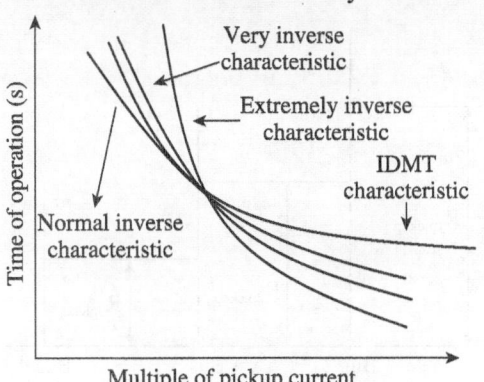

Fig. 2.3 Characteristic of inverse time overcurrent relay

Table 2.1 Values of constants for different characteristics of overcurrent relay

Relay characteristic	Electromechanical relay			Static relay		
	A	**B**	**C**	**a**	**b**	**n**
Normal inverse	0.092	0.02	0.149	5.4	0.18	2.0
Very inverse	18.92	2.0	0.492	5.4	0.11	2.0
Extremely inverse	28.08	2.0	0.13	5.4	0.03	2.0
IDMT	0.14	0.02	0.0	0.14	0.0	0.02

2.2.4 Inverse Definite Minimum Time Overcurrent Relay

This type of relay is widely used by the utilities in the field. Initially, the characteristic of the relay follows inverse law, and thereafter, when the current becomes very high, it follows definite minimum operating time pattern. This is because of the constant operating torque due to the saturation of flux at a high value of current in the electromechanical relay. The characteristic of IDMT relay is shown in Fig. 2.3. The values of constants from which the characteristics of the IDMT relay are obtained are given in Table 2.1.

Different characteristics of overcurrent relays used in the field are given in Appendix C.

2.3 Application of Overcurrent Relay Using Different Relay Characteristics

In order to understand the application of overcurrent relays, consider a radial system fed by utilities from one side. Figure 2.4 shows a radial feeder containing three sections protected by the relays R_1, R_2, and R_3, respectively.

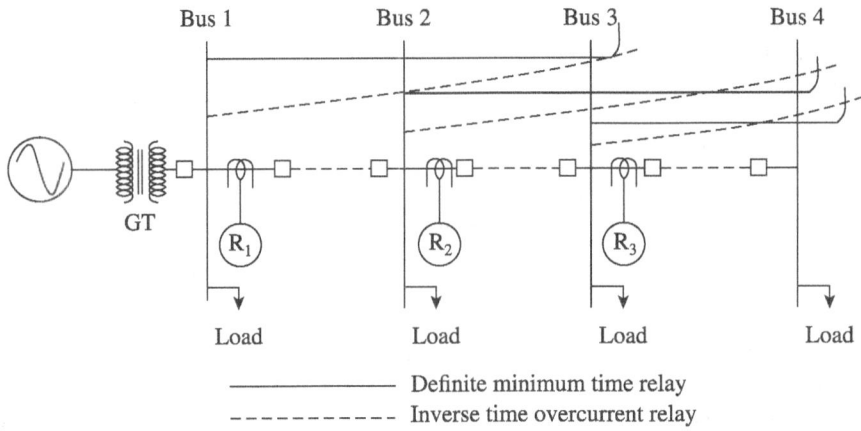

Bus 1   Bus 2   Bus 3   Bus 4

GT

R_1 R_2 R_3

Load   Load   Load   Load

———————— Definite minimum time relay

‑ ‑ ‑ ‑ ‑ ‑ ‑ ‑ ‑ ‑ Inverse time overcurrent relay

Fig. 2.4   Single line diagram of a radial system with relay characteristics

2.3.1 Instantaneous Overcurrent Relay

If the relays R_1, R_2, and R_3 are instantaneous overcurrent relays, then each relay (R_1, R_2, and R_3) is set in such a way that it does not operate beyond the reach of its own section. They are adjusted to operate progressively in decreasing order from source to load. Their current–distance characteristics are shown in Fig. 2.5.

Advantages

1. Settings of these relays are independent of load.
2. They operate instantaneously in all sections.

Disadvantages

1. It is not possible to achieve backup protection using instantaneous relays.
2. Instantaneous overcurrent relays are affected by the ratio of the source impedance to the load impedance (Z_s/Z_l). *Source impedance* is the

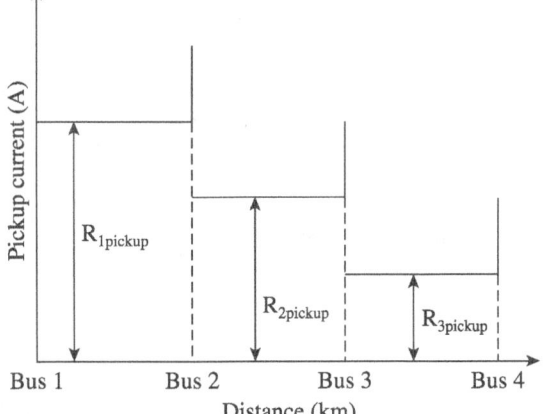

Fig. 2.5   Current–distance characteristics of instantaneous relays

impedance from the source to the relaying point, whereas *load impedance* is the impedance from the relaying point to the fault point. In this condition, the relay R_1 is not able to discriminate between the remote end fault in its own section and the close-in fault in the next section.

3. Instantaneous overcurrent relays suffer from the problem of transient overreach. *Transient overreach* is defined as the tendency of a relay to operate instantaneously for faults beyond its section. This is related to the time constant of the decaying DC component of fault current. Fault current is always asymmetric in nature. Whenever a fault occurs, this asymmetry depends on the instant of fault occurrence. If a fault occurs at a maximum voltage ($V = V_m$), there is no asymmetry, whereas for a fault at zero voltage

($V = 0$), the asymmetry is maximum. The transient overreach is more if the decay of the DC component of fault current is slow. Time delay relays are not affected by the transient overreach phenomena.

2.3.2 Definite Minimum Time Relay

If the relays R_1, R_2, and R_3 are definite minimum time relays, then each relay is set in such a way that it must operate for all faults in its own zone. Further, it also provides backup protection to the adjoining line section. They are adjusted to operate progressively in decreasing order from source to load. Their time–distance characteristics are shown in Fig. 2.6.

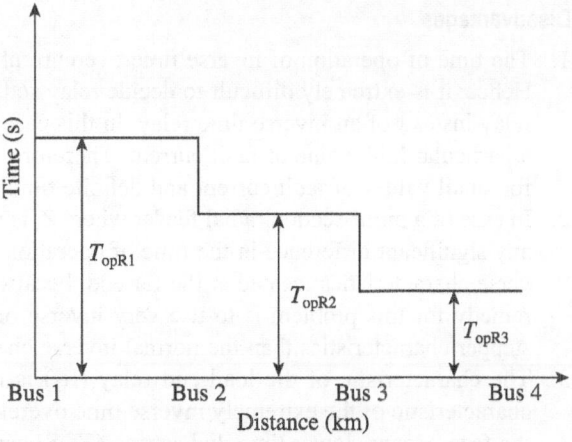

Fig. 2.6 Time–distance characteristics of definite time relays

Advantages

1. It provides backup protection.
2. It is immune to the ratio of source impedance to load impedance (Z_s/Z_l).

Disadvantage It is to be noted from Fig. 2.6 that the time of operation (T_{op}) of relay for a fault near the generator can be dangerously high. This is obviously undesirable since the magnitude of such faults is very high. If these faults persist for a longer period of time, then they produce destructive effects.

The solution to this problem is to use instantaneous high set unit along with definite time delay unit. Another solution is to use inverse time overcurrent relay, which is explained in Section 2.3.3.

2.3.3 Inverse Time Overcurrent Relay

If the relays R_1, R_2, and R_3 are inverse time overcurrent relays, then each relay is set in such a way that it operates for all faults in its own zone, and at the same time, provides backup protection to the adjoining line section. They are adjusted to operate progressively in decreasing order from source to load. Their time–current characteristics are shown in Fig. 2.7.

The time–distance characteristics of definite minimum time relays and inverse time overcurrent relays are shown in Fig. 2.4. It has been observed from Fig. 2.4 that the time of operation of inverse time overcurrent relays is lower than that of definite minimum time relays.

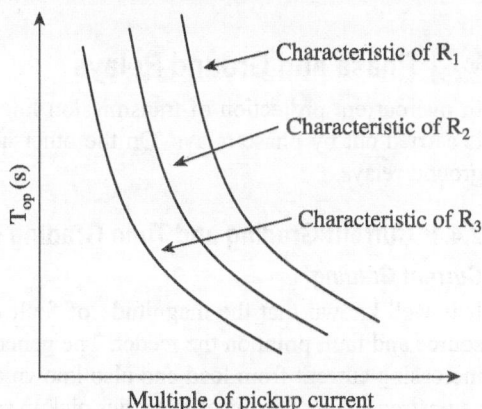

Fig. 2.7 Time–current characteristics of inverse time overcurrent relays

Advantages

1. Inverse time overcurrent relays operate faster for faults near the generator (source).
2. They provide backup protection.
3. They maintain selectivity criteria of the protection system.
4. In order to reduce the time of operation of inverse time overcurrent relays near the generator, an instantaneous high set unit is used along with inverse time unit. The instantaneous unit is set to operate for the faults near the generator, whereas for the faults in the remaining section, the inverse time overcurrent relay operates.

Disadvantages

1. The time of operation of inverse time overcurrent relays is very small for high values of fault current. Hence, it is extremely difficult to decide relay settings. The remedy for this problem is to use an IDMT relay instead of an inverse time relay. In this case, the operating time of the relay remains constant for a particular high value of fault current. Therefore, the IDMT relay follows inverse time–current pattern for small values of fault current and definite time characteristic for large values of fault current.

2. In case of a multi-section radial feeder where Z_s is very large compared to Z_1, it is not possible to achieve any significant difference in the time of operation of inverse time overcurrent relays having normal inverse characteristics located at the far end, because of the insignificant magnitude of fault current. The remedy for this problem is to use very inverse or extremely inverse characteristics, which give more stepper characteristics than the normal inverse characteristics.

3. The characteristic of the load-end relay (R_3) is coordinated with the characteristic of the fuse. The characteristic of the extremely inverse time overcurrent relay closely matches with the characteristic of the fuse. Hence, lower time dial setting (TDS) can be selected for load-end relay, and similarly, lower TDSs are selected for relays located towards the source.

4. The settings of inverse time overcurrent relays are highly affected by the change in source impedance. This is because of the change in loading conditions. There is a possibility of reduction of magnitude of the fault current below the full load current of the feeder. Therefore, it is not possible to set the overcurrent relay because if it is set to operate at minimum fault current, it would not allow the feeder to draw full load current during an increase in load; and if it is set considering the full load current, then it would not operate in case of fault current that is lower than the full load current. Though the magnitude of fault current is lower than the full load current, it is harmful to the system and equipment as it produces voltage dip and negative- and zero-sequence currents during unsymmetrical faults. The remedy for this problem is to monitor the inverse time overcurrent relay using an under-voltage relay.

2.4 Phase and Ground Relays

In overcurrent protection of transmission lines, protection against phase faults that do not involve ground is carried out by phase relays. On the other hand, protection against ground faults is provided by separate ground relays.

2.4.1 Current Grading and Time Grading of Overcurrent Relay

Current Grading

It is well known that the magnitude of fault current is inversely proportional to the distance between the source and fault point on the feeder. The general practice is to set the relay such that it operates at gradually increasing current from load end also known as *downstream end* of the network to source end, also known as *upstream end*. The lower to higher pickup setting of the relay from load end towards source end is known as current grading. With reference to Fig. 2.4, relay R_3 located at bus-3 would operate for any fault in the third section. The pickup setting of relay R_2 is graded at a higher current level in coordination with the pickup setting of relay R_3. Similarly, relays R_1 and R_2 would pickup for any fault in its respective section (zone) but at a progressively higher current. Moreover, if there is sufficient margin in the current ratio of the CTs located on bus-3 and bus-2 (Fig. 2.4), equal pickup setting of relays R_3 and R_2 may be selected.

However, relay settings based on current grading fail in case of a close-in fault in section-3 which has equal current magnitude for a fault on the other side of bus-3 (Fig. 2.4). Thus, relay R_2 may mal-operate considering a close-in fault of section-3 as a remote end fault in its own section. Moreover, correct relay operation is not guaranteed if only current grading based relay setting is done. Thus, current grading based relay setting alone cannot be used for feeder protection. Hence, it is necessary to extend the relay setting by incorporating the time grading.

Time Grading

As stated previously in Chapter 1 (Section 1.3) the selectivity criteria of protection scheme should not be lost in any situation. The time of operation of protective relays is gradually increased starting from load end in the direction of the source. The time margin permitted between two successive relays to satisfy selectivity criteria is called the time grading. If adequate time margin is not provided, the precise location of fault is not determined and this leads to redundant supply failure in the healthy section of the feeder. Hence, in order to prevent simultaneous relay operations during common fault on radial feeder, sufficient interval of operating time is provided to consecutive relays.

Figure 2.6 illustrates the approach of definite time graded protection of radial feeder (Fig. 2.4). In order to achieve discrimination, the time margin between relays R_3 and R_2 is set such that relay R_3 operates for any fault in section-3 and R_2 provides backup protection. The least time interval which is decided during coordination of two successive relays (i.e., R_3 and R_2) is known as *minimum coordination time* (MCT).

The following factors are considered while deciding the MCT:

1. Operating time of the circuit breaker (t_{cb})
2. Error in relay operating time (t_e)
3. Overshoot or over-travel time of the relay (t_o)
4. CT ratio error (t_{ct})
5. Safety margin (t_{sm})

In time grading, the breaker operating time is included as the complete interruption of circuit is ensured before selective relay–relay coordination. However, the time margin for breaker operation (t_{cb}) depends on the magnitude of current and the type of circuit breaker. It normally varies in the range of 2.5 cycles to 5 cycles.

Operating time of the relay cannot be ideal as per the characteristic adopted. Hence, while grading relay, error in the operating time of the relay (t_e) is taken into consideration as per the error index cited by manufacturer.

The overshoot time (t_o) of the relay is integrated in MCT while grading electromagnetic relays. This is due to the momentum of induction disc that continues to travel after the relay is de-energized abruptly by reducing the operating current below the threshold value. Sometimes, this overshoot or over-travel of induction disc causes unnecessary operation of the relay during transient condition. This is not applicable to digital/numerical relays.

Generally, protection class CTs are used for relaying purpose, but they are not ideal during normal operation. The magnetizing current required to excite CT core leads to ratio and phase errors in secondary signals. Thus, CT error (t_{ct}) has a direct effect on relay operating time while coordinating it with other relays.

For any possible error after adding all the above factors, safety margin (t_{sm}) is incorporated to ensure that an acceptable contact gap is maintained and coordinated relay does not operate.

While coordinating electromagnetic induction disc type relays, as per ANSI/IEEE Std-242: 1986, the minimum time interval for 5-cycle breaker usually varies from 0.3 to 0.4 seconds. This includes:

1. 4-cycle breaker: 0.08 seconds
2. Relay error: 0.1
3. Relay over-travel: 0.10 seconds
4. CT ratio and safety margin: 0.12 seconds

On the other hand, digital relays get rid of the overshoot. Thus by removing overshoot time, 0.2 to 0.25 second time interval can be considered for coordination of digital relays.

However, when coordinating more number of relays in a long feeder, the operating time of relay progressively increases towards the utility supply. As a result, the relay at source end has longer time delay compared to the relay located downstream. Thus, the severe fault occurring next to the source is isolated after a comprehensive interval of time compared to the light fault at the end of the section of the feeder.

This is the main limitation of distinct time graded protective scheme and to overcome it, a combined time and current graded protection is brought into action.

Combined Time and Current Grading Protection

With reference to the drawbacks of only current graded and time alone graded protection, a combined time and current grading system can be employed for the protection of feeder. This is accomplished with the use of inverse time overcurrent relays. Different operating times can be achieved for dissimilar values of fault current using inverse time characteristic. Thus, the required time of operation can be obtained in any direction by grading the relay for particular fault in terms of magnitude and location. It means, if the relay characteristic has both 'time' and 'current' settings, the time of operation (T_{op}) of relay is inversely proportional to the magnitude of the fault current. The relay characteristics described, in Section 2.3.3 are the best example of combined time and current grading protection. In IDMT characteristic, current grading is possible in a wide range and relay can be set to operate with the required definite minimum time. In some applications, very inverse and extremely inverse characteristics are used by considering both time and current grading. The grading is done by means of adjusting either the *time dial setting* (TDS) or the current setting known as *pickup setting or plug setting*. These two settings are available on any inverse time relay and are used for fuse–relay or relay–relay coordination.

2.4.2 Setting Rules for Phase and Ground Relays

The following are the points required to be considered during the coordination of consecutive phase relays.

Plug Setting (PS)

1. The phase relay shall reach at least up to the end of the next substation for double-line fault with maximum source impedance (minimum generation). Similarly, the ground relay shall reach at least up to the end of the next substation for single line-to-ground fault with minimum generation. For example, as shown in Fig. 2.4, the relay R_1 must reach up to bus 3.
2. The value of PS is always greater than the maximum full load current of the line. However, this rule is not applicable to two successive ground relays where star–delta transformer is situated.
3. While deciding the PS of any relay with reference to another relay, the relay pickup varies from 105% to 130% of the PS of the relay.
4. Plug settings of ground relays are lower than that of phase relays. This is because of the fact that the magnitude of earth-fault current is reduced owing to the tower footing resistance, fault resistance, ground resistance, and zero-sequence impedance of the system. Further, ground relays are usually connected in the residual circuit of three line CTs. Hence, while deciding the PS of ground relays, we have to consider the excitation current of CT.

Time Dial Setting (TDS)

1. The TDS of a relay is selected in such a way that the downstream relay (relay located near load) achieves the lowest possible time of operation for overload or fault near the load. As we move towards the upstream relay (relay located near source), the value of TDS progressively increases. The TDS is chosen such that it gives the desired selective interval from the downstream relay at maximum fault conditions (for phase relays, the triple-line fault just beyond the next relay is considered, whereas for ground relays, single-line-to-ground fault beyond the next relay is considered).
2. Fault current calculations are usually carried out by considering the impedances of all associated equipment in per unit.
3. While deciding the TDS of an upstream relay with reference to the downstream relay, proper minimum coordination time (MCT) interval must be considered. This MCT contains errors in relay, operating time of breaker, and safety margin. Considering the fast-acting breakers having two-cycle operating time, a fixed selective interval of 0.2 s between successive relays is used by the utilities.

4. While deciding the TDS of ground relays, we have to consider the excitation current of the CT. If a delta–star transformer is involved between two successive ground relays, there is no need to coordinate the primary side relay with the relay located on the secondary side of the star–delta transformer. They can be set independently.

2.4.3 Scheme Used in Practice

Figure 2.8 shows the power and control circuit of three overcurrent and one earth-fault scheme used in the field. With reference to Fig. 2.4, R_1 relay means it is a group of three overcurrent and one earth-fault relays. Similarly, R_2 and R_3 also have a group of relays. All the phase relays carry equivalent value of full load current of the feeder as transferred by the CT. Conversely, the ground relay does not carry any current during normal condition as well as in case of double-line and triple-line faults. However, due to non-identical CT saturation characteristics of the three-line CTs, the ground relay may carry spill current. In case of faults involving ground (L–G/L–L–G/L–L–L–G), the ground relay carries current depending upon the magnitude of fault current.

Now, in case of double-line fault (say, R–Y), relays 51A and 51B operate and give commands to 51A-1 and 51B-1, which further energize the trip coil of the auxiliary relay (86). The contact of the auxiliary relay (86-1) energizes the trip coil of the CB and hence, the contacts of the CB open. Similarly, in case of triple-line fault (R–Y–B), all three-phase relays operate, and the contact of the CB opens. On the other hand, during a single-line-to-ground fault (say R–G), the ground relay operates because of the increase in the residual current with reference to its PS and hence, the contacts of the CB open. In case of double-line-to-ground fault (say, R–Y–G), both phase relays and ground relay operate and open the contacts of the CB. This scheme is widely used in the distribution system, particularly when a star–delta transformer is involved between two successive relays. If a transformer is not present between the two successive relaying points, then two overcurrent schemes and one earth-fault scheme can be used. Here, one of the phase relays in Fig. 2.8 (say, 51B) can be removed. In this situation, in case of any type of fault, either of the two relays (51A/51C) operates and opens the contact of the CB. Therefore, the cost of one relay can be saved.

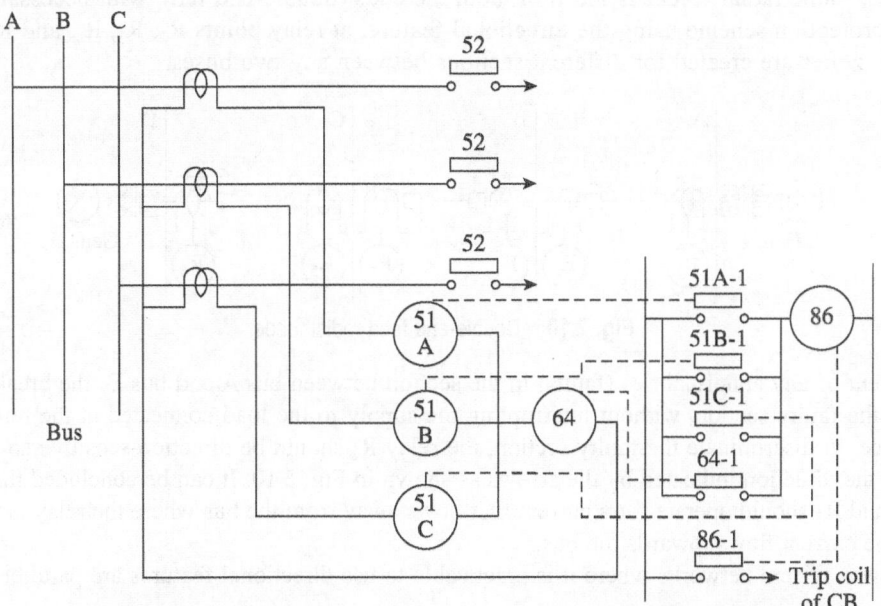

Fig. 2.8 Three overcurrent schemes and one earth-fault scheme of protection

2.5 Directional Protection

In plain radial feeders, the non-directional relays are used as they operate when the CT secondary current exceeds the threshold value of pickup setting in relays. This type of relay operates irrespective of the direction of current flow.

The feeders other than plain radial feeders are not protected by the non-directional overcurrent relays as they require the creation of zones. The protection of such parallel feeders or double-end-fed feeders is achieved by the directional feature.

2.5.1 Necessity

The non-directional relays used for plain radial feeders offer tuning of time and pickup setting, which is not sufficient to discriminate the faulted zone in parallel feeders. To obtain fault zone discrimination in case of the protection of parallel feeders and ring main systems, the directional feature is incorporated. By introducing the directional feature in relays, uninterrupted supply can be made possible at all load points connected in the parallel/ring system.

In the plain radial feeder shown in Fig. 2.9, if the breaker 1 trips because of any abnormalities in the section between bus A and bus B, it will interrupt the power supply at the buses B, C, and D. Thus, because of the tripping of the first breaker, the load connected to the other buses will not receive power supply.

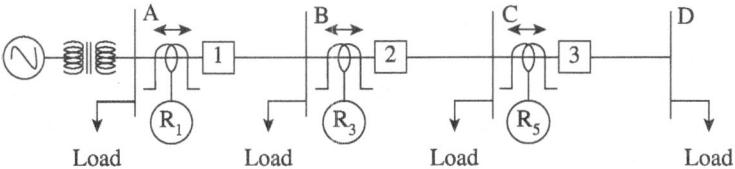

Fig. 2.9   Single-line diagram of a radial system

In case the same radial feeder is fed from both the ends (double-end fed), with necessary modification in the protection scheme using the directional feature, at relay points R_2, R_3, R_4, and R_5, as shown in Fig. 2.10, zones are created for different sections between any two buses.

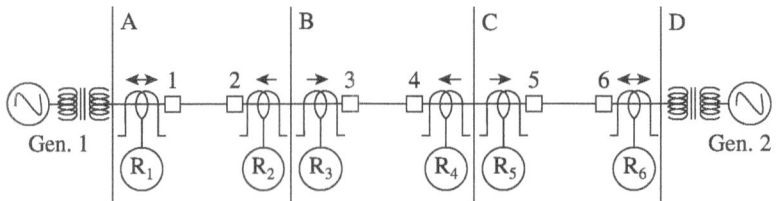

Fig. 2.10   Double-end feed radial feeder

In the event of any abnormalities (faults) in the section between bus A and bus B, the breakers 1 and 2 will isolate the faulty section, without interrupting the supply to the load connected at the buses A, B, C, and D. Hence, to discriminate the faulty section, the relay R_2 should be direction sensitive so that it operates only in the direction indicated by the arrows as shown in Fig. 2.10. It can be concluded that the relays R_2, R_3, R_4, and R_5 should operate for a current that flows away from the bus where the relay is located, and restrain if the current flows towards the bus.

Other power system networks where it is practicable to use directional features are parallel feeders and ring mains.

Figure 2.11 shows the single-end feed supply to parallel feeders (line 1 and line 2).

In case a fault occurs on line 1 at point F, the fault is fed from both the buses (A and B) because line 2 is in healthy condition. If the directional feature is provided to the relay R_3 (and R_4), only the relays R_1 and R_3 trip the respective breakers of line 1 for a fault at F. If the directional feature is not provided to the relay R_3, both line 1 and line 2 trip for a fault at F on line 1 even when line 2 is in healthy condition. Thus, to ensure proper fault discrimination, directional overcurrent relays are used at the load end of the parallel feeders. The relays on bus A, R_1 and R_2, are non-directional overcurrent relays. The relay R_2 is graded with the relay R_3 in such way that R_2 provides backup to R_3, if the relay R_3 fails to clear the fault on line 1. Similarly, R_1 is to be graded with the relay R_4.

In ring main feeders, it is desired to maintain supply to all the load buses irrespective of whether any line section is under outage. Referring to Fig. 2.12, if a fault occurs on the line section between bus A and bus B, only the breakers of that line section should trip. Thus, it maintains the supply to all the buses from an alternative path. This is achieved by applying directional features to the relay R_3. Similarly, the fault in any section causes the CBs of that section to be tripped out.

It is to be noted that the directional feature is essential where fault current

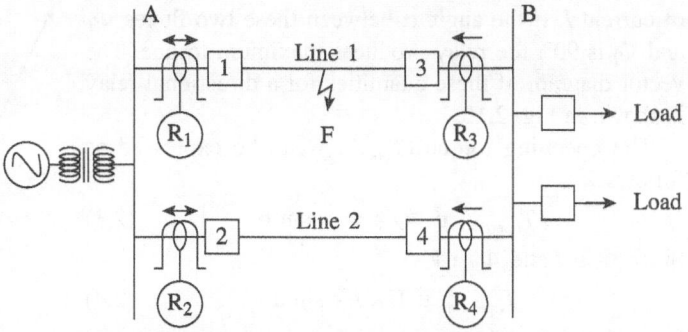

Fig. 2.11   Single-end feed parallel feeder

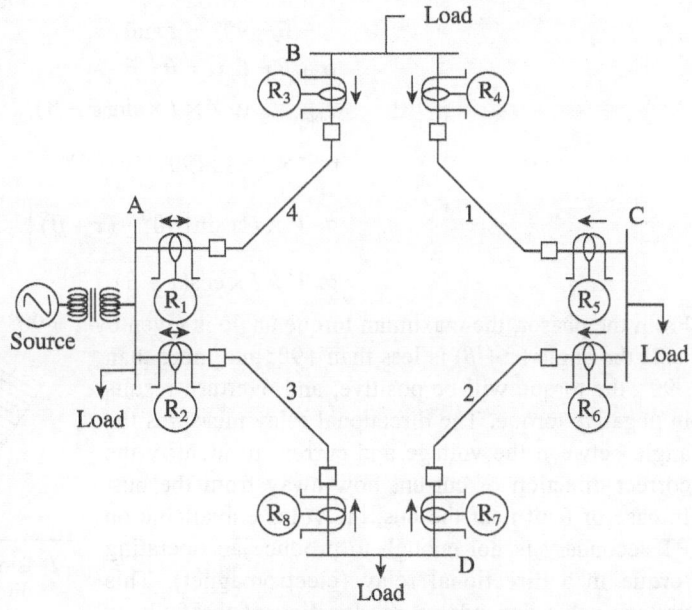

Fig. 2.12   Ring main system

can flow in both directions from the bus. The directional relays have their tripping direction away from the bus where the relays are located.

2.5.2 Directional Relay Characteristics

The *directional relay* takes two input quantities, namely line current and bus voltage. The relay compares the direction of the current flow with reference to the bus voltage by measuring the phase angle between line current and bus voltage. The directional relay operates on watt metric principle, where the voltage coils (VCs) receive voltage from the bus potential transformer (PT) and the current coils (CC) receive the current from the line CT secondary. A maximum positive torque is produced when the current and voltage supplied to the CCs and VCs are in phase. Hence, the angle between current and voltage at which the relay develops maximum torque is defined as the *maximum torque angle* (MTA).

In a directional relay, if V is the voltage given to the VC of the directional relay, then a current I_V lags the voltage V by very large angle θ (because of inductive nature of VC). The flux produced by this current I_V is Φ_V. I is the current given to the current coil of the directional relay, which sets up a flux Φ_I because of current I. If the angle α between these two fluxes Φ_V and Φ_I is 90°, the relay produces maximum torque. The vector diagram of these quantities for a directional relay is shown in Fig. 2.13.

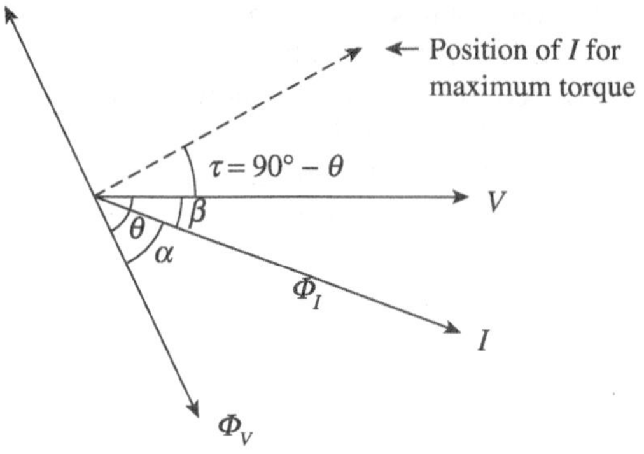

The operating torque ($T_{operating}$) can be expressed as follows:

$$T_{operating} \,\mu\, \Phi_V \times \Phi_I \times \sin\alpha \qquad (2.3)$$

Since $\Phi_I \,\mu\, I$ and $\Phi_V \,\mu\, V$

$$T_{operating} \,\mu\, V \times I \times \sin a \qquad (2.4)$$

From the vector diagram, it can be seen that

$$\theta + \tau = 90°$$
$$\theta = 90° - \tau \text{ and}$$
$$\alpha + \beta = \theta, \alpha = \theta - \beta$$
$$T_{operating} \,\mu\, V \times I \times \sin(\theta - \beta)$$

$$\propto V \times I \times \sin(90° - \tau - \beta)$$

$$\propto V \times I \times \sin\left[90° - (\tau + \beta)\right]$$

$$\propto V \times I \times \cos(\tau + \beta) \qquad (2.5)$$

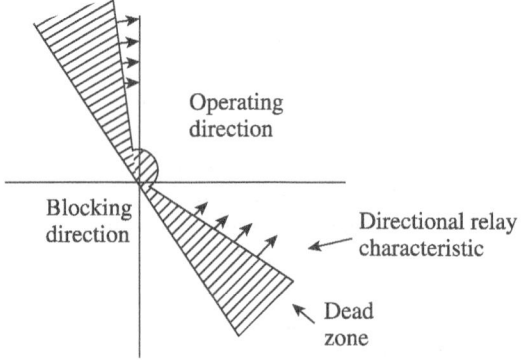

Fig. 2.13   Vector diagram for directional relay

From the phasor, the maximum torque angle is given by $\tau = 90° - \theta$.

If the angle $(\tau + \beta)$ is less than +90° and more than −90°, the torque will be positive, and overturn results in negative torque. The directional relay measures the angle between the voltage and current to identify the correct direction of current flow away from the bus. In case of fault near the bus, the voltage available on PT secondary is not enough to produce an operating torque in a directional relay (electromagnet). This voltage value depends on the location of the fault on the line from the relaying point. The minimum fault distance from the relay point for which the relay fails to operate is known as *dead zone*. Figure 2.14 shows the characteristic of directional relay with dead zone.

Fig. 2.14   Directional relay characteristic

2.5.3 Polarizing Quantity

A *directional relay* is a two-quantity relay, which compares the phase angle of the input voltage and current quantities. The directional overcurrent relay operates only when the magnitudes of current become higher than the set value of the threshold, and the current flows in its correct operating direction (forward direction). The torque produced in the directional overcurrent relay is maximum when $\tau = -\beta$. During fault, the power factor angle is large, that is, of the order of 80° to 90°, depending on the location of fault. Hence, the maximum torque angle $\tau = 90° - \theta$ should be of the same order to achieve maximum torque in the relay during fault.

The maximum torque angle can be set to 30°, 60°, and 90° by suitable connection of CTs and PTs in the relaying circuit. Connections of 30° offer negative torque and maloperation of the directional relaying scheme for certain types of faults. Connections of 60° produce low torque for certain types of faults. Hence, 30° and 60° connections are not widely used for directional relaying scheme. In 90° connections, the polarizing voltage is fed to phase element in such way that it produces maximum torque.

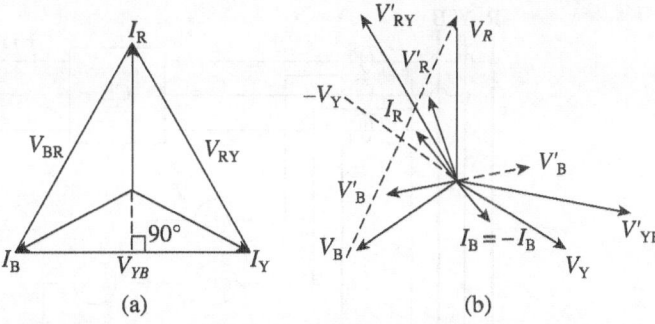

In case of R–B (L–L) fault, the voltage across the R element is V_{YB} and across the B element is V_{RY}. Thus, the required maximum torque is produced by providing the polarizing voltage of healthy phase to the voltage coil of active (faulted) phase.

Fig. 2.15   R–B fault (a) 90° connection (b) Vector diagram

Figures 2.15(a) and 2.15(b) show the 90° connection and the vector diagram for R–B fault, respectively.

In case of unity power factor, the position of faulted phase current (I_R) leads the polarizing voltage (V_{YB}) by 90°. In the event of high resistance fault, the directional overcurrent relay with 90° connections produces less torque. Hence, to achieve maximum torque, the maximum torque angle can be adjusted to any desired value by inserting a resistance or capacitance in series with the voltage coil of the directional relay.

Table 2.2 shows the various combinations of voltages and current fed to the directional relays for 30°, 60°, and 90° connections.

Table 2.2   Quantities fed to phase element of directional relay

Types of connections	Fault involving phase R		Fault involving phase Y		Fault involving phase B	
	Current	Voltage	Current	Voltage	Current	Voltage
30°	I_R	V_{RB}	I_Y	V_{YR}	I_B	V_{BY}
60°	$I_R - I_Y$	V_{RB}	$I_Y - I_B$	V_{YR}	$I_B - I_R$	V_{BY}
90°	I_R	V_{YB}	I_Y	V_{BR}	I_B	V_{RY}

2.5.4 Directional Ground-fault Relays

A residual current and a residual voltage are fed to the directional ground-fault relay. The value of residual voltage ($V_R + V_Y + V_B$) is zero for normal operating condition as well as during phase faults. During ground fault, the residual voltage of open delta PT secondary operates the directional relay. Figure 2.16 shows the vector diagram for L–G (R–G) ground fault.

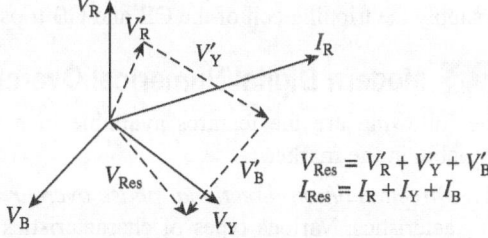

$V_{Res} = V'_R + V'_Y + V'_B$
$I_{Res} = I_R + I_Y + I_B$

Fig. 2.16   Vector diagram for L–G (R–G) ground fault

2.5.5 Directional Overcurrent Protection Scheme for Transmission Line

Figure 2.17 shows the complete protective scheme for directional overcurrent protection for a high voltage line (67). Figure 2.17(a) shows a power circuit where the current coil of each phase element (67R, 67Y, and 67B) of the directional relay is connected across the relevant line CT secondary. The potential coil of any particular phase element is connected across the bus PT secondary of remaining two phases, that is, the 67R element is connected across YB phase of bus PT secondary. The current coil of ground-fault detection element (67N) is connected residually, and the voltage coil is connected across the open delta of intermediate voltage transformer (IVT).

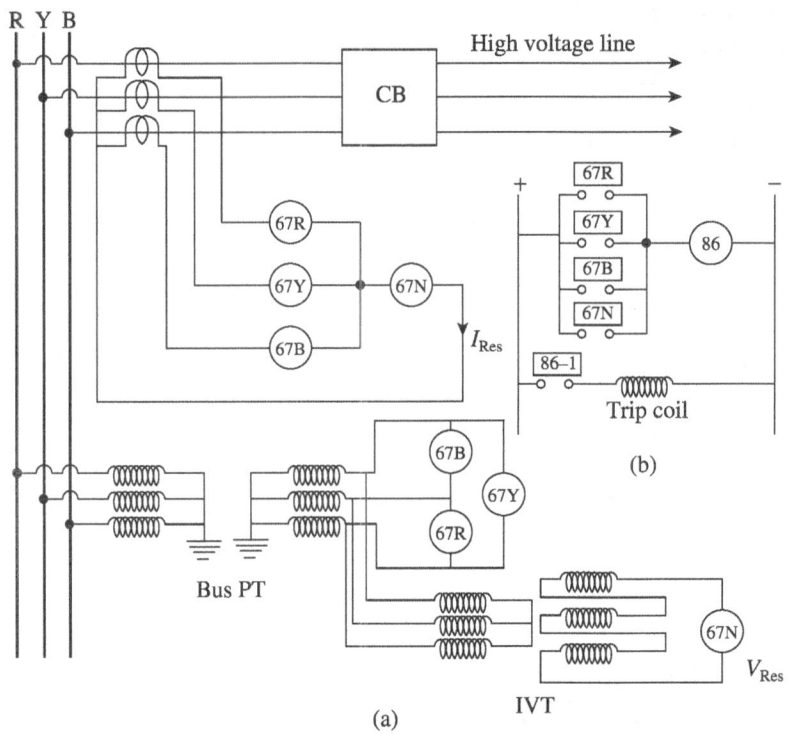

Fig. 2.17 Complete directional overcurrent protective scheme for a line
(a) Protective scheme of directional O/C relay (b) Tripping circuit

Figure 2.17(b) shows the DC tripping circuit where the relay contacts 67R, 67Y, 67B, and 67N are connected parallel and in series with the auxiliary relay coil (86). On operation of directional overcurrent element, this contact closes and the auxiliary relay element is energized. Auxiliary relay (86) closes one of its contacts, 86-1, to supply the tripping coil of the CB and CB trips as required.

2.6 Modern Digital/Numerical Overcurrent and Earth-fault Relay

The following are the features available in modern digital/numerical overcurrent and earth-fault relays available in the market.

1. *Directional/non-directional phase overcurrent and earth fault*: This includes two or three stage characteristics. Various types of characteristics are definite time, IDMT, IEC standard inverse, IEC very inverse, IEC extremely inverse, IEEE medium inverse, IEEE very inverse, and rarely inverse.

2. *Thermal overload protection*: Thermal overload protection can be used to prevent the plant from operating at a temperature in excess of its designed maximum withstand capacity. Prolonged overloading causes excessive heating, which may result in premature deterioration of the insulation or insulation failure. Numerical relays incorporate current-based thermal replica using load current, to model heating and cooling of the protected plant. The element can be set with both alarm and trip stages.

3. *Undercurrent protection*: The undercurrent function makes it possible to detect a loss of load using definite time undercurrent protection. The function is also used for circuit breaker failure detection.

4. *Negative phase sequence overcurrent*: Any unbalanced fault condition will produce negative sequence current of some magnitude. Thus, a negative phase sequence overcurrent element can operate for both phase-to-phase and phase-to-earth faults.

5. *Undervoltage/Overvoltage*: Undervoltage condition may arise due to increased system loading, fault occurring on power system resulting in a reduction in voltage, and complete loss of busbar voltage due to a fault on the busbar itself. On the other hand, as a result of load shedding, the supply voltage will increase in magnitude. Numerical relays are facilitated with detection of such voltage variation in the system, based on which they take a decision to control the system voltage. In the worst condition of undervoltage or overvoltage, the numerical relay operates and protects the system insulation and the system itself.

6. *Residual overvoltage*: This feature gives protection against neutral voltage displacement. It offers earth-fault protection to a system in which the residual voltage is produced. It measures this situation at the secondary terminals of a voltage transformer having a broken delta secondary connection.

7. *Broken conductor detection*: On a lightly loaded line, negative sequence current protection fails to detect the series fault because in this situation, the negative sequence current may be very close to or less than the full load steady state unbalance arising from CT errors, load unbalance, etc. This situation is solved using the element that measures the ratio of negative to positive sequence current. Most of the modern relays contain this feature.

8. *Auto-reclose*: This function allows instantaneous protection for time graded circuit. It gives a high-speed first trip. In other words, it reduces the chances of damage to the line, which might otherwise cause a transient fault to develop into a permanent fault.

9. *Circuit breaker failure*: This feature monitors whether the CB has opened with a reasonable time or not. This detection is extremely important as slow fault clearance can endanger system stability.

10. *Trip circuit supervision*: The trip circuit needs supervision as it includes many components such as fuse, links, relay contacts, and auxiliary switch contact. In modern digital relays, this is achieved with the help of binary input.

11. *Wattmetric protection*: This feature is used for zero-sequence power measurement.

12. *Cold load pickup*: This feature is required to cater to temporary overload conditions that may occur during cold starts. This situation may often arise at the time of switching on large heating loads after a sufficient cooling period, or loads that draw high initial starting currents.

2.7 Overcurrent Relay Coordination in Interconnected Power System

When two or more protective apparatus installed in series have characteristics that provide a specified operating sequence, they are said to be *coordinated* or *selective*. Here, the device set to operate first to isolate the fault (or interrupt the fault current), provides primary protection. The operating device that is set to operate only when the primary protection fails to operate to clear the fault furnishes backup protection.

Certain time intervals must be maintained between the operating times of various protective devices to ensure correct sequential operation of the devices. These are guided by the principle that primary protection should get an adequate chance to protect the zone under its primary protection. Only if the primary protection does not clear the fault, the backup protection should initiate tripping. Thus, as soon as the fault occurs, it is sensed by the primary protection and the backup protection. Naturally, primary protection is the first to operate. The time interval essential for maintaining selectivity between primary and backup protections is known as *coordination time interval* (CTI) or *coordination margin*. This is sometimes referred to as *selective time interval* (STI) also.

2.7.1 Introduction

The coordination of protective relays is an important aspect of the protection system design. It has always been very difficult to coordinate the protective relays on the transmission lines of power networks. A relay operates as a primary relay (PRI) for a fault in its primary protection zone and the same relay operates as a backup to some other PRIs in the system. Each relay must be set so that it not only identifies a fault in its zone of protection, known as the *primary protection zone*, and acts very quickly, but also operates discriminately in a proper time sequence with the relays on the neighbouring lines. By discrimination, it means the ability of a relay to recognize a fault outside its intended zone of protection, and act sufficiently slower than the other relays, which are meant to provide primary protection for this fault.

Relaying schemes and setting procedures vary from utility to utility as they reflect different company philosophies and practices. Changing a single relay setting or any structural/operational change taking place in the system is likely to affect the system relaying.

Many experts have attempted to define relay coordination. In general, *relay coordination* is defined as the problem of coordinating protective relays in an electric power system and consists of selecting their fundamental protective function under the requirements of sensitivity, selectivity, reliability, and speed.

Historically, it has always been a difficult task to set overcurrent relays in large systems. This subject has always attracted the interest of the investigators.

The coordination of overcurrent relays involves the determination of the following:

1. Time dial setting
2. Pickup current (I_{pickup}) or plug setting

The TDS adjusts the time delay before the relay operates whenever the fault current reaches a value equal to or greater than the relay current setting. Pickup current is the minimum current for which the relay operates. Pickup current is determined by selecting one of the PS taps available on the relay. The most vital task when installing overcurrent relays on the system is selecting their settings such that all backup/PRI pairs operate as planned and at the same time fulfil the major requirements of the protection. Relays in different locations detect greatly different currents during the same fault; this makes the relay-setting calculations difficult. The main problem that arises with the protection of overcurrent relays is the difficulty in performing the relay coordination, especially in the multiloop, multisource networks.

In the initial phase of the development of proper coordination procedures, the time-consuming and tiring relay-setting calculations used to be done manually. This scenario changed in the early 60s, when computers were introduced in this field. Relay-setting calculations and the way that the analysis, design, and control of power systems were done were revolutionized. Usually, finding the coordinated settings takes several iterations of settings before a satisfactory solution is achieved. Traditionally, a trial and error procedure is employed for setting relays in multiloop networks. In the early days of the coordination of overcurrent relays, elaborate and complex topological analysis programs had been prepared to determine the breakpoints set, relative sequence matrix, set of sequential pairs, and facility for data management. Proper mathematical modelling of overcurrent relays was a necessity because if the relay characteristics were not modelled properly, errors would appear with the setting of relays obtained from the coordination program. With the introduction of computers in the early 60s, mathematical implementation of relay characteristic curves was easy. It took a while for interactive graphical and analytical approaches to develop. Optimization techniques were first applied in this area in 1988. One of the main advantages of optimization techniques is that there is no need to determine the breakpoints set and hence no need to employ elaborate and complex topological analysis programs. The other equally important advantage is that the settings obtained are optimal. Though optimization techniques have gained popularity, yet there are experts who have been working with traditional trial and error techniques, and much work is still performed using trial and error, more or less based on hand calculations.

2.7.2 LINKNET Structure

The first task of the relay coordination process is to store the network information optimally in the computer memory. LINKNET structure is used to store the topology of the power system network. The topological operations of power networks are required in coordination studies to identify the relay pairs corresponding to each possible location of faults and for storing as well as retrieving the various data, whenever required in the relay coordination algorithm. In this structure, the bus numbers are assigned manually, but the computer assigns relay numbers (branch-end numbers) sequentially. In other words, in coordination studies, the relays are assigned numbers by LINKNET, commensurate with the bus numbers, which are assigned by the user. A procedure to constitute the LINKNET structure vectors for any power network is described in the flowchart shown in Fig. 2.18. The topological properties are represented by specifying the connections between the nodes and the branches, assuming that the ends of each branch are numbered as follows:

end A = f(branch) = (2 × branch) − 1
end B = g (branch) = (2 × branch)

Conversely, a branch number may be derived from either of its end numbers using,

branch = h (end) = (end + 1)/2

In this relationship, the integer round off is used to obtain the two to one mapping between ends and branches. The topology of the network can be defined by constructing a linked-list of the branch ends that are connected to each node. Three one-dimensional vectors that are used in the LINKNET structure are described here.

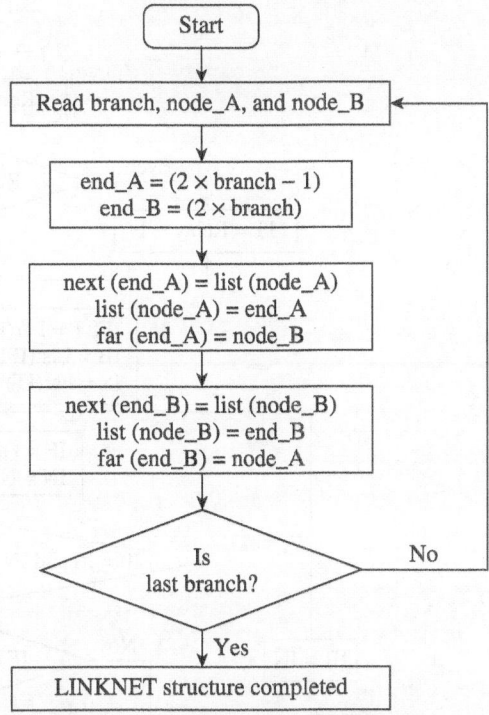

Fig. 2.18 Flowchart of LINKNET structure

Bus links *List* (*node*) or *list* (*bus*) is a vector that stores all the bus links. It has the dimension of the number of buses in the system. For any given bus, the element *list* (*bus*) points to the first branch end on the list from bus.

Directional relay links To store the links between the directional relays incident at a given bus, next vector is used. For any given relay incident at a bus, the element *next* (*end*) or *next* (*relay*) points to the next branch end on the list after *end*. The last relay incident at the bus is indicated when the element *next* (*relay*) assumes a zero value. The dimension of the vector *next* is equal to the number of branch ends, that is, the relays in the system.

Remote relay links *Far* (*end*) vector is used to link any relay with its remote bus. It points to the node at the far or opposite end of the branch.

Figure 2.18 shows the flowchart that is used to constitute the LINKNET structure by simply adding each branch to the network.

2.7.3 Determination of Primary/Backup Relay Pairs

Once the LINKNET structure is established, the next step is to determine the primary/backup relay pairs. Before finding the primary/backup relay pairs, load-flow studies and short-circuit studies are essential. These are not explained here.

The flowchart for the determination of primary/backup relay pairs is shown in Fig. 2.19. It is applied to any network such as radial feeder, parallel feeder, and ring networks. The steps of the algorithm are explained here.

1. Initially, the relay is considered as odd or even.
2. Afterwards, IFLT (the bus number near which the relay under consideration is located), IB (first directional relay looking towards this bus), and IS (bus on which opposite end relay is placed) are obtained.

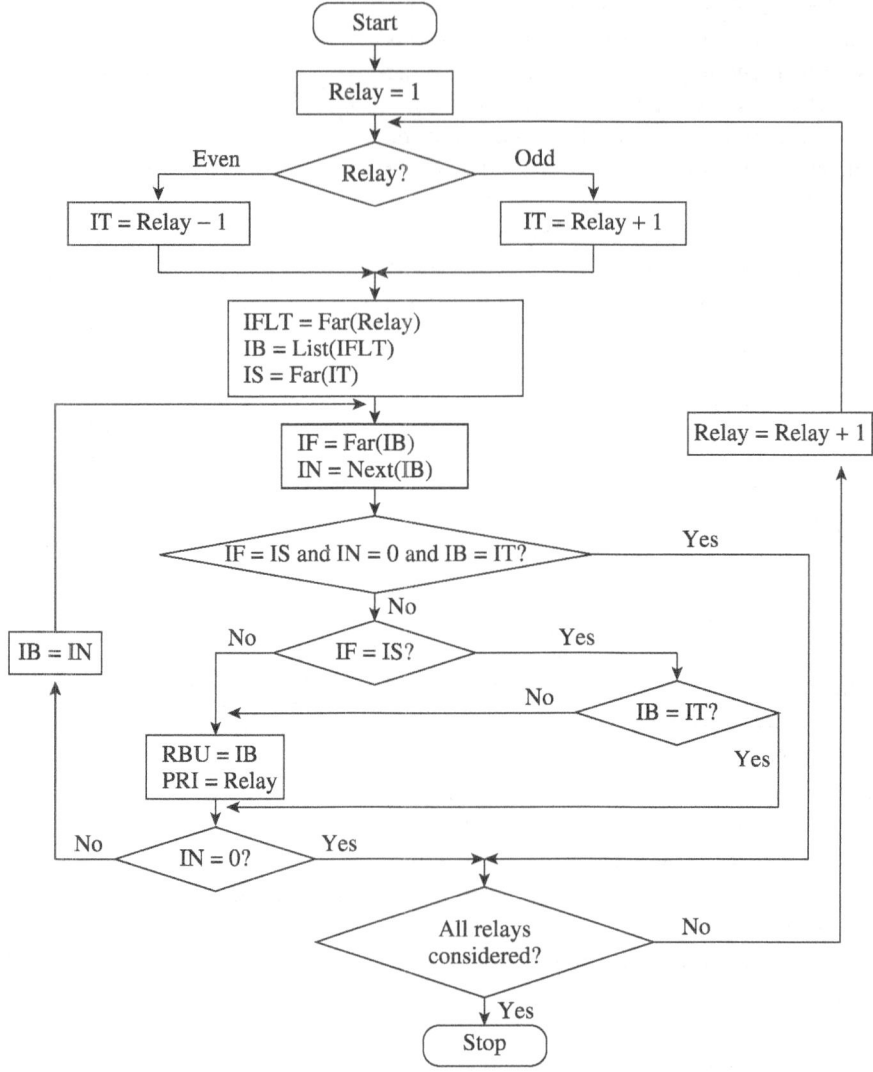

Fig. 2.19 Flowchart for primary-backup relay pairs

3. Subsequently, IF (bus near which the backup relay to the relay under consideration is placed) and IN (the next directional relay incident at bus IFLT) are calculated.
4. After two/three decision logic blocks, primary/backup relay pairs are found out.
5. Once the primary/backup relay pairs are obtained, the next step is to determine the PS (pickup current) and TDSs of all the relays. These are not shown in this book.

2.7.4 MATLAB Code for LINKNET Structure and PRI-BACKUP Pairs

In this sub-section, MATLAB codes for LINKNET structure and primary-backup relay pairs are described. Moreover, a case study for 9-bus system as shown, in Fig. 2.20, is also given along with its execution and results in terms of primary backup relay pairs. Relay numbers and branch numbers associated with relays

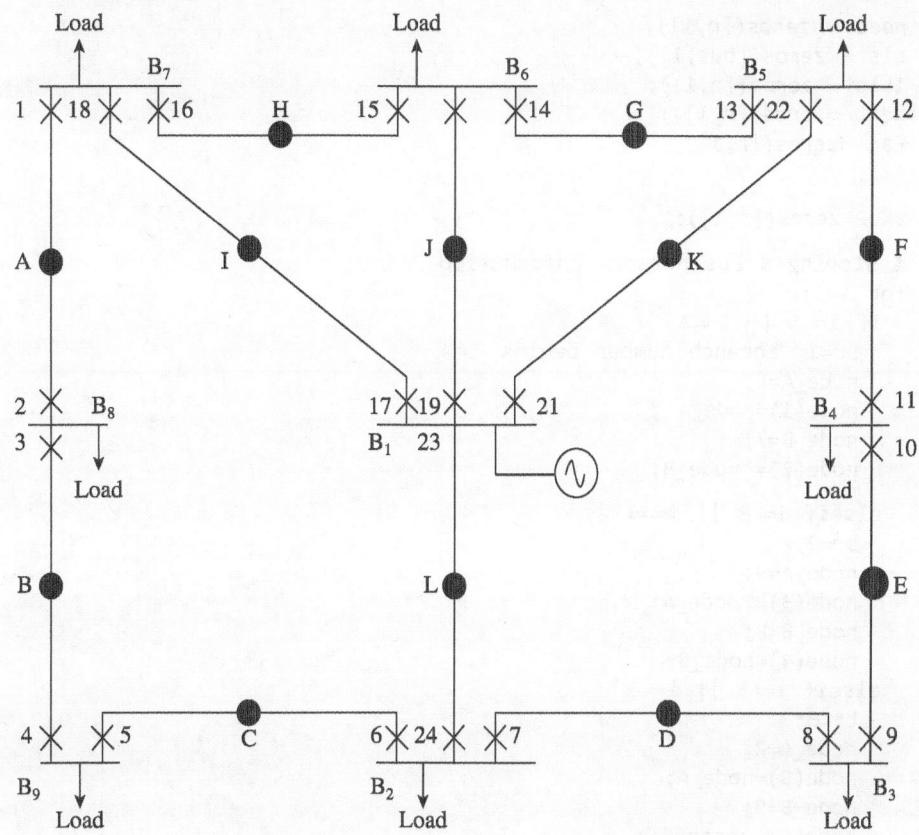

Fig 2.20 Single line diagram of a 9-bus system

are to be entered manually according to the system configuration such as 3-bus system, 9-bus system, 14-bus system, 30-bus system, and so on. Accordingly, bus numbers are assigned by LINKNET structure. The below steps describe the execution of the program for a sample case, that is, a 9-bus system.

Step-1: Enter system data (as per your selected system, e.g., 9-bus system):

The following data is to be entered before program execution.

1. Assign variable "r" in the program equal to the total number of relays in the selected system.
2. Enter the total number of buses in the system.
3. Enter the system configuration details.

The data of 9-bus system is entered for generation of various vectors in LINKNET structure.

Section of MATLAB code showing 9-bus system data

```
clear all
close all

NoOfBranches=12;
r=24; % number of relays in a system
bus=9;
End =zeros([r,1]);
```

```
node = zeros([r,1]);
List= zeros([bus,1]);
IList= zeros([r,1]);
Next =zeros([r,1]);
Far =zeros([r,1]);
F=1;

BKUP=zeros([r,1]);

% storing 9 bus network information
for i=1:r
  if i==1 || i==2
    br=1; %branch number begins
    node_A=8;
    node(1)=node_A ;
    node_B=7;
    node(2)= node_B;

  elseif i==3 || i==4
    br=2;
    node_A=9;
    node(3)= node_A;
    node_B=8;
    node(4)=node_B;
  elseif i==5 || i==6
    br=3;
    node_A=2;
    node(5)=node_A;
    node_B=9;
    node(6) =node_B;
  elseif i==7 || i==8
    br=4;
    node_A=3;
    node(7)=node_A;
    node_B=2;
    node(8)=node_B ;
  elseif i==9 || i==10
    br=5;
    node_A=4;
    node(9)=node_A;
    node_B=3;
    node(10)=node_B ;
  elseif i==11 || i==12
    br=6;
    node_A=5;
    node(11)=node_A;
    node_B=4;
    node(12)=node_B ;
  elseif i==13 || i==14
    br=7;
    node_A=6;
    node(13)=node_A;
```

```
            node_B=5;
            node(14)=node_B ;
        elseif i==15 || i==16
            br=8;
            node_A=7;
            node(15)=node_A;
            node_B=6;
            node(16)=node_B ;
        elseif i==17 || i==18
            br=9;
            node_A=7;
            node(17)=node_A;
            node_B=1;
            node(18)=node_B ;
        elseif i==19|| i==20
            br=10;
            node_A=6;
            node(19)=node_A;
            node_B=1;
            node(20)=node_B ;
        elseif i==21|| i==22
            br=11;
            node_A=5;
            node(21)=node_A;
            node_B=1;
            node(22)=node_B ;
        elseif i==23|| i==24
            br=12;
            node_A=2;
            node(23)=node_A;
            node_B=1;
            node(24)=node_B ;
            end
    % storing 9 bus network information ends
```

Step-2: Execute the program

The system configuration, as entered above, helps in storing network information in computer memory, that is, forming LINKNET structure as shown in Table 2.3. This results in the following output vectors:

NEXT, LIST, and FAR.

```
    %%LINKNET STRUCTURE FORMATION BEGINS
        if mod(i,2)~=0
            end_A=(2*br)-1;
            End(i)=end_A;
            Next(end_A)=IList(node_A);
            IList(node_A)=end_A;
            List(i)= IList(node_A);
            Far(end_A)=node_B;

        elseif (mod(i,2)==0)
            end_B=2*br;
            End(i,1)=end_B;
```

Table 2.3: Formation of LINKNET structure for 9-bus system

NODE	END	NEXT	LIST	FAR
8	1	0	1	7
7	2	0	2	8
9	3	0	3	8
8	4	1	4	9
2	5	0	5	9
9	6	3	6	2
3	7	0	7	2
2	8	5	8	3
4	9	0	9	3
3	10	7	10	4
5	11	0	11	4
4	12	9	12	5
6	13	0	13	5
7	14	11	14	6
1	15	2	15	6
6	16	13	16	7
7	17	15	17	1
1	18	0	18	7
6	19	16	19	1
1	20	18	20	6
5	21	14	21	1
1	22	20	22	5
2	23	8	23	1
1	24	22	24	2

```
Next(end_B,1)=IList(node_B,1);
IList(node_B)=end_B;
List(i)=IList(node_B);
Far(end_B)=node_A;
end
end
%%LINKNET STRUCTURE FORMATION ENDS
```

Step-3: Obtain primary backup relay pairs:

Once LINKNET structure is formed, the next task is to enter the relay number whose backup pairs are to be calculated. Any primary relay number can be entered depending upon the total number of relays in the system.

As a sample case, for the 9-bus system, one can enter any primary relay number (between 1 and 24) as the total number of relays in the system.

After providing all relay numbers one by one to the program, the backup pairs of the relay number are calculated and finally stored in the vector **"BKUP"** corresponding to the relay number entered. Here, row number is equal to relay number. A relay can have any number of backups depending on system configuration as shown in Table 2.4.

```
%% CALCULATION OF PRI-BACKUP PAIRS BEGIN
R=input('Relay number');
k=1;

if mod(R,2)~=0
    IT=R+1;
else
    IT=R-1;
end

    IFLT=Far(R);
    IB= IList(IFLT);
    IS=Far(IT);
    Flag=1;

while(Flag==1)
IF=Far(IB);
IN=Next(IB);
Flag=0;

if IF==IS && IN==0 && IB==IT
    Flag=3;

elseif IF==IS
    if IB==IT
        if IN==0;
            Flag=2;

        else
            IB=IN;
            Flag=1;
        end

    else
        RBU=IB;
        PRI=R;
        BKUP(R,k)=RBU;
        k=k+1;
        if IN==0;
            Flag=2;
        else
            IB=IN;
            Flag=1;
        end

    end
```

Table 2.4: Primary backup relay pairs for 9-bus system

PRI	BACKUP 1	BACKUP 2	BACKUP 3
1	17	15	
2	4		
3	1		
4	6		
5	3		
6	23	8	
7	23	5	
8	10		
9	7		
10	12		
11	9		
12	21	14	
13	21	11	
14	19	16	
15	19	13	
16	17	2	
17	24	22	20
18	15	2	
19	24	22	18
20	16	13	
21	24	20	18
22	14	11	
23	22	20	18
24	8	5	

Note: In the 9-bus system, as per the system connections relay numbers 17, 19, 21, and 23 are located near the generating bus and practically no backup exists for these relays. Hence, these relays are not considered in coordination studies.

```
        else
                RBU=IB;
                PRI=R;
                BKUP(R,k)=RBU;
                k=k+1;

        if IN==0;
            Flag=2;
        else
                IB=IN;
                Flag=1;
        end

    end

        if (Flag==3 || Flag==2)
            break,
        end
    end
end

%%CALCULATION OF PRI-BACKUP PAIRS ENDS
```

2.8 Examples

Examples 2.1   Figure 2.21 shows the single line diagram of a portion of a radial distribution system. The PS of $R_3 = 75\%$ of CT secondary. The TDS of $R_3 = 0.1$. Determine the settings of the relays R_1 and R_2. The normal range of PS is 50–200% of 1 A in seven equal steps, whereas the TDS setting range is 0.1 to 1 in steps of 0.05.

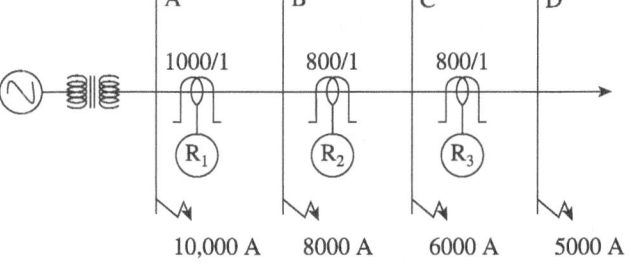

Fig. 2.21   Single line diagram of a portion of radial distribution system

Solution:

(i) Plug setting

The PS of relay R_3 is given as 75% of CT secondary = 600 (primary side).

$$\text{PS of } R_2 > \frac{1.3}{1.05} \times \text{PS of } R_3 > \frac{1.3}{1.05} \times 600 > 742.85 \text{ A (primary current)}$$

$$> \frac{742.85}{800} > 92.85\% \text{ of CT rating.}$$

Hence, the PS of R_2 is 100% of relay rating.

$$\text{PS of } R_1 > \frac{1.3}{1.05} \times \text{PS of } R_2 > \frac{1.3}{1.05} \times 100\% \text{ of } 800 > \frac{1.3}{1.05} \times 800$$

$$> 990.47 \text{ A (primary current)} > \frac{990.47}{1000} > 99.04\% \text{ of CT rating.}$$

Thus, the PS of relay R_1 is selected as 100% of relay rating.

(ii) Time dial setting

To achieve the TDS of relay R_2, the fault at bus C is considered.

With a fault current of 6000 A at bus C, the plug setting multiplier (PSM) of R_3 is given by

$$\text{PSM of } R_3 = \frac{6000}{75\% \text{ of } 800} = \frac{6000}{600} = 10$$

Hence, the time of operation of relay R_3 can be obtained by,

$$T_{op} = \frac{0.14}{\left(\text{PSM of } R_3\right)^{0.02} - 1} \times \text{TDS of } R_3 = \frac{0.14}{\left(10\right)^{0.02} - 1} \times 0.1 = 0.297s$$

Thus, the required time of operation at relay R_2 is 0.547 s by considering 0.25 s coordination time interval. Now, the PSM of R_2, for the fault at bus C (6000 A) is given by,

$$\text{PSM of } R_2 = \frac{6000}{100\% \text{ of } 800} = \frac{6000}{800} = 7.5$$

Hence, the time of operation of relay R_2 can be obtained by,

$$T_{op} \text{ of } R_2 = \frac{0.14}{\left(\text{PSM of } R_2\right)^{0.02} - 1} \times \text{TDS of } R_2$$

$$0.547 = \frac{0.14}{\left(7.5\right)^{0.02} - 1} \times \text{TDS of } R_2$$

Therefore, the TDS of $R_2 = 0.16 = 0.2$ (selected) from the available setting range.

To set the TDS of relay R_1, the fault at bus B is considered.

With a fault current of 8000 A at bus B,

$$\text{PSM of } R_2 = \frac{8000}{100\% \text{ of } 800} = 10$$

Hence, the time of operation of the relay R_2 can be obtained by,

$$\text{Required } T_{op} \text{ of } R_2 = \frac{0.14}{\left(\text{PSM of } R_2\right)^{0.02} - 1} \times \text{TDS of } R_2 = \frac{0.14}{\left(10\right)^{0.02} - 1} \times 0.2 = 0.594 \text{ s}$$

Now, the required time of operation of the relay R_1 is 0.845 by considering 0.25 s coordination time interval between R_2 and R_1.

PSM of R_1 for the same fault at bus B is given by

$$\text{PSM of } R_1 = \frac{8000}{100\% \text{ of } 1000} = 8$$

The TDS of R_1 can be found by,

$$\text{Required } T_{op} \text{ of } R_1 = \frac{0.14}{\left(\text{PSM of } R_1\right)^{0.02} - 1} \times \text{TDS of } R_1$$

$$\text{That is, } 0.845 = \frac{0.14}{\left(8\right)^{0.02} - 1} \times \text{TDS}$$

TDS of $R_1 = 0.256 = 0.3$ (selected) from available setting range.

The calculated relay settings are given in Table 2.5.

Table 2.5 Relay settings

Relay setting	Relays		
	R_1	R_2	R_3
PS	100%	100%	75%
TDS	0.3	0.2	0.1

Example 2.2 Figure 2.22 shows three overcurrent and one earth-fault schemes. An earth-fault relay 64 is set to operate at 20% of relay rating. The setting range of earth-fault relay is $10 - 40\%$ of 1 A in seven equal steps. The excitation current of CT core is 30 mA. Find out the percentage of the set current at which the relay will pick up. Find out the PSM of relay 64 for fault currents of 50 A, 100 A, and 200 A.

Solution:

$$I_{\text{pickup}} = 0.2 + (0.03 \times 3) = 0.29 \text{ A}$$

(a) For $I_f = 50$ A, $i_f(\text{secondary}) = 0.5$ A,
    Current through ground relay is $0.5 - (0.03 \times 3) = 0.41$ A.

$$\text{PSM(50)} = \frac{0.41}{0.2} = 2.05$$

(b) For $I_f = 100$ A, $i_f(\text{secondary}) = 1$ A,

$$\text{PSM(100)} = \frac{0.91}{0.2} = 4.55$$

(c) For $I_f = 200$ A, $i_f(\text{secondary}) = 2$ A,

$$\text{PSM(200)} = \frac{1.91}{0.2} = 9.55$$

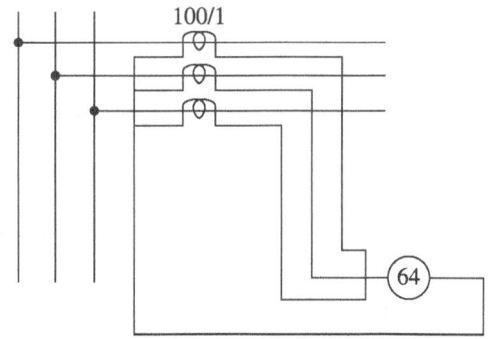

Fig. 2.22 Three overcurrent and one earth-fault schemes

Example 2.3 Figure 2.23 shows the single line diagram of a power system. A ground relay is connected across a 100/1 A CT. Find out the pickup current in primary ampere for PS of relay equal to 5%, 10%, 20%, 40%, and 80%. In addition, find out the current in primary ampere for PSM of relay equal to 2, 4, 5, and 10 with PS = 10%. Give your comments on the results obtained. The excitation current of each CT core is 50 mA.

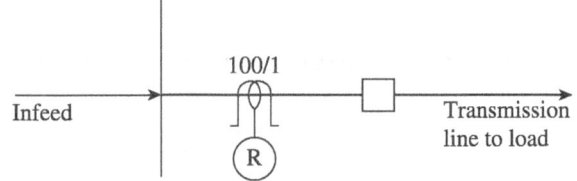

Fig. 2.23 Single line diagram of power system

Solution:
For PS = 5%, relay pickup current is 5% of 1 A = 0.05 A.

Considering the excitation current of CT, the actual relay pickup current = $0.05 + (3 \times 0.05) = 0.2$ A. Similarly, for other values of PS, the relay pickup currents are shown in Table 2.6.

For PSM = 2 and PS = 10%, $\text{PSM} = \dfrac{I(\text{secondary})}{\text{PS}}$

$$I(\text{secondary}) = 2 \times 0.1 = 0.2 \text{ A}, \quad I(\text{secondary}) = 0.2 + (0.05 \times 3) = 0.35 \text{ A}$$

$$I(\text{primary}) = 35 \text{ A}$$

Table 2.6 Different relay pickup currents for different values of PS

Plug setting (%)	Pickup current (A)
5	0.2
10	0.25
20	0.35
40	0.55
80	0.95

Table 2.7 Primary current values for different values of PSM

PSM	Primary current (A)
2	35
4	55
5	65
10	115

Similarly, for other values of PSMs, the currents in primary ampere are shown in Table 2.7.

Comments: The reduction in relay sensitivity does not provide any improvement because of the excitation current of three line CTs. If the relay settings are 5%, 10%, 20%, 40%, and 80%, the relay actually picks up at 20%, 25%, 35%, 55%, and 95%, respectively.

Example 2.4 Figure 2.24 shows the single line diagram of a portion of a power system network. Determine the PS and TDS of ground relays R_1, R_2, and R_3. The PS and TDS of R_4 are 10% and 0.1, respectively. The relays have the setting range of 10 – 40% of 1 A in seven equal steps. The excitation current of each CT core is 50 mA. The relevant current for single-line-to-ground fault is given in Fig. 2.24.

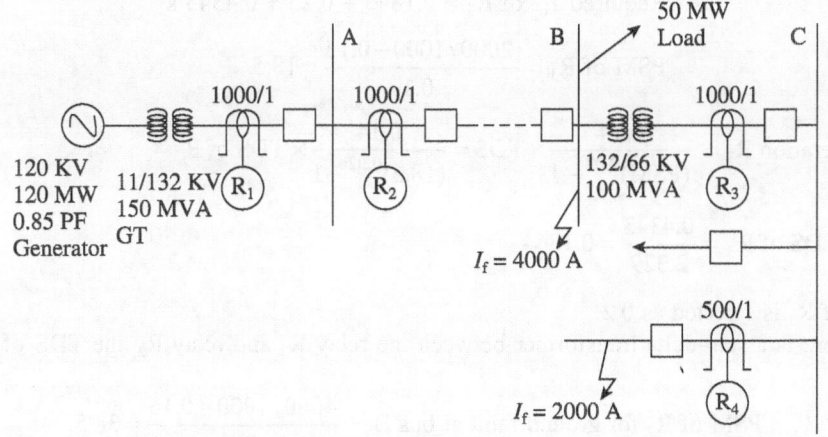

Fig. 2.24 Single line diagram of a portion of power system

Solution:

PS of R_4 (secondary) = 10% of 1 A = 0.1 A

Considering the excitation current of three line CTs, the PS of R_4 (secondary) = 0.1 + 0.15 = 0.25 A

PS of R_4 (primary) = 125 A

$$\text{PS of } R_3 > \frac{1.3}{1.05} \times \text{PS of } R_4 > \frac{1.3}{1.05} \times 125 > 154.76 \text{ (primary)}$$

PS of R_3 (secondary) = 0.1548 A

The current through ground relay $(R_3) = 0.15476 - 0.15$

$$= 0.004762 \text{ A} = 0.4762\%.$$

Therefore, the selected value of PS of $R_3 = 10\%$.

Now, as there is star–delta transformer between the relay R_3 and relay R_2, the relay R_2 can be set independently, and there is no need to coordinate relay R_2 with relay R_3.

Assume PS of R_2 (secondary) = 10% of 1 A = 0.1 A

Considering the excitation current of three line CTs, the PS of R_2 (secondary) = 0.1 + 0.15 = 0.25 A.

$$\text{PS of } R_2 \text{ (primary)} = 250 \text{ A}$$

$$\text{PS of } R_1 > \frac{1.3}{1.05} \times \text{PS of } R_2 > \frac{1.3}{1.05} \times 250 > 309.52 \text{ (primary)}$$

$$\text{PS of } R_1 \text{ (secondary)} = 0.3095$$

Current through ground relay (R_1) = 0.3095 − 0.15 = 0.1595 A = 15.95%.

Therefore, the selected value of PS of $R_1 = 20\%$.

$$\text{PSM of } R_4 = \frac{2000/500 - 0.15}{0.1} = \frac{3.85}{0.1} = 38.5$$

$$\text{Time of operation } R_4 = \frac{0.14}{(\text{PSM})^{0.02} - 1} \times \text{TDS} = \frac{0.14}{(38.5)^{0.02} - 1} \times 0.1 = 0.1843$$

$$\text{Required } T_{\text{op}} \text{ of } R_3 = 0.1843 + 0.25 = 0.4343 \text{ s}$$

$$\text{PSM of } R_3 = \frac{2000/1000 - 0.15}{0.1} = 18.5$$

$$\text{Time of operation } R_3 = \frac{0.14}{(\text{PSM})^{0.02} - 1} \times \text{TDS} = \frac{0.14}{(18.5)^{0.02} - 1} \times \text{TDS of } R_3$$

$$\text{TDS of } R_3 = \frac{0.4343}{2.329} = 0.1864$$

Hence, TDS of R_3 is selected as 0.2.

Now, as there is a star–delta transformer between the relay R_3 and relay R_2, the TDS of R_2 can be assumed as 0.2.

$$\text{PSM of } R_2 \text{ for ground fault at bus B} = \frac{4000/1000 - 0.15}{0.1} = 38.5$$

$$T_{\text{op}} \text{ of } R_2 = \frac{0.14}{(38.5)^{0.02} - 1} \times 0.2 = 0.3696 \text{ s}$$

$$\text{Required } T_{\text{op}} \text{ of } R_1 = 0.3696 + 0.25 = 0.6196 \text{ s}$$

$$\text{PSM of } R_1 = \frac{4000/1000 - 0.15}{0.2} = 19.25$$

$$\text{Time of operation } R_1 = \frac{0.14}{(19.25)^{0.02} - 1} \times \text{TDS of } R_1$$

$$\text{TDS of } R_1 = \frac{0.6196}{2.297} = 0.2696 = 0.3 \text{ (selected)}$$

Example 2.5 Figure 2.25 shows the single line diagram of a portion of radial distribution system. The TDS of R_5 is given as 0.1. Assume suitable discrimination time and determine the settings of relays R_1, R_2, R_3, and R_4. The normal range of PS is 50–200% of 1 A in seven equal steps, whereas the TDS setting range is 0.1 to 1 in steps of 0.05.

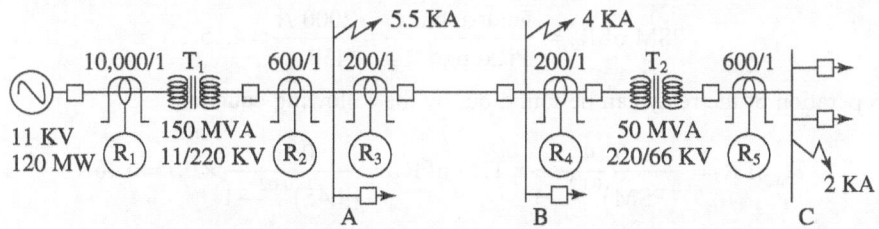

Fig. 2.25 Single line diagram of a portion of power system

Solution:

(i) Plug setting

The PS of relay R_5 is decided on the basis of rated secondary current of transformer T_2.

The rated secondary current of transformer $T_2 = \dfrac{50 \times 10^6}{66 \times 10^3 \times \sqrt{3}} = 437.38$ A

$$\text{PS of R}_5 > \frac{437.38}{600} = 72.89\% \text{ of } 600 \text{ A}$$

Thus, the setting R_5 is selected as 75% of CT rating.

Now, for finding out the PS of R_4, the following calculations are performed.

$$\text{PS of the relay R}_4 > \frac{1.3}{1.05} \times \frac{66}{220} \times 75\% \text{ of } 600$$

$$> 0.3714 \times 450 > 167.14 > 167.14/200$$

$$\text{PS of the relay R}_4 > 0.8357$$

Hence, the PS of R_4 is selected as 100% of CT rating.

Now, the PS of relay R_3 can be decided using the PS of the relay R_4.

$$\text{PS of R}_3 > \frac{1.3}{1.05} \times \text{PS of the relay R}_4 > \frac{1.3}{1.05} \times 100\% \text{ of } 200$$

$$> \frac{1.3}{1.05} \times 200 > 247.6 > 1.23\% \text{ of } 200 \text{ A}$$

Hence, the PS of R_3 is selected as 125% of CT rating.

PS of R_2 can be decided with respect to the rated secondary current of the transformer T_1.

The rated secondary of transformer T_1 is $= \dfrac{50 \times 10^6}{220 \times 10^3 \times \sqrt{3}} = 393.64$

$$= 65.60\% \text{ of } 600 \text{ A}$$

Hence, the PS of R_2 is selected as 75% of CT rating.

To find the PS of R_1, the value of PS of the relay R_2 can be used with the transformation ratio.

$$\text{PS of R}_1 > \frac{1.3}{1.05} \times \frac{220}{11} \times 75\% \text{ of } 600 > 24.76 \times 450 > 11142.85$$

$$> 111.42\% \text{ of } 10{,}000 \text{ A}$$

Hence, the PS of R_1 is selected as 125% of the CT rating.

(ii) Time dial setting

The TDS of the relay R_5 is given as 0.1.

$$\text{PSM of } R_5 = \frac{\text{Fault at bus C}}{\text{Pickup of } R_5} = \frac{2000 \text{ A}}{450 \text{ A}} = 4.45$$

The time of operation of the relay can be found out by the following equation:

$$T_{op} \text{ of } R_5 = \frac{0.14}{(\text{PSM})^{0.02} - 1} \times \text{TDS of } R_5 = \frac{0.14}{(4.45)^{0.02} - 1} \times 0.1 = 0.46 \text{ s}$$

Thus, the required T_{op} of relay R_4 is $0.46 + 0.25 = 0.71$ s, by considering 0.25 s time margin between two relays.

The PSM of R_4 for a fault on bus C can be given by

$$\text{PSM} = \frac{2000 \text{ A}}{200 \text{ A}} \times \frac{66}{220} = 3$$

$$T_{op} \text{ of } R_4 = \frac{0.14}{(\text{PSM})^{0.02} - 1} \times \text{TDS} = \frac{0.14}{(3)^{0.02} - 1} \times \text{TDS of } R_4 = 0.71 \text{ s}$$

$$\text{TDS of } R_4 = 0.112$$

Therefore, the TDS of R_4 is selected as 0.15, which is the next higher step to 0.112 in the multiples of 0.05.

Now, for deciding the TDS for relay R_3, the fault at bus B is considered.

$$\text{PSM of } R_3 = \frac{4000}{125\% \text{ of } 200} = \frac{4000 \text{ A}}{250 \text{ A}} = 16$$

Now, the T_{op} of relay R_4 for a fault at bus B is calculated as

$$\text{PSM of } R_4 = \frac{4000}{200} = 20$$

$$T_{op} \text{ of } R_4 = \frac{0.14}{(\text{PSM})^{0.02} - 1} \times \text{TDS of } R_4 = \frac{0.14}{(20)^{0.02} - 1} \times 0.15 = 0.34 \text{ s}$$

Therefore, the required T_{op} of R_3 is $0.34 + 0.25 = 0.59$ s.

Now, with PSM of $R_3 = 16$,

$$T_{op} \text{ of } R_3 = \frac{0.14}{(16)^{0.02} - 1} \times \text{TDS of } R_3$$

$$0.59 = 2.455 \times \text{TDS of } R_3$$

$$\text{TDS of } R_3 = 0.24$$

Hence, the TDS of R_3 is selected as 0.25.

Similarly, the TDS of R_2 and R_1 is selected by grading R_2 with R_3 and R_1 with R_2. The respective fault levels are to be considered as 4 KA and 5.5 KA.

Thus, the TDS of R_2 is 0.3, and the TDS of R_1 is 0.35.

Example 2.6 Calculate the relay settings of the relays used for the protection of the radial feeder shown in Fig. 2.26. The relays used are standard IDMT, having a rated current of 1 A. The PS range is of 50–200% of 1 A in seven equal steps, whereas the TDS setting range is 0.1 to 1 in steps of 0.05. The PS of relay R_5 is given as 75% of CT secondary rating, and the TDS setting is given as 0.1.

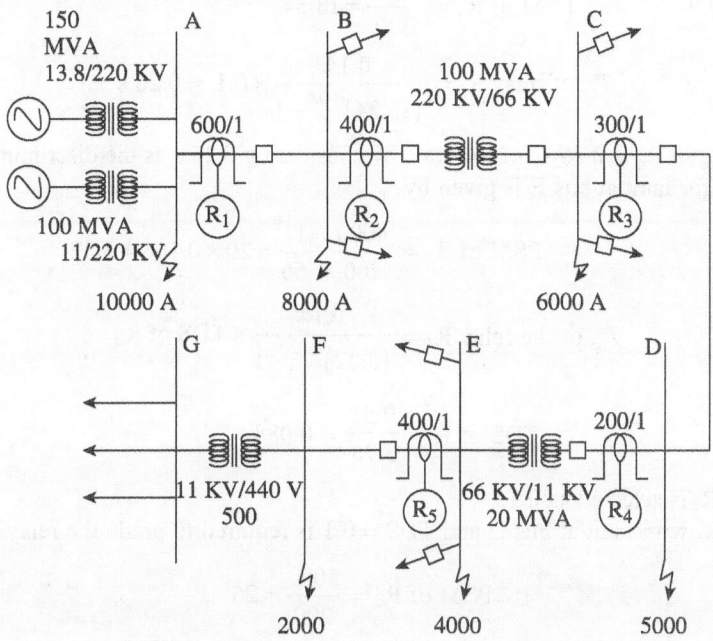

Fig. 2.26 Single line diagram of power system

Solution:
(i) Plug setting

$$\text{PS of relay } R_4 > \frac{1.3}{1.05} \times \text{PS of } R_5 \times \frac{11}{66} > \frac{1.3}{1.05} \times 300 \times \frac{11}{66}$$

$$> 61.90 \text{ A (primary side)} > 30.95\% \text{ of } 200 \text{ A}$$

Thus, the PS of the relay R_4 can be selected as 50% of CT rating, but the rated primary current of 20 MVA transformer is 174.95 A. Therefore, the PS of the relay R_4 is selected 100% of CT rating.

The relay R_3 gives the backup to the relay R_4.

$$\text{PS of } R_3 > \frac{1.3}{1.05} \times \text{PS of } R_4 > \frac{1.3}{1.05} \times 200 \text{ A} > 247.61 \text{ A (primary side)}$$

$$> 82.53\% \text{ of } 300 \text{ A}$$

Hence, the PS of R_3 is selected as 100% of the CT rating.

The PS of R_2 depends on rated primary current of 100 MVA transformer, and it is 262.43 A, which is 65.60% of 400 A. Hence, the PS of relay R_2 is selected as 75% of CT rating.

$$\text{PS of } R_1 > \frac{1.3}{1.05} \times \text{PS of } R_2 > \frac{1.3}{1.05} \times 300 > 371.42 \text{ A (primary)}$$

$$> 61.90\% \text{ of CT primary rating}$$

Hence, the PS of R_1 is selected as 75% of the CT rating.

(ii) Time dial setting

To decide the TDS of relay R_4, the TDS of R_5 is used with fault at the bus E.

$$\text{PSM of } R_5 = \frac{4000}{300} = 13.34$$

$$T_{op} \text{ of relay } R_5 = \frac{0.14}{(13.34)^{0.02} - 1} \times 0.1 = 0.26 \text{ s}$$

Thus, the required T_{op} of R_4 is $0.26 + 0.25 = 0.51$, by considering 0.25 s as the discrimination time.

PSM of relay R_4 for fault at bus E is given by

$$\text{PSM of } R_4 = \frac{4000}{200} \times \frac{11}{66} = 20 \times 0.166 = 3.32$$

$$T_{op} \text{ of the relay } R_4 = \frac{0.14}{(3.32)^{0.02} - 1} \times \text{TDS of } R_4$$

$$\text{TDS of } R_4 = \frac{0.51}{5.76} = 0.088$$

Hence, the TDS of R_4 is selected as 0.1.

Now, the T_{op} of R_4 with fault at bus D and TDS = 0.1 is required to grade the relay R_3 with R_4.

$$\text{PSM of } R_4 = \frac{5000}{200} = 25$$

PSM = 20 is selected because the T_{op} of the electromagnetic relay beyond PSM = 20 becomes constant.

$$T_{op} \text{ of the relay } R_4 = \frac{0.14}{(20)^{0.02} - 1} \times 0.10 = 0.226 \text{ s.}$$

Hence, the required T_{op} of R_3 is selected as $0.226 + 0.25 = 0.476$ s.

$$\text{PSM of } R_3 = \frac{5000}{300} = 16.67$$

$$T_{op} \text{ of the relay } R_3 = \frac{0.14}{(16)^{0.02} - 1} \times \text{TDS of } R_3$$

$$\text{TDS of } R_3 = \frac{0.476}{2.42} = 0.197$$

Hence, the TDS of R_3 is selected as 0.2.

Now, the T_{op} of R_3 with fault at bus C and TDS = 0.2 is given by

$$\text{PSM of } R_3 = \frac{6000}{300} = 20$$

$$T_{op} \text{ of the relay } R_3 = \frac{0.14}{(20)^{0.02} - 1} \times 0.2 = 0.453$$

Hence, the required T_{op} of R_2 is $0.453 + 0.25 = 0.703$ s.

$$\text{PSM of } R_2 = \frac{6000}{300} \times \frac{66}{220} = 6$$

$$T_{op} \text{ of the relay } R_2 = \frac{0.14}{(6)^{0.02} - 1} \times \text{TDS of } R_2$$

$$\text{TDS of } R_2 = \frac{0.703}{3.84} = 0.183$$

Hence, the TDS of R_2 is selected as 0.2.

Similarly, by considering the fault level at bus B, the TDS of relay R_1 is graded with TDS of relay R_2, and it is 0.3.

Example 2.7 Figure 2.27 shows the single line diagram of a portion of power system network. Find out the time of operation of relays R_1 and R_2 for a high resistance fault immediately after relaying point R_2 with a magnitude of 500 A. Relay R_1 is voltage monitored IDMT overcurrent relay and its plug setting reduces to 40% of the set value if voltage collapses below 70% of the rated voltage. The set value of plug setting of relay R_1 is 100% and TDS is 0.25. The plug setting of R_2 is 75% and TDS is 0.2. Other parameters are shown in Fig. 2.27.

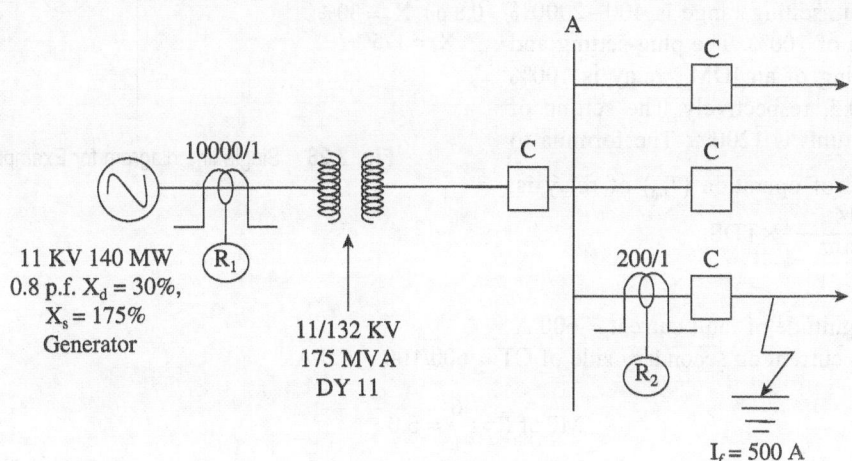

Fig. 2.27 Circuit diagram for Example 2.7

Solution:
The fault current magnitude on 132 kV side of the transformer, $I_f = 500$ A.
PS of $R_2 = 75\%$ of 1 A $= 0.75 \times 1 = 0.75$ A.

$$\text{MP of } R_2 = \frac{I_f \text{ (CT secondary)}}{\text{PS of } R_2} = \frac{500/200}{0.75} = 3.33$$

$$T_{op} (R_2) = \frac{0.14}{(\text{MP})^{0.02} - 1} \times \text{TDS} = \frac{0.14}{(3.33)^{0.02} - 1} \times 0.2 = 1.149 \text{ s}$$

Now, PS of R_1 will reduce to 40% from 100% as the voltage will collapse below 70%.

$$\text{PS of } R_1 = 0.4 \text{ A}.$$

The magnitude of fault current on 11 kV side,

$$I_f = \frac{500 \times 132 \times 10^3}{11 \times 10^3} = 6000 \text{ A}.$$

$$\text{MP of } R_1 = \frac{I_f \text{ (CT secondary)}}{\text{PS of } R_1} = \frac{6000/10000}{0.4} = 1.5$$

$$T_{op} (R_1) = \frac{0.14}{1.5^{0.02} - 1} \times 0.25 = 4.3 \text{s}.$$

Example 2.8 Figure 2.28 shows the single line diagram of a portion of power system. Determine the time of operation of an IDMT relay (R) having normal inverse characteristic for two different values of fault current (i) 600 A and (ii) 1500 A. The setting range of the relay is 50–200% of 1 A in steps of 25%. The high set instantaneous unit of the said relay is enabled and its setting range is 400–2000% of 1 A in step of 100%. The plug-setting and time dial setting of an IDMT relay is 100% of 1 A and 0.5, respectively. The setting of instantaneous unit is 1200%. The formula to calculate time of operation (T_{op}) of relay is,

$$T_{op} = \frac{0.14}{(\text{PSM})^{0.02} - 1} \times \text{TDS}.$$

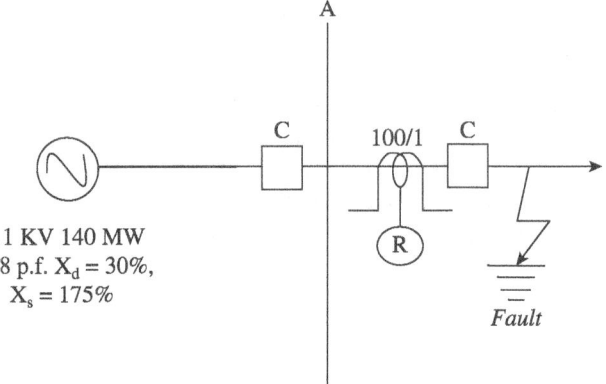

11 KV 140 MW
0.8 p.f. $X_d = 30\%$,
$X_s = 175\%$

Fig. 2.28 Single line diagram for Example 2.8

Solution:
(i) When magnitude of fault current = 600 A
Referring this current on secondary side of CT = 600/100 = 6.0 A.

$$\text{MP of } R = \frac{6}{1} = 6.0$$

$$T_{op} = \frac{0.14}{(6)^{0.02} - 1} \times 0.5 = 1.918 \text{s}.$$

(ii) When magnitude of fault current = 1500 A
Now, in case of fault current is 1500 A which is higher than 1200% of 1 A = 1200 A, the relay will operate instantaneously (two to three cycles). The time of operation of relay is obtained using the above formula only for $I_f < 1200$ A.

Example 2.9 The transmission line, as shown in Fig. 2.29, takes maximum full load current of 300 A when all the generators are connected in the network. The high resistance single line to ground fault occurs on the transmission line with a magnitude of 250 A when minimum number of generator is connected in the network. The transmission line is protected by the relay R which comprises two units. The first unit is of overcurrent relay (51) whereas the second unit is of under-voltage relay (27). Suggest the settings of the overcurrent and under-voltage relays with respect to Fig. 2.29 and Fig. 2.30. Moreover, write down the sequence of all the contacts shown in Fig. 2.30 when the transmission line takes a full load current of 300 A during normal/pre-fault condition.

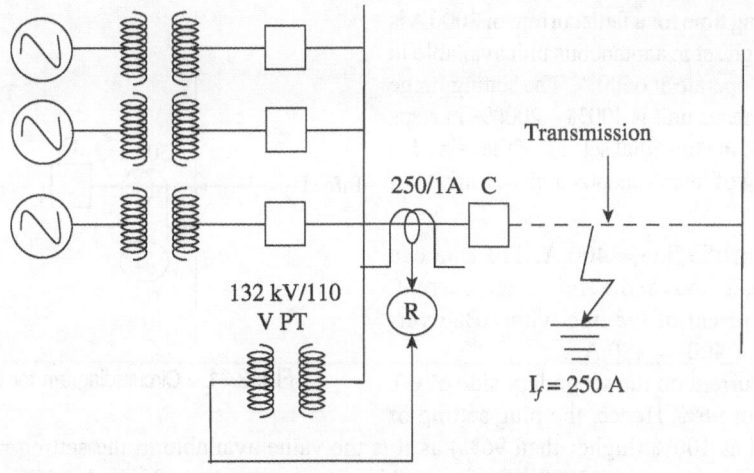

Fig. 2.29 Power circuit for Example 2.9

Solution:

The relay R, as shown in Fig. 2.29, is an overcurrent relay monitored by under-voltage relay. The plug setting of an overcurrent relay is done on maximum full load current of the transmission line with maximum number of generators connected in the network. Hence, it is done on 300 A current. Therefore,

$$\text{PS of R} = \frac{I_{\text{full load}}}{\text{CT Ratio}} = \frac{250}{250} = 1.0 \text{ or } 100\%$$

Hence, the plug setting of 100% is selected.

The setting of under-voltage relay is carried out on 70% of PT secondary voltage. Hence, it comes out to be $0.7 \times 110 = 77$ V.

The sequence of contacts with respect to Fig. 2.30 is as below.

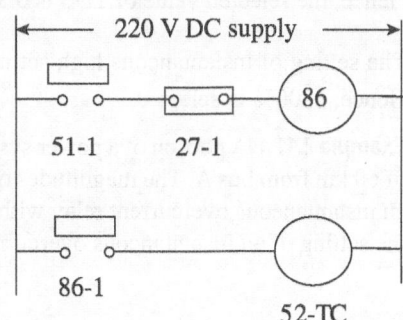

Fig. 2.30 Control circuit for Example 2.9

(i) The normal practice related to control circuit is that all relays (relay coils) are shown in de-energized condition and all circuit breakers are shown in open condition.

(ii) During normal/pre-fault condition, the voltage available to relay 27 is normal and hence, its coil is energized. Therefore, its contact (27-1) will remain in open condition. The setting of relay 51 is carried out on maximum full load current with maximum number of generators connected in the network (300 A). Hence, during normal/pre-fault condition, the contact of relay 51 (51-1) will remain in close condition. However, the coil of auxiliary relay (86) is not energized and tripping of CB (52) is not initiated as 27-1 will be in open condition.

(iii) In case of a fault on transmission line, voltage reduces and hence, relay 27 drops off (de-energized) and its contact (27-1) resumes its original (normally closed) position and final tripping is initiated.

Example 2.10 Figure 2.31 shows the single line diagram of a portion of a power system. The CT ratio is 500/1 A. The minimum current at which the relay 51 is desired to operate is 400 A. The relay 51 is a standard IDMT overcurrent relay having normal inverse characteristic. The rated current of relay 51 is 1 A and its PS range is 50%–200% of 1 A in steps of 25% whereas TDS range is 0–1 in steps of 0.05. The overload withstand is 20% above the normal current.

The desired operating time for a fault current of 3000 A is 1.0 seconds. The high set instantaneous unit available in the relay 51 should operate at 6500 A. The setting range of high set instantaneous unit is 400% – 2000% in steps of 100%. Determine the time dial setting of the relay 51. Also find out setting of instantaneous high set unit.

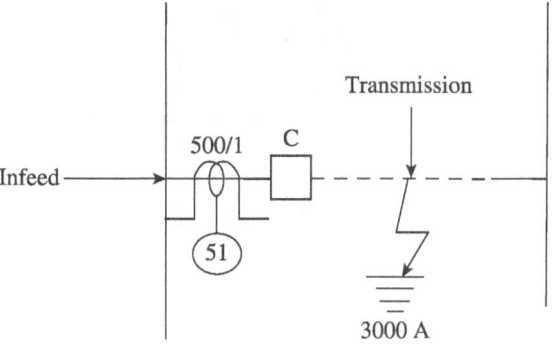

Fig. 2.31   Circuit diagram for Example 2.10

Solution:

Full load current of the line = 400 A. The line can take 20% overload above the full load current. Hence, full load current of the line with 20% overload = 400 + (0.2 × 400) = 480 A.

Referring this current on the secondary side of CT = 480/500 = 0.96 or 96%. Hence, the plug setting of relay R is selected as 100% (higher than 96%) as it is the value available in the setting range of the relay.

The magnitude of fault current = 3000 A. Referring this current on the secondary side of CT = 3000/500 = 6.0 A.

$$\text{MP of R} = \frac{6}{1} = 6.0$$

$$T_{op} = \frac{0.14}{(\text{MP})^{0.02} - 1} \times \text{TDS}$$

$$1.0 = \frac{0.14}{(6)^{0.02} - 1} \times \text{TDS}, \text{TDS} = 0.261.$$

Hence, the selected value of TDS is 0.30.

The setting of instantaneous high set unit = $\frac{6500}{500} = 13.0$ or 1300% of 1 A.
Hence, 1300% is selected.

Example 2.11   A portion of a power system network is shown in Fig. 2.32 in which a fault occurs at a distance of 60 km from bus A. The magnitudes of fault currents at the remote bus are given in Table 2.8. The relay R is an instantaneous overcurrent relay with a setting range of 400% –2000% of 1 A in steps of 100%. Determine the setting of an instantaneous overcurrent relay R in terms of its rated current, i.e., 1 A.

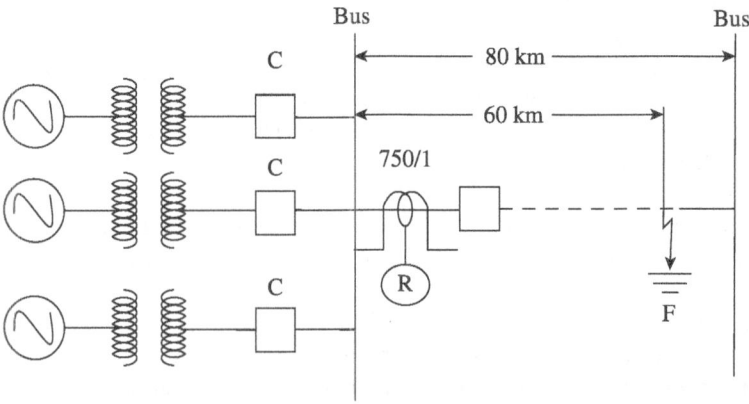

Fig. 2.32   Circuit diagram for Example 2.11

Table 2.8 Magnitude of fault currents for different generation

Faults with reference to generation	Fault currents (kA) for following faults	
	L-L	L-L-L-G
Minimum generation	5.5	8.5
Maximum generation	7.5	11.5

Solution:

The relay used in this example is an instantaneous overcurrent relay. Therefore, in order to decide the setting of this relay, we have to consider maximum value of fault current with maximum generating capacity. This situation is valid for L-L-L-G fault at the remote bus. Hence, considering this value, the magnitude of fault current on secondary of CT is,

$$i_f = \frac{11500}{750} = 15.33 \text{ or } 1533.33\% \text{ of } 1A.$$

Therefore, taking higher range, i.e., 1600% of 1 A, the setting of an instantaneous overcurrent relay is selected as 1600% of 1 A.

Example 2.12 Figure 2.33 shows the single line diagram of a portion of the power system network. Relays R_1, R_2, R_3, and R_4 are standard IDMT relays having normal inverse characteristic. At each bus, the fault level in MVA is mentioned. Th PS and TDS of relay R_4 are 100% and 0.1, respectively. The rated current of each relay is 1 A and its PS range is 50% –200% of 1 A in step of 25% whereas TDS range is 0 –1 in steps of 0.05. Calculate PS and TDS of relays R_1, R_2, and R_3. Assume 0.25s as the MCT between two successive relays.

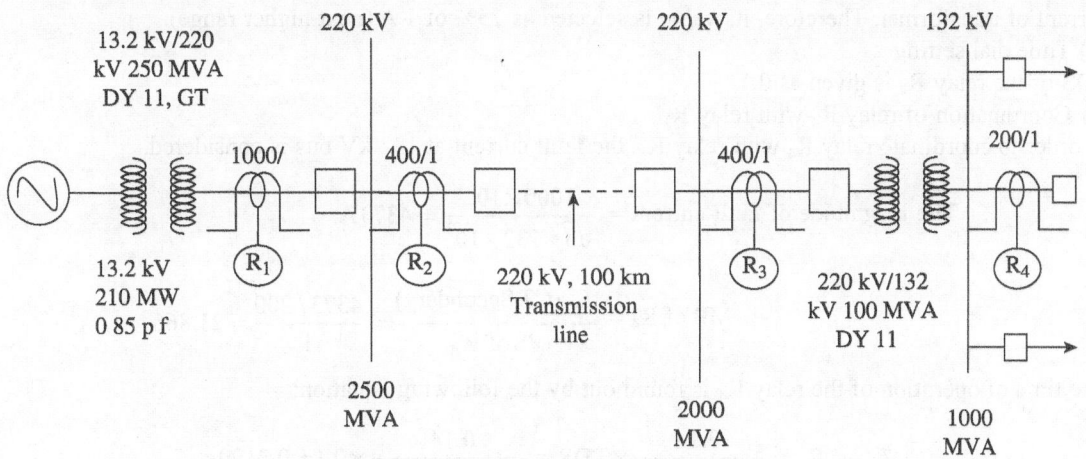

Fig. 2.33 Power system network for Example 2.12

Solution:

(i) Plug setting

The PS of relay R_4 is given as 100% of 1 A, i.e., 200 A on primary side of the CT on which relay R_4 is connected.

$$\text{PS of } R_3 > \frac{1.3}{1.05} \times \text{PS of } R_4, \Rightarrow \frac{1.3}{1.05} \times 200 \times \frac{132 \times 10^3}{220 \times 10^3}$$

$$\text{PS of } R_3 > 148.57 \text{ (primary side)} \Rightarrow \frac{148.57}{400} > 37.14\%$$

Hence, PS of R_3 can be selected as 50%.

However, if this setting is selected then relay R_3 may pick-up for rated current of 220 kV/132 kV transformer.

$$\text{The rated current of transformer on 220 kV side} = \frac{100 \times 10^6}{\sqrt{3} \times 220 \times 10^3} = 262.43 \text{ A}$$

Referring this current on the secondary side of CT on which relay R_3 is connected is 262.43/400 = 65.61% of 1 A. Hence, plug setting of R_3 is selected in such a way that it should not operate due to full load current of transformer. Therefore, its value is selected as 75% of 1 A (next higher range).

$$\text{PS of } R_2 > \frac{1.3}{1.05} \times \text{PS of } R_3, \Rightarrow \frac{1.3}{1.05} \times 0.75 \times 400$$

$$\text{PS of } R_2 > 371.42 \text{ (primary side)} \Rightarrow \frac{371.42}{400} > 92.85\% \text{ of 1A}$$

Hence, PS of R_2 can be selected as 100% of 1A.

Now, plug setting of R_1 can be selected by two ways (i) by coordinating R1 with R2 (ii) by considering rated current of the transformer on 220 kV side. Finally, the higher setting out of the two methods is selected.

$$\text{Following the above rule, the rated current of transformer on 220 kV side} = \frac{250 \times 10^6}{\sqrt{3} \times 220 \times 10^3} = 656.08 \text{ A}$$

Referring this current on the secondary side of CT on which relay R_1 is connected is 656.08/1000 = 65.61% of 1 A. Hence, plug setting of R_1 is selected in such a way that it should not operate due to full load current of transformer. Therefore, its value is selected as 75% of 1 A (next higher range).

(ii) Time dial setting

TDS of the relay R_4 is given as 0.1.

(a) Coordination of relay R_4 with relay R_3

In order to coordinate relay R_4 with relay R_3, the fault current at 132 kV bus is considered.

$$\text{The magnitude of fault current} = \frac{1000 \times 10^6}{\sqrt{3} \times 132 \times 10^3} = 4373 \text{ A}$$

$$\text{MP of } R_4 = \frac{I_f \text{ (CT Secondary)}}{\text{PS of } R_4} = \frac{4373/200}{1} = 21.865$$

The time of operation of the relay R_4 is found out by the following equation:

$$T_{op} \text{ of } R_4 = \frac{0.14}{(\text{MP})^{0.02} - 1} \times \text{TDS} = \frac{0.14}{(21.865)^{0.02} - 1} \times 0.1 = 0.2199 \text{ s.}$$

Thus, the required time of operation of relay R_3 is 0.2199 + 0.25 = 0.4699 s, by considering 0.25 s as the MCT between two relays.

Now, for the same magnitude of fault current,

$$\text{MP of } R_3 = \frac{I_f \text{ (CT Secondary)}}{\text{PS of } R_3} = \frac{\dfrac{4373}{400} \times \dfrac{132}{220}}{0.75} = 8.746$$

$$\text{Required } T_{op} \text{ of } R_3 = 0.4699 = \frac{0.14}{(MP)^{0.02} - 1} \times TDS = \frac{0.14}{(8.746)^{0.02} - 1} \times TDS$$

TDS=0.1487

Hence, TDS of R_3 is selected as 0.15 (next higher value from the available range).
(b) Coordination of relay R_3 with relay R_2
In order to coordinate relay R_3 with relay R_2, the fault current at 220 kV bus is considered.

$$\text{The magnitude of fault current} = \frac{2000 \times 10^6}{\sqrt{3} \times 220 \times 10^3} = 5248.63 \text{ A}$$

$$\text{MP of } R_3 = \frac{I_f \text{ (CT Secondary)}}{PS \text{ of } R_3} = \frac{5248.63/400}{0.75} = 17.49$$

The time of operation of the relay R_3 is found out by the following equation:

$$T_{op} \text{ of } R_3 = \frac{0.14}{(MP)^{0.02} - 1} \times TDS = \frac{0.14}{(17.49)^{0.02} - 1} \times 0.15 = 0.3565 \text{ s.}$$

Thus, the required time of operation of relay R_2 is 0.3565 + 0.25 = 0.6065 s, by considering 0.25 s as the *MCT* between two relays.

Now, for the same magnitude of fault current,

$$\text{MP of } R_2 = \frac{I_f \text{ (CT Secondary)}}{PS \text{ of } R_2} = \frac{5248.63/400}{1.0} = 13.12$$

$$\text{Required } T_{op} \text{ of } R_2 = 0.6065 = \frac{0.14}{(MP)^{0.02} - 1} \times TDS = \frac{0.14}{(13.12)^{0.02} - 1} \times TDS$$

TDS=0.2289

Hence, TDS of R_2 is selected as 0.25 (next higher value from the available range).
(c) Coordination of relay R_2 with relay R_1
In order to coordinate relay R_2 with relay R_1, the fault current at 220 kV bus between R_2 and R_1 is considered.

$$\text{The magnitude of fault current} = \frac{2500 \times 10^6}{\sqrt{3} \times 220 \times 10^3} = 6560.79 \text{ A.}$$

$$\text{MP of } R_2 = \frac{I_f \text{ (CT Secondary)}}{PS \text{ of } R_2} = \frac{6560.79/400}{1.0} = 16.4$$

The time of operation of the relay R_2 is found out by the following equation:

$$T_{op} \text{ of } R_2 = \frac{0.14}{(MP)^{0.02} - 1} \times TDS = \frac{0.14}{(16.4)^{0.02} - 1} \times 0.25 = 0.61 \text{ s.}$$

Thus, the required time of operation of relay R_1 is 0.61 + 0.25 = 0.86 s, by considering 0.25 s as the MCT between the two relays.
Now, for the same magnitude of fault current,

$$\text{MP of } R_1 = \frac{I_f \text{ (CT Secondary)}}{PS \text{ of } R_1} = \frac{\dfrac{6560.79}{1000}}{0.75} = 8.7477$$

$$\text{Required } T_{op} \text{ of } R_1 = 0.86 = \frac{0.14}{(MP)^{0.02} - 1} \times TDS = \frac{0.14}{(8.7477)^{0.02} - 1} \times TDS$$

$$TDS = 0.2723$$

Hence, TDS of R_1 is selected as 0.30 (next higher value from the available range).

Example 2.13 Figure 2.34 shows the single line diagram of a portion of the power system network. Relays R_1, R_2, R_3, and R_4 are standard IDMT relays having normal inverse characteristic. The maximum full load current for each line along with other lines on each bus is also shown in Fig. 2.34. The TDS of relay R_4 is 0.1. The relay rated current is 5 A and PS range of each relay is 50%–200% of 5 A in steps of 25% whereas TDS range is 0–1 in steps of 0.05. Calculate the PS of four relays along with TDS of relays R_1, R_2 and R_3. Assume 0.25s as the MCT between two successive relays. The magnitudes of fault current at each bus for 3-phase and single line-to-ground fault are shown in Table 2.9.

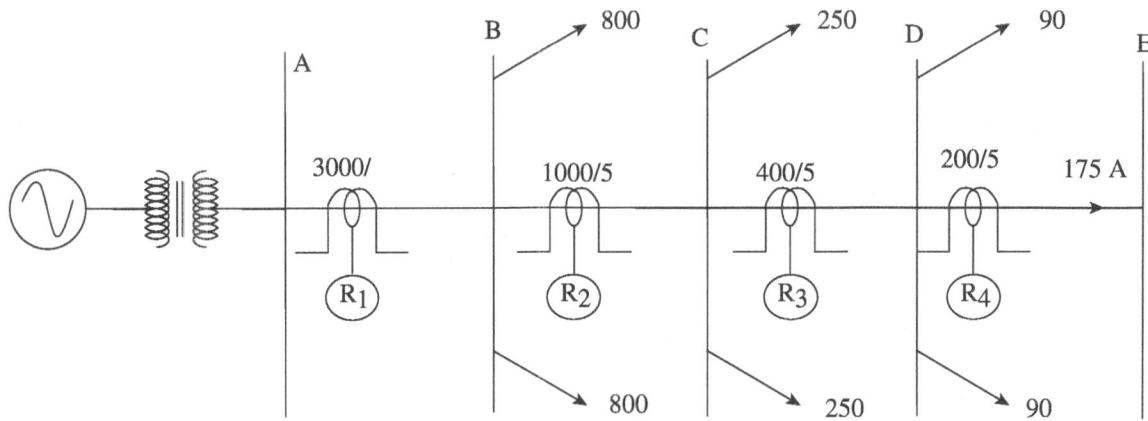

Fig. 2.34 Power system network for Example 2.13

Table 2.9 Magnitude of fault current at different substations

Magnitude of fault current	Bus-A	Bus-B	Bus-C	Bus-D	Bus-E
3-phase fault	12000	8000	4000	2000	1500
Single line-to-ground fault	10000	5000	3000	1200	650

Solution:

(i) Plug setting

The magnitude of current for the line in which relay R_4 is connected with 10% overload = 175 + 17.5 (10% of 175) = 192.5 A.

Referring this current on the secondary side of CT (200/5), the current is $\frac{192.5}{200} = 0.9625$ or 96.25% of 5 A. Hence, PS of R_4 is selected as 100% of 5 A.

The magnitude of current for the line in which relay R_3 is connected = 175 + 90 + 90 = 355 A.

Considering 10% overload, the magnitude of current = 355 + 35.5 = 390.5 A.

Referring this current on the secondary side of CT (400/5), the current is $\frac{390.5}{400} = 0.9763$ or 97.63% of 5 A. Hence, PS of R_3 is selected as 100% of 5 A.

The magnitude of current for the line in which relay R_2 is connected = 355 + 250 + 250 = 855 A. Considering 10% overload, the magnitude of current = 855 + 85.5 = 940.5 A.

Referring this current on the secondary side of CT (1000/5), the current is $\dfrac{940.5}{1000} = 0.9405$ or 94.05% of 5 A. Hence, PS of R_2 is selected as 100% of 5 A.

The magnitude of current for the line in which relay R_1 is connected = 800 + 855 + 800 = 2455 A. Considering 10% overload, the magnitude of current = 2455 + 245.5 = 2700.5 A.

Referring this current on the secondary side of CT (3000/5), the current is $\dfrac{2700.5}{3000} = 0.9001$ or 90.01% of 5 A.

Hence, PS of R_1 is selected as 100% of 5 A.

(ii) Time dial setting

(a) Coordination between relay R_3 with relay R_4

During coordination of relay R_4 with relay R_3, we have to consider maximum magnitude of fault current at bus D for phase relays and minimum magnitude of fault current at bus D for ground relays. Assuming all relays are phase relays, we consider 2000 A current at bus D.

$$\text{MP of } R_4 = \frac{I_f \text{ (CT Secondary)}}{\text{PS of } R_4} = \frac{\dfrac{2000}{200}}{1} = 10.0$$

The time of operation of the relay R_4 is found out by the following equation:

$$T_{\text{op}} \text{ of } R_4 = \frac{0.14}{(\text{MP})^{0.02} - 1} \times \text{TDS} = \frac{0.14}{(10.0)^{0.02} - 1} \times 0.1 = 0.2971 \text{ s}$$

Thus, the required time of operating of relay R_3 is 0.2971 + 0.25 = 0.5471 s, by considering 0.25 s as the *MCT* between two relays.

Now, for the same magnitude of fault current,

$$\text{MP of } R_3 = \frac{I_f \text{ (CT Secondary)}}{\text{PS of } R_3} = \frac{\dfrac{2000}{400}}{1} = 5.0$$

$$\text{Required } T_{\text{op}} \text{ of } R_3 = 0.5471 = \frac{0.14}{(\text{MP})^{0.02} - 1} \times \text{TDS} = \frac{0.14}{(5.0)^{0.02} - 1} \times \text{TDS}$$

$$\text{TDS} = 0.1278$$

Hence, TDS of R_3 is selected as 0.15 (next higher value from the available range).

(b) Coordination between relay R_2 with relay R_3

During coordination of relay R_3 with relay R_2, we have to take 4000 A current at bus C.

$$\text{MP of } R_3 = \frac{I_f \text{ (CT Secondary)}}{\text{PS of } R_3} = \frac{\dfrac{4000}{400}}{1.0} = 10.0$$

The time of operation of the relay R_3 is found out by the following equation:

$$T_{\text{op}} \text{ of } R_3 = \frac{0.14}{(\text{MP})^{0.02} - 1} \times \text{TDS} = \frac{0.14}{(10.0)^{0.02} - 1} \times 0.15 = 0.4456 \text{ s}$$

Thus, the required time of operation of relay R_2 is 0.4456 + 0.25 = 0.6956 s, by considering 0.25 s as the MCT between two relays.

Now for the same magnitude of fault current,

$$\text{MP of } R_2 = \frac{I_f \text{ (CT Secondary)}}{\text{PS of } R_2} = \frac{4000/1000}{1.0} = 4.0$$

$$\text{Required } T_{op} \text{ of } R_2 = 0.6956 = \frac{0.14}{(\text{MP})^{0.02} - 1} \times \text{TDS} = \frac{0.14}{(4.0)^{0.02} - 1} \times \text{TDS}$$

$$\text{TDS} = 0.1397$$

Hence, TDS of R_2 is selected as 0.15 (next higher value from the available range).

(c) Coordination of relay R_1 with relay R_2

While coordinating relay R_1 with relay R_2, we have to take 8000 A current at bus-B.

$$\text{MP of } R_2 = \frac{I_f \text{ (CT Secondary)}}{\text{PS of } R_2} = \frac{\dfrac{8000}{1000}}{1.0} = 8.0$$

The time of operation of the relay R_2 is found out by the following equation:

$$T_{op} \text{ of } R_2 = \frac{0.14}{(\text{MP})^{0.02} - 1} \times \text{TDS} = \frac{0.14}{(8.0)^{0.02} - 1} \times 0.15 = 0.4945 \text{ s.}$$

Thus, the required time of operating of relay R_1 is $0.4945 + 0.25 = 0.7445$ s, by considering 0.25 s as the MCT between two relays.

Now, for the same magnitude of fault current,

$$\text{MP of } R_1 = \frac{I_f \text{ (CT Secondary)}}{\text{PS of } R_1} = \frac{\dfrac{8000}{3000}}{1.0} = 2.67$$

$$\text{Required } T_{op} \text{ of } R_1 = 0.7445 = \frac{0.14}{(\text{MP})^{0.02} - 1} \times \text{TDS} = \frac{0.14}{(2.67)^{0.02} - 1} \times \text{TDS}$$

$$\text{TDS} = 0.11$$

Hence, TDS of R_1 is selected as 0.15 (next higher value from the available range).

Example 2.14   Figure 2.35 shows the single line diagram of a portion of the power system network. Relays R_1, R_2, and R_3 are standard IDMT relays having normal inverse characteristic. The magnitudes of fault current (for L-L and L-L-L fault) are shown at each bus. The TDS of relay R_3 is 0.1. The relay rated current is 1 A and PS range of each relay is 50% – 200% of 1 A in steps of 25% whereas TDS range is 0 – 1 in steps of 0.05. All three IDMT relays have instantaneous high set units each having a setting range of 400% to 2000% in steps of 100%. The percentage overreach of each instantaneous high set unit is 10%. Considering 10% overload withstand, calculate PS and TDS of all three overcurrent relays. Also, calculate the pickup setting of all the instantaneous high set units. Assume 0.25s as the MCT between two successive relays. Also assume TDS of relay $R_3 = 0.1$.

Solution:

(i) Plug setting

The magnitude of current for the line connected between bus C and bus D with 10% overload = 100 + 10 (10% of 100) = 110 A.

Referring this current on the secondary side of CT (200/1), the current is $\dfrac{110}{200} = 0.55$ or 55% of 1 A. Hence, PS of R_3 is selected as 75% of 1 A.

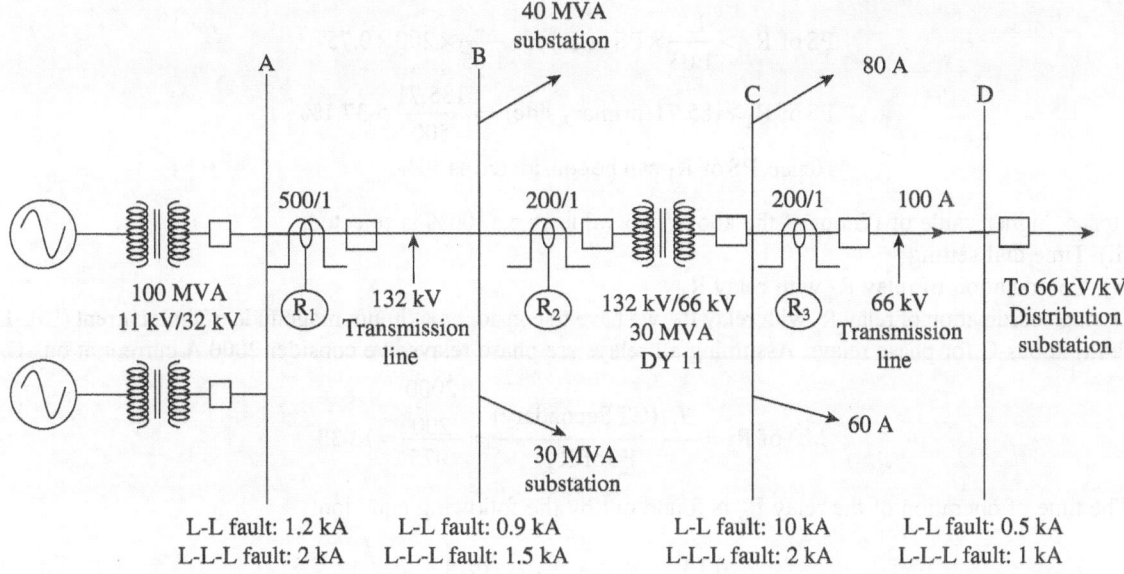

Fig. 2.35 Power system network for Example 2.14

Now, the PS of relay R_2 is decided by two ways (i) by coordinating rule between relay R_3 and relay R_2 and (ii) by considering full load current of 30 MVA transformer on 132 kV side.

The magnitude of full load current of 30 MVA transformer on 132 kV side is given by,

$$I_{FL} = \frac{30 \times 10^6}{\sqrt{3} \times 132 \times 10^3} = 131.21 \text{ A}.$$

Referring this current on the secondary side of CT (200/1), the current is $\frac{131.21}{200} = 0.6561$ or 65.61% of 1 A.

Hence, PS of R_2 can be considered as 75% of 1 A.

Now using coordinating rule between relay R_3 and relay R_2,

$$\text{PS of } R_2 > \frac{1.3}{1.05} \times \text{PS of } R_3, \Rightarrow \frac{1.3}{1.05} \times 200 \times 0.75 \times \frac{66 \times 10^3}{132 \times 10^3}$$

$$\text{PS of } R_2 > 92.857 \,(primary\,side) \Rightarrow \frac{92.857}{200} > 46.43\%$$

Hence, PS of R_2 can be considered as 50%.

Hence, higher value of PS out of the above two values, i.e., 75% is selected.

Similarly, the PS of relay R_1 is decided by two ways (i) by coordinating rule between relay R_2 and relay R_1 and (ii) by considering full load current of 100 MVA transformer on 132 kV side.

The magnitude of full load current of 100 MVA transformer on 132 kV side is given by,

$$I_{FL} = \frac{100 \times 10^6}{\sqrt{3} \times 132 \times 10^3} = 437.39 \text{ A}.$$

Referring this current on the secondary side of CT (500/1), the current is $\frac{437.39}{500} = 0.8748$ or 87.48% of 1 A.

Hence, PS of R_1 can be considered as 100% of 1 A.

Now using coordinating rule between relay R_2 and relay R_1,

$$\text{PS of R}_1 > \frac{1.3}{1.05} \times \text{PS of R}_2, \Rightarrow \frac{1.3}{1.05} \times 200 \times 0.75$$

$$\text{PS of R}_1 > 185.71 \, (\text{primary side}) \Rightarrow \frac{185.71}{500} > 37.1\%$$

Hence, PS of R_1 can be considered as 50%.

Hence, higher value of PS out of the above two values, i.e., 100% is selected.

(ii) Time dial setting

(a) Coordination of relay R_2 with relay R_3

During coordination of relay R_2 with relay R_3, we have to consider maximum magnitude of fault current (L-L-L fault) at bus-C for phase relays. Assuming all relays are phase relays, we consider 2000 A current at bus-C.

$$\text{MP of R}_3 = \frac{I_f \, (\text{CT Secondary})}{\text{PS of R}_3} = \frac{\dfrac{2000}{200}}{.075} = 13.33$$

The time of operation of the relay R_3 is found out by the following equation:

$$T_{op} \text{ of R}_3 = \frac{0.14}{(\text{MP})^{0.02} - 1} \times \text{TDS} = \frac{0.14}{(13.33)^{0.02} - 1} \times 0.1 = 0.2633 \, \text{s}$$

Thus, the required time of operation of relay R_2 is 0.2633 + 0.25 = 0.5133 s, by considering 0.25 s as the *MCT* between two relays.

Now, for the same magnitude of fault current,

$$\text{MP of R}_2 = \frac{I_f \, (\text{CT Secondary})}{\text{PS of R}_2} = \frac{\dfrac{2000}{200} \times \dfrac{66 \times 10^3}{132 \times 10^3}}{0.75} = 6.66$$

$$\text{Required } T_{op} \text{ of R}_2 = 0.5133 = \frac{0.14}{(\text{MP})^{0.02} - 1} \times \text{TDS} = \frac{0.14}{(6.66)^{0.02} - 1} \times \text{TDS}$$

$$\text{TDS} = 0.1418$$

Hence, TDS of R_2 is selected as 0.15 (next higher value from the available range).

(b) Coordination of relay R_1 with relay R_2

During coordination of relay R_1 with relay R_2, we have to consider maximum magnitude of fault current (L-L-L fault) at bus-B, i.e., 1.5 kA.

$$\text{MP of R}_2 = \frac{I_f \, (\text{CT Secondary})}{\text{PS of R}_2} = \frac{1500 / 200}{0.75} = 10.0$$

The time of operation of the relay R_2 is found out by the following equation:

$$T_{op} \text{ of R}_2 = \frac{0.14}{(\text{MP})^{0.02} - 1} \times \text{TDS} = \frac{0.14}{(10.0)^{0.02} - 1} \times 0.15 = 0.4456 \, \text{s}$$

Thus, the required time of operation of relay R_1 is 0.4456 + 0.25 = 0.6956 s, by considering 0.25 s as the MCT between two relays.

Now, for the same magnitude of fault current,

$$\text{MP of R}_1 = \frac{I_f \text{ (CT Secondary)}}{\text{PS of R}_1} = \frac{\dfrac{1500}{500}}{1.0} = 3.0$$

$$\text{Required } T_{op} \text{ of R}_1 = 0.6956 = \frac{0.14}{(\text{MP})^{0.02} - 1} \times \text{TDS} = \frac{0.14}{(3.0)^{0.02} - 1} \times \text{TDS}$$

$$TDS = 0.1104$$

Hence, TDS of R_1 is selected as 0.15 (next higher value from the available range).

(c) Pick-up value of instantaneous high set unit

Now, in order to decide the pick-up of instantaneous high set unit, it is required to consider phase-to-phase fault with minimum generation at bus-C, bus-B, and bus-A for relays R_3, R_2, and R_1, respectively.

Hence, for relay R_3, the fault current for L-L fault at bus-C, i.e., 1000 A is considered. By considering 10 % overreach of instantaneous highest unit, the current magnitude = 1000 + 100(10% of 1000) = 1100.

Referring this current on the secondary of CT (200/1), we get 1100/200 = 5.1 or 510% of 1A.

Hence, pick-up value of 600 % of 1A is selected for relay R_3.

For relay R_2, the fault current for L-L fault at bus-B, i.e., 900 A is considered.

By considering 10% overreach of instantaneous highest unit, the current magnitude = 900 + 90(10% of 900) = 990.

Referring this current on the secondary of CT (200/1), we get 990/200 = 4.95 or 495% of 1A.

Hence, pick-up value of 500 % of 1A is selected for relay R_2.

For relay R_1, the fault current for L-L fault at bus-A, i.e., 1200 A is considered.

By considering 10% overreach of instantaneous highest unit, the current magnitude = 1200 + 120(10% of 1200) = 1320.

Referring this current on the secondary of CT (500/1), we get 1320/500 = 2.64 or 240% of 1A.

Hence, minimum pick-up value of 400% of 1A is selected for relay R_1.

Example 2.15 The relays shown in Fig. 2.36 are directional and non-directional relays. Identify the directional relays. The breaking capacities of the breakers are given in Table 2.10. The PS of relays are given in Fig. 2.36; the TDS value of the relays R_2 and R_5 is set at 0.1. Determine the TDS of other relays.

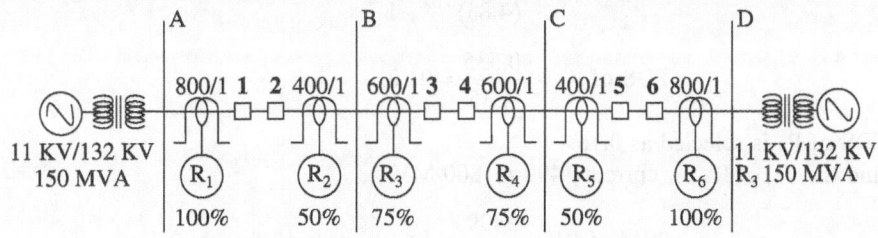

Fig. 2.36 Single line diagram of power system

Solution:

The relays R_1 and R_6 are non-directional, whereas all others relays are directional because the fault current direction can change on bus B and bus C.

We will split the network into two radial networks.

The TDS value of relays R_2 and R_5 is given as 0.1. The TDS of the other relay can be calculated on the basis of its operation in the

Table 2.10 Breaking capacities of breakers

Breaker no.	Capacities (MVA)
1, 6	1000
3, 4	800
2, 5	500

correct direction of the current flow as shown in Fig. 2.37. The PSM of relay R_5 for the current equivalent to the breaking capacity of breaker 5 (500 MVA):

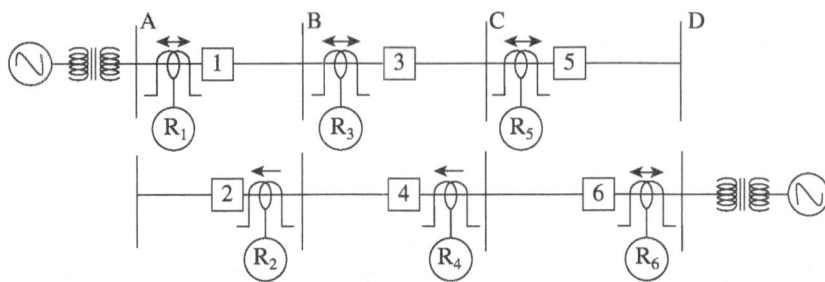

Fig. 2.37 Single line diagram of two-radial network of Fig. 2.36

$$\text{Current} = \frac{500 \times 10^6}{132 \times 10^3 \times \sqrt{3}} = 2186.93 \text{ A}$$

$$\text{PSM of } R_5 = \frac{2186.93}{200} = 10.93$$

$$T_{op} \text{ of } R_5 = \frac{0.14}{(10.93)^{0.02} - 1} \times 0.1 = 0.285 \text{ s}$$

Hence, the required T_{op} of relay R_3 is 0.535 s with the consideration of 0.25 s discrimination time. PSM of R_3 for the same breaking capacity is given by

$$\text{PSM of } R_3 = \frac{2186.93}{450} = 4.85$$

$$T_{op} \text{ of } R_3 = \frac{0.14}{(4.85)^{0.02} - 1} \times \text{TDS of } R_3$$

$$\text{TDS of } R_3 = \frac{0.535}{4.35} = 0.122$$

Hence, the TDS of R_3 is selected as 0.15.

Now, to find the T_{op} of R_3 for current 3499 A (800 MVA),

$$\text{PSM of } R_3 = \frac{3499}{450} = 7.8$$

$$T_{op} \text{ of } R_3 = \frac{0.14}{(7.8)^{0.02} - 1} \times 0.15 = 0.5 \text{ s}$$

The required T_{op} of relay R_1 is 0.5 + 0.25 = 0.75 s.

$$\text{PSM of } R_1 = \frac{3499}{800} = 4.38$$

$$T_{op} \text{ of } R_1 = \frac{0.14}{(4.38)^{0.02} - 1} \times TDS \text{ of } R_1$$

$$TDS \text{ of } R_1 = \frac{0.75}{4.67} = 0.160$$

Hence, the TDS of R_1 is selected as 0.2.

The TDS of relays R_2, R_4, and R_6 are calculated in a similar manner, considering the second network of Fig. 2.37 with the same data.

Example 2.16 Figure 2.38 shows the single line diagram of a power system. The relays are directional and non-directional. The PS and TDS of relay R_5 are given as 75% and 0.15, respectively. Determine the PS and TDS of the other relays by considering suitable discrimination time.

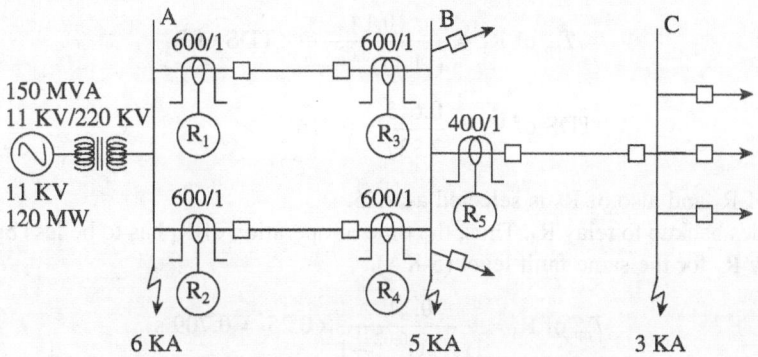

Fig. 2.38 Single line diagram of power system

Solution:
(i) Plug setting
The PS and TDS of relay R_5 is given as 75% of CT rating and 0.15.

Here, the relays R_3 and R_4 are directional relays, and the relays R_1 and R_2 are non-directional relays. The relays R_1 and R_2 are required to be coordinated with the relay R_5.

$$PS \text{ of } R_1 > \frac{1.3}{1.05} \times PS \text{ of } R_5 > \frac{1.3}{1.05} \times 300 > 371.42 \text{ A (primary)}$$

$$> 61.90\% \text{ of } 600 \text{ A}$$

Hence, the PS of the relays R_1 and R_2 is selected as 75% of the CT rating.

Now, the relay R_1 gives backup to the relay R_4, and the relay R_2 gives the backup to the relay R_3.

$$PS \text{ of } R_1 > \frac{1.3}{1.05} \times PS \text{ of } R_4$$

$$PS \text{ of } R_4 < \frac{1.05}{1.3} \times PS \text{ of } R_1 < \frac{1.05}{1.3} \times 450 < 363.46 \text{ A (primary)}$$

$$< 60.57\% \text{ of } 600 \text{ A}$$

Hence, the PS of the relays R_4 and R_3 is selected as 50% of the CT rating.
(ii) Time dial setting

Here, R_1 and R_2 are graded with the relay R_5.

PSM of R_5 for a fault at bus B (5 KA) is given by

$$\text{PSM of } R_5 = \frac{5000}{300} = 16.67$$

T_{op} of R_5 with TDS set at 0.15 is given by

$$T_{op} \text{ of } R_5 = \frac{0.14}{(16.67)^{0.02} - 1} \times 0.15 = 0.363 \text{ s}$$

Hence, the required T_{op} of the relay R_1 and R_2 is $0.363 + 0.25 = 0.613$, by considering 0.25 time discrimination.
Now, the PSM of the relay R_1 for the same fault level is

$$\text{PSM of } R_1 = \frac{5000}{450} = 11.11$$

$$T_{op} \text{ of } R_1 = \frac{0.14}{(11.11)^{0.02} - 1} \times \text{TDS of } R_1$$

$$\text{TDS of } R_1 = \frac{0.613}{2.837} = 0.216$$

Hence, the TDS of R_1 and also of R_2 is selected as 0.25.

Now, R_1 provides backup to relay R_4. Thus, the time of operation of R_4 has to be less by 0.25 s compared to that of the relay R_1 for the same fault level (5 KA).

$$T_{op} \text{ of } R_1 = \frac{0.14}{(11.11)^{0.02} - 1} \times 0.25 = 0.709 \text{ s}$$

Thus, the required T_{op} of the relay R_4 is $0.709 - 0.25 = 0.459$ s.

$$\text{PSM of relay } R_4 = \frac{5000}{300} = 16.67$$

$$T_{op} \text{ of } R_4 \text{ (and } R_3) = \frac{0.14}{(16.67)^{0.02} - 1} \times \text{TDS of } R_4$$

$$0.459 = 2.418 \times \text{TDS of } R_4$$

$$\text{TDS of } R_4 = \frac{0.459}{2.418} = 0.189$$

Hence, the TDS of the relays R_4 and R_3 is selected as 0.15.

Example 2.17 Figure 2.39 shows the single line diagram of a power system. When fault occurs at F on line 2 just beyond the relaying point R_2, the fault current through breakers and respective CTs is given in Table 2.11.

All relays are directional overcurrent relays. Assume the TDS of the relays R_1 and R_2 as 0.1. Calculate the relay settings for the relays R_3 and R_4.

Solution:
(i) Plug setting
The PS of any relay is decided in the worst case. To decide the PS of relay R_1 the load of 100 MW on bus B with one of the lines in outage condition is considered along with generator 2, which is under service outage.

Table 2.11 Fault current through breaker

Breakers	Fault current (A)
1	1000
2	3000
3	1000
4	1000

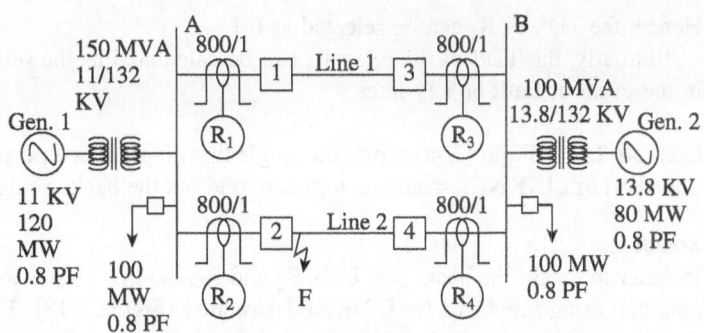

Fig. 2.39 Single line diagram of power system

$$\text{Hence, the PS of } R_1 \text{ (and } R_2) = \frac{100 \times 10^6}{\sqrt{3} \times 132 \times 800 \times 0.8 \times 10^3}$$

$$= 0.6834 \text{ A (secondary side of CT)}$$

$$= 68.34\% \text{ of relay rated current}$$

Hence, the PS of R_1 (and R_2) is selected as 75% of the CT rating.

In the same way, the PS of R_3 (and R_4) is decided on the basis of maximum loading condition of 100 MW on bus A, with one line under service. However, the generating capacity of generator 2 is 80 MW. Thus,

$$\text{PS of } R_3 \text{ (and } R_4) = \frac{80 \times 10^6}{\sqrt{3} \times 132 \times 800 \times 0.8 \times 10^3}$$

$$= 0.5467 \text{ A (secondary side of CT)}$$

$$= 54.67\% \text{ of relay rated current}$$

Hence, the PS of R_3 (and R_4) is decided as 75% of the CT rating.

(ii) Time dial setting

The TDS of the relays R_1 and R_2 is assumed as 0.1. The relay R_3 gives backup to relay R_2. At the time of fault at F, the current through relay R_2 is 3000 A.

$$\text{PSM of relay } R_2 = \frac{3000}{0.75 \times 800} = 5$$

$$T_{op} \text{ of } R_2 = \frac{0.14}{(5)^{0.02} - 1} \times 0.1 = 0.43 \text{ s}$$

Hence, the required T_{op} of R_3 is selected as $0.43 + 0.25 = 0.68$ s considering 0.25 s time discrimination.

$$\text{PSM of } R_3 = \frac{1000}{0.75 \times 800} = 1.67$$

The current through the breaker 3 for the said fault at F is 1000 A, on line 2.

$$T_{op} \text{ of } R_3 = \frac{0.14}{(1.67)^{0.02} - 1} \times \text{TDS of } R_3$$

$$\text{TDS of } R_3 = \frac{0.68}{13.58} = 0.05$$

Hence, the TDS of R_3 can be selected as 0.1.

Similarly, the TDS of the relay R_4 can be calculated as the relays R_1 and R_4 give backup to each other in the event of fault on any lines.

Example 2.18 Figure 2.40 shows the single line diagram of a portion of a power system network. Using the flowchart of LINKNET structure, logically find out the backup relay for the PRIs R_2 and R_6.

Solution:
In order to obtain the backup of PRIs R_2 and R_6, the first step is to prepare the branch–node table, which is obtained using flowchart for LINKNET structure (Figure 2.18). This is shown in Table 2.12.

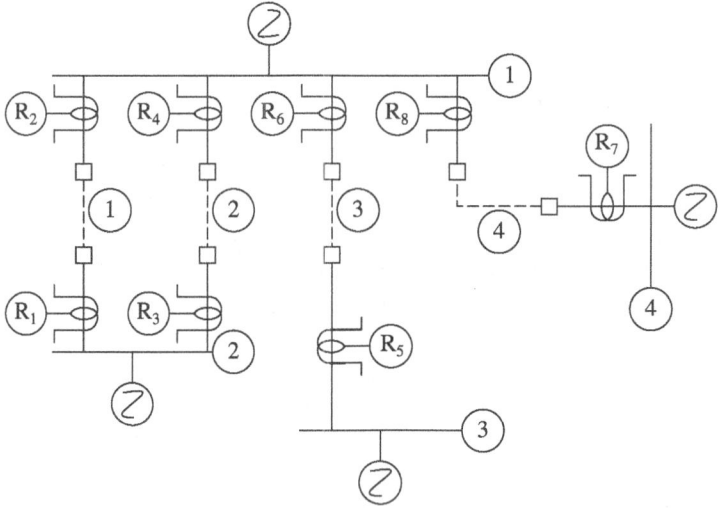

Fig. 2.40 Single line diagram of power system network

The second column in Table 2.12 is obtained based on the fact that branch 1 is connected between bus 1 and bus 2. Hence, the opposite bus of the lowest relay (R1 for branch 1) is node_A = 1(indicates bus number). Conversely, node_B = 2 (indicates bus number). Now, the other arrays such as *end*, *next*, *list*, and *far* are obtained using Fig. 2.18.

Table 2.12 Branch–Node obtained using flowchart for LINKNET structure

Branch	Node	End	Next (N)	List (L)	Far (F)
1	Node_A = 1	End_A = 1	$N(1) = L(1) = 0$	$L(1) = 1$	$F(1) = 2$
	Node _B = 2	End_B = 2	$N(2) = L(2) = 0$	$L(2) = 2$	$F(2) = 1$
2	Node_A = 1	End_A = 3	$N(3) = L(1) = 1$	$L(1) = 3$	$F(3) = 2$
	Node_B = 2	End_B = 4	$N(4) = L(2) = 2$	$L(2) = 4$	$F(4) = 1$
3	Node_A = 1	End_A = 5	$N(5) = L(1) = 3$	$L(1) = 5$	$F(5) = 3$
	Node_B = 3	End_B = 6	$N(6) = L(3) = 0$	$L(3) = 6$	$F(6) = 1$
4	Node_A = 1	End_A = 7	$N(7) = L(1) = 5$	$L(1) = 7$	$F(7) = 4$
	Node_B = 4	End_B = 8	$N(8) = L(4) = 0$	$L(4) = 8$	$F(8) = 1$

After preparing branch–node in Table 2.12, the primary backup relay pairs are obtained using the flowchart for primary/backup relay pair (Fig. 2.19). This is shown in Table 2.13.

Table 2.13 Primary/backup relay pair determination

Primary relay number	Primary relay number
PRI = 2	PRI = 6
Even relay and hence, IT = 1	Even relay and hence, IT = 5
IFLT = $F(2)$ = 1, IB = $L(1)$ = 7, IS = FFR(1) = 2	IFLT = $F(6)$ = 1, IB = $L(1)$ = 7, IS = FFR(5) = 3
IF = $F(7)$ = 4, IN = $N(7)$ = 5	IF = $F(7)$ = 4, IN = $N(7)$ = 5
(IF = IS, IN = 0, and IB = IT) is not true	(IF = IS, IN = 0, and IB = IT) is not true
(IF = IS) is not true	(IF = IS) is not true
Remote backup (RBU) relay = 7, PRI = 2	RBU relay = 7, PRI = 6
IN = 0 is not true and hence, IB = IN = 5	IN = 0 is not true and hence, IB = IN = 5
Repeating the same procedure twice, we obtain RBU relay = 5, PRI = 2 RBU relay = 3, PRI = 2	Repeating the same procedure, we obtain RBU relay = 3, PRI = 6
Stop	Stop

Example 2.19 Figure 2.41 shows the single line diagram of a power system network. Using the flow chart of LINKNET structure, logically find out the backup relay for the PRI R_8.

Solution:

In order to obtain the backup of the PRI R_8, the first step is to prepare the branch–node table, which is obtained using the flowchart for LINKNET structure (Fig. 2.18). This is shown in Table 2.14.

The second column in Table 2.14 is obtained based on the fact that branch 1 is connected between bus 1 and bus 2. Hence, the opposite bus of the lowest relay (R_1 for branch 1) is node_A = 1 (indicates bus number). Conversely, node_B = 2 (indicate bus number). Now, the other arrays such as *end*, *next*, *list*, and *far* are obtained using Fig. 2.18.

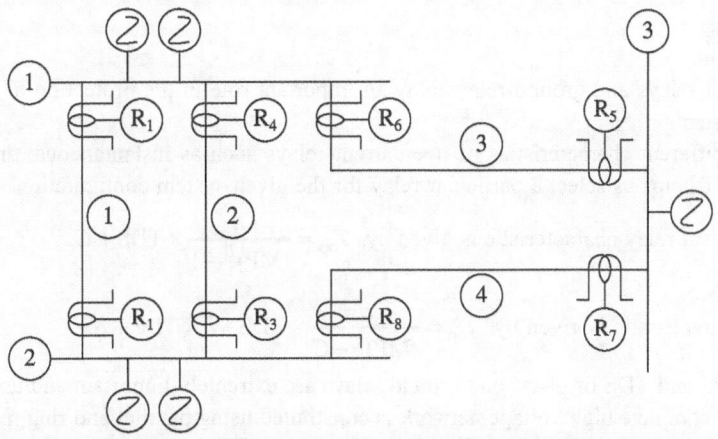

Fig. 2.41 Single line diagram of power system network

Table 2.14 Branch–Node obtained using flowchart for LINKNET structure

Branch	Node	End	Next (N)	List (L)	Far (F)
1	Node_A = 1	End_A = 1	N(1) = L(1) = 0	L(1) = 1	F(1) = 2
	Node_B = 2	End_B = 2	N(2) = L(2) = 0	L(2) = 2	F(2) = 1
2	Node_A = 1	End_A = 3	N(3) = L(1) = 1	L(1) = 3	F(3) = 2
	Node_B = 2	End_B = 4	N(4) = L(2) = 2	L(2) = 4	F(4) = 1
3	Node_A = 1	End_A = 5	N(5) = L(1) = 3	L(1) = 5	F(5) = 3
	Node_B = 3	End_B = 6	N(6) = L(3) = 0	L(3) = 6	F(6) = 1
4	Node_A = 2	End_A = 7	N(7) = L(2) = 4	L(1) = 7	F(7) = 3
	Node_B = 3	End_B = 8	N(8) = L(3) = 6	L(4) = 8	F(8) = 2

After preparing the branch–node Table 2.14, the primary/backup relay pairs are obtained using the flowchart for primary/backup relay pairs. This is shown in Table 2.15.

Table 2.15 Primary/Backup relay pair determination

Primary relay number
PRI = 8
Even relay and hence, IT = 7
IFLT = F(8) = 2, IB = L(2) = 7, IS = FFR(7) = 3
IF = F(7) = 3, IN = N(7) = 4
(IF = IS, IN = 0, and IB = IT) is not true
(IF = IS) is true, (IB = IT) is not true
RBU relay = 4, PRI = 8
IN = 0 is not true and hence, IB = IN = 5
Repeating the same procedure once, we obtain RBU relay = 2, PRI = 8
Stop

Recapitulation

- The overcurrent relays and ground relays play an important role in the protection of transmission and distribution systems.
- The study of different characteristics of overcurrent relays such as instantaneous, time delay, inverse time, and IDMT helps us select a particular relay for the given system configuration.
- Electromechanical relay characteristic is given by, $T_{op} = \dfrac{A}{(MP)^B - 1} \times TDS + C$.
- Static relay characteristic is given by, $T_{op} = \dfrac{a}{(MP)^n - C} \times TDS + b \times TDS + K$.
- The rules for PS and TDS of phase and ground relays are extremely important and useful.
- Nowadays, the complete high voltage network is constituted using parallel and ring mains transmission line. The evaluation of the directional feature is a must for the protection of such a network.
- Operating torque for directional relay is given by, $T_{operating} \propto V \times I \times \cos(\tau + \beta)$.

- It is important to study the different features of modern digital/numerical overcurrent and ground relays.
- Extensive coordination of overcurrent and ground relays is a demanding task in an interconnected power system network.

Multiple Choice Questions

1. The operating time of definite minimum time relay
   (a) varies with reference to current
   (b) is independent of current magnitude
   (c) is dependent of current magnitude
   (d) none of the above

2. Instantaneous overcurrent relays suffer from the problem of
   (a) transient overreach
   (b) transient underreach
   (c) both (a) and (b)
   (d) none of the above

3. The characteristic of an overcurrent relay having very inverse characteristic is more
   (a) stepper than normal inverse overcurrent relay
   (b) stepper than extremely inverse overcurrent relay
   (c) both (a) and (b)
   (d) none of the above

4. Which is the closest relay characteristic to the characteristic of fuse/MCCB?
   (a) normal inverse
   (c) extremely inverse
   (b) very inverse
   (d) all (a), (b), and (c)

5. The application of overcurrent relay monitored by an under-voltage relay is preferred in case of
   (a) varying load conditions
   (b) varying fault conditions
   (c) varying fault resistances
   (d) varying generating conditions

6. The value of coordination time interval between the successive relays is
   (a) 0.6 s
   (c) 0.25 s
   (b) 0.5 s
   (d) none of the above

7. The PS range of phase relays is
   (a) lower than ground relays
   (b) higher than ground relays
   (c) the same as ground relays
   (d) none of the above

8. While protecting a distribution feeder having transformer, the preferred relaying scheme is
   (a) two overcurrent and one earth fault
   (b) three overcurrent and one earth fault
   (c) both (a) and (b)
   (d) none of the above

9. The connection most widely used in case of directional relays for protection against phase faults is
   (a) 90° connection
   (b) 30° connection
   (c) 60° connection
   (d) none of the above

10. The overcurrent relays are widely used as PRIs up to
    (a) 11 kV
    (b) 132 kV
    (c) 400 kV
    (d) all of the above

Review Questions

1. Give the classification of overcurrent relays based on the characteristic used.

2. Discuss various advantages and disadvantages of definite minimum time relay.

3. Why are instantaneous relays alone not capable to protect transmission line?

4. Give the relative merits and demerits of instantaneous overcurrent relays.

5. Compare overcurrent relays giving definite time characteristic, inverse time characteristic, and IDMT characteristic.

6. Why is inverse time overcurrent relay having normal inverse characteristic not suitable for higher Z_s/Z_i ratio?

7. Why are extremely inverse and very inverse time overcurrent relays more suitable to the fuse/MCCB (moulded case circuit breaker) characteristic?

8. Why do IDMT relays provide better discrimination than inverse time overcurrent relay?

9. Why are the settings of ground relays lower than the phase relays?

10. Is it possible to design an extremely sensitive ground relay? If yes, what are the factors to be considered?

11. Justify the statement, 'current grading based relay setting alone cannot be used for feeder protection'.

12. List the factors to be considered while grading two consecutive overcurrent relays.

13. Discuss the PS and TDS rules for phase relays and ground relays.

14. For the protection of distribution feeders containing transformer, why is the three overcurrent and one earth-fault scheme preferred instead of the two overcurrent and one earth-fault scheme?

15. Why is the application of overcurrent relays difficult in the case of varying generating conditions?

16. Why are directional overcurrent relays used at the load end of the parallel feeders?

17. What do you mean by dead zone with reference to directional relay?

18. Draw the characteristic of directional relays and clearly show tripping zone, blocking zone, maximum torque angle, maximum torque line, and minimum polarizing voltage.

19. What is polarizing quantity? Why are 30°, 60°, and 90° connections used in directional relay?

20. Draw the vector diagram of the 90° connection for double line fault showing maximum torque angle.

21. Explain LINKNET structure for coordination of overcurrent relay.

Numerical Exercises

1. Calculate the relay setting of the single line diagram of a radial feeder shown in Fig. 2.42. The relays used are standard IDMT phase overcurrent relay, having rated current of 1 A. The normal range of PS is 50 – 200% of 1 A in seven equal steps, whereas the TDS range is 0.1 to 1 in steps of 0.05. The PS of the relay R_2 is given as 50% of CT secondary rating, and the TDS is given as 0.2.

[PS of R1 = 50%, TDS of R1 = 0.3]

2. Figure 2.43 shows the single line diagram of a radial feeder. The relays used are standard IDMT, having rated current of 1 A. The PS range is 50 – 200% of 1 A in seven equal steps, whereas the TDS range is 0.1 to 1 in step of 0.05. The PS of the relay R_3 is given as 75% of CT secondary rating, and the TDS is given as 0.2. Calculate the relay setting for relay R_1 and R_2.

[PS of R_2 = 75%, PS of R_1 = 150%, TDS of R_2 = 0.35, TDS of R_1 = 0.5]

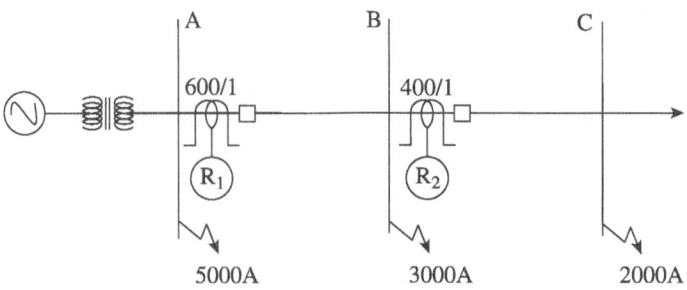

Fig. 2.42

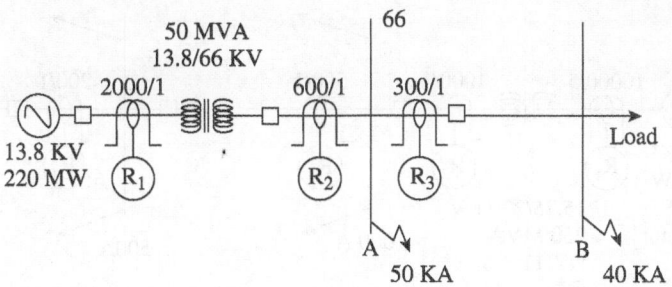

Fig. 2.43

3. In Fig. 2.44, the relays used are standard IDMT, having rated current of 5 A. The PS range is 50 to 200% of 1 A in seven equal steps, whereas TDS setting range is 0.1 to 1 in step of 0.05. The PS of the relay R_4 is given as 50% of CT secondary rating, and the TDS is given as 0.15. Determine the PS and TDS settings of other relays.

[PS of R_3 = 75%, PS of R_2 = 50%, PS of R_1 = 50%, TDS of R_3 = 0.25, TDS of R_2 = 0.3, TDS of R_1 = 0.4]

4. Figure 2.45 shows a single line diagram of a radial feeder. The relays used are standard IDMT, having rated current of 1 A. The PS range is 50 to 200% of 1 A in seven equal steps, whereas the TDS range is 0.1 to 1 in step of 0.05. The PS of the relay R_3 is given as 100% of CT secondary rating, and the TDS is given as 0.15.

[PS of R_2 = 75%, PS of R_1 = 75%, TDS of R_2 = 0.2, TDS of R_1 = 0.3]

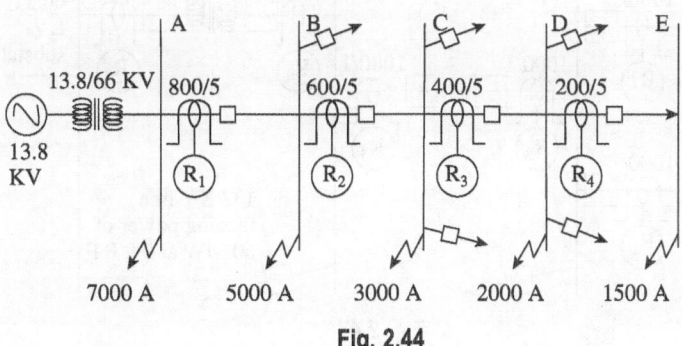

Fig. 2.44

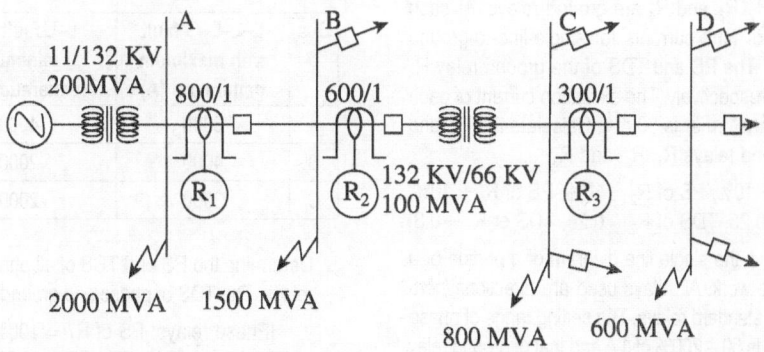

Fig. 2.45

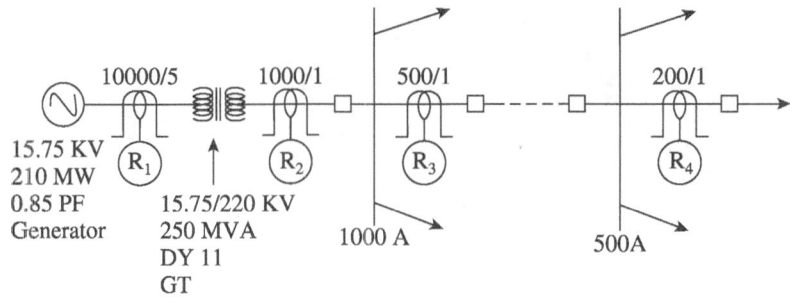

Fig. 2.46

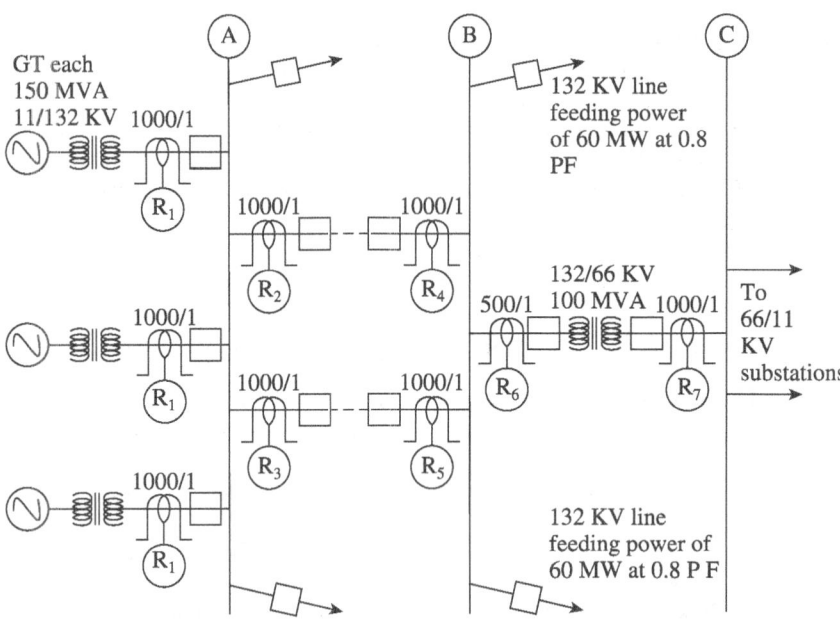

Fig. 2.47

5. Figure 2.46 shows a portion of the power system network. The relays R_1, R_2, R_3, and R_4 are ground relays. At each bus, the levels of fault currents for single-line-to-ground faults are given. The PS and TDS of the ground relay R_4 is 20% and 0.2, respectively. The excitation current of each of the current transformer is 50 mA. Calculate the PS and TDS of the ground relays R_1, R_2, and R_3.

[PS of $R_3 = 10\%$, PS of $R_2 = 10\%$, PS of $R_1 = 10\%$, TDS of $R_3 = 0.25$, TDS of $R_2 = 0.26$, TDS of $R_1 = 0.3$]

6. Figure 2.47 shows the single line diagram of a portion of a power system network. All relays used are directional/non-directional IDMT standard relays. The setting range of phase overcurrent relay is 50 – 200% of 1 A and that of ground relay is 10 – 40% of 1 A. The excitation current of all CT cores is 0.05 A. The fault currents are tabulated in Table 2.16.

Table 2.16   Fault currents

Bus	L–L–L–G fault with maximum generation (A)	L–L fault with minimum generation (A)	L–G fault (A)
A	6000	4000	3000
B	4000	2000	2000
C	3000	2000	2000

Determine the PS and TDS of all phase relays and ground relays. The TDS of phase and ground relay R7 is set at 0.1.

[Phase relays: PS of R7 = 100%, PS of R6 = 125%, PS of R5 = 75%, PS of R4 = 75%, PS of R3 = 125%, PS of R2 = 125%, PS of R1 = 150%, TDS of R6 = 0.15, TDS of R5 = 0.15, TDS of R4 = 0.15, TDS of R3 = 0.1,

TDS of R2 = 0.1, TDS of R1 = 0.1
Ground relays: PS of R7 = 10%, PS of R6 = 10%,
PS of R5 = 10%, PS of R4 = 10%, PS of R3 = 20%,
PS of R2 = 20%, PS of R1 = 30%, TDS of R6 = 0.1,
TDS of R5 = 0.1, TDS of R4 = 0.1, TDS of R3 = 0.15,
TDS of R2 = 0.15, TDS of R1 = 0.15]

7. Identify the directional relays in Fig. 2.48. The TDS of the relays R_9 and R_{10} is given as 0.1 (refer to Table 2.17). Considering suitable time discrimination, find the PS and TDS of all others relays. Figure 2.48 shows the fault levels at the given buses.

Table 2.17 TDS of relays

Relays	PS (% of CT rating)	TDS
R_1 and R_2	125	0.2
R_3 and R_4	100	0.15
R_5 and R_6	100	0.15
R_7 and R_8	75	0.1
R_9 and R_{10}	50	0.1

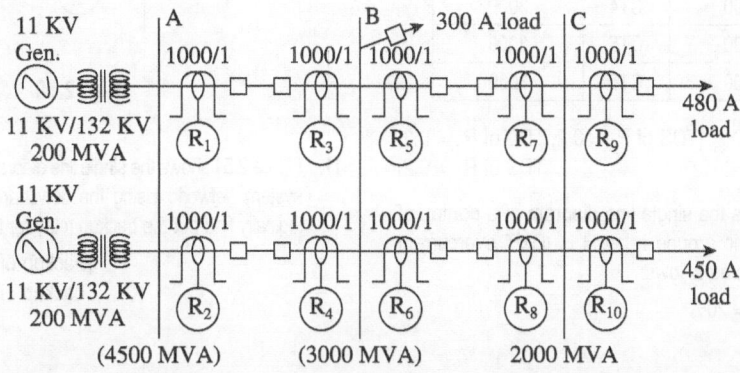

Fig. 2.48

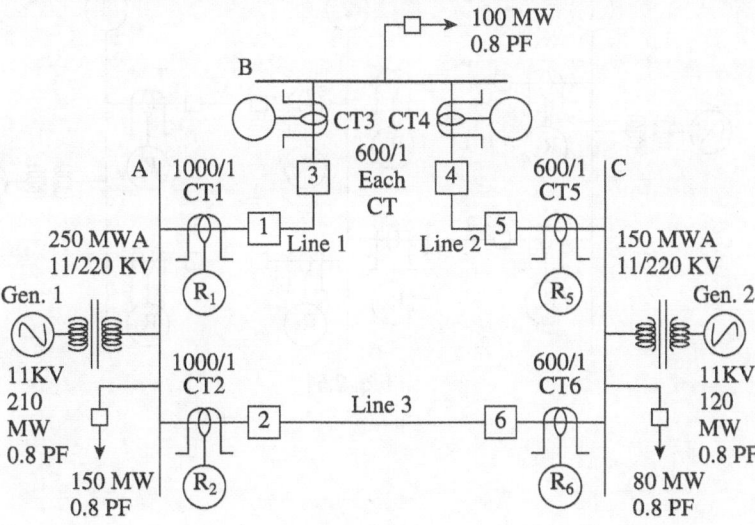

Fig. 2.49

8. Figure 2.49 shows the single line diagram of a power system. Identify the directional and non-directional relays and also find the backup relay of each PRI. Calculate the PS of each relay with the given load condition.

[PS of R_1 = 75%, PS of R_2 = 75%, PS of R_3 = 75%, PS of R_4 = 50%, PS of R_5 = 75%, PS of R_6 = 75%]

9. In Numerical Exercise 8, a fault occurs just beyond the relay location R_1 on line 1 with fault current through different CTs. This is given in Table 2.18. The TDS of the relays R_1 and R_3 is assumed to be 0.1. Find the TDS of the other relays in conjunction with the said fault by considering suitable time discrimination.

Table 2.18 Fault currents through different CTs

CT	Current (A)	CT	Current (A)
CT1	5000	CT4	3000
CT2	3000	CT5	3000
CT3	3000	CT6	3000

[TDS of R_2 = 0.2, TDS of R_5 = 0.2, TDS of R_6 = 0.2]

10. Figure 2.50 shows the single line diagram of a portion of power system. The ground relay R is IDMT overcurrent relay with settings as follows:

 (a) Plug setting = 20%

 (b) TDS = 0.2

The excitation current of three line CTs is 0.05 A each.

Determine the fault current for single-line-to-ground for which the relay will just pickup. In addition, find out the time of operation of the relay if the fault current is equal to 60 A and 100 A, respectively.

[Fault current = 35 A, T_{op} of R (60 A) = 1.7125 s, T_{op} of R (100 A) = 0.9536 s]

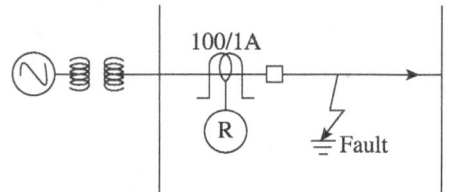

Fig. 2.50

11. Figure 2.51 shows the single line diagram of a portion of a power system network. Using the flowchart of LINKNET structure, logically find out the backup relay for the PRIs R_2 and R_6.

[Backup of relay R_2 = R_3, R_5, R_7; backup of relay R_6 = R_2, R_4, R_7]

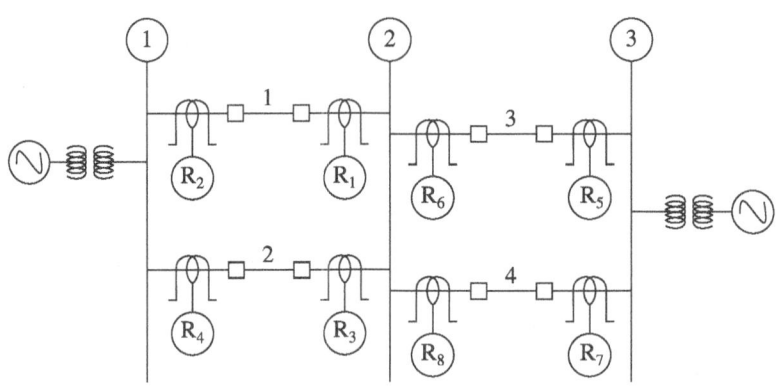

Fig. 2.51

Answers to Multiple Choice Questions

1. (b)   2. (a)   3. (a)   4. (c)   5. (d)   6. (c)   7. (b)   8. (b)   9. (a)   10. (a)

Distance Relaying Scheme for Transmission Line

3

Learning Objectives

After going through this chapter, the students will be able to:

- Explain the different schemes used to protect transmission lines
- Discuss the principle underlying distance protection
- Understand the factors influencing the reach of distance relay
- Explain the connections for phase-fault and earth-fault relays
- Explain the working of different types of static comparators
- List the various problems in distance protection and discuss the ways to overcome them
- Explain digital distance relay and protection

3.1 Introduction

Transmission lines connect the various parts of an electric power system. As these lines are located in areas that encounter harsh environmental conditions, they are vulnerable to numerous faults. Extra high voltage (EHV) transmission lines are important links in the bulk power transfer system, and the faithful operation of these lines is vital for system stability, power transfer capability, and voltage control. Hence, the protection of transmission lines is a task of fundamental importance in modern power systems. Owing to the network configurations and complexities, added with the special restrictions on the system operation, such as single-phase switching, transmission line protection is becoming a great challenge, and presents a multifaceted problem for protection engineers. Faults in the transmission system must be identified, and the faulted line must be isolated from the network with minimum delay. A distance relay is an economic option that can be widely used for the protection of EHV and ultra high voltage (UHV) transmission lines.

3.2 Transmission Line Protection

Depending on the requirements, transmission lines are protected by overcurrent, distance, and pilot relaying systems. The current-based scheme fails in many situations and does not provide instantaneous protection to the entire line. Moreover, the coordination of overcurrent relays is very difficult, particularly for a large, interconnected power system. Pilot relaying is used to protect a line when high-speed protection of an entire line is desirable.

In addition to these three relaying schemes, differential protection, phase comparison, and directional comparison methods are also used.

Differential protection The differential protection scheme operates on the principle of the comparison of the currents at both ends of the protected line. If the difference in these currents is not equal to zero, then there is an internal fault, and if the difference in these current equals zero, then the fault is external or the system is healthy. The measured current at one end of the line must be transmitted via a communication channel to the other end of the line. Communication channels are cables, microwaves, power line carriers, or fibre optics. In order to minimize the amount of transmitted data, the three-phase currents are usually mixed, and only one resulting current is transmitted (non-segregated differential protection). Unless phase selectors are installed, the single-pole tripping capability is lost. Current transformer (CT) mismatches and security problems caused by CT saturation during external faults with high short-circuit currents or DC components are the major limitations of differential protection schemes.

Phase comparison The above mentioned drawbacks can be avoided by using the phase comparison scheme that utilizes the phases of the currents at both ends of the line. The relay trips if the phases of both currents are essentially in phase and blocks if they are out of phase. The differential protection and phase comparisons are only current-based schemes. They do not provide any protection during the failure of communication link and cannot be used as a backup protection for adjacent lines.

Directional comparison In directional comparison scheme, the directional elements connected at each end of the line categorize faults as backward or forward faults by evaluating the phase difference of voltages and currents at their relay locations. However, like differential protection and phase comparison, directional comparison depends on the communication channels. The relay tripping time includes the time required for the communication exchange. Reliability and tripping times of directional comparison schemes are negatively affected during communication channel failure. Distance protection, which we will discuss in detail in this chapter, is independent of communication channels.

3.3 Distance Protection

Distance relaying scheme has been widely used for the protection of transmission lines for the last five decades. The philosophy of distance protection is explained in Sections 3.3.1 and 3.3.2.

3.3.1 Fault Distance Measurement

The operation of distance relays is mainly based on the impedance measured at the relaying point. The voltage-to-current ratio of the fundamental frequency components seen at the relaying point is an indicator of the system's normal or fault condition. Figure 3.1 shows a single line diagram of a faulty power network.

A distance relay is connected at bus A and bus B to protect the transmission line having an impedance $Z_L = R_L + jX_L$. KS is the set value of impedance. Figure 3.2 illustrates the impedance measurement at the relaying point for different faults. It has been observed from Fig. 3.2 that for an internal fault at F1, the measured impedance at bus A lies in the first quadrant of the impedance plane and is less than the set value of impedance ($Z_1 <$ KS). For an external fault at F2, it lies in the first quadrant of the impedance plane, but it is greater than the set value of impedance ($Z_2 >$ KS). For a backward or reverse fault at F3, it lies in the third quadrant of the impedance plane (negative impedance). The faults F1–F3 are assumed not to have any fault resistance value. For an internal fault, the fault trajectory of the measured impedance moves from its pre-fault condition (load area) to its post-fault steady state measurement. As soon as the fault trajectory

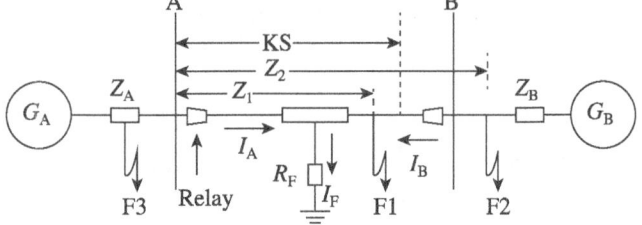

Fig. 3.1 Transmission line protected by distance relay

enters the trip area and settles down on a line defined by the transmission line angle ($\phi_L = \tan^{-1}(X_L/R_L)$), the relay (local and remote ends) sends a trip signal to the respective breakers (local and remote ends).

3.3.2 Three Stepped Distance Characteristics

Backup protection is made possible in distance protection by stepped distance characteristics. It is one of the most commonly used backup protection schemes on high voltage (HV) and EHV transmission lines. In such a scheme of protection, a distance relay has three zones of protection as shown in Fig. 3.3.

The first zone or the high-speed zone, designated as zone I, is set to trip without any intentional time delay and provides primary protection for the line

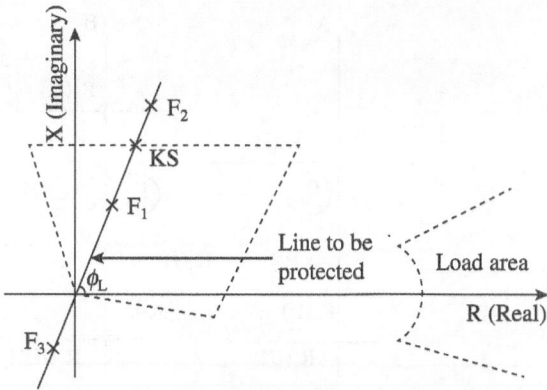

Fig. 3.2 Distance measurement at relay location for different faults

section to be protected. To avoid unnecessary operation on faults beyond the remote end due to the transient overreach and errors in the instrument (current and voltage) transformers (CTs and VTs), zone I is adjusted to reach 80–90% of the length of the line section.

The second zone, designated zone II, is used to provide high-speed protection for the remainder of the line with an adequate margin. It gives instantaneous protection for the line section that is uncovered by zone 1 and time-delayed protection for half of the following line section. It should be adjusted so that it will be able to operate even for arcing faults at the end of the line. Even if the arcing faults did not have to be considered, the underreaching tendency of the relay because of the effect of intermediate current sources and the errors in the data, CTs, potential transformers (PTs), and relays, should be taken into account. It is customary to have the second-zone unit reach 50% of an adjoining line section. The second-zone relays have to be time delayed to coordinate with relays at the remote bus, and the typical time delays are of the order of 0.3–0.6 s.

As shown in Fig. 3.3(a) the relay R_1 will provide backup protection in the second zone if the primary protection of relay R_3 fails to operate for a fault in 50% of section CD.

The third zone, designated as zone III, is used to provide remote backup to the first and second zone of an adjacent line sections when a relay or breaker fails to clear the fault locally. The usual practice is to extend its reach beyond the end of the largest adjoining line section or more than double the line section to be protected. The third-zone operation is usually delayed by about 1.4–1.8 s. The third-zone reach setting is a more complex problem. It has been observed that the zone III unit trips under heavy and unusual loading conditions, and thereby leads to the cascade tripping of the power system. The third-zone setting must be blocked in case of extreme loading conditions. In certain conditions, its reach can be modified. There are certain critical locations where zone III protection can be removed if alternative protection functions in other forms are available.

Figure 3.3(b) shows the operating characteristic of mho relay with the three zones. The control circuit of distance protection is also shown in Fig. 3.3(c) including contacts of directional element (D-1), contacts of three zones of distance relay (Z_1-1, Z_2-1, and Z_3-1), auxiliary relay (86), and contacts of a timer (T). If the three relay units are of impedance or reactance type, then separate directional measurement is required. On the contrary, Mho relay does not require any directional element as it is an inherently directional unit. As shown in Fig. 3.3(b), the zone I circle is nominal and covers 80% of line length. If any fault occurs within the reach of zone-I, contact of unit Z_1, that is, Z_1-1 closes and energizes the contact of directional element (D-1) which further energizes the coil of auxiliary relay (86) and finally trips the circuit breaker (CB). The seal-in contact (86-1) of auxiliary relay provides a hold on path to the trip coil of CB. The zone II of distance relay covers 50% of the next line section. In case when the fault impedance lies within the reach of zone II but outside the reach of zone I, both relay units zone II and zone III operate because zone III relay unit

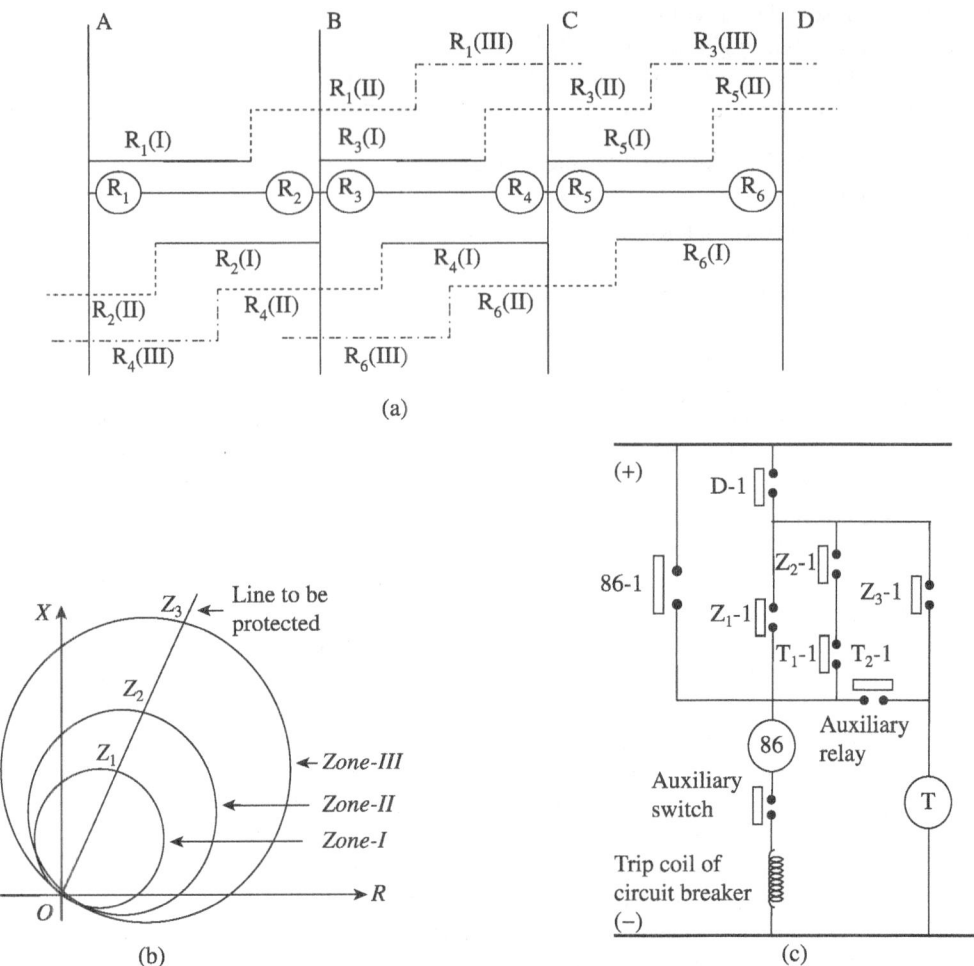

Fig. 3.3   Distance protection scheme (a) Stepped distance characteristics, (b) Mho relay characteristic, (c) control circuit of three zone distance protection

covers the complete region with the largest diameter. Thus, for second-zone fault, contacts Z_2-1 and Z_3-1 operate immediately. Whenever the contact of directional element (D-1) and Z_3-1 operates for second-zone fault, the timer (T) is energized. Whenever a definite time of timer (T) elapses, the closing of contact T_1-1 energizes the trip coil of CB. If the fault impedance is measured outside zone II, that is, in zone III region, after a delayed time, the contact T_2-1 closes and subsequent tripping is issued. It is to be noted that both timers of zone II and zone III are individually adjustable.

3.4 Reach of Distance Relay

The settings of phase distance relays are done on the basis of the positive-sequence impedance between the relaying point and the fault point. On the other hand, the settings of ground distance relays are carried out on the basis of the zero-phase-sequence impedance. Hence, the corresponding distance or impedance is known as the *reach of the relay*. Now, the relay is always connected on the secondary side of the CT and PT. Hence, to transfer the impedance of the line referred to the primary of CTs and PTs to the line impedance referred to the relay side, the following equation is to be used.

$$Z_{sec} = Z_{pri} \times \frac{CTr}{PTr} \tag{3.1}$$

where, CTr is the ratio of the CT primary current to the CT secondary current, and PTr is the ratio of the primary phase-to-phase voltage to the secondary phase-to-phase voltage. These values are under balanced three-phase conditions.

The phenomenon when a distance relay operates beyond its zone of protection or for impedances greater than its set value is known as *overreaching of the relay*. Similarly, the tendency of a distance relay not to operate within its zone of protection or lower than its set value of impedance is known as *underreaching of the relay*.

The value of DC offset in the fault current is responsible for the overreaching of the distance relay. The ratio of DC offset to the fundamental frequency component depends on the instant at which the fault occurs. However, the instant of fault is not predicated, and hence, it is not in the user's hand. The rate of decay of DC offset depends on the *X/R* ratio of the system, which is very high for modern power systems. However, this DC offset is present only up to the first few cycles after the inception of fault. This overreach is transient in nature as the DC offset decays rapidly. Hence, this overreach is widely known as *transient overreach*. As the transient overreach disappears after a few cycles, the second and the third zones of distance relays are not affected by this phenomenon. Conversely, the first zone or high-speed protection zone is affected by the transient overreach.

The percentage transient overreach is defined as

$$\text{Percentage transient overreach} = \frac{Z_x - Z_y}{Z_y} \times 100 \tag{3.2}$$

where, Z_x is the maximum impedance for which the relay will operate with an offset current wave, for a given adjustment. Z_y is the maximum impedance for which the relay will operate for symmetrical currents, for the same adjustment as for Z_y.

As the angle of the system ($\phi = \tan^{-1}(X/R)$) increases, the transient overreach increases. It also increases for long EHV and UHV lines as the value of inductive reactance is very high because of the bundling of conductors and greater spacing between them.

3.5 Types of Distance Relay

The discrimination of fault condition against heavy load and such other conditions when the relay is not required to operate requires the measurement of not only the magnitude, but also the angle of the impedance of the line up to the fault point. Many types of distance relays have therefore been developed and applied for line protection. These are as follows:

1. Impedance relay
2. Reactance relay
3. Mho or admittance relay
4. Ohm or angle impedance relay
5. Offset mho relay
6. Quadrilateral and other special characteristics

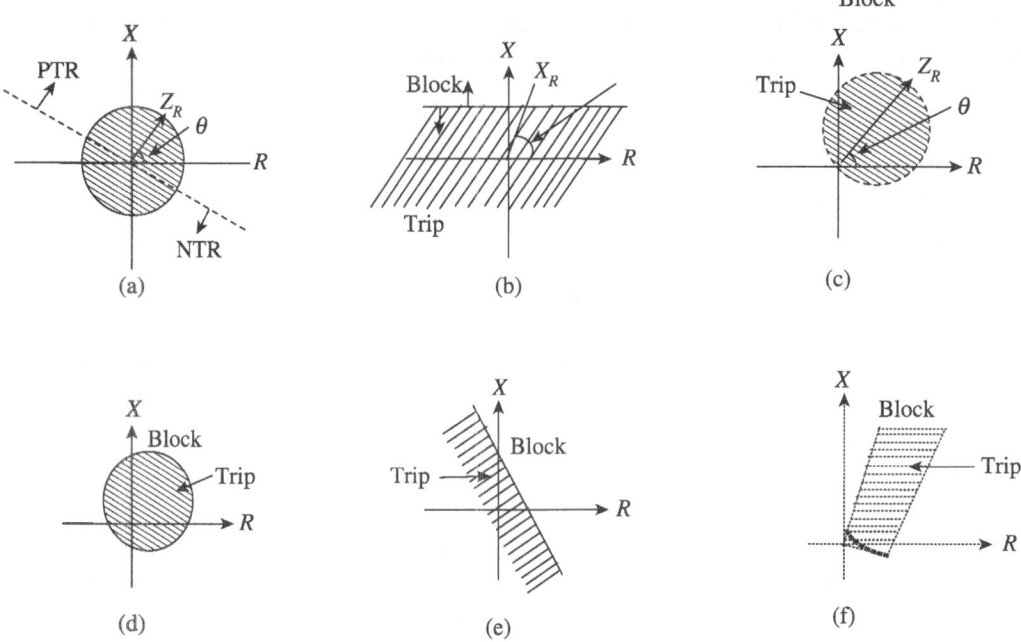

Fig. 3.4 Different characteristics of distance relays (a) Impedance (b) Reactance (c) Mho (d) Offset mho (e) Ohm (f) Quadrilateral

R-X Diagram

Any single or two-input quantity relay requires plotting of a characteristic using voltage (V), current (I), and their phase angle difference (θ) or a combination of these measurements. For plotting distance relay characteristic, the above-mentioned system parameters (V, I, and θ) are transformed into variables, R and X. Thus, only two values R and X (or Z and θ) are utilized instead of using the above three measurements. Moreover, the R-X diagram permits to plot both relay characteristic and system parameters (line to be protected) on the same plane. The measured rms values of CT secondary current (I_{CTS}) and PT secondary voltage (V_{PTS}) are used to derive the operating quantities such as impedance, reactance, and admittance. The conventional representation of the operating characteristics of these relays on R–X (impedance) plane (polar characteristic) along with their positive torque region (PTR), and negative torque region (NTR) is given in Fig. 3.4.

Impedance Characteristic

Relay which computes the ratio of voltage to current, that is, V/I is known as impedance relay. The impedance characteristic, as shown in Fig. 3.4(a), is plotted with a circle that has centre as the origin of R-X plane and radius as Z. In impedance relay, the current coil provides actuating force whereas the voltage coil produces restrictive force. Thus, this relay operates only when the calculated impedance (also known as measured value of impedance, Z_{mes}) falls below the pre-set value of impedance (Z_{set}) in relay. Here, Z_{mes} is the impedance measured between the relaying location and the fault point on line. As we know that the measured value of impedance Z_{mes} is related to the distance between the relaying point and fault position, it is sometimes labelled as distance relay. Plain impedance relay is bidirectional in nature and operates for any fault on either side of the relaying location. Hence, impedance relay requires a separate directional unit for line protection. To reduce the effect of the resistance component of an arc on reach of the relay, the relay characteristic angle of directional element is made smaller than the line angle.

Reactance Characteristic

As shown in Fig. 3.4(b), the characteristic of reactance relay is a straight line having only reactance component at a fixed distance from R-axis. Here, X_R is the reactance of protected line between the relay location and the fault point. This relay possess an overcurrent measuring unit which produces positive torque along with a directional unit which either allows or restrains the operation of an overcurrent element. The direction of operation is decided based on the angle difference (θ) between the voltage and current. If the current lags the voltage by 90°, maximum negative force is developed in the reactance measuring unit. This relay senses only reactance component and operation is not affected by the resistance component of the line. There is a possibility of mal operation of the reactance element, because of power swings and loading condition. Hence, to overcome mal operation, a fault detector unit is used in conjunction with the reactance element. Moreover, separate directional element is also required to prevent the maloperation of this unit in case of reverse faults.

Mho Characteristic

A relay which measures admittance of the line is called mho relay or admittance relay (Fig. 3.4(c)). This relay can overcome the limitation of impedance relay by providing a unidirectional feature. The mho characteristic of distance relay is the boundary of a circle passing through the origin of R-X plane. This is achieved by polarizing the impedance relay with additional voltage which provides current biasing to offset the characteristics and makes it directional. Hence, the quantity $Y\angle\theta$ is measured by mho unit, where θ is the characteristic angle of relay. Mho elements are inherently directional, and no separate directional unit is required. However, mho element cannot incorporate the value of the fault resistance. Hence, in practical field, a quadrilateral element is widely used, which incorporates reasonable value of fault resistance. Normally, the first- and the second-zone elements are of mho type and the third-zone element is of offset mho type (Fig. 3.4(d)). The function of zone 3 is to provide backup protection in case a relay or a breaker in the next bus section fails to clear a fault. Inaccuracy due to the loads and power infeeds is worse with zone 3 because there are two buses between the relay and the cut-off point of the relay.

Overreaching can cause the third-zone unit to trip undesirably for a fault on the other side of a large distribution transformer. Further, if the setting is too long, the third-zone unit becomes susceptible to operation on power swings. In addition, the discrimination between faults and loads becomes poor. Therefore, at certain lines, the recent practice is to omit the third zone of the distance relay and use only the two zones of the distance element.

Quadrilateral Characteristic

As stated above, the prime limitation of distance relay having MHO characteristic is that it measures incorrect value of impedance during fault which involves arc resistance or fault resistance. Recent practice is to use a distance element which provides better resistive reach than any other type of characteristic (impedance, mho). Quadrilateral type distance characteristic has an independent setting of resistive and reactive reach. Hence, modern digital distance relay utilizes quadrilateral or polygonal characteristic (Fig. 3.4(f)) having flexible fault impedance coverage for both phase and earth faults unit. Implementation of quadrilateral characteristic on the R-X diagram requires (i) directional element, (ii) reactance element, and (iii) left and right resistance blinder elements. Directional element distinguishes between forward and reverse fault and is set to operate for forward direction only. Reactance and resistance elements are the most important as the performance of the quadrilateral characteristic mainly depends on them. The reactance element determines the impedance reach of relay (load flow) while two resistive blinders are adjusted to detect high resistance fault so that distance relay does not overreach or underreach unnecessarily.

3.6 Current and Voltage Connections

If the measurement of fault distance is to be made with reasonable accuracy, the voltage at the relay terminals must be proportional to the voltage drop to the fault. This cannot be achieved with a single measuring relay, and it is therefore an usual practice to employ three units of phase-fault measuring relays and three units of earth-fault measuring relays. However, in a switched scheme, one relay for phase faults and one for earth faults (or only one in all) is used, and the connections to appropriate phase pair are made by the selectors or starters.

3.6.1 Connections for Phase-fault Relays

A phase relay must have the same balance point (i.e., should measure the same distance of the fault) on three-phase, double-line, and double-line-to-ground fault involving a given phase pair. This is achieved by supplying the relay with the delta voltage and delta current from the phase pair in question. Thus, the three-phase fault relays should be supplied with $(V_b - V_a)$ and $(I_b - I_a)$, $(V_c - V_b)$ and $(I_c - I_b)$, and $(V_a - V_c)$ and $(I_a - I_c)$. With these signals, the relay measures positive-sequence impedance of the line section upto the fault location for any type of the phase faults. This is proved as follows:

A double-line-to-ground fault is represented by the three sequence networks in parallel, whereas a double-line fault is represented by the positive- and negative-sequence networks in parallel. So for both types of faults on a given phase pair, say b–c, positive- and negative-sequence voltage at fault point V_{f1} (phase b) and V_{f2} (phase c) are equal. At the relay location, the positive- and negative-sequence voltages are given by

$$V_{r1} = V_{f1} + I_1 Z_1$$
$$V_{r2} = V_{f2} + I_2 Z_2 \tag{3.3}$$

where I_1 and I_2 are sequence currents and Z_1 and Z_2 are the sequence impedances of the line. Since $Z_1 = Z_2$ for long transmission lines and $V_{f1} = V_{f2}$ as described earlier,

$$V_{r1} - V_{r2} = (I_1 - I_2) Z_1$$

Or
$$\frac{V_{r1} - V_{r2}}{I_1 - I_2} = Z_1 \tag{3.4}$$

Next,
$$V_b = a^2 V_{r1} + a V_{r2} + V_{r0}$$

and
$$V_c = a V_{r1} + a^2 V_{r2} + V_{r0}$$

So,
$$V_b - V_c = (a^2 - a) V_{r1} + (a - a^2) V_{r2}$$
$$= (a^2 - a)(V_{r1} - V_{r2}) \tag{3.5}$$

Similarly,
$$I_b - I_c = (a^2 - a)(I_1 - I_2) \tag{3.6}$$

From Eqs (3.5) and (3.6)
$$\frac{V_b - V_c}{I_b - I_c} = \frac{V_{r1} - V_{r2}}{I_1 - I_2} = Z_1,$$

which is the same as the value obtained from Eq. (3.4).

The basic connection with delta currents and delta voltages to phase-fault relays is given in Fig. 3.5.

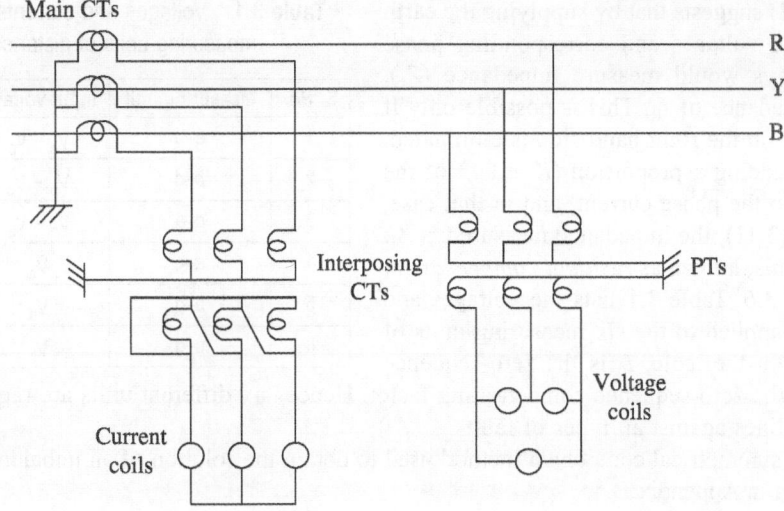

Fig. 3.5 Connections for phase-fault relays

3.6.2 Connections for Earth-fault Relays

On the occurrence of an earth fault, the phase-to-earth voltage at the fault location is zero. The voltage drop to the fault is not the product of the phase current and the line impedance but depends on the number of earthing points, the method of earthing, and the sequence impedance of the fault loop. The voltage drop to the fault is the sum of the sequence voltage drops between the relaying point and the fault. For fault on phase 'a',

$$V_a = V_{r1} + V_{r2} + V_{r0}$$

$$V_a = I_1 Z_1 + I_2 Z_2 + I_0 Z_0 \tag{3.7}$$

During line-to-ground fault,
$$I_1 = I_2 = I_0 \tag{3.8}$$

In addition,
$$I_a = I_1 + I_2 + I_0, \ I_b = 0, \ \text{and} \ I_c = 0 \tag{3.9}$$

From Eqs (3.8) and (3.9),
$$I_R = I_a + I_b + I_c = 3I_0 \tag{3.10}$$

where I_R is the residual current and I_a, I_b, I_c are the phase currents at the relaying point.

Substituting $Z_1 = Z_2$ and $Z_0 = KZ_1$ in Eq. (3.7), where K is the zero-sequence compensating factor, gives the following expression.

$$V_a = Z_1(I_1 + I_2 + KI_0)$$
$$= Z_1(I_1 + I_2 + I_0 + KI_0 - I_0)$$

Hence,
$$Z_1 = V_a \{I_a + I_0(K-1)\}$$

From Eqs (3.9) and (3.10) we get,

$$V_a = Z_1 \left(I_a + \frac{(K-1)}{3} I_R \right) \tag{3.11}$$

Therefore,
$$\frac{V_a}{I_a} = Z_1 + \left(Z_1 \frac{(K-1)}{3} \times \frac{I_R}{I_a} \right) \tag{3.12}$$

Equation (3.12) suggests that by supplying the earth relays with phase voltages and corresponding phase currents, the relays would measure impedance (Z_1), which is the impedance of I_R. This is possible only if the second factor on the right hand side is eliminated. This is done by adding a proportion $(K - 1)/3$ of the residual current to the phase current, and in that case, as shown in Eq. (3.11), the impedance measured is Z_1. The scheme for this, known as *residual compensation*, is shown in Fig. 3.6. Table 3.1 lists the voltages and currents that are applied to the six measuring units of a distance relay. In the table, I_0 is the zero-sequence current and K is the zero-sequence compensating factor. Hence, six different units are required to protect the transmission lines against all types of faults.

Table 3.1 Voltages and currents applied to the measuring units of distance relays

S. No.	Measuring unit	Input voltage	Input current
1	a–b	$V_a - V_b$	$I_a - I_b$
2	b–c	$V_b - V_c$	$I_b - I_c$
3	c–a	$V_c - V_a$	$I_c - I_a$
4	a–g	V_a	$I_a + I_0 (K - 1)$
5	b–g	V_b	$I_b + I_0 (K - 1)$
6	c–g	V_c	$I_c + I_0 (K - 1)$

The details of symmetrical component method used to obtain the solution of an unbalanced three-phase network are given in Appendix B.

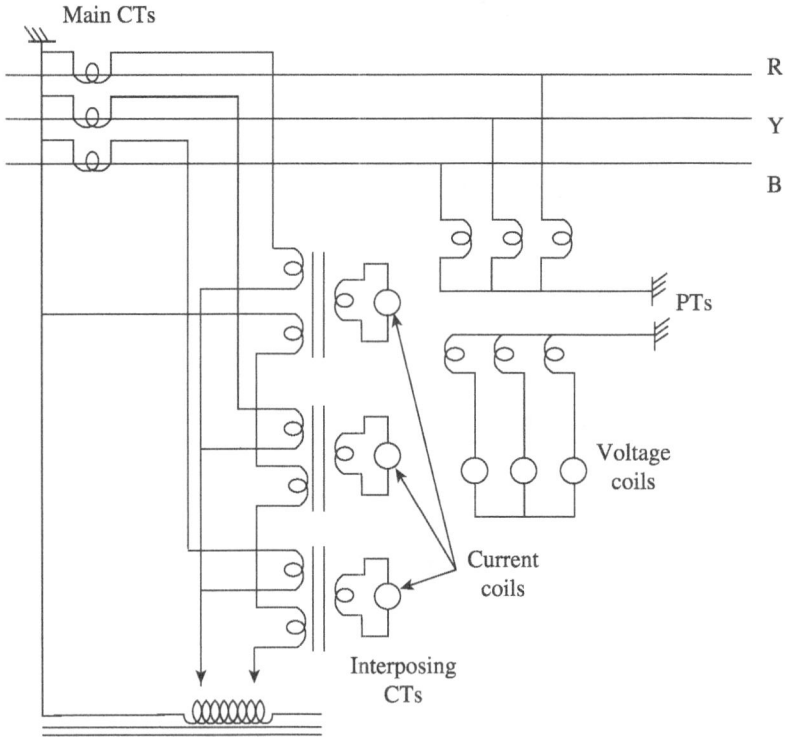

Fig. 3.6 Connections for earth-fault relays

3.7 Distance Protection Using Static Comparators

The two input relays, such as directional and distance, operate by comparing the magnitude and phase of the given two input signals. When the phase relationship or magnitude relationship act in accordance with the predetermined threshold condition, the relay initiates tripping signal. This relationship can be achieved by some form of mixing devices known as *comparators*. There are two types of comparators used in the field, namely amplitude comparator and phase comparator. An amplitude comparator can be converted into phase comparator by changing the inputs with the help of addition and subtraction of two original input quantities.

In order to understand the operation of a comparator, assume two input signals X and Y. For an amplitude comparator, the following condition must be satisfied.

$$|X| > |Y| \tag{3.13}$$

If the inputs are changed to $(X + Y)$ and $(X - Y)$, then the amplitude comparator becomes the phase comparator subject to the fulfilment of the following condition.

$$|X + Y| > |X - Y| \tag{3.14}$$

In this condition, an amplitude comparator inherently compares the phase angles of the original quantities X and Y. The phase comparators have a threshold at a phase difference between two quantities equal to 90°, as shown in Fig. 3.7.

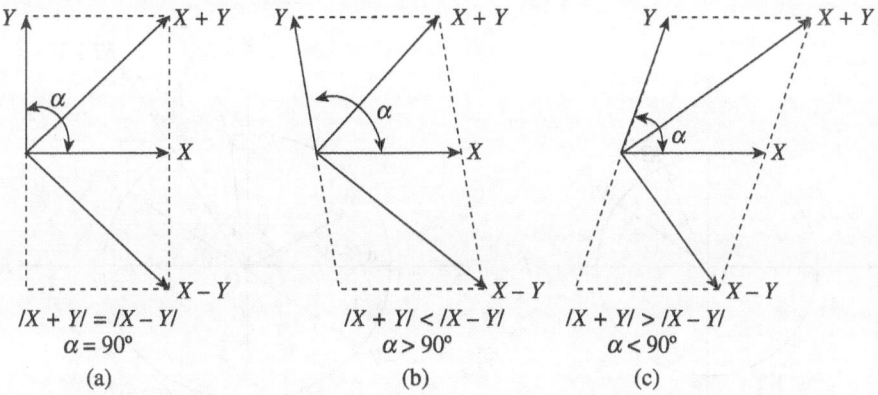

Fig. 3.7 Amplitude comparator (a) Threshold (b) No operation (c) Relay operation

Similarly, the phase comparator having two inputs X and Y can be used as an amplitude comparator by changing the input to $(X + Y)$ and $(X - Y)$. In this condition, it compares the magnitude of X and Y by sensing the angle between $(X + Y)$ and $(X - Y)$ as shown in Fig. 3.8.

Depending on the requirements, different relay elements can be used as an amplitude comparator or a phase comparator. The comparator for impedance relay is explained subsequently. However, comparators for the rest of the relay elements such as reactance, mho, and quadrilateral can be obtained in a similar manner.

Impedance relay receives the local voltage and local current, and compares them in such way that an operation is activated if the ratio of voltage to current is less than the set impedance K. Here, K is a threshold value and used as the radius of the circle for the impedance characteristic on the R–X plane.

As shown in Fig. 3.9(a), for an amplitude comparator, the inputs to the impedance relay are two quantities, namely operating quantity $(X = KI)$ and restraining quantity $(Y = V)$. If the condition mentioned in Eq. (3.13) is satisfied, the relay operates; otherwise, it blocks.

Similarly, as shown in Fig. 3.9(b), for a phase comparator, the impedance relay compares the phase angle between the two quantities $KI + V$ and $KI - V$. If the measured value of angle is less than 90° $(KI > V)$, the relay operates; otherwise, it blocks.

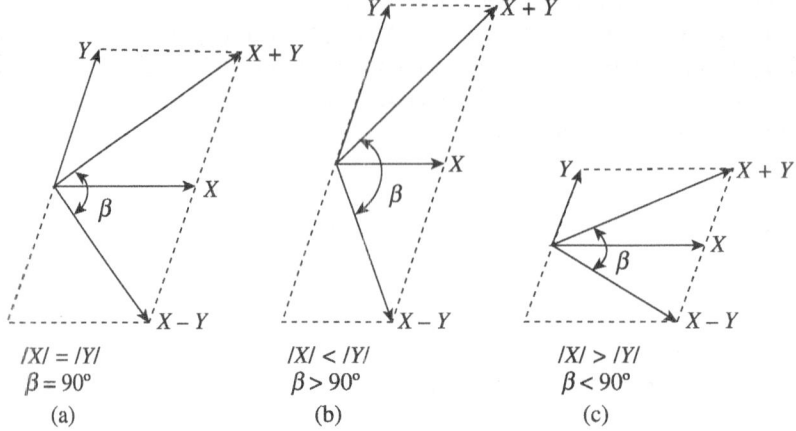

Fig. 3.8 Phase comparator (a) Threshold (b) No operation (c) Relay operation

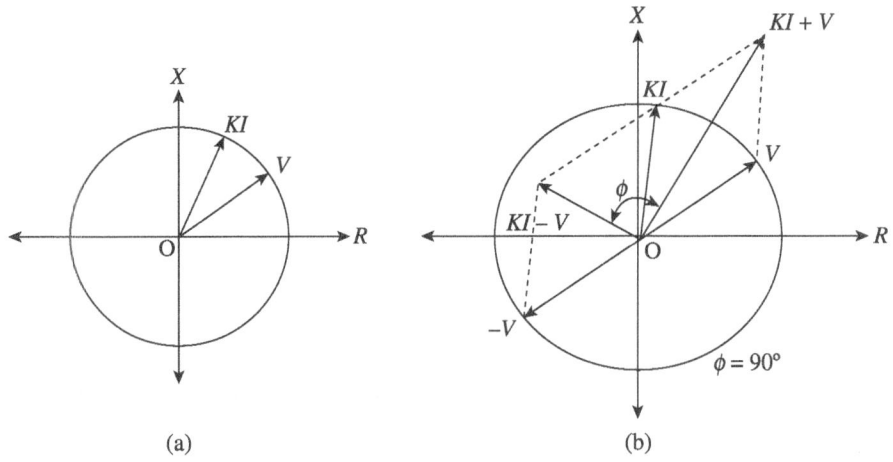

Fig. 3.9 Comparators for impedance relay (a) Amplitude comparator
(b) Phase comparator

3.8 Comparisons between Impedance, Reactance, and Mho Relays

Day by day power system becomes more multifaceted and the protection requirement to such complex system varies depending on system configuration. The simple overcurrent relay based protection fails to protect the line having unforeseen event. On the other hand, the distance relaying scheme is cooperatively used for wide variation in system changes due to its consistent setting. Hence, distance relays have replaced the directional overcurrent protection of high voltage AC transmission lines and distribution lines.

There are three general distance relay types which are used for short, medium, and long transmission lines. The choice of a particular relay unit for a given line depends on line length, line configurations, and system behaviour. Table 3.2 shows the general comparison of impedance, reactance, and Mho type distance relays for the protection of transmission line with respect to its application and operating characteristic.

Table 3.2 Comparison of impedance, reactance, and Mho relay for line protection

Impedance relay	Reactance relay	MHO relay
In this relay, current coil provides operating torque whereas voltage coil produces restraining torque.	In this relay, an over current type measuring unit produces positive force along with a directional unit which either allows or restrains the operation of over current element.	By providing additional polarizing voltage which gives current biasing to offset the impedance characteristics and makes it directional. Thus, in Mho relay, directional current element produces operating torque whereas voltage coil restrains the operation.
This relay can operate in forward as well as reverse direction if additional directional feature is not incorporated.	The trip characteristic of reactance relay engages desired area in forward direction and larger distance in reverse direction. Thus, reactance relay operates for reverse zone fault if directional feature is not provided.	Mho relay is inherently directional and hence, it remains stable for any fault in reverse zone.
During normal load condition, the impedance locus of load is well outside the impedance characteristic (trip region). Thus, this relay does not operate during normal loading condition.	The load impedance during normal load situation falls well below the large trip region of reactance relay. Hence, this relay operates during normal load condition.	Like impedance relay, the load impedance measured by Mho relay moves outside the tripping characteristic. Hence, Mho relay restrains during normal load condition.
Plain impedance relay under reaches during arcing fault or high resistance fault. This is due to the impedance measured by relay reclines towards the boundary of characteristic in the direction of *R*-axis rather than falling on line to be protected.	The reach of reactance relay is not affected by the fault resistance and is, therefore, suitable for protecting the short transmission lines where fault resistance could be comparable to the line reactance.	Behaviour of Mho relay during arcing fault is similar to that of impedance relay. However, Mho relay can incorporate more fault resistance than the impedance relay. Hence, for long transmission lines, mho relay is preferred.
In case of power swing, the apparent impedance of system slowly moves towards the trip region and it enters into the relay characteristic. The effect of power swing on plain impedance relay is considerable as the characteristic covers significant area on *R-X* diagram.	As the area covered by reactance relay is very large on *R-X* diagram, it is highly prone to power swing and mal-operate during normal load flow/power swing situation. This is due to the fact that the apparent impedance comes well within the trip area of reactance relay.	This relay has small area on *R-X* plane. Thus, operation of Mho relay is restrained to some extent during low power swing. However, for large power swing situation, the apparent impedance enters into the trip characteristic of Mho relay and it mal-operates.
The impedance of heavy loads can actually be less than the impedance of some faults. For particular value of load, the locus of apparent impedance enters the third zone of a relay and the relay may operate.	Reactance relays are more susceptible to load encroachment phenomenon as the apparent impedance is already in trip region during full load and overload conditions.	Mho relay performs better than impedance relay during load encroachment phenomenon because the loadability limit of Mho relay is more than impedance relay.

3.9 Problems in Distance Protection

Distance relaying is one of the most commonly used and preferred schemes in transmission line protection. According to a survey conducted by the Institute of Electrical and Electronics Engineers (IEEE), more than 61% of utility companies preferred to use distance protection for primary and backup protection. However, there are various parameters that cause problems in the implementation and operation of distance protection schemes:

1. Loss of potential due to blowing of fuse in the PT circuit
2. Direction
3. Effects of faults on relays with unfaulted phases
4. Fault resistance
5. Effect of Δ–Y transformer between a relay and a fault
6. Series capacitor (SC)
7. Close-in faults
8. Overloads
9. Power swings
10. Transient conditions
11. Remote infeed
12. Mutual coupling

The present discussion is confined only to the eight major problems, namely close-in fault, fault resistance, remote infeed, mutual coupling, SC, power swing, overload, and transient condition which have drawn a lot of interest from the research community.

3.9.1 Close-in Fault

Distance relays, which use voltage signal as the polarizing or reference quantity for their operation, suffer from the problem that on occurrence of close-in faults their accuracy drops and sometimes they even fail to operate. This is due to the collapse of the reference (polarizing) signal (voltage). On the other hand, the fault currents are very high, and protection is most needed on close-in faults. Mho characteristic, elliptical characteristic, and quadrilateral characteristic relays are used on heavily loaded lines and fall in this category. As illustrated in Fig. 3.4(c) and 3.4(f), if the voltage falls sharply, the characteristics of the relays get distorted near the origin and, in some cases, the origin may get excluded from its operating zones as shown by the dotted lines in Fig. 3.4(c) and 3.4(f). Noting that the origin represents a solid fault at the relay location, the problem is quite serious. It is present in static, electromagnetic, as well as numerical relays. All these relays are adversely affected because of their lower sensitivity.

One technique used to ensure accuracy of the relay operation on close-up fault is to offset its characteristic such as to bring the origin well inside its operating zone. However, this makes the relay directionally insensitive near the origin leading to a possible tripping on the busbar faults behind the relay. Another technique consists in providing the polarizing signal wholly or partly from a healthy phase, popularly known as *cross-polarization*. The limitation of this technique is that in the event of a three-phase close-up fault, no polarization signal would be available to the relay as the voltages on all the phases collapse. An yet another mitigation feature has been in use in these relays and is called *memory action*. It uses a tuned circuit (in electromagnetic and analog static relays) or semiconductor memory (in digital static relays) to retain (remember) the pre-fault voltage information for a few cycles to enable the relay to complete its operation accurately. The limitation of the memory action is that it becomes ineffective in case the voltage transformer is connected on the busbar side of the circuit breaker (CB) and the line is energized on a fault.

A new technique, which is applicable to both analog and digital distance relays to improve their sensitivity on very low values of reference (polarizing) voltages, consists of employing an adaptive digital distance relay. It uses software-programmable gain amplifier that is adapted to the value of the said voltage. However, this is beyond the scope of this book and not discussed here.

3.9.2 Fault Resistance/Arc Resistance

Whenever a fault occurs, there is a possibility of involvement of resistance, which varies depending upon the types of fault. For double-line and triple-line faults, the value of resistance is negligible and contains only arc resistance. This arc resistance changes with reference to time as the fault current continues to flow. However, for faults that involve ground, the value of fault resistance is significant. In this situation, the fault path consists of arc resistance in series with tower footing resistance and resistance of ground. The value of tower footing resistance depends on the resistivity of soil (earth). For all practical purposes, it is considered as a constant parameter in case of fault and varies between 5 and 50 Ω. The resistance of ground depends on the type of the surface.

Whenever a fault occurs, the value of arc resistance is very small and can be neglected, particularly during the first few cycles. It increases as the fault current prolongs. However, for all practical relaying calculations, the value of arc resistance is assumed to be constant and is given by,

$$R_{\text{arc}} = \frac{76V^2}{S_{\text{sc}}}$$

(3.15)

where V is the system voltage in kV and S_{sc} is the short-circuit in kVA at the fault location.

Since ground resistance may vary considerably, a ground distance relay must not be practically unaffected by large variations in fault resistance. Figure 3.10 shows the comparison of various distance relay characteristics. It is noted that the reactance relay is not affected by the fault resistance and is, therefore, suitable for protecting the short transmission lines where fault resistance could be comparable to the line reactance. The mho relay can incorporate more fault resistance than the impedance relay. Hence, for long transmission lines, mho relay is preferred. The distance relay having quadrilateral characteristic is even better than the mho relay.

With reference to Fig. 3.10, for a fault at the middle of the line, having fault resistance R_F, the apparent impedance seen by the relay at bus A is given by,

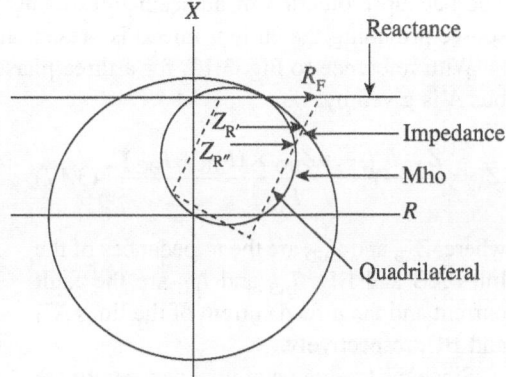

Fig. 3.10   Effect of fault resistance

$$Z_{\text{a}} = pZ_{\text{L}} + 3R_{\text{F}} \times \frac{I_{\text{F}}}{I_{\text{A}}}$$

(3.16)

Due to the pre-fault power flow, the phase angles of the two sources (A and B) connected through the transmission line differ, resulting in a phase shift of the currents flowing from each side of the line to the fault. Thus, not only does the resistance measured at the relay location differ from the actual resistance from the relaying point to the fault, but also the measured inductive reactance is affected. The error in the inductive reactance due to fault resistance can be positive or negative depending on the pre-fault power flow direction. It becomes more pronounced with the increase in pre-fault power flow and fault resistance. If the pre-fault power flows from bus A to bus B, then the measured reactance at the relay location is less than the inductive reactance from the relay location to the fault, causing an eventual tripping for an external fault. On the other hand, if the pre-fault power flows from bus B to bus A, then the measured reactance is greater than the actual reactance from the relay location to the fault, and an internal fault within the first zone of protection may not be detected, resulting in a delayed fault clearance. This is illustrated graphically in Fig. 3.11 for different pre-fault power flow directions.

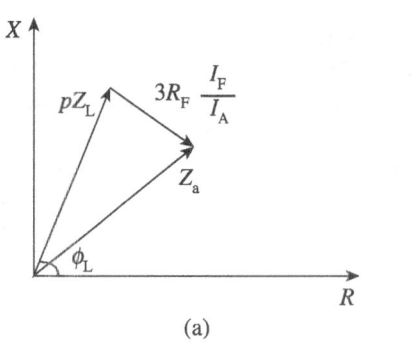 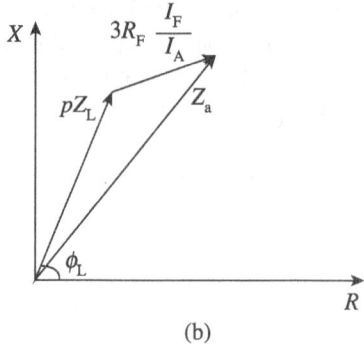

Fig. 3.11 Effect of fault resistance on *R–X* diagram (phasor representation) (a) Pre-fault power flow from bus A to bus B (b) Pre-fault power flow from bus B to bus A

3.9.3 Remote Infeed

The backup protection of adjacent lines using the zone-II and zone-III elements are rarely achieved if a source providing the current infeed is present at the remote bus.

With reference to Fig. 3.12, for a three-phase fault at location F on line BD, the apparent impedance at bus A is given by,

$$Z_a = \frac{Z_{AB}I_{AB} + pZ_{BD} \times (I_{AB} + I_{BC})}{I_{AB}} \quad (3.17)$$

where, Z_{AB} and Z_{BD} are the impedances of the lines AB and BD; I_{AB} and I_{BC} are the fault current and the infeed current of the lines AB and BC, respectively.

Since the true value of impedance upto the fault point is $(Z_{AB} + pZ_{BD})$, it is obvious that the infeed has increased the apparent impedance of the line by $(I_{BC}/I_{AB}) \times (pZ_{BD})$. As a result, the relay 1 underreaches.

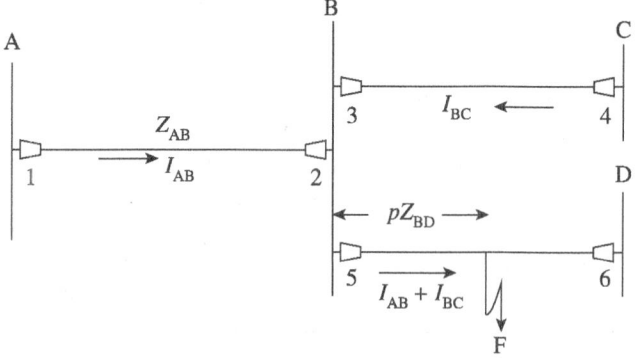

Fig. 3.12 Effect of infeed at a remote bus

3.9.4 Mutual Coupling

Independent operation of parallel transmission lines that run close to each other is not possible because of the presence of mutual inductance and mutual capacitance between them. If parallel lines run on the same double circuit towers, then the effect of the mutual coupling is very high because of the close proximities of the different conductors. The positive- and negative-sequence coupling between the two feeders is usually less than 5–7% and, hence, has negligible effect on protection. The zero-sequence coupling, on the other hand, can be strong and its effect cannot be ignored. The mutual zero-sequence impedance can be as high as 50–70% of the self-impedance.

Due to mutual inductances and capacitances, and the currents in the conductors of line 1, electromagnetic fields (EMFs) are induced in the conductors of line 2 as shown in Fig. 3.13. As the value of mutual inductances and capacitances is different between pairs of conductors, the induced EMF is also different. However, it is to be noted that the induced EMF is negligible in case of balanced condition, that is, during equal line currents. Conversely, it is maximum during unsymmetrical faults as in this situation, the zero-sequence currents flow in the line. Therefore, the apparent impedance seen by the ground distance relays is affected.

Hence, the zero-sequence coupling between the lines that are parallel for a part, or all along their length, can induce false information in the unfaulted circuit and cause protection problems in both faulted and unfaulted circuits.

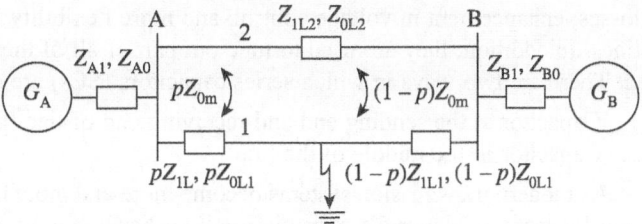

Fig. 3.13 Parallel line model with one faulted circuit

Quite often, parallel lines are used at either one or both terminals. With currents flowing in the same direction, the impedance with mutual coupling is higher than without.

For a single-line-to-ground fault on a single circuit line, as shown in Fig. 3.13, the voltage of the faulted phase at bus A is given by,

$$V_{am} = pZ_{1L_1} \times (I_{am} + k_{0L} \times I_{0m}) + pZ_{0m} \times I_{0L_2} \tag{3.18}$$

where,

$$k_{0L} = \frac{Z_{0L_1} - Z_{1L_1}}{Z_{1L_1}}$$

k_{0L} is the residual compensation factor; Z_{0L_1} and Z_{1L_1} are the zero-sequence and positive-sequence impedances of line 1, respectively; Z_{0m} is the total zero-sequence mutual coupling impedance; V_{am} and I_{am} are the post-fault phase voltage and current at the relaying point; I_{0m} is the post-fault zero-sequence current at the relay location; and I_{0L_1} and I_{1L_1} are the zero- and positive-sequence currents of the line 1, respectively.

The relaying current and the calculated impedance is given by,

$$I_R = I_{am} + k_{0L} \times I_{0m} \tag{3.19}$$

$$Z_{mes} = \frac{V_{am}}{I_{am} + k_{0L} \times I_{0m}} = pZ_{1L_1} + \alpha Z_{1L_1} \tag{3.20}$$

where,

$$\alpha = \frac{p \times (Z_{0m}/Z_{1L_1}) \times I_{0L_2}}{I_{am} + k_{0L} \times I_{0m}}$$

Here, α is known as *error factor*. The error may cause the relay to overreach or underreach depending upon the relative direction of the parallel line's zero-sequence current with respect to the compensated current given to the relay. If they are in the opposite direction, the relay will overreach and vice versa. As the power system condition changes, α varies accordingly.

Mutual coupling can cause incorrect tripping of distance relays. The problem is compounded by the composite voltage circuits and the remote source infeed to the fault branch; these factors can significantly modify the apparent impedance presented to the relay at the local end. The aforementioned problems are particularly endemic when there is a cross-country fault on parallel transmission lines. In particular, overreach/underreach can cause sympathy trips, which can lead to major power system disturbances.

3.9.5 Series-compensated Transmission Lines

The series-compensated transmission lines are employed in modern power systems as they offer many advantages such as increase in transmittable power, improvement in the system stability, reduction in transmission losses, enhancement in voltage control, and more flexibility in power flow control over the uncompensated lines. In addition, they are used to tune out part or all of the transmission line inductance.

There are two ways in which series capacitors (SCs) are connected to the line:

1. Capacitor at the sending end and receiving end of line (half of compensation at each end)
2. Capacitor in the middle of the line

As modern transmission systems become more and more heavily loaded, the benefits of series compensation for many of the grid's transmission lines become more obvious. Unfortunately, the SC can undermine the effectiveness of many of the protection schemes used for long distance transmission lines. The introduction of capacitance in series with the line reactance adds certain complexities to the effective application of impedance-based distance relays. It is of course possible to correct the settings of the relay when it is known that the capacitor will always be part of the fault circuit. However, this is not always known. By compensating for some of the line's series inductance, the SC can make the remote forward faults look as if they are in zone 1 of the relay when the capacitor is switched into the transmission line circuit and the relay setting rules are based on *no capacitor in the fault loop* (i.e., they can cause the relay to 'overreach'). Under these conditions, the close-in faults can appear to be reverse faults due to voltage reversal (voltage inversion).

Basic Components of Series-compensated Transmission Lines

Figure 3.14 shows the model of an SC.

The basic components used along with an SC are as follows:

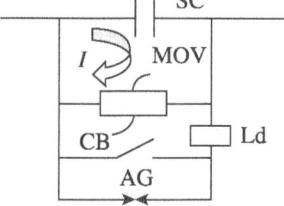

1. Series capacitor (SC)
2. Air gap (AG)
3. Metal oxide varistor (MOV)
4. CB along with damping circuit

The function of AG is to protect the compensating capacitor against overvoltages. The gap flashes over when the voltage across the gap increases beyond the spark over limit of the gap protecting the capacitor.

Fig. 3.14 Model representation of a series capacitor

In recent years, MOVs are widely used as main protection for SCs. As shown in Fig. 3.14, the MOV along with the bypass CB and AG is placed across the SC for protection purpose. The damping circuit 'Ld' limits the oscillatory transient current through AG when it fires.

Whenever short-circuits occur on the line, the voltage across the SC increases. It is to be noted at this juncture that the MOV is able to limit the voltage only upto a certain predetermined level. This level is typically 2 pu assuming that 1 pu is the voltage across SC during the passage of normal-rated current. Hence, there is an upper limit below which the MOV dissipates the energy in the unfaulted line for the worst-case external fault. Beyond this limit, it is not capable to dissipate the energy during some of the internal faults. Hence, to protect the MOV under these conditions, AG diverts the current away from the MOV when the energy limit of MOV is crossed. The AG also has some finite energy limit and therefore, a bypass CB is required, which acts as a diverter.

The key advantage of MOV is its small reinsertion time. When MOV is not used along with SC, it is mandatory that the bypass CB of healthy line must be closed before the SC could be put back into service. On the other hand, when the MOV is used along with SC, the overvoltage across SC disappears as soon as the MOV stops conducting. Hence, the SC is connected back in service without the bypass CB.

Protection Problems

Series capacitors, MOVs, and AGs create several problems for protective relays and fault locators used for transmission lines. This is due to different impedance seen by the protective relays for faults occurring before and after SC.

The fault signals captured under such conditions contain different frequency components such as odd harmonics because of the conduction of MOV, high-frequency components because of the resonance between line capacitance and line inductance, non-fundamental decaying frequency components because of the resonance between the system inductance and SC, and fundamental frequency components of the steady state fault current.

Therefore, the operating conditions for protective relays become unfavourable and include different phenomena as follows:

1. Voltage inversion
2. Current inversion
3. Subharmonic oscillations and additional transients
4. Overreaching of distance elements
5. Underreaching of distance elements

Some of the problems are discussed briefly in this section.

Overreaching of Distance Element

Normally, three single-phase banks of capacitors are used for series compensation of transmission line. Each capacitor must be protected against over-voltages by AG or MOVs or both. The problem of overreaching of distance element is explained by the concept of equivalent impedance known as *Gold-sworthy's equivalent impedance*. This equivalent impedance has been obtained using the voltage drop across SC and the current passing through it. It is to be noted that this equivalent impedance depends on

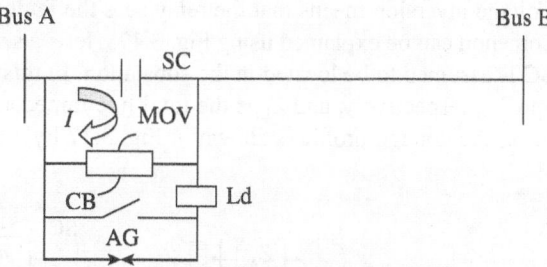

Fig. 3.15   Single line diagram of transmission line with series capacitor

the location of the SC on the transmission line. Figure 3.15 shows the single line diagram of transmission line along with SC.

If a fault occurs after the SC, then the equivalent impedance is made up of the line impedance between the relaying point and the fault point, fault resistance (if any), and the equivalent impedance of SC and MOV. Owing to this non-linearity, the apparent impedance seen by the relay at bus A moves down to the right side and transverses between regions as shown in Fig. 3.16.

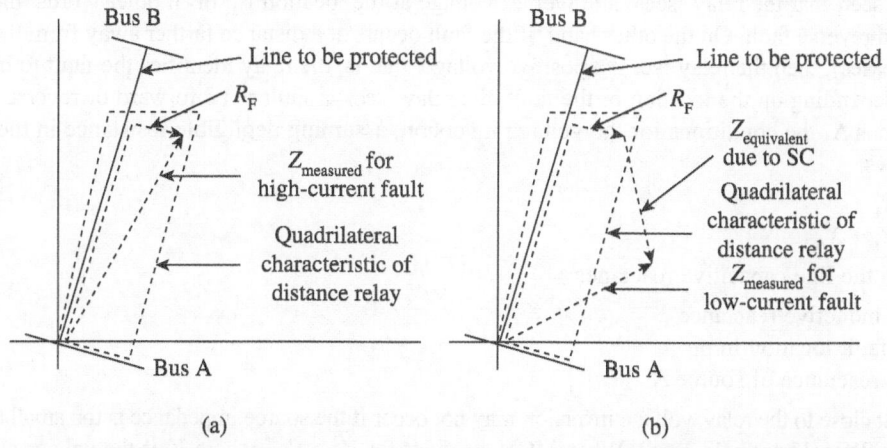

Fig. 3.16   Impedance seen by relay at bus A (a) High-current fault (b) Low-current fault

For high-current faults, as the SC is bypassed because of the conduction of AG and/or MOV, the equivalent impedance of the SC bank is a small resistance. Hence, the apparent impedance as seen by the relay at bus A slightly shifts to the right as shown in Fig. 3.16(b), and there is no danger of overreaching.

For low-current faults, the AG or MOV does not conduct any current because of the reduced voltage drop (below the voltage protection level) across the SC. In this situation, the equivalent impedance of the SC is equal to the reactance of the actual physical capacitor. Hence, the apparent impedance seen by the relay at bus A reduces drastically as shown in Fig. 3.16. Because of this, it is quite possible that the distance element may not able to operate particularly for close-in low-current fault. Sometimes, the apparent impedance seen by the relay may move into the fourth quadrant and create a problem of directional discrimination.

For medium-current faults, the equivalent impedance of the SC is such that the apparent impedance seen by the relay at bus A shifts to the right, and hence, it falls outside the operating characteristic of the relay.

Voltage Inversions

Voltage inversion means that the relay sees the fault on the protected line in the reverse direction. This phenomenon can be explained using Fig. 3.17. Here, a series-compensated transmission line is shown, where the SC is assumed to be located in the substation. In this figure, Z_A and Z_B are the impedances of the sources V_A and V_B, respectively, and Z_L is the total line impedance. Now, for a fault F_1 occurring at a per-unit distance of m, the voltage profile is shown in Fig. 3.17 by a dotted line.

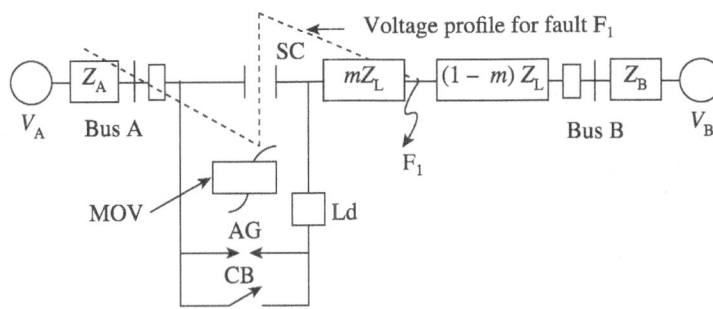

Fig. 3.17 Sample power system for voltage inversion phenomenon

It can be seen that the relay 'sees' an inverted voltage at the location F_1, or in other words, the relay sees this fault as a reverse fault. On the other hand, if the fault occurs at a distance farther away from the substation (i.e., m increases), then the relay 'sees' a positive voltage, that is, the relay identifies the fault to be a forward one. Thus, depending on the location of the fault, the relay 'sees' a fault to be forward or reverse fault.

Now, at bus A, the conditions for the voltage inversion, assuming negligible resistance in the fault loop, are as follows:

1. $X_C > mX_L$
2. $X_C < mX_L + X_A$

where, X_C is the line capacitive reactance

X_L is the inductive reactance

m is the fault location in pu

X_A is the reactance of source A

For a fault close to the relay, voltage inversion may not occur if the source impedance is too small (in this case, the condition 2 would not hold good). When MOV conducts for $V_C > V_{PR}$, where V_C is the voltage across the capacitor and V_{PR} is the predetermined voltage level of the MOV, the combined reactance of MOV and SC becomes less than that of X_C. As a result, the possibility of condition 1 getting satisfied reduces. This in turn reduces the possibility of voltage inversion. Thus, it can be said that a voltage inversion may occur for faults located within certain distance on the line depending on source and line impedance, fault resistance, and capacitor protective circuit.

Current Inversions

Current inversion means that the relay sees fault current in the reverse direction because of large capacitive reactance in the fault loop. The voltage and current inversion cannot happen simultaneously. From Fig. 3.17, for fault F_1, the conditions for current inversion, assuming only reactance in the fault loop, are given as follows:

1. $X_C > mX_L + X_A$
2. $V_C = (X_C/\{X_C - (mX_L + X_A)\}) \times V_A < V_{max}$

where V_{max} is voltage that causes the gap to flash.

Current inversion may occur for the faults closer to the relay and for systems having small source impedances. The possibility of current inversion reduces under the conduction of MOV ($V_C > V_{PR}$) due to the reduction of capacitive reactance to X_{CMOV}, which is less than X_C. The current inversion and voltage inversion depend on the location of SC installation on the line. However, voltage and current inversion are rare phenomena.

Solution to the Problem of Series-compensated Lines

The protection scheme of the series-compensated transmission line comprises three stages: fault classification, selection of faulted section (whether the fault is in front or behind the SC), and distance calculation. Various techniques for the fault classification, fault zone detection, and distance calculation are used by different researchers. However, the solution is beyond the scope of this book.

3.9.6 Power Swing

Power swing is a phenomenon that causes large fluctuations in the three-phase power flow between two areas of a power system due to the variation of rotor angles in response to changes in load magnitude and direction, line switching, loss of generation, faults, and other system disturbances. These large fluctuations die down if the swing is stable. On the other hand, if the swing is unstable, the fluctuations cannot die down; it causes large fluctuations of voltages and current, and may finally result in a loss of synchronism. Under steady state conditions, the trajectory of load impedance is well outside the operating zone of the relay. Conversely, with the presence of power swing, the locus of load impedance can enter the operating zone of the distance relay. Moreover, the directional or bidirectional relay may mal-operate because of the unstable power swings, which can generate high currents. Furthermore, the unwanted operation of short time undervoltage relays also occurs because of the reduction in voltages. During large frequency excursions, digital distance relays measure a wrong value of phasors and may mal-operate. Figure 3.18 shows the behaviour of mho and quadrilateral relays on power swings. Mho relays are likely to operate undesirably on large power swings, even if the power swing blocking feature is provided to prevent such operation. This tendency of undesired operation is less with relays having quadrilateral characteristics.

As shown in Fig. 3.18, in situations where the ratio of the local end to far end source magnitude (r) is unity, the impedance trajectory passes through the

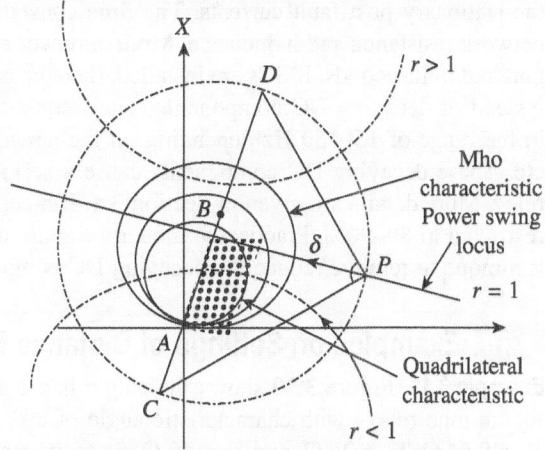

Fig. 3.18 Behaviour of relays on power swings

centre of the line to be protected. When the power angle difference is close to 180°, the apparent impedance seen by a distance relay can be within the operating zone of the relay. Hence, the relay sees this condition as a three-phase fault. A blocking function is provided, which blocks the operation of a distance relay in case of a power swing. However, if a fault occurs, particularly symmetrical in nature, during a power swing, then this function is not useful. The solution to the said problem is to use a method that tracks the rate of change of impedance using blinders and a timer.

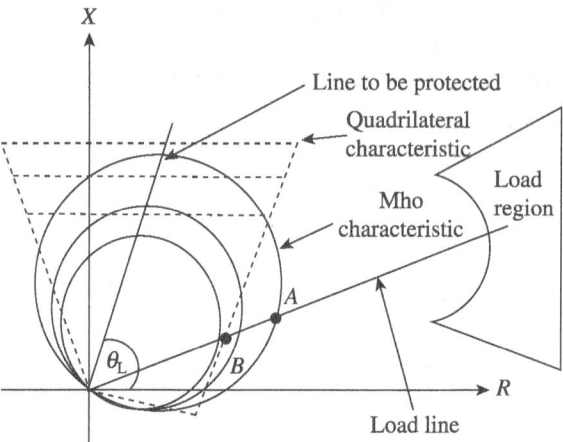

Fig. 3.19 Effect of overload on mho and quadrilateral distance relay characteristics

3.9.7 Overload

Figure 3.19 shows the behaviour of distance relay having mho and quadrilateral characteristics during overload condition. The locus of apparent impedance moves towards the origin of the R–X plane. For a particular value of load, the locus of apparent impedance enters the third zone of a relay (point A and point B), and the relay will operate. Hence, the setting of the third zone of distance relay is decided by the probable overload. The value of load at which the relay is on the verge of operation is known as the *loadability limit of the relay*. It is to be noted from Fig. 3.19 that the loadability limit of the quadrilateral distance relay (point B) is significantly greater than that of a mho distance relay (point B). Quadrilateral characteristic distance relay is commonly used for transmission line protection because of its ability to accommodate the predetermined arc resistance while maintaining sufficient margin from the load region.

3.9.8 Transient Condition

Whenever a fault occurs on the transmission line, the magnitude of the fault current is very high and asymmetrical in nature. Hence, the energy stored in the line inductance generates transients. The asymmetrical fault current contains decaying DC components, whose maximum values can be almost twice as high as the stationary post-fault currents. The time constant, which determines the rate of decay, depends on the network resistance and inductance. Modern power system networks can have time constants of upto several hundred milliseconds. If SCs are installed, then the relay voltage and currents contain decaying subharmonics instead of decaying DC components. The frequency of the decaying component for three-phase faults is in the range of 15–150 Hz depending on the network parameters, degree of compensation, fault distance, etc. These decaying DC components cause a serious problem of overreach for the first zone of distance relay. More details are given in Section 3.4. Hence, to avoid overreaching of distance relay, its first zone is restricted to 80–90%. Practically, modern digital/numerical relays use modified discrete Fourier transform technique to remove/reduce the decaying DC components.

3.10 Examples on Settings of Distance Protection Relays

Example 3.1 Figure 3.20 shows the single line diagram of a portion of power system. The relays R_1 and R_2 are mho relays with characteristic angle of 60°. The secondary ohms of both the relays are as follows. $K_1 = 9.64 \ \Omega$, $K_2 = 26 \ \Omega$, and $K_3 = 40 \ \Omega$. State the reach of the relay R_1 in all three zones. Your answer should be supported by relevant calculations.

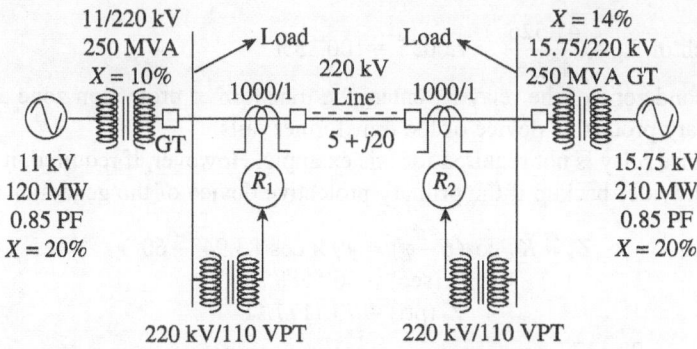

Fig. 3.20   Single line diagram of a portion of power system network

Solution:

Let us assume that X_1 is the reach of transmission line in the first zone.

Now, the impedance of line on primary side, Z_1 (pri) $= 5 + j20 \ \Omega = 20.61\angle75.96°\Omega$.

$$Z_1 \text{ (pri)} = X_1 \times 20.61\angle75.96°$$

$$Z_1 \text{ (sec)} = \frac{\text{CTR}}{\text{PTR}} \times Z_1 \text{ (pri)} = \frac{1000/1}{220 \times 10^3 / 110} \times X_1 \times 20.61\angle75.96°\Omega$$

$$Z_1 \text{ (sec)} = X_1 \times 10.305\angle75.96°\Omega$$

$$K_1 = \frac{Z_1(\text{sec})}{\cos(\theta - \varphi)}$$

$$9.64 = \frac{Z_1(\text{sec})}{\cos(75.96° - 60°)}$$

$$X_1 \times 10.305 = 9.64 \times \cos(75.96° - 60°)$$

$$X_1 = 0.8994 = 0.8994$$

Hence, the first zone of the relay is 89.94%.

To find out the reach of the relay R_1 in the second zone, we have to consider the impedance of a transformer.

$$Z_{\text{pu}} = \frac{Z_{\text{act}}}{Z_{\text{base}}} \Rightarrow Z_{\text{act}} = Z_{\text{pu}} \times Z_{\text{base}}$$

$$Z_{\text{act}} = 0.14 \times \frac{(\text{KV})^2}{\text{MVA}} = 0.14 \times \frac{(220)^2}{250} \ \Omega$$

$$Z_{\text{act}} = 27.104 \ \Omega$$

$$Z_{\text{total}} = 5 + j20 + j27.104 \ \Omega = 5 + j47.104 \ \Omega,$$

$$Z_{\text{total}} = 47.36\angle83.94°\Omega$$

$$Z_2(\text{sec}) = K_2 \times \cos(\theta - \varphi) = 26 \times \cos(83.94° - 60°)$$

$$Z_2 \text{ (sec)} = 23.76 \ \Omega$$

$$Z_2 \text{ (pri)} = 47.526 \ \Omega$$

The second-zone reach of $R_1 = \dfrac{47.526}{47.36} = 1.0035 = 100.35\%$

It is clear that the second zone of the relay R_1 enters the transformer protection zone and provides backup protection if the primary protective device of the transformer fails.

Third zone of distance relay is not required for this example. However, if required, it enters the generator protection zone and provides backup if the primary protective device of the generator fails.

$$Z_3 = K_3 \cos(\theta - \varphi) = 40 \times \cos(83.94° - 60°)$$
$$Z_3 \text{ (sec)} = 36.5588 \ \Omega$$
$$Z_3 \text{ (pri)} = 73.1177 \ \Omega$$

Third zone reach of $R_1 = \dfrac{73.1177}{47.36} = 1.5438 = 154.38\%$

Example 3.2 Decide the three zone settings of mho distance relay R, as shown in Fig. 3.21, having a characteristic angle of 60°.

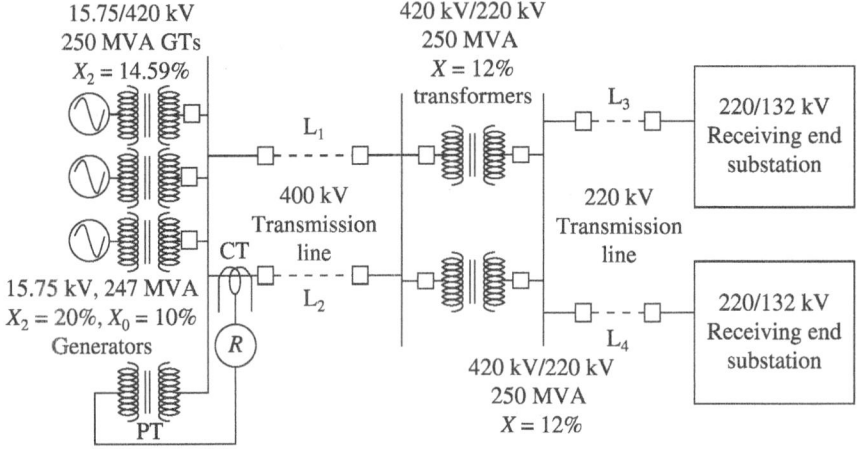

Fig. 3.21 Single line diagram of a portion of power system network

The required data is as follows:

400 kV lines L_1 and L_2:
Impedance: $0.01 + j \times 0.1 \ \Omega/\text{km/phase}$
Line length: 200 km
220 kV lines L_3 and L_4:
Impedance: $0.02 + j \times 0.1 \ \Omega/\text{km/phase}$
Line length: L_3 is 100 km long, whereas L_4 is 125 km long.
CT ratio: 400/1 A
PT ratio: 420 kV/110 V

Solution:

Impedance of $L_1 = (0.01 + j \times 0.1) \times 200 = 2 + j \times 20 \ \text{W} = 20.09\angle 84.29° \ \Omega$

Assuming that the reach of mho relay R in first zone is 80%, the impedance of zone 1 of relay R is given by,

$$Z_1 \text{ (pri)} = 0.8 \times 20.09\angle 84.29° = 16.072\angle 84.29° \ \Omega$$

Transferring this primary impedance to the relay side (secondary side of CT),

$$Z_1 \text{ (sec)} = \frac{\text{CTR}}{\text{PTR}} \times Z_1 \text{ (pri)} = \frac{400/1}{420 \times 10^3/110} \times 16.072 \angle 84.29°$$

$$Z_1 \text{ (sec)} = 1.6837 \angle 84.29° \ \Omega$$

Hence, the first-zone setting of mho relay R is given by,

$$K_1 = \frac{|Z_1|}{\cos(\theta - \varphi)} = \frac{1.6837}{0.9115} = 1.8472 \ \Omega$$

Assume that the second-zone reach of mho relay R covers line L_1 and two parallel transformers. Parallel impedance of two transformers is given by,

$$Z_T \text{ (referred on 400 kV side)} = \frac{(420)^2}{250} \times \frac{0.12}{2} = j \times 42.336 \ \Omega$$

$$Z_2 \text{ (pri)} = Z_{L_1} + Z_T \text{ (referred on 400 kV side)}$$

$$Z_2 \text{ (pri)} = 2 + j \times 20 + j \times 42.336 = 2 + j \times 62.336 \ \Omega$$

$$Z_2 \text{ (pri)} = 62.368 \angle 88.16° \ \Omega$$

Transferring this primary impedance to the relay side (secondary side of CT),

$$Z_2 \text{ (sec)} = \frac{\text{CTR}}{\text{PTR}} \times Z_2 \text{ (pri)} = \frac{400/1}{420 \times 10^3/110} \times 62.368 \angle 88.16°$$

$$Z_2 \text{ (sec)} = 6.5337 \angle 88.16° \ \Omega$$

Hence, the second-zone setting of mho relay R is given by,

$$K_2 = \frac{|Z_2|}{\cos(\theta - \varphi)} = \frac{6.5337}{0.882} = 7.4078 \ \Omega$$

$$\text{Impedance of } L_3 = (0.02 + j \times 0.1) \times 100 = 2 + j \times 10 \ \Omega = 10.198 \angle 78.69° \ \Omega$$

$$\text{Impedance of } L_4 = (0.02 + j \times 0.1) \times 125 = 2.5 + j \times 12.5 \ \Omega = 12.7475 \angle 78.69° \ \Omega$$

Considering the third-zone of mho relay R cover lines L_3 and L_4, its setting is to be done with respect to the impedance of line L_4 to achieve maximum coverage. If the setting of relay R is carried out with respect to the impedance of line L_3, then the relay R underreaches for line L_4.

$$Z_3 \text{ (pri)} = Z_{L_1} + Z_T + Z_{L_4} \text{ (referred on 400 kV side)}$$

$$Z_3 \text{ (pri)} = 2 + j \times 20 + j \times 42.336 + \left(\frac{420}{220}\right)^2 \times 12.7475 \angle 78.69°$$

$$Z_3 \text{ (pri)} = 108.46 \angle 84.12° \ \Omega$$

Transferring this primary impedance to relay side (secondary side of CT),

$$Z_3 \text{ (sec)} = \frac{\text{CTR}}{\text{PTR}} \times Z_3 \text{ (pri)} = \frac{400/1}{420 \times 10^3/110} \times 108.846 \angle 84.12°$$

$$Z_3 \text{ (sec)} = 11.36 \angle 84.12° \ \Omega$$

Hence, the third-zone setting of mho relay R is given by,

$$K_3 = \frac{|Z_3|}{\cos(\theta - \varphi)} = \frac{11.36}{0.913} = 12.45\,\Omega$$

Example 3.3 Figure 3.22 shows the single line diagram of a portion of power system network. The line parameters of L_1, L_2, and L_3 are as shown in the table.

Line	Impedance (Ω/km)	Distance (km)
L_1	0.0316 + j × 0.1265	95
L_2	0.04 + j × 0.16	100
L_3	0.0175 + j × 0.075	75

The relay R is an mho relay with characteristic angle of 65°. The CT ratio is 1000/1 A and PT ratio is 220 kV/110 V. Three zone settings of relay R is $K_1 = 5.2$, $K_2 = 17.32$, and $K_3 = 22.72$, respectively. Determine (a) zone 1 reach of relay R from bus 1 for the line L_1 in km, (b) zone 2 reach of relay R form bus 2 for line L_2 in km, (c) zone 3 reach of relay R from bus 2 for line L_2.

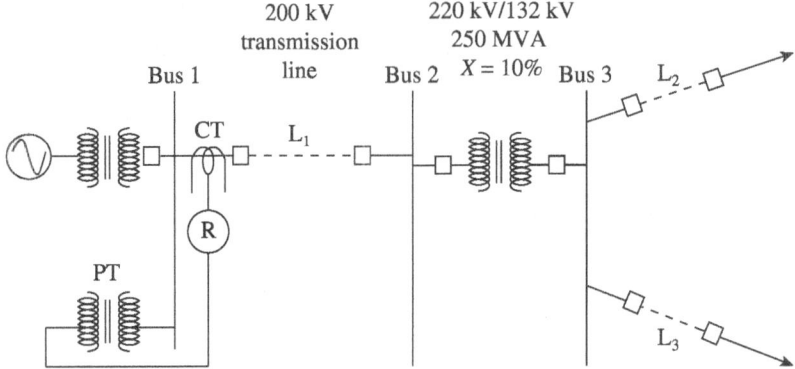

Fig. 3.22   Single line diagram of a portion of power system network

Solution:
(a) Impedance of $L_1 = (0.0316 + j \times 0.1265) \times 95 = 3 + j \times 12\,\Omega = 12.37\angle 75.96°\,\Omega$

$$Z_1\,(\text{sec}) = K_1 \times \cos(\phi - \theta)° = 5.2 \times \cos(75.96 - 65) = 5.1052\,\Omega$$

Transferring this secondary impedance to primary side,

$$Z_1\,(\text{pri}) = \frac{\text{PTR}}{\text{CTR}} \times Z_1\,(\text{sec}) = \frac{220 \times 10^3/110}{1000/1} \times 5.1052 = 10.2103\,\Omega$$

Hence, zone 1 reach of mho relay R from bus 1 is

$$= \frac{Z_1\,(\text{pri})}{Z_{L_1}/\text{km}} = \frac{10.2103}{0.0316 + j \times 0.1265} = 78.54\,\text{km}$$

(b) Now, the second-zone reach of mho relay R covers the line L_1 and the transformer reactance.

$$Z_T\,(\text{referred on 220 kV side}) = \frac{(220)^2}{250} \times 0.10 = j \times 19.36\,\Omega$$

$$Z_{L_1} + Z_T = 3 + j \times 12 + j \times 19.36 = 3 + j \times 31.36 = 31.50\angle 84.53° \; \Omega$$

$$Z_2 \text{ (sec)} = K_2 \times \cos(\phi - \theta)° = 17.32 \times \cos(84.53 - 65)° = 16.32 \; \Omega$$

Transferring this secondary impedance to primary side,

$$Z_1 \text{ (pri)} = \frac{\text{PTR}}{\text{CTR}} \times Z_1 \text{ (sec)} = \frac{220 \times 10^3/110}{1000/1} \times 16.32 = 32.64 \; \Omega$$

Hence, the zone 2 reach of mho relay R from bus 2 is

$$= \frac{Z_2 \text{(pri)} - Z_{L_1} - Z_T}{Z_{L_2}/\text{km}} = \frac{32.64 - 31.5}{0.04 + j \times 0.16} = 6.91 \; \text{km}$$

(c) Now, the impedance of L_2 as seen by relay R in the third zone is given by,

$$Z_{L_2} \text{ (referred on 220 kV side)} = Z_{L_1} + Z_T + \left(\frac{220}{132}\right)^2 \times (4 + j \times 16)$$

$$= 77.1\angle 79.46° \; \Omega$$

$$Z_3 \text{ (sec)} = K_3 \times \cos(\phi - \theta)° = 22.72 \times \cos(79.46 - 65)° = 22.0 \; \Omega$$

Transferring this secondary impedance to primary side,

$$Z_3 \text{ (pri)} = \frac{\text{PTR}}{\text{CTR}} \times Z_3 \text{ (sec)} = \frac{220 \times 10^3/110}{1000/1} \times 22.0 = 44.0 \; \Omega$$

Hence, the zone 3 reach of mho relay R from bus 2 is

$$= \frac{Z_3 \text{(pri)} - Z_{L_1} - Z_T}{Z_{L_2}/\text{km}} = \frac{44.0 - 31.5}{0.04 + j \times 0.16} = 75.8 \; \text{km}$$

As $Z_3 \text{(pri)} < Z_{L_2}$ (referred on 220 kV side), the line L_2 will not be covered fully by zone 3 of the relay R.

Example 3.4 Figure 3.23 shows the single line diagram of a portion of power system network. The line parameters of L_1 and L_2 are shown in the table.

Line	Voltage (kV)	Impedance (Ω/km)	Distance (km)
L_1	220	$0.04 + j \times 0.16$	100
L_2	132	$0.03 + j \times 0.09$	50

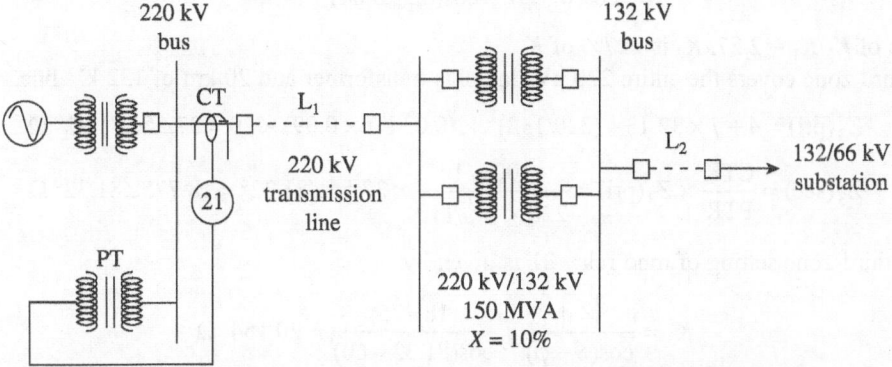

Fig. 3.23 Single line diagram of a portion of power system network

The mho relay 21 with characteristic angle of 60° is used to protect the 220 kV line. The CT ratio is 1000/1 A and the PT ratio is 220 kV/110 V. The first zone of relay 21 covers 90% of 220 kV line. The second zone covers the entire 220 kV line plus the transformers upto 132 kV bus. The third zone covers 20 km of 132 kV bus. Determine the relay settings K_1, K_2, and K_3. Note that K_2 and K_3 should be in percentage of K_1.

Solution:
Impedance of $L_1 = (0.04 + j \times 0.16) \times 100 = 4 + j \times 16\ \Omega = 16.49\angle75.96°\ \Omega$.
As the first zone covers 90% of line L_1,

$$Z_1(\text{pri}) = 0.9 \times 16.49\angle75.96° = 14.84\angle75.96°\ \Omega$$

$$Z_1(\text{sec}) = \frac{\text{CTR}}{\text{PTR}} \times Z_1(\text{pri}) = \frac{1000/1}{220 \times 10^3/110} \times 14.84 = 7.42\angle75.96°\ \Omega$$

Hence, the first-zone setting of mho relay 21 is given by,

$$K_1 = \frac{|Z_1|}{\cos(\theta - \varphi)} = \frac{7.42}{0.9615} = 7.72\ \Omega$$

Now, the second zone covers the complete 220 kV line along with the transformer. The equivalent impedance of two transformers is given by,

$$Z_T(\text{referred on 220 kV side}) = \frac{(220)^2}{150} \times \frac{0.1}{2} = j \times 16.13\ \Omega$$

So, the impedance of second zone on 220 kV side is given by,

$$Z_2(\text{pri}) = 4 + j \times 16 + j \times 16.13 = 32.38\angle82.9°\ \Omega$$

$$Z_2(\text{sec}) = \frac{\text{CTR}}{\text{PTR}} \times Z_2(\text{pri}) = \frac{1000/1}{220 \times 10^3/110} \times 32.38\angle82.9°$$

$$= 16.19\angle82.9°\ \Omega$$

Hence, the second-zone setting of mho relay 21 is given by,

$$K_2 = \frac{|Z_2|}{\cos(\theta - \varphi)} = \frac{16.19}{\cos(82.9 - 60)} = 17.57\ \Omega$$

As the ratio of $K_2/K_1 = 2.27$, K_2 is 227% of K_1.
Now, the third zone covers the entire 220 kV line plus transformer and 20 km of 132 kV line.

$$Z_3(\text{pri}) = 4 + j \times 32.13 + (220/132)^2 \times (0.03 + j \times 0.09) \times 20 = 37.55\angle81.32°\ \Omega$$

$$Z_3(\text{sec}) = \frac{\text{CTR}}{\text{PTR}} \times Z_3(\text{pri}) = \frac{1000/1}{220 \times 10^3/110} \times 37.55\angle81.32° = 18.775\angle81.32°\ \Omega$$

Hence, the third-zone setting of mho relay 21 is given by,

$$K_3 = \frac{|Z_3|}{\cos(\theta - \varphi)} = \frac{18.775}{\cos(81.32 - 60)} = 20.154\ \Omega$$

As the ratio of $K_3/K_1 = 2.61$, K_3 is 261% of K_1.

Example 3.5 Figure 3.24 shows the single line diagram of a portion of power system. The relay R (mho and reactance type relay) is used to protect the first section (primary protection) and second section (secondary protection) of the transmission line. The characteristic angle of mho relay is 60°. Determine the three zone settings of relay R by assuming its characteristic as (i) Reactance (ii) Mho. The first zone covers 80% of the first section. The second zone covers the first section plus 30% of the second section and the third zone covers the first section plus 120% of the second section.

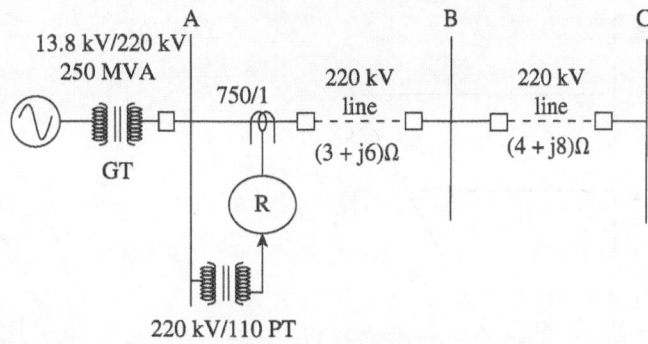

Fig. 3.24 Single line diagram of a portion of power system network

Also draw the characteristics of relay R having reactance and mho characteristic for all the three zones.

Solution:

An mho relay R is used to protect the long 220 KV transmission line.

The characteristic angle of R is = 60°

Impedance of first section = $3 + j6 = 6.71\angle63.43°$ Ω

Impedance of second section = $4 + j8 = 8.94\angle63.43°$ Ω

The primary impedance for all zones:-

Impedance of first zone $Z_1 = 0.8\times6.71\angle63.43° = 5.368\angle63.43°$ Ω

Impedance of second zone $Z_2 = (3+j6)+(0.3\times8.94\angle63.43) = 4.2 + j8.4$

$\qquad\qquad\qquad\qquad\qquad = 9.392\angle63.43°$ Ω

Impedance of third zone $Z_3 = (3 + j6) + 1.2 (4 + j8) = 7.7998 + j15.596$

$\qquad\qquad\qquad\qquad = 17.438\angle63.43°$ Ω

The secondary impedance for all zones, as relay receives the secondary quantities of CT and PT.

Z_1 (Secondary) = $\dfrac{CTR}{PTR}\times Z_1$ (Primary) = $\dfrac{750/1}{220\times10^3\big/110}\times5.368\angle63.43° = 2.013\angle63.43°$ Ω

For mho relay,

The first zone setting of relay R is

$$K_1 = \frac{|Z_1|}{Cos(\theta-\varphi)} = \frac{2.013}{Cos(63.43°-60°)} = 2.02\ \Omega$$

Similarly, Z_2 (Secondary) = $\dfrac{CTR}{PTR}$ × Z_2 (Primary) = 0.375 × $9.392\angle63.43° = 3.52\angle63.43°$ Ω

The second zone setting of relay R is

$$K_2 = \frac{|Z_2|}{Cos(\theta-\varphi)} = \frac{3.522}{Cos(63.43°-60°)} \quad K_2 = 3.53\ \Omega$$

Similarly, Z_3 (Secondary) = $6.5393\angle63.43°$ Ω

The third zone setting of relay R is $K_3 = 6.55$ Ω.

$X_1 = Z_1 \times \sin(63.43) = 2.013 \times \sin(63.43) = 1.8$ Ω.

Similarly, $X_2 = 3.15$ Ω, and $X_3 = 5.85$ Ω.

The characteristic of reactance type distance relay and mho type distance relay is shown in Fig. 3.25(a) and (b), respectively.

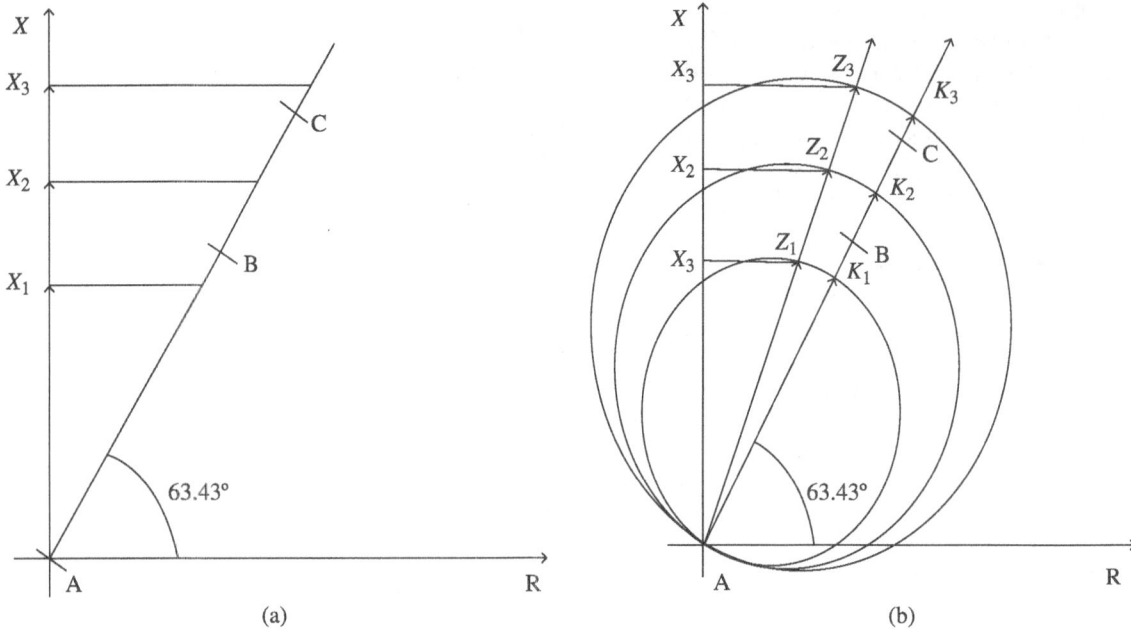

Fig. 3.25 (a) Characteristic of reactance distance relay (b) Characteristic of mho distance relay

Example 3.6 Figure 3.26 shows the single line diagram of a portion of power system. Impedance of all the three line sections is shown in Fig. 3.26. The relay R is used to protect the line sections and it is a mho type distance relay having a characteristic angle of 60°. The transient overreach of the relay R is 15%. The rated current of the line section is 1049.72 A at 0.9 lagging power factor. The probable overloading of the line is 200% of the rated current and the probable voltage dip is 15%. The CT ratio and PT ratio are given in Fig. 3.26. The primary impedance of line section 1, 2, and 3 is $5 + j20\ \Omega$, $4 + j16\ \Omega$, and $3 + j12\ \Omega$, respectively. Determine the three zone settings of the distance relay R. Also, find out the reach of all the three zones for the line to be protected in terms of percentage of impedance of the first line section.

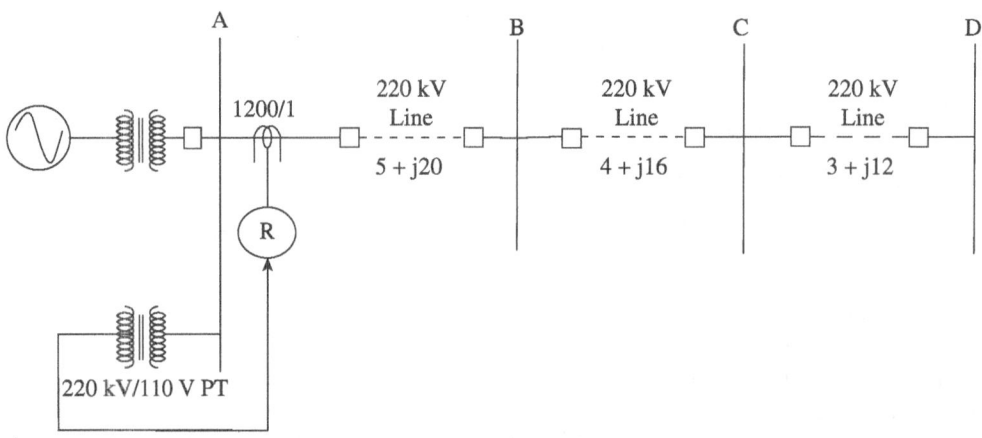

Fig. 3.26 Single line diagram for Example 3.6

Solution:

The reach of the third zone of the relay R is decided based on overloading and voltage dip conditions on the transmission line to be protected. This is due to the fact that the locus of the third zone of the relay is near to the locus of the transmission line when it carries full load current during normal situation. Hence, in this situation, voltage of the 220 kV transmission line is taken as 15% less (85%) and current carrying capacity of the line is taken as double the full load current of the given value (1049.72A). The impedance of load under the said situation is given by,

$$Z_{Load} = \frac{\frac{220 \times 10^3}{\sqrt{3}} \times 0.85}{2099.46 \angle -25.84°} = 51.42 \angle 25.84° \, \Omega$$

Considering 10% margin for error in the relay R, the load impedance is given by,

$$Z'_{Load} = 0.9 \times 51.42 \angle 25.84° \, \Omega = 46.28 \angle 25.84° \, \Omega$$

Referring this load impedance on the secondary side (relay side),

$$Z'_{Load} = 46.28 \angle 25.84° \times \frac{1200/1}{220 \times 10^3/110} = 27.77 \angle 25.84° \Omega$$

Hence, the setting of the third zone of the relay is given by,

$$K_3 = \frac{|Z_{Load}'|}{\cos(60° - 25.84°)} = \frac{27.77}{0.8275} = 33.56 \ \Omega$$

$$Z_3 = K_3 \times \cos(\varphi - \theta) = 33.56 \times \cos(75.96° - 25.84°) = 32.27 \Omega.$$

Here, the value of φ is obtained from the inverse of the tangent of the ratio of inductive reactance and resistance of the line section.

Impedance of section 1 = $5 + j20 = 20.62 \angle 75.96° \Omega$.

Considering 15% transient overreach of the relay R, the primary impedance of the first zone of the relay R is given by,

$$Z_1(primary) = \frac{1}{1.15} \times 20.62 \angle 75.96° = 17.93 \angle 75.96° \Omega$$

$$Z_1(secondary) = 10.76 \angle 75.96° \Omega.$$

$$K_1 = \frac{|Z_1|}{\cos(75.96 - 60)} = \frac{10.76}{\cos(75.96 - 60)} = 11.19 \Omega.$$

Now, impedance of sections 1 and 2 is given by,

$$Z_{12}(primary) = 5 + j20 + 4 + j16 = 37.1 \angle 75.96°$$

$$Z_{12}(secondary) = 22.26 \angle 75.96° \ \Omega.$$

The impedance of zone-3 of the relay R (32.27 Ω) is more than the impedance of sections 1 and 2 (22.26 Ω). Hence, the relay R will cover section 1 and section 2 completely as desired by the three-step distance relay.

Now, considering the impedance of zone-2 of the relay R is the impedance of the whole first section along with 50% of the impedance of the second section.

$$Z_2(primary) = 5 + j20 + 0.5 \times (4 + j16) = 28.86 \angle 75.96°$$

$$Z_2(secondary) = 17.32 \angle 75.96° \Omega$$

$$K_2 = \frac{|Z_2|}{\cos(75.96 - 60)} = \frac{17.32}{\cos(75.96 - 60)} = 18.01 \Omega$$

Example 3.7 The 220 kV transmission line having an impedance of 2.5 + j6 Ω is protected by distance relay R as shown in Fig. 3.27. If the distance relay R is assumed to be (i) reactance relay (ii) ohm relay and (iii) Mho relay, then draw the characteristic of all these relays on *R-X* diagram. All these relays are set to operate in their first zone which covers 80% of the line section.

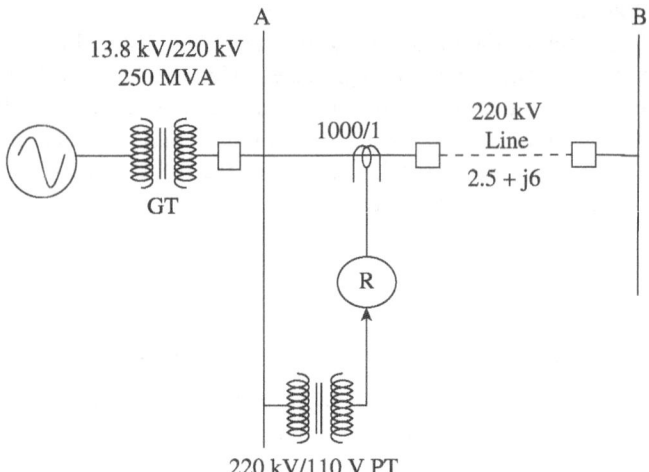

Fig. 3.27 Single line diagram for Example 3.7

Solution:
Impedance of line, $Z_L = 2.5 + j6$
$$= 6.5\angle 67.38° \ \Omega.$$
Impedance of the line for the first zone
$$= 0.8 \times (2.5 + j6) = 2 + j4.8 \ \Omega.$$
(i) Reactance relay
Considering 80% setting for zone-1 of reactance relay which has only imaginary part of the impedance, the set value of X is = $0.8 \times j6 = j4.8 \ \Omega$. The characteristic of reactance relay is shown in Fig. 3.28. With reference to Fig. 3.28, when the measured value of impedance by reactance relay R is less than the set value, i.e., 4.8 Ω, the relay R operates; otherwise, it blocks. The reactance relay has a characteristic angle of 90°.
(ii) Ohm relay
The ohm relay has a modified reactance characteristic and hence, its characteristic angle is not 90°. It has a characteristic angle equal to the inverse tangent of the ratio of inductive reactance and resistance of the line section, respectively. With reference to Fig. 3.28 characteristic of ohm relay forms 90° angle with reference to the angle of the line to be protected, i.e., 67.38°.
(iii) Mho relay
The characteristic of mho relay is represented by a circle having radius of 2.6 ($\sqrt{1^2 + 2.4^2} = 2.6$) and the centre of circle comes out to be at (1, 2.4). With reference to Fig. 3.28, k_m represents 80% of the line impedance vector.

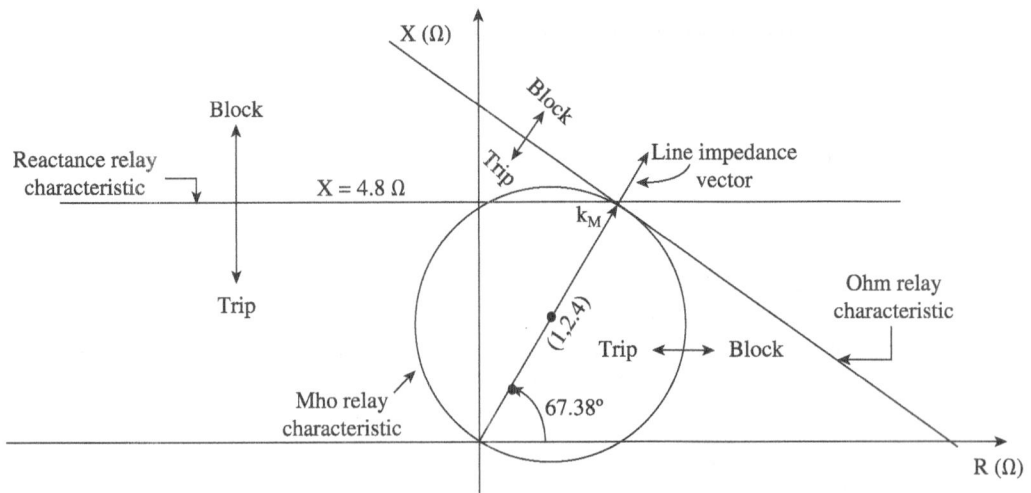

Fig. 3.28 Characteristics of all three relays on *R-X* diagram

Example 3.8   Figure 3.29 shows the single line diagram of a two terminal 220 kV transmission line. The impedance of the whole line is $5 + j20 \ \Omega$. The fault occurs at point F on the line and the impedance from the relaying point to the fault point, i.e., $Z_{AF} = 2.5 + j10 \ \Omega$. The fault resistance $R_F = 15 \ \Omega$. The fault current contribution from bus A to the fault, i.e. $I_A = 200\angle - 76.36° \ A$. The contribution of fault current from bus B to the fault, i.e., $I_B = 300\angle - 84.29° \ A$. Determine the apparent impedance seen by the relay R with and without considering the contribution of current from the remote bus. Also give your comments.

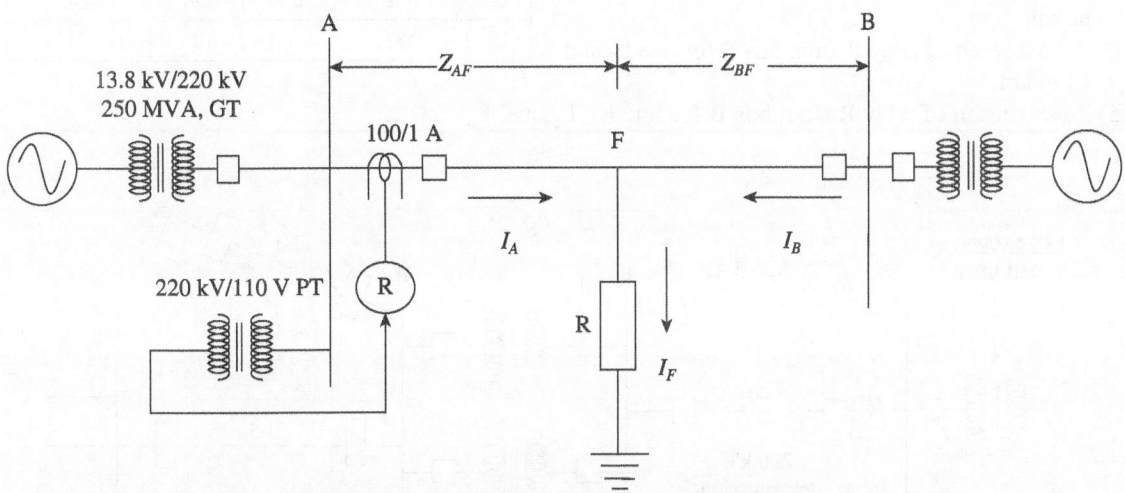

Fig. 3.29   Single line diagram for Example 3.8

Solution:
Considering the contribution of current from remote bus, the apparent impedance seen by the relay R is given by the following equations.

$$Z_A = Z_{AF} + R_F \times \left(1 + \frac{I_B}{I_A}\right)$$

$$Z_A = 2.5 + j10 + 15 \times \left(1 + \frac{300\angle -84.29°}{200\angle -76.36°}\right)$$

$$Z_A = 39.789 + j9.9985 \Omega = 41.02\angle 14.11° \Omega$$

Now, neglecting the contribution of current from remote bus, the apparent impedance seen by the relay R is given by

$$2.5 + j \times 10 + 15 = 17.5 + j \times 10 \Omega = 20.16\angle 29.74° \Omega$$

By observing both values of Z_A, it is to be noted that the apparent impedance seen by the relay R is greatly affected by the current from remote end. If this current is neglected then the relay R will measure the lower value of impedance and underreaches. Moreover, while considering the current from the remote bus, the error in the apparent impedance seen by the relay will be in real as well as imaginary part as I_A and I_B are not in phase.

Example 3.9   Figure 3.30 shows the single line diagram of a portion of power system network. Line parameters of L_1, L_2, and L_3 are as shown in the following table.

Mho relay R with characteristic angle of 60° is used to protect the power system network. CT ratio is 1000/1 A and PT ratio is 220 kV/110 V. Three-zone settings of relay R is $K_1 = 9\ \Omega$, $K_2 = 160\%$ of K_1 and $K_3 = 270\%$ of K_1.

Find out in terms of distance in km

(i) Zone-1 reach of relay R from bus A for the line L_1 in km.
(ii) Zone-2 reach of relay R from bus B for line L_2 and L_3 in km.
(iii) Zone-3 reach of relay R from bus B for line L_4, L_5, and L_6.

Line	Voltage (kV)	Impedance (Ω/km)	Distance (km)
L_1	220	$0.025 + j \times 0.10$	200
L_2	220	$0.03 + j \times 0.12$	100
L_3	220	$0.02 + j \times 0.1$	90
L_4	132	$0.025 + j \times 0.1$	30
L_5	132	$0.08 + j \times 0.24$	80
L_6	132	$0.1 + j \times 0.3$	50

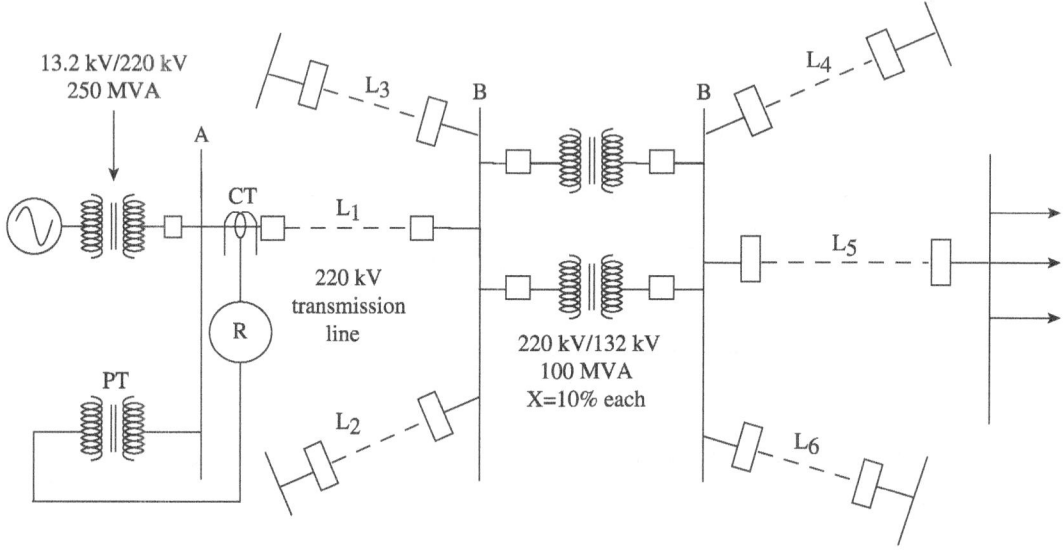

Fig. 3.30 Single line diagram for Example 3.9

Solution:

(i)
$$Z_{L1} = 0.025 + j0.1(\Omega/\text{km}) \quad = 0.1031\angle 75.96°(\Omega/\text{km})$$

$$K_1 = 9.0$$

$$Z_1 (\text{sec}) = K_1 \cos(\theta - \varphi) = 15 \times \cos(75.96 - 60)$$

$$Z_1 (\text{sec}) = 8.65\ \Omega$$

$$Z_1 (\text{prim}) = Z_1 (\text{sec}) \times \frac{PTr}{CTr} = 8.65 \times \frac{220 \times 10^3}{1000 \times 110}$$

$$Z_1 (\text{prim}) = 17.31\ \Omega$$

Zone-1 reach of R from bus A for line L_1 is

$$= \frac{Z_1(\text{prim})}{Z_{L1}/\text{km}} = \frac{17.31}{0.1031} = 167.85\,\text{km}$$

(ii)
$$K_2 = 1.6 \times 9 = 14.4\,\Omega$$
$$Z_2\,(\text{sec}) = K_2 \cos(\theta - \phi) = 14.4 \times \cos(75.96 - 60)$$
$$Z_2\,(\text{sec}) = 13.85\ \Omega$$
$$Z_2\,(\text{prim}) = 13.85 \times 2.0 = 27.69\,\Omega$$
$$Z_{L1} = 200 \times (0.025 + j \times 0.1) = 5 + j \times 20 = 20.62\angle 75.96°\,\Omega.$$

Now,
$$Z_2\,(\text{prim}) - Z_{L1} = 27.69 - 20.62.$$
$$Z_{L2} = (0.03 + j \times 0.12) = 0.1236\angle 75.96°\,\Omega.$$

Therefore, zone-2 reach of relay R from bus B for line L_2

$$= \frac{Z_2 - Z_{L1}}{Z_{L2}} = \frac{27.69 - 20.62}{0.1237} = \frac{7.07}{0.1237} = 57.15\,\text{km}$$
$$Z_{L3} = (0.02 + j \times 0.1) = 0.1019\angle 78.66°\,\Omega.$$

Therefore, zone-2 reach of relay R from bus B for line L_3

$$= \frac{Z_2 - Z_{L1}}{Z_{L3}} = \frac{27.69 - 20.62}{0.1019} = \frac{7.07}{0.1019} = 69.38\,\text{km}$$

(iii)
$$Z_T = \frac{(220)^2}{100} \times \frac{0.1}{2} = j24.2\,\Omega$$
$$Z_{L1} + Z_T = 5 + j20 + j24.2 = 5 + j44.2 = 44.48\angle 83.55°\,\Omega$$
$$= \text{Impedance seen by relay R for } L_4$$
$$= Z_{L1} + Z_T + Z_{L4}\ (\text{Referred on 220 kV side})$$
$$Z_{L4} = (0.025 + j \times 0.1) \times 30 = 0.75 + j \times 3\,\Omega$$
$$= 5 + j44.2 + \left(\frac{220}{132}\right)^2 \times (0.75 + j3)$$
$$= 5 + j44.2 + 2.08 + j8.33 = 7.08 + j52.53 = 53.0\angle 82.33°\ \Omega$$
$$Z_3\,(\text{sec}) = K_3 \cos(\theta - \varphi) = 2.7 \times 9 \times \cos(82.33 - 60)$$
$$Z_3\,(\text{sec}) = 22.48\ \Omega$$
$$Z_3\,(\text{prim}) = Z_3\,(\text{sec}) \times \frac{PTr}{CTr} = 22.48 \times 2.0 = 44.96\ \Omega$$

Hence, reach of relay R in zone-3 from bus B for line L_4

$$= \frac{Z_3\,(\text{prim}) - (Z_{L1} + Z_T)}{Z_{L4/km}} = \frac{44.96 - 44.48}{0.1031} = 4.6\,\text{km}$$

Similarly, the reach of relay R in zone-3 from bus B for L_5 and L_6 can be calculated.

Example 3.10   Figure 3.31 shows the single line diagram of a multi-terminal 220 kV transmission line network. This line is protected by mho type distance relays R_A, R_B, and R_C at bus A, bus B, and bus C, respectively. The source and line parameters at each bus are as under.

Source parameters: $Z_{1S} = 0.55 + j7.98\ \Omega$, $Z_{0S} = 1.5 + j23.98\ \Omega$
Line parameters: $Z_{1L} = 0.032 + j0.318\ \Omega/\text{km}$, $Z_{0L} = 0.2586 + j1.174\ \Omega/\text{km}$

Assuming positive sequence impedance of the line is same as negative sequence impedance of the line, determine zone-1 and zone-2 settings of distance relays R_A, R_B, and R_C.

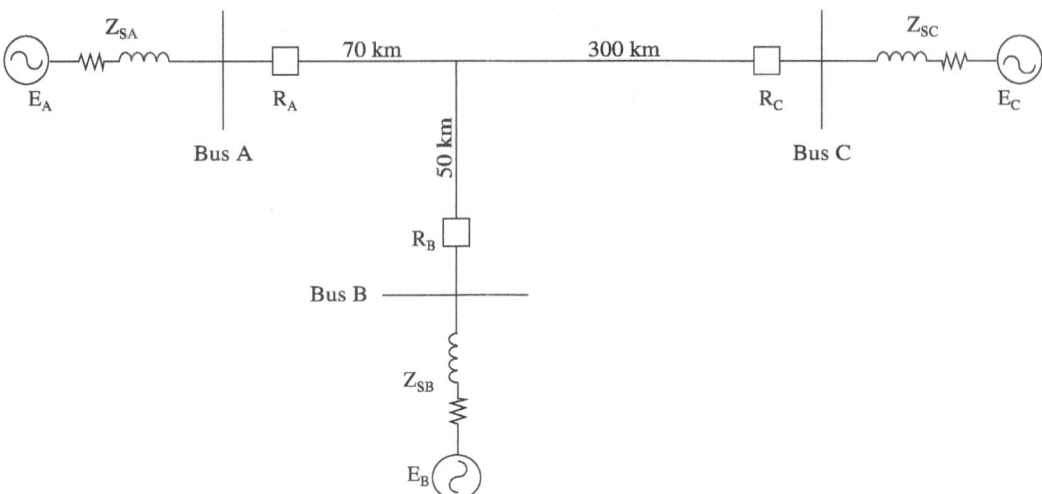

Fig. 3.31 Single line diagram of multi-terminal transmission line network

Solution:

Let us assume that the first zone of all three distance relays cover 80% of the positive sequence impedance of the line.

Hence, the zone-1 setting of distance relay R_A is obtained by considering 80 % of line impedance. Here, in order to calculate total line impedance for relay R_A, shortest line length is taken as otherwise relay may overreach.

$$Z_{1RA} = 0.8 \times (0.032 + j0.3184) \times (70 + 50) = 3.072 + j30.5664 \ \Omega$$

Similarly, zone-1 setting of relay R_B and R_C are calculated as given below.

$$Z_{1RB} = 0.8 \times (0.032 + j0.3184) \times (50 + 70) = 3.072 + j30.5664 \ \Omega$$
$$Z_{1RC} = 0.8 \times (0.032 + j0.3184) \times (300 + 50) = 8.96 + j89.152 \ \Omega$$

Let us assume that zone-2 of all distance relays cover 125% of positive sequence impedance to farthest breaker terminal i.e. longest adjacent line. This is required to achieve back-up protection for faults on the adjoining line section.

Therefore, zone-2 setting of all three distance relays are calculated considering the longest adjoining line section as given below.

$$Z_{2RA} = 1.25 \times (0.032 + j0.3184) \times (70 + 300) = 14.8 + j147.26 \ \Omega$$
$$Z_{2RB} = 1.25 \times (0.032 + j0.3184) \times (50 + 300) = 14 + j139.30 \ \Omega$$
$$Z_{2RC} = 1.25 \times (0.032 + j0.3184) \times (70 + 300) = 14.8 + j147.26 \ \Omega$$

The time distance characteristic of all the three relays for zone-1 and zone-2 is shown in Fig. 3.32.

Example 3.11 Figure 3.33 shows the single line diagram of a 400 kV three terminal transmission lines system. A three phase fault occurs on the line section connected between T point and bus C. The magnitude and direction of contribution of currents from each bus are shown in Fig. 3.33. In addition, impedance of line section connected between bus A and T-point, between bus B and T-point and between bus C and T-point is $25\angle80° \ \Omega$, $15\angle85° \ \Omega$ and $12\angle75° \ \Omega$, respectively. Determine the impedance seen by relays located at bus A and bus B.

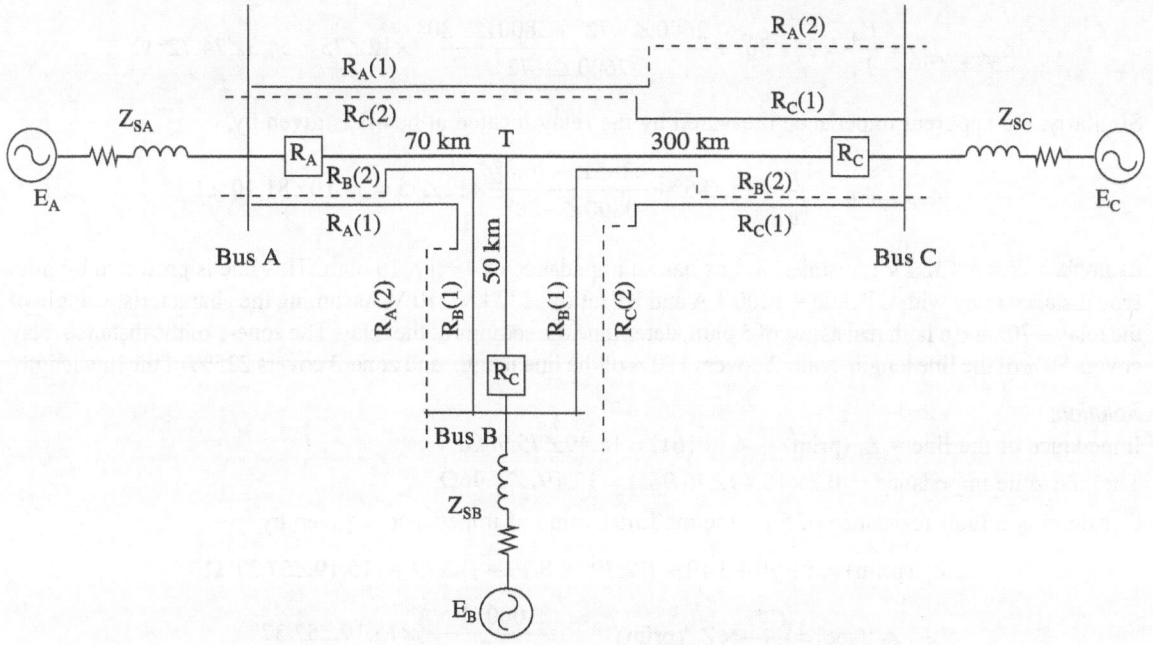

Fig. 3.32 Time distance characteristic of all three distance relays

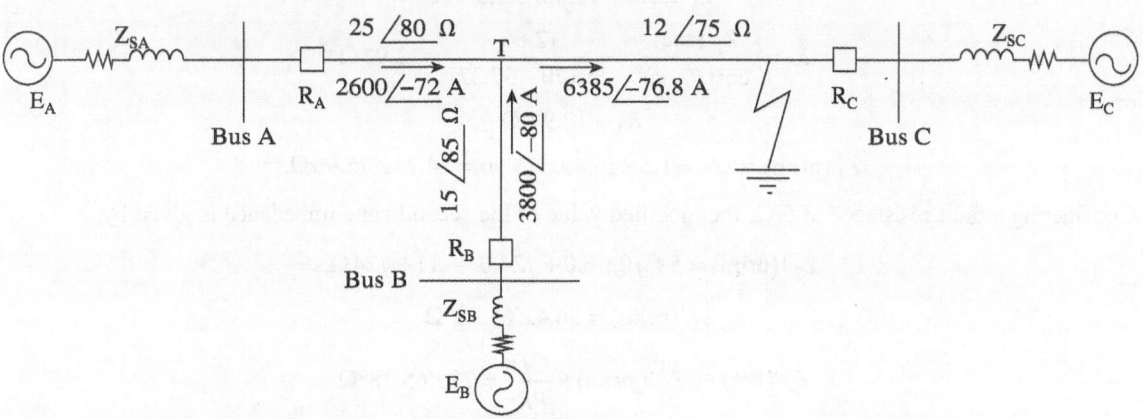

Fig. 3.33 Multi-terminal transmission line network showing current and impedance values

Solution:

Let us assume that V_A and V_B are the voltages measured at bus A and bus B, respectively.

Moreover, I_A and I_B are the currents measured at bus A and bus B, respectively.

Hence, the voltage measured at bus A for the said fault is given by,

$$V_A = I_A \times Z_{AT} + (I_A + I_B) \times Z_{TC}$$

Therefore, the apparent impedance measured by the relay located at bus A is given by,

$$Z_{apparent, A} = \frac{V_A}{I_A} = Z_{AT} + \frac{(I_A + I_B) \times Z_{TC}}{I_A}$$

$$Z_{apparent,A} = \frac{V_A}{I_A} = 25\angle 80° + \frac{2600\angle -72° + 3800\angle -80°}{2600\angle -72°} \times 12\angle 75 = 54.3\angle 74.72° \ \Omega$$

Similarly, the apparent impedance measured by the relay located at bus B is given by,

$$Z_{apparent,B} = \frac{V_B}{I_B} = 15\angle 85° + \frac{6385\angle -76.8°}{3800\angle -80°} \times 12\angle 75 = 35.10\angle 81.10° \ \Omega$$

Example 3.12   A 132 kV transmission line has an impedance of $4 + j \times 16$ ohm. This line is protected by mho type distance relay with CT ratio = 1000/1 A and PT ratio = 132 kV/110 V. Assuming the characteristic angle of the relay = 70° and a fault resistance of 5 ohm, determine the settings of the relay. The zone-1 of the distance relay covers 80% of the line length, zone-2 covers 150% of the line length, and zone-3 covers 225% of the line length.

Solution:

Impedance of the line = Z_1 (prim) = $4 + j16\Omega = 16.49\angle 75.96\Omega$.

The first zone impedance = $0.8 \times 16.49\angle 75.96\Omega. = 13.19\angle 75.96\Omega$.

Considering a fault resistance of 5 Ω, the modified value of impedance is given by,

$$Z_1'(\text{prim}) = 5 + j0 + 3.19 + j12.79 = 8.19 + j12.79 = 15.19\angle 57.37°\Omega$$

$$Z_1'(\text{sec}) = \frac{CTr}{PTr} \times Z_1'(\text{prim}) = \frac{1000/1}{132\times 10^3 /110} \times 15.19\angle 57.37°$$

$$Z_1'(\text{sec}) = 12.66\angle 57.37°\Omega$$

$$K_1 = \frac{Z_1'(\text{sec})}{\cos(\theta - \varphi)} = \frac{12.66}{\cos(70 - 57.37)} = 12.97\Omega$$

$$K_1 = 12.97 \ \Omega$$

$$Z_2 \ (\text{prim}) = Z_2 = 1.5 \times 16.49\angle 75.96 = 24.74\angle 75.96\Omega.$$

Considering a fault resistance of 5 Ω, the modified value of the second zone impedance is given by,

$$Z_2'(\text{prim}) = 5 + j0 + 6.0 + j24.0 = 11 + j24\Omega.$$

$$Z_2'(\text{prim}) = 26.4\angle 65.38°\Omega$$

$$Z_2'(\text{sec}) = Z_2'(\text{prim}) \times \frac{CTr}{PRr} = 22\angle 65.38°\Omega$$

$$K_2 = \frac{22}{\cos(70 - 65.38)} = 22.07\Omega$$

$$K_2 = 22.07 \ \Omega$$

$$Z_3 \ (\text{prim}) = 2.25 \times 16.49\angle 75.96 = 37.1\angle 75.96\Omega$$

Considering a fault resistance of 5 Ω, the modified value of the second zone impedance is given by,

$$Z_3'(\text{prim}) = 5 + j0 + 9.0 + j35.99 = 14 + j35.99 \ \Omega$$

$$Z_3'(\text{prim}) = 38.62\angle 68.74°\Omega$$

$$Z_3'(\text{sec}) = Z_3'(\text{prim}) \times \frac{CTr}{PRr} = 32.18\angle 68.74\Omega$$

$$K_3 = \frac{32.18}{\cos(70 - 68.74)} = 32.19 \ \Omega$$

$$K_3 = 32.19 \ \Omega$$

Example 3.13 An mho relay is used to protect a 132 kV transmission line having an impedance of $5 + j \times 15$. The first zone of a distance relay is set to $K_1 = 14.5 \ \Omega$. The other data is as under:

CT ratio: 1000/1 A
PT ratio: 132kV/110V
Characteristic angle (θ): 86°
Line angle (Φ): 71°

Determine the value of fault resistance which the mho relay incorporates for a short circuit at 80% of the line length.

Solution:
The first zone setting of distance relay is given by $K_1 = \dfrac{Z_1}{\cos(\theta - \varphi)}$.

$$Z_1 = K_1 \times \cos(\theta - \varphi) = 14.5 \times \cos(86 - 71) = 14.0 \ \Omega$$

$$Z_1 \ (\text{prim}) = (Z_1 \ \text{sec}) \times \frac{PTR}{CTR} = 14 \times \frac{132}{110} \times \frac{1000}{1000} = 16.8 \angle 71° \Omega. = 5.47 + j \times 15.88 \Omega$$

Now, the primary impedance value of Z_1 for a short circuit at 80% of the line length is given by,

$$Z_1(\text{prim}) = 16.8 \angle 71 \times \frac{1}{0.8} = 21.0 \angle 71 \Omega. = 6.83 + j \times 19.86 \Omega$$

Now, the impedance of the line section is $= Z_L = 5 + j \times 15 \Omega. = 15.81 \angle 71.56 \Omega.$
Therefore, the value of fault resistance accommodated by the mho distance relay is given by,

$$R_F = Z_1(\text{prim})\_\text{Real} - Z_L(\text{Real}) = 6.83 - 5 = 1.83 \Omega$$

3.11 Symmetrical Component-based Digital Distance Relay

Figure 3.34 shows the functional block diagram of the symmetrical component-based digital distance relay.

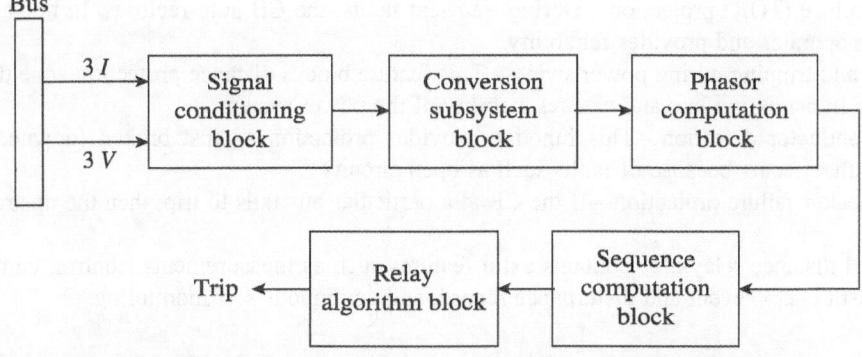

Fig. 3.34 Block diagram of symmetrical component-based digital distance relay

The voltage and current inputs ($3\ V + 3\ I$) are fed into the signal conditioning block that contains the surge protection device, isolation transformer, and anti-aliasing filter. The signal conditioning block removes unwanted frequency components from the input signals. Further, it also avoids aliasing error. Thereafter, the signals are given to the conversion subsystem block, which converts analog signal into the digital form. Moreover, it also captures instantaneous samples of currents and voltage signals. The phasor computation block contains a cosine filter, based on a discrete Fourier transform, which calculates the phasor values (magnitude and phase angle) of the fundamental component of voltages and currents. The sequence computation block converts the phasor values of currents and voltages into zero-sequence and negative-sequence values using symmetrical component theory. The sequence computation block contains two filters, namely negative-sequence filter and zero-sequence filter. The zero-sequence components and negative-sequence components are both measurable indications of abnormal/fault conditions. Negative-sequence filters require phase shifting by the operator 'a', which can be achieved using operational amplifiers or using other solid state components such as capacitors and resistors. The relay algorithm block computes the ratio of the negative-sequence voltage to the negative-sequence current and the ratio of the zero-sequence voltage to the zero-sequence current. If these two ratios exceed the predetermined value of threshold, then the tripping command is issued; otherwise, it is blocked.

3.12 Digital Distance Protection

The digital distance relay is a multifunction relay that supports different protection, control, and communication features. It offers a comprehensive range of protection functions for application in many overhead line and underground cable circuits. The protection features available in the digital distance relay are described here. It is to be noted that these features may vary from one company to another. However, after the execution of IEC 61850 rules for substation and automation, all manufacturers now design relays that are compatible to each other. Protection features include the following:

1. Phase- and ground-fault distance protection—This is usually given with five independent zones.
2. Instantaneous and time delayed overcurrent protection—For protection against short-circuit, distance relays are used as a primary protection. As a second line of defence, instantaneous and time delayed overcurrent protection is provided.
3. Undervoltage/overvoltage protection—This protection is provided against sudden increase or decrease in voltage with reference to the base voltage of transmission line.
4. Voltage and current transformer supervision—This is a unique feature of modern digital distance relay. It continuously monitors the current and voltage transformers. Monitoring of instrument transformers is highly essential as they provide current and voltage juice to the distance relay.
5. Trip on reclose (TOR) protection—During transient faults, the CB auto-recloses. In this situation, this protection operates and provides reliability.
6. Blocking and tripping during power swing—This feature blocks distance protection zone during power swings on transmission line and ensures stability of the power system.
7. Broken conductor detection—This function provides protection against broken (downed) conductor condition that occurs because of faults such as open circuits.
8. Circuit breaker failure protection—If the CB at a particular bus fails to trip, then the upstream CB will operate.
9. The digital distance relay also contains extra features such as measurements, control, communication, synchronism check, event and disturbance record, and continuous self-monitoring.

Recapitulation

- Most long EHV and UHV transmission lines are protected by the distance relaying scheme.
- The protection scheme is based on the stepped distance characteristics, and it measures the fault distance from the relaying point.
- Proper selection of measuring unit such as impedance relay, reactance relay, mho or admittance relay, ohm or angle impedance relay, offset mho relay, quadrilateral, and other special characteristics provides exhaustive protection to any long transmission line.
- The first zone setting of mho relay is given by $K_1 = \dfrac{|Z_1|}{\cos(\theta - \varphi)} \, \Omega$.
- In a distance relay, the phase and ground relay units should be properly connected to the available CTs and PTs, which is based on the combination of voltages and currents applied to the measuring units.
- The percentage transient overreach is given by $\dfrac{Z_x - Z_y}{Z_y} \times 100$.
- Different problems such as close-in fault, fault resistance, remote infeed, mutual coupling, power swing, and overload should be considered while implementing the distance relay in practice.
- Nowadays, all the long EHV transmission lines utilize the digital distance relaying scheme, which provides several useful features.

Multiple Choice Questions

1. Distance relays are widely used for the protection of
   (a) long EHV and UHV transmission lines
   (b) distribution feeders
   (c) induction motors
   (d) none of the above

2. The distance characteristic that is best suitable as far as the incorporation of fault resistance is concerned is
   (a) reactance
   (b) impedance
   (c) Mho
   (d) quadrilateral

3. The zone of distance relay that is affected by the transient overreach phenomenon is
   (a) first zone
   (b) second zone
   (c) third zone
   (d) none of the above

4. Transient overreach increases as
   (a) the length of line increases
   (b) the load increases
   (c) the spacing between conductors increases
   (d) all (a), (b), and (c) increase

5. The characteristic in which separate directional unit is not required is
   (a) impedance
   (b) Mho
   (c) reactance
   (d) all of the above

6. For a single circuit one-terminal transmission line (three conductors), the number of distance units required is
   (a) one
   (b) two
   (c) three
   (d) six

7. The problem of reduction in sensitivity in case of a close-in fault is rectified with the help of
   (a) electromechanical relays
   (b) static relays
   (c) numerical relays
   (d) adaptive relays

8. The value of mutual zero-sequence impedance is of the order of
   (a) 5–10% of zero-sequence impedance of the line
   (b) 1–2% of zero-sequence impedance of the line
   (c) 50–70% of zero-sequence impedance of the line
   (d) None of the above

9. The tendency of mal operation of distance relay during power swing condition is less in case of

   (a) reactance element

   (b) quadrilateral element

   (c) mho element

   (d) none of the above

10. Overreaching of distance relay due to the decaying DC component is avoided by

    (a) electromechanical relay

    (b) solid state relay

    (c) digital relay

    (d) none of the above

Review Questions

1. Explain how backup protection is achieved in a distance relay using step distance characteristics.

2. What do you mean by reach of the relay? Explain over-reach and underreach phenomena with reference to distance relay.

3. Explain the characteristic of different types of distance relays.

4. Compare reactance, impedance, mho, and quadrilateral characteristic used in distance relay.

5. How many distance relay units are required at a particular bus for a three-phase transmission line?

6. How is protection of transmission line achieved using static distance comparators?

7. What is the impact of close-in fault on distance relay? Suggest various remedies to overcome the problem of close-in fault in a distance relay.

8. What is the behaviour of distance relay during a fault with a considerable value of fault resistance? Explain this effect with vector diagram.

9. Which characteristic do you prefer for distance relay that reduces the impact of fault resistance?

10. How does distance relay behave when the current from a remote bus is considered?

11. How does a distance relay overreach and underreach when mutual coupling effect is considered?

12. What different problems occur in distance relay used for the protection of long series-compensated transmission line?

13. Explain the voltage and current inversion phenomenon that occurs during the protection of series-compensated transmission line.

14. How does a distance relay overreach during the protection of series-compensated transmission line?

15. What is the impact of power swing on various characteristics of distance relay? Discuss this phenomenon for various characteristics with the relative merits and demerits.

16. Discuss the various features available in the modern digital distance relay.

Numerical Exercises

1. In Example 3.1 (Fig. 3.20), find out the reach of relay R_2 in all three zones.

   [First-zone reach = 88.27%, second-zone reach = 121.17%, third-zone reach = 186.41%]

2. Figure 3.35 shows the single line diagram of a portion of power system. The mho relay R is used to protect the portion of the line with a characteristic angle of 40°.

   Determine the three zone settings of mho relay. The first zone covers 85% of the first section. The second zone covers the first section plus 30% of the second section, and the third zone covers the first section plus 120% of the second section.

   [K_1 = 4.43 Ω, K_2 = 6.31 Ω, and K_3 = 9.64 Ω]

3. Decide the three zones setting of an mho relay R as shown in Fig. 3.36. The relay R has the characteristic angle of 60°.

The required data is as follows:

(a) 220 kV line (L_1) having the length of 150 km and impedance of 0.012 + j0.05 Ω/km/phase

(b) 132 kV line (L_2) having the length of 70 km and impedance of 0.05 + j0.10 Ω/km/phase

(c) Power transformer: 220/132 kV, 250 MVA having the reactance X of 10%

(d) CT ratio: 1000/1 A, PT ratio: 220 kV/110 V

   [K_1 = 3.21 Ω, K_2 = 14.92 Ω, and K_3 = 18.31 Ω]

4. Figure 3.37 shows the single line diagram of a portion of power system. The distance relay with mho characteristic having the characteristic angle 75° is used to protect the 400 kV transmission line. The required data is as follows:

   CT ratio: 400/1 A, PT ratio: 420 kV/110 V.

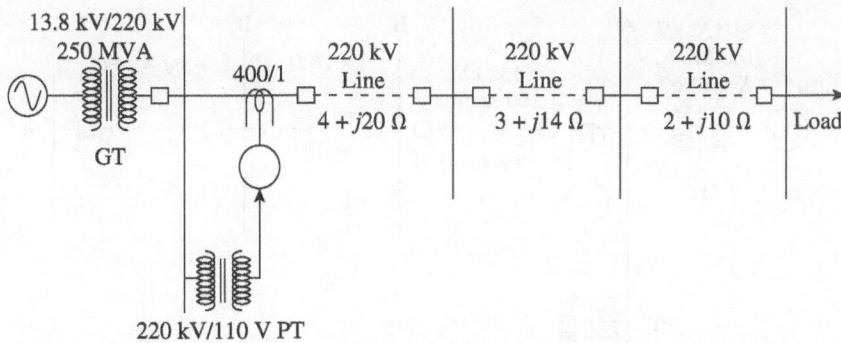

Fig. 3.35

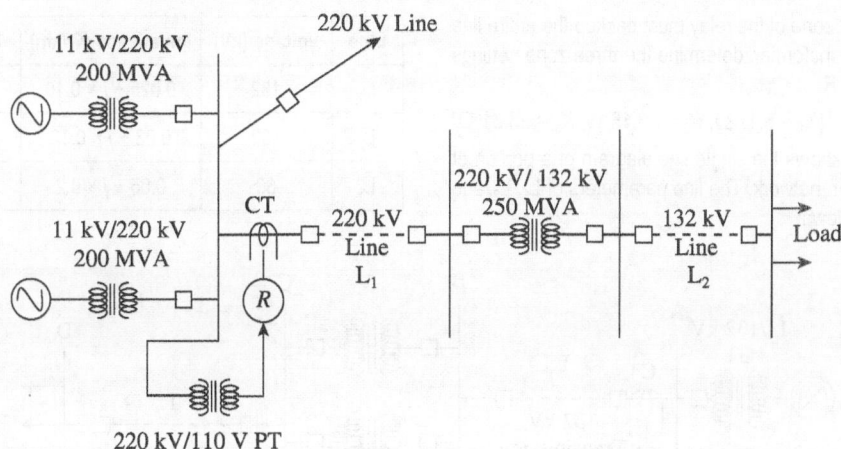

Fig. 3.36

Determine the relay settings for all three zones assuming that zone 1 covers 90% of the line length, zone 2 covers 140% of line length, and zone 3 covers 220% of the line length.

$[K_1 = 9.08\ \Omega, K_2 = 14.13\ \Omega,$ and $K_3 = 22.20\ \Omega]$

5. In Fig. 3.36, if the value of fault resistance of 10 Ω is included in the mho relay characteristic, then find out the new values of the three zone settings of the relay. The rest of the data is the same as given in Numerical Exercise 3.

$[K_1 = 9.11\ \Omega, K_2 = 14.113\ \Omega,$ and $K_3 = 22.16\ \Omega]$

6. Figure 3.38 shows the single line diagram of a portion of power system network. All the sequence impedances shown in Fig. 3.38 are in ohms as viewed from 132 kV side. The other data is as follows. The CT ratio and PT ratio are 500/1 A and 132 kV/110 V, respectively. The available three zone settings ranges of the distance relay are as follows:

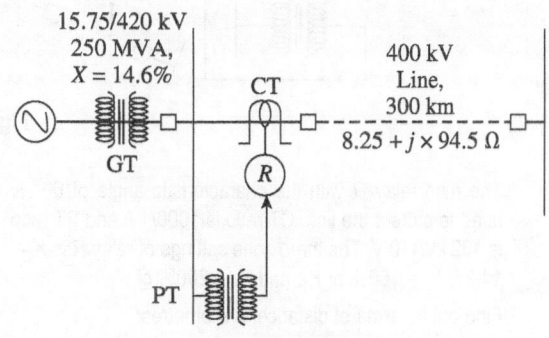

Fig. 3.37

Zone 1: 0.1–10 Ω in steps of 0.1 Ω

Zone 2: 0.15–15 Ω in steps of 0.1 Ω

Zone 3: 0.5 W–50 Ω in steps of 0.1 Ω

The characteristic angle of mho relay R is 75°. Considering

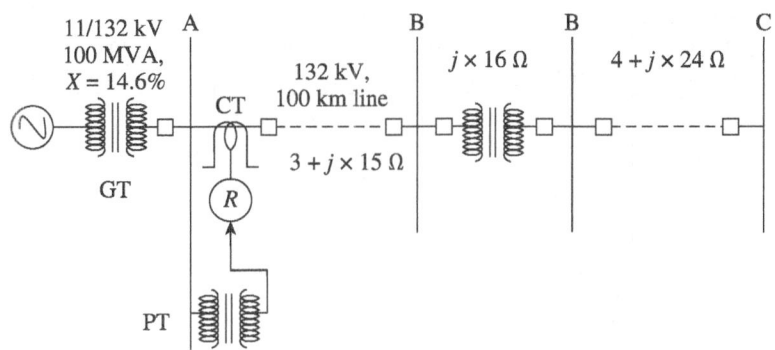

Fig. 3.38

that the third zone of the relay must backup the entire line along with transformer, determine the three zone settings for the relay R.

$[K_1 = 5.11 \ \Omega, \ K_2 = 13.15 \ W, \ K_3 = 23.31 \ \Omega]$

7. Figure 3.39 shows the single line diagram of a portion of power system network. The line parameters of L_1, L_2, and L_3 are as follows:

Line	Voltage (kV)	Impedance (Ω/km)	Distance (km)
L_1	132	$0.025 + j \times 0.10$	200
L_2	132	$0.03 + j \times 0.12$	100
L_3	66	$0.05 + j \times 0.2$	50

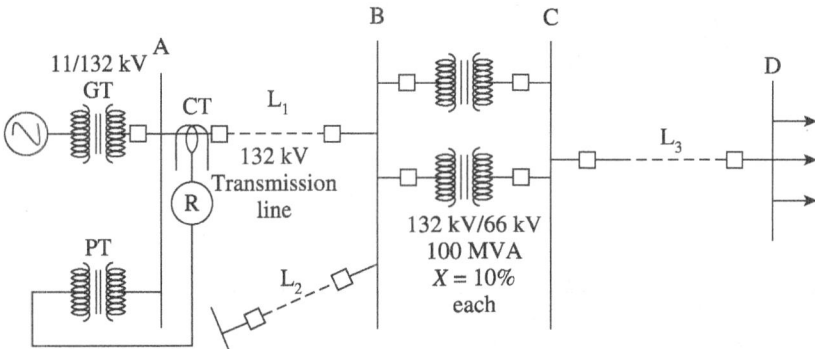

Fig. 3.39

The mho relay R with the characteristic angle of 60° is used to protect the line. CT ratio is 1000/1 A and PT ratio is 132 kV/110 V. The three zone settings of relay R is K_1 = 14.21, K_2 = 160% of K_1, and K_3 = 200% of K_1.

Find out in terms of distance in kilometres:

(a) Zone 1 reach of relay R from bus A for the line L_1

(b) Zone 2 reach of relay R form bus B for line L_2

(c) Zone-3 reach of relay R from bus B for line L_3

[(a) 159.02 km, (b) 45.37 km, (c) 16.25 km]

8. Figure 3.40 shows a single line diagram of a portion of power system. The distance relay having reactance

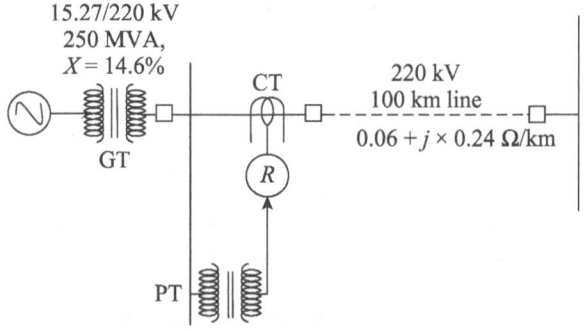

Fig. 3.40

characteristic is used to protect 220 kV transmission line. The required data is as follows:

CT ratio: 1000/1 A, PT ratio: 220 kV/110 V

The first zone covers 85% of the line length. The second zone covers 160% of the line length, and the third zone covers 225% of the line length. Determine the three zone settings of reactance relay R.

$$[X_1 = 10.21 \ \Omega, X_2 = 19.999 \ \Omega, \text{ and } X_3 = 27.0 \ \Omega]$$

9. A 132 kV transmission line with an impedance of $4 + j \times 16 \ \Omega$ is protected by a reactance relay. On R–X plane, draw the operating characteristic of a reactance relay. Furthermore, determine the reach of the reactance relay in secondary ohms for all three zones. The CT ratio and PT ratio are 600/1 A and 132 kV/110 V, respectively. First zone, second zone, and third zone cover 90%, 140%, and

225% of the line section, respectively.

$$[X_1 = 7.199 \ \Omega, X_2 = 11.198 \ \Omega, X_3 = 17.99 \ \Omega]$$

10. An mho relay is used to protect a line having an impedance of $7.28 + j \times 16.94 \ \Omega$. The first zone of a distance relay is set to $K_1 = 9.11 \ \Omega$.

The other data is as follows:

CT ratio: 1000/1 A

PT ratio: 220 kV/110 V

Characteristic angle (θ): 75°

Line angle (ϕ): 60°

Determine the value of the fault resistance that the mho relay incorporates for a short-circuit at 90% of the line length.

[Fault resistance $R_f = 2.5 \ \Omega$]

Carrier Aided Distance Scheme for Transmission Lines

4

Learning Objectives

After going through this chapter, the students will be able to:

- Explain the basic block diagram of a pilot protection scheme and justify the need for the same
- Classify the different types of pilot communication
- Explain the different types of wire pilot relaying schemes
- Understand the concept of carrier blocking and unblocking schemes
- Analyse the differences between carrier tripping and blocking schemes
- Explain overreach and underreach transfer tripping scheme

4.1 Introduction

Owing to the installation of compensation devices, distribution reforms, and deregulated environment, modern power systems operate close to their stability limits. Hence, fault clearing time is very important in such systems. Failure to comply with the requirements may result in instability of the system, and in the worst case, lead to the shutdown of large parts of the network or complete blackout. Various protection methods are used to protect long extra high voltage (EHV)/ultra high voltage (UHV) transmission lines.

Current-based protection schemes cannot be used as they do not provide instantaneous operation throughout the entire length. Moreover, they also suffer from the problem of transient overreach. Distance protection is commonly used for the protection of transmission lines. However, it has several limitations. The two important limitations are the following:

1. As 80–90% of the faults on long overhead transmission lines are single-line-to-ground faults, all UHV and EHV transmission line protection devices are equipped with single-pole tripping facility.
2. As 85% of the faults are transient in nature, most protective devices have an auto-reclosing feature.

Due to these two reasons, distance relays cannot be used to protect important long EHV and UHV lines. Furthermore, they provide delayed clearing of the fault, at about 20% from each side of the bus. This is explained in detail in this section.

Figure 4.1 shows the stepped distance protection of a transmission line. The first zone of both relays R_1 and R_2 is set to cover 80% of the line. It is to be noted from Fig. 4.1 that both relays provide instantaneous operation for internal faults occurring within approximately 60% (from the midpoint) of the transmission line, whereas the remaining 40% (20% from bus A and 20% from bus B) will be cleared by the time de-layed zone 2 protection of both the relays. Conversely, by measuring the signals at all line terminals of the intended zone, instantaneous operation can be achieved over the entire line by pilot protection. However, distance relays are still used by the utilities because of the backup protection they provide.

In order to achieve instantaneous fault clearing at both ends, pilot protection requires a communication channel to exchange information from one end of the line to the other. The type and effectiveness of the pilot protection scheme depends on the capabilities of the applied communication channel.

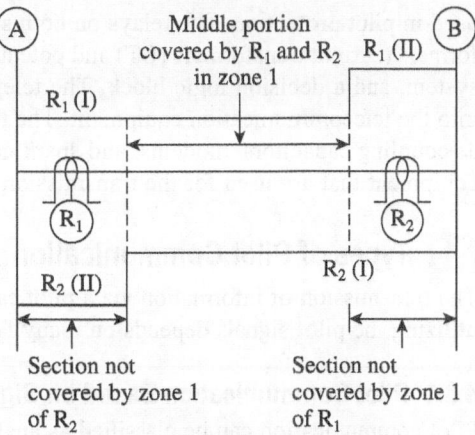

Fig. 4.1 Stepped distance protection of transmission line

4.2 Pilot Protection System and its Need

Figure 4.2 shows the basic block diagram of a pilot protection system. It is to be noted from this figure that the relay at each side (R1 and R2) measures local voltages and local currents. Therefore, these signals are supplied to the respective communication equipment, which transmits them to the other end of the line. As the information from both ends of the line is available, the relay at each end operates instantaneously for a fault within the zone of the protected line. The signalling system used to transfer the signal between the two ends of a transmission line is known as the pilot system. The term 'pilot' refers to the communication medium, such as wire, microwave, power line carrier, fibre optic cable, or satellite. Figure 4.3 shows various types of equipment

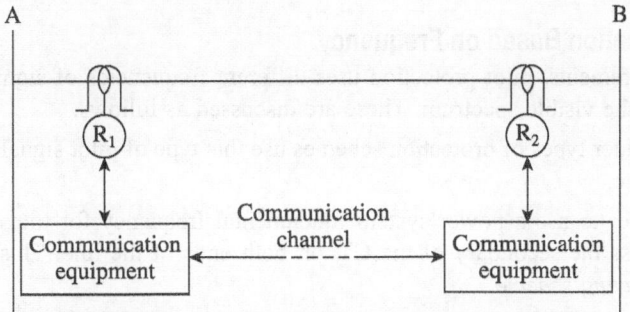

Fig. 4.2 Basic block diagram of pilot protection system

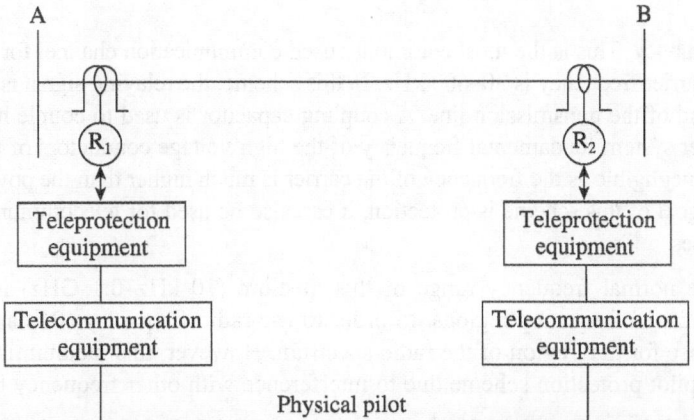

Fig. 4.3 Various types of equipment used in pilot protection

used in pilot protection. The relays on both sides work as protective devices and contain instrument transformers (current transformers (CT) and potential transformers (PT)). They also contain transducers, a logical system, and a decision logic block. The teleprotection equipment takes data from the relay and injects it into the telecommunication equipment. The teleprotection block contains various types of equipment such as coupling capacitors, modems, and spark gaps. The telecommunication block includes different types of equipment that are used for the transmission of data in text, voice, or binary form.

4.3 Types of Pilot Communication

The transmission of information via a pilot can be achieved in several ways. However, the exact method of utilizing the pilot signals depends on many factors.

4.3.1 Pilot Communication Based on Signal

Pilot communication can be classified as analog and digital, based on the type of signal used.

Continuous (analog) signal In this signal, all the information such as amplitude of a quantity, system frequency, phase shift, and width of the pulse of infinite number of levels are considered at regular intervals between specified minimum and maximum levels.

Discrete (digital) signal In this signal, all the information of a limited number of levels is passed in digital form. The communication of information in digital form offers high channel density. Moreover, one can connect different types of devices through programming using digital signals.

4.3.2 Pilot Communication Based on Frequency

Depending on the requirements, pilot protection uses different frequencies of signals, comprising a broad bandwidth from DC to the visible spectrum. These are discussed as follows.

Direct current Certain older types of protection schemes use this type of pilot signal for transmission. However, it is almost obsolete.

Power frequency In order to use a power system fundamental frequency for transmission, a pair of pilot wires is connected across the secondary of the CT, on both ends of the line. This arrangement is widely known as *pilot wire relaying scheme*.

Audio frequency The normal audio frequency range is 0.02–20 kHz. In this scheme, tone generators and receivers are used to transmit and receive the audio frequency carrier. However, this scheme has also become obsolete.

Power line carrier frequency This is the most commonly used communication channel for relaying. The normal range of power line carrier frequency is 30–600 kHz. In this scheme, the relaying signal is modulated and transmitted to the other end of the transmission line. A coupling capacitor is used to couple high frequency carrier signals with the power system fundamental frequency of the high voltage conductors of the line. The problem of induced voltage is negligible as the frequency of the carrier is much higher than the power system frequency. Though the primary goal of this scheme is protection, it can also be used for telecommunication, data transfer, and signalling purposes.

Radio frequency The normal frequency range of this medium (10 kHz–0.1 GHz) is between the audio frequency and the infrared frequency regions. In order to use radio frequency (RF) as a carrier, the utility has to acquire a license for that region of the radio spectrum. However, this medium is not used because of maloperation of the pilot protection scheme due to interference with other frequency bands.

Microwave frequency This medium has the frequency range of 0.3–3 GHz. As the electromagnetic wave propagates in a straight line, this medium is free from refraction by the ionosphere and has little interference from lightning. Furthermore, as its bandwidth is much wider than that of the carrier channel, it can carry more information.

Fibre optics The wavelength of a signal in a fibre optic link is about 0.85–1.6 μm. As the frequency of a signal in a fibre optic link is much higher than that of the microwave signal, the communication capacity of a fibre optic link is very high. Moreover, a fibre optic link is immune to electromagnetic interference because of its insulating property. Furthermore, it is free from induced voltage. In addition, there is no need for repeaters to transmit information from one end to the other of long EHV/UHV transmission lines.

4.4 Wire Pilot Relaying Scheme

The wire pilot relaying scheme is divided into three basic groups, as discussed in Sections 4.4.1 – 4.4.3.

4.4.1 Circulating Current-based Wire Pilot Relaying Scheme

Figure 4.4 shows the schematic diagram of a circulating current-based wire pilot relaying scheme. It is to be noted that under pre-fault/normal conditions, the CT secondary current flows through the restraining coil (RC) of each current balance relay, whereas no current is present in the operating coil (OC). Conversely, in case of a fault on the line, the CT secondary current for the far end CT reverses, and hence, the current will flow through the OC of both relays.

There are several methods to combine the three-phase currents into one signal. One of the methods is to use a summing transformer, which is shown in Fig. 4.4.

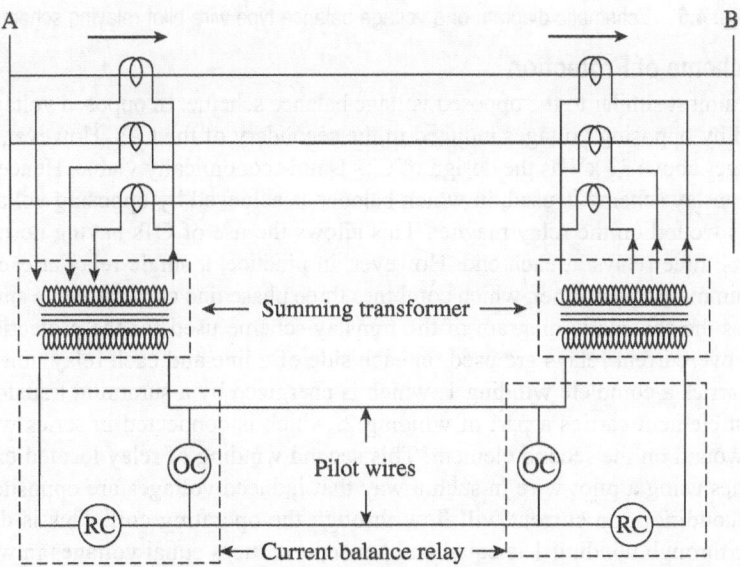

Fig. 4.4   Schematic diagram of a circulating current-based wire pilot relaying scheme

4.4.2 Voltage Balance Type Wire Pilot Relaying Scheme

Figure 4.5 shows the schematic block diagram of a voltage balance type wire pilot relaying scheme. It is to be noted that under normal conditions, no current flows through the pilot wires as they cross each other. The working of this scheme is similar to the previous scheme except for the reversal of roles of the two windings.

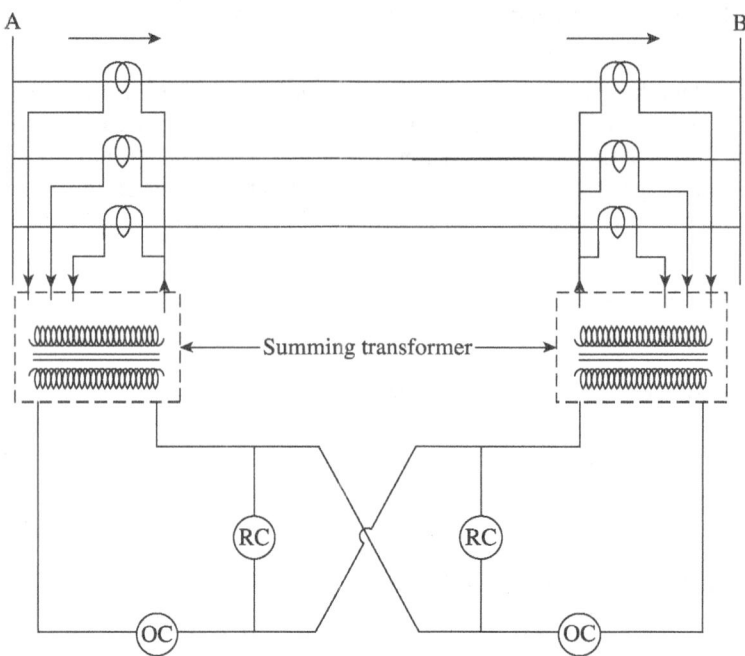

Fig. 4.5 Schematic diagram of a voltage balance type wire pilot relaying scheme

4.4.3 Translay Scheme of Protection

Translay scheme is almost similar to the opposed voltage balance scheme. In opposed voltage balance scheme, balance is achieved by opposing voltages induced in the secondary of the CTs. However, this scheme is not preferable for voltages above 33 kV as the design of CTs is not economically viable. Hence, in order to rectify this problem, the translay scheme is used, in which balance is achieved by opposing voltages induced in the secondary windings wound on the relay magnet. This allows the use of CTs having normal design. Ideally, this scheme requires three relays at each end. However, in practice, a single relay at each end can be used with the help of a summing transformer, which combines three phase line currents into a single phase quantity.

Figure 4.6 shows the schematic diagram of the translay scheme used for the protection of transmission lines. Inverse time overcurrent relays are used on each side of a line and each relay unit has two elements. The first element carries a complete winding 1, which is energized by a summing transformer from the CT secondary. The first element carries a part of winding 2, which is connected in series with the winding of the operating coil wound on the second element. This second winding of relay located on both ends of line is connected in series using a pilot wire in such a way that induced voltages are opposite to each other.

During normal condition, no current will flow through the operating coil. This is due to the fact that the current flowing through winding 1 is equal and hence, it induces equal voltages in winding 2. As these windings are connected in opposition, no current flows through the operating coil. Conversely, at the time of fault on the feeder, voltage induced in winding 2 is different due to imbalance of currents at each end of the line. Therefore, current will flow through the operating coil and the pilot wires. In this situation, both relay elements on each side of the line are energized and open the CB of the respective end of the line.

Limitations of wire pilot relaying scheme

1. It can be applied only to short transmission lines because of the high cost of installation.
2. Sensitivity of this scheme is reduced because of the charging current between the pilot wires.

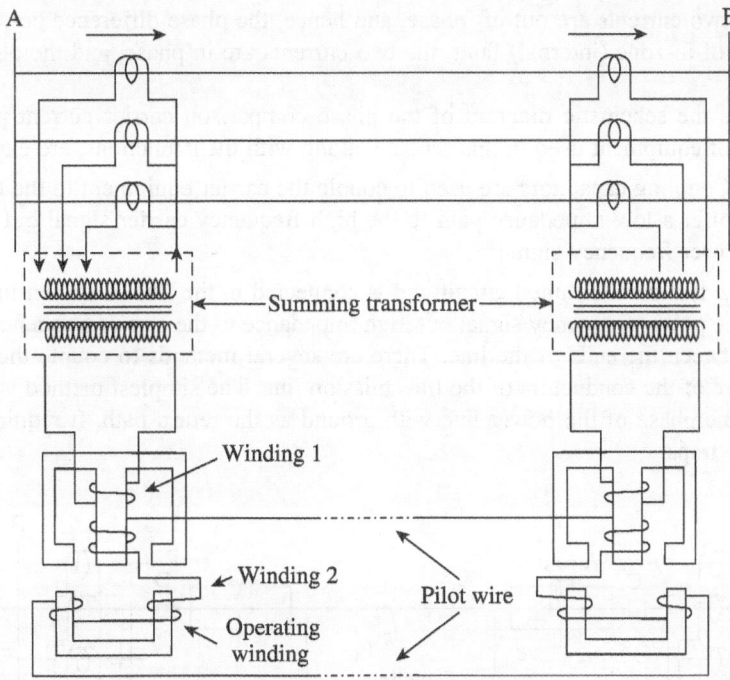

Fig. 4.6 Translay scheme of protection

3. This scheme requires special tuning to optimize signal transmission.
4. It suffers from the problem of induced voltage from parallel power transmission lines.
5. Difference in ground potentials between the two ends causes connection problems in the metallic link.

4.5 Carrier Current Protection Scheme

The pilot wire relaying scheme is economical only for short transmission lines (15–20 km). Hence, for long transmission lines, the carrier current relaying scheme is used. Moreover, this scheme is also used to achieve simultaneous tripping at both ends of the transmission line.

In this scheme, a carrier signal is used either to initiate the tripping or to block the tripping of the relay. When the carrier signal is used to initiate the tripping of the relay, it is known as *carrier tripping scheme*. On the other hand, when the carrier signal is used to block the tripping of the relay, it is known as *carrier blocking scheme*.

The carrier current protection scheme can be classified as follows:

1. Phase comparison carrier protection scheme
2. Directional comparison carrier protection scheme

The phase comparison scheme is explained in detail in Section 4.5.1. The directional comparison scheme works in a similar manner and is hence explained in brief, in Section 4.5.2.

4.5.1 Phase Comparison Carrier Protection Scheme

The phase comparison scheme compares the phase angle between currents at the two ends of a line (current entering at one end and current leaving the other end). During normal condition or out-of-zone (external) fault condition, the two currents are out of phase, and hence, the phase difference between them is 180°. Conversely, in case of in-zone (internal) fault, the two currents are in phase, and the phase difference between them is 0°.

Figure 4.7 shows the schematic diagram of the phase comparison carrier current protection scheme. The different types of equipment used in this scheme, along with their functions, are explained as follows.

Coupling capacitor Coupling capacitors are used to couple the carrier equipment to the high voltage transmission line. They offer a low impedance path to the high frequency carrier signal but a high impedance path to the 50 Hz power frequency signal.

Wave trap *Wave trap* is a parallel tuned circuit and is connected in the line as shown in Fig. 4.7. It offers low impedance to the power frequency signal but high impedance to the carrier frequency signal. Thus, the signal is trapped between the ends of the line. There are several methods to couple the carrier frequency signal to one or more of the conductors of the transmission line. The simplest method is single-phase coupling, which uses one phase of the power line with ground as the return path. It requires fewer coupling capacitors and wave traps.

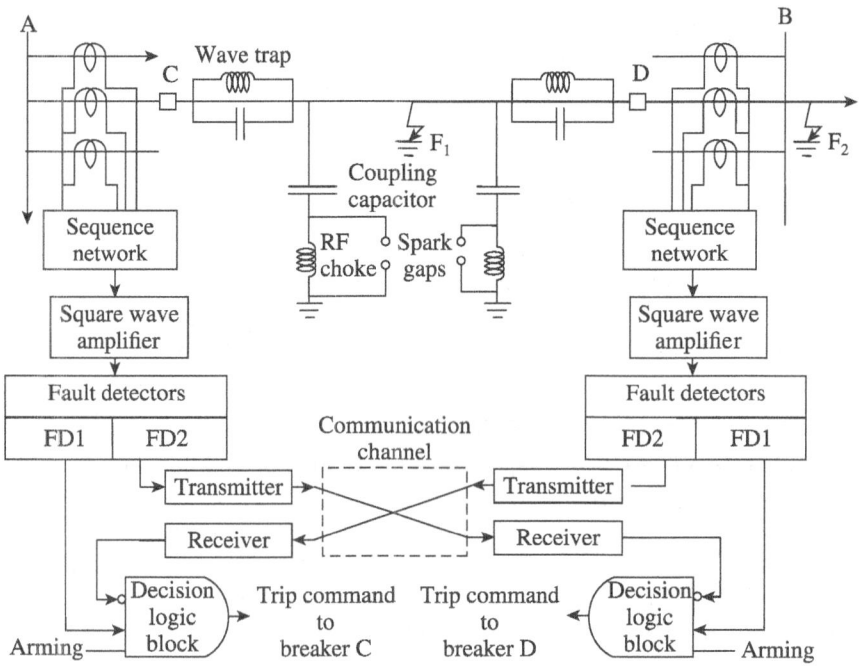

Fig. 4.7 Schematic diagram of phase comparison carrier current protection scheme

Spark gaps Spark gaps provide protection to the coupling equipment against overvoltage.

Transmitter and receiver unit A transmitter unit contains an oscillator block, which generates a high frequency carrier signal. It also contains an amplifier block, which amplifies the signal. The receiver unit contains an attenuator block, a matching element, and a filtering circuit.

The phase comparison scheme is a unit protection scheme. It measures the phase angle of the current entering the protected line at the local and far ends, and by comparing these two phase values of currents, it issues a trip decision or blocking command.

Let us assume that this scheme works on the blocking principle. As shown in Fig. 4.7, the measured currents are processed through a summation block or a composite sequence network block. The function of this block is to combine three sequence currents and convert them into the equivalent single-phase quantity. The square wave amplifier blocks convert and amplify the single-phase voltage to a square wave. Thereafter, this signal is given to the fault detectors. Two overcurrent fault detectors, namely FD_1 and FD_2, are used. The pickup setting of FD_1 is low, and is used to start the carrier. It is more sensitive than FD_2 and must be set on the basis of the maximum load on the line. FD_2 is a high set relay. It waits for the comparison of the transmitted signal and then arms the tripping circuit. The usual setting of FD_2 is 125–200% of FD_1. A scheme in which the signal is transmitted on the positive half of the square wave and no signal is transmitted on the negative half of the square wave is known as *single-phase comparison scheme*. A decision logic block (either flip-flop or comparator) at each end compares the received signal with the transmitted signal.

Figure 4.8 shows the wave shapes for internal and external fault conditions. During an internal fault at F_1, both overcurrent fault detectors operate as the current exceeds their pickup settings. They issue a local input signal to the decision logic block, which contains the flip-flop circuit. If the input to the flip-flop continues for a preset time (4–5 ms), then it is energized. In this situation, the square wave input from the local end and the receiver input to the decision logic block are in phase. Hence, the trip signal is initiated as both signals are present.

On the other hand, for an external fault at F_2, the phase of the receiver input to the decision logic block is out of phase with the local input square wave. As the signal is transmitted from the far end (bus B) to the local end (bus A), the receipt of a carrier signal from the channel prevents tripping. It is to be noted that the local square wave is used to turn the carrier signal on and off. Since a single-phase comparison scheme transmits the signal only during the positive half cycle, there is a delay of one half cycle in tripping. A remedy is to use a dual phase comparison scheme, which transmits the signal on both halves of the square wave. Therefore, it issues a trip signal on each half cycle.

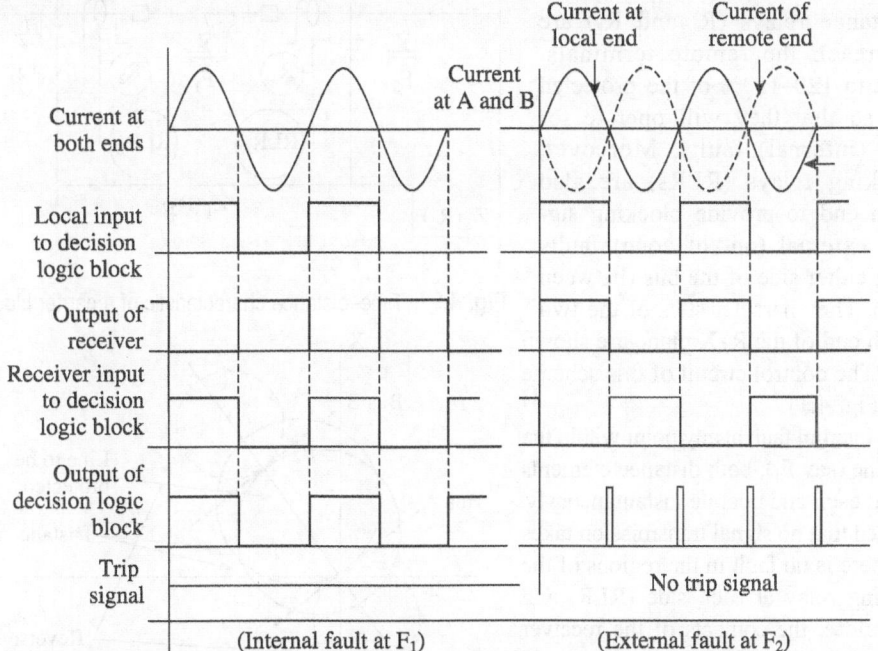

Fig. 4.8 Wave shapes for internal and external fault conditions

4.5.2 Directional Comparison Scheme

In the directional comparison scheme, the direction of power flow at the local and far ends of the line is compared. The relays located at both ends of the line are directional featured which continuously check the direction of current flow away from the bus (i.e., towards the line). Like, phase comparison, the power line carrier (PLC) is used to communicate information regarding the fault direction sensed by relay from one end to the other end of the line. An AND logic is used at each end to compare the local as well as remote end fault direction status transferred through PLC. During normal operating condition, if the flow of current at the local end of line is away from the bus, that is, correct direction, the relay generates signal "1" whereas if the flow of the current at the remote end is towards the bus, that is, incorrect direction, the relay generates signal "0". In case of an internal fault, the direction of the power flow reverses at remote end, which initiates the tripping signal "1". Thus, by combining the local end relay generated signal "1" and the signal conveyed by remote end relay "1" during internal fault, a trip signal is issued to associate line breakers. During an external fault, the direction of power flow is outwards at any one bus, similar to normal operating condition. Thus, a combination of signals generated by local and remote end relays at either end during external fault does not issue trip signal through AND logic.

4.6 Blocking and Unblocking Carrier-aided Distance Scheme

When a carrier signal is used to block the tripping of relays, the scheme is known as *carrier blocking scheme*. In this scheme, the relays located at each end monitor the region within as well as behind the protected line. A blocking signal is to be sent for any fault behind the relay at either end of the line and tripping is prevented.

4.6.1 Carrier Blocking Scheme

Figure 4.9 shows the time–distance characteristic of the distance elements (relays) installed at each end.

Both distance relays (R_1 and R_2) are set to overreach the remote terminals. They are set to 120–150% of the protected line length so that they will operate for all in-zone (internal) faults. Moreover, reverse looking relays (RLRs) are also used at each end to provide blocking signals for all external (out of zone) faults occurring on either side of the bus (between AC and BD). The characteristics of the two relays at each end of the R–X plane are shown in Fig. 4.10. The control circuit of this scheme is shown in Fig. 4.11.

During an internal fault at any point within the zone of the line (say F_1), both distance elements (R_1 and R_2) at each end operate instantaneously. It is to be noted that no signal transmission takes place since there is no fault in the regions of the reverse looking relay at each side (RLR$_A$ and RLR$_B$). Therefore, the contacts of the receiver relay (RR$_1$-1, RR$_2$-1) at both ends remain in closed condition.

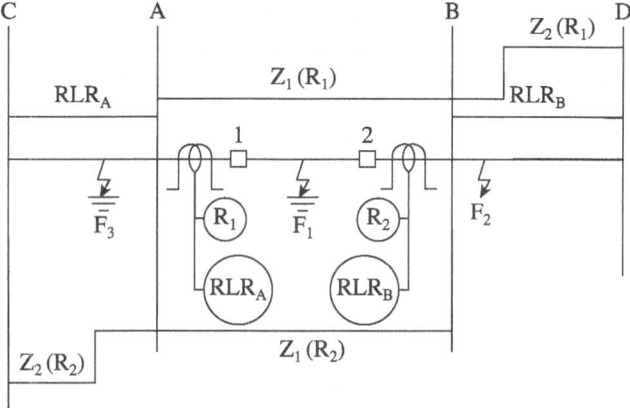

Fig. 4.9 Time–distance characteristic of a carrier blocking scheme

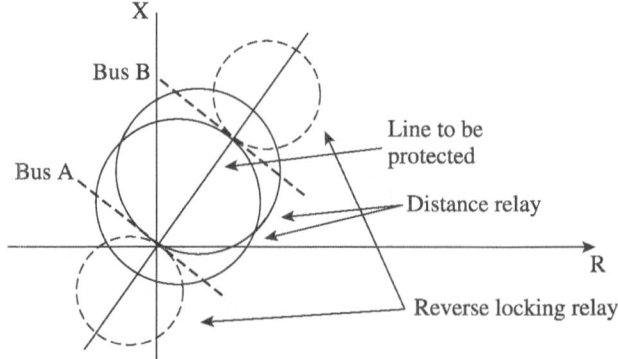

Fig. 4.10 Characteristics of relays on R–X plane

On the other hand, during an external fault beyond the bus B (say F_2), the distance relay R_2 at bus B will not operate. However, the reverse looking relay RLR_B at bus B picks up and sends a blocking signal at bus A. As the blocking signal is present at bus A, even after the operation of relay R_1 at bus A the CB will not trip due to opening of RR_1-1 contact. The CB will trip only when the blocking signal is not available (as in this situation, it sees the fault in the first zone). It is extremely important to note that the blocking signal must be received at each end before the local distance element operates for faults beyond bus A (say F_3) or bus B (say F_2). In the field, the normal practice is to use a timer that provides a time delay to ensure that no unwanted tripping would occur for a fault on the adjoining line on each side of the bus.

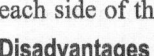

Fig. 4.11 Control circuit of carrier blocking scheme

Disadvantages

1. There is a possibility of relay maloperation in case of failure of the carrier blocking signal or the communication equipment.
2. During an internal fault, there is a possibility of attenuation of the carrier signal due to which a trip signal would not be received at the other end.

4.6.2 Carrier Unblocking Scheme

In order to avoid maloperation in the directional blocking scheme due to the failure of carrier signal during an external fault, a carrier unblocking scheme is used. In this scheme, a low-energy carrier signal is continuously transmitted as a check on the communication link. During an in-zone fault, the signal frequency is shifted to the unblock frequency, and hence, tripping is initiated. Figure 4.12 shows the time–distance characteristic of a directional comparison unblocking scheme.

In the case of an internal fault at F1, the relay R_1 at bus A and the relay R_2 at bus B operate in their first zone. Owing to the shifting of channel 1 and channel 2 into the unblock condition, the receiver relay at bus A and bus B operates, and tripping is initiated at each bus. During an external fault at F_2, the relay

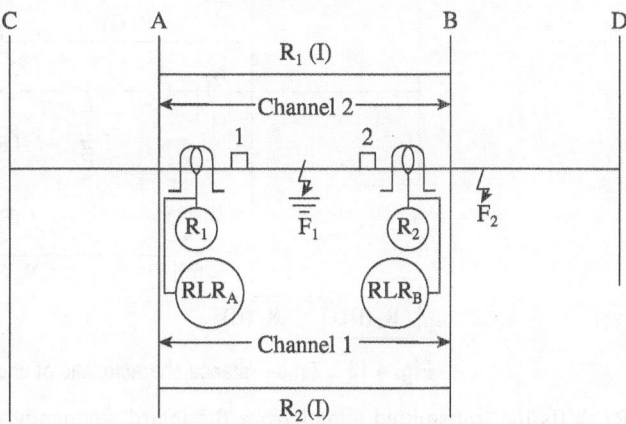

Fig. 4.12 Time–distance characteristic of directional comparison unblocking scheme

R_1 operates and channel 1 shifts from block condition to unblock condition. However, no tripping is initiated as channel 2 remains in blocking condition. On the other hand, at bus B, the relay R_2 does not operate, and hence, tripping is prevented.

4.7 Transfer Tripping Carrier-aided Distance Scheme

Transfer tripping is a viable scheme, particularly when the communication channel is independent of the power line. Tripping schemes are more advantageous than blocking schemes because of the following reasons:

1. In the tripping scheme, there is no need to use an additional time delay or coordination time, which is mandatory for the blocking scheme to avoid mal-trip during external faults.
2. The setting of the blocking relay is lower than that of the tripping relay. There is a possibility of mal-operation of the blocking relay in case of heavy loading or highly unbalanced condition.

Therefore, to avoid these problems, the transfer tripping scheme is used. In this scheme, a continuously transmitted guard signal is provided, which must shift to a trip frequency. Transfer tripping can be achieved in two ways.

1. Underreach transfer tripping scheme
2. Overreach transfer tripping scheme

4.7.1 Underreach Transfer Tripping Scheme

Figure 4.13 shows the time–distance characteristic of the underreach transfer tripping scheme. The control circuit of the same is depicted in Fig. 4.14. In this scheme, zone 1 of both relays, R_1 and R_2, is set to reach only the original 80% of the total line length. In case of a fault within the first zone of both relays (say at F_1), they, along with the fault detectors (FD_1 and FD_2), operate instantaneously and trip the respective breaker at each end.

In the case of a fault in the remaining 20% of the line section on each side (say at F_2), the distance relay R_2 operates instantaneously, and at the same time, sends a transfer tripping carrier signal at bus A. The relay

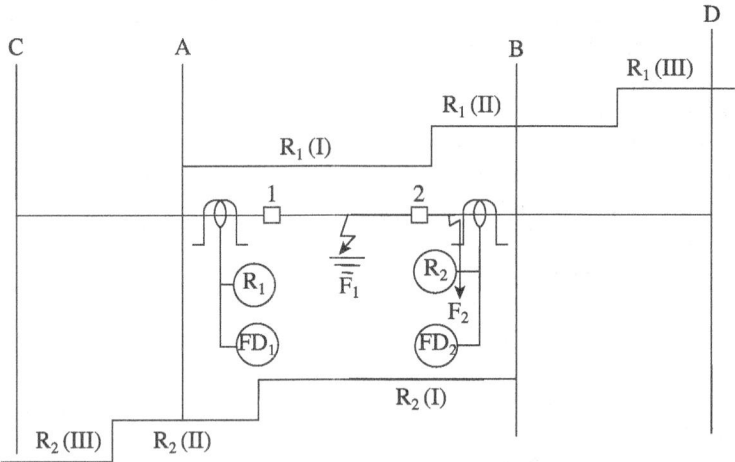

Fig. 4.13 Time–distance characteristic of underreach transfer tripping scheme

R_2 shifts the transmitted signal from the guard frequency to the trip frequency. Instantaneous tripping is initiated at bus A after the closure of the guard relay contact and the receiver relay contact (RR_1-1) as shown in Fig 4.14. Here, the incoming carrier signal is used to trip the breaker directly, and hence, this scheme is known as *direct underreach transfer tripping scheme*.

Disadvantages

1. The prime limitation of direct underreach transfer tripping scheme is its maloperation due to inadvertent closing of the receiver relay's contact. This can happen during maintenance or calibration, or because of the noise initiated by switching in the substation or transients during relay operations.

2. As this scheme requires phase selection at each end, it cannot be used when single-phase auto-reclosing is involved.

The remedy to the aforementioned problem is to use the direct underreach transfer tripping scheme with dual transmitter–receiver sets. However, the dependability of this solution is limited since there are twice as many components involved. Moreover, it is also costlier than the conventional direct underreach transfer tripping scheme.

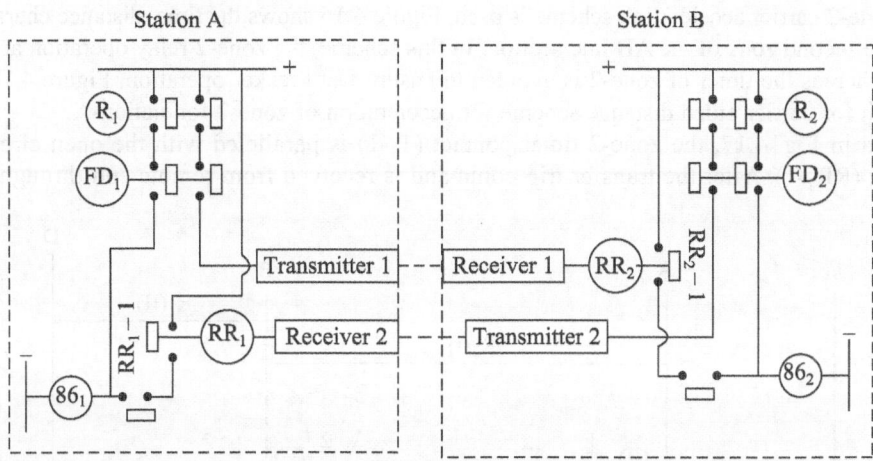

Fig. 4.14   Control circuit of underreach transfer tripping scheme

4.7.2 Permissive Inter Tripping or Permissive Underreach Transfer Tripping Scheme

As discussed in Section 4.7.1 the remote end receiver relay (RR) operates as and when it receives tripping carrier signal and instantly trips the associated CB directly for fault in the remaining 20% of the line section on each side. However, to avoid maloperation due to noise and to ensure the actual internal fault on line section, a permissive underreach transfer tripping scheme is used. In this scheme, the received carrier signal from the far end is utilized to trip the breaker after conforming that the local fault detector unit is operated. Thus contact of local fault detector (FD_1-1) and receiver relay contact (RR_1-1) are connected in series with the auxiliary trip coil of CB as shown in Fig. 4.15. Here, no purposeful time delay is added in breaker

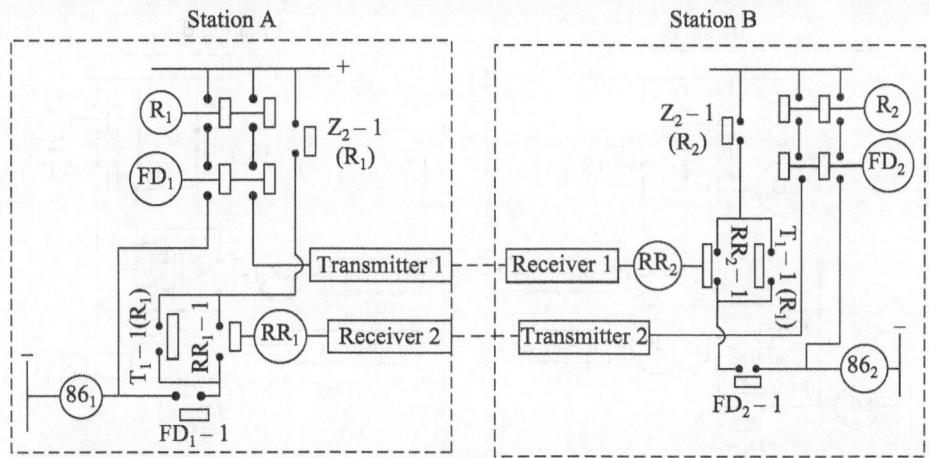

Fig. 4.15   Control circuit of permissive underreach transfer (inter) tripping scheme

operation at each end; hence, associated line breakers are tripped within same time for fault within first zone or in the remaining 20% of the line section. With reference to Fig. 4.15 if the local fault detector contact (FD_1-1) is bypassed, the scheme becomes direct underreach transfer tripping scheme as stated previously.

4.7.3 Carrier-aided Distance Scheme for Acceleration of Zone-2

To protect 20% of line section beyond the first zone (assumed to be 80%) from each end and to achieve fast operation, zone-2 carrier acceleration scheme is used. Figure 4.16 shows the time–distance characteristic for fault F_2 in the second zone of the AB line section. In this scheme, the zone-2 relay operation at local end is considered whereas the timer of zone-2 is avoided to ensure fast breaker operation. Figure 4.17 shows the control circuit for carrier aided distance scheme for acceleration of zone-2 for station-A.

As shown in Fig. 4.17, the zone-2 timer contact (T_1-1) is paralleled with the open circuit receiver relay contact (RR_1-1). After the transfer trip command is received from remote end through power line

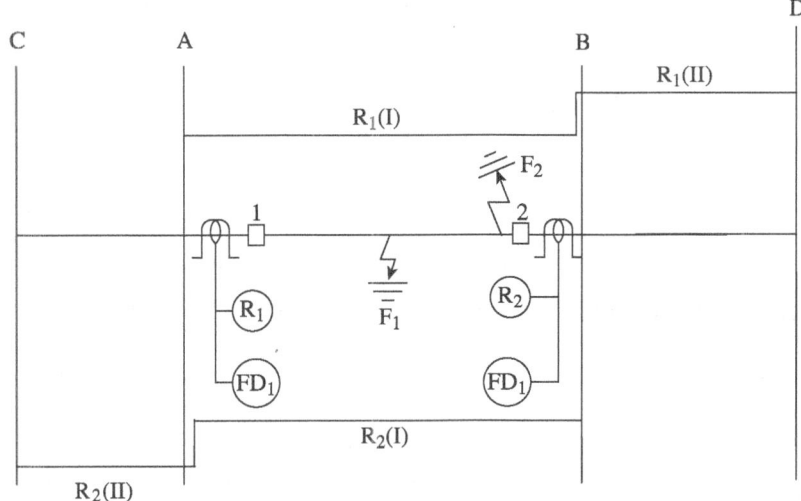

Fig. 4.16 Time–distance characteristic of acceleration of zone-2

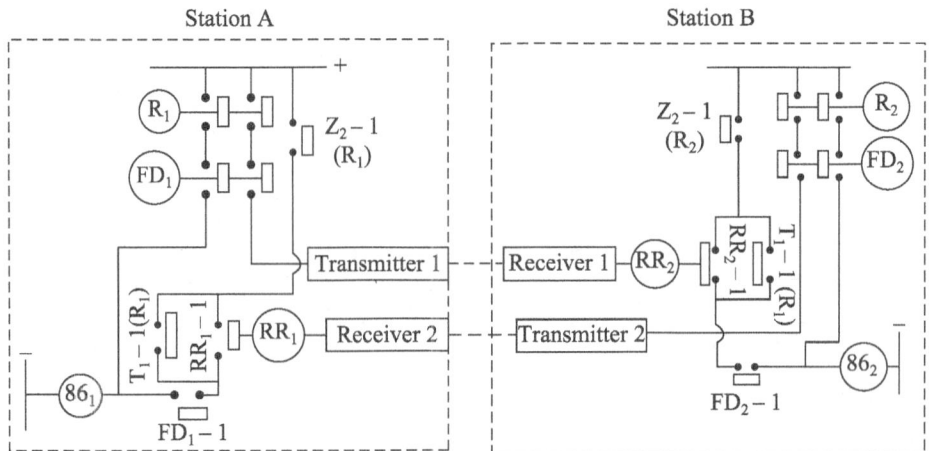

Fig. 4.17 Control circuit of acceleration of zone-2

carrier, the zone-2 timer contact is bypassed by closing of receiver relay contact. As a result, the zone-1 reach of relay is extended upto the end of line section by adding extra measurement of zone-2 (Z_2-1). Thus, all the faults in the remaining 20% of the protected zone can be cleared in the order of zone-1 time plus time lag due to carrier propagation delay. The usual operating time of zone-1 is around 25 ms and by adding the time of 15 ms for carrier transmitted, the overall operating time may be 40 ms for accelerated zone-2 protection. Conversely, this scheme provides enhanced security compared to direct underreach and permissive inter tripping scheme against noise and unwanted tripping. However, this scheme fails to protect a line which is being energized from one end and if fault occurs beyond the reach of zone-1. It is to be noted that in the case of failure of carrier channel or carrier equipment, the carrier aided tripping scheme goes back to a normal three-stepped distance protection as mentioned in Chapter 3.

4.7.4 Carrier-aided Distance Scheme for Pre-acceleration of Zone-2

The protection scheme explained earlier depends on carrier tripping signal transmitted through carrier channel. However, a carrier blocking scheme is used to prevent the instantaneous operation of zone-2 external fault as discussed in Section 4.6.1. This scheme is known as pre-acceleration of zone-2 for fault (F_2) beyond the line section to be protected as shown in Fig. 4.18. In this scheme, the zone-2 timer is pre-accelerated by connecting its contact T_1-1(Z_2) in parallel with the closed contact of receiver relay (RR_1-1) at each end. As can be seen from Fig. 4.19, the RR_1 is energized by the reverse looking relay (RLR_B) located at the other end of the line. In the event of fault (F_2) beyond the bus B (in second zone), the relay R_1 at bus A operates in zone-2 and close one of its contacts R_1-1(Z_2). At the same time the RLR_B relay senses the fault and sends a blocking signal at bus A to pickup RR_1 relay. Thus, RR_1 relay at bus A changes the status of RR_1-1 contact from normally CLOSED to OPEN and allows the zone-2 timer circuit (T_1-1(Z_2)) to decide the delayed operating time. Here, the action of relay R_1 in zone-2 for a fault at F_2 (external to line section) was pre-accelerated by (1) operating time for reverse looking directional relay, (2) time for carrier transmission over the channel and (3) operating time of the receiver relay.

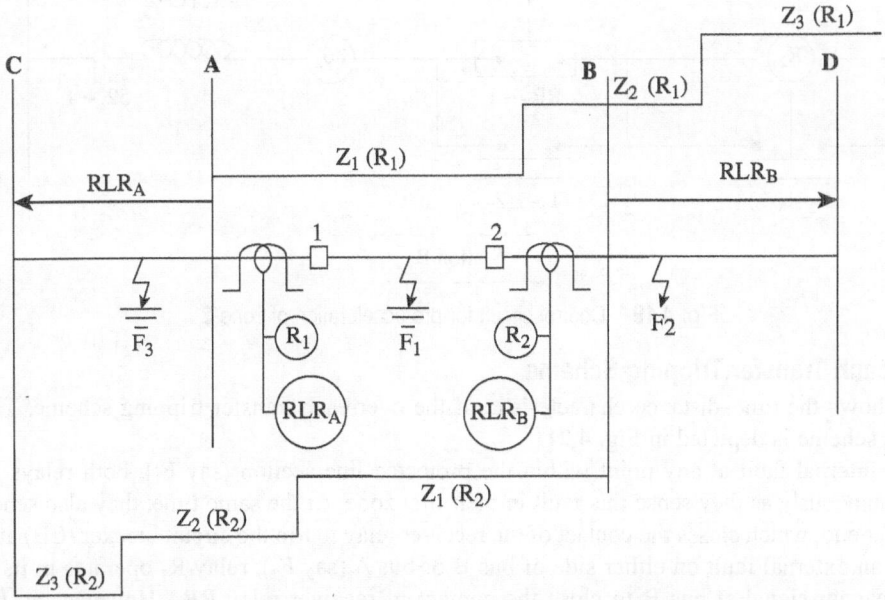

Fig. 4.18 Time–distance characteristic for pre-acceleration of zone-2

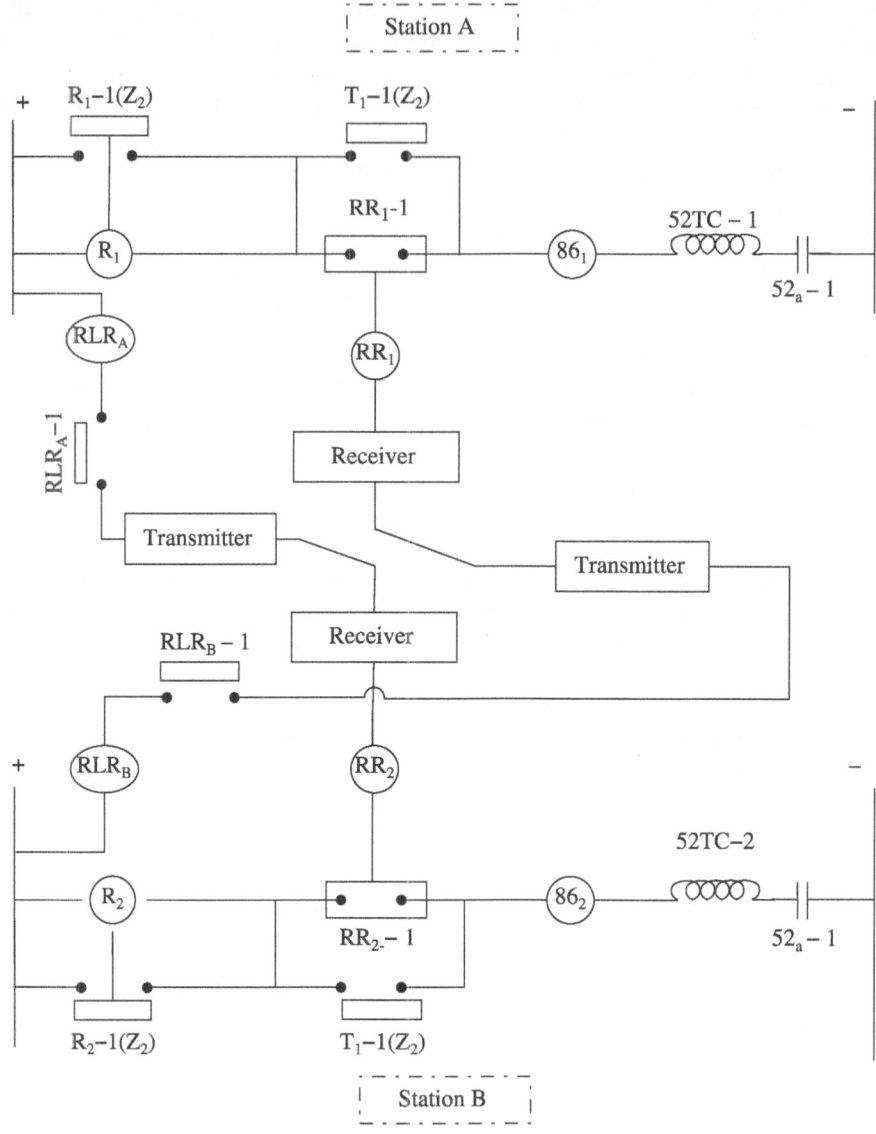

Fig. 4.19 Control circuit for pre-acceleration of zone-2

4.7.5 Overreach Transfer Tripping Scheme

Figure 4.20 shows the time–distance characteristic of the overreach transfer tripping scheme. The control circuit of this scheme is depicted in Fig. 4.21.

During an internal fault at any point within the protected line section (say F_1), both relays R_1 and R_2 operate instantaneously as they sense this fault in their first zone. At the same time, they also send a carrier signal to the far end, which closes the contact of the receiver relay to trip the circuit breaker (CB) at each end.

In case of an external fault on either side of bus B or bus A (say F_2), relay R_1 operates in its first zone and sends a carrier signal at bus B to close the contact of receiver relay RR_2. However, no tripping is initiated at bus B as relay R_2 does not operate. Since relay R_2 does not operate at bus B, no carrier signal is transmitted to bus A. Therefore, tripping is prevented at bus A even though the local fault detector unit operates.

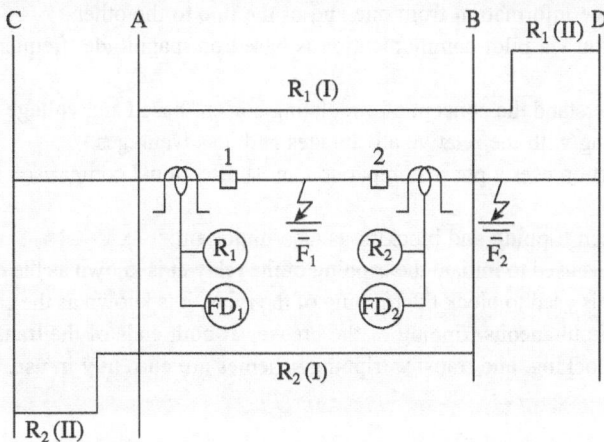

Fig. 4.20 Time–distance characteristic of overreach transfer tripping scheme

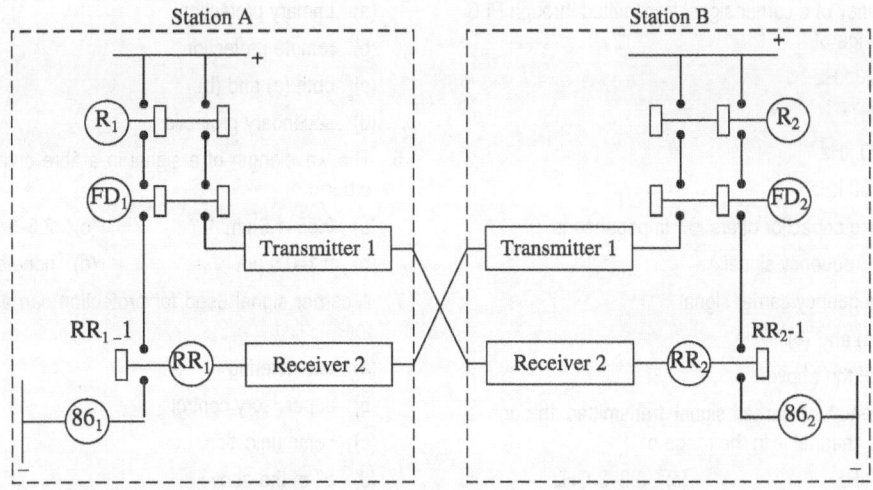

Fig. 4.21 Control circuit of overreach transfer tripping scheme

Disadvantages

1. High speed tripping in the overreach transfer tripping scheme is entirely dependent on the pilot channel. Failure of this pilot channel will lead to non-isolation of the fault.
2. If Power Line Carrier (PLC) is used as a pilot channel, then it is difficult to send a signal in case of a fault.

Recapitulation

- In pilot relaying scheme, to achieve instantaneous fault clearing at both ends, a communication channel is required to exchange information from one end of the line to the other.
- Transfer of information via pilot communication is based on magnitude, frequency, and phase shifting of the signal.
- It is important to understand the concept of circulating current-based and voltage balance type wire pilot relaying schemes along with the relative advantages and disadvantages.
- Carrier current protection uses a phase comparison and a directional comparison scheme for the transfer of tripping signal.
- The difference between tripping and blocking is also important:
- When a carrier signal is used to initiate the tripping of the relay, it is known as the carrier tripping scheme.
- When a carrier signal is used to block the tripping of the relay, it is known as the carrier blocking scheme.
- In order to achieve simultaneous tripping of the breaker at both ends of the transmission line, different carrier blocking, unblocking, and transfer tripping schemes are currently in use.

Multiple Choice Questions

1. The frequency of a carrier signal transmitted through PLC is in the range of
   (a) 0.02–20 kHz
   (b) 1–10 kHz
   (c) 50–100 kHz
   (d) 100–150 kHz

2. The coupling capacitor offers low impedance to
   (a) power frequency signal
   (b) high frequency carrier signal
   (c) both (a) and (b)
   (d) none of the above

3. The frequency of carrier signal transmitted through a microwave channel is in the range of
   (a) 0.3–3 kHz
   (b) 0.3–3 MHz
   (c) 0.3–3 GHz
   (d) none of the above

4. The RF choke offers high impedance to
   (a) carrier frequency currents
   (b) power frequency currents
   (c) both (a) and (b)
   (d) none of the above

5. A pilot protection scheme based on unit protection principle does not provide

   (a) primary protection
   (b) remote protection
   (c) both (a) and (b)
   (d) secondary protection

6. The wavelength of a signal in a fibre optic link is of the order of
   (a) 0.85–1.6 μm
   (b) 0.1–0.5 μm
   (c) 2.5–5 μm
   (d) none of the above

7. A carrier signal used for protection can also be utilized for
   (a) telemetering
   (b) supervisory control
   (c) communication
   (d) all of the above

8. Simultaneous tripping of the circuit breaker at both ends of a transmission line is achieved by
   (a) pilot protection
   (b) distance protection
   (c) overcurrent protection
   (d) none of the above

9. The frequency of a carrier signal transmitted through an RF channel is in the range of

(a) 1 kHz–0.1 MHz

(b) 0.1 kHz–1 MHz

(c) 10 kHz–0.1 GHz

(d) none of the above

10. Which transfer tripping scheme is widely used by the utili-

ties in the field?

(a) Overreach transfer tripping scheme

(b) Underreach transfer tripping scheme

(c) Both (a) and (b)

(d) None of the above

Review Questions

1. What is the need for a pilot protection scheme even though distance protection schemes are widely used in the field?

2. Draw and explain the basic block diagram of a pilot protection system.

3. Explain the various types of equipment used in the pilot protection scheme.

4. How is pilot communication carried out on the basis of various types of signal?

5. Explain the transmission of information on the basis of different types of frequency.

6. Explain with a schematic diagram the circulating current wire pilot relaying scheme.

7. Explain with a schematic diagram an opposed voltage wire pilot relaying scheme.

8. Draw a schematic diagram of a phase comparison carrier current protection scheme and explain the various types of carrier equipment used in this scheme.

9. Explain with a neat schematic diagram the concept of carrier blocking scheme.

10. Explain the concept of carrier unblocking scheme.

11. What is the difference between blocking and transfer tripping?

12. Explain the concept of underreach transfer tripping scheme.

13. What is the difference between direct underreach and permissive underreach transfer tripping schemes?

14. Explain the concept of overreach transfer tripping scheme.

15. Compare underreach and overreach transfer tripping schemes.

Generator Protection 5

5.1 Introduction

A generator is the heart of an electrical power system, as it converts mechanical energy into its electrical equivalent, which is further distributed at various voltages. The capacity of generators has increased in recent times from 50 MW to 500 MW, with the result that the loss of any single unit may cause system instability. Power-generating plants contribute to approximately 50% of the capital cost in an electrical power system, and the generator is the most expensive electrical equipment in a power plant. The protection of generators requires more consideration, to reduce the outage period by rapid clearance of faults.

A modern generator unit is a complex system comprising the generator stator winding, associated transformer, rotor with its field winding and excitation system, and prime mover with its associated auxiliaries. Faults of many kinds can occur within this system, for which different forms of electrical and mechanical protection are required. The generator outage caused by faults, abnormal operating conditions, or maloperation of the generator protection scheme is very costly in terms of repair or replacement of any part and amounts to loss of revenue because of the loss of generation. Abnormal operating conditions, other than short circuits, do not necessarily require the immediate removal of a machine from service, and that might be left to the control of an assistant.

As generators are exposed to more harmful operating conditions than any other power system element, more sophisticated and innovative protection schemes are required. Over the last two decades, protection relay technology has evolved from single-function electromechanical relays to static relays and finally to multifunction digital/numerical relays. Multifunction numerical relays are capable of providing complete protection, including differential protection, stator ground fault protection, generator field winding fault protection, stator winding turn-to-turn fault protection, out-of-step protection, and loss-of-excitation protection, to generators of all sizes at low cost.

This chapter includes a detailed study of the various types of faults that occur in generators and the appropriate protection schemes with their fundamental concepts, including fault detection and clearance.

5.2 Differential Protection

Differential protection (87) is the protection of selected equipment based on the current measuring principle at the input and output sides of the equipment being protected. Depending on the earthing method used, the neutral can also be incorporated in measurement and balancing. The area between the input and output connections of current transformers (CTs) for the protection of the equipment is classified as the protection zone supervised by the relay unit. Protection against the stator phase faults is normally covered by a high-speed differential relay covering the three phases separately. All types of phase faults (phase–phase) and phase-to-ground faults are covered by this type of protection scheme.

Differential protection is generally applied to generators with ratings more than 1 MVA. It provides protection to the stator windings of a generator against various types of phase faults. If a generator is solidly grounded, then the magnitude of ground fault current is very high, which can be easily detected by the differential relaying scheme. However, if the generator is grounded through high impedance, then the ground fault currents may not have sufficient magnitude to operate the differential relay, particularly when the fault occurs closer to the neutral point of the generator stator winding. For this purpose, other types of protective schemes are used, which are discussed in Sections 5.2.1–5.2.3.

5.2.1 Merz–Price Differential Protection (Circulating Current Differential Protection)

Figure 5.1 shows the principle of the circulating current differential protection scheme. In this scheme, the primaries of the CTs are connected in series on both sides of each phase winding of the generator. The secondaries of the CTs are connected in an additive manner to pass the circulating currents through a closed path. The differential relay constantly checks on the secondary sides of the CTs as to whether the incoming current of a phase winding is equal to the respective outgoing current of the same winding.

The directions of currents passing through the secondary side of the CTs are shown in Fig. 5.1. If the currents on the primary sides, that is, I_{A1} and I_{A2}, have the same magnitude, then the secondary side currents, that is, i_{a1} and i_{a2}, will also have the same magnitude, considering that the CTs on both sides have the same transformation ratio and have identical characteristics. If there is a significant difference between the

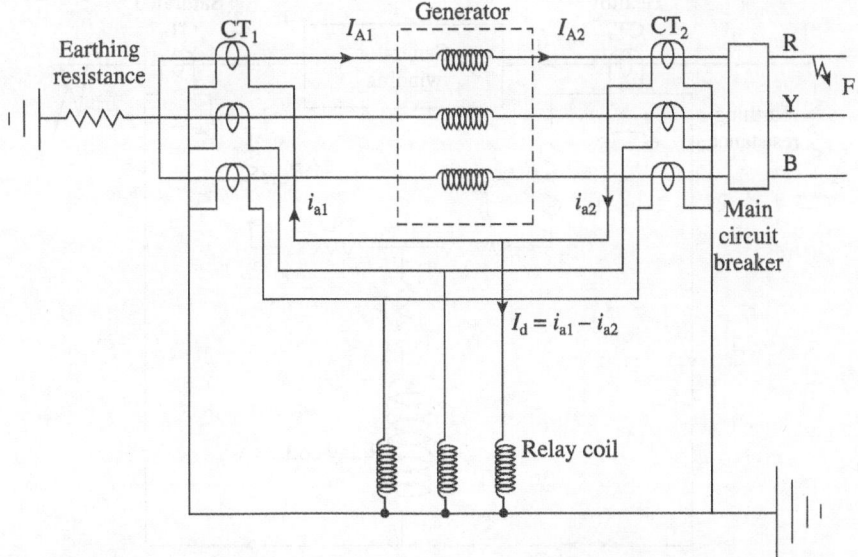

Fig. 5.1 Circulating current differential protection

CT secondary currents (I_d) on both sides of the windings, then this indicates a fault within the protection zone of the stator winding of the generator. Hence, the differential relay trips if the value of I_d exceeds a predetermined threshold value (relay setting). On the other hand, during external faults, the differential relay remains stable and does not initiate a trip signal.

If the CTs are identical in nature, then the functioning of the differential relay is straightforward. However, in practice, it is impossible to achieve CTs with identical saturation characteristics. Hence, the secondary currents of the CTs are unequal even though the primary currents are the same. This current is widely known as *spill current*. The spill current passes through the relay and may maloperate the relay if its value exceeds the setting of the relay. This is possible particularly in case of heavy through-fault conditions. Moreover, if the length of the connecting wires (also known as *pilot wires*) is unequal, then the value of the spill current increases. In order to avoid maloperation of the differential relay in these situations, a stabilizing resistance is connected in series with the relay. However, incorporation of the stabilizing resistance reduces the sensitivity of the relay during an internal fault.

5.2.2 High Impedance Differential Protection

The high impedance differential protection scheme differs from the biased differential protection scheme (explained in Section 5.2.3) in the manner in which the relay stability is accomplished in case of external faults and by the fact that the differential current must be managed through the electrical connections of the CT secondary circuits. In the event of a heavy through-fault, it is quite possible that the CT may saturate, and hence, current flows through the saturated CT rather than through the relay. The remedy for this is to use a tuned relay element that provides the required stability. The necessary high impedance is gained by adding an external resistance in the relay circuit.

The application of the principle of the high impedance protection scheme is illustrated in Fig. 5.2. In some applications, protection may be required to limit voltages across the CT secondary circuits when the differential secondary current for an internal phase fault flows through the high impedance relay circuit(s). However, this is not a common requirement for generator differential applications unless very

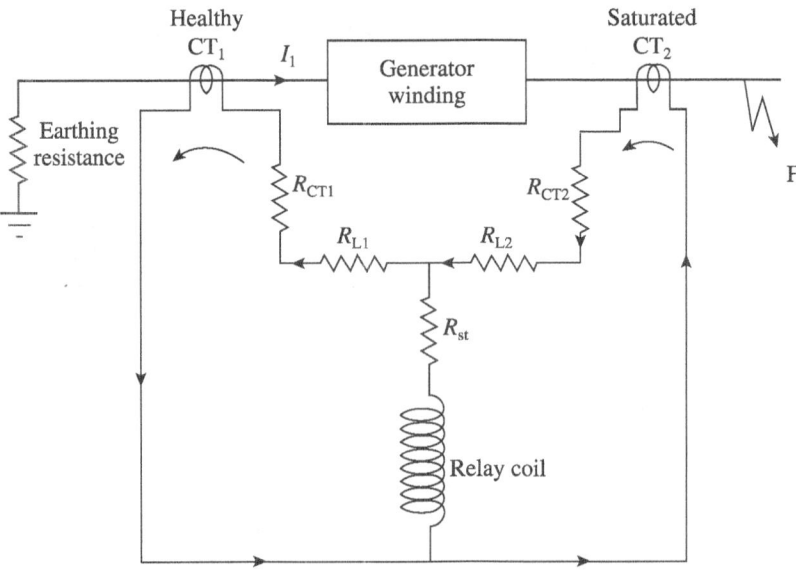

Fig. 5.2 High impedance differential protection scheme

high impedance relays are applied. Shunt-connected, non-linear resistors should be deployed to fulfil the necessary requirement of the high impedance relaying scheme, as shown in Fig. 5.2.

The primary operating current is given by

$$I_R = \text{CTR} \times (I_s + n \times I_e) \tag{5.1}$$

where,

I_R = primary operating current
CTR = CT ratio
n = *the number of CTs in parallel with the relay element*
I_e = *CT magnetizing current*
I_s = *relay setting*

Voltage across the relay circuit is given by

$$V_r = I_f \times (R_{CT} + 2 \times R_L) \tag{5.2}$$

where, R_{CT} and R_L are the CT secondary winding resistances and CT lead resistance, respectively. I_f is the fault current on the CT secondary side.

The knee-point voltage (KPV) of a CT decides the working range of the CT, and it should be high for higher saturation flux density. The KPV of the CT is given by

$$V_K \geq 2V_r \tag{5.3}$$

The stabilizing resistor, R_{ST}, limits the spill current below the relay setting.

$$R_{ST} = (V_r/I_s) - R_R \tag{5.4}$$

where R_R is the relay burden resistance given by

$$R_R = \textit{Relay burden/(Relay setting)}^2$$

If the value of the stabilizing resistance (R_{ST}) as calculated using Eq. (5.4) is added to the circuit, the sensitivity of the relay to internal faults is reduced. Further, it produces a high voltage across the CT during heavy external faults. Hence, to avoid this problem, the value of R_{ST} can be considered as nearly about one-third of the calculated value, in practice. Relay setting is typically set to 5–10% of the generator-rated secondary current.

It follows from these calculations that the application of the high impedance differential protection scheme is much more complex than that of the biased differential protection scheme. However, it is still used because of its simplicity and stability during heavy through-faults and external switching events.

5.2.3 Biased Differential Protection

The main drawback of Merz–Price differential protection is the reduction in sensitivity of the relay due to the incorporation of the stabilizing resistance. Hence, to minimize this effect and also to increase the sensitivity of the differential relay, biased differential protection scheme is used. Figure 5.3 shows the principle of the biased percentage differential protection scheme where no additional stabilizing resistance is connected in series with the relay.

The biased percentage differential relay has two settings, namely basic setting and bias setting. *Basic setting* is the difference between two CT secondary currents $(i_1 - i_2)$. *Bias setting* is the ratio of the difference between two secondary currents to the average value of those two currents, that is, $\dfrac{(i_1 - i_2)}{\left(\dfrac{i_1 + i_2}{2}\right)}$.

In the case of a normal condition/external fault condition, some amount of spill current $(i_1 - i_2)$ is already available because of the non-identical CTs and unequal lead length. However, at the same time, the average

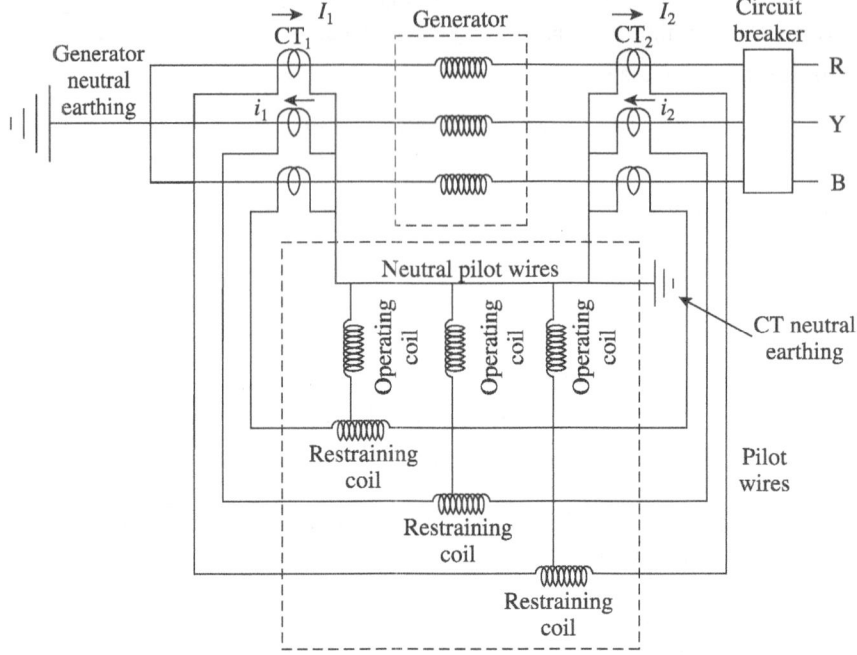

Fig. 5.3 Biased percentage differential relay

value of the two currents $\left(\dfrac{i_1+i_2}{2}\right)$ also increases. Hence, the fault current will not cross the bias setting of

the relay even though the basic setting is exceeded. Therefore, maloperation of the relay can be avoided.

On the other hand, during an internal fault, two currents on the primary sides of the CTs become unequal. Hence, a high spill current (i_1-i_2) is produced on the secondary side of the two CTs. This crosses not only the basic setting limit of the relay, but also the bias setting limit because of the increase in spill current and decrease in the average of the two currents. Finally, the relay operates and trips the generator as per the requirements.

Figure 5.4 shows the typical biased differential protection characteristic of the modern digital/numerical differential protection relay used for generator protection. The first slope gives the basic setting of the relay, and usually, it is set to 5% of the rated current. Bias setting is shown in the second slope of the characteristic, which is set to 120% of the rated current with a slope of about 30%. The third slope is designed considering the mismatch between the two CTs (non-identical CT saturation characteristics of two CTs), which is set to about 70%.

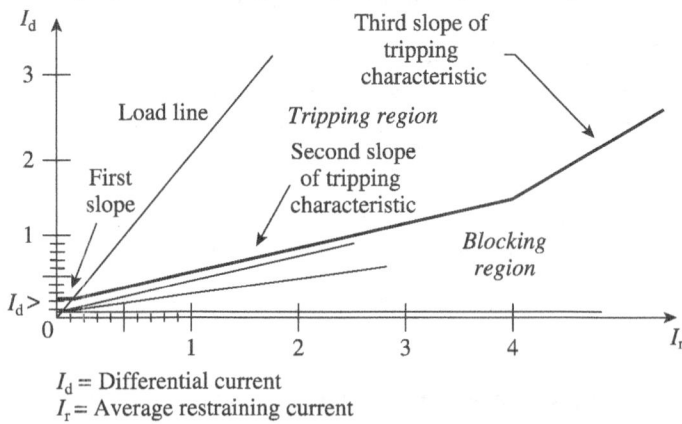

I_d = Differential current
I_r = Average restraining current

Fig. 5.4 Biased differential protection characteristic

The fundamental advantage of this scheme is that it makes the scheme more permissive towards CT mismatch, thus reducing the relay's sensitivity to external faults. Moreover, it provides high sensitivity in case of internal faults.

5.2.4 Relay Setting

Table 5.1 shows the typical relay setting range for phase-to-phase and phase-to-frame short circuits for high impedance and biased differential protection schemes.

Table 5.1 Relay setting for differential protection scheme

Differential protection scheme	Internal phase-to-phase short circuits	Internal phase-to-frame short circuits
High impedance differential protection	I_{set} = 5 – 15% of I_n*, without any time delay	I_{set} = 20% of I_n, without any time delay
Biased differential protection	1. Percentage characteristic = 37.5% 2. Minimum threshold = 30% of I_n, without any time delay	—
* I_n is the nominal current of the generator.		

Example 5.1 A 200 MW, 13.8 kV, 0.9 power factor (PF), 50 Hz, 3-phase, Y-connected generator is protected by differential protection with CTs with ratio 10,000/5 A, CT secondary resistance of 1.5 Ω, and lead resistance of 0.15 Ω. The rated current of the differential relay is 5 A, and its setting range is 5–20% of the relay rated current. The relay burden is 1 VA. If a through-fault occurs, with the fault current 12 times the full load current of the generator, then determine the value of the stabilizing resistance. In addition, suggest the suitable value of KPV for the CT.

Solution:
CTR = 10,000/5, R_{CT} = 1.5 Ω, R_L = 0.15 Ω
The full load current (I_{fl}) of the generator is given by

$$I_{fl} = \frac{MW \times 10^6}{\sqrt{3} \times kV \times p.f. \times 10^3} = \frac{200 \times 10^6}{\sqrt{3} \times 13.8 \times 0.9 \times 10^3} = 9297 \text{ A}$$

Now, the fault current is 12 times the full load current of the generator.

Hence, $\qquad I_F = 12 \times 9297 = 111{,}565 \text{ A}$

The CT secondary current for this fault current is given by

$$i_f = \frac{I_F}{CTR} = \frac{111{,}565 \times 5}{10{,}000} = 55.782 \text{ A}$$

A generator differential relay is set to pickup at 5–20% of the CT secondary current. Therefore, we select the relay setting (I_s) = 10% of the relay rated current, that is, 0.5 A.

$$\text{Relay resisitance } (R_R) = \frac{\text{Relay burden}}{(\text{Pickup})^2} = \frac{1VA}{I_s^2} = \frac{1VA}{(0.5)^2} = 4 \ \Omega$$

Voltage across the relay circuit is given by

$$V_r = i_f (R_{CT} + 2 \times R_L) = 55.782 (1.5 + 2 \times 0.15) = 100.4076 \text{ V}$$

Stabilizing resistance is given by

$$R_{ST} = \frac{V_r}{I_s} - R_R = \frac{100.4076}{0.5} - 4 = 196.8152 \ \Omega$$

The actual value of R_{ST} is one-third of the calculated value. Hence, $R_{ST} = 65.6 \cong 66 \ \Omega$ can be selected. This resistance is connected in series with the relay.

The KPV of CT is given by $V_K > 2 \times V_r$

$$> 2 \times 100.4067$$

$$> 200.8 \ V$$

Hence, considering some safety margin, the KPV of the CT used for differential protection should not be less than 250 V.

5.3 Stator Earth-fault Protection

Due to failure of insulation between generator stator and core, the earth fault occurs in the stator of the generator. An arc will be formed between the conductor and the core due to breakdown of insulation between the conductor which is at high potential and the core which is at low potential (earthed). The intensity of the arc will be very high due to the high magnitude of ground fault current which flows through the arc. This in turn increases temperature of the arc which results in failure of insulation between laminations and core. Finally, a large portion of the core will be damaged due to increase in eddy current losses. This happens in case when the earth fault occurs near the neutral of the generator. The severity and further damage due to the fault current is higher in case of fault occurring near the terminal of a generator.

Therefore, in order to avoid the destructive effect of large ground fault current, large generators are always grounded through a resistor/reactor in the neutral circuit of the generator. The selection of the value of resistor is extremely important not only to restrict the magnitude of ground fault, but also to avoid transient overvoltages. Hence, the value of resistor (R_n) should not be greater than

$$R_n = \frac{10^6}{6 \times \pi \times f \times C}$$

where C is the capacitance of the generator stator circuit to earth on per phase basis (μF). If its value exceeds the value obtained by the above equation then high transient overvoltages are developed due to ferro-resonance effect. For a value of C say 0.25 μF, the value of R_n is given by

$$R_n = \frac{10^6}{6 \times \pi \times 50 \times 0.25} = 4246 \ \Omega$$

Hence, for a standard rating of generator, say 15.75 kV, the magnitude of ground fault current will be given by,

$$I_{fg} = \frac{15.75 \times 10^3}{\sqrt{3} \times 4246} = 2.14 \ A$$

Therefore, the potential of the neutral of the generator which was previously zero will increase to a value of 4246 × 2.14 = 9.0864 kV and the peak value will be around 12.85 kV. Hence, a high value of resistor having high voltage rating is required which is not economically viable. Hence, in actual field, Neutral Grounding Transformer (NGT) is used.

The utilization of NGT will reduce the value of resistor to $4246 \times \dfrac{(240)^2}{(15750)^2} = 0.985\ \Omega$. The main advantage of NGT, as clearly visible from Fig. 5.5 (a), is that the value of resistor is reduced to a great extent and its voltage rating is also reduced to 240 V.

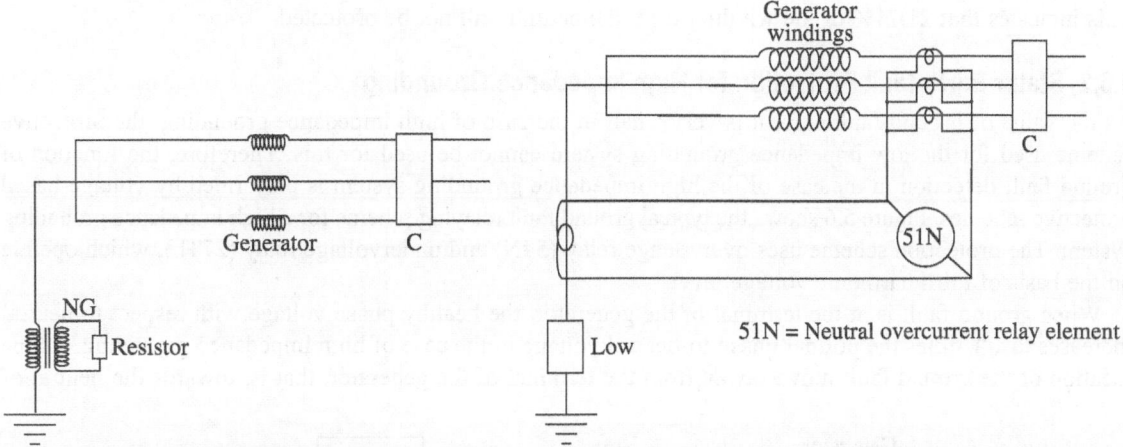

Fig. 5.5 (a)   NGT connected in the neutral circuit of generator

Fig. 5.5 (b)   Relaying scheme for low impedance grounding

However, grounding of the neutral of the generator through NGT produces two negative effects also.

(i) In case of earth fault on one of the phases, the potential of the other two healthy phases with respect to earth rises from its normal phase-to-neutral value to the potential of the neutral point ($V_{ph-n} + I_{fg} \times Z_n$). Hence, the insulation between stator conductor and core must withstand this value. This additional cost of insulation extends up to the primary of a generator transformer (GT) only. However, this cost is not that much high compared to the reduction of the magnitude of ground fault current.

(ii) Larger the neutral impedance connected in the neutral circuit of the generator, more will be the portion of the generator stator winding that remains unprotected in case of earth fault.

5.3.1 Stator Earth-fault Protection during Low Impedance Grounding

In case where the neutral of the small generators is earthed through small resistor/reactor, a restricted earth-fault protection, as shown in Fig. 5.5 (b), is used.

Here, the magnitude of fault current in case of an earth fault on any of the phases of the generator winding is given by,

$$I_{fg} = \frac{10 \times V_{P-P} \times x}{\sqrt{3} \times Z_n}$$

where V_{P-P} is the phase-to-phase voltage rating of a generator (kV)

x is the location of ground fault in terms of percentage of winding from neutral end

Z_n is the value of neutral impedance

Therefore, percentage of winding remaining unprotected is obtained by,

$$x = \frac{I_{pickup} \times CT_{primary} \times \sqrt{3} \times Z_n}{1000 \times V_{P-P}}$$

where, I_{pickup} is the pickup current of the relay in terms of percentage of the CT rating.

$CT_{primary}$ is primary rating of the CT.

For example, for a generator of 13.8 kV having a CT ratio of 10,000/5 A, pickup current of the relay = 10% and Z_n=1.73 ohm, the percentage of winding that remains unprotected on the generator stator winding will be

$$x = \frac{10 \times 10000 \times \sqrt{3} \times 1.73}{1000 \times 13.8} = 21.71\% \; .$$

This indicates that 21.71% of the winding from the neutral will not be protected.

5.3.2 Stator Earth-fault Protection for High Impedance Grounding

As the value of ground fault current is very small in the case of high impedance grounding, the protective scheme used for the low impedance grounding system cannot be used for this. Therefore, the function of ground fault detection in the case of the high impedance grounding system is performed by voltage-based protective schemes. Figure 5.6 shows the typical ground fault relaying scheme for a high impedance grounding system. The protection scheme uses overvoltage relay (59N) and undervoltage relay (27Th), which operate on the basis of third harmonic voltage level.

When ground fault is at the terminal of the generator, the healthy phase voltage with respect to neutral increases to $\sqrt{3}$ times the normal phase-to-neutral voltage in the case of high impedance grounding. As the location of the ground fault moves away from the terminal of the generator, that is, towards the neutral of

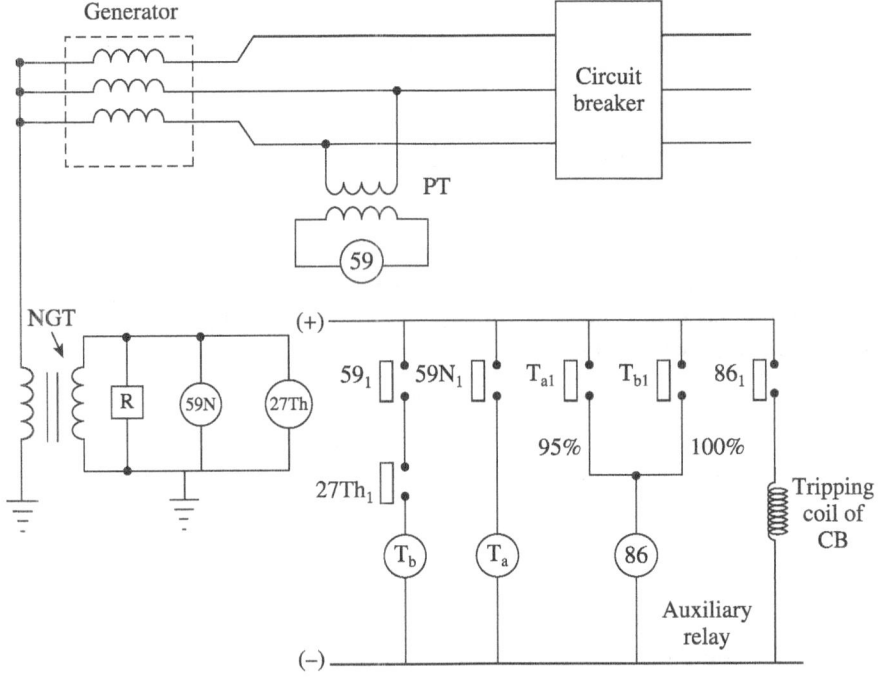

59 = Overvoltage relay
59N = Overvoltage relay tuned to fundamental frequency
27Th = Undervoltage relay tuned to third harmonic frequency
T_a and T_b = Timers; T_{a1} and T_{b1} = Contacts of timers
86 = Auxiliary relay; 86_1 = Contacts of auxiliary relay
NGT = Neutral grounding transformer

Fig. 5.6 Stator earth-fault relaying scheme for medium/high impedance grounding

the generator, the value of healthy phase voltages with respect to the neutral goes down. Eventually, for the ground fault at the neutral of the generator, the level of healthy phase voltages becomes the same as that of the normal phase-to-neutral voltage. During normal running conditions, the generator produces typically 1–5% of third harmonic voltages with respect to the fundamental voltages. At the time of the ground fault close to the terminal of the generator, healthy phase voltages are high enough to operate the overvoltage relay (59N) because of shifting of the neutral of generator. Hence, the relay (59N) will protect about 90–95% of the generator winding from the terminal of the generator.

Now, if the ground fault is present at the neutral of the generator, it gives a bypass path for the third harmonic current to pass through the faulty path to ground. Hence, the undervoltage relay (27Th) operates when the third harmonic voltage level goes below its set value. One precaution has also been taken to avoid maloperation of this scheme at the time of synchronization of the generator and when it is taken out of service. As shown in Fig. 5.6, one contact of overvoltage relay (59) is connected in series with the contact of undervoltage relay (27Th). This arrangement avoids the possible maloperation of the generator due to improper variation of fundamental and third harmonic voltages.

As the proposed scheme provides complete protection to the high impedance grounded generator against ground faults at any point on the stator windings, it is extensively known as *100% stator earth-fault protection scheme*.

Example 5.2   A 200 MW, 13.8 kV, 0.9 PF, 50 Hz, three-ϕ, Y-connected generator is protected by an earth-fault relay. The relay is set to operate at 10%. The CT ratio is 10,000/1 A. A resistor is used in the neutral circuit of the generator to limit the earth-fault current to 50% of the normal load current. Determine the value of the resistor and the percentage of stator winding protected.

Solution:

The full load current of the generator is $I_{fl} = \dfrac{200 \times 10^6}{\sqrt{3} \times 13.8 \times 10^3 \times 0.9}$

$$= 9297.1 \text{ A}$$

When earth fault occurs on the terminal of the generator, the maximum earth-fault current is 50% of the full load current.

$$I_{fmax} = 9297.5 \times 0.5$$

$$= 4648.5 \text{ A} \tag{5.5}$$

The value of the earth-fault current is given by

$$I_{fmax} = \frac{3E_{ph}}{Z_1 + Z_2 + Z_0 + 3Z_n}$$

where E_{ph} is the line-to-neutral voltage of the generator. Z_1, Z_2, and Z_0 are sequence impedances, and Z_n is the impedance of the neutral circuit. Z_n is relatively higher than Z_1, Z_2, and Z_0 and hence we can neglect them.

Hence, $$I_{fmax} = \frac{E_{ph}}{Z_n} = \frac{13.8 \times 10^3}{\sqrt{3} \times Z_n} \tag{5.6}$$

Equating Eqs (5.5) and (5.6), the neutral resistor is given by

$$Z_n = R_n = \frac{13.8 \times 10^3}{\sqrt{3} \times 4648.5} = 1.71 \ \Omega$$

Instead of a terminal fault, if the earth fault occurs at $x\%$ of the generator winding from the neutral end, the fault current is given by

$$I_F = \frac{13.8 \times 10^3}{\sqrt{3} \times R_n} \times \frac{x}{100}$$

$$= \frac{13.8 \times 10^3}{\sqrt{3} \times 1.71} \times \frac{x}{100}$$

$$= 46.59x \text{ A}$$

This value of the fault current is compared with the sensitivity of the relay as

$$46.59x = 10\% \text{ of } 10000 \text{ A}$$
$$46.59x = 1000$$
$$x = 21.46$$

Hence, 21.46% of the generator winding is unprotected from the neutral end and 78.54% winding is protected from the generator terminal.

Example 5.3 A 50 MVA, 11 kV, 50 Hz, Y-connected generator is protected by an earth-fault relay with 5% setting. A neutral resistor is required to protect the generator against the earth fault such that only 5% of the winding is left unprotected from the neutral end. The CT ratio is 3000/1 A. Determine the value of the earth resistor and earth-fault current with this resistor in neutral circuit.

Solution:
The full load current of the generator is given by

$$I_{fl} = \frac{50 \times 10^6}{\sqrt{3} \times 11 \times 10^3}$$
$$= 2624.3 \text{ A} \qquad (5.7)$$

The value of earth-fault current is given by

$$I_{fmax} = \frac{3E_{ph}}{Z_1 + Z_2 + Z_0 + 3Z_n}$$

where E_{ph} is the line-to-neutral voltage of the generator. Z_1, Z_2, and Z_0 are sequence impedances, and Z_n is the impedance of neutral circuit. Z_n is relatively higher than Z_1, Z_2, and Z_0.

Hence,

$$I_{fmax} = \frac{E_{ph}}{R_n} = \frac{11 \times 10^3}{\sqrt{3} \times R_n}$$

where $R_n = Z_n$

Earth-fault current for earth fault at $x\%$ of generator winding from neutral end is given by

$$\text{Fault current } I_F = \frac{11 \times 10^3}{\sqrt{3} \times R_n} \times \frac{x}{100}$$

Considering that only 5% of the winding is unprotected and equating the earth-fault current to the sensitivity of the relay, we get

$$\text{Sensitivity of relay (5\% of 3000 A)} = \frac{11 \times 10^3}{\sqrt{3} \times R_n} \times \frac{5}{100}$$

$$150 = \frac{317.54}{R_n}$$

$$R_n = 2.11 \text{ }\Omega$$

With this value of earth resistor in neutral circuit, the maximum earth-fault current is

$$I_{\text{fmax}} = \frac{11 \times 10^3}{\sqrt{3} \times R_n}$$

$$= \frac{11 \times 10^3}{\sqrt{3} \times 2.11}$$

$$= 3009.88 \text{ A} \qquad (5.8)$$

Comparing the values in Eqs (5.7) and (5.8), the maximum earth-fault current is 114.69% of the full load current of the generator.

Example 5.4 An 11 kV, 25 MW, three-phase, 0.9 lagging power factor, star-connected generator is protected by an earth-fault relay. The pick-up setting of the relay is 10%. A resistor is connected in the neutral circuit of the generator which restricts the maximum earth-fault current to 40% of the full load current of the generator. Determine the value of resistor and the percentage of winding protected by the relay. The CT ratio is 2000/1 A. Also find out that value of resistor so that only 7% of the winding remains unprotected.

Solution:

The full load current of the generator is given by, $I_{\text{FL}} = \dfrac{25 \times 10^6}{\sqrt{3} \times 11 \times 10^3 \times 0.9} = 1457.95 \text{ A}$

The maximum value of earth-fault current is $I_{\text{fmax}} = 1457.95 \times 0.4 = 583.18 \text{ A}$

Now, the value of earth-fault current is given by, $I_{\text{fmax}} = \dfrac{3 \times E_{ph}}{Z_1 + Z_2 + Z_0 + 3 \times Z_n}$

Normally, the value of $Z_n >>>> Z_1/Z_2/Z_0$ and hence, $I_{\text{fmax}} = \dfrac{E_{ph}}{Z_n}$

Hence, $Z_n = \dfrac{E_{ph}}{I_{\text{fmax}}} = \dfrac{11 \times 10^3}{\sqrt{3} \times 583.18} = 10.89 \, \Omega$

Now, the value of location of fault expressed as a percentage of winding from the neutral end (x) is given by the equation as under

$$x = \frac{I_{pickup} \times CT_{primary} \times \sqrt{3} \times Z_n}{1000 \times V_{PP}} = \frac{10 \times 2000 \times \sqrt{3} \times 10.89}{1000 \times 11} = 34.29\%$$

Here, I_{pickup} is the pick-up current of the relay, $CT_{primary}$ is the primary rating of the CT, and V_{PP} is the line voltage of the generator in kV.

This indicates that 34.29% of the winding will remain unprotected or 65.07% of the winding will remain protected.

Now, for allowing only 7% of the winding to remain unprotected, the magnitude of earth-fault current will be,

$$I_{\text{fmax}} = \frac{10 \times V_{PP} \times x}{\sqrt{3} \times Z_n}$$

But the value of I_{fmax} is 10% of 2000 A = 200 A.

Hence, $200 = \dfrac{10 \times 11 \times 7}{\sqrt{3} \times Z_n}$ and $Z_n = 2.22\ \Omega$

With this value of Z_n, the magnitude of earth-fault current is

$$I_{f\,max} = \frac{11 \times 10^3}{\sqrt{3} \times 2.22} = 2860.75\,\text{A}$$

The above calculation clearly indicates that the maximum value of earth-fault current will increase if we want to increase more percentage of winding that will remain protected.

Example 5.5 A three-phase, 11 kV, 120 MW, 0.85 lagging power factor generator is protected by a circulating current differential protection scheme. The CT ratio is 7500/5 A. The CT secondary resistance is 1.5 Ω and lead resistance is 1 Ω. The relay rated current is 5 A and its setting range is 5 – 20% of 5 A. Burden of the relay is 1 VA. Calculate the value of stabilizing resistance (R_{stab}) for an external fault having a magnitude of 10 pu. Also suggest the KPV required for CTs. The relay is set at 10% of 5 A.

Solution:

The stabilizing resistance is given by, $R_{stab} = \dfrac{i_f \times (R_{CT} + R_L)}{I_S} - R_R$

I_S = 10% of 5 A = 0.5 A

Now, relay resistance is given by, $R_R = \dfrac{Burden}{(I_S)^2} = \dfrac{1}{(0.5)^2} = 4\,\Omega$

The full load current of the generator is calculated as under

$$I_{FL} = \frac{120 \times 10^6}{\sqrt{3} \times 11 \times 10^3 \times 0.85} = 7409.84\,\text{A}$$

Referring this current on the CT secondary side,

$$i_{FL} = \frac{7409.84}{7500} \times 5 = 4.9399\,\text{A}$$

Hence, the magnitude of fault current is,

$$i_f = 10 \times 4.9399 = 49.399\,\text{A}$$

$$R_{stab} = \frac{49.399 \times (1.5 + 1)}{0.5} - 4 = 242.99\,\Omega$$

Hence, the actual value of R_{stab} = 242.99/3 = 80.99 Ω.
Now, knee point voltage is given by,

$$KPV > 2 \times V_R$$
$$> 2 \times i_f \times (R_{CT} + R_L)$$
$$> 2 \times 49.399 \times (1.5 + 1)$$
$$> 246.99\,\text{V}$$

Example 5.6 A 13.8 kV, three-phase, 50 Hz, 100 MVA star-connected generator is protected by the circulating current differential protection scheme. The CT ratio is 5000/5 A. The relay is set to operate at 100 mA. The voltage at one end of CT is 90% of that at the other end during an external fault (heavy through fault) with a magnitude of 12 times full load current of the generator. The relay with a resistance of 90 Ω is connected to the physical mid-point of the pilot wires. The pilot wire has a resistance of 0.5 Ω per 100 m. The distance between two sets of CTs is 300 m. Determine the value of stabilizing resistance required to be connected in series with the relay to have a stability factor of 3 for this fault condition. Stability factor is defined as the ratio of pickup current of the relay to the current through the relay during normal condition (I_S / I_R).

Solution:

The total resistance of pilot wires is given by $2 \times R_L = 2 \times \dfrac{0.5}{100} \times 300 = 3\,\Omega$

The full load current of the generator is calculated as under

$$I_{FL} \frac{100 \times 10^6}{\sqrt{3} \times 13.8 \times 10^3} = 4183.69\,\text{A}$$

Referring this current on the CT secondary side,

$$i_{FL} = \frac{4183.69}{5000} \times 5 = 4.1837\,\text{A}$$

Hence, the magnitude of fault current is $i_f = 12 \times 4.1837 = 50.20\,\text{A}$

Now, the voltage required to circulate current in the pilot wires is given by,

$$E + 0.9 \times E = i_f \times 2 \times R_L = 50.20 \times 3 = 150.61$$
$$E = 79.27\,\text{V}$$

With reference to Fig. 5.7, from similar triangles TAB and TPQ, we can write the equation as shown below.

$$\frac{AB}{PQ} = \frac{AT}{TP}\text{ i.e., } \frac{E}{0.9 \times E} = \frac{X}{300 - X}$$

$$X = 157.89\,\text{m}$$

Similary, using triangles TAB and TCD,

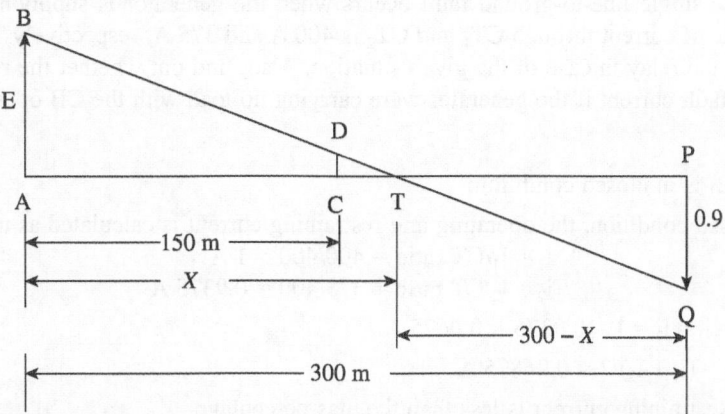

Fig. 5.7   Voltage distribution across secondary of two CTs

$$\frac{CD}{AB} = \frac{TC}{TA} \text{ or } CD = AB \times \frac{TC}{TA}$$

$$= 79.27 \times \frac{7.89}{157.89} = 3.96 \text{ V}$$

$$CD = Voltage\ across\ relay = V_R = I_R \times (R_R + R_{stab})$$

$$Stability\ factor = \frac{I_S}{I_R}$$

$$I_R = \frac{I_S}{3} = \frac{0.1}{3} = 0.033 \text{ A}$$

$$R_{stab} = \frac{V_R}{I_R} - R_R = \frac{3.96}{0.033} - 90 = 28.8 \Omega$$

Example 5.7 Figure 5.8 shows the single line diagram of a generator winding which is protected by a percentage differential protection scheme. The relay settings for the said scheme are as under.
  (i) Pick-up current = 0.05 A
  (ii) Slope = 10 %

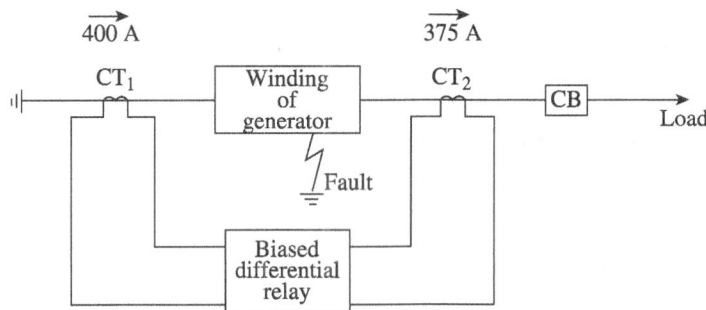

Fig. 5.8 Percentage differential relay for Example 5.7

A high resistance single line-to-ground fault occurs when the generator is supplying the power to the load. The magnitude of current through CT$_1$ and CT$_2$ is 400 A and 375 A, respectively. Determine whether the CB will trip by the relay in case of the given situation. Also, find out whether the relay will operate at the given value of fault current if the generator were carrying no load with the CB open.

Solution:

(i) Case: 1: CB (52) is in closed condition:

When CB is in closed condition, the operating and restraining current is calculated as under.
$$i_1 = I_1/CT \text{ ratio} = 400/400 = 1 \text{ A}$$
$$i_2 = I_2/CT \text{ ratio} = 375/400 = 0.9375 \text{ A}$$

Operating current = $i_1 - i_2 = 1 - 0.9375 = 0.0625$ A

Restraining current = $(i_1 + i_2)/2 = 0.96875$ A

Operating current/Restraining current is less than the bias percentage

$$0.0625/0.96875 < 0.1$$

So the relay will not operate.

(ii) Case: 2: CB (52) is in open condition:

The operating and restraining current in the case when CB is in open condition are calculated as under.

$$i_1 = I_1/CT \text{ ratio} = 400/400 = 1A$$
$$i_2 = I_2/CT \text{ ratio} = 0 \text{ A (due to opening of CB)}$$

Operating current = $i_1 - i_2 = 1$ A

Restraining current = $(i_1 + i_2)/2 = 0.5$ A

Operating current/Restraining current is more than the bias percentage

$$1/0.5 > 0.1$$

Relay current $(i_1 - i_2)$ is also more than the minimum pick-up current of 0.05 A.

So, the relay will operate.

Example 5.8 A generator is provided with restricted earth-fault protection. The ratings are 11 kV, 5000 KVA. The percentage of winding protected against phase-to-ground fault is 80%. The relay is set such that it trips for 25% out of balance. Calculate the resistance to be added in neutral to ground connection.

Solution:

The value of phase voltage and full load current of the generator is given by,

$$V_P = \frac{11000}{\sqrt{3}} = 6340 \text{ V}$$

$$I = \frac{5000 \times 10^3}{\sqrt{3} \times 11 \times 10^3} = 262 \text{ A}$$

Now, the percentage of winding that remains unprotected is given by,

$$X = \frac{R \times I_0 \times 100}{V}$$

where, I_0 = Minimum operating current in primary of CT

R = Ohmic value of resistor connected in the neutral circuit of the generator

V = Line-to-neutral voltage

$$I_0 = 262 \times \frac{25}{100} = 65.5 \text{ A}$$

Now, the percentage of winding that remains protected = 80% (given)

Hence, the percentage of winding that remains unprotected = 20%

$$\text{i.e., } 20 = \frac{R \times 65.5 \times 100}{6340}$$

$$R = 1.94 \ \Omega$$

Example 5.9 The neutral point of a 10,000 V alternator is earthed through a resistance of 10 Ω. The relay is set to operate when there is an out of balance current of 1 A. The CT ratio is 1000/5 A. What percentage of the winding is protected against earth fault and what must be the minimum value of earthing resistance in order to give 90% protection to each phase winding?

Solution:

Out of balance current = 1 A

Corresponding current in CT primary $= 1 \times \dfrac{1000}{5} = 200$ A

Hence, current I_0 for which the relay operates = 200 A

Percentage of winding that remains unprotected $= \dfrac{R \times I_0}{V} \times 100$

$= \dfrac{10 \times 200 \times 100}{10000 / \sqrt{3}} = 34.8\%$

Percentage of winding that remains protected = 100 − 34.8 = 65.2%

Resistance to get 90% of winding protection is given by

$$10 = \dfrac{R \times 200 \times 100}{10000 / \sqrt{3}}$$

$$R = 2.88 \ \Omega$$

Example 5.10 A three-phase, two-pole, 11 kV, 10,000 KVA alternator has neutral earthed through a resistance of 7 Ω. The machine has current balance protection which operates upon out of balance current that exceeds 20% of the full load. Determine the percentage of winding protected against earth fault.

Solution:

The value of phase voltage and full load current of the generator is given by,

$$V_P = \dfrac{11 \times 10^3}{\sqrt{3}} = 6.35 \text{ kV}$$

$$I = \dfrac{10000}{\sqrt{3} \times 11} = 525 \text{ A}$$

Out of balance current = 20% of full load $= \dfrac{20}{100} \times 525 = 105$ A

Now, the percentage of winding remains unprotected $= \dfrac{R \times I_0}{V} \times 100$

$$= \dfrac{7 \times 105 \times 100}{6.35 \times 1000} = 11.6\%$$

Subsequently, the percentage of winding remains protected = 100 − 11.6 = 88.4%

Example 5.11 A star-connected three-phase, 10 MVA, 6.6 kV alternator has a per phase reactance of 10%. It is protected by Merz-Price circulating current principle which is set to operate for fault currents not less than 175 A. Calculate the value of earthing resistance to be provided in order to ensure that only 10% of the alternator winding remains unprotected.

Solution:

Let r ohm be the earthing resistance required to leave 10% of winding unprotected.

Now, the value of phase voltage and full load current of the generator is given by,

$$V_{Pn} = \frac{6.6 \times 10^3}{\sqrt{3}} = 3810 \text{ V}$$

$$I = \frac{10 \times 10^6}{\sqrt{3} \times 6.6 \times 10^3} = 875 \text{ A}$$

Let the reactance/phase be x ohm. Its value is obtained by,

$$10 = \frac{\sqrt{3} \times x \times 875 \times 100}{6600} \quad \Rightarrow x = 0.436 \ \Omega$$

Reactance of 10% winding = $0.436 \times 0.1 = 0.0436 \ \Omega$

Emf induced in 10% winding = $V_{Pn} \times 0.1$

$$= 3810 \times 0.1$$
$$= 381 \text{ V}$$

$$\text{Impedance } (Z_f) = \sqrt{0.0436^2 + r^2}$$

$$I_f = \frac{V_P}{Z_f} = \frac{381}{\sqrt{0.0436^2 + r^2}}$$

Now, the value of fault current is given as 175 A. Hence, the value of r is calculated by putting the value of fault current in the above equation.

$$175 = \frac{381}{\sqrt{0.0436^2 + r^2}} \quad \text{or, } 0.0436^2 + r^2 = \left(\frac{381}{175}\right)^2$$

$$0.0436^2 + r^2 = 4.715 \quad \text{or, } r^2 = 4.713$$

Hence, $\qquad r = 2.171 \ \Omega$

Example 5.12 A star-connected three-phase, 10 MVA, 6.6 kV alternator is protected by Merz–Price circulating current principle using 1000/5 A current transformers. The star point of the alternator is earthed through a resistance of 7.5 Ω. If the minimum operating current for the relay is 0.5 A, calculate the percentage of each phase of the stator winding which is unprotected against earth faults when the machine is operating at normal voltage.

Solution:

The phase voltage is given by,

$$V_{Pn} = \frac{6.6 \times 10^3}{\sqrt{3}} = 3810 \text{ V}$$

Now, the minimum fault current for the relay is given by,

$$I_{fmin} = 0.5 \times \frac{1000}{5} = 100 \text{ A}$$

Emf induced in x% of winding = $V_{Pn} \times \dfrac{x}{100} = 38.1x$

Since $\qquad \dfrac{38.1x}{r} = \dfrac{38.1x}{7.5}$

Therefore, the percentage of winding that remains unprotected is calculated by,

$$100 = \frac{38.1x}{7.5} \text{ or } x = 19.69\%$$

Example 5.13 A star-connected three-phase, 10 MVA, 6.6 kV alternator is protected by circulating current protection, the star point being earthed via a resistance r. Estimate the value of earthing resistor if 85% of the stator winding is protected against earth faults. Assume an earth-fault setting of 20%. Neglect the impedance of alternator winding.

Solution:

Since 85% of winding is protected, 15% would remain unprotected.

The full load current is given by $I_{FL} = \dfrac{10 \times 10^6}{\sqrt{3} \times 6.6 \times 10^3} = 876$ A

The minimum value of fault current for relay is given by,

$$I_{f\min} = 20\% \text{ of full load current}$$

$$= \frac{20}{100} \times 876 = 175 \text{ A}$$

Now, the voltage induced in 15% of winding is given by,

$$= \frac{15}{100} \times \frac{6.6 \times 10^3}{\sqrt{3}} = 330\sqrt{3} \text{ V}$$

The minimum value of fault current = 175 A and is given by $\dfrac{330\sqrt{3}}{r}$

$$175 = \frac{330\sqrt{3}}{r} \text{ or } r = 3.27 \text{ }\Omega$$

Hence, earthing resistance $r = 3.27$ Ω.

5.4 Stator Winding Turn-to-turn Fault Protection (Transverse Differential Scheme)

The differential protection discussed in Section 5.2 cannot protect against the turn-to-turn faults on the same phase of the stator winding. This is because of the fact that it does not produce any current unbalance in the secondary of CTs. In general, large-size generators have two or more windings per phase. Hence, a split-phase protection scheme is used to detect the turn-to-turn fault. Figure 5.9 shows the typical split-phase protection scheme used for the protection of a two-winding generator. The total current produced by each phase is divided into two equal halves in the two parallel windings of the generator. In the case of a turn-to-turn fault, the currents of the two windings get unbalanced because of

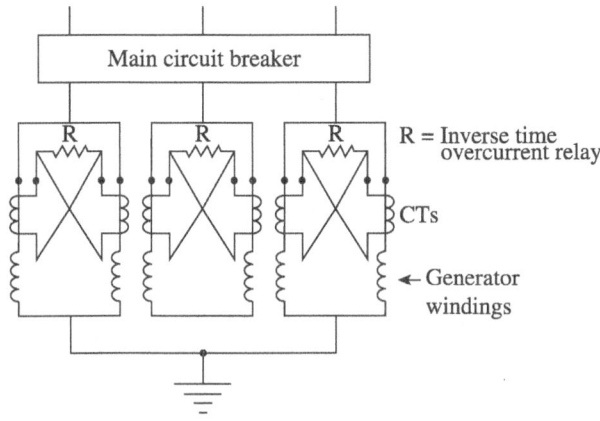

Fig. 5.9 Turn-to-turn split-phase differential protection scheme

the change in the impedance of the faulted winding. This unbalanced current is measured by an inverse time overcurrent relay that is connected in the secondary side of the CTs in a transverse form. A split-phase relaying scheme operates for any type of short circuit in the generator windings, whereas it does not provide protection against the short circuit on the primary or secondary winding of the generator. The split-phase relays should operate the same hand reset auxiliary tripping relay that is operated by the differential relays. The split relay is set above the normal unbalanced currents of the two windings of the generator winding. At the same time, it is set below the unbalanced currents caused by the turn-to-turn fault. In order to avoid

maloperation of the relay during transient fault conditions, an intentional time delay is provided. In this type of faulty conditions, it is not necessary to trip the generator immediately for repair purpose.

For generators containing only one winding per phase, the split-phase differential protection scheme cannot be applied. To detect turn-to-turn fault in this case, a neutral voltage measurement method, as shown in Fig. 5.10, is used. The primary sides of potential transformers (PTs) are connected in a star pattern, and this star point is connected to the neutral. The secondary sides of PTs are connected in an *open delta* manner across which the 59G relay is connected, which measures the zero-sequence component of the voltage caused by the reduction of the electromagnetic force (emf) in the faulted phase. If the turn-to-turn fault exists, it will be sensed by the relay (59G) connected on the secondary sides of the PTs, which finally operates the protection system and triggers the alarm.

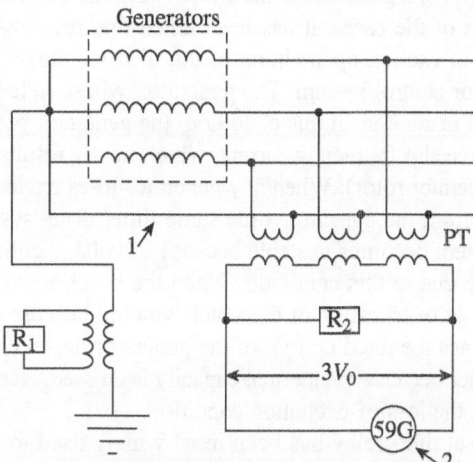

1 = Insulated cable for generator line to neutral voltage
2 = 59G Overvoltage relay needs to be tuned to
     fundamental frequency

Fig. 5.10   Turn-to-turn protection by neutral overvoltage scheme

5.5 Rotor Ground Fault Protection (Field Earth Fault)

Ungrounded DC supply is provided to the field winding of the generator through a brush assembly fitted over the rotor periphery. Therefore, in the case of the first rotor ground fault (64), no current will flow to the ground as no return path is available for the fault current. However, the existence of a single ground fault increases the stress at

the other points in the field winding when voltages are induced in the field by stator transients. If the second ground fault occurs, then the part of the field winding will be bypassed, and the circulating current will pass through the faulted paths. Hence, the distribution of flux in the air gap between the stator and the rotor becomes uneven, which results in increased heating of the rotor, and in the worst case, may damage the rotor.

In all unattended stations, the protective-relaying equipment is organized to trip the generator's main and field breakers immediately when the first ground fault occurs. The existing practice in the industry is to use a very sensitive DC voltage relay to protect the generator against the field ground fault. Figure 5.11 shows a field ground fault protection scheme which

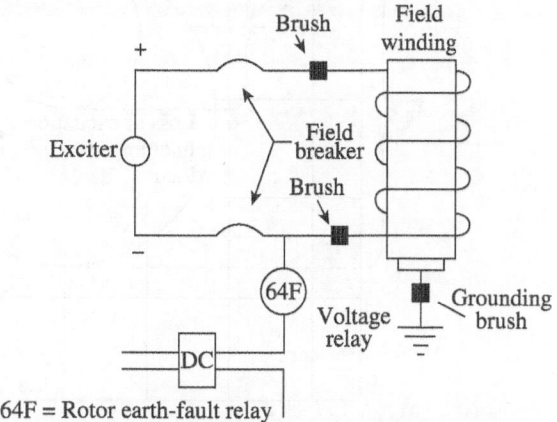

64F = Rotor earth-fault relay

Fig. 5.11   Ground fault detection using DC source

uses a DC source. The rotor earth-fault relay (64F) is connected between one end of the field winding and ground. Whenever a ground fault occurs anywhere in the field circuit, the potential difference across the 64F relay is disturbed, and hence the relay operates as soon as the fault current crosses the set limit.

5.6 Loss-of-excitation (Field Failure) Protection

In general, all the generators are connected to the infinite busbar for supplying power to the utilities. Loss of excitation of a generator is harmful to both the generator and the power system network to which it is connected. In most of the cases, it has been noted that the loss-of-excitation condition is caused by faulty field breaker, failure of exciter, open circuit of the filed windings, accidental tripping of the field breaker, and failure of the regulator control system. The generator, whose field has failed, takes up most of the load and works as an induction generator. In this condition, the generator runs above the synchronous speed, and the rotor induces an excessive slip frequency current, which finally results in excessive iron losses in the rotor (excessive heating of the generator rotor). When any generator loses excitation power, it begins to absorb a large amount of reactive power from the system. At the same time, if the system is not able to feed such a large reactive power, then the system becomes unstable because of voltage collapse. The generator must be isolated quickly to avoid any damage due to this condition. When the synchronous generator runs as an induction generator, there is a possibility of overheating of the stator winding because of overcurrent. The overcurrent may be as high as two to four times the rated current of the generator depending on the slip. This finally results in severe damage to the generator because the thermal capacity is crossed. Hence, adequate protection is required to protect the generator against the loss-of-excitation condition.

Offset mho relay has been most widely used to protect the generator against the loss-of-excitation condition by sensing the variation of impedance as seen from the generator terminals. Figure 5.12 shows a

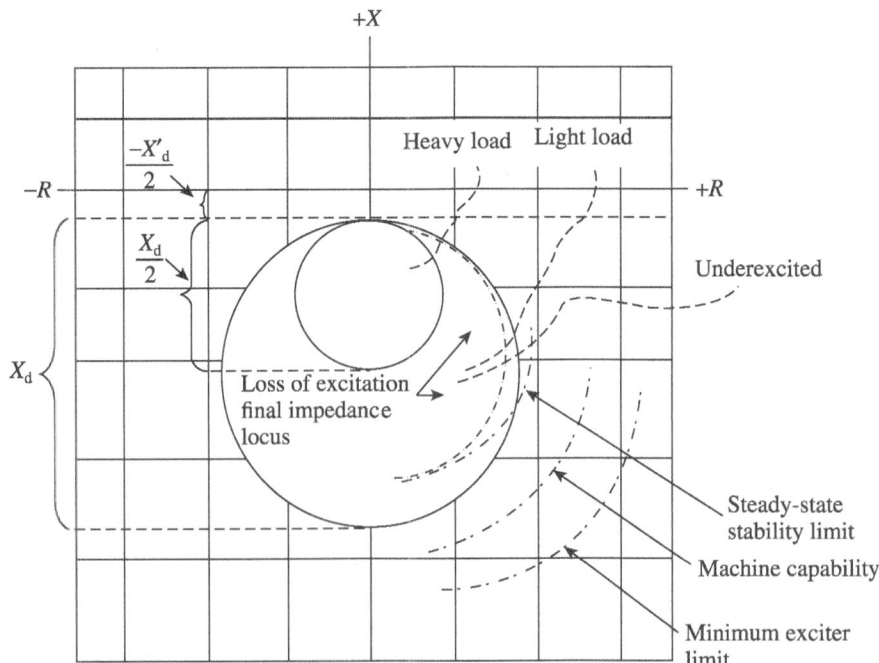

Fig. 5.12 Loss-of-excitation using two-zone offset mho relay

variation in impedance as seen from the generator terminals when it loses its excitation. If the impedance as seen from the generator terminals falls inside the circular characteristic, the offset mho relay operates.

The recommended offset is equal to one-half of the direct axis transient reactance ($X'_d/2$) from R-axis. This offset is provided to offer selectivity between the loss-of-excitation condition and other normal or abnormal operating conditions such as an out-of-step situation or power swing. The loss-of-excitation characteristics will be plotted with respect to two settings, namely high-speed setting and time-delayed setting. High-speed setting has a circle of diameter equal to 1.0 per unit impedance of the generator base and is set to operate without any time delay (instantaneous operation). This setting provides fast protection to the generator against more severe loss-of-excitation conditions. The time-delayed setting has a circle of diameter equal to the direct axis synchronous reactance (X_d) of the generator and is set to operate after some time delay (0.5–3 s) to avoid maloperation of the relay against transient fault conditions. This setting is useful to detect a loss-of-excitation condition when the generator is lightly loaded.

5.7 Negative Phase Sequence (Unbalance Loading) Protection

The main causes for the generation of negative phase sequence (NPS) currents are asymmetrical faults, load unbalance, open conductors, or other unsymmetrical operating conditions. The phenomena that increase NPS currents are line-to-ground faults, line-to-line faults, the widespread use of non-transposed lines, series capacitor compensation, single phase railroad electrification loads, arc furnace loads, and independent pole switching of circuit breakers (CBs). These NPS components of current produce double-frequency currents in the rotor core, which rapidly causes rotor overheating. In the case of the solid rotor, the very less resistance of the rotor core leads to very high negative-sequence current. This may eventually overheat the rotor core because of high iron losses. In general, the generator is designed to withstand 5–10% of the negative-sequence currents of the full load current generator continuously, depending on its size and design. The generator has continuous and short-time unbalanced current-carrying capabilities, which are expressed in terms of negative-sequence currents in Table 5.2.

The continuous negative-sequence current-carrying capability is also covered by the new standards for cylindrical pole type and salient pole type generators, which is given in Table 5.3.

The generator is protected against the negative-sequence currents by using the negative-sequence overcurrent relay (46) whose characteristic equation is given by

$$I_2^2 t = K$$

where,

I_2 = root mean square (RMS) value of negative-sequence current (per unit)

t = time (seconds)

K = constant, depending on the design and size of the generator

Table 5.2 Short-time unbalanced current-carrying capabilities

Type of generator	Permissible $I_2^2 t$ (percentage of stator rating)
Salient pole generator	40
Synchronous condenser	30
Cylindrical rotor generators:	
Indirectly cooled	30
Directly cooled (0–800 MVA)	10

Table 5.3 Continuous negative-sequence current-carrying capability

Type of generator	Permissible I_2 (percentage of stator rating)
Salient pole rotor: with amortisseur windings	10
without amortisseur windings	5
Cylindrical rotor:	10
Indirectly cooled	8
Directly cooled to 960 MVA	6
961–1200 MVA	6
1201–1500 MVA	5

Figure 5.13 shows the protection of the generator against unbalanced current. The NPS filter separates out the negative-sequence current from the positive- and zero-sequence current. The relay calculates the percentage of the negative-sequence current with respect to the rated stator current. When it crosses the set limit, it gives an alarm during low values of negative-sequence current and trips the generator during high values.

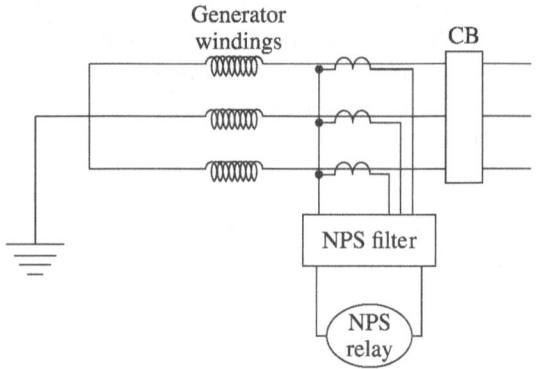

Fig. 5.13 Negative phase sequence protection

5.8 Out-of-step Protection

The main causes of out-of-step condition (78) of the generator are delayed clearance of a heavy fault on the power system, high load angle close to the stability limit, and loss-of-excitation. When an out-of-step situation occurs, variation in the internal angle of the generator is noticed and the generator runs out of synchronism with respect to the rest of the network.

When the generator pulls out of synchronism, a relatively high peak current is produced because of variation in the internal angle of the generator. Further, off-frequency operation of the generator causes increase in winding stress, uneven torque, and mechanical resonance. This situation is dangerous to the generator as well as to the turbine. Therefore, to minimize the possibility of damage, the turbine–generator set is required to isolate as fast as possible, preferably within half a cycle from the rest of the power system network. The best way to detect an out-of-step condition of the generator is to analyse the variation in the apparent impedance with respect to time. This apparent impedance locus depends on the type of governing system and the excitation system of the generator. The level of severity of the fault also affects the locus of the apparent impedance.

The conventional impedance relaying scheme detects the loss of synchronism condition by analysing the variation in the apparent impedance as seen from the terminals of the generator. In most of the cases, the generator can be separated from the system before completion of the one slip cycle.

Typical impedance loci obtained with this procedure are illustrated in Figs 5.14(a) and 5.14(b). These three loci represent the variation in impedance seen from the terminals of the generator. They are plotted as a function of the ratio of the two equivalent system voltages E_A/E_B, which is generally assumed to remain constant during the power swing condition.

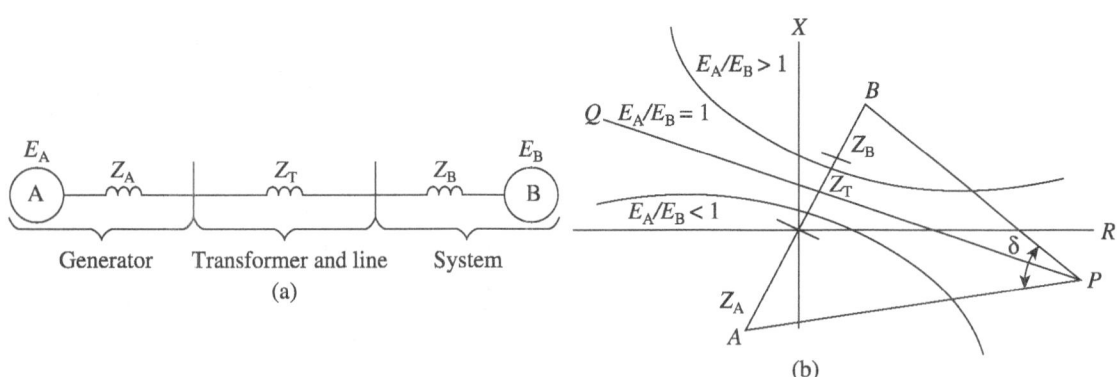

Fig. 5.14 Loci of impedance during out-of-step protection (a) Reactance diagram of a sample system (b) Loci of impedance

When the voltage ratio $E_A/E_B = 1$, the impedance locus is a straight line PQ, which is a perpendicular bisector of the total system impedance between A and B. When the voltage ratio $E_A/E_B > 1$, the centre of the circle is in the positive reactance region, and the electrical centre is above the impedance centre of the system. For $E_A/E_B < 1$, the centre of the circle is in the negative reactance region, and the electrical centre is below the impedance centre of the system. Thus, the diameters and the location of the centres of these circles are a function of the voltage ratios E_A/E_B. Generally, the equivalent internal voltage of the generator is less than 1.0 per unit and less than the equivalent system voltage. Hence, during the loss of synchronism, the locus of impedance as seen from the terminals of the generator follows swing characteristic where the voltage ratio is less than 1 ($E_A/E_B < 1$).

Figure 5.15 shows the two-zone out-of-step protection of the generator. As the impedance moves from zone 1 to zone 2, the out-of-step relay sees an apparent load impedance swing. The relay measures the time needed for the impedance locus to pass from zone 1 to zone 2. If a generator pulls out of synchronism with the system, the current will rise at a relatively slow rate compared to the drastic change in current in the case of a severe fault. Therefore, the impedance locus takes more time to pass from zone 1 to zone 2. During abnormal conditions other than the out-of-step condition, the impedance locus takes very less time to pass from zone 1 to zone 2. Thus whether or not the out-of-step relay trips the generator is based on the time required. If the required time is more than the set time (time taken by the impedance locus during out-of-step condition), then the relay operates; otherwise, it blocks.

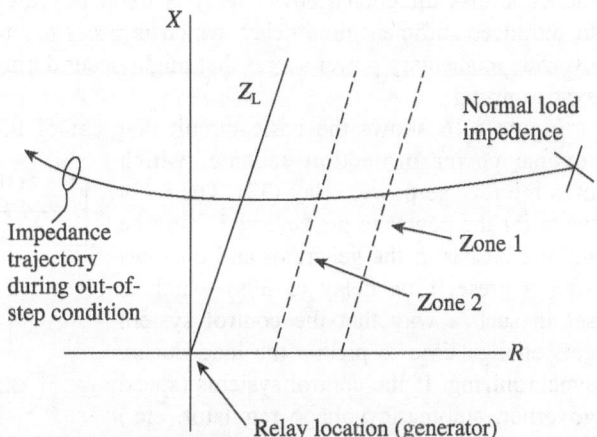

Fig. 5.15   Two-zone out-of-step protection

5.9 Reverse Power Protection (Prime Mover Failure)

Reverse power protection (32) scheme is applied to a generator against the condition of reversal of power, which may occur when the generator loses its prime mover (turbine) input. Reverse power relay is used to identify the direction of flow of power. Due to the unavailability of the prime mover input, the generator is not able to generate active power, and hence, it does not work as a synchronous generator. In this condition, the generator works as an overexcited synchronous motor drawing active power from the infinite busbar, and the prime mover works as a load. The consequences of reversal of power on various types of turbines are discussed here.

For a steam turbine, the low pressure blades will overheat because of the trapped steam inside the turbine, whereas in the case of a hydro-turbine, cavitations will be created because of lack of water flow. For diesel and gas turbines, there is a possibility of hazards of fire and/or explosion due to the presence of unburnt fuel.

Therefore, to avoid the damage to the prime mover, it is essential to isolate the generator from the infinite busbar. Now, before we apply the reverse power protection scheme to the generator, some important points should be considered.

1. Damage of the prime mover will not occur instantaneously. It takes some time, and hence, reverse power protection is a time-delayed method.
2. The reverse power protection scheme should not operate during the synchronizing of the generator with an infinite busbar in case of an internal fault in the generator or during system disturbances.

3. Therefore, the relay is set to operate after some time delay, for about 5–30s, depending on the severity of the system.

The magnitude of reverse power varies considerably depending on the types of prime movers used. Table 5.4 shows the maximum motoring power for different prime movers. Therefore, the reverse power relay should have sufficient sensitivity such that the motoring power provides 5–10 times the minimum pick-up power of the relay. An induction disk directional power relay is frequently used to introduce sufficient time delay, which is necessary to override momentary power surges that might occur during synchronizing.

Table 5.4 Maximum motoring power for different prime movers

Different types of prime movers	Maximum motoring power for prime movers (%)
Steam turbine	3
Water wheel turbine	0.2
Gas turbine	50
Diesel engine	25

Figure 5.16 shows the basic circuit diagram of the reverse power protection scheme, which uses the reverse power relay (32). The relay waits for the power to get reversed from the infinite busbar to the generator and operates after a preset time delay (5–30s) which is set in such a way that the control system gets enough time to pickup the load during synchronizing. If the control systems (speed governor, automatic voltage regulator, etc.) of a generator fail to control the voltage and frequency of the generator or if the supply system is unable to feed active power, the relay is required to trip within a short time.

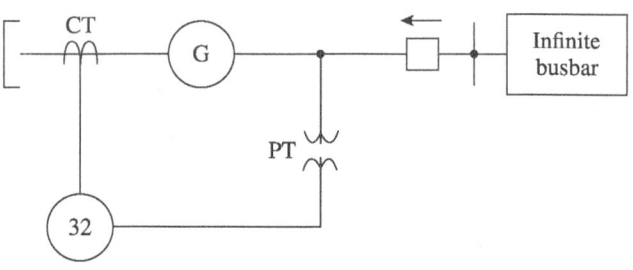

32 = Directional reverse power relay

Fig. 5.16 Reverse power protection scheme

5.10 Abnormal Frequency and Voltage Protection

In order to protect the generator against sudden changes in frequency and voltage within the prescribed limit, over/under frequency (81) and over/undervoltage protection is provided by the manufacturers.

5.10.1 Overfrequency Protection

Overfrequency in a generator occurs when the mechanical power input to the generator is more than the electrical load. The most common reason for the occurrence of overfrequency is the sudden loss of load. When the speed of the turbine increases, the governor responds quickly to reduce the input to it, as otherwise severe overfrequency operation of a large size high-speed generator could result in plant damage.

In general, two setting limits are provided by the overfrequency relay. The lower limit is meant for giving an alarm, whereas the higher limit is for tripping the generator. The relay is equipped with both instantaneous (for higher limit) and time-delayed (for lower limit) operations. In general, a small time delay is provided to avoid maloperation of the relay due to transient faults.

5.10.2 Underfrequency Protection

Underfrequency occurs because of the excessive loading of a generator. In this condition, the generator is forced to produce more power than its rated capacity, and therefore, the frequency gets reduced. The power system remains in a stable condition if the load demand is decreased below the maximum generating capacity of the generator. If the load demand exceeds the generating capacity of the generator, frequency reduces and if proper load shedding is not done, the system frequency falls below its specified limit. The rate at which the frequency drops depends on the duration for which this condition exists, percentage of overload, and load and generator variations with respect to the change in frequency. The frequency decay that occurs within a short period cannot be corrected manually. Therefore, automatic load shedding techniques are widely used.

In order to avoid this problem, the current practice is to use underfrequency relays at various load stations (substations and power generating plants). The setting of the underfrequency relay is based on the maximum overloading condition. During overload conditions, load shedding must be done before the underfrequency relay operates. In other words, the load must be shed before the generators are tripped.

The setting of the underfrequency relay is the same as that of the overfrequency relay. The only difference is the limit in the frequency setting. The underfrequency relay gives an alarm or tripping signal, depending on the reduction in frequency.

5.10.3 Overvoltage Protection

If the generator is running independently (not connected to a power system) or when the generator is providing power to an islanded power system, there is a possibility of increase in the generator terminal voltage (which is known as *overvoltage*). Overvoltage can also occur because of increase in the speed of the prime mover owing to the sudden loss in the load on the generator or to the failure of the automatic voltage regulator. Generally, overvoltage does not occur in the turbo-generator because of its highly sensitive speed governing systems. However, this protection is required for hydro generators or gas turbine generators to prevent possible damage of generator insulation or prolonged overfluxing of the generating plant.

A two-stage overvoltage element is used in the overvoltage protection (59) scheme. The element can be set to operate from phase–phase or phase–neutral voltages. Each stage has an independent time delay that can be set to zero for instantaneous operation. The first stage, which has an instantaneous overcurrent characteristic, is set to pickup at 130 – 150% of the rated voltage. The second stage with an inverse definite minimum time (IDMT) characteristic is set to pickup at 110% of the rated voltage.

5.10.4 Undervoltage Protection

In case of parallel operation of generators, the total load is shared according to the capacity of the generating plant. If any accidental tripping of one or more generators occurs, then the other generators try to deliver the load.

Each of these generators will experience a sudden increase in current. Therefore, the terminal voltage of the generator decreases. Undervoltage protection (27) is not a commonly specified requirement for the generator protection scheme. However, undervoltage elements are sometimes used as interlocking elements for other types of protection such as field failure protection.

In the relay, this type of interlocking can be arranged via the relay logic scheme. Undervoltage protection may also be used as a backup protection where it is difficult to provide adequate sensitivity with voltage dependent/underimpedance/NPS elements. The automatic voltage regulator connected to the system tries to restore the voltage.

5.10.5 Voltage-dependent Overcurrent Protection

Whenever a fault occurs, the terminal voltage of the generator reduces. Therefore, a voltage measuring element can be used to control the current setting of this element. On detection of a fault, the current setting is reduced by a factor K. This ensures that the faults are cleared in spite of the presence of the generator decrement characteristic.

A single-stage, non-directional overcurrent element is provided. The element has a time-delayed characteristic that can be set as either IDMT or definite time (DT). The element can be selectively enabled or disabled and can be blocked via a relay input so that it can be integrated into a blocked overcurrent protection scheme. The element is fed from the CTs at the terminal or neutral end of the generator. If a voltage dependent overcurrent operation is selected, the element can be set in one of two modes, that is, voltage controlled overcurrent or voltage restrained overcurrent.

5.10.6 Neutral Voltage Displacement Protection (59N)

On a healthy three-phase power system, the summation of each of the three phase-to-earth voltages is nominally zero, as it is the vector addition of the three balanced vectors at 120° to one another. However, when an earth fault occurs on the primary system, this balance gets disturbed, and a *residual voltage* is produced. This could be measured at the secondary terminals of a PT connected in a *broken delta* manner. Hence, a residual voltage measuring relay is used to provide earth-fault protection to such a system. In general, two-stage settings are provided by the relay, namely high setting and low setting, with and without time delay. The relay gives an alarm when the residual voltage crosses a lower threshold, whereas it issues a trip signal when the residual voltage crosses a higher threshold.

5.10.7 Protection Against Unintentional Energization at Standstill

Accidental energization of the generator from a standstill condition can cause severe damage to the generator. If the breaker is closed, then the generator will begin to act as an induction motor. In this condition, the rotor draws an abnormal current that can cause an arcing between the rotor components, for example, slot wedge to core, which finally results in rapid overheating.

To provide fast protection against this abnormal operating condition, an instantaneous overcurrent relay is used in conjunction with a three-phase undervoltage detector. The undervoltage detector gets the voltage signal from the secondary side of the PT. The relay is enabled by the undervoltage detector, with the help of a logic scheme, when the generator is not generating any voltage (generator is not running). Therefore, the instantaneous overcurrent relay is set below the rated current to increase the sensitivity and speed of the protection scheme.

5.10.8 Thermal (Overheating) Protection Using Resistance Temperature Detector

If the generator is overloaded for a long duration, then there is a possibility of overheating of the windings. This situation results in the premature ageing of the insulation, and in the worst case, the insulation may fail. Worn or unlubricated bearings can also generate localized heating within the bearing housing. To protect the generator against any general or localized overheating, continuous monitoring of the temperature is required for various parts of the generator. If at any of the point, the temperature rise is excessive, it may result in damage to that part of the generator. The temperature rise depends on I^2Rt and also on the cooling. Hence, to avoid these situations, the generator is provided with resistance temperature detectors (RTDs) (49) to measure the temperature of various parts of the generator. The output of each RTD reaches the temperature relay and the relay gives out an alarm or trip signal according to the level of temperature rise.

5.10.9 ROCOF Protection to Generator

Generally, the generator is operated close to the fundamental frequency which indicates the balancing of power generated and load demand. However, sometimes low-to-high variation of system frequency is observed due to unbalance. The rate at which the frequency deviates depends on the nature and severity of disturbance in system. Thus, the rate of change of frequency (ROCOF) protection of generator is normally applied to re-establish the balance between the generated power and the consumed power. Though the underfrequency (UF) protection is used for protecting the generator against blackout, ROCOF (df/dt) protection of generator operates slightly earlier compared to UF protection to detect unusual changes in the system frequency. In this way, ROCOF relay takes the correct action by shedding appropriate load from the system in order to protect the generator from overloading. Usually, the setting of ROCOF relay is done in Hz/s, for example, 0.25 Hz/s. Whenever, the measured value of the rate of change of frequency crosses the set value in ROCOF relay, it generates a trip signal which is transferred to the respective breaker(s) to shed the load. The factors to be considered while designing load-shedding scheme are discussed in Chapter 12.

5.11 Numerical Protection of Generator

Using the latest numerical technology, the numerical generator protection relays are designed to cater to the protection of a wide range of generators. These can range from small machines providing standby power on industrial sites to large machines in power stations providing for the base load on the grid transmission network. These illustrate the use of individual relays for different fault conditions. However, the modern numerical relays combine most of these functions in a single relay with programming features that make them useful for any large rating generator. The latest technological rise in generator protection has been the release of digital multifunction relays by various manufacturers. The various protection functions available in a typical numerical relay are inverse time overcurrent, voltage restrained phase overcurrent, negative-sequence overcurrent, ground overcurrent, phase differential, ground directional, high-set phase overcurrent, undervoltage, overvoltage, underfrequency, overfrequency, neutral overvoltage (fundamental), neutral undervoltage (third harmonic), loss of excitation, distance elements, and low forward power.

5.11.1 Functions of Modern Numerical/Digital Generator Protection Relay

In this section, we will discuss in detail the functions of the modern numerical/digital generator protection relay.

Protection Functions

Generator differential protection Three slope differential characteristics are provided to detect the generator stator winding fault.

Complete stator earth-fault protection Protection logic based on the third harmonic voltage measurement technique is provided.

Non-directional phase overcurrent protection This protection is provided in two stages where the first stage contains IDMT and DT overcurrent scheme, whereas the second stage includes only the DT overcurrent scheme.

Non-directional earth-fault protection This protection is the same as the non-directional phase overcurrent protection. However, the setting is generally done at lower values of current and time.

Neutral displacement/residual overvoltage protection This type of protection is generally provided to detect earth faults in high impedance grounding systems. The PTs are connected in an open delta manner to detect the residual voltage produced because of the earth fault.

Sensitive directional earth-fault protection The direction of the earth fault is found by using residual voltage and earth-fault current. In general, a core balance CT is used to improve the sensitivity and accuracy of the protection scheme.

Restricted earth-fault protection This protection scheme is generally provided for low impedance grounding systems where a sufficient value of the fault current is available for sensing purpose.

Voltage dependent overcurrent/underimpedance protection This protection function performs logical operations such as AND logic and OR logic between voltages and currents of the generator.

Undervoltage/overvoltage protection Protection is provided in multiple stages with time delay or IDMT characteristics.

Underfrequency/overfrequency protection A multistage protection scheme with time delay facility is provided.

Reverse power/low forward power/overpower protection This protection is provided by the relay that senses the direction and magnitude of the power of the generator with respect to the busbar.

Loss of field protection Offset mho relay is provided to sense a loss of field condition. Instantaneous operation is provided for severe conditions, while the time-delayed operation is performed during less severe conditions.

Negative phase sequence thermal protection Multistage protection is provided to protect the generator against the unbalanced stator faults or external fault conditions.

Overfluxing protection Multistage overfluxing protection based on volts per Hertz measurement logic is used.

Unintentional energization at standstill protection If a generator CB is closed accidentally when it is not running, very high current will flow from infinite busbar to the generator. Protection against this type of condition is provided by an instantaneous overcurrent element. Time delay is provided to make the protection scheme stable for normal voltage dips that could occur for system faults or generator reconnection.

Thermal protection by RTDs Temperature is measured by RTDs at various points in the generator. Instantaneous and time-delayed functions are provided for giving out alarm and tripping signals, respectively.

Circuit breaker failure protection Multistage CB failure protection schemes are used for tripping upstream CBs and/or the local secondary trip coil of the same breaker.

Potential transformer supervision function (PTS) This is provided to detect the loss of one or more phases of PT signals. It is achieved by providing an indication and inhibition of voltage-dependent protection elements.

Current transformer supervision function (CTS) This is provided to detect loss of one or more phases of CT signals and inhibit the operation of current-dependent protection elements.

Control Functions

Programmable logic scheme (AND, OR) and general purpose timers allow the user to customize the protection and control functions. They are also used to program the functionality of the optically isolated inputs, relay outputs, and LED indications. The relay possesses multiple setting groups, where independent setting groups are available to set the relay for different requirements.

Measurements

The relay can measure instantaneous and integrated values of phase currents, phase voltages, active power, reactive power, apparent powers, and temperatures of generator parts.

Post Fault Analysis

In addition to the various protection functions, numerical relays are capable of recording a few events, disturbances, and faults that are stored in non-volatile memory of the relay.

Monitoring

Trip circuit monitoring This function monitors the status (open and close) of the breaker. This information is used in conjunction with the programmable logic scheme.

Breaker state monitoring This function is used to check the presence of any discrepancy between the status contacts of the CB.

Breaker condition monitoring This function monitors the number of trip operations, operating time, etc., of the generator breaker.

Temperature monitoring This is used to know the variation of temperature of the generator parts with respect to the change in load and other external conditions.

Communications

All these features can be accessed remotely from one of the relay's remote serial communication options. Front communication is done by the RS 232 port, and rear communication is carried out by the RS 485 port.

Diagnostics

Certain automatic tests are performed, such as power-on diagnostics and continuous self-monitoring, to ensure a high degree of reliability.

User-friendly Interface

Liquid crystal display with backlight is provided to operate the relay from its front plate. Fixed and programmable LED indications are provided to give indications such as relay status and fault status. Password protection is given to the relay to provide access to controls and settings.

5.11.2 Numerical Differential Protection of Generator

In analog Merz–Price or biased differential protection scheme, the secondary sides of two CTs are physically connected with each other for circulation of current. While in numerical differential protection scheme (87), the secondary side of each CT is connected directly to the relay. It means that the two CTs are isolated electrically. The protection program resides permanently inside the relay memory (ROM), derives the basic and bias currents mathematically, and checks the variation in data sent by the digital signal processor. As soon as the processor gives a new differential current, the protection task checks whether it is within the tripping limit or outside the limit. If it is within the tripping limit, then the relay issues the tripping command. The total tripping time of the numerical relay is less than 35 ms.

Recapitulation

- This chapter focusses on the effect of the abnormal or faulty operating conditions of the generator and various types of protection systems applied to minimize their adverse effects.
- The protection functions applied on generators are differential protection, stator and rotor earth-fault protection, out-of-step protection, loss-of-excitation protection, and reverse power protection.
- The CTs used for the generator differential protection should have the KPV as $V_K \geq 2V_r$, where $V_r = I_f \times (R_{CT} + 2 \times R_L)$ is the voltage across the relay circuit.
- In differential protection, the basic setting is given by $(i_1 - i_2)$ and the bias setting is given by

$$(i_1 - i_2)\Big/\left(\frac{i_1 + i_2}{2}\right).$$

- Different abnormal overfrequency/underfrequency, overvoltage/undervoltage protection, and thermal protection schemes are also applied to generators.

- The latest trend of using digital technology in protection systems has helped a lot to provide complete protection to generators of various capacities.
- Multifunction numerical/digital relays help the generator work smoothly by providing facilities such as measurement of operating quantities, control functions, operational logics, faults, and disturbances recording.

Multiple Choice Questions

1. Which of the following protections applied to a generator is time delayed?
   (a) Differential protection
   (b) Unbalance protection
   (c) Stator earth-fault protection
   (d) Rotor second earth-fault protection

2. Which of the following protections applied to a generator is instantaneous?
   (a) Turn-to-turn fault protection
   (b) Unbalance protection
   (c) Field failure protection
   (d) Underfrequency protection

3. In differential protection scheme, a stabilizing resistance is required to
   (a) block the relay current
   (b) increase the relay sensitivity
   (c) reduce the effect of non-identical CTs
   (d) none of the above

4. The percentage bias relay setting should be
   (a) 5–10%   (c) 30–35%
   (b) 20–30%   (d) 35–40%

5. In case of a generator, the differential relay operates for
   (a) rotor earth fault   (c) internal fault
   (b) external fault   (d) mechanical fault

6. NPS current produces
   (a) overvoltage   (c) rotor heating
   (b) vibration   (d) over speed

7. Failure of the generator field breaker causes
   (a) overspeed   (c) overheating
   (b) voltage drop   (d) none of the above

8. The magnitude of the generator earth-fault current is limited by
   (a) inserting stabilizing resistance in series with the relay
   (b) using higher rating CTs
   (c) controlling field exciter
   (d) inserting suitable resistance in neutral grounding circuit

9. The reverse power protection is applied for
   (a) overspeed   (c) turbine failure
   (b) excitation failure (d) stator earth fault

10. Rotor first earth-fault protection
    (a) is instantaneous in operation
    (b) is a time-delayed operation
    (c) gives an alarm to the operator
    (d) none of the above

Review Questions

1. Explain the Merz–Price differential protection for generator. Why does the relay sometimes tend to operate in case of heavy external fault?

2. Explain the biased differential protection of a generator with percentage characteristic.

3. Explain with diagram how the generator is protected against turn-to-turn faults.

4. Explain with the help of a relevant diagram the complete (100%) stator earth-fault protective scheme.

5. Explain the loss-of-excitation and out-of-step protection for a generator.

6. Why is a time-delayed relay required for reverse power protection of a generator?

7. Explain abnormal frequency and voltage protection applied to a large generator.

8. Enumerate the different features of numerical relays used for the protection of a generator.

Numerical Exercises

1. A 13.8 kV, 100 MVA generator is protected by a percentage differential relay. The relay is set to operate with minimum pickup of 0.2 A and bias setting of 10%. A high resistance fault occurs close to the grounded neutral while the generator is carrying a load with the fault current of 450 A. Suggest suitable CT ratio, and show with relevant calculation whether the relay will operate or not.

   [Operating current = 0.4506 A, Pickup ratio = 8.78%, Relay will not operate.]

2. Figure 5.17 shows the single line diagram of a system where the generator is protected by a biased differential relay. The settings are as follows:

   Pickup setting: 20% of relay rated current (5 A)

   Bias setting: 30%

   CT ratio: 7500/5 A

   Show with relevant calculation that the relay will operate for internal fault (F_{int}). All reactance are in per unit and on common 100 MVA, 11 kV base.

   [Operating current = 63.75 A, Pickup ratio = 58.32%, Relay will operate.]

3. The following is the data obtained for the generator differential protection:

   KVA = 19,500, voltage = 12.47 kV, X''_d = 10.7%, X'_d = 15.4%, X_d = 154%, and the through-fault withstand current is 10 times the rated current. The CT ratio is 1000/5 A with R_{CT} = 1.2 Ω, lead resistance R_L = 0.1 Ω. The relay with the burden of 1 VA with rated current is 5 A, and the setting range of 5–20% of rated current is used for differential protection. Suggest a suitable stabilizing resistance and KPV for the CTs.

   [R_{ST} = 41 Ω, KPV > 150 V]

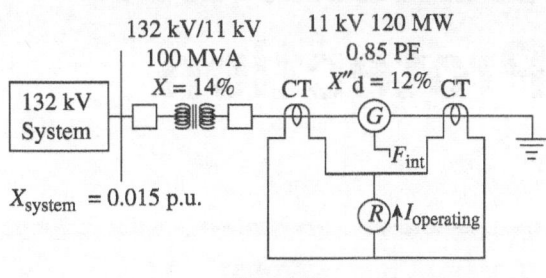

$$X_{system} = 0.015 \text{ p.u.}$$

Fig. 5.17

4. A 120 MW, 15.75 kV, 0.85 PF, 50-Hz, 3-φ, Y-connected generator is protected by an earth-fault relay. The relay is set to operate at 5%. The CT ratio is 7500/1 A. A resistor is used in neutral circuit of the generator to limit the earth-fault current equal to 75% of normal load current. Determine the value of the resistor and the percentage of stator winding protected.

   [R_n = 3.24 Ω, Percentage of winding protected = 90.33%]

5. A 10 MW, 6.6 kV, 0.9 PF, Y-connected generator is protected by an earth-fault relay. The star point is earthed through a resistor to limit the maximum earth-fault current to 60% of full load current. The earth-fault relay with the sensitivity of 10% is connected to the secondary of CT with the ratio of 1000/1 A. Determine the value of the resistor and percentage of stator winding protected. In addition, find the earth resistor to protect 92% of the winding from generator terminal and fault current with this value of the resistor.

   [R_n = 6.53 Ω, Percentage of winding protected = 82.87%, R_n = 3.04 Ω for 92% Winding to be protected, I_F = 1253.45 A]

Transformer Protection

After going through this chapter, the students will be able to:

- List and explain the different types of abnormal conditions in a transformer
- Explain the different protection schemes to provide protection against faults in a power transformer
- Analyse the effects of magnetizing inrush and overfluxing on transformers and the feasible protection
- Explain the working of Buchholz relay
- Discuss the overcurrent and restricted earth fault protection
- List the various features of modern digital relays

6.1 Introduction

Ever since the AC supply system was developed, the transformer has been a part of transmission and distribution systems. Since it is part of every network, it is important to provide proper protection to it. In general, a well-designed transformer with proper maintenance provides uninterrupted service for many years. However, it is almost always subjected to undue stress during its operation because of the faults in the power system. Since transformers are expensive and unwieldy to transport in the event of a fault, it is advisable to provide proactive protection.

In general, faults in a transformer occur because of the weakening/failure of insulation. Insulation failure may cause the following faults:

1. Turn-to-core fault (earth fault)
2. Turn-to-turn fault (inter-turn fault)

In both situations, the temperature of the transformer oil will increase, and this may lead to mechanical damage to the transformer tank and thus to the transformer. The transformer is submerged in insulating oil. The insulation may weaken because of ageing, contaminated oil, and corona discharge in the insulating oil. Transient overvoltage due to switching/lightning surges also causes overheating of the insulating oil. Increase in the temperature of transformer coils gives rise to the formation of gas in the tank, and if this condition is left unnoticed for some time, the tank pressure may rise, causing mechanical damage.

The transformer is also subjected to some abnormal operations that are quite similar to the fault conditions, but do not require the protective gear meant for the protection of the transformer. These conditions are as follows:

1. Magnetizing inrush

2. Overfluxing
3. Low oil level

Although these abnormal conditions are not faults in a transformer, prolonged exposure to these conditions results in overheating of the transformer winding and damage to the insulation.

6.2 Abnormal Conditions in Transformers

As discussed earlier, there are different abnormal conditions in a transformer. Some of them require the immediate operation of transformer protective relays and some others require other relays in the system to clear the abnormality, while some require either no operation or delayed operation of transformer protective relays. The abnormal conditions that require delayed operation of the transformer protective system are not internal faults of the transformer, but if allowed to persist longer result in a fault in the transformer. These abnormal conditions are magnetizing inrush, overfluxing, and low oil level in the transformer tank. In the following subsections, the reasons for the abnormalities and the method to handle each are explained.

6.2.1 Magnetizing Inrush

Magnetizing inrush is a condition when the transformer draws a very large current from the supply while the load current is either zero or of nominal magnitude. To understand the magnetizing inrush, the basic operation of a transformer needs to be understood.

During energization of a power transformer, with secondary winding open circuited or lighted loaded, the steady state value of flux in the core, as shown in Fig. 6.1, lags the voltage waveshape by angle 'θ'.

This flux is symmetrical or asymmetrical in nature which depends on switching instant of a voltage signal given to the primary winding of a transformer. If switching of a transformer is carried out at the instant when the voltage wave is at its peak value, then the flux is symmetrical in nature and is given by

$$\phi = \phi_{max} \pm \phi_R$$

where, ϕ_{max} is the maximum value of steady state flux in the core of a transformer and ϕ_R is the residual flux present in the core of the transformer.

In this situation, the B-H curve operates in the linear region. Hence, the magnitude of magnetizing inrush current is restricted to a normal value as shown in Fig. 6.2.

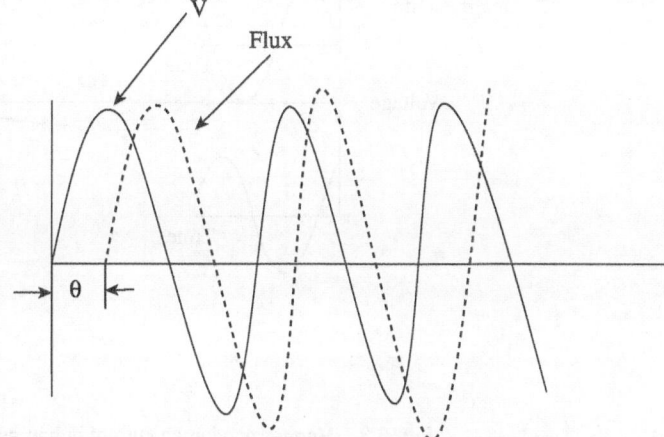

Fig. 6.1 Waveform of voltage and flux

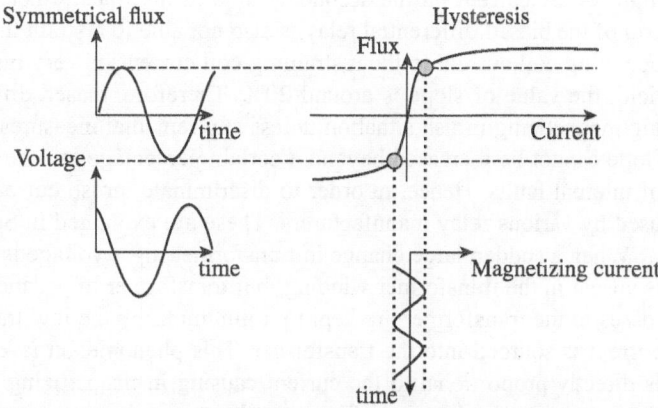

Fig. 6.2 Magnetizing inrush current during symmetrical flux

Conversely, if the switching of a transformer is done at an instant when voltage wave is passing through zero, then the flux is asymmetrical in nature and it is given by

$$\phi = 2 \times \phi_{max} \pm \phi_R$$

In this condition, upper point of B-H curve gets elongated (it does not remain in linear region) as shown in Fig. 6.3. Hence, in order to produce the same amount of flux, the magnitude of current becomes very high. This is due to the fact that the flux is going to link through air core instead of iron core for the first 5 – 7 cycles. As the inductance of air is very low and hence, according to the equation $I = \dfrac{N \times \varphi}{L}$, the magnitude of current becomes very high. Here, N is the number of turns of the transformer winding, ϕ is the flux in the core of the transformer, and L is the inductance of air path. This current is of the order of 6–10 times the rated current of the transformer. The magnetizing inrush current is going to die out after a few (5–7) cycles as more and more flux is going to link through iron core which has higher inductance than the air path.

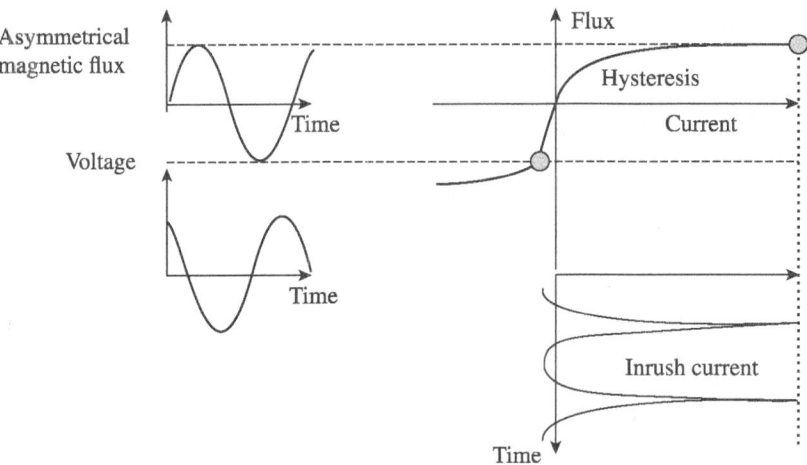

Fig. 6.3   Magnetizing inrush current during asymmetrical magnetic flux

Now, the magnetising inrush current through the operating coil of the biased differential relay is very high, as the current on the secondary side of the transformer would be zero or very less. The restraining coil of the biased differential relay is also not able to restrain the operation of the relay as the slope (ratio of operating coil current to the restraining coil current) is very high (more than 100%). However, in practical field, the value of slope is around 20%. Therefore, biased differential relay gives unwanted tripping during magnetizing inrush situation unless any remedial measures are taken. It is to be noted that the value of slope cannot be increased beyond a certain percentage as otherwise sensitivity of the relay reduces in case of internal faults. Hence, in order to discriminate inrush current with internal faults, different methods are used by various relay manufacturers. These are explained in Section 6.8.2.

When a sudden large change in transformer input voltage is made, the induced back electromotive force is absent in the transformer winding, but transformer impedance is present to restrict the input current. The losses in the transformer are kept to a minimum by the low transformer winding impedance. Thus, a large current is sourced into the transformer. This phenomenon is known as *magnetizing inrush*. Since the flux is directly proportional to the current causing it, magnetizing inrush results in a large flux/flux density in the core with the objective of keeping the core volume minimal. Transformers are designed to operate just below the knee region of the B–H curve of the core material. During normal operation, the flux is within the linear region of the B–H curve, but during magnetizing inrush, a large flux results in saturation of the transformer core. Thus, the flux waveform will be asymmetrical with respect to the time axis and will contain a DC value (Figs 6.4–6.6). The nature and magnitude of the inrush current will be decided by the direction and magnitude of the residual magnetization flux in the core and switching instant. On harmonics analysis, it is found that this flux consists of a significant amount of second and odd harmonics. These higher order harmonics, out of the second harmonics, is predominant, generally in the range of 15–20% of the fundamental frequency component. The amplitudes of other odd harmonics gradually decrease with the order of harmonics. Therefore, under the magnetizing inrush condition, the current in the transformer winding, which may be as high as 12–15 times of the rated current, will be rich in harmonics content.

This discussion indicates that the magnetizing inrush will occur not only when the transformer is switched on, but also when a fault outside the transformer is cleared. In this case also, there is a sudden large change in the transformer input terminal voltage.

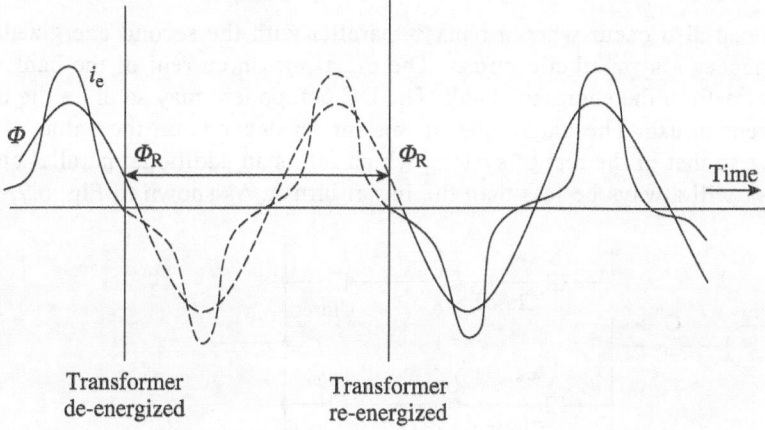

Fig. 6.4 Magnetizing current when transformers are re-energized at the instant of the voltage wave corresponding to the residual magnetic density within the core

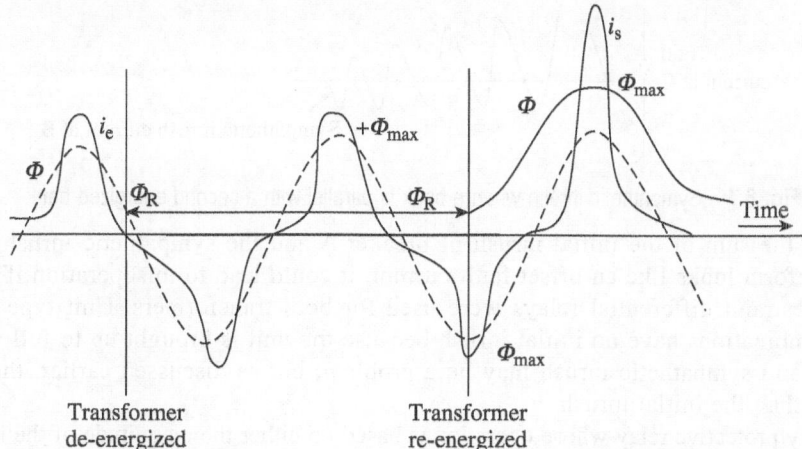

Fig. 6.5 Magnetizing current when transformers are re-energized at the instant when the flux is normally at its negative maximum value

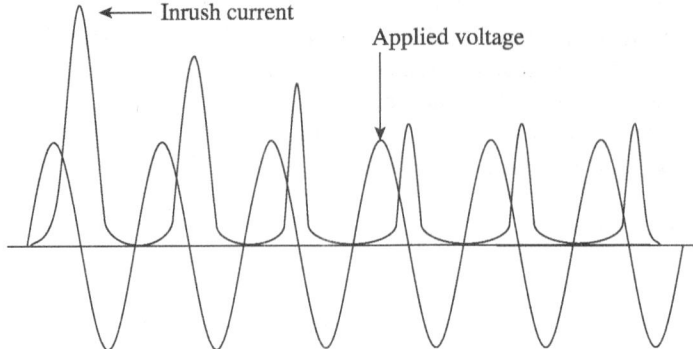

Fig. 6.6 A typical magnetizing inrush current wave

Further inrush can also occur when a bank is parallel with the second energized bank; the energized bank experiences a sympathetic inrush. The offset inrush current of the bank being energized will find a parallel path in the energized bank. The DC component may saturate the transformer iron, creating an apparent inrush. The magnitude of this inrush depends on the value of the transformer impedance relative to that of the rest of system, which forms an additional parallel circuit. Again, the sympathetic inrush will always be less than the initial inrush. As shown in Fig. 6.7, the total current

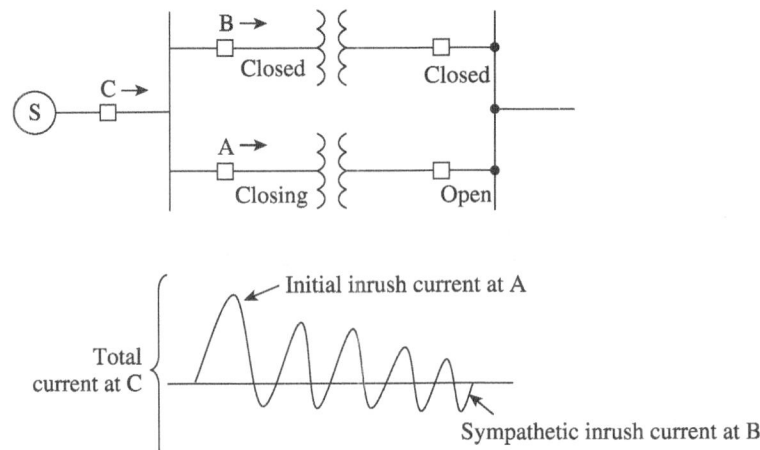

Fig. 6.7 Sympathetic inrush when a bank is parallel with a second energized bank

at breaker C is the sum of the initial inrush of breaker A and the sympathetic inrush of breaker B. Since this waveform looks like an offset fault current, it could lead to misoperation if a common set of harmonics restraint differential relays were used for both transformers. Unit type generator and transformer combinations have no initial inrush because the unit is brought up to full voltage gradually. Recovery and sympathetic inrush may be a problem, but as discussed earlier, these conditions are less severe than the initial inrush.

Therefore, any protective relay whose operation is based on either the magnitude of the input current or the difference in the input or output current will operate. Conventionally, during this phase (when a transformer is energized) current-magnitude-based protective relays are blocked for the initial few cycles till the input current settles down around the rated current, based on which current-based relay settings are done.

6.2.2 Overfluxing

The steady state value of flux in a transformer core is directly proportional to the ratio V/f. As discussed in Section 6.2.1, a transformer operates just below the knee region of the B–H curve. Therefore, either an increase in voltage or decrease in frequency will result in an increased value of the core flux, driving it into the saturation region. Because of this core saturation, the induced voltage in the two windings and their corresponding currents will become flat topped. In other words, it consists of a higher order odd harmonics. Because of the transformer connections, generally the third harmonics are cancelled out (star/delta, delta/star), but other harmonics currents will circulate in the transformer windings. When a transformer core is overexcited, the core operates in a non-linear magnetic region, and creates a significant amount of the fifth harmonic component in the exciting current. Usually, 10% overfluxing can be allowed without damage. This will result in overheating or heating of the transformer winding. If this condition is allowed to continue for long, it will cause damage to/weakening of the transformer insulation and gas formation in the transformer oil. Although this condition is not any abnormality within the transformer, it requires proper protection of the transformer against it. Since the heating will be gradual, the relay can wait for some time before it operates, but the transformer should be isolated in 1 or 2 minutes if overfluxing persists.

In such cases V/f relays are used to detect overfluxing in the transformer. For a 50 Hz AC supply, the setting of the relay should be 1.0 to 1.3 on 110 V. The relay is provided with an adjustable time delay to prevent the transient operation due to momentary disturbances. The transformer-overfluxing protection is implemented for both sides of the interconnecting transformer. This is to cover all possible operating conditions. For other transformers, the overfluxing relay should be provided on the untapped winding of the transformer.

6.2.3 Low Oil Level in Transformer Tank

The transformer oil continuously absorbs heat from the transformer's winding either because of copper losses occurring in the winding or because of overheating of the transformer coils owing to faults internal or external to the transformer. This results in regular decomposition of the transformer oil in the tank. Furthermore, transformer oil samples are drawn at regular intervals to check their quality and to proactively detect faults in the transformer windings through gas content analysis of the transformer oil. These two factors can result in reduction in the transformer tank oil up to a level when some part of the winding does not remain submerged in the transformer oil. The conditions are worsened if either the stopcock meant for taking the transformer oil samples goes bad or is not properly closed post taking of the samples, resulting in a sudden loss of transformer oil.

Whatever may be the reason, if the level of the transformer tank oil is below the critical level, causing in part, dry winding, it will result in overheating of the exposed part. This overheating of the transformer winding will result in weakening of the solid insulation, thus leading to a hot spot development. If not attended to in a timely manner, this will precipitate a solid internal fault in the transformer, requiring compromise with power supply and major maintenance of the transformers.

6.3 Non-electrical Protection

In order to protect a transformer against incipient faults, special types of protection are required. Incipient faults are those which occur in the transformer due to overheating, overfluxing, or sudden increase in pressure. Usually, these types of faults develop due to deterioration of insulation. Though these faults grow slowly, they can cause a major arcing fault which will be detected by the protective devices used in the transformer. Early detection of these faults can reduce repair time and hence, the transformer can be placed back into service without a prolonged outage. Protection against these faults is explained in Section 6.3.1.

6.3.1 Buchholz Relay

Buchholz relay belongs to a category of relays that responds to pressure (oil or gas), fluid flow, liquid level, and so on. The most widely used member of this category, the Buchholz relay is used to protect the oil-immersed transformers. The Buchholz relay may consist of one or two floats as shown in Fig. 6.8. These floats are contained in a closed housing located in the pipeline from the transformer tank to the conservator tank. Any fault in the transformer causes the heating of transformer oil, resulting in its decomposition. The decomposition process causes the formation of gases, which travel from the hot spot in the pipe towards the conservator tank. During this process, these gases are trapped in the gas-operated relay. In case of a heavy fault, a bulk displacement of oil takes place. In a two-float relay, the upper float responds to the slow accumulation of gas due to mild or incipient faults, while the lower float is being deflected by the oil surge caused by a major fault. The float control contact, in the first case, actuates an alarm so that the operator is informed about the abnormal condition prevailing in the transformer. If this gas formation continues causing a large displacement of oil, the second float causes the actuation of the contacts, which in turn energizes

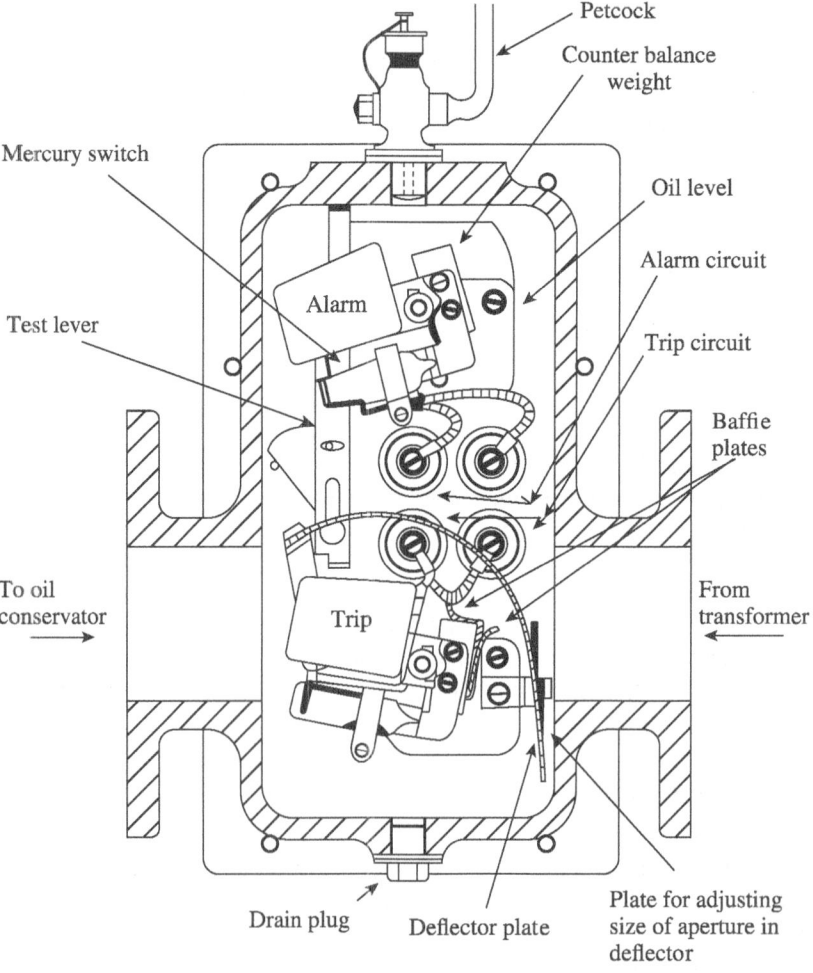

Fig. 6.8 Typical Buchholz relay

the normally open contacts of the tripping relay. This closure results in the actuation of the trip coil of the circuit breaker meant to isolate the transformer from service. Thus, the Buchholz relay, which is a mechanical relay, protects the transformer from severe damage.

A typical Buchholz relay comprises two pivoted aluminium floats of buckets. These two floats counterbalance each other so that with the device empty or completely full of oil the bucket or float is in a high position. Each pivoted assembly carries a mercury switch as seen in Fig. 6.8.

In normal operating conditions, the casing is filled with the oil so that the bucket floats are high, causing the mercury switches to remain open. If the gas bubbles formed (because of the formation of the hot spot due to insulation weakening) pass up in the piping, they will be trapped in the relay casing. The trapping of gases in the bucket manifests in the displacement of oil. As the oil level falls, the upper float will also go down as the weight of the bucket filled with the oil exceeds the counterbalance when the buoyancy from the surrounding oil is lost. As the float falls, the mercury switch tilts, causing closure of the alarm circuit. A similar operation will occur if a tank leak causes the oil level to fall below the minimum threshold level. The device will therefore give an alarm for the prevailing fault conditions. In general, these conditions are of relatively less significance; however, if left unnoticed for a longer duration they may result in a severe internal fault in the transformer. The conditions that cause activation of alarm circuit from Buchholz relay may be one or more of the following.

1. Formation of hot spots on the core due to the short circuit of laminated insulation
2. Core belt insulation weakening/failure
3. Faulty joints
4. High resistance inter-turn faults or other winding faults involving only lower power infeeds
5. Loss of oil due to leakage

Whenever a major internal winding fault occurs, it causes a surge of oil. This surge of oil displaces the lower float and thus isolates the transformer. This operation of Buchholz relay occurs during the following.

1. All severe winding faults (solid inter-turn or turn to earth fault).
2. Fall in the transformer tank oil resulting from either continuous decomposition of oil due to higher operating temperatures or a faulty sample collection tap. This loss of oil if allowed to continue will bring down the transformer oil level to a dangerous degree.

This relay is provided with an inspection window on either side of the gas collection space, which is meant to monitor the oil level in the transformer tank. The colour of the gas collected is useful in diagnosing the fault in the transformer. For example, white or yellow gas collected is indicative of an insulation burnout; on the other hand, dissociated oil is indicated by black or grey colour gas. In most of the cases, the gas so collected is inflammable, and hence, due care should be taken to release the gas into the atmosphere. To this end, a petcock is provided on the top of the housing for the safe release of the gas into the atmosphere. Nowadays, gas analysis techniques for the proactive protection of a transformer are gaining popularity in conjunction with electrical means. For such an application, rubber tubing over the nozzle is used to collect gases for analysis. An important issue to be addressed during the mounting of relays is ensuring that any gas formed should travel up to the pipeline freely without causing much turbulence in the oil stream. An arrow is provided on the relay housing indicating the direction of the gas flow. As a matter of practice, the relay should be mounted such that the arrow points towards the conservator and is at an angle of 5°. To ensure linear gas flow, so that the relay sensitivity is not adversely affected, the pipe should be straight for at least six times the pipe diameter on the transformer side and three times the pipe diameter on the conservator side.

For large transformers consisting of separate radiators and forced circulation, oil pressure surges while starting the pump result in expansion of radiator tubes; thus, oil flows into the conservator pipe. The arrangement should be made in the relay to ensure its stability under this condition.

The oil is regularly cleaned by centrifuging or filtering to maintain its original characteristics. During this process, aeration takes place. The trapped air is released and collected by the Buchholz relay. It may cause false tripping of the relay; therefore, during this process only an alarm function is enabled while the tripping function is blocked. The blocking is achieved by disconnecting the tripping contacts. In general, this is left disconnected for about two days; hence, during this period, a mature decision must be taken whenever the alarm goes off. The utility engineers in such a situation generally collaborate their decision on the basis of electrical parameters.

Since the Buchholz relay operation is based on the formation of gas in the transformer tank, in many of the cases, it can detect faults that are otherwise not possible or difficult to be detected. Thus, the Buchholz relay is indispensable for the protection of a transformer. Some consider it as the main protection for the transformer while others consider it as supplementary protection, but either way transformer protection is not complete without the Buchholz relay. Literature has reported operating times of the Buchholz relay of the order of 0.05–0.1 s. In every case, electrical protection is provided for faster operation in the eventuality of a heavy fault. For the smaller transformers (generally less than 1000 KVA), some countries do not recommend the use of Buchholz relay. This is because in such transformers, only heavy faults can initiate gas formation, and for such faults, electrical relays will definitely operate faster. Thus, by the time the Buchholz relay can operate, major damage would have been done to the equipment.

6.3.2 Sudden Pressure Relay

The gas pressure relay facilitates the protection of a transformer merely in conjunction with a simple differential relay. This relay has to be made stable against the inrush current. In addition to the Buchholz relay, another gas-operated relay is used for the protection of the transformer. This relay operates in the eventuality of a sudden large formation of gas in the transformer tank. Owing to its operating characteristic, it is known as *sudden pressure relay* (SPR). The operation of this relay is based on the rate of rise of the gas in the transformer. Thus, it can be used for the protection of any transformer having a sealed gas chamber at the top of the tank. In general, to facilitate its maintenance, this relay is fastened to the manhole cover of the transformer tank. Care should be taken to mount the manhole above the oil level. Figure 6.9 shows the construction of gas-operated SPR. The design of this relay is such that it only operates

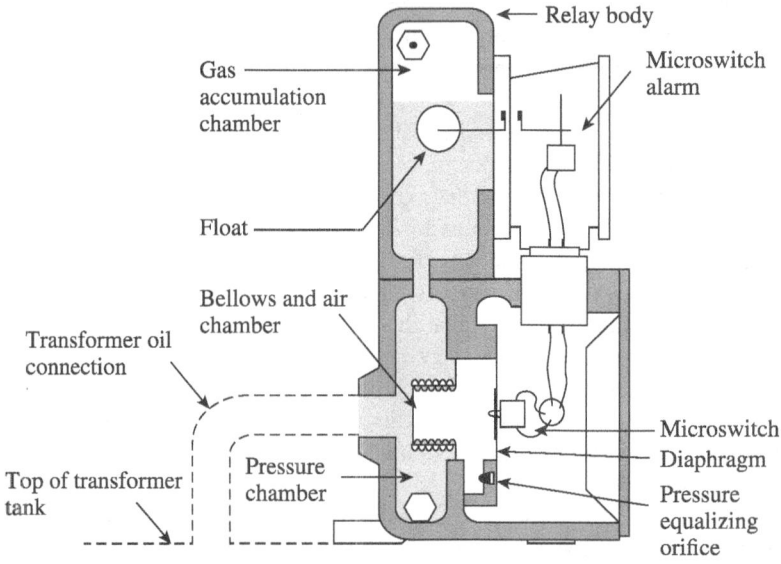

Fig. 6.9   Construction of gas-operated sudden pressure relay

on dynamic pressure changes; thus, under static pressure, it remains stable. Utility engineers recommend SPR for all transformers greater than a size of 5 MVA. In contrast to the Buchholz relay, which operates even for light internal faults, this relay operates only for heavy internal faults of the transformer. Although for light internal faults it is more sensitive than the differential relay, it cannot dispose off the differential relay. The reason is that the differential relay not only takes care of the transformer internal faults, but also provides protection against faults outside the transformer tank (e.g., bushing faults). The operating time of SPR varies over a large span depending on the rate of gas formation.

Utility engineers have experienced false operation of this relay; hence, there is reluctance in using it to energize a tripping coil, while there is no dispute about its use for alarm purposes. In general, protection schemes are designed in such a way that the SPR tripping threshold is set just below the cut in threshold of the differential relay; this can be ensured with a high-speed current blocking type relay such as a simple overcurrent relay to supervise the SPR trip.

6.3.3 Oil and Winding Temperature Relay

Transformers are designed to operate within a specified temperature range depending on the type of insulation being used within them. In order to take away heat from the transformer, cooling arrangements are provided as part of the design of the transformer tank. Any abnormal temperature rise of the winding is indicative of weakening of insulation either between the turns or from turn to the core. These hot spots should be carefully monitored and timely corrective measures should be taken so that they do not precipitate a major fault in the transformer.

For monitoring the formation of hot spots in transformers, resistance temperature detectors (RTDs) are embedded in the transformer windings. Since oil carries away heat from the transformer winding, some RTDs are also deployed for the monitoring of transformer oil temperature. The prevailing temperature information from all of these RTDs is given to a temperature scanner. In the normal course of action, this temperature scanner cyclically displays the temperature measured by each RTD. Their corresponding locations, in most of the cases, are decoded by the operator corresponding to the RTD number. This continuous monitoring is very useful in the detection of the hot spot and its approximate location.

The winding temperature can also be measured by inserting a small current transformer (CT) in series with the main winding of the transformer. The secondary of this CT is connected to the heaters inside the transformer tank. The sensing bulb (RTD) is surrounded by the heater and measures the temperature proportional to the current flow through the windings. The output leads of the sensing bulb are connected to the winding temperature inductor (WTI) and alarm/protective unit. Figure 6.10 shows the connection of the WTI with the transformer. When the temperature of any RTD (corresponding voltage input from the RTD) crosses the threshold value, an alarm is actuated. This alarm is used by the operator to take a decision regarding the continuation of service of the transformer under consideration. Furthermore, if the temperature continues to rise beyond the alarm limit, the signal is used to close the contacts of all or one of the auxiliary relays that are transferred to the master trip relay to disconnect the transformer.

Some relay engineers advise a threshold for temperature duration integration instead of a threshold for temperature. This philosophy

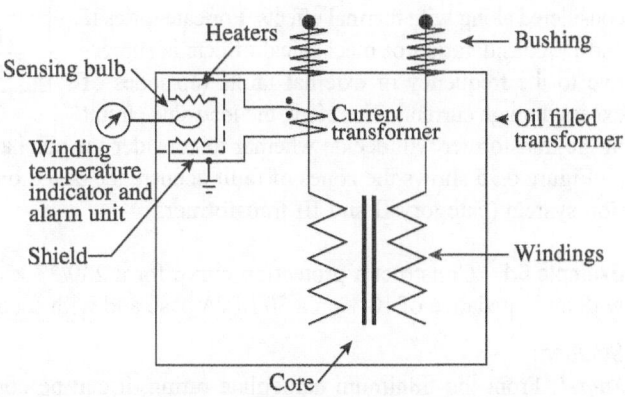

Fig. 6.10 Connection of WTI and alarm unit with transformer

rules out the isolation of transformers because of the transient temperature rise. Thus, the reliability of supply is improved.

6.4 Overcurrent Protection

For smaller transformers, generally, protection is achieved by the overcurrent relays. A transformer size of 100–500 kVA is protected by overcurrent relays. These relays are also used as backup protection for the large transformers.

In general, overcurrent relays are set above the normal rated current so that a short time overload is allowed by the relay. Since during low level internal fault the current may not exceed the pickup value of the relay, in such cases, the instantaneous overcurrent relay will let this fault continue indefinitely. To address this problem, inverse time overcurrent relays are used. Although in this case the operation will be delayed to allow short time overloading, in case of an internal fault, the relay will definitely operate. The operating time will depend on the level of the fault. An added advantage with the inverse time overcurrent relays is that they provide better coordination with all other relays in the system that overreaches. In general, these relays are set at 2–3 times the rating of the transformer. To provide primary protection against heavy internal faults, instantaneous overcurrent relays are still used. These relays are usually set at 1.25 times the maximum through-fault current. Care must be taken so that this setting is always greater than the maximum inrush current. Thus, in a nutshell, it can be concluded that primary protection is provided by instantaneous overcurrent relays, and secondary protection is provided by inverse time overcurrent relays. Separate overcurrent protection must be provided for tertiary windings.

With increasing size of power transformers during overloaded operation, not only electrical but mechanical forces are also generated. These mechanical forces are in the form of rise of winding temperature and movement of transformer coils. This movement of transformer coils results in insulation damage, which results in hot spot development. Therefore, a log must be maintained for such overloads and faults. Efforts should be made to isolate the transformer in minimum time by using proper protective devices to maximize the life of the transformer.

Table 6.1 provides different protection curves to be used depending on the size of the transformer as per ANSI/IEEE Standard C57.109-1985 (Figs 6.11–6.14).

For small transformers (category I), only thermal effects are taken into consideration, whereas for large transformers (category IV), mechanical effects are also considered along with thermal effects. For categories II and IV consideration of mechanical effects is subjective to the frequency of external faults (episodes of excess through current). Therefore, the load side circuit of the transformer will decide whether to consider the mechanical effects.

Table 6.1 Transformer category (ANSI/IEEE Standard C57.109-1985 curves)

Category	Minimum nameplate (kVA)		Reference protective curve
	Single-phase	Three-phase	
I	5–500	15–500	Fig. 6.8
II	501–1667	501–5000	Fig. 6.9
III	1668–10,000	5001–30,000	Fig. 6.10
IV	Above 10,000	Above 30,000	Fig. 6.11

Figure 6.15 shows the zones of fault occurrence based on the fault frequency for a typical radial distribution system (category II and III transformers).

Example 6.1 Construct a protection curve for a 230/33 kV, 30/50 MVA transformer (thermal/mechanical) with an impedance of 10% on a 30 MVA base and with secondary side overhead lines.

Solution:

Step 1: From the minimum nameplate rating, it can be concluded that this transformer is of category III. Therefore, the curve of Fig. 6.13 will be applicable.

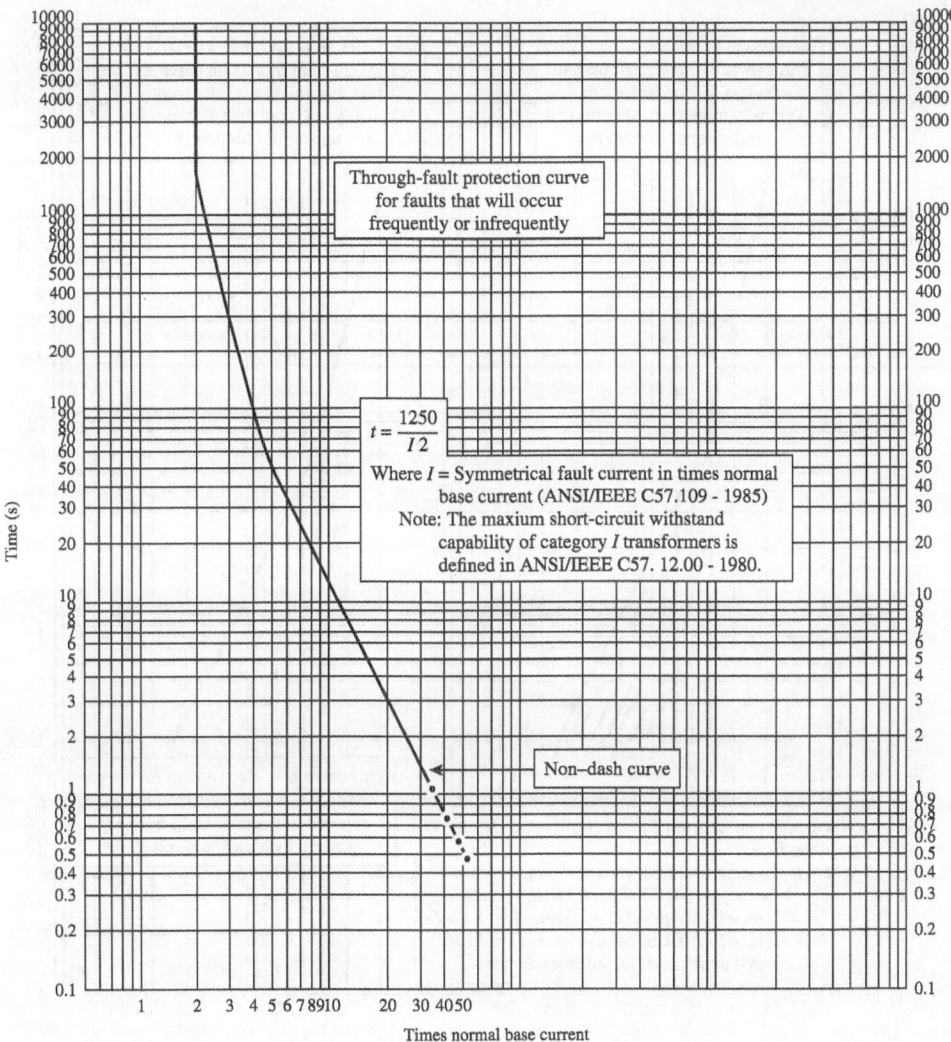

Through-fault protection curve
for faults that will occur
frequently or infrequently

$$t = \frac{1250}{I\,2}$$

Where I = Symmetrical fault current in times normal
base current (ANSI/IEEE C57.109 - 1985)
Note: The maxium short-circuit withstand
capability of category I transformers is
defined in ANSI/IEEE C57. 12.00 - 1980.

Non–dash curve

Times normal base current

Fig. 6.11 Through-fault protection curve for category I transformer 5–500 kVA single-phase, 15–500 kVA three-phase

Step 2: From Fig. 6.13, plot the infrequent through-fault curve of category III. This is shown by curve A in Fig. 6.16.

Step 3: Using the following procedure, determine the protection curve:

(a) Calculate the maximum per unit through-fault current:

$$I = (1/0.10) = 10 \times \text{base current at 2 s}$$

This is the point 1 in Fig. 6.16.

(b) Calculate the constant $K = I^2t$ at 2 s:

$$K = (1/0.10)^2 \times 2 = 200$$

(c) Calculate the time at 50% (note: use 50% for category III and IV, 70% for category II) of the maximum per unit through-fault current:

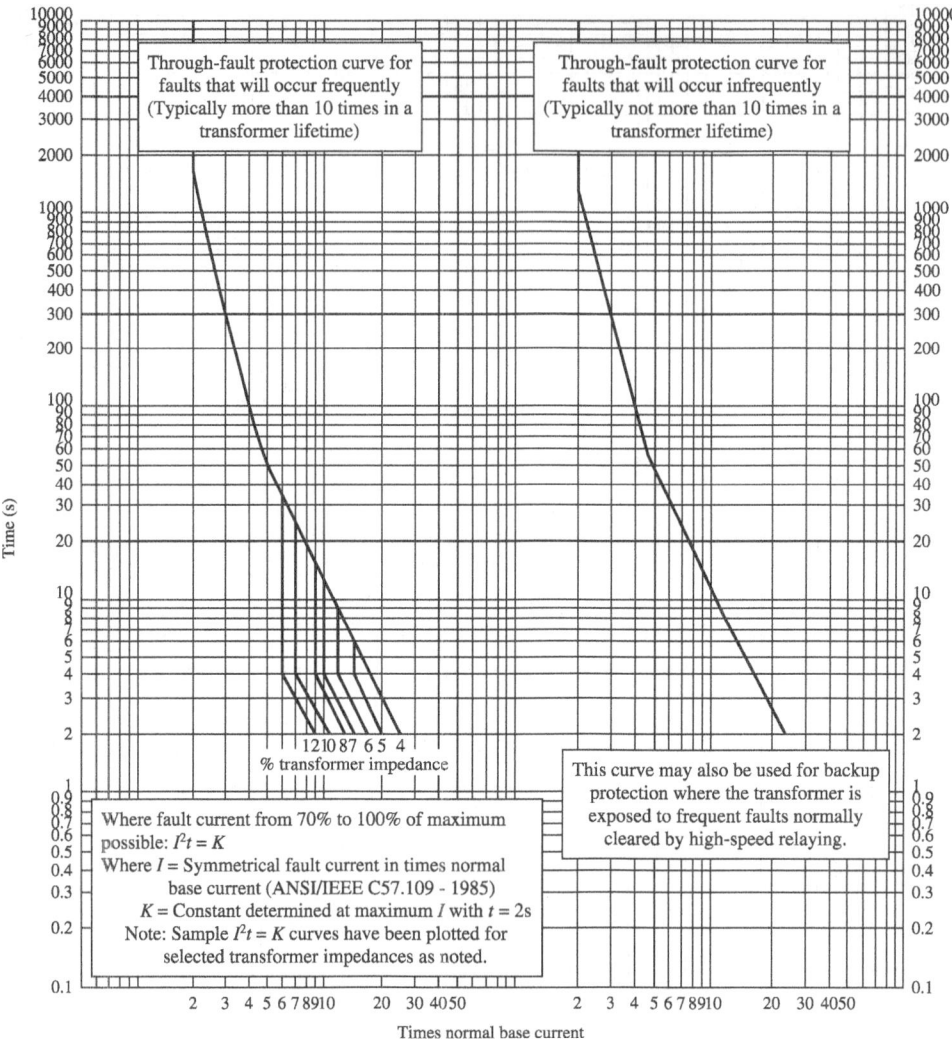

Fig. 6.12 Through-fault protection curve for category II transformer 501–1667 kVA single-phase, 501–5000 kVA three-phase

$$T = (K/I^2) = (200/[0.5 \times (10)]^2 = 8 \text{ s}$$

This is point 2 in Fig. 6.16.

(d) Connect points 1 and 2 and draw a vertical line from point 2 to the infrequent curve to complete the dog leg portion curve as shown in Fig. 6.16.

6.4.1 Overcurrent Relay with Harmonic Restrain Unit Supplement

Owing to the large value of inrush current, the transformer protection relays are blocked for the initial few cycles till the inrush settles down. During this period, no protection is available to the transformer; in case the transformer is switched during a fault, it will cause a heavy damage to the transformer as well as to the source side system. To overcome this, three single-phase harmonic restrain units (HRUs) with instantaneous

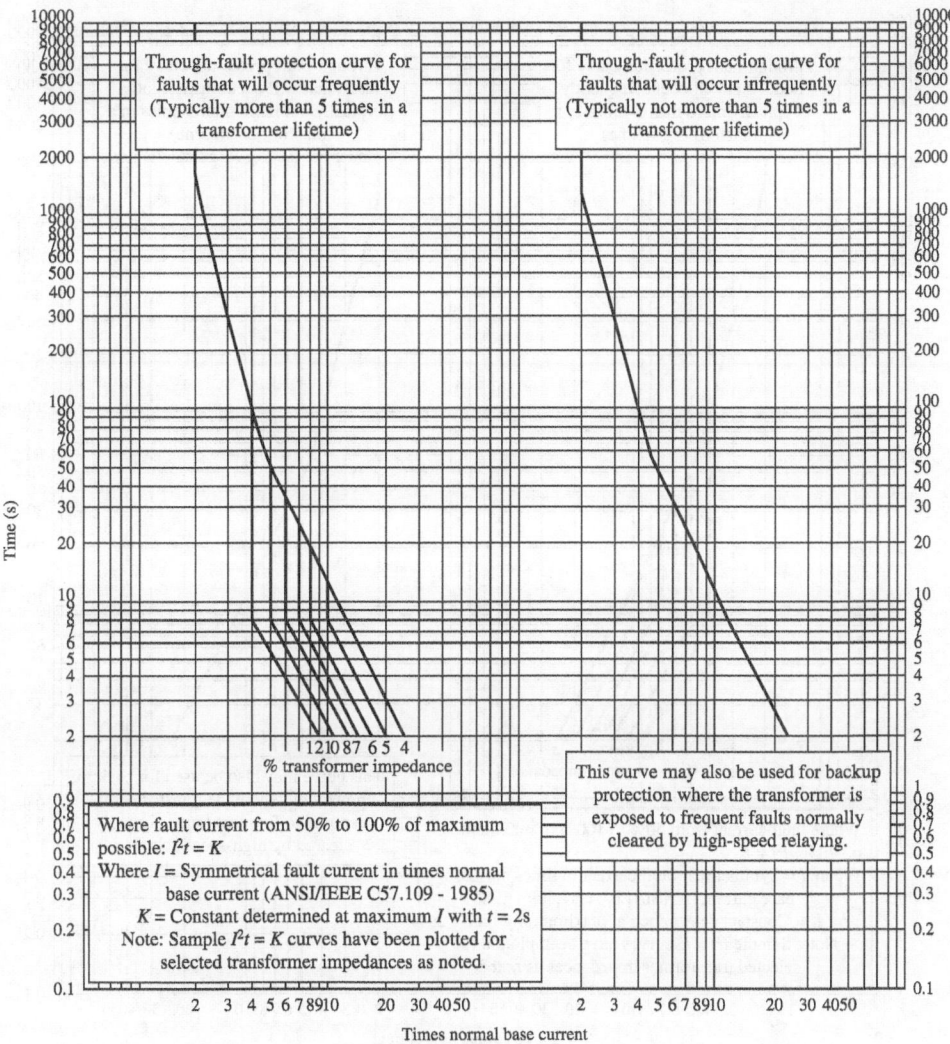

Fig. 6.13 Through-fault protection curve for category III transformer 1668–10,000 kVA single-phase, 5001–30,000 kVA three-phase

tripping arrangement are used to supplement the inverse time overcurrent relay. The arrangement is shown in Fig. 6.17. Thus, whenever switching in on fault occurs, these devices provide high-speed tripping. Care should be taken not to allow load pickups during switching in and to ensure that all the feeders originating from the transformer bus are monitored by local breakers.

6.5 Earthing/Grounding Transformer

Generally, at least one neutral point is available at every system voltage level starting from the generating station upto the consumer point. However, neutral point is not present in a delta-connected winding of three-phase power transformer. In case of earth fault which occurs in the part of the system where the neutral point is not available, the relay does not operate due to insufficient magnitude of fault current. Hence, artificial

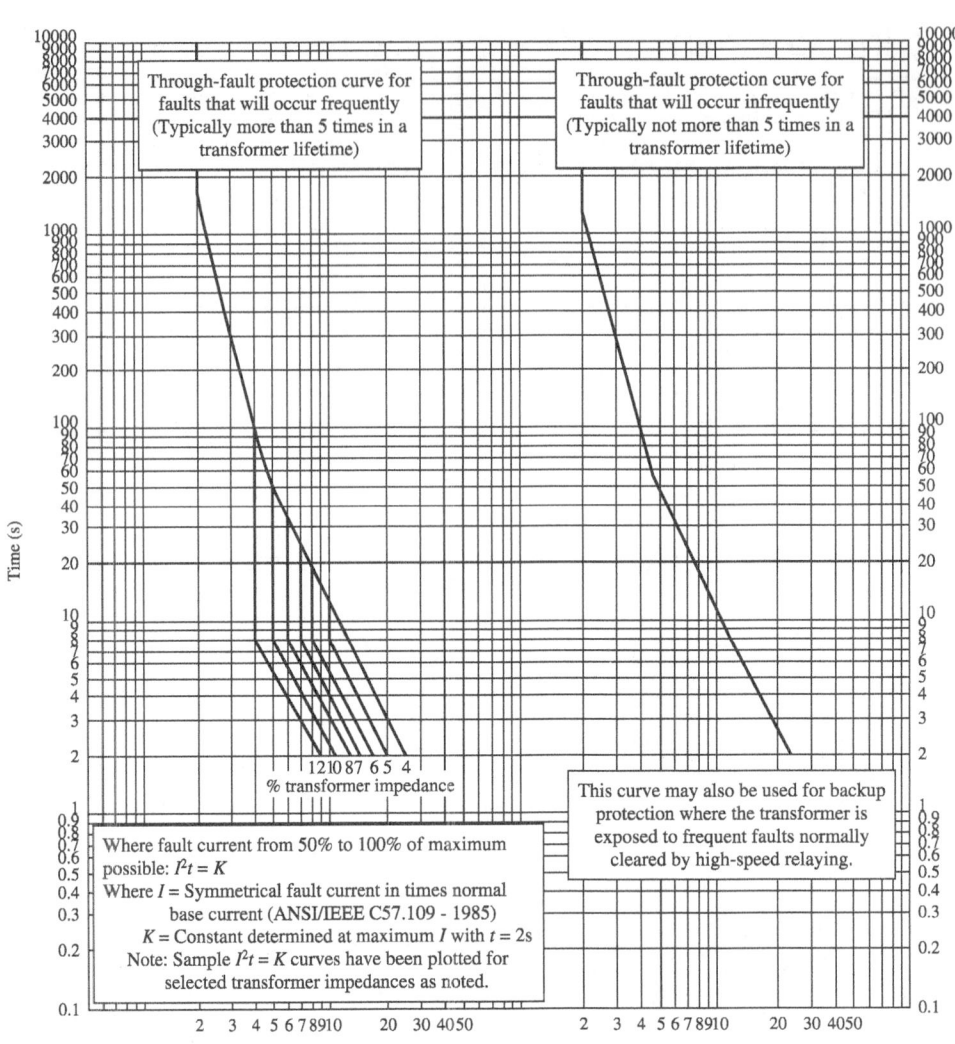

Through-fault protection curve for faults that will occur frequently (Typically more than 5 times in a transformer lifetime)

Through-fault protection curve for faults that will occur infrequently (Typically not more than 5 times in a transformer lifetime)

% transformer impedance

This curve may also be used for backup protection where the transformer is exposed to frequent faults normally cleared by high-speed relaying.

Where fault current from 50% to 100% of maximum possible: $I^2t = K$
Where I = Symmetrical fault current in times normal base current (ANSI/IEEE C57.109 - 1985)
K = Constant determined at maximum I with $t = 2s$
Note: Sample $I^2t = K$ curves have been plotted for selected transformer impedances as noted.

Time (s)

Times normal base current

Fig. 6.14 Through-fault protection curve for category IV transformer above 10,000 kVA single-phase, above 30,000 kVA three-phase

arrangement is done to provide a neutral point for grounding purpose using earthing transformer or grounding transformer. Grounding transformer provides easy return path to sense earth fault and it also limits transient overvoltages (due to arcing ground fault). Winding of earthing transformer is connected in zig-zag style which takes a very small magnetizing current during normal operations. In a zig-zag connection, each phase winding is divided in to two equal parts. One-half of the winding is wound on one limb and the remaining part is wound on the other limb of the core type transformer. One end of all three windings is connected together to form a neutral point. Thus, flux distributed in concentric connected coil opposes each other, and offers high impedance (takes a very small magnetizing current) during normal conditions. Conversely, it provides very low impedance to zero sequence current during earth fault by cancelling the ampere turns on the same limb. Fig. 6.18 shows L-g fault on B-phase of a delta-connected system in the presence of a grounding transformer.

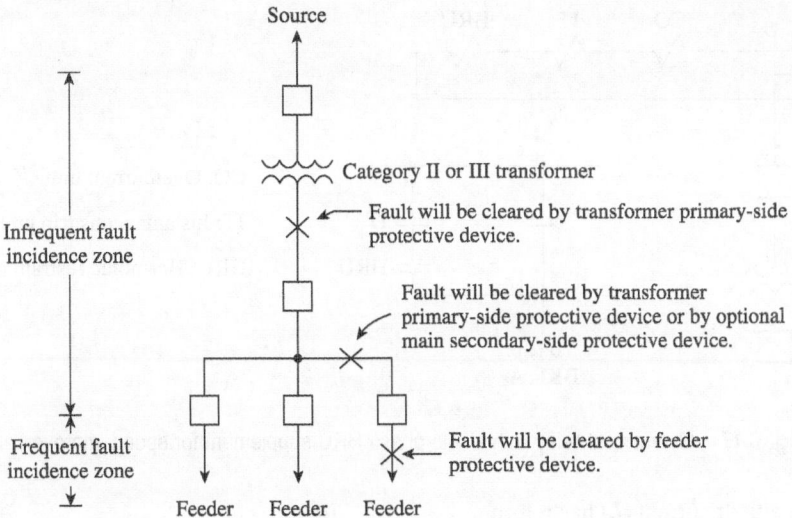

Fig. 6.15　Infrequent–frequent fault incidence zones for category II and category III transformers

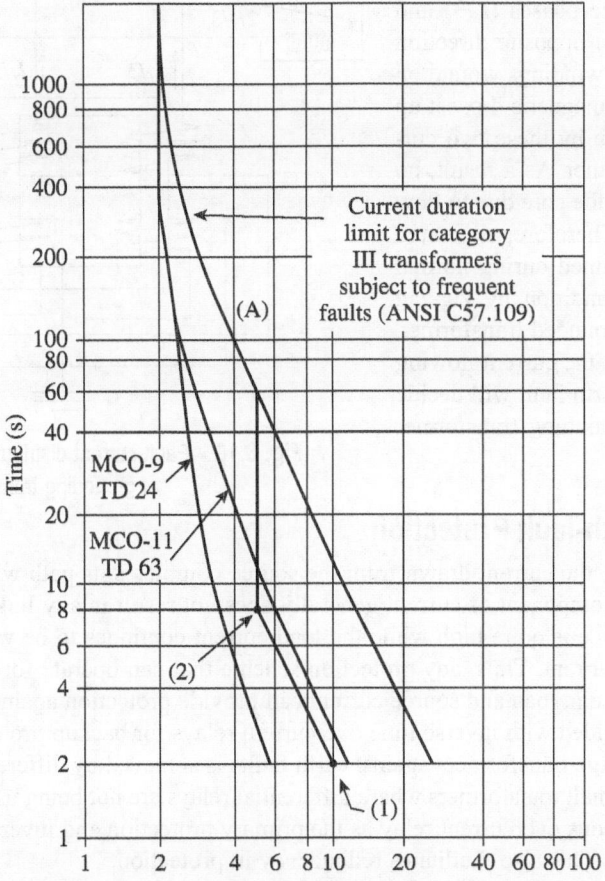

Fig. 6.16　Multiple of transformer full load (per unit)

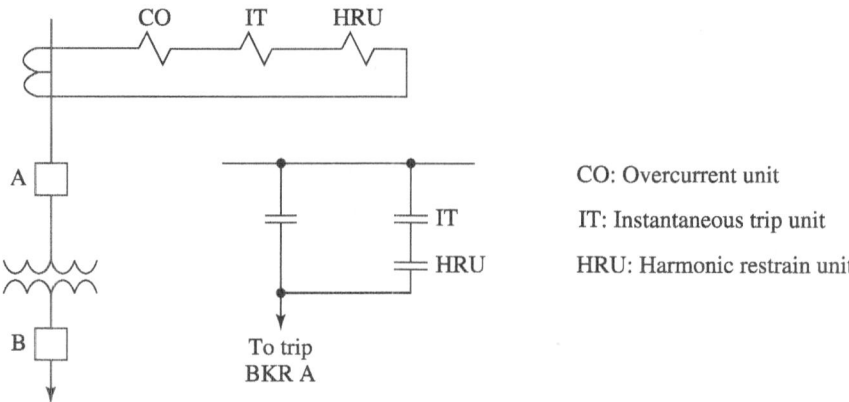

Fig. 6.17 Overcurrent relay with single-phase HRU supplement for speed improvement

The zero sequence fault current (I_f) has return path through earth and grounding transformer. As shown in Fig. 6.18, the current I_f divides uniformly in all the three phases ($I_f/3$) and these currents flow in the opposite direction in two different bisected windings wound on the same limb. Thus, the magnetic flux set up on the same limb of core by these two currents will oppose each other. As a result, no additional flux retains in the core due to flow of earth-fault current. Therefore, the rated supply voltage is maintained during normal as well as earth-fault condition by zig-zag or interconnected star grounded transformer. For a given definite time, the current flowing through neutral during earth fault will decide the current rating of grounding transformer in ampere.

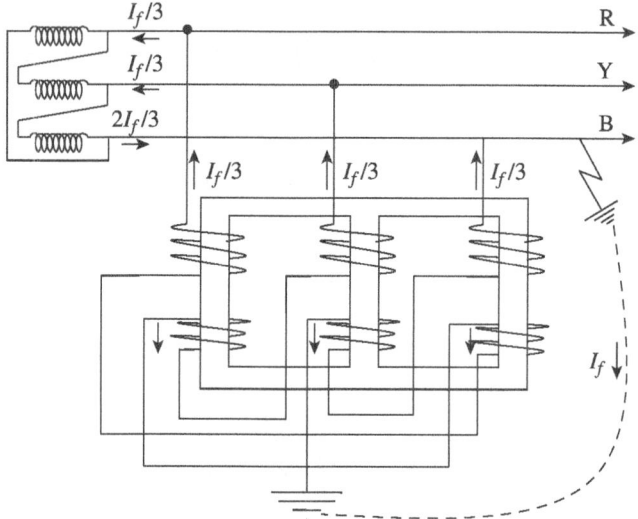

Fig. 6.18 Fault-current distribution in zig-zag winding of grounding transformer

6.6 Restricted Earth-fault Protection

In the case of earth fault, the current drawn from the source completes its path without contributing to the load current. Thus, this component of current generally does not result in any linkage flux; hence, the current drawn from the source is quite high while the load current continues to be within normal limits. This situation results in overcurrent. Thus, any protection scheme that can operate for abnormally high current or for abnormal difference in load and source current can provide protection against earth fault. In practice, the transformers are provided with inverse time overcurrent relays for backup protection against earth fault. Primary protection for large transformers against earth faults is achieved by differential relays (discussed in Section 6.8). In case of small transformers where differential relays are not being used, earth-fault protection is provided by instantaneous overcurrent relay as the primary protection and inverse time overcurrent relay as the backup protection. This also facilitates redundancy in protection.

As such, for star connected primary winding, a simple overcurrent earth-fault relay will not provide proper protection, especially if the neutral is earthed through neutral grounding impedance. The situation can be handled more efficiently with the use of restricted earth-fault (REF) relay. A single line diagram employing REF relay is shown in Fig. 6.19. In this case, three line side CTs are connected in parallel with the same polarity; thus, only residual current will be available at the output. This residual current is balanced against the output of neutral CT current. The operating element is a high impendence relay, and the region of operation is the star winding of the transformer. The quality of protection is enhanced not only because of the use of the instantaneous element, but also because the entire fault current is used as the operating quantity. Thus,

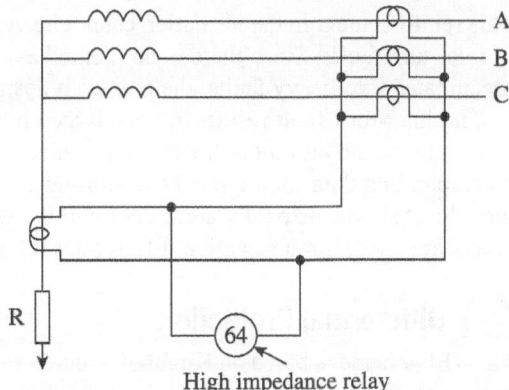

Fig. 6.19 Restricted earth-fault protection for a star winding

the scheme is capable of detecting faults closer to the neutral end of the winding. In this case the level of fault current decreases. This arrangement can provide protection to a significant part of the winding. Since the operating quantity is residual current, the relay will remain stable for any fault outside its zone.

The scheme can be equally applied for protection of the transformer with solid grounding of neutral. In this case, because of the absence of neutral grounding impedance, the current is quite large. Since neutral current is also measured and used as restraining quantity, it is possible to provide protection cover to the entire winding. The delta connected system, since no zero-sequence components can be transformed to secondary circuit, is inherently restricted. A high impedance relay can be used to provide high-speed protection for winding faults and stability for outside faults.

The REF scheme is quite simpler and is used independently on either side of the transformer for providing high-speed earth-fault protection.

6.7 Inter-turn Fault

As we have already discussed in Section 6.4 for a low voltage transformer, it is very unlikely that insulation breakdown will occur until and unless a heavy external fault has caused mechanical forces to develop. This will cause damage to the insulation. Another situation is when an inter-turn fault may occur when moisture is admitted to the transformer coil either due to oversaturation of silica gel or missing silica gel.

From the preceding discussion, it is clear that a high voltage transformer connected to an over-head transmission line is prone to skip impulse voltages, which are many times the rated voltage. These surges, due to their high equivalent frequency, most seriously affect the end turns of the winding. To overcome this, better inter-turn insulation at the end of the winding is provided.

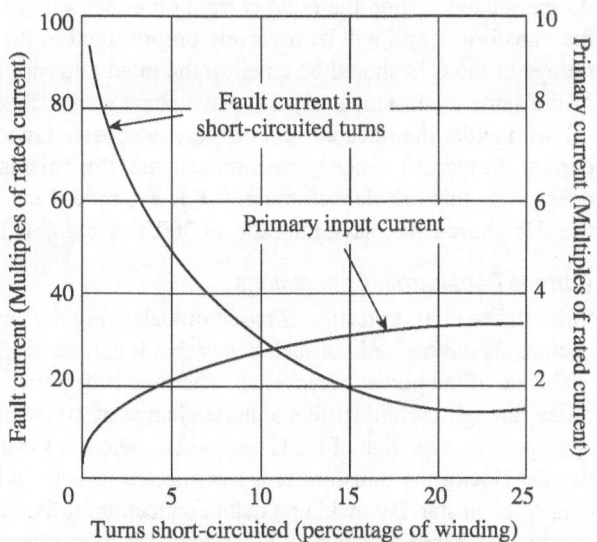

Fig. 6.20 Inter-turn fault current/number of turns short-circuited

This reinforcement in the insulation cannot be very high; thus, maximum transformer failures occur because of inter-turn faults. Since there is no method available to detect the failures during their inception, they get precipitated into heavy faults, thereby even destroying the location of their actual occurrence.

The inter-turn fault results in very large circulating currents at the local level. Owing to the transformation ratio, the current at the transformer terminal will be very small. The graph in Fig. 6.20 shows the corresponding data for a typical transformer of 3.25% normal through impedance with the short-circuited turns located symmetrically at the centre of the winding. In general, for such faults, thermal or gas-operated relays are used, but a sensitive differential relaying scheme can also be used for this purpose.

6.8 Differential Protection

The REF scheme is based on Kirchhoff's current law, that is, the sum of all the incoming and outgoing currents is equal to zero. Since the transformer is a high efficiency device, a differential scheme can be designed to protect the entire transformer. In this case, barring some small difference, the volt-ampere on either side of the transformer is equal. Thus, if the CTs on either side are properly chosen, the difference in current will be almost zero. The CTs on either side are collected to form a circulating current system. Figure 6.21 shows the basic single line diagram to illustrate the principle of transformer differential protection.

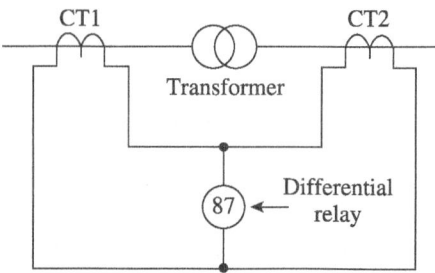

Fig. 6.21 Principle of transformer differential protection

6.8.1 Basic Considerations for Transformer Differential Protection

Before applying differential protection philosophy to the transformer, one has to consider several factors. These factors are discussed in the following subsections.

Line Current Transformer Primary Ratings

As mentioned earlier, the rated current on either side of the transformer will depend on the MVA rating of the transformer and will be inversely proportional to the corresponding voltages. Thus, ideally, the primary ratings of the CTs should be equal to the rated current on either side, whereas the secondary current can be of the same level. However, CTs have some standard primary and secondary current ratings. Therefore, a CT with either the rated current or more is chosen. Owing to this fact, there is a mismatch in the secondary current. In the difference protection scheme, this mismatch is to be taken care of and the relay should not operate for this unbalanced current. For example, for a 100 MVA, 220 kV/132 kV, delta/star transformer, the CTs chosen will have the ratio of 300/1 A and 450/1 A.

Current Transformer Connections

Since differential protection is based on balancing the currents on either side of the transformer, the CT connection should be made in such a way that it compensates the phase differences between the line currents on each side of the power transformer wherever it becomes necessary. As shown in Fig. 6.22, a balanced three-phase through-current suffers a phase change of 30°, which must be corrected in the CT secondary leads by appropriate connection of the CT secondary windings if the transformer is connected in delta/star. The CTs on the star side of the transformer are connected in delta, whereas on the delta side of the transformer, they are connected in star. By making a delta connection of the CTs on the star side, a phase shift of the required 30° is achieved. Moreover, this will eliminate the zero-sequence current on the star side while the zero-sequence component on the delta side will not produce current outside the delta on the other side. This is a universal rule; the CTs on both sides would need to be connected in delta if the transformer were connected star/star.

To provide current balance with the secondary current of the star connected transformers, the secondary ratings of the delta connected CTs must be reduced to $1/\sqrt{3}$ times the secondary rating of the star connected CTs. When the line CT ratios provide adequate matching between the currents supplied to the differential relay under through-load and through-fault conditions, the necessary phase shift can be obtained by suitable connection of the line CTs. Figure 6.23 shows the required connections for various power transformer winding arrangements. When the delta connected CTs are required, it is a common practice to use the star connected line CTs and to obtain the delta connection by means of star/delta interposing CTs (ICTs), as exemplified in the section on ICTs.

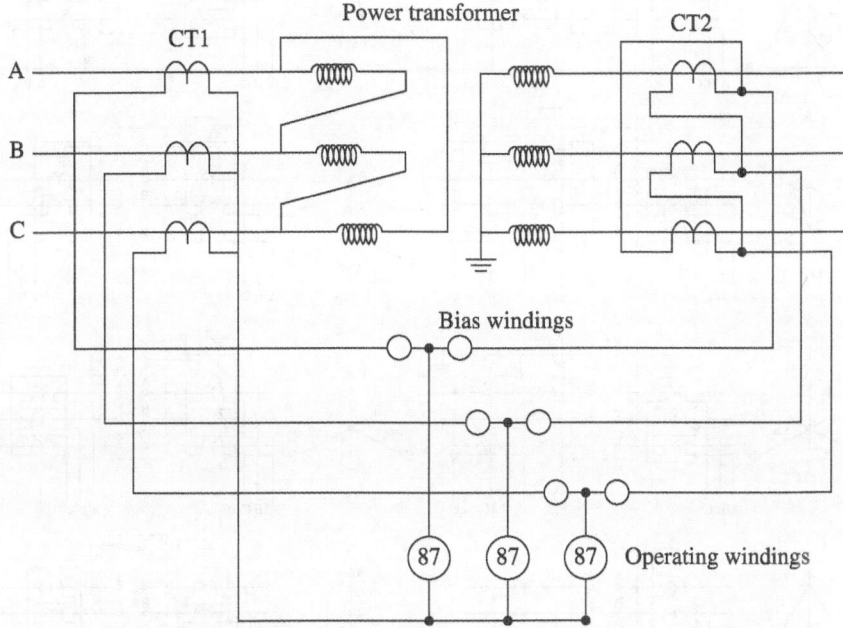

Fig. 6.22 Biased differential protection for two-winding delta/star transformer

Note on relay connections

1. The connection 'k' or 'I' is omitted when the corresponding CTs are delta connected. When both sets of CTs are delta connected, 'k', 'I', and 'm' are omitted.
2. It is essential that the CT connections are earthed at one point only.
3. When the power transformer to be protected has three windings, it may be helpful to first consider the relative connections for the CTs associated with the windings 1 and 2, and then both with windings 1 and 3. As a final check, the relation between windings 2 and 3 of the CTs may be considered.
4. The power transformer vector group references correspond to those specified in IEC 76: 1967 and BS 171: 1970.

Bias to Cover Tap Changing Facility and Current Transformer Mismatch

In most of the cases, tap changing transformers are used, and thus, the CTs can be chosen on the basis of a fixed tap (nominal tap). Transformer operation on any tap other than this will result in mismatch of the CT secondary current. This mismatch of current will be over and above the mismatch due to the use of standard ratio CTs. This unbalanced current may be sufficient to cause the differential relay to operate. In practice, the

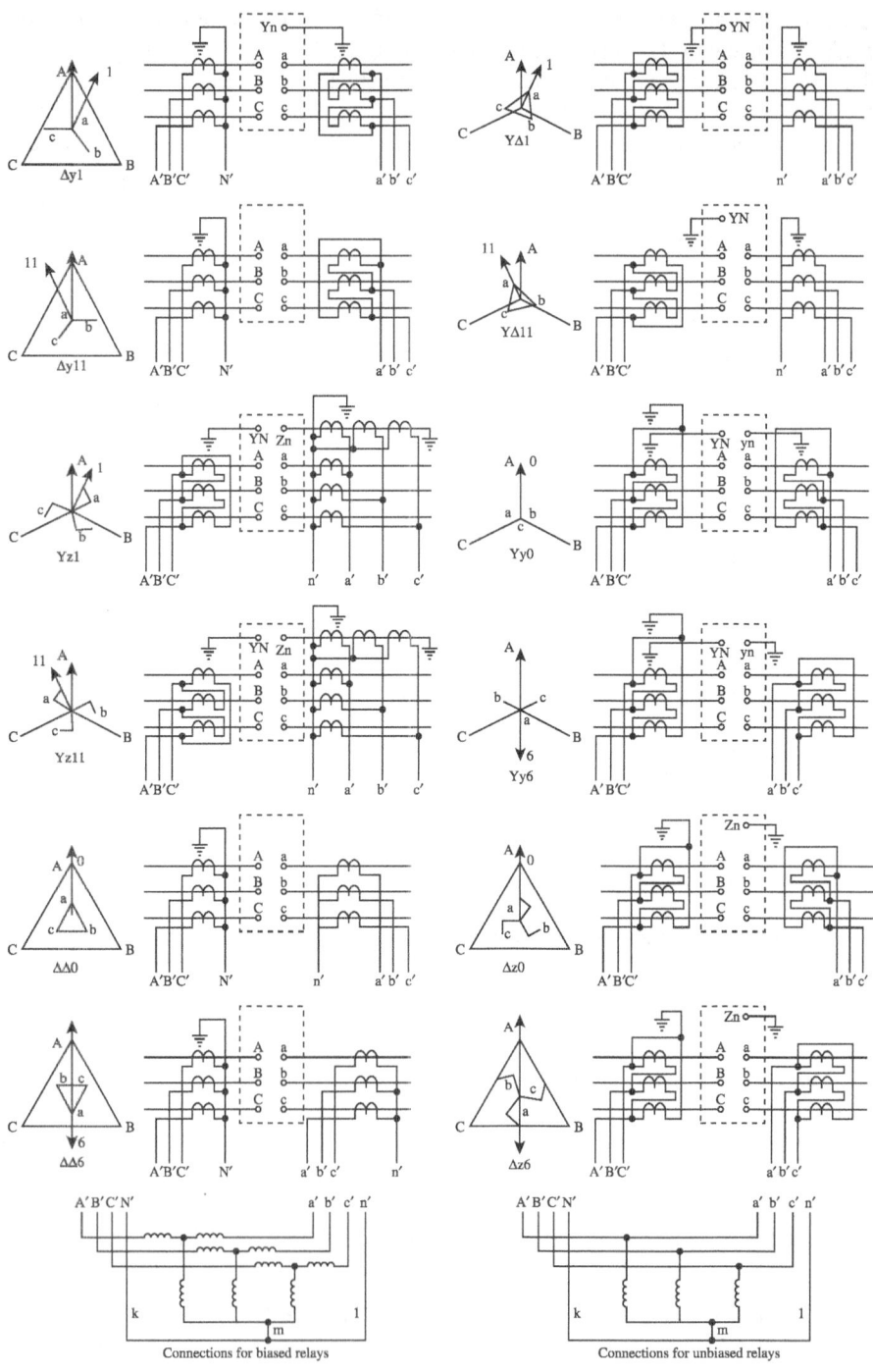

Fig. 6.23 Current transformer connections for power transformers at various vector groups

mean tap is taken as the nominal tap. The differential protection scheme is provided a bias such that the relay does not operate even for the maximum CT mismatch resulting from tap changing. This bias stabilizes the relay operation due to the CT mismatch and continues to provide good sensitivity under through-fault conditions.

Figure 6.24 shows a typical bias characteristics for differential protection of the transformer. An increase in restraining quantity (through bias current), also results in an increase in the operating quantity (differential current). Since the increase in bias makes the relay less sensitive, care must be taken that the current inputs to the differential relay should match at the midpoint of the tap range.

Figure 6.18 shows the percentage bias differential protection scheme for a two-winding transformer. The two bias windings per phase

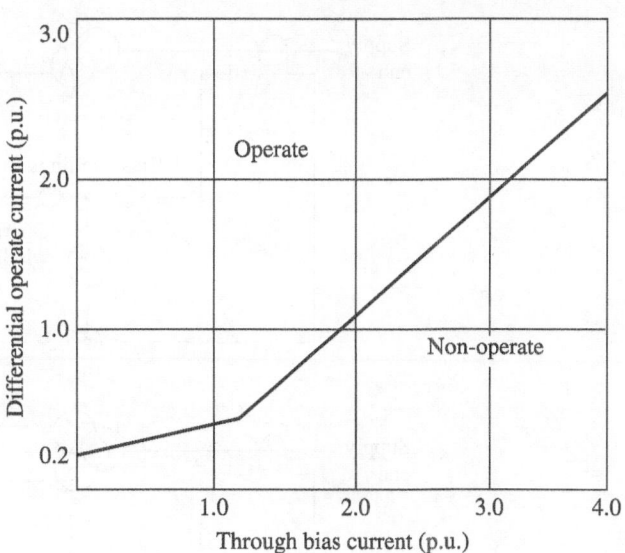

Fig. 6.24 Typical bias characteristics

are commonly provided on the same core. The bias differential protection scheme can be used for the protection of the transformer with three or more windings. This is because the Merz-Price principle remains valid for a system with more than two connections. Similarly, the bias differential protection scheme can also be used for protection of the transformer with two secondary windings at different voltage levels. The connection diagram for this situation is shown in Fig. 6.25(a). In this case, the two secondary side CT currents will balance out the current drawn from the supply (the infeed current). In the event of more than one winding being connected to the source, the current in the individual windings cannot be estimated, and in some cases it may result in a circulating current between the two CTs without producing any bias. In all such conditions, bias is obtained separately from each set of the CTs. The two bias windings are arranged in such a way that the total bias is obtained by the scalar addition of individual bias magnitudes in place of vector addition. The scheme is shown in Fig. 6.25(b). For the delta-connected tertiary winding, when no connection is brought out, these considerations are not required and the scheme as shown in Fig. 6.25(c) is used.

Interposing Current Transformers to Compensate for Mismatch of Line Current Transformers

Under ideal conditions, the relay through-currents should be perfectly matched. However, because of the availability of CTs for specific ratios, there is always some mismatch in the two relay currents. If this gap is not much, the bias can take care of this and the relay remains stable against this mismatch. However, the situation is more serious once the gap in the two currents is more either because of the standard CT ratios or because of the tap changing transformer. For the tap changing transformer, the CTs are selected on the basis of the nominal tap, which is generally a central tap. For any other tap position, the CT mismatch will be more. To address this, the bias should be either dynamically changed on the basis of the operating tap or set corresponding to the worst case scenario. The higher bias will adversely affect the relay sensitivity of the transformer operating at other taps.

To address these issues, interposing CTs (ICTs) are used. Interposing CTs are used to match the relay currents under through-load conditions corresponding to the ratings of the transformer. For a tap changing transformer, this is achieved corresponding to the nominal tap. For any other tap position, the bias can take care of CT mismatch.

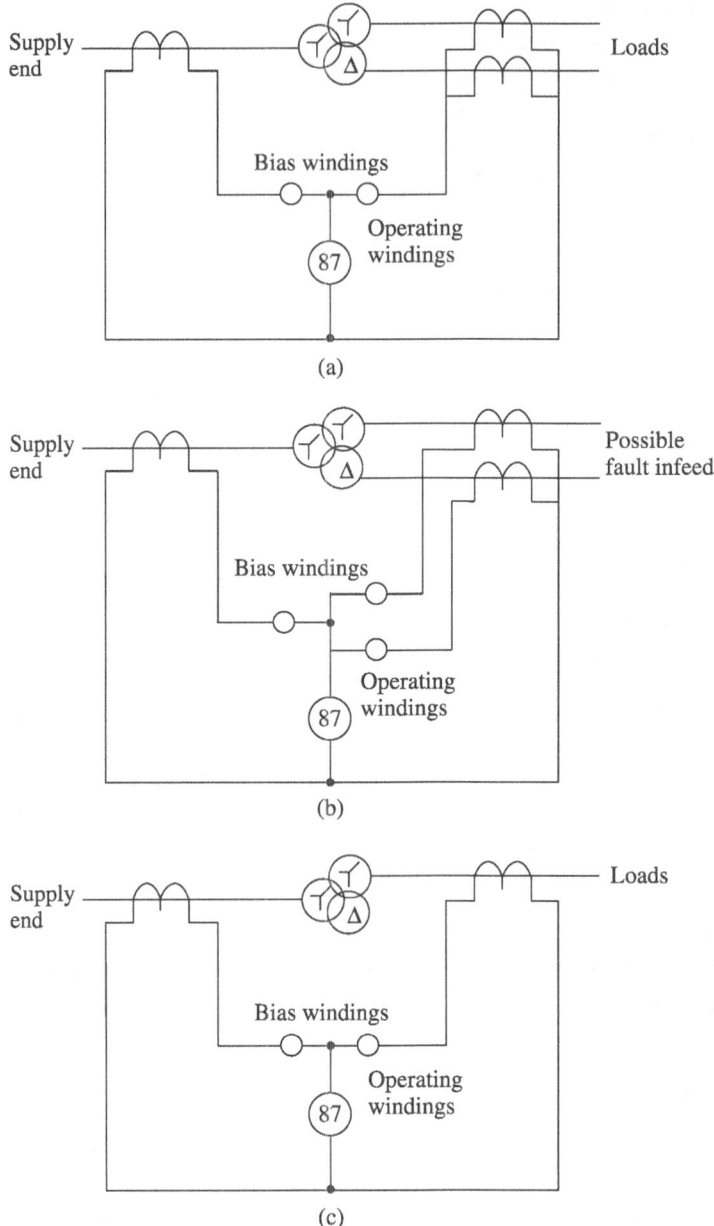

Fig. 6.25 Biased differential protection arrangements, shown as single-phase diagrams for simplicity (a) Three-winding transformers (one power source) (b) Three-winding transformers (three power sources) (c) Three-winding transformers with unloaded tertiary delta

Example 6.2 Find the ratio of the star/delta ICTs to achieve ideal matching of CT secondaries in case of differential protection of an 11 kV/132 kV, 30 MVA, delta/star transformer as shown in Fig. 6.26.

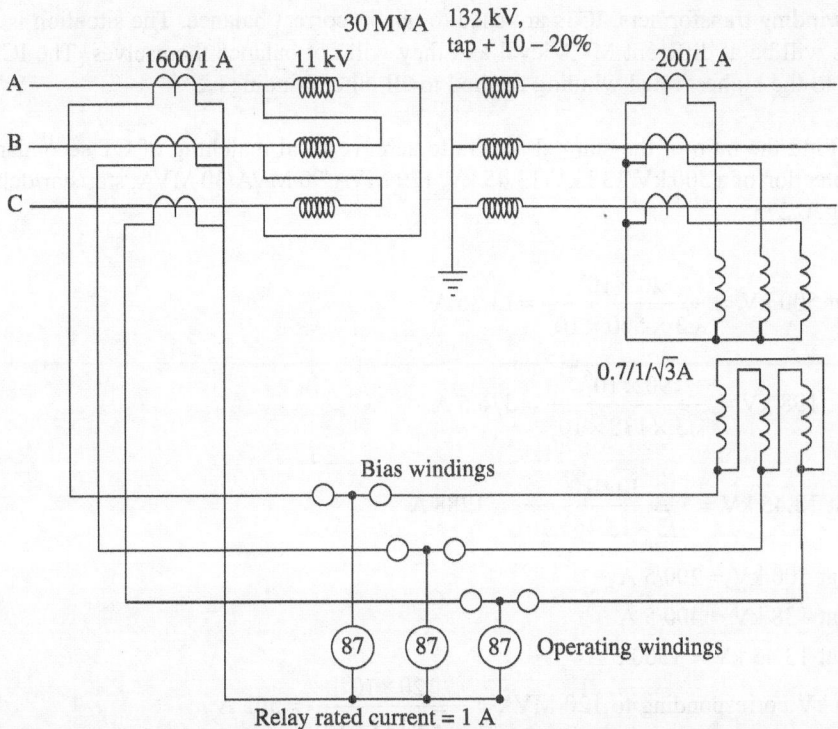

Fig. 6.26 Example of differential protection applied to a two-winding transformer showing the use of interposing current transformers

Solution:

First, the primary ratings of 1600 A and 200 A chosen for the main CTs should not be less than the maximum full load currents in each winding. These are derived as follows:

$$\frac{30 \times 10^6}{\sqrt{3} \times 11 \times 10^3} = 1575 \text{ A} \text{ for the 11 kV winding and}$$

$$\frac{30 \times 10^6}{\sqrt{3} \times 0.8 \times 132 \times 10^3} = 164 \text{ A} \text{ for the 132 kV winding, considering 20% tap.}$$

For the 11 kV winding, this is also the normal full load current, but for the 132 kV winding with 5% tap, the latter is given as

$$\frac{30 \times 10^6}{\sqrt{3} \times 0.95 \times 132 \times 10^3} = 138 \text{ A}$$

Equivalent secondary currents in the line CTs are 0.984 A and 0.69 A. Thus, the ratio of the star/delta ICTs to achieve ideal matching is given by

$$\frac{0.69}{0.984/\sqrt{3}} = \frac{0.70}{1/\sqrt{3}} \text{ or } 0.70/0.577 \text{ A}$$

For the three-winding transformers, ICTs are used to obtain correct balance. The situation is a bit tricky as three windings will be at different MVA level and they will not balance themselves. The ICTs are chosen corresponding to the highest rated winding applied to all other windings.

Example 6.3   Find the ratio of the star/delta ICTs to achieve ideal matching of CT secondaries in case of differential protection of a 500 kV/138 kV/13.45 kV, 120 MVA/90 MVA/30 MVA, star/star/delta transformer as shown in Fig. 6.27.

Solution:

Load current at 500 kV = $\dfrac{120 \times 10^6}{\sqrt{3} \times 500 \times 10^3}$ = 138.6 A

Load current at 138 kV = $\dfrac{90 \times 10^6}{\sqrt{3} \times 138 \times 10^3}$ = 376.5 A

Load current at 13.45 kV = $\dfrac{30 \times 10^6}{\sqrt{3} \times 13.45 \times 10^3}$ = 1288 A

Line CT ratio at 500 kV = 200/5 A

Line CT ratio at 138 kV = 400/5 A

Line CT ratio at 13.45 kV = 1500/5 A

Current at 138 kV corresponding to 120 MVA = $\dfrac{120 \times 10^6}{\sqrt{3} \times 138 \times 10^3}$ = 502 A

Current at 13.45 kV corresponding to 120 MVA = $\dfrac{120 \times 10^6}{\sqrt{3} \times 13.45 \times 10^3}$ = 5151 A

Secondary current from 500 kV line CTs corresponding to 120 MVA = $\dfrac{138.6 \times 5}{200}$ = 3.46 A

Therefore, the ratio of the required star/delta ICTs = $3.46 \Big/ \dfrac{5}{\sqrt{3}}$ A or 3.46/2.89 A

Secondary current from 138 kV line CTs corresponding to 120 MVA = $\dfrac{502 \times 5}{400}$ = 6.28 A

Therefore, the ratio of the required star/delta ICTs = $6.28 \Big/ \dfrac{5}{\sqrt{3}}$ A or 6.28/2.89 A

Secondary current from 13.45 kV line CTs corresponding to 120 MVA = $\dfrac{5151 \times 5}{1500}$ = 17.17 A

Therefore, the ratio of the required star/delta ICTs = 17.17/5 A

Under full load conditions of 30 MVA for the 13.45 kV delta winding, the current appearing in the primary of the 17.17/5 A ICT will be only 4.29 A, with the corresponding secondary current being 1.25 A. However, the ratios of the primary and secondary windings should ideally be 17.17 A and 5 A, respectively, to minimize winding resistances.

Example 6.4   Draw a detailed protection scheme for biased differential protection of a 66/132 KV, 250 MVA, DY1 power transformer. Suggest suitable CT ratios. In addition, suggest the proper ICT for the scheme.

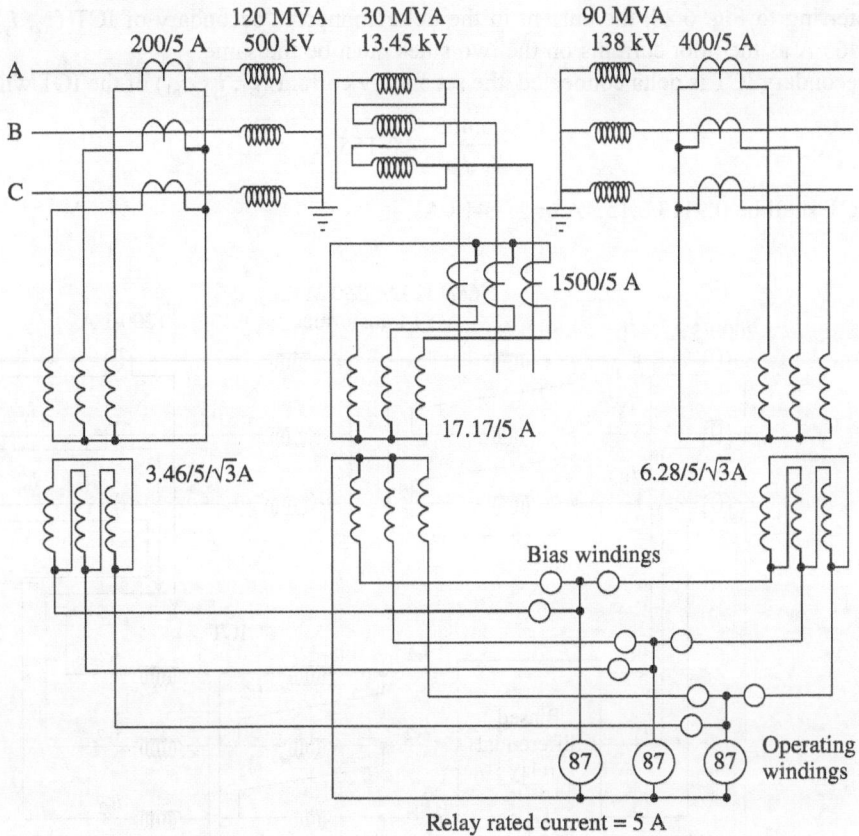

Fig. 6.27 Example of differential protection applied to a three-winding transformer showing the use of interposing current transformers

Solution:

Figure 6.28 shows the detailed biased differential protection scheme of the given power transformer. The rated full load current on the primary of the power transformer is given by

$$I_p = \frac{250 \text{ MVA}}{\sqrt{3} \times 66 \text{ kV}} = 2186.933 \text{ A}$$

Hence, the CTs on the primary side shall have the ratio 2000/5 A.

Now, the secondary full load current is given by

$$I_s = 2186.933 \times \frac{66}{132} = 1093.5 \text{ A}$$

Hence, the CTs on the secondary side will have the ratio 1200/1 A.

Now, the CTs on both sides are star connected.

The CT secondary equivalent of primary and secondary full load currents calculated here will be as follows:

$$i_p = 5.465 \text{ A}$$

$$i_s = 0.9113 \text{ A}$$

Now, referring to Fig. 6.28, the current in the delta connected secondary of ICT (i_{r2}, i_{y2}, i_{b2}) has to be equal to 5.465 A as the pilot currents on the two sides must be the same.

As the secondary ICT is delta connected, the secondary current (i_{r1}, i_{y1}, i_{b1}) of the ICT will be as follows:

$$\frac{5.465}{\sqrt{3}} = 3.155 \text{ A}$$

Thus, the ICT shall be 0.9113/3.155 A or 1:3.418 A.

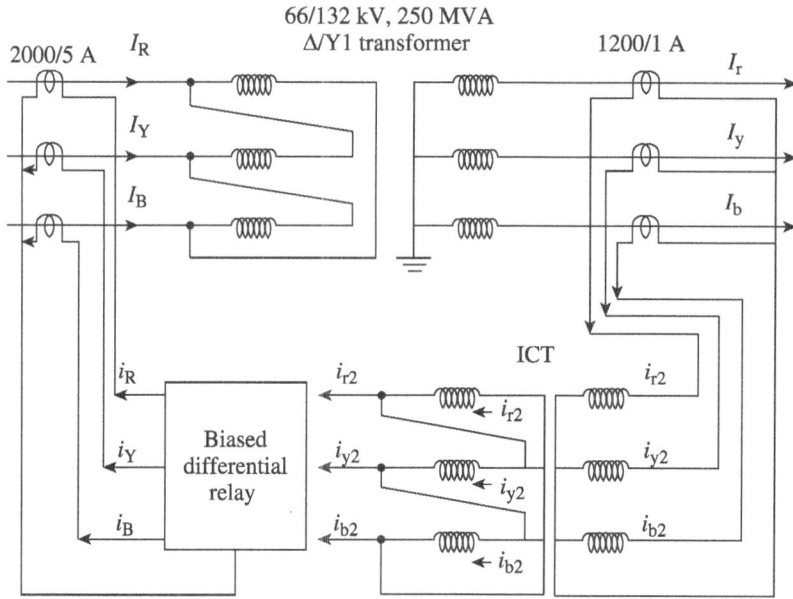

Fig. 6.28 Detailed biased differential protection scheme
of power transformer for Example 6.4

Example 6.5 A 33/6.6 kV, three-phase, 50 Hz, Y/Δ1 connected transformer is protected by a differential scheme. Draw the detailed connection diagram for the differential protection of the transformer. If the CTs employed on the high tension (HT) side have the ratio of 6.93:1 A, determine the CT ratio on the low tension (LT) side. Let the current on LT lines be 300 A.

Solution:

Figure 6.29 shows the detailed biased differential protection scheme of a given power transformer.

On LT lines, current = 300 A.

The current on HT lines of the transformer is given by,

$$\frac{6.6 \times 300}{33} = 60 \text{ A}$$

Hence, the HT line CT ratio = 6.93:1.

The secondary current of the CT on the HT line for 60 A line current will be

$$60/6.93 = 8.66 \text{ A}$$

Now, the CTs on the HT side are delta connected; hence, their secondary current will be $\frac{1}{\sqrt{3}}$ times the secondary current of the CTs on the LT side, which are star connected.

Thus, the current on the secondary of the CTs on the LT side will be $\frac{8.66}{\sqrt{3}} = 5$ A.

The ratio of the CTs on the LT side will be 300:5 A.

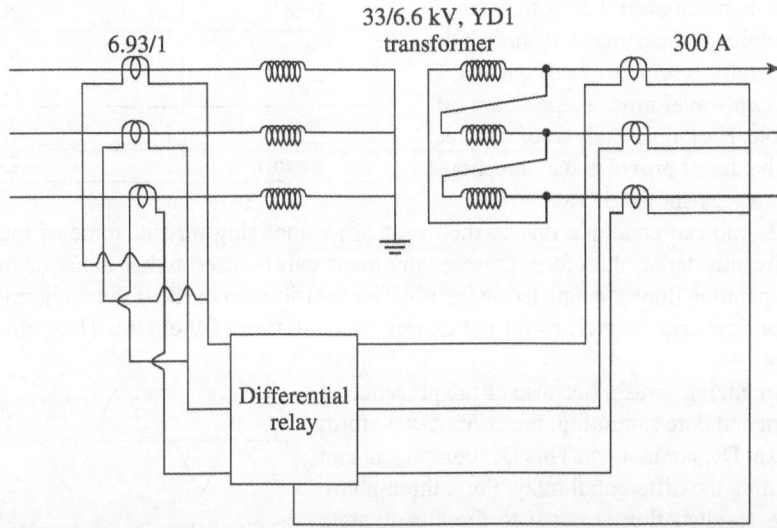

Fig. 6.29 Detailed biased differential protection scheme of power transformer for Example 6.5

6.8.2 Stabilization of Differential Protection during Magnetizing Inrush Conditions

When a transformer is energized or recovers from a fault, there is a large change in voltage. Since there is no back electromotive force in the transformer corresponding to this voltage, large current flows into the transformer. Corresponding to this current, there is no equivalent to the other side winding. Because of this, the entire current drawn from the source appears as an unbalanced current to the differential relay. Since this magnitude will be too large to be handled by a normal bias, the relay will operate. To avoid this situation, if the bias value is set too high, the protection will be of no use. To overcome this, either time delay or harmonic restrain is used.

Time Delay

Post energization, the heavy inrush current causes a voltage to get induced in the source winding of the transformer, which in turn will gradually reduce the inrush current. Therefore, the magnetizing inrush is a transient phenomenon lasting only a few cycles. To provide stabilization against the magnetizing inrush, the easiest approach can be to block the relay or provide a time delay to the relay immediately after the transformer is energized. The other approach can be to provide a fuse link in parallel to the instantaneous relay so that the inrush current is diverted through the fuse link. Inverse definite minimum time (IDMT) induction cup relays, because of their sluggish response, can also be considered for providing stabilization against magnetizing inrush.

Harmonic Restraint

In the eventuality of a transformer being energized on fault or occurrence of fault during the period when the transformer protective relay is time delayed, damages to transformers will occur. Therefore, some other method is required to retain the protection function in place even during the energization process to minimize the damage. Although the inrush current generally resembles fault current, if one compares the two waveforms, a significant difference is observable. Harmonic analysis of a typical magnetizing inrush current provides the percentage of different harmonics as shown in Table 6.2.

Table 6.2 Harmonic analysis of a typical magnetizing inrush current signal

Harmonic component	Amplitude in per cent of fundamental
Second	63.0
Third	26.8
Fourth	5.1
Fifth	4.1
Sixth	3.7
Seventh	2.4

From Table 6.2, one can conclude that in the event of magnetizing inrush, some of the harmonic components are significantly large; therefore, these components can be used either alone or in combination to restrain the relay operation during magnetizing inrush. The contribution of the different harmonic components depends on the core material as well as on the degree of saturation of the core. Thus, this aspect requires further studies.

During the magnetizing inrush, because of the presence of residual magnetism and core saturation, the current waveform contains a significant DC component. This DC component can be used for restraining the differential relay. For a three-phase transformer, if the residual flux is equal to the steady state flux, then no transient will be there for that phase while, for the other two phases, the inrush current will be very large.

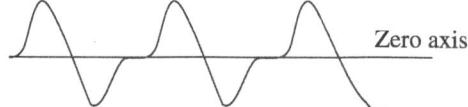

Fig. 6.30 Transformer inrush without offset because of yoke saturation

This current, which will be rich in DC, can be used for cross phase biasing. Figure 6.30 shows distorted current waveform under this condition. Since some fault currents may also have initial offset, tripping will be delayed in that situation.

The second harmonic component being significantly large during magnetizing inrush can be utilized for providing stabilization bias to the relay during switching in. To this end, the differential current is passed through a suitable filter to extract the second harmonic component. This component is used to restrain the relay.

The third harmonic is the only other harmonic component whose magnitude is comparable to the second harmonic. However, in case of delta connected CTs, it will only result in circulating current within the CTs. This circulating current may result in core saturation under heavy fault. Owing to these reasons, the third harmonic is not used for providing stabilization for the differential relay.

From Table 6.2, it is clear that except the second and third harmonics, all other harmonic magnitudes are less than 5% of that of the fundamental frequency component. Therefore, none of them can be used to provide stabilization against the magnetizing inrush. However, their combination with the DC component or the second harmonic component can be utilized very efficiently to provide stabilization for the relay.

Limitations of Harmonic Restraint

Harmonic-restrained differential relay is based on the fact that the magnetizing inrush current has a large second harmonic component, and nowadays the above technique is widely applied. But this technique must be modified, because harmonics occur in a normal state of power system and the quantity of second harmonic frequency component in inrush state has been decreased because of the improvement in core steel. There are cases in which the presence of differential currents cannot make a clear distinction between fault and inrush. Therefore, a new technique must be required based on aritificial intelligence or based on advanced signal processing techniques.

6.9 Digital/Numerical Protection of Transformer

The transformer is a major and very important equipment in the power system, and requires highly reliable protective devices. The protective scheme depends on the size of the transformer. For small transformers, overcurrent relays are used, whereas for large transformers, differential protection is recommended. A *protective relay* is a device that detects the faults and initiates the operation of the circuit breaker to isolate the defective element from the rest of the system. A high-performance numerical protective relay is used to minimize the operating and maintenance costs of power systems. These relays not only handle fault detection and location tasks but also control metering and monitoring functions. These additional functions were not possible before the advent of numerical technology, which offers major cost-cutting potential.

6.9.1 Numerical Relays

Numerical differential protection is a fast and selective short circuit protection scheme for transformers of all voltage levels. In addition to differential function, a backup overcurrent protection for one winding/star point is integrated in the relay. Optionally, a low or high impedance REF protection system, a negative-sequence protection system, or a breaker failure protection system can be used. Measurement and supervision of temperature can also be done in the relay. Therefore, complete thermal supervision of a transformer is possible.

6.9.2 Features of Digital/Numerical Relays

Modern digital transformer protection relays are capable of performing many functions. These are explained in detail in this section.

Protection Functions

1. Differential protection with phase-segregated measurement
2. Sensitive measuring stage for low fault currents
3. Restraint against the inrush of transformer
4. Phase/earth overcurrent protection
5. Overload protection with or without temperature measurement
6. Negative-sequence protection
7. Breaker failure protection
8. Low or high impedance REF
9. Overexcitation protection
10. Thermal monitoring of the transformer via temperature measurement with external thermo-box upto maximum measuring points

Control Functions

These include commands for controlling CBs and isolators.

Monitoring Functions

1. Self-supervision of the relay
2. Trip circuit supervision
3. Oscillographic fault recording
4. Permanent differential and restraint current measurement

Communication Interfaces

1. Communication with PC through the front and rear ports
2. IEC recommended system interface protocols
3. Service interface for temperature monitoring (thermo-box)
4. Time synchronization

   Apart from processing the measured values, the numerical relays perform the following functions:

1. Filtering and conditioning of measured signals
2. Continuous supervision of measured signals
3. Monitoring of the pickup conditions of each protection function
4. Conditioning of the measured signals, that is, conversion of currents according to the connection group of the protected transformer (when used for transformer differential protection) and matching of the current amplitudes
5. Formation of the differential and restraint quantities
6. Frequency analysis of the phase currents and restraint quantities
7. Calculation of the root mean square (RMS) values of the currents for thermal replica and scanning of the temperature rise of the protected object
8. Interrogation of threshold values and time sequences
9. Processing of signals for the logic functions
10. Reaching trip command decisions
11. Storage of fault messages, fault annunciations, as well as oscillographic fault data for system fault analysis
12. Operating system and related function management such as data recording, real time clock, communication, and interfaces

Example 6.6 Figure 6.31 shows the 3-phase diagram of a 250 MVA, 15.75/440 kV, DY-1 power transformer having a reactance of 12% and its neutral is solidly grounded. The CT ratio of line and neutral is 500/1 A on HV side of transformer. The knee point voltage (KPV) is greater than 300 V. CT secondary resistance is 5 Ω and lead resistance is 2.5 Ω. The relay connected on HV (Y) side of the transformer is instantaneous overcurrent relay having setting range 20–80% of 1 A. Burden of the relay is 1 VA. If an earth-fault occurs on HV side of the transformer having magnitude of fault current (I_F) 3.3 kA, determine the value of stabilizing resistance (R_{stab}). Assume pick-up setting of the relay is 50% of the relay rated current.

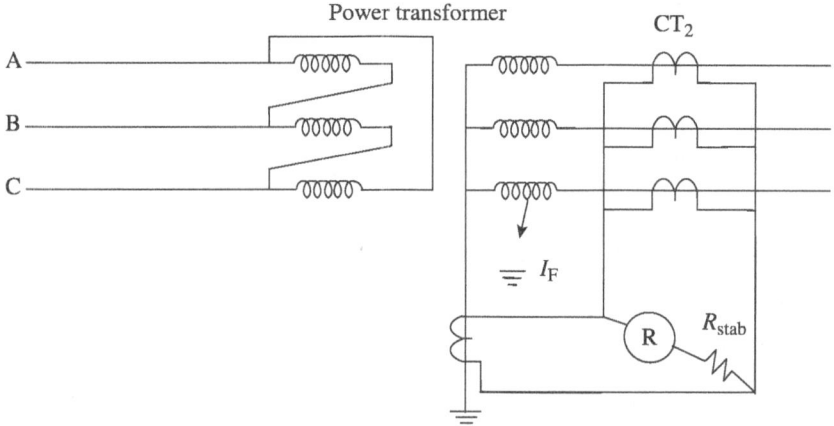

Fig. 6.31 Power transformer protected by restricted earth-fault scheme

Solution:

Rated current of the transformer on HV side = $I_R = \dfrac{250 \times 10^3}{\sqrt{3} \times 440} = 328.03\,A$

The CT secondary equivalent current for this rated current = 328.03/500 = 0.65 A

Now the pick-up current of the relay is given as 50% of 1 A , i.e., 0.5 A. Hence, 50% of 0.65 A is 0.328 A which is lower than 0.5 A and hence, relay does not operate in normal condition.

Now, the value of fault current = 3300 A.

The fault current on secondary side of CT is = 3300/500 = 6.6 A. Therefore, the voltage across the relay (V_r) is given by,

$$V_r = i_f \times (R_{CT} + R_L) = 6.6 \times (5 + 2.5) = 49.5\ V.$$

This clearly indicates that KPV of the CTs should not be less than 49.5 × 2 = 99 V.

Now, $V_r = I_r \times (R_r + R_{stab})$, where, $R_r = \dfrac{(Burden)}{(pick-up)^2} = \dfrac{1.0}{(0.5)} = 4\Omega$

$R_{stab} = \dfrac{V_r}{I_r} - R_r = 99.0 - 4.0 = 95\Omega$, where I_r is the current through relay coil.

Hence, the selected value of the stabilizing resistance is one-third of the calculated value, i.e., 95/3 = 31.66 Ω. Hence, 30 Ω is selected.

Example 6.7 Draw a detailed protection scheme for a power transformer protected by a biased differential relay. Also, suggest CT ratio on both sides of the transformer along with the ratio of ICT. Furthermore, provide proper verification of the operation/non-operation of a biased differential relay during normal/pre-fault situation. The relevant data is as under.

 (a) Power transformer is of 250 MVA, 15.75 kV/440 kV, DY-11 connection, with tappings of –5% to +7.5% on HV side having percentage reactance of 14.59%.
 (b) Assume ratio error of all CTs is ± 3%.
 (c) Biased differential relay has a fixed sensitivity (differential) setting of 15% of 5 A and three variable settings, that is, 10%, 20%, and 30%. It has instantaneous high-set unit setting of 10 times the rated current. The relay has a feature of second harmonic restraint equal to 15% (relay operation is blocked when second harmonic content in the operating coil of the relay exceeds 15%). In addition, the relay has also a feature of 5th harmonic bypass which avoids unwanted operation of relay under overexcited conditions.

Solution:

Detailed protection scheme for a power transformer protected by a biased differential relay is shown in Fig. 6.32.

Primary rated current of the transformer = $I_R / I_Y / I_B = \dfrac{250 \times 10^6}{\sqrt{3} \times 15.75 \times 10^3} = 9164.28\,A$

Hence, CTs on the primary side of the transformer are selected as 10,000/5 A. Reflecting this current on the secondary of CTs,

$$i_R / i_Y / i_B = \dfrac{9164.28 \times 5}{10000} = 4.582\,A$$

This is the pilot current that flows on the delta side of the transformer.

Now, secondary rated current of the transformer = $I_r / I_y / I_b = \dfrac{250 \times 10^6}{\sqrt{3} \times 440 \times 10^3} = 328.04\,A$

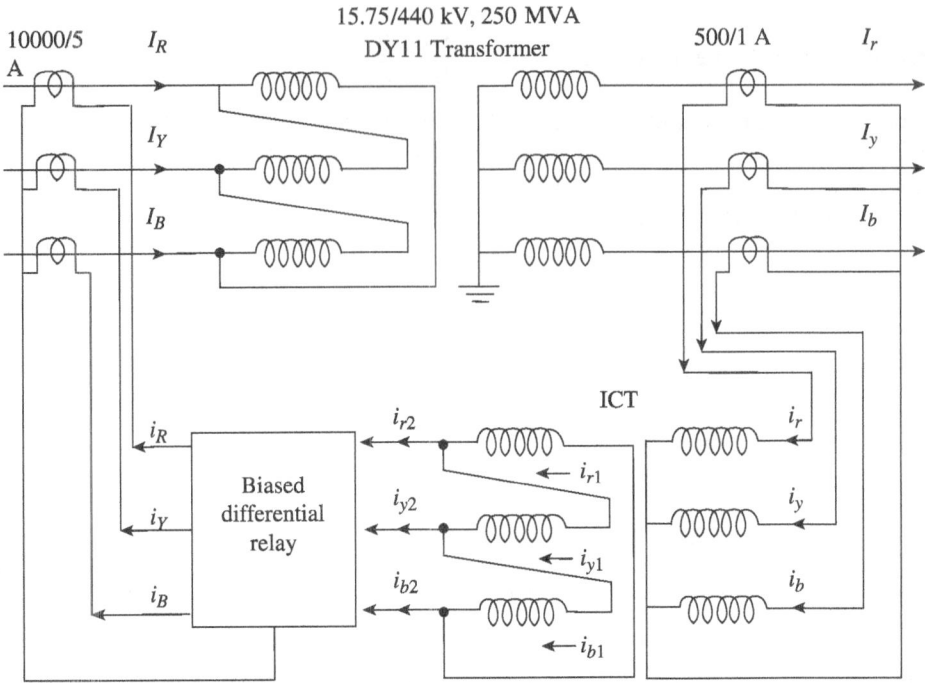

Fig. 6.32 Detailed biased differential protection scheme of power transformer for Example 6.7

Hence, CTs on the secondary side of the transformer are selected as 500/1 A. Reflecting this current on the secondary of CTs, $i_r / i_y / i_b = \dfrac{328.04}{500} = 0.6561\,\text{A}$.

Now, the pilot current on right hand side of biased differential relay, i.e., $i_{r2}/i_{y2}/i_{b2}$ must be equal to 4.582 A. So $i_{r1}/i_{y1}/i_{b1} = \dfrac{4.582}{\sqrt{3}} = 2.645$ A. Therefore, the ratio of ICT should be 0.6561/2.645 or 1: 4.1.

Now, HV side of the transformer is provided with tappings. Hence, if relay setting is carried out on nominal tap, then at the highest tap (+7.5%), relay may maloperate. Therefore, the highest tap must be considered while deciding relay settings. At the highest tap, the turns ratio of the transformer

$$= \frac{440 \times 1.075 \times 1000}{15750} = \frac{473000}{15750}$$

Moreover, percentage bias setting is selected/decided in such a way that the relay does not give unwanted tripping for a three-phase short-circuit (heavy through fault) just after the CTs on the secondary side of the transformer. Considering the highest tap (+7.5%), the fault current magnitude on the secondary side of the transformer for a three-phase short-circuit is calculated as under.

$$I_{sf} = \frac{328.04}{0.1459} = 2248.39\,\text{A}$$

The fault current on primary side of the transformer is given by,

$$I_{Pf} = \frac{2248.39 \times 473000}{15750} = 67.52\,\text{kA}$$

The CT secondary equivalent currents are as given by,

$$i_{pf} = 33.67 \text{ A and } i_{sf} = 4.4968 \text{ A}$$

Considering +3% error in the CT connected on primary side of the transformer, $i_{pf} = 34.192$ A. Similarly, considering −3% error in the CT connected on secondary side of the transformer,

$$i_{sf} = 4.3619 \text{ A}$$

Reflecting i_{sf} on the secondary of ICT,

$$i_{sf1} = 4.3619 \times 4.1 = 17.8838 \text{ A}$$

Once again considering −3% error in ICT, $i_{sf1} = 17.3473$ A

Thus,

$$i_{sf2} = \sqrt{3} \times 17.3473 = 30.05 \text{ A}$$

Hence, the differential current = $i_{pf} - i_{sf2} = 34.192 - 30.05 = 4.142$ A

The restraining current = $\dfrac{i_{pf} - i_{sf2}}{(i_{pf} + i_{sf2})/2} \times 100 = 8.42\%$

Considering CT saturation and CT ratio mis-matches, percentage bias can be selected as 20%.

Example 6.8   A 160 MVA, 132/66 kV, DY-1, 3-phase transformer is to be protected against short-circuit. An instantaneous overcurrent relay is used to achieve the above task. The magnetizing inrush current of the transformer is 8 times the rated current. The setting range of an instantaneous overcurrent relay is 400–2000% of 1 A in step of 50%. The CT ratio is 1000/1 A. Suggest the setting of an instantaneous overcurrent relay.

Solution:
Rated current of the transformer on 132 kV side, $I_R = \dfrac{160 \times 10^6}{\sqrt{3} \times 132 \times 10^3} = 699.82$ A

The instantaneous overcurrent relay should not operate due to magnetizing inrush current during energization of transformer. Hence, magnetizing current is given by,

$$I_M = 699.82 \times 8 = 5598.55 \text{ A}$$

Referring this current on the secondary side of CT, $I_r = \dfrac{5598.55}{1000} = 5.5986 \ or \ 559.8\%$

Hence, 600% is suggested.

Example 6.9   A 3-ϕ transformer of 220/11000 V is connected in star/delta. The protective transformers on 220 V side have a current ratio of 600/5 A. What should be the CT ratio on 11000 V side?

Solution:
Line voltage connected in star/delta = 220/11000 V_t
For star/delta power transformer, CTs will be connected in delta on 220 V side, i.e., star side of power transformer and in star on 11000 V side (i.e. delta side of power transformer).
Line current of 220 V side is 600 A.
So phase current of delta connected CT on 220 V side = 5 A
Line current of delta connected CTs on 220V side = $5\sqrt{3}$ A
This current (i.e., $5\sqrt{3}$ A) will flow through pilot wires. Also this current flows through the secondary of CT on the 11000 V side.

So phase current of star-connected CTs on 11000 V = $5\sqrt{3}$ A

If I is the line current on 11000 V side, then

$$\text{Primary power} = \text{Secondary power}$$
$$\sqrt{3} \times 220 \times 600 = \sqrt{3} \times 11000\ I$$
$$I = 12\ A$$

Turn ratio of CT on 11000 V side = $12 : 5\sqrt{3}$ A

Example 6.10   A 30 MVA, 11.5 kV/69 kV, star/delta power transformer is to be protected by differential protection. The high voltage side phase lags behind low voltage side phase by 300. The continuous current carrying capacity to restraining coils of the differential relay should not exceed 5 A. CT ratio is 3000/5 A on 11.5 kV side. Determine CT ratio on 69 kV side. Also calculate the full load current of transformer on HV and LV side.

Solution:

30 MVA 11.5 kV/69 kV, $\gamma - \Delta$

CT ratio = 3000/5 A on 11.5 kV side

On 11.5 kV side,

$$I_{fL} = \frac{30,000}{\sqrt{3} \times 11.5} = 1505\ A$$

$$\text{CT ratio} = \frac{3,000}{5} = 600$$

$$I_s = \frac{1505}{600} = 2.51\ A$$

Since 11.5 kV is star connected and CT secondary will be connected in Δ,. Hence current fed into pilot wires from 11.5 kV side CT of secondary is

$$\sqrt{3} \times 2.51 = 4.35\ A$$

On 69 kV side, $I = \dfrac{30,000}{\sqrt{3} \times 69} = 251\ A$

Current in secondary of CT's = Current in pilot wires since 69 kV side CT secondary are connected in star = 4.35 A

$$\text{CT ratio} = \frac{251}{4.35} = 57.7$$

Take CT ratio = 60, Secondary current = 5 A, Primary current = 60 × 5 = 300

So CT ratio on 69 kV side = 300/5 A.

Example 6.11   A 3 - ϕ, 33kV/6.6 kV star/delta connected transformer is protected by a differential scheme. The CTs on LT side have a ratio of 300/5 A. Show that the CTs on HT side will have a ratio $60 : 5\sqrt{3}$ A.

Solution:

CTs on delta side are star connected; hence the secondary phase currents are equal to currents in pilot wires. CTs on star connected side are delta connected; hence current in secondary is equal to current in pilot wires divided by $\sqrt{3}$.

$$\sqrt{3} \times 6.6 \times 300 = \sqrt{3} \times 33 \times I$$
$$I = 60\ A = \text{current in HT line}$$
$$= \text{Primary current of CT on HT side}$$

Secondary current is 5 A. Hence current fed in pilot wires from LT side is 5 A and the same current is fed from CT connections on HT side which are delta connected.

Secondary current of CT on HT side = $5/\sqrt{3}$ A

So CT ratio on HT = 60 : $5/\sqrt{3}$ A

Example 6.12 A 3 - ϕ transformer having line voltage ratio of 0.4 kV/ 11 kV is connected in star/delta and protective transformers on the 400 V side have a current ratio of 500/5 A. What must be the ratio of the protective current transformers on the 11 kV side?

Solution:

Line current on 400 V side is 500 A.

So phase current of delta-connected CTs on 400 V side = 5 A.

Line current of delta-connected CTs on 400 V side = $5/\sqrt{3}$ A.

This current ($5/\sqrt{3}$ A) will flow through the pilot wires, so this will be the current which flows through the secondary of CTs on 11000 V side.

Phase current of star-connected CTs on 11000 V side = $5/\sqrt{3}$ A.

If I is the line current on 11000 V side, then

$$\text{Primary apparent power} = \text{secondary apparent power}$$

$$\sqrt{3}\times 400 \times 500 = \sqrt{3}\times 11000 \times I$$

$$I = \frac{200}{11}\ \text{A}$$

So CT ratio of CTs on 11000 V side = $\frac{200}{11} : 5\sqrt{3}$ A

<h2>Recapitulation</h2>

- The power transformer is one of the most important equipment in the power system network. Thus, it is worthwhile reviewing the different protection schemes applicable to the transformer against faults.
- In addition to the faults in a transformer, it is subjected to some abnormal operations such as magnetizing inrush and overfluxing, which are quite similar to the fault conditions but do not require the protective gear meant for the protection of the transformer to operate.
- The relays that respond to pressure (oil or gas), fluid flow, liquid level, and temperature are Buchholz relay, SPR, and oil and winding temperature relays.
- The main protections applied to transformers are overcurrent protection, REF protection, and differential protection.
- The selection of CT ratings on both sides of transformer winding, CT connections, CT mismatch, CT saturation, magnetizing inrush conditions, and tap changing are a few problems encountered during differential protection of the transformer.
- Interposing CTs are used to match the relay currents under through-load conditions corresponding to the ratings of the transformer.
- To overcome the magnetizing inrush conditions and the harmonic generation caused to it during differential protection, either time delay or harmonic restrain method is used.
- The modern practice of protecting a power transformer is to use a wide-ranging digital/numerical relay.
- Protection of transformer using digital/numerical relay offers control and metering and monitoring functions in addition to fault detection and location tasks.

Multiple Choice Questions

1. For the protection of inter-turn fault in transformer, the relay preferred is
   - (a) overcurrent relay
   - (b) Buchholz relay
   - (c) differential relay
   - (d) none of the above

2. Switching of unloaded transformer during zero crossing of a voltage wave generates
   - (a) maximum amount of magnetizing inrush current
   - (b) minimum amount of magnetizing inrush current
   - (c) zero amount of magnetizing inrush current
   - (d) none of the above

3. Overfluxing protection of a transformer is
   - (a) inverse time protection
   - (b) instantaneous protection
   - (c) time delayed protection
   - (d) none of the above

4. The earth-fault relay used for the protection of delta connected winding of transformer is
   - (a) inverse time protection
   - (b) instantaneous protection
   - (c) time delayed protection
   - (d) none of the above

5. The purpose of the restraining coil in a biased differential relay is
   - (a) to reduce CT saturation during fault
   - (b) to increase the sensitivity of the relay
   - (c) to match the transformation ratio
   - (d) to limit the spill current through the relay during heavy external fault

6. A line-to-ground fault occurs on the star side of a feeder transformer; the same fault appears on the delta side as
   - (a) a line-to-ground fault
   - (b) a line-to-line fault
   - (c) a double-line-to-ground fault
   - (d) a three-phase fault

7. The magnetizing inrush current in a transformer is rich in the
   - (a) second harmonic component
   - (b) third harmonic component
   - (c) fifth harmonic component
   - (d) seventh harmonic component

8. Interposing CT is used in differential protection of a transformer
   - (a) to limit the main CT secondary current
   - (b) to increase the sensitivity of the relay
   - (c) to minimize the effect of saturation of the main CTs
   - (d) to match the standard ratio of CTs used on both sides of the transformer winding

9. A harmonic restrain feature is used in a relay for transformer protection to
   - (a) increase the relay speed
   - (b) detect external fault
   - (c) stabilize the magnetizing inrush current
   - (d) none of the above

10. The percentage bias setting of differential relay used for transformer protection should be
    - (a) below 10%
    - (b) between 10% and 50%
    - (c) above 50%
    - (d) none of the above

Review Questions

1. Explain the different abnormal conditions in a transformer.

2. Explain the concept of magnetizing inrush current in a power transformer.

3. Why is time delayed protection provided for overfluxing in a transformer?

4. Explain the construction and operation of Buchholz relay protection in a transformer.

5. Explain the purpose of sudden pressure relay and WTI in transformer protection.

6. Explain the overcurrent and restricted earth fault protection of power transformer using a neat diagram.

7. Discuss the problems that arise in the application of differential protection of a transformer.

8. Why are ICTs used in differential protection of transformer?

9. Enlist the different features of digital relays used for transformer protection.

Numerical Problems

1. Draw a detailed protection scheme for biased differential protection of a 66/220 kV, 100 MVA, DY1 power transformer. Suggest suitable CT ratios. In addition, suggest the proper ICT for the scheme.

   [Primary CT ratio = 1000:5 A, Secondary CT ratio = 300:1 A, ICT ratio = 1:2.886 A]

2. A 220 kV/66 kV, 250 MVA, DY1 three-phase power transformer is connected in star solidly earthed on the LV side,

   and in delta on the HV side. Draw a detailed connection diagram for the differential protection of the transformer. If the CTs employed on the LV side have a transformation ratio 2000/1 A, determine the CT ratio on the HV side of the transformer. Suggest the proper ICT ratio.

   [Primary CT ratio = 750:5 A, Secondary CT ratio = 2000:1 A, ICT ratio = 1:2.76 A]

Protection of Induction Motor

7

Learning Objectives

After going through this chapter, the students will be able to:

- Discuss the different types of faults in induction motor
- Analyse the various protection schemes available to tackle faults in induction motor
- List the additional features supported by modern digital/numerical induction motor protection scheme

7.1 Introduction

Induction motors are an important component in the energy conversion process for electric utilities all over the world. Large high voltage motors are used to drive water pumps, fans, and other auxiliary items in industries and power generating plants. Therefore, losses in the motor due to any type of fault results in loss of production, which finally results in a loss of revenue. Hence, protection of the induction motor is very essential for its satisfactory working. The type of protection scheme applied to the motor depends mainly on the size of the motor, its importance, and the load connected to the motor. For example, fuses are adequate to protect small motors that drive unimportant loads, whereas a comprehensive motor protection scheme using sophisticated digital/numerical relays is required to protect large motors used in industries.

This chapter deals with various types of protection schemes including modern numerical/digital relays used in large induction motors. A numerical motor protection relay uses phase currents, negative sequence currents, and ground currents to protect the motor from various types of abnormalities/faults such as short circuit, earth fault, thermal overload, locked rotor, negative sequence current, unbalanced current, loss of load, and underload conditions. Owing to the use of solid-state technology in numerical relays, it is now possible to reduce the size of devices and also to add new functions/features. One of the hidden benefits is the improved simplicity and accuracy of calculation of the symmetrical (positive, negative, and zero sequences) components from phase quantities, which is comparatively difficult in electromechanical relays.

7.2 Faults/Abnormal Conditions in Induction Motor

In an AC motor, different types of faults/abnormal conditions may occur. The majority of the faults are due to insulation failure and mechanical failure. The main cause of failure of motor is excessive heating. If it is sustained for a long period, the motor will finally burn out. Overheating also reduces the lifetime of the motor. For example, if a motor is continuously overheated by 10 degrees above its specified rated temperature limit, its life can be reduced by almost 50%. In general, overheating occurs because of the overcurrent, which may be due to many abnormal/faulty conditions such as overload, locked rotor, low supply voltage, single phasing, repeated starts, or voltage/current unbalance. The reasons for the presence of various types of faults/abnormal conditions in the motor and the related protection schemes are explained in the following sections.

7.2.1 Overloading

All motors can withstand overload within their designated limit for a specified duration. In general, an induction motor is designed in such a manner that it can continuously take 10–20% overload with reference to its rated current. However, as the degree of the overload increases, the heating effect also increases, which finally results in damage of the insulation of the motor winding. One can easily find the thermal withstand limit of the motor from its thermal withstand characteristics (current vs time curve). This graph gives the values of time limits for different values of overload currents.

7.2.2 Single Phasing

Single phasing is the worst kind of current/voltage unbalance condition. When a load using a three-phase power source is subjected to a loss of one of the three phases from the power distribution system, single phasing occurs. This condition may occur because of a downed line or the blowing out of the fuse of any one of the three phases of the supply system or of the motor itself. The loss of one phase of a three-phase line causes serious problems for the induction motor. The three-phase motor may continue to run by taking more currents in the remaining healthy phases, but it is not capable of starting during single phasing. The motor windings get overheated because of the flow of negative sequence current. It also creates an 'unbalance' in the three-phase voltages.

7.2.3 Phase Unbalance

When the three-phase voltage supplied to the motor is not balanced with respect to magnitude or phase, the currents are 'unbalanced' in the three phases. A condition of unbalance in a three-phase system is less dangerous than single phasing but may have similar consequences. An unbalance of just 5% can reduce the output by 25% even though the motor continues to draw the same current as during the balanced condition. This means that the motor current increases under unbalanced condition to deliver the same power, which eventually increases the heating of the motor. This excessive heating is mainly because of the negative sequence currents which cause the motor to turn in the direction opposite to its normal direction of rotation. This higher temperature results in degradation of the motor insulation and shortened lifetime of the motor. The percentage rise in temperature of the highest current winding is nearly twice the square of the voltage unbalance.

7.2.4 Phase Reversal

Phase reversal in a motor is also a very dangerous situation. If the motor is not designed to run in the reverse direction, but tries to run in the opposite direction then there is a possibility of severe damage to the gear boxes, material flow problems, and hazards to the operating personnel as well. If the motor is connected in reverse polarity, it produces a negative sequence of nearly 100%, which is detected by a relay with negative sequence protection.

7.2.5 Short Circuit

Insulation failure due to the shorting of any two or all three phases of the motor is known as *short circuit*. Thermal overload relays protect a motor against overload situations when the temperature of the motor rises at a very slow rate. However, they do not provide adequate protection during short circuit of stator windings of the motor because the rate of rise of temperature is very high. Hence, a separate short circuit protection scheme which operates instantaneously is required.

7.2.6 Earth-fault

The shorting of any one or more phases of the motor with the earth because of failure of insulation is known as an *earth-fault condition*. Similar to the short circuit, in this case too the rate of rise of temperature is very high. In such situations, a separate earth-fault protection scheme is required.

7.2.7 Stalling or Locked Rotor

If the load on the motor is increased abruptly, it will fail to start or will run very slowly. This condition is known as *stalling* of the motor. If the rotor is stalled (which would inevitably lead to a thermal trip) while running or when starting the motor, it is necessary to disconnect the motor immediately using a protective device and not wait until the motor becomes hot. This reduces motor stress and even protects its machinery.

7.2.8 Underload

A sudden loss of mechanical load and a loss in the connection of the conductor at the terminal box create the condition of underload. Induction motors are often cooled using media such as air (in case of fans) or water (in case of submersible pumps). Inadequacy of the cooling medium (because of an obstruction of air or water/liquid flow) due to underload would result in excessive motor heating. For this reason, underload protection should be provided to the motors connected to such loads. The tripping device triggers a warning signal or a trip signal before damage occurs.

7.3 Protection Schemes of Induction Motor

Many protective schemes (ranging from basic overload protection using fuse to comprehensive digital protective device for performing a number of protection, measurement, and control functions) are available to protect the motor against various types of faults/abnormalities. Many protective schemes that adopt the latest techniques used by modern digital relays are explained in Sections 7.3.1–7.3.8. These schemes protect the motor against faulty/abnormal conditions explained earlier.

7.3.1 Thermal Overload Protection

The primary protective element of motor protection relays is the *thermal overload element*, which is accomplished through motor thermal image modelling. This model accounts for all thermal processes in the motor while it starts and runs at normal, overload, and standstill (stop) conditions.

Thermal protection (49) is the simplest type of protection against overloading for the induction motor. Its basic function is to operate the motor within its predefined temperature withstand limit (thermal limit). For this purpose, usually two philosophies are used in the relays:

1. Measurement of temperature by a direct method using resistance temperature detectors (RTDs) or thermistors
2. Measurement of temperature by an indirect method using the data of phase currents and deriving an equation that gives an equivalent thermal state of the motor

The relay takes the decision to disconnect the motor based on data available from one or both of these methods as well as the settings made by the user.

Figure 7.1 shows the acceleration curves which give an indication of the amount of current and the associated time for the motor to accelerate from a stop condition to a normal running condition. Usually, for large motors, there are two acceleration curves: the first is the acceleration curve at the rated stator voltage (100%), and the second is the acceleration curve at 80% of the rated stator voltage (soft

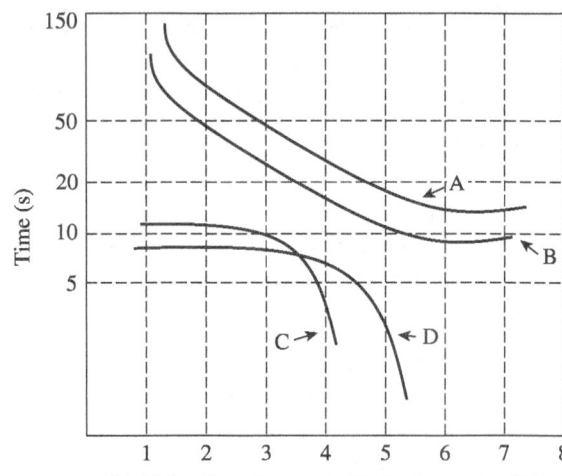

A. Cold running overload
B. Hot running overload
C. Acceleration curve at 80% of rated voltage
D. Acceleration curve at 100% of rated voltage

Fig. 7.1 Motor thermal limits and acceleration curves

starters are commonly used to reduce the amount of inrush current during starting).

Three-phase motors are designed in such a manner that the overloads are below the thermal withstand limit of the machine. The motor thermal limit curves consist of three distinct segments, which are based on the three running conditions of the motor: the locked rotor or stall condition, motor acceleration, and motor running overload. Ideally, the curves should be provided for both hot and cold running conditions of the motor.

Figure 7.2 shows the starting curve of an induction motor and the suggested relay characteristics. The thermal relay (49) characteristic has to match with the thermal limit curve of the motor as shown in Fig. 7.2. Therefore the settings of the relays have to match with the thermal curve.

The inverse time overcurrent relay (51) is used for locked rotor protection, whereas the instantaneous overcurrent relay (50) is used for the heavy short circuit protection of induction motor.

7.3.2 Protection Against Unbalanced Currents

Unsymmetrical faults within the motor or on the feeder may produce unbalanced voltages and hence, unbalanced currents (46). The typical conditions that can give rise to unbalanced currents are as follows:

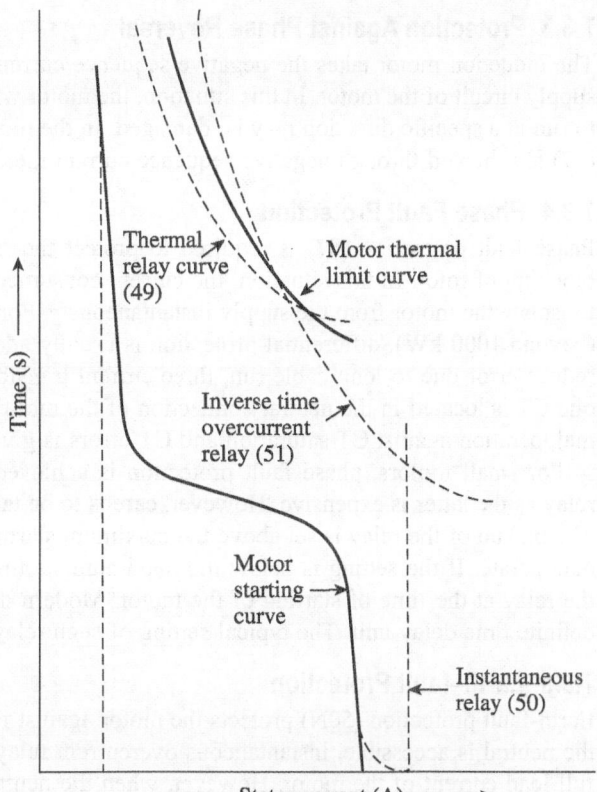

Fig. 7.2 Starting curve and relay curve for induction motor

1. Single phasing due to the disconnection of fuse for large motors
2. Unbalanced fault conditions and failure of breakers

The motor will draw excess current because of single phasing. Hence, its winding will be overheated, which leads to deterioration/damage of insulation. In case of unbalancing of currents, negative sequence current is produced. This may overheat/damage the rotor structure. To protect the large motor against such unbalance, negative sequence relay (46) is used. However, for small motors, phase unbalanced relay is usually used.

Each motor is designed for specific permissible unbalance according to the industry standards. The general equation for allowable negative sequence current is

$$I_2^2 \times t = 40 \qquad (7.1)$$

where I_2 is the negative sequence current in per unit and t is the time in seconds.

Modern digital relays measure the negative sequence component of current and provide protection against unbalanced conditions, broken conductor, and phase inversions. They use two relays, namely definite time relay and inverse time relay. Definite time relay is used in the first stage of negative sequence protection, whereas inverse time relay is used for the second stage of protection. The typical setting range of this relay is $I_2 = 10$–50% in steps of 5%. This setting is decided on the basis of the ratio of the negative sequence impedance (Z_2) to the positive sequence impedance (Z_1).

7.3.3 Protection Against Phase Reversal

The induction motor takes the negative sequence current if there is a change in the phase sequence in the supply circuit of the motor. In this situation, the motor will run in the opposite direction. The loads designed to run in a specific direction may be damaged. In the modern digital relay, protection against phase reversal (47) is achieved through negative sequence current measurement, as explained in Section 7.3.2.

7.3.4 Phase Fault Protection

Phase fault protection (87) is required to protect the motor against short circuit in motor windings and blocking of rotor. In this situation, the current consumed by the motor is very high. Hence, there is a need to isolate the motor from the supply instantaneously. For short circuit in the stator windings of large motors (beyond 1000 kW), differential protection is usually adopted. In order to minimize the burden and also to reduce error due to long cable run, three current transformers (CTs) are placed within the switchgear and one CT is located in the neutral connection of the motor. The importance of stabilizing resistance to avoid maloperation against CT saturation and CT errors is given in Chapter 5.

For small motors, phase fault protection is achieved through overcurrent relay instead of differential relay as the latter is expensive. However, care is to be taken at the time of deciding the setting of this relay. The pickup of the relay is set above the maximum starting current of the motor as otherwise the relay may maloperate. If the setting is below the maximum starting current, then an interlock is used, which blocks the relay at the time of starting of the motor. Modern digital relays use instantaneous relays along with a definite time delay unit. The typical setting of such relays is 400–2000% of the relay rated current.

7.3.5 Earth–fault Protection

Earth-fault protection (50N) protects the motor against faults between one or more phases and earth. When the neutral is accessible, instantaneous overcurrent relay provides this protection with a setting of 20% of full load current of the motor. However, when the neutral is not accessible, residual ground relay is used. This relay considers the vector sum of the secondary current of all the three phases. In order to increase the sensitivity of the relay, it is sometimes connected in the core balance CT. In this case, the relay can be set instantaneously (3–10% of the rated current).

7.3.6 Stalling (Locked Rotor) Protection

Stalling protection (51) is required in a motor at the time of starting or running. Any motor can dissipate more heat during the running condition than when it is in a standstill condition. If a motor fails to start after the energization, then the heat produced in the rotor and stator windings is 10 to 15 times that produced during the rated conditions. The motor can withstand this type of extreme heating condition only for a limited time period. This time limit depends on the applied voltage and the I^2t limit. A relay with I^2t characteristic can be set for any permissible overload. With respect to the starting or running condition of the motor, protection is applied in two different ways.

Stalling at the time of starting This function is activated only during the starting of the motor, that is, during the course of the starting time delay (t_{start}). It uses a speed signal from the motor and the time delay (t_{stall}). On detection of a start this function is activated, and the time delay provided for safe stalling (t_{stall}) begins. At the end of this time delay, the motor should gain the acquired speed. If the motor does not gain the acquired speed, it is an indication of a locked rotor situation, and the relay generates a tripping command.

Stalling at the time of running This function is activated immediately after the starting period (t_{start}). Two parameters are required to be set in the relay, that is, the stall rotor current threshold (I_{stall}) and the stalled rotor time (t_{stall}). The relay detects the overcurrent caused by stalling and generates a tripping command if the phase current exceeds the I_{stall} value.

A typical setting of the stalling relay is given by two settings: the current setting range which is usually 150–600% of the relay rated current and the time setting range which is usually 6–60 s. The normal practice for current setting is 1/3–1/4 of the starting current of the motor. On the other hand, time setting is done on the basis of the accelerating time of the motor and safe stalling time of the motor. The time setting is higher than the accelerating time and lower than the safe stalling time of the motor.

7.3.7 Loss of Load Protection

The sudden reduction of load from the motor is known as *loss of load*. The reasons for the loss of load are breakage of a conveyor belt, prime failure on a pump, or shearing of a drive pin. It is mandatory to trip the motor in this situation. The usual practice is to use definite time overcurrent relay, which recognizes the difference between no-load before the application of load and no-load after the application of load. In this relay, three parameters are required to be set, namely undercurrent threshold, time associated with this threshold, and inhibit start time delay. After the starting of the motor, this function is activated at the end of the inhibit time delay (T_{inhibit}), which is useful for motors to start at no-load and also to increase the load gradually at the end of starting. During the running condition of the motor, a loss of load signal occurs in case of reduction of one of the phase currents below the threshold value for a period longer than or equal to the set time delay.

7.3.8 Undervoltage

When there is reduction in the voltage below the rated voltage in a running motor, the current drawn by the motor increases beyond the rated current of the motor. Hence, the insulation of the motor windings is damaged because of heating. Thus, the undervoltage (27) relay gives protection against this phenomenon. The typical setting range of an undervoltage relay is 70–100% of the rated voltage.

Figure 7.3 shows the detailed protection scheme of a three-phase induction motor, where the thermal relay module (49), phase unbalance relay (46), overcurrent relay unit (51), instantaneous overcurrent relay (50), earth-fault relay unit (50N), and undervoltage relay unit (27) are incorporated. Measurement of temperature by direct method using RTDs or thermistors (H) is also illustrated. The residual current measurement is carried out using core balance current transformer (CBCT).

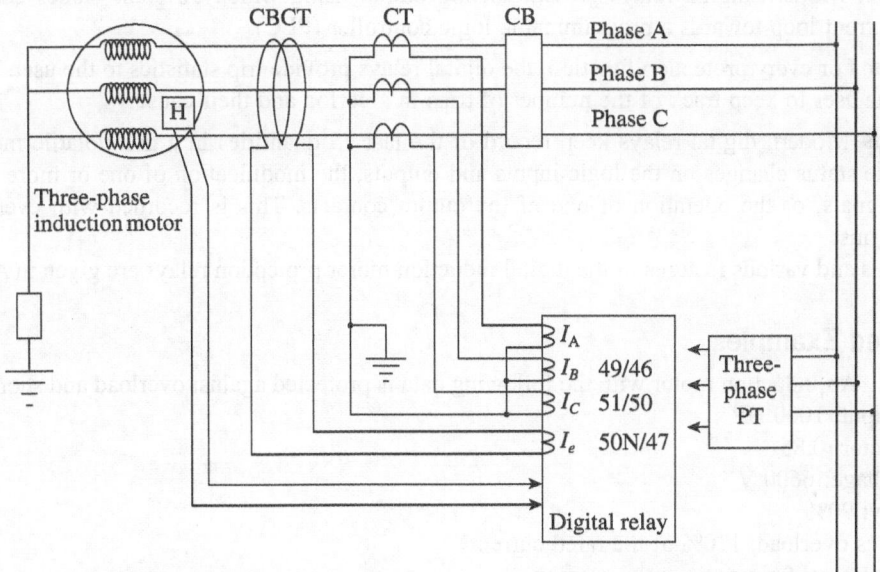

Fig. 7.3 Detailed protection scheme of induction motor

7.4 Numerical Protection of Induction Motor

The modern digital induction motor protection relay contains all the protection features mentioned in Section 7.3. Further, it includes the following additional features.

Number of starts The setting for the number of starts is included in motors, to avoid temperature rise due to several start attempts. For any motor, only a few start attempts are permitted for a specific period. If the number of such attempts is exceeded, then the starting of the motor is blocked. Different parameters, such as monitoring period, number of hot starts, number of cold starts, and start inhibit time delay, are adjusted.

Excessive start time The setting for excessive start time is provided to avoid a long starting period of the motor. In this situation, the motor draws very high current, and hence the insulation deteriorates because of the temperature rise. Long starting periods may result due to many conditions such as excessive load torque, jamming of bearing and gear drives, single phasing, or low voltage supply. In order to avoid the situation, the digital relay uses a starting current threshold and a starting time delay. Both values can be adjusted in such a way that the starting current passes only for a specific period. If it is passed for a longer duration, then the relay will trip the motor, in the specified time period.

Minimum time between two starts A time delay is required between two consecutive starts of the motor to provide proper cooling between the two starts. This time delay is initiated on detection of the motor start by the relay. If the minimum time between two starts of an induction motor has not lapsed and the motor stops, then the start inhibit signal is generated until the specified time delay is over.

Programmable logic scheme In modern digital relays, the facility of custom design is included. Hence, four logical equations can be achieved by combining the internal and external information. This results in saving in the external relaying and in the relay/process interactivity.

Measurements Modern digital relays provide continuous measurements of a wide range of data such as current, voltage, frequency, speed, and status of the motor.

Analog output Modern digital relays provide analog output using which different values can be driven through a current loop towards a programmable logic controller (PLC).

Trip statistics For every protection function, the digital relays provide trip statistics to the user. This will be helpful to the user to keep track of the number of trips in a period and their cause.

Event records Modern digital relays keep record of the last 75 quantities in a non-volatile memory. This covers all the status changes on the logic inputs and outputs, the modification of one or more parameters, the alarm signals, or the operation of one of the output contacts. This is recorded with every sampling interval of 1 ms.

The details and various features of the digital induction motor protection relays are given in Appendix C.

7.5 Solved Examples

Example 7.1   An induction motor with the following data is protected against overload and short circuit.
   Rated output: 1000 HP
   Power factor: 0.85
   Rated voltage: 6600 V
   Efficiency: 90%
   Continuous overload: 110% of the rated current
   Starting current: 5 times the rated current
   Pickup setting: 100% of 1 A

Suggest a suitable CT ratio and calculate the overload and instantaneous relay settings.

Solution:

Rated current of the motor is given by $(I_R) = I_R = \dfrac{1000 \times 746}{6600 \times 0.85 \times \sqrt{3} \times 0.9} = 85.3$ A

Considering 10% overload, the input current of the motor = $85.3 \times 1.1 = 93.83$ A

Thus, the CT of 100/1 A is selected.

Thermal overload protection:

Pickup setting $< \dfrac{93.83}{100} < 93.84\%$. Hence, the setting of thermal overload relay is selected as 90% of 1 A.

Short circuit protection:

The starting current at 75% of rated voltage $(I_{start}) = \dfrac{5 \times 85.3}{0.75} = 568.67$ A

The secondary equivalent of 568.67 A for 100/1 A CT is $= \dfrac{568.67}{100} = 5.68$ A

Thus, 600% of 1 A setting is selected for instantaneous overcurrent relay.

Example 7.2   The details of a 2000 HP, three-phase, 50 Hz induction motor are as follows:

    Rated output: 2000 HP
    Power factor: 0.85
    Rated voltage: 6600 V
    Efficiency: 90%
    Continuous overload: 110% of the rated current
    Starting current: 6 times the rated current
    Starting time
       (i)   at 100% voltage: 12 s
       (ii) at 80% voltage: 16 s
    Safe stalling time: 22 s
    Safe stalling current: 1/3 of starting current of motor
    CT ratio: 200/1 A
    Negative sequence impedance: 20%
    Positive sequence impedance: 80%
    Setting range of thermal relay: 70–130% of 1 A in steps of 5%
    Setting range of negative sequence relay: 10–40% of 1 A
    Setting range of stalling relay
       (i)   current range: 150–600% of 1 A in steps of 30%
       (ii) time range: 6–60 s
    Pickup setting: 100% of 1 A

Calculate the overload and instantaneous relay settings. In addition, suggest the relay setting for thermal overload relay, instantaneous overcurrent relay, negative phase sequence relay, and stalling relay with their timer settings.

Solution:

Rated current of the motor is given by $(I_R) = I_R = \dfrac{2000 \times 746}{6600 \times 0.85 \times \sqrt{3} \times 0.9}$

$= 170.61$ A

Considering 10% overload, the input current of motor = $170.61 \times 1.1 = 187.67$ A

Thermal overload relay:

Pickup setting of thermal relay is $< \dfrac{187.67}{200} < 93.83\%$. Hence, the setting of thermal overload relay is selected as 90% of 1 A.

Instantaneous overcurrent relay:

Starting current at 80% of rated voltage $(I_{start}) = \dfrac{6 \times 170.61}{0.8} = 1279.56$ A

Secondary equivalent of 1279.56 A for 200/1 A CT is $= \dfrac{1279.56}{200} = 6.39$ A $= 639\%$

Thus, 650% of 1 A setting is selected for instantaneous overcurrent relay.

Negative phase sequence relay:

The setting of such a relay can be decided on the basis of the ratio of the negative phase sequence imped-ance to the positive phase sequence impedance. Hence, $\dfrac{Z_2}{Z_1} = \dfrac{0.2}{0.8} = 0.25$

Thus, the setting of negative phase sequence relay is selected as 30% of 1 A.

Stalling relay:

$$I_{stall} = \dfrac{I_{start}}{3} = \dfrac{1279.5}{3} = 426.5 \text{ A}$$

Secondary equivalent of this current is 2.13 A.

Hence, a setting of 210% is selected for stalling relay. The time setting of the stalling relay has to be higher than the accelerating time (12 s) and lower than the safe stalling time. Hence, the stalling timer set-ting is taken as 15 s.

Recapitulation

- The induction motor undergoes a few abnormal conditions such as overloading, single phasing, phase unbalance, phase reversal, short circuit, earth-fault, stalling or locked rotor, and underload.
- The protection functions employed for the induction motor against the abnormalities are thermal overload protection, protection against unbalanced current, protection against phase reversal, phase fault protec-tion, earth-fault protection, stalling protection (locked rotor), loss of load protection, and undervoltage protection.
- The latest trend of using digital technology in protection systems has helped to provide comprehensive protection to large and small induction motors.

Multiple Choice Questions

1. Which of the following protections is not applied to small induction motors?

   (a) Differential protection

   (b) Overcurrent protection

   (c) Short circuit protection

   (d) Thermal overload protection

2. The starting current of an induction motor is of the order of

   (a) 1–2 times the rated current of the motor

   (b) 2–4 times the rated current of the motor

   (c) 4–6 times the rated current of the motor

   (d) None of the above

3. The characteristic of thermal overload relay must be coordinated with
   (a) the thermal withstand characteristic of the motor
   (b) the starting characteristic of the motor
   (c) both (a) and (b)
   (d) none of the above

4. The typical setting range of overcurrent relay used for the protection of small motors against short circuit is
   (a) 5–10%
   (b) 50–100%
   (c) 100–200%
   (d) 400–2000%

5. The detection of unbalance in currents in large induction motors is done by
   (a) negative sequence current
   (b) positive sequence current
   (c) zero sequence current
   (d) none of the above

6. The time setting in stalling protection of an induction motor is
   (a) higher than the accelerating time and lower than the safe stalling time of the motor
   (b) lower than the accelerating time and higher than the safe stalling time of the motor
   (c) higher than both accelerating time and safe stalling time of the motor

7. Extremely sensitive ground protection is provided in an induction motor when
   (a) a relay is connected in the residual circuit of three line CTs
   (b) a relay is connected in the core balance CT
   (c) both (a) and (b)
   (d) none of the above

8. The typical setting range of negative sequence relay used for the protection of motors against unbalanced currents is
   (a) 10–50%
   (b) 1–5%
   (c) 100–500%
   (d) none of the above

9. Which relay operates during the removal of one phase of the induction motor?
   (a) Short circuit
   (b) Earth fault
   (c) Negative phase sequence
   (d) None of the above

10. The majority of faults in an induction motor are due to
    (a) insulation and mechanical failure
    (b) bearing failure
    (c) breakage of conveyer belt
    (d) none of the above

Review Questions

1. Discuss various abnormal conditions of an induction motor and their causes and consequences.

2. Draw both hot and cold characteristics of thermal overload relay used for induction motor protection.

3. Explain the loss of load protection and its requirement for an induction motor.

4. What is the need for stalling protection in an induction motor?

5. Explain the differential protection used for phase fault protection in a large induction motor.

6. Justify the use of differential protection for a large induction motor.

7. Explain the abnormal frequency and voltage protection applied to an induction motor.

8. What is the need for protection against the reversal of phase in an induction motor?

9. Explain the protection against unbalanced currents in an induction motor.

10. Enumerate different features of the numerical relay used for the protection of an induction motor.

Numerical Exercises

1. An induction motor is protected with overload and short circuit. The details of the motor are as follows.

   Rated output: 750 kW
   Rated power factor: 0.85

Rated voltage: 6.6 kV

Efficiency of motor: 92%

Continuous overload: 110% of rated current

Starting current: 6 times the rated current at 75% the rated voltage

Suggest a suitable CT ratio and calculate the relay setting for the overload and short circuit protection of the motor.

[CT ratio: 100/1 A, Thermal overload relay setting = 90% of 1 A, and instantaneous overcurrent relay setting = 700% of 1 A]

2. The following is the data for a three-phase, 50 Hz induction motor.

Rated output: 1500 HP

Rated power factor: 0.85

Rated voltage: 6.6 kV

Efficiency of motor: 90%

Continuous overload: 110% of rated current

Starting current: 6 times the rated current at 75% the rated voltage

Acceleration time: 8 s

Safe stalling time: 18 s

Safe stalling current: one-third of starting current

Suggest a suitable CT ratio used for protection. Calculate the setting of the relay for the following types of protection.

(a) Overload protection (setting range = 70–130% of 1 A)

(b) Short circuit protection (setting range = 400–2000% of 1 A)

(c) Stalling protection with timer setting (setting range = 150–600% of 1 A, time setting = 6–30 s)

[CT ratio: 150/1 A, thermal overload relay setting = 90% of 1 A, instantaneous overcurrent relay setting = 700% of 1 A, and definite time overcurrent relay = 240% of 1 A with timer setting = 13 s]

Busbar Protection

8 ⚡

Learning Objectives

After going through this chapter, the students will be able to:

- List and explain the different types of busbar arrangements
- Classify the various busbar protection schemes based on the type of relay used
- Discuss the effect of CT saturation on busbar protection and the remedies to reduce it
- Explain high impedance bus differential protection scheme
- Differentiate linear coupler from ordinary CT
- Explain double busbar protection scheme

8.1 Introduction

The *busbar* is the junction of an electrical network where many lines are connected together. It is the area in the power system where the magnitude of fault current is very high. The busbar protection scheme demands smaller time of operation by quick operation of breakers, better reliability during internal faults, and high stability in case of external faults. Whenever a fault occurs, the busbar protection scheme operates and trips all the transmission lines connected to that bus. Hence, it isolates the bus from the rest of the system. In case of heavy through-faults on some of the circuits connected to the bus, the insufficiently rated current transformers (CTs) will result in saturation and may maloperate the busbar protection scheme.

Whenever a bus fault occurs, it results in severe disturbances as clearance of this fault requires tripping of all the breakers of the lines connected to the faulted bus. Moreover, the damage caused by an unprotected bus during bus fault is found to be very severe.

Conversely, the magnitude of fault current is very high during bus fault. Therefore, a high-speed busbar protection scheme is required to achieve maximum security, as otherwise uncleared internal faults generate severe dynamic and thermal forces that imperil the entire substation.

8.2 Busbar Arrangements

There are several types of busbar arrangements. The choice of arrangement depends on many factors such as system voltage, reliability of supply, position of substation in the system, flexibility, and cost. The other factors are as follows:

1. Simplicity of busbar arrangement
2. Easy maintenance with interrupting power supply
3. Economic viability with reference to the continuity of supply

4. Availability of backup of busbar arrangement in case of any outage
5. Flexibility in expansion or augmentation with reference to future load growth

The different busbar arrangements are explained in Sections 8.2.1–8.2.5.

8.2.1 Single Busbar Arrangement

Figure 8.1 shows the single line diagram of a single busbar arrangement scheme. It is the simplest busbar arrangement scheme, and consists of a single busbar and other associated equipment.

Advantages

1. It has low initial cost.
2. Maintenance is low.
3. It can be used for small and medium-sized stations where shutdown can be permitted.

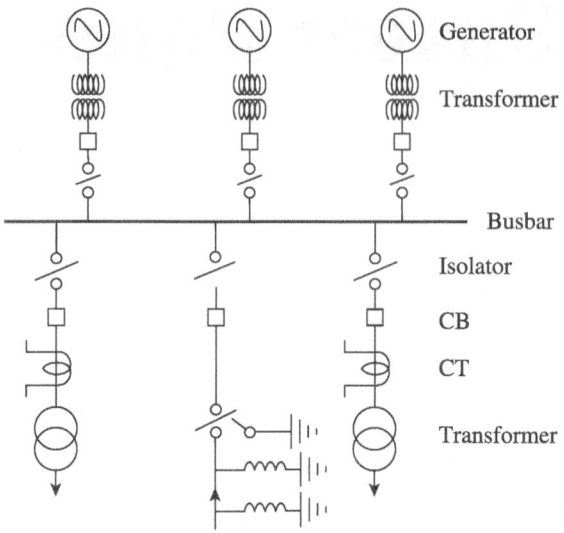

Fig. 8.1 Single busbar arrangement

Disadvantages

1. During a busbar fault, all the healthy feeders are also disconnected.
2. Maintenance on any of the station feeders is not possible without interruption of supply.

8.2.2 Single Busbar Arrangement with Sectionalization

Complete shutdown during a fault on a busbar can be avoided by using sectionalization in a single busbar arrangement. Sectionalization is carried out using circuit breakers and isolators. The maximum advantage of this scheme is achieved by even distribution of incoming and outgoing feeders on the sections. Figure 8.2 shows the single line diagram of a single busbar arrangement with sectionalization.

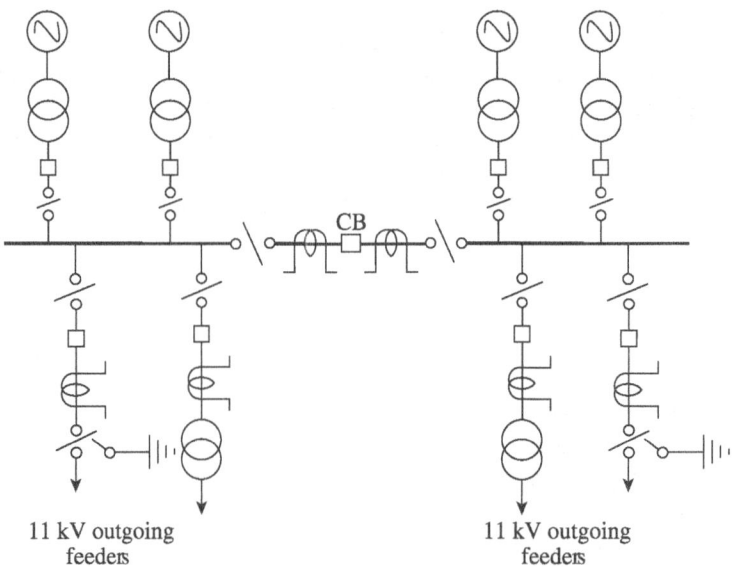

Fig. 8.2 Single busbar arrangement with sectionalizer under off-load conditions

Advantages

1. Maintenance of one section is possible without interrupting the supply of the other section.
2. The breaking capacity of circuit breakers can be reduced by inserting a reactor between the sections.

Disadvantages

1. In case of a fault on any one of the sections of the scheme, healthy feeders may be affected.
2. If air break isolators are used as a sectionalizer, then isolation is not carried out automatically at the time of fault.

8.2.3 Main and Transfer Busbar Arrangement

Main and transfer busbar arrangement is used when continuity from the power supply to the load is the main consideration. By providing additional flexibility, continuity of supply, and periodic maintenance, this scheme justifies additional costs. This arrangement is an alternative to the double busbar scheme and is suitable for highly interconnected power system networks. This scheme contains two buses, namely main bus and transfer bus. Each generator and feeder is connected to the main bus as well as to the transfer bus through a bus coupler. Figure 8.3 shows the single line diagram of a main and transfer busbar arrangement.

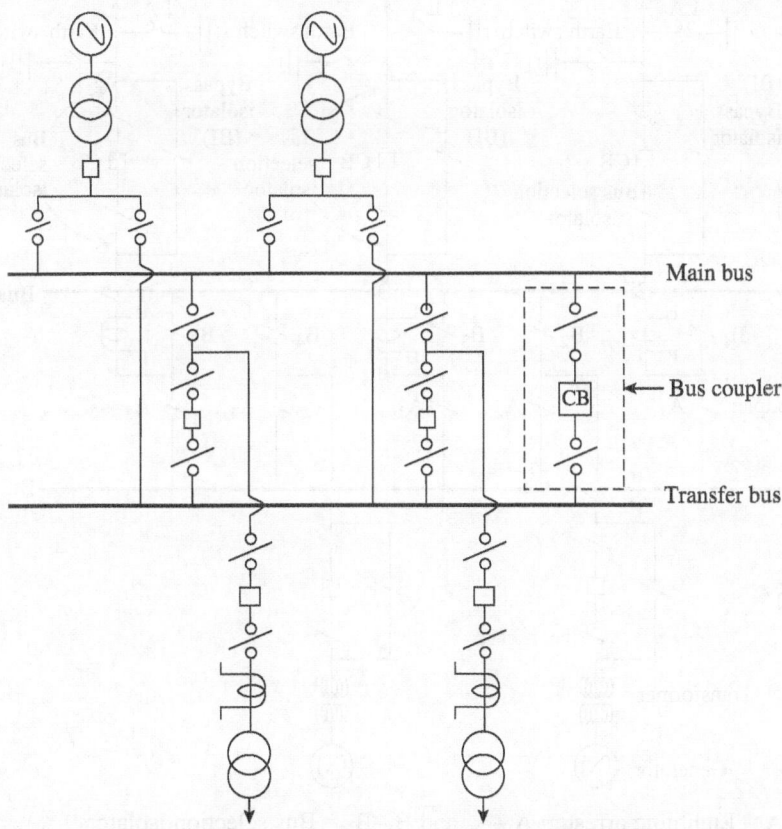

Fig. 8.3 Main and transfer busbar arrangement

Advantages

1. It provides continuity of supply during a fault on the busbar.
2. Repair and maintenance can be carried out on either of the buses (main or transfer) by transferring the entire load to either of the buses.
3. The bus potential can be used for protective devices.

Disadvantages

1. Cost is higher than that of the single busbar arrangement.
2. After the transfer of all loads on to the transfer bus, a fault on any of the circuits connected to the transfer bus would shut down the entire station.

8.2.4 Double Busbar Arrangement

In this scheme, two identical busbars are used so that each load may be fed to either of the buses. This arrangement is used where the loads and continuity of the power supply justify additional costs. The working of this arrangement during different situations is explained here. Figure 8.4 shows the single line diagram of a double busbar arrangement.

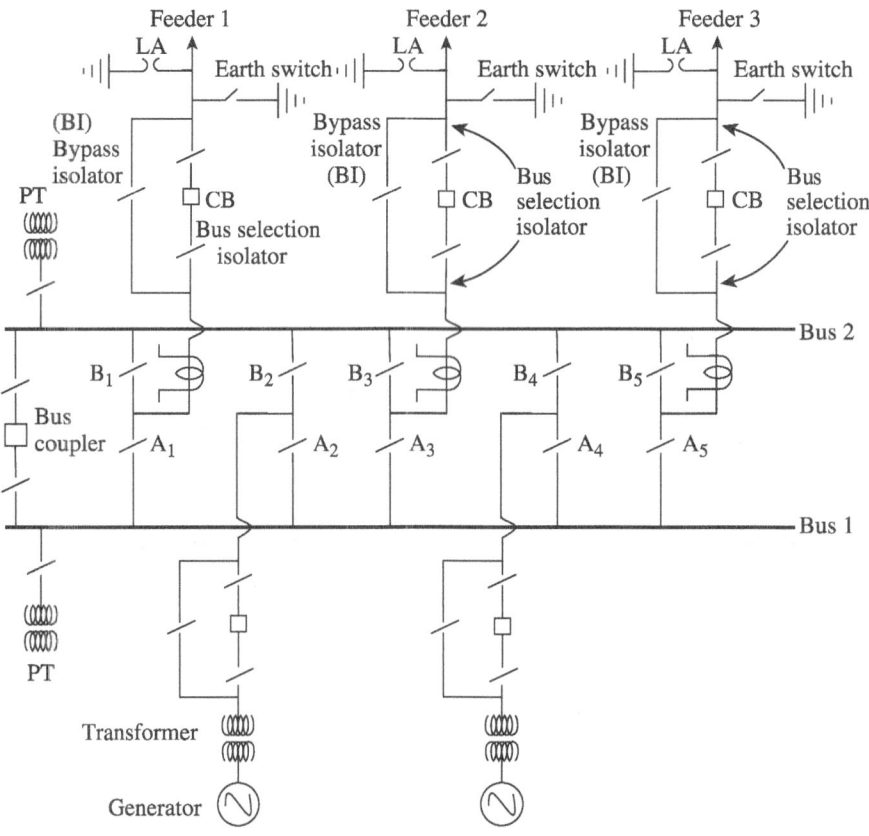

LA = Lightning arrester; A_1–A_5 and B_1–B_5 = Bus selection isolators

Fig. 8.4 Single line diagram of double busbar arrangement

Normal Condition

Let us assume that under normal conditions, power is fed through bus 1. Further,

1. All the isolators and circuit breakers are on.
2. All earthing switches and bypass isolators are off.
3. Bus coupler and associated isolators are on.
4. All bus selection isolators of bus 1 are on (A_1 to A_5).
5. All bus selection isolators of bus 2 are off (B_1 to B_5). Hence, bus 2 is live, but no power is fed through it in normal condition.

Maintenance of Bus 1

The following operations are to be performed during the maintenance of bus 1.

1. Switch on all the isolators of bus 2 (B_1 to B_5). Now, the power will be fed by both the buses.
2. Switch off all the isolators of bus 1 (A_1 to A_5). Now, the power is fed through bus 2 alone. In this situation, although heavy currents will flow because of the closing of isolators during these two steps, there will be no sparking.
3. Switch off the bus coupler and the respective isolators. Now, you can maintain bus 1 by closing the respective earthing switches of bus 1, as the total power is now transferred to bus 2 without any power interruption.

Fault on Bus 1

Let us assume that power is fed through bus 1 during normal condition. If a fault occurs on bus 1, then all feeder breakers will trip because of which momentary interruption of power occurs. Power can be re-established by performing the following steps.

1. Switch off the respective bus selection isolators of bus 1.
2. Switch off the bus coupler and the associated isolators.
3. Switch on all the isolators of bus 2.
4. Switch on all line isolators and breakers. It is to be noted at this juncture that first the isolators should be switched on and thereafter the breakers should be switched on.

Maintenance of Line Breaker

Let us assume that we want to carry out maintenance of the breaker of feeder 1. The following steps are to be performed:

1. Switch on the bus selection isolator of feeder 1 from bus 1.
2. Switch on the bypass isolator of feeder 1.
3. Switch off the breaker and then the respective isolators of feeder 1. Now, the power of feeder 1 will flow through the bypass isolator to bus 2, the bus coupler, bus 1, and finally to load.

The advantages of double bus arrangement are as follows:

1. It provides continuity of supply during a fault on any of the feeders.
2. Repair and maintenance can be carried out on either of the buses by transferring the entire feeders onto the other bus.
3. The bus potential can be used for protective devices.
4. Double busbar arrangement is more reliable than single busbar and sectionalizer busbar arrangements.

The disadvantages of double bus arrangement are as follows:

1. The installation cost is very high compared to the single bus arrangement.
2. Only one line breaker can be maintained at a time.
3. During a fault on one of the busbars, there is a momentary interruption of supply.

8.2.5 One-and-half Breaker Arrangement

As there is a saving in the number of circuit breakers, this arrangement is much more popular and better than the double busbar arrangement. In this scheme, the number of circuit breakers is one and half per circuit, and hence, it is known as *one-and-half breaker arrangement*. Figure 8.5 shows the single line diagram of one-and-half breaker arrangement.

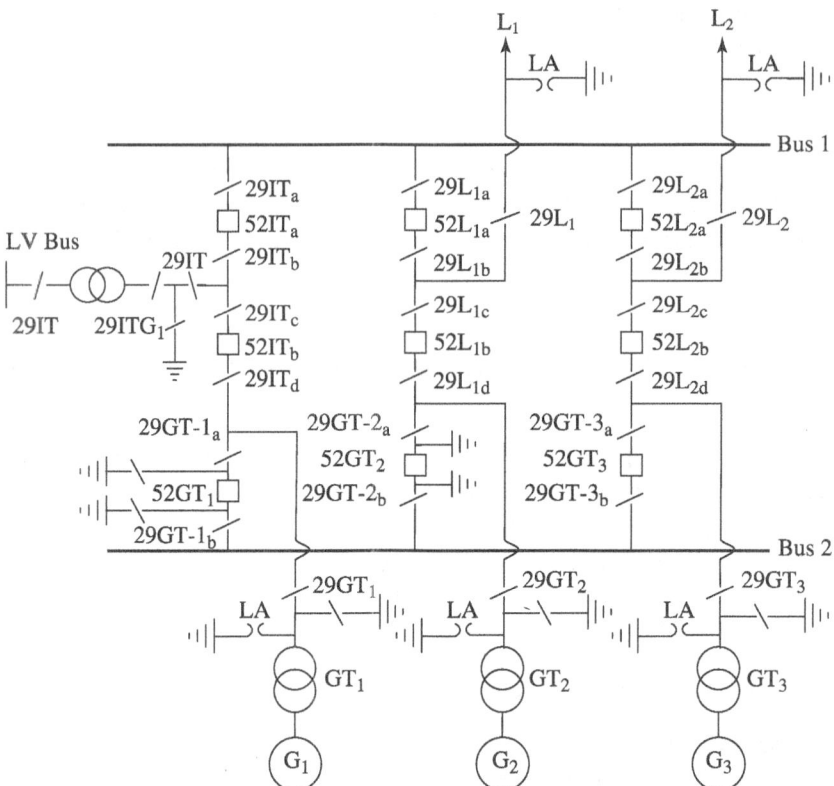

LA = Lightning arrester; 52 = Circuit breaker, 29 = Isolator,
IT stands for interconnected transformer

Fig. 8.5 Single line diagram of one-and-half breaker arrangement

The working of this arrangement during different conditions is explained here.

Normal Condition

Let us assume that bus 1 and bus 2 are live.

1. All the circuit breakers and isolators ($52GT_1$, $29GT-1_a$, $29G-1_b$, $52GT_2$, $29GT-2_a$, $29GT-2_b$, $52GT_3$, $29GT-3_a$, and $29GT-3_b$) of GT_1, GT_2, and GT_3 are in a closed condition.
2. In order to feed power to transmission lines L_1 and L_2, breakers $52L_{1b}$ and $52L_{2b}$ and their associated isolators, that is, $29L_{1c}$, $29L_{1d}$, $29L_{2c}$, and $29L_{2d}$, are in a closed condition.
3. To energize bus 1, breakers $52L_{1a}$ and $52L_{2a}$ and their associated isolators, that is, $29L_{1a}$, $29L_{1b}$, $29L_{2a}$, and $29L_{2b}$, are also in a closed condition.

Maintenance of Busbar 1

Let us assume that we want to carry out maintenance of busbar 1. The following steps are to be performed.

1. Switch off all the breakers, that is, 52 L_{1a}, 52 L_{2a}, and 52 IT_a, connected to bus 1.
2. Switch off all the associated isolators, that is, $29L_{1a}$, $29L_{1b}$, $29L_{2a}$, $29L_{2b}$, $29IT_a$, and $29IT_b$, connected to these breakers.
3. Now busbar 1 is maintained without interrupting power to L_1, L_2, and the interconnected transformer (ICT).

Fault on Busbar 1

Let us assume that the fault occurs on busbar 1. The following steps are to be performed.

1. All the breakers, that is, $52L_{1a}$, $52L_{2a}$, and $52IT_a$, connected to bus 1 will be tripped by the bus zone protective device.
2. In this situation, the power to the transmission lines L_1, L_2, and the ICT will be continued through $52GT_1$, $52GT_2$, $52GT_3$, $52L_{1b}$, $52L_{2b}$, and $52IT_b$.

Maintenance of any Breaker

The following steps are to be performed for the maintenance of any of the two breakers.

1. If $52GT_2$ is to be maintained, then the power will be fed through $52L_{1b}$. $52L_{1a}$ will also be on.
2. If $52L_{1a}$ is to be maintained, then the power will be fed through $52GT_2$ and $52L_{1b}$.
3. If $52L_{1b}$ is to be maintained, then the line L_1 will be fed through $52L_{1a}$. $52GT_2$ will also be on. Bus 1 will always remain live.

Advantage More than one line breaker can be maintained at a time.

Disadvantage More number of breakers is required for the bus as well as the lines; hence cost is higher than the double busbar arrangement.

8.3 Busbar Faults and Protection Requirement

Protection engineers try to implement a dedicated bus zone protection scheme using the various schemes invented by many researchers. If it is not implemented perfectly, the clearing of bus fault will be performed by the backup protection provided to the lines terminating at the bus. The high fault levels associated with a busbar require fast protection. Typical fault clearing time should be less than 100 ms; with fast breakers, this means that the measuring time should be about 20–30 ms.

The prime requirement of the busbar protection scheme is the identification of fault in a particular region, and the disconnecting of breakers associated with that particular region so that the healthy section of the power system network remains unaffected. This discriminating feature of the busbar protection scheme minimizes interruption to the plant. Hence, it is justified that discrimination on the basis of time graded relays is not a good choice, and hence, there is need for unit protection scheme for the busbar.

Busbar protection scheme should not operate for any external (through) fault, as otherwise it unnecessarily trips other healthy lines connected to the bus. This is a very important feature of the busbar protection scheme with respect to the stability of the power system.

It is to be noted that all these requirements cannot be achieved completely by any of the schemes without affecting the other requirements. Therefore, there is always a compromise between many factors such as speed vs selectivity and stability vs dependability.

The bus fault occurs infrequently. It is likely to cause extensive damage, and may destroy an entire power system. Bus faults have been observed to be relatively rare, around 6–7% of all faults compared with line faults, which are over 60%. It has been observed from the widely published literature that most of the bus faults are ground faults with 67% being single-line-to-ground faults, 15% being double-line-to-ground faults, and 19% being triple-line-to-ground faults. The causes of bus faults are classified as follows:

1. Insulation failure due to the deterioration of the material
2. Flashover caused by prolonged and excessive overvoltages

3. Failure of circuit breakers or other switchgears
4. Human errors in operating and maintaining switchgear
5. Foreign objects falling across busbar
6. Contact by animals

Damage to the equipment depends on the fault type, fault duration, fault level, and the withstand capability of the switchgear. The isolation of a busbar disrupts all the circuits connected to the busbar. The busbar protection system, therefore, must be carefully monitored to prevent inadvertent operations. A scheme that is simple in design and easy to apply is most likely to provide a reliable service.

To meet these requirements, a bus protection system must satisfy the following criteria.

1. The bus protection system gives high-speed protection during in-zone faults to minimize damage and maintain system stability.
2. It remains stable for all external faults to avoid unnecessary interruption of supply.
3. It provides proper discrimination between two zones (in-zone and out of zone), that is, tripping a minimum number of circuit breakers.
4. It gives reliable operation to avoid extensive damage to the equipment, danger to personnel, and disruption of service.

8.4 Impact of Current Transformer Saturation on Busbar Protection

While discussing the busbar protection scheme, it is necessary to evaluate the characteristic of CTs and certain parameters related to their saturation. A few issues related to the saturation of CTs are discussed here.

In general, a small amount of current always flows in the operating coil because of the mismatch of CT ratios and differences in characteristics of the CTs. The phenomenon of CT saturation and CT ratio mismatch, which causes the differential relay to operate incorrectly, is described in the following sections.

8.4.1 Current Transformer Saturation

Whenever the required flux density to produce the CT secondary current exceeds the core limit, CT saturation may occur. In this situation, the CT secondary current is distorted and does not follow the CT ratio. CT saturation depends on many factors such as burden, ratio of the CT, core material, cross-sectional area of the core, level of remnant flux, and DC offset in the fault current.

A simplified circuit of a CT is shown in Fig. 8.6. L_m and R_m represent the non-linear magnetizing inductance and the iron loss equivalent resistance, respectively.

I_m and I_r are reactive and active components of the magnetizing current, respectively. R_b is the load, containing impedances of all leads and the relay coils, connected to the CT. R_p and L_p represent resistance and leakage inductance of the primary winding, whereas R_s and L_s represent

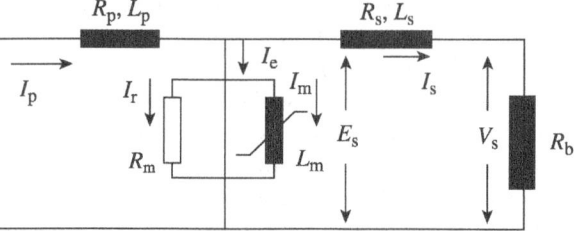

Fig. 8.6 Simplified circuit of a current transformer

the resistance and the inductance of the secondary winding. E_s and V_s are the induced electromotive force (emf) of the CT secondary winding and the voltage at the CT terminals, respectively.

The primary winding of a CT is in series with the line and, therefore, carries the current that flows in the line. When short circuit occurs, the line current becomes large and flows in the CT primary winding (I_p). The secondary current (I_s) also increases, and ideally, it should be proportional to the primary current. The CT must develop sufficient voltage (E_s) to make this current flow in the secondary circuit. To generate this voltage, a part of the primary current becomes the magnetizing current (I_e), which produces a flux in the core of the CT.

Normally, the magnetizing current is small, and the secondary current remains proportional to the primary current for all practical purposes. If a CT has to develop a large voltage to overcome the voltage drop in the secondary circuit, the level of core flux must be high. If the flux approaches the saturation level, the exciting current (I_e) becomes large and the secondary current decreases. The secondary current of the CT in this case is less than what it was when the CT was not saturated. As the primary current increases beyond the saturation level, the core saturates during a part of the cycle only. The result is that the secondary current becomes distorted.

As the CTs are connected in series with line, they carry high magnitudes of current at the time of in-zone and out-of-zone faults. A high level of external fault current can cause a particular CT to saturate, resulting in different secondary currents out of the many CTs. This results in the flow of a differential current in the operating element of the relay, and hence, the differential relay operates during external faults. Therefore, it is essential that corrective steps be taken to detect CT saturation and block the relay operation.

8.4.2 Ratio Mismatch

The ratio of various CTs used in a differential protection scheme should be selected in such a way that it provides no differential current during normal operation and through-fault conditions. Any mismatch in the ratio of CTs will result in maloperation of the relay due to differential current during through-faults. The realization of exactly matching CT ratios is difficult in practice. As such, there always are some currents in the differential coils of the relays. To avoid incorrect operations of differential relays, additional features are employed in the relay.

8.4.3 Remedies

Figure 8.7(a) shows the concept of the differential protection scheme applied to a busbar during external fault. Figure 8.7(b) shows the same scheme during an internal fault.

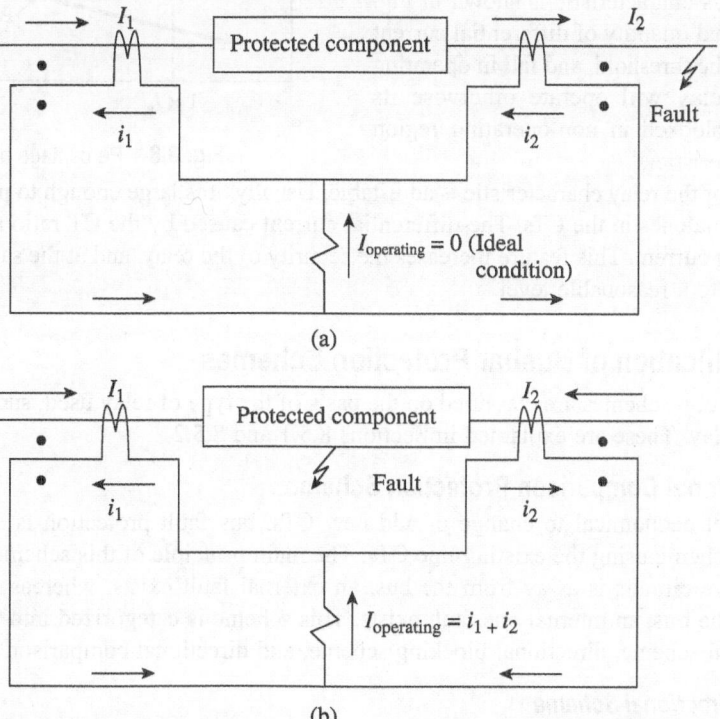

(a)

(b)

Fig. 8.7 Principle of a differential protection scheme (a) External fault
(b) Internal fault

The lengths of the leads, connecting the CTs to the relay, are not usually equal. The burdens (in VA) on the CTs, therefore, are not equal. This causes the CTs to produce different current outputs for the same level of input currents. Hence, during normal and through-fault conditions, the differential current, of course small in magnitude, flows in the operating element of the differential relay. Therefore, the application of the simple differential protection scheme is adversely affected by the characteristics of the CTs.

There are many ways to reduce the effect of CT saturation phenomena. Increasing the cross-sectional area of the CT core increases the sustainability of flux density in core and hence, reduces the saturation of CT. In order to improve the stability of unit protection scheme during heavy through-fault, a stabilizing resistance is connected in series with the operating coil of the relay. This increases the burden on the relay and also reduces the differential current. However, insertion of stabilizing resistance reduces the sensitivity of the differential protection scheme during the internal fault.

Most of the differential relays are provided with a percentage bias feature to avoid tripping due to CT ratio mismatch. The differential relays include restraints derived from the arithmetical sum of all the currents. Hence, the operating current in this relay is the vector sum of all the currents, whereas the restraining current is their scalar sum.

The required differential current to operate the relay must exceed a set percentage of the total restraint. The ratio of the operating current to the restraining current is expressed in percentage, and it is usually known as the *slope of the relay characteristic*. A typical percentage bias characteristic is shown in Fig. 8.8. If the actual quantity of differential current (I_{diff}) crosses the threshold and fall in operating region, the relay will operate otherwise its operation is blocked in non-operating region of the characteristic.

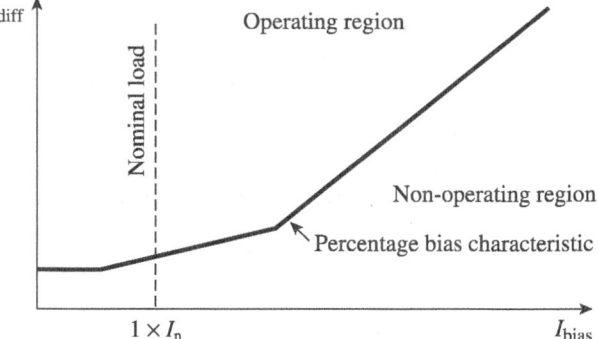

Fig. 8.8 Percentage bias characteristic

The slope of the relay characteristic is adjustable. Usually, it is large enough to prevent the relay operation owing to mismatches in the CTs. The differential current caused by the CT ratio mismatch is overcome by the restraining current. This feature increases the security of the relay, and at the same time, it also maintains its sensitivity to a reasonable level.

8.5 Classification of Busbar Protection Schemes

Busbar protection schemes are classified on the basis of the type of relay used, such as directional relay and differential relay. These are explained in Sections 8.5.1 and 8.5.2.

8.5.1 Directional Comparison Protection Scheme

When it is not economical to change or add new CTs, bus fault protection is achieved by a directional comparison scheme using the existing line CTs. The main principle of this scheme is that if the power flow in one or more circuits is away from the bus, an external fault exists, whereas for power flow in all the circuits into the bus, an internal bus fault exists. This scheme is categorized into three parts, namely series trip directional scheme, directional blocking scheme, and directional comparison scheme.

Series Trip Directional Scheme

In this scheme, all the directional relay trip contacts are connected in series. Figure 8.9(a) shows two directional relays, R_1 and R_2, of line 1 and line 2 for bus protection. Figure 8.9(b) shows its control circuit where the contacts of all line relays are connected in series. In case of an internal fault, the contacts R_1 and R_2 close simultaneously and energize the auxiliary relay (Aux). As a result, auxiliary relay contacts AX_1 and AX_2 close. The mechanical switches A_1 and A_2 are replicated with breaker position and are normally in closed position. Hence, whenever AX_1 and AX_2 close, it energizes the tripping coil (TC) of circuit breaker and circuit breakers of respective line trips.

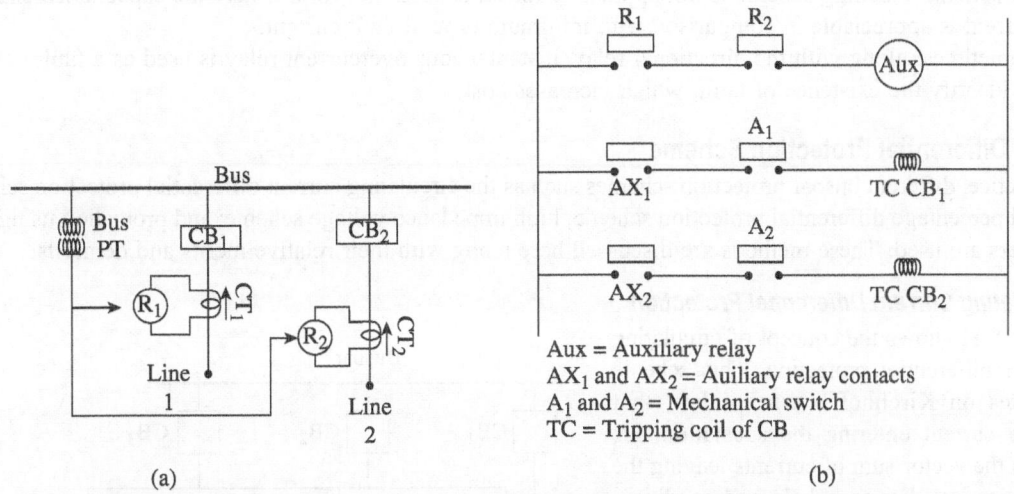

Aux = Auxiliary relay
AX_1 and AX_2 = Auiliary relay contacts
A_1 and A_2 = Mechanical switch
TC = Tripping coil of CB

(a) (b)

Fig. 8.9 Series trip directional comparison scheme (a) Power circuit (b) Control circuit

Directional Blocking Scheme

Figure 8.10 shows the control circuit of the directional blocking scheme for busbar protection. The power circuit is the same as that shown in Fig. 8.9(a). In this scheme, all the tripping contacts (R_1 and R_2) are connected in parallel and then to the tripping relay (Aux), whereas all the blocking contacts are connected in parallel and then to the blocking relay (B). During an external fault, the blocking relay (B) operates and blocks the operation of bus protection as per requirement by opening one of its contacts, B1, which is in series with the tripping relay.

Directional Comparison Scheme

The use of directional blocking scheme can be avoided by the directional comparison scheme, which uses a voltage restraint relay. However, the main problem with this scheme is that the relay settings must be reviewed and changed whenever system changes are made near the protected bus.

Advantage The fundamental advantage of the directional protection scheme is that it is not affected by CT saturation as it compares the direction of the current and not the magnitude.

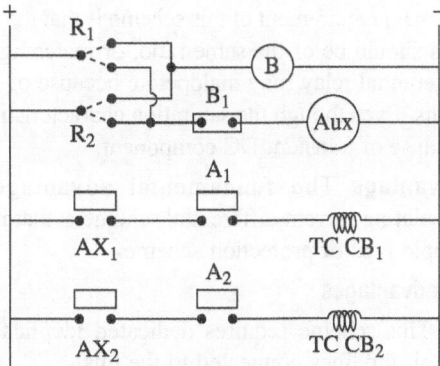

Aux = Auxiliary relay
AX_1 and AX_2 = Auxiliary relay contacts
A_1 and A_2 = Mechanical swictch
TC = Tripping coil of CB
B = Blocking relay
B_1 = Contact of blocking relay

Fig. 8.10 Control circuit of directional blocking scheme

Disadvantages

1. The reliability of series trip directional scheme can be compromised by too many series contacts. Further, its circuitry is too complex, and hence, it requires careful and periodic review. Moreover, this scheme requires more time to provide coordination, which initiates all series-connected contacts of directional relays to clear a bus fault.
2. Directional blocking scheme is not applicable for large cable networks where the capacitance charging current is appreciable in comparison with minimum ground fault current.
3. Sometimes, along with the directional relay, instantaneous overcurrent relay is used as a fault detector to identify the existence of fault, which increases cost.

8.5.2 Differential Protection Scheme

In practice, different busbar protection schemes such as the circulating current differential protection scheme, biased percentage differential protection scheme, high impedance voltage scheme, and protection using liner couplers are used. These methods are discussed here along with their relative merits and demerits.

Circulating Current Differential Protection

Figure 8.11 shows the concept of circulating current differential protection. This scheme operates on Kirchhoff's current law, that is, the current entering the substation bus equals the vector sum of currents leaving the substation bus. For an unbalanced condition, owing to internal fault at bus, the differential relay (R) operates whereas in case of external fault, the differential relay does not operate as no current is fed to the relay coil. However, the main requirement of this scheme is that the

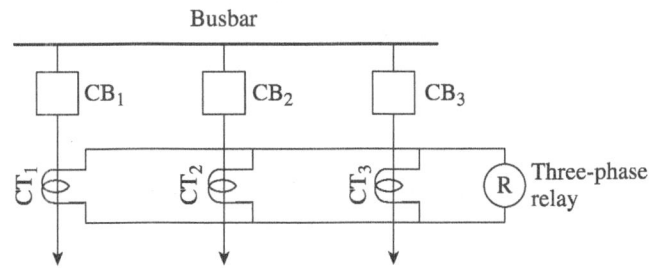

Fig. 8.11 Circulating current differential protection

CTs should be of the same ratio, or matching CTs are required. If these conditions are not satisfied, then the differential relay may maloperate because of the presence of spill current, particularly during heavy through-faults. Even though the saturation characteristic of all the CTs is identical, the differential relay may maloperate because of transient DC component.

Advantage The fundamental advantage of the circulating current differential scheme is that it is a very simple type of protection scheme.

Disadvantages

1. This scheme requires dedicated identical CTs for all the lines connected to the bus.
2. This scheme may maloperate in presence of a transient DC component in the fault.

Biased Percentage Differential Protection

In order to avoid maloperation of relays due to CT saturation and transient DC current, biased percentage differential protection scheme is used. Figure 8.12 shows the connection logic of the biased percentage differential protection scheme. The differential relay

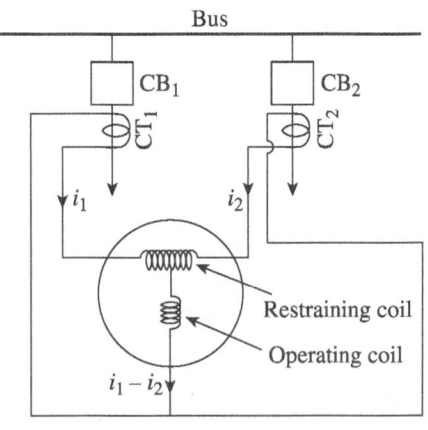

Fig. 8.12 Biased percentage differential protection

has two coils, namely restraining coil and operating coil. The percentage bias characteristic is shown in Fig. 8.8. Maximum security for external faults is obtained when all the CTs have identical CT ratio. In order to operate the percentage differential relay, the differential current ($i_1 - i_2$) in the operating coil must exceed a fixed percentage of the restrain current (($i_1 + i_2$)/2) in the restraining coil. The *restraining current* is defined as the average of the incoming and outgoing line currents at the bus.

Advantage The fundamental advantages of the biased percentage differential scheme are high tolerance against substantial CT saturation, reduced requirement of dedicated CTs, and its applicability where comparatively high-speed tripping is required.

Disadvantage The fundamental limitation of this scheme is that the relay may maloperate in case of close-in external fault due to complete saturation of the CT.

High Impedance Voltage Differential Protection

To overcome spill current due to CT saturation in case of external fault, the high impedance voltage relaying scheme, as shown in Fig. 8.13, is widely used. The effect of saturation is controlled by keeping the CT's secondary and lead resistance low and by adding resistance to the relay circuit. Here, full wave bridge rectifier adds substantial resistance to the circuit. The series L–C circuit is turned to 50 Hz fundamental frequency to respond to only the fundamental component of current and make the overvoltage relay immune to DC offset and harmonics. This scheme discriminates between internal and external faults through the relative magnitudes of the voltage across the differential junction points.

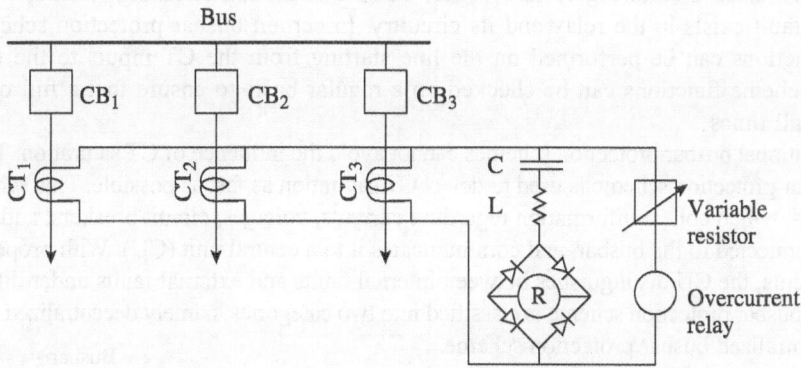

Fig. 8.13 High impedance voltage differential protection

Advantage The main merits of this scheme are its stability against transient DC component due to the tuned circuit, improved CT saturation characteristics because of stabilizing resistors, and faster operating time.

Disadvantage The disadvantages of this scheme include the need for dedicated CTs (cost increases), maloperation of relay when the secondary leakage reactance is present, and inapplicability of the scheme to the reconfigurable busbar.

Protection using Linear Couplers

CT saturation in case of iron core CT is rectified using linear couplers (air core mutual reactors) as they use air core. The secondary of all linear couplers are connected in series as shown in Fig. 8.14. The output voltage of the linear coupler is proportional to the derivative of the input current because of its

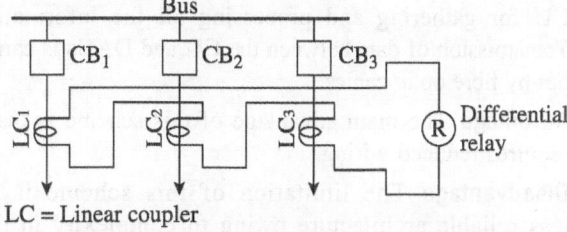

LC = Linear coupler

Fig. 8.14 Bus differential relay with linear coupler

linear characteristics. If the voltage sum across the relay is zero, then the input current is equal to the output current at the bus. During an internal fault, all the line currents flow towards the bus, and thus, the induced voltage appears across the relay.

Advantage A linear coupler is a special device that can operate low energy relays compared to conventional CTs.

Disadvantage This scheme needs extra equipment to realise the benefits of microprocessor-based relays, which increases the overall cost of the scheme.

8.6 Digital/Numerical Protection of Busbar

Digital/numerical relays provide significant benefits for industrial as well as commercial systems where an economical but effective busbar protection scheme is required. These techniques are increasingly being applied to major busbars and continuously improve power system operation. Further, they provide many advantages such as reduction in lifetime management costs and maintenance costs, and the ability to extend the protection to cater to primary system changes with minimal hardware alterations and transmission system changes.

The digital technique allows a fast and easy connection to substation automation systems, providing fast fault analysis and monitoring. Modern numerical busbar relays incorporate modelling of the CT's response to eliminate errors caused by effects such as CT saturation. Numerical busbar protection schemes include a sophisticated monitoring feature widely known as the *self-checking feature*, which initiates an alarm when a fault exists in the relay and its circuitry. In certain busbar protection schemes, continuous monitoring functions can be performed on the line starting from the CT inputs to the tripping outputs, and thus, the scheme functions can be checked on a regular basis to ensure that a full operational mode is available at all times.

Most conventional busbar protection schemes cannot avoid the influence of CT saturation. Therefore, digital/numerical busbar protection scheme is used to detect CT saturation as fast as possible. This scheme uses several processing units, which collect information regarding currents, voltages, circuit breakers, and isolator status of lines that are connected to the busbar, and communicates it to a central unit (CU). With proper algorithms and logical judgements, the CU distinguishes between internal faults and external faults under different situations. Modern digital busbar protection scheme is classified into two categories, namely decentralized busbar protection scheme and centralized busbar protection scheme.

8.6.1 Decentralized Busbar Protection

The decentralized busbar protection scheme, as shown in Fig. 8.15, uses data acquisition units (DAUs) installed in each line to sample and pre-processes the signals. Further, it also provides tripping signals to the circuit breakers. The decentralized busbar protection scheme uses a separate CU for gathering and processing all the information. Transmission of data between the CU and DAUs is carried out by fibre optic cables.

Advantage The main advantage of this scheme is that it requires reduced wiring.

Disadvantage The limitation of this scheme is the less reliable architecture owing to complexity in data transfer.

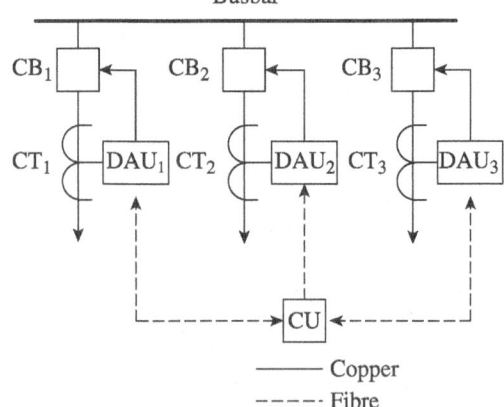

Fig. 8.15 Schematic block diagram of decentralized busbar protection

8.6.2 Centralized Busbar Protection

Figure 8.16 shows the structure of centralized busbar protection scheme. In this scheme, it is mandatory to connect all the signals at a central location, where a single *relay* performs all the functions. Further, in this scheme, all the preprocessings and computations such as sampling, filtering, analog to digital and digital to analog conversion, and relay logic are performed by the CU. Hence, this scheme imposes more computational burden on the CU.

Advantages

1. This scheme is more reliable than the decentralized busbar protection scheme as it performs all computations in a single relay module.
2. This scheme is more suitable for retrofit applications.

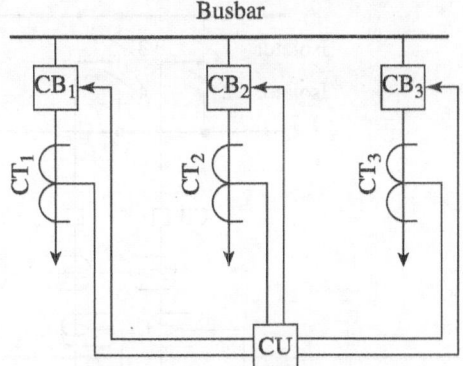

Fig. 8.16 Schematic block diagram of centralized busbar protection

8.7 Recent Trends in Busbar Protection

In certain busbar protection schemes, continuous monitoring of the line starting from the CT inputs to the tripping outputs can be performed. The IEC 61850 international standard for communications in substations brings in a new era in the development of substations and future application for bus protection. This new protocol affects not only the design of the substation protection, monitoring, and control system but also the design of the substation secondary circuits. High-speed peer-to-peer communications using generic object oriented substation events (GOOSE) messages and sampled analogue values allow the development of distributed applications based on current and voltage values communicated between devices connected to the substation local area network.

8.8 Commercially Used Technique

It is clear from our earlier discussions that there is a need to develop a protection scheme for the double busbar system, which is widely used in the industry. Figure 8.17 shows the block diagram of the double busbar protection scheme. To distribute the load equally, some of the feeders are connected at bus 1, whereas the others are connected at bus 2. Using the bypass isolator, all the lines can be switched over to either bus 1 or bus 2 during fault or maintenance on either of the buses. The main digital relay contains three modules, namely bus 1 zone relay, bus 2 zone relay, and check relay in which the relay algorithm is located. During an internal fault, the relay operates according to its tripping logic.

Figure 8.18 shows the tripping logic of this scheme. The check zone relay works as an overall bus zone protection, where all the line signals are fed to detect the internal fault either on bus 1 or on bus 2. Here, the check zone relay contacts are in series with the contacts of the main relay of bus 1 zone relay and bus 2 zone relay. If the fault occurs on bus 1, the contacts of the check zone and bus 1 zone relays close and give the tripping signal to the respective breakers of lines connected to bus 1. The CT secondary signals fed to the main relay of the respective bus zones can be selected through an isolator selector switch.

Maloperation of busbar protection can result in widespread system failure. Therefore, its operation is monitored by some form of check relay. In the case of high impedance relay, the setting calculations are quite high, and sometimes, low settings can be adopted to provide more safety. Hence, there may be a possibility of maloperation from the design point of view. The provision of a check feature is therefore purely a measure against maloperation caused by external agencies.

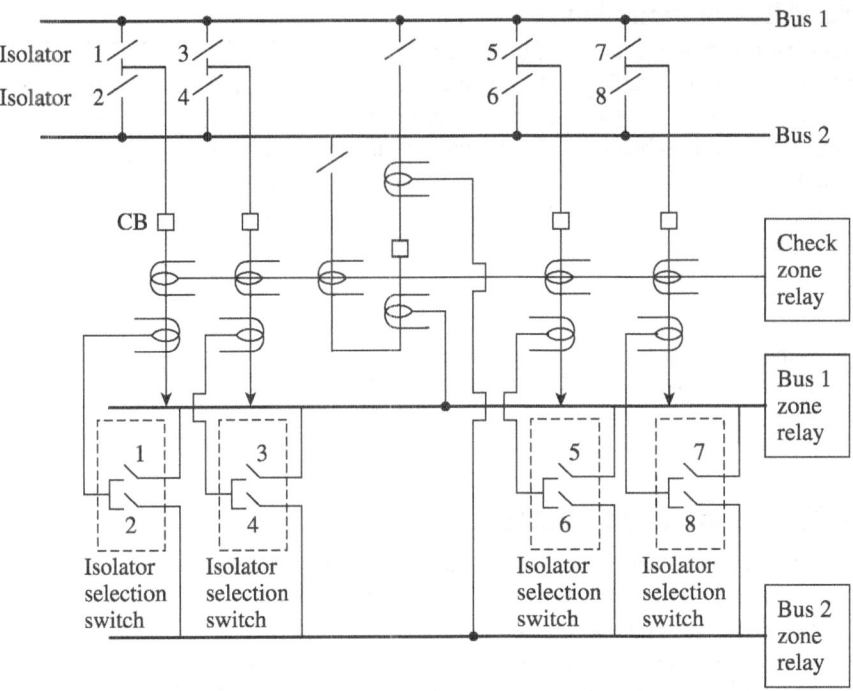

Fig. 8.17 Schematic block diagram of double busbar protection scheme

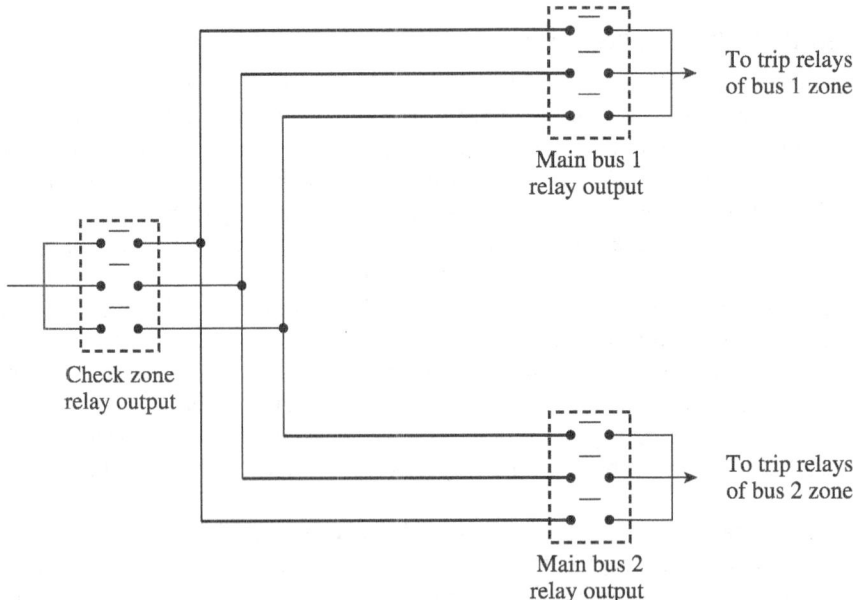

Fig. 8.18 Tripping logic of double busbar protection scheme

The ideal check feature should possess the following characteristics:

1. Check feature should be provided in the main relay module, which is independent of each bus zone relay.
2. The check relay should be able to detect all types of faults in the zone of bus 1 or bus 2.
3. The check relay must remain stable for all out of zone faults of bus 1 and bus 2.
4. As the contact of the check relay is in series with the contact of bus 1 or bus 2 relay, the check relay must operate before bus 1 or bus 2 relay.
5. All the line current signals should be available to the check relay irrespective of the connection of line to bus 1 or bus 2.

Recapitulation

- Busbar is the intersection of an electrical network where many lines are connected together and have very high fault levels (fault MVA).
- The various configurations of busbar arrangement used in substations are single busbar arrangement, single busbar arrangement with sectionalizer, main and transfer busbar arrangement, double busbar arrangement, and one-and-half breaker arrangement.
- The protection schemes used for busbar are directional relaying scheme, differential relaying scheme, and high impedance voltage differential scheme.
- Further, differential relaying schemes can be divided into circulating current differential scheme, biased differential scheme, and opposed voltage balance scheme.
- It is necessary to evaluate the problems and remedies of CT saturation during implementation of the differential relaying scheme of busbar protection.
- Currently, a study of centralized and decentralized type digital busbar protection schemes is going on.
- In practice, digital protection along with check zone feature is the commercially used busbar protection scheme for double bus arrangement.

Multiple Choice Questions

1. In the case of busbar fault, the bus zone relay must
   (a) trip all the breakers connected to the bus
   (b) give an alarm for bus fault
   (c) trip one breaker connected to the bus
   (d) trip some breakers connected to the bus

2. In order to avoid CT ratio mismatch,
   (a) circulating current differential protection scheme is the most suitable
   (b) biased percentage differential protection scheme is the most appropriate
   (c) directional comparison scheme is the most suitable
   (d) directional blocking scheme is the best suitable

3. The effect of CT saturation can be reduced by
   (a) decreasing the cross-section of the CT core
   (b) increasing the cross-section of the CT core
   (c) changing the CT ratio
   (d) using identical CTs

4. In order to avoid CT saturation during heavy through-fault,
   (a) high impedance voltage differential relay is suitable
   (b) biased percentage differential protection scheme is most appropriate
   (c) directional comparison scheme is most suitable
   (d) modern digital/numerical relay is best suitable

5. The advantage of linear coupler is that it
   - (a) requires high energy relay
   - (b) requires low energy relay
   - (c) requires extra equipments
   - (d) none of the above

6. Bus coupler connects the
   - (a) transmission line with the busbar
   - (b) generator with the busbar
   - (c) transformer with the busbar
   - (d) main bus and the transfer bus

7. With reference to computational requirements,
   - (a) centralized busbar protection scheme is most suitable
   - (b) decentralized busbar protection scheme is most suitable
   - (c) both (a) and (b) are suitable
   - (d) none of the above

8. The recent trend is to use
   - (a) single bus arrangement scheme
   - (b) one-and-half breaker bus arrangement scheme
   - (c) double bus arrangement scheme
   - (d) none of the above

9. The advantage of double bus arrangement over single bus arrangement is
   - (a) low cost
   - (b) better reliability and flexibility
   - (c) that of complete shutdown when fault occurs on one bus
   - (d) that it requires a simple protection scheme

10. The digital/numerical bus protection scheme should operate within
    - (a) 1–3 cycles
    - (c) 30–40 cycles
    - (b) 10–20 cycles
    - (d) none of the above

Review Questions

1. Explain the requirements of busbar protection.
2. Explain the effect of CT saturation on busbar protection.
3. Discuss various remedies to reduce the saturation of CT.
4. Explain the biased differential protection of busbar with percentage characteristic.
5. Explain with diagram the high impedance bus differential protection scheme.

6. How does the linear coupler differ from ordinary CT? Explain the bus protection scheme using linear couplers.
7. Explain the numerical protection of busbar.
8. Explain double busbar protection scheme with check zone feature.

Answers to Multiple Choice Questions

1. (a)   2. (b)   3. (b)   4. (d)   5. (b)   6. (d)   7. (a)   8. (c)   9. (b)   10. (a)

Current and Potential Transformers for Relaying Schemes 9

9.1 Introduction

The current and voltage magnitudes in a primary circuit are required to be stepped down to such a level that they can be easily utilized by the metering circuit and protective relay coil. Current transformers (CTs) and potential transformers (PTs) are required to insulate the secondary circuit used for metering and protective purposes. These instrument transformers produce secondary quantities that are proportional to the primary quantities. The secondary quantities are of standard value, that is, 1 A or 5 A for CT secondary and 110 V or 220 V for PT secondary. Thus, the relay and meter designers can design standard relays and meters to match the secondaries of the CTs and PTs, irrespective of the actual primary current and voltage. Ideal CTs and PTs transform the primary quantities without any errors. However, in practice, there are always some errors, known as *ratio errors* and *phase angle errors*.

The design of CTs and PTs mainly depends on their application in the power system. A metering CT may be of low class, that is, it will saturate in the case of fault, whereas a protective CT must be of a higher class, that is, it faithfully transforms the proportional primary current at the time of a fault.

9.2 Operating Principle of Current Transformer and Potential Transformer

The operation of a CT and PT is the same as that of an ordinary transformer. When an alternating current flows through the primary winding, an alternating flux is set up in the core because of the production of magnetomotive force (mmf). This flux induces an electromotive force (emf) in the primary as well as the secondary windings wound on the same core. For an ordinary transformer, the primary winding current is divided into two components, namely exciting current and secondary current. The primary winding of the CT is connected in series with the transmission line/distribution feeder having a relatively high current magnitude. A small component of this current excites the CT core with a flux density just enough to induce an emf in the secondary winding to drive the secondary circuit. Thus, the flux density in the core of a CT mainly depends on the amount of primary current and the loading condition (impedance) of the secondary circuit.

Conversely, the primary winding of a PT is connected in parallel with the power system with relatively high voltage. The core flux, and hence, the flux density in the core, depends on the emf induced in the primary winding, which is proportional to the applied voltage at the primary of the PT. When the PT is in open circuit condition, the current drawn by the primary winding is just sufficient to excite the core. With impedance (load) connected to the secondary circuit of the PT, the primary current of the PT depends on the secondary current and the exciting currents, by maintaining almost constant flux in the core. Thus, the flux density in the core of the PT is constant during normal operating conditions.

As the impedance offered by the secondary circuit of the CT is very low, the number of secondary ampere turns (AT) required is very small. It is around 1% of primary ATs. On the other hand, the impedance offered by the secondary circuit of the PT is very high. Thus, the primary ATs of the PT are equal to the secondary ATs plus the amount necessary to excite the core.

9.3 Construction and Performance of Current Transformer

Current transformers are usually constructed in two ways. The primary conductor itself forms a one-turn primary winding and becomes part of the CT assembly, known as *bar-primary CT*. This single-turn primary winding is properly insulated to withstand high system voltage. The secondary winding of such CTs can be located inside the bushing.

A multiturn primary winding is known as *wound CT*. The primary and secondary windings of such CTs are placed concentrically over the core. The primary and secondary winding turns are separated by cast resin insulation. The whole assembly is placed in a porcelain housing in which oil is filled.

9.3.1 Equivalent Circuit and Vector Diagrams of Current Transformer

Figure 9.1 shows an equivalent circuit of a CT. In a CT, the main feature of interest is the relationship between the primary current I_p and the secondary current I_s. These two currents are of different magnitudes and phases because of the exciting current I_e, which is a component of the primary current. Thus, for the circuit in Fig. 9.1, the secondary equivalent current on the primary side of the CT with turns ratio N is $NI_s = I_p - I_e$. Here, I_e depends on the secondary induced emf, E_s, and the exciting impedance, $(R_m + jX_m)$.

The impedances $R_p + jX_p = Z_p$ and $R_s + jX_s = Z_s$ are primary and secondary winding impedances, respectively, and Z_B is the secondary burden impedance. The error in the CTs is due to the component of primary current being utilized to excite the core and the balance of the primary current being utilized to generate secondary currents in the secondary circuit.

The corresponding vector diagram is shown in Fig. 9.2. It shows the phase angle difference (θ) between I_p and NI_s, which is due to the exciting current I_e. The iron loss component (I_r) of the exciting current (I_e)

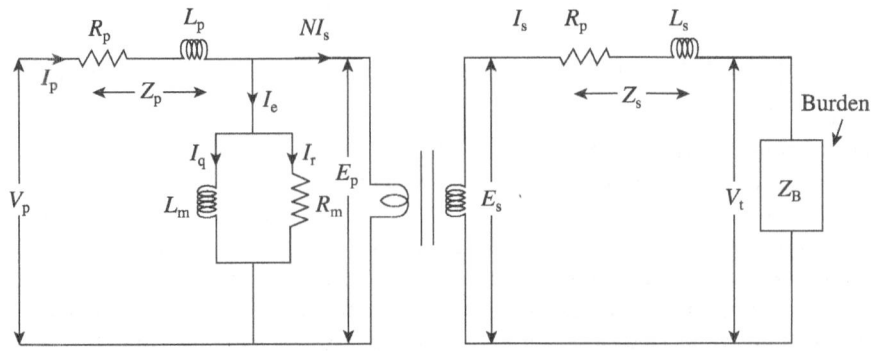

Fig. 9.1 Equivalent circuit of CT

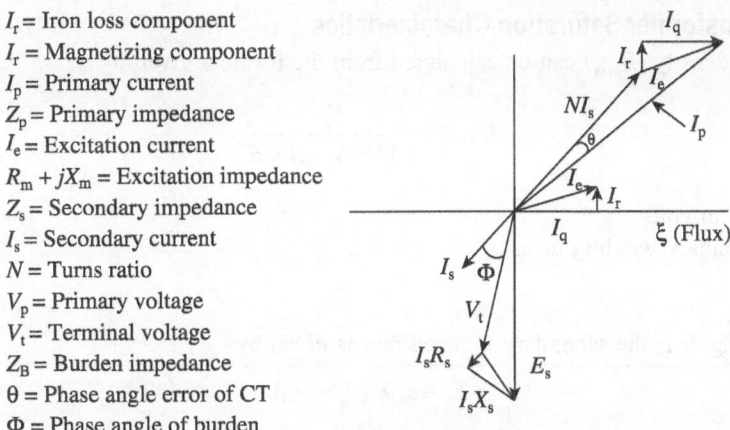

I_r = Iron loss component
I_r = Magnetizing component
I_p = Primary current
Z_p = Primary impedance
I_e = Excitation current
$R_m + jX_m$ = Excitation impedance
Z_s = Secondary impedance
I_s = Secondary current
N = Turns ratio
V_p = Primary voltage
V_t = Terminal voltage
Z_B = Burden impedance
θ = Phase angle error of CT
Φ = Phase angle of burden

Fig. 9.2   Vector diagram of CT

constitutes the current error, and the magnetizing component (I_q) of exciting current (I_e) results in the phase angle error θ.

The relative values of the current ratio error component I_r and the phase angle error component I_q depends on the phase displacement of NI_s and I_e. Thus,

$$\text{Ratio error} \propto I_r/NI_s \tag{9.1}$$

$$\text{Phase angle error} \propto \frac{I_q}{NI_s} \tag{9.2}$$

9.3.2 Magnetization (Excitation) Curves

It is clear from Section 9.3.1 that the errors in the CT results are due to the primary component of the current (exciting current) required to excite the core. Thus, it is necessary to find out the amount of exciting current required at the time of performance of the CT.

The excitation or magnetization curve of a CT depends on the magnetic properties of the core material, cross-sectional area, length of the magnetic path of the core, and number of turns in the windings. The shape of the curve changes for different core materials. Figure 9.3 shows the magnetization curves of widely used core materials of CTs. These are hot rolled non-oriented silicon steel (a), cold rolled oriented silicon steel (b), and nickel iron (c).

The curves are drawn between the exciting force H (mmf) in AT/cm and the maximum flux density B_{max} in tesla. In Fig. 9.3, at low flux density, curve (a) has the lowest permeability, whereas curve (c) has the highest permeability. Further, curve (b) has an extremely high permeability at high flux densities. It indicates that material (b) is used in CT manufacturing as it gives good accuracy at high current magnitude (10–15 times the rated current). However, the accuracy of the core material of curve (b) is not maintained as well as that of a CT with core material (c), for currents less than five times the rated current.

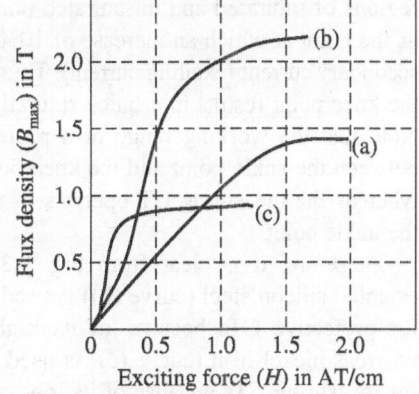

Fig. 9.3   Magnetization curve for different CT core materials (a) Hot rolled non-oriented silicon steel (b) Cold rolled oriented silicon steel (c) Nickel iron (80% nickel)

9.3.3 Current Transformer Saturation Characteristics

The core peak flux density (B_{max}) can be calculated from the formula given by

$$B_{max} = \frac{E_s}{4.44 \times N \times A \times F} \qquad (9.3)$$

where,

E_s = secondary emf in volts
N = number of secondary winding turns
A = core area (m^2)
F = frequency (Hz)
With reference to Fig. 9.1, the secondary induced emf is given by

$$E_s = I_s \times (Z_s + Z_B)$$

where Z_s and Z_B are the secondary impedance and burden impedance, respectively.

From Eq. (9.3), $$B_{max} = \frac{I_s \times (Z_s + Z_B)}{4.44 \times N \times A \times F}$$

Thus, $$B_{max} \propto I_s (Z_s + Z_B) \qquad (9.4)$$

It is clear from Eq. (9.4) that the flux density is directly proportional to the secondary current for a constant value of secondary burden impedance (Z_B). Thus, an increase in primary and secondary currents leads to higher flux density in the core material. Hence, the core material starts saturating, which further leads to a reduction in secondary current. This will result in a high current error.

During the performance of a CT, the exciting current (I_e) can be obtained at different values of the secondary induced emf (E_s). This is achieved by applying variable voltage to the secondary winding of the CT. In this situation, the primary winding is kept open. Figure 9.4 shows such a characteristic drawn between the secondary emf (E_s) and the exciting current (I_e).

This characteristic is divided into three regions, separated by the ankle point (A) and knee point (K). The regions of saturated and unsaturated portions are distinguished by the knee point. The *knee point* is defined as the point at which an increase of 10% in the secondary voltage (exciting emf) produces a 50% increase in secondary current (exciting current). Thus, the voltage above the knee point results in a quick saturation of the CT core. Note that the working range of a protective CT must be between the ankle point and the knee point (linear region), whereas the measuring CT operates in the region close to the ankle point.

Therefore, it is clear from Fig. 9.3 that cold rolled-oriented silicon steel (curve (b)) is used as a core material for protective CTs because of its high saturation level, whereas nickel iron (curve (c)) is used as a core material for measuring CTs because of its low saturation level.

9.3.4 Current Transformer Burden

The load connected to the secondary winding of a CT is called *burden*. It is usually expressed as VA or as impedance. The VA rating of the burden is defined at the nominal

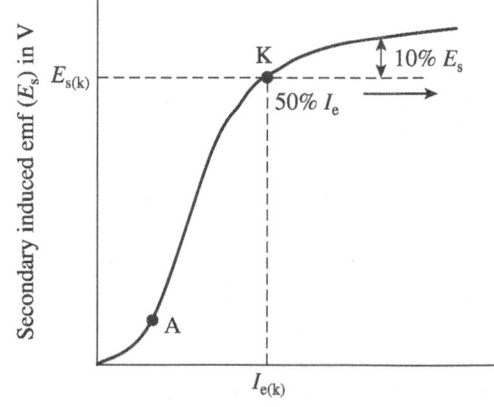

Fig. 9.4 CT saturation curve

CT secondary current. For example, if a burden of 10 VA is connected to a CT with a secondary rating of 5 A, then it offers an impedance of 2 Ω. Conversely, the same burden, when connected to a CT with a secondary current of 1 A, offers an impedance of 10 Ω. If the secondary of the CT is short-circuited, then the VA burden is zero due to zero voltage. The size of the CT is directly affected by the burden and the current rating of the secondary winding. Moreover, the errors in a CT depend on the type and amount of burden connected to the secondary of the CT. This is because the exciting current (I_e) is proportional to E_s, which drives the secondary current through the connected burden impedance (Z_B). The burden on the CT is not only due to the connection of the external load (relay, meter, etc.) to the secondary of the CT but also due to the resistance of the interconnecting lead.

9.3.5 Current Transformer Accuracy

The proper functioning of a CT depends on the accuracy and selection (of the class and rating) of the CT because these factors directly affect the performance of protective relaying. The accuracy requirements of relaying schemes are not the same. This requires the determination of inaccuracies of the CT under different operating conditions. In order to determine the effect of inaccuracies of CTs, evaluation of ratio error and phase angle error is extremely important. The accuracy of the CT at high current magnitude depends on the cross section of the iron core and the number of turns in the secondary winding. The greater the cross section of the iron core, the more flux is developed before saturation occurs. The saturation of the CT core results in a rapid increase in ratio error. The greater the number of secondary turns, the lower the flux required to force the secondary current through the relay.

The *accuracy of a CT* is defined as the difference between the actual ratio and the true ratio. Hence, in terms of error, it is given by

$$\% \text{ error} = \frac{(N \times I_s) - I_p}{I_p} \times 100 \tag{9.5}$$

For most protective relays, an accuracy variation of $\pm 10\%$ to $\pm 15\%$ is acceptable. Furthermore, the ratio of error should not exceed 10% at any current ranging from 1 to 20 times the rated current at standard burden, as otherwise the secondary current may be distorted because of CT saturation. The main function of the protective CT is to faithfully transform the maximum possible current under normal conditions as well as during faulty conditions. The *accuracy limit factor* (ALF) indicates the overcurrent as the product of a multiple and the rated current up to which the rated accuracy is fulfilled (with the rated burden connected). For most protective schemes, accuracy class of a CT is specified in the format 5P150 F15, where the first number 5 indicates composite error, the letter P indicates protection class, the number 150 indicates rated knee point voltage (KPV) (150) in volts, and F15 indicates the rated ALF.

9.3.6 Open Circuit Current Transformer Secondary Voltage

During normal operation of a CT, with load burden connected to the secondary circuit, the AT required on the primary side differs by a small value from that required on the secondary side. That is because of the fact that a part of the primary AT produces magnetic flux, which further induces the voltage on the secondary side. In the event of an open-circuited secondary winding, there are no secondary ATs present to compensate the ATs produced by the primary current. Thus, the primary ATs are entirely used to produce maximum exciting flux in the core, which drives the core into the saturation region. When the primary current is at the zero crossing in every half cycle, the high rate of change of flux induces a very high secondary emf. In case of a protective CT with a large cross section, the value of such an induced emf may be in kV. Such high voltages are harmful to the insulation of the CT and may damage it completely. It is therefore necessary to take precautions against the open circuit of the CT secondary terminal by connecting a conductor capable of carrying the CT secondary current. However, in the case of modern digital/numerical relays, this problem will not occur because of the presence of the CT shorting switch.

9.4 Performance of Potential Transformer

The voltage coil of the protective device cannot be connected directly to the high voltage systems. Therefore, it is necessary to step down the high system voltage at a suitable level. Moreover, the protective device must be isolated from the primary power circuit. This is achieved by a voltage transformer (VT), which is also known as *potential transformer* (PT). The performance of a PT is the same as that of the conventional power transformer operating at a very light load. PTs are rated in terms of the VA burden they can deliver to protective equipment without exceeding the specified limit of error. The output (secondary) voltage is usually 110 V between the phases. Most of the protective relaying schemes mainly use two types of PTs:

1. Electromagnetic type PT
2. Capacitive voltage transformer (CVT)

9.4.1 Equivalent Circuit of Electromagnetic Type Potential Transformer

In a PT, it is necessary to evaluate the magnitude and phase difference between the primary voltage V_p and the secondary terminal voltage V_t. This difference is analysed in terms of ratio error and phase angle error. Ratio error and phase angle error occur because of voltage drops in both primary and secondary windings, due to burden current I_s and the exciting current I_e. Figure 9.5 represents the equivalent circuit of a PT, and the corresponding vector diagram with reference to voltages is illustrated in Fig. 9.6. The difference between V_p and V_t' is shown in terms of magnitude and phase. The angle difference between V_p and V_t' is shown by θ (phase angle error of PT).

The ratio error is defined as

$$\frac{(N \times V_t) - V_p}{V_p} \times 100\% \tag{9.6}$$

where N is the turns ratio. The permissible errors for a PT are of the order of $\pm 3\%$ (ratio error) and $\pm 2\%$ (phase angle error).

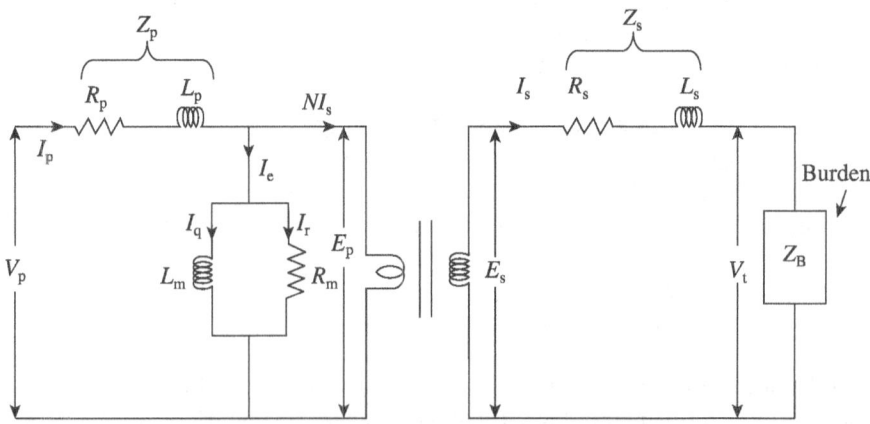

Fig. 9.5 Equivalent circuit of electromagnetic type PT

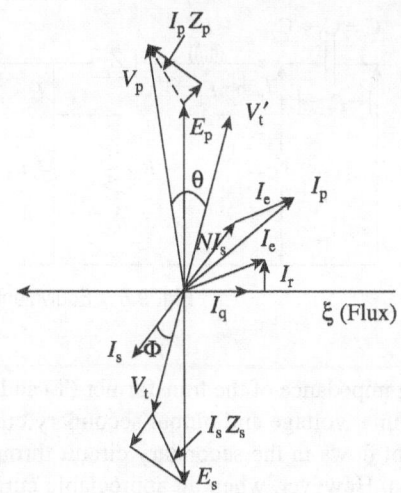

I_r = Iron loss component
I_q = Magnetizing component
I_p = Primary current
Z_p = Primary impedance
I_e = Excitation current
$R_m + jX_m$ = Excitation impedance
Z_s = Secondary impedance
I_s = Secondary current
N = Turns ratio
V_p = Primary voltage
V_t = Terminal voltage
Z_B = Burden impedance
θ = Phase angle error of CT
Φ = Phase angle of burden
E_p = Primary induced emf
E_s = Secondary induced emf

Fig. 9.6 Vector diagram of PT

9.4.2 Capacitive Voltage Transformer

The electromagnetic type PT is used for system voltages up to 132 kV. However, beyond 132 kV, the size of such PTs becomes large, and hence, it is not economical. On the other hand, the CVT, which is cost effective and also serves for carrier current coupling, is widely used for voltages beyond 132 kV.

The CVT is basically a capacitance potential divider where the primary high line-to-ground voltage is applied across a series-connected capacitor, whereas the low voltage end is connected to the primary winding of the electromagnetic PT. Figure 9.7 shows the basic circuit of a

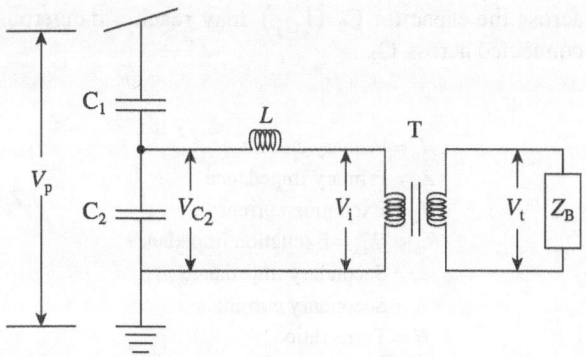

Fig. 9.7 Basic circuit of CVT

CVT where the line-to-ground voltage (Vp) is applied across C1 and C2, and the intermediate voltage is given to the wound type PT (T) through inductance L. The value of inductance L is chosen in such a way that it resonates with capacitors C1 and C2 at the system frequency. The transformer T steps down the input voltage Vi to the secondary voltage Vt.

Equivalent Circuit

Figure 9.8 shows the equivalent circuit of the CVT. In Fig. 9.8, V_i is the nominal intermediate voltage and its value is given by $V_i = V_p \times \dfrac{C_1}{C_1 + C_2}$. L and Z_p are the tuning inductance and primary impedance, respectively. $R_m + jX_m$, that is, Z_e, is the impedance offered by the intermediate transformer T, which takes the exciting current (I_e) from the primary current (I_p). On the secondary side of the CVT, Z_s and Z_B represent the

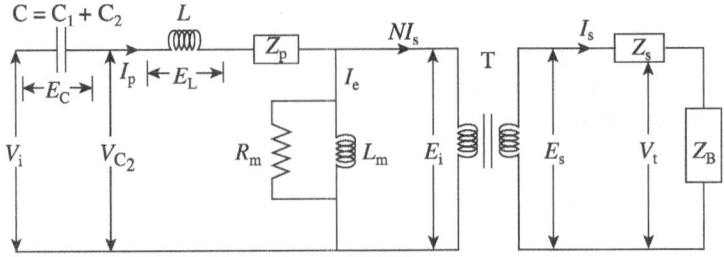

Fig. 9.8 Equivalent circuit of CVT

secondary winding impedance of the transformer (T) and burden impedance, respectively. V_t and I_s represent the secondary terminal voltage and output secondary current, respectively.

When no current flows in the secondary circuit through the burden impedance, the voltage drop on the primary side is less. However, when an appreciable current flows through the burden impedance, the ratio error and phase error become visible because of the passage of load current through the capacitor C and inductor L. Therefore, at a resonant frequency, the effect of the voltage drop across capacitance C can be compensated by the voltage drop across inductor L.

Figure 9.9 shows the vector diagram of the equivalent circuit of a CVT. It is to be noted from Fig. 9.9 that the voltage error in this diagram is the difference between V_i and V_t', whereas the phase angle error is the angle (θ) between vector V_i and vector V_t'.

During the short circuit condition of the secondary circuit (zero burden impedance), the circuit voltage across the capacitor C_2 (V_{C_2}) may reach a dangerously high value. Hence, to protect C_2, a spark gap is connected across C_2.

I_p = Primary current
Z_p = Primary impedance
I_e = Excitation current
$R_m + jX_m$ = Excitation impedance
Z_s = Secondary impedance
I_s = Secondary current
N = Turns ratio
Z_B = Burden impedance
θ = Phase angle error of PT
Φ = Phase angle of burden
E_i = Primary induced emf
E_s = Secondary induced emf
V_i = Input voltage
V_{C_2} = Voltage across capacitor C_2
V_t = Terminal voltage
E_C = Voltage across capacitor C
E_L = Voltage across inductor L

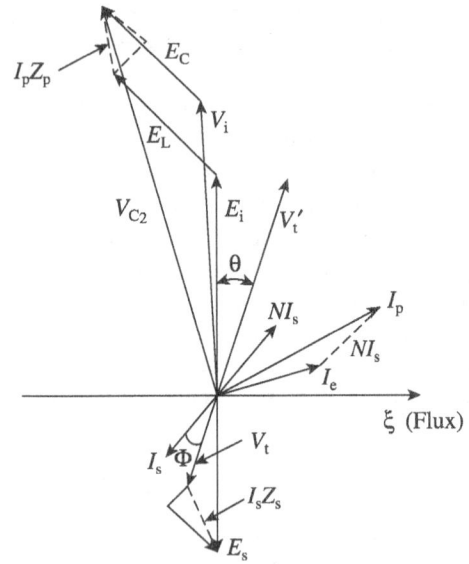

Fig. 9.9 Vector diagram of CVT

Capacitive Voltage Transformer Transient Response and Ferroresonance

The transient response of the CVT is inferior to that of the electromagnetic PT. During fault, the reproduction of secondary voltage never follows the sudden change in the primary voltage. Hence, the interaction of series tuning reactance, magnetic inductance, and capacitance of the intermediate transformer winding leads to decaying oscillations.

It is possible to induce transient oscillations during the energization of the CVT. This is because ferroresonance occurs between the inductance of the exciting impedance and the capacitance of the potential divider. The frequency of such oscillations is around one-third of the supply frequency. This oscillation will result in a rise in the output voltage, upto 25–50% of the normal value. To reduce such oscillations, a resistive burden is preferred. An intermediate transformer with a large core is another option.

9.5 Specifications of Current and Potential Transformers

At the time of designing a transmission line or part of an electric network, proper selection of measuring and protective CTs and PTs are very essential. This exercise is carried out during the design stage in which the designer has to choose from the standard CTs and PTs available in the market. The following are the factors to be considered while tendering orders for CTs and PTs.

9.5.1 Current Transformer

1. Type: Bushing CT or oil filled CT or SF6 filled CT
2. Installation: Outdoor/indoor
3. Standard: IS 2705/IS 4201
4. Rated maximum system voltage: 7.2 kV, 12.5 kV, 72.5 kV, 145 kV, etc.
5. Frequency: 50 Hz
6. Rated continuous primary current: 100 A, 150 A, 200 A, ..., 10,000 A
7. Rated continuous secondary current: 1 A or 5 A
8. Rated CT ratio: 100/1, 200/1 or 100/5, 200/5, etc.
9. Accuracy class: Class P5 (differential and distance protection)
   Class 5P20 (overcurrent protection)
   Class 0.5 (metering)
10. Burden: 5 VA, 10 VA, 15 VA, etc.
11. Knee point voltage: As per the rating
12. Accuracy limit factor: As per the rating
13. CT secondary resistance in ohm: As per the rating
14. Insulation level and its flashover voltage: As per the rating

9.5.2 Potential Transformer

1. Type: Oil filled or dry type
2. Installation: Outdoor/indoor
3. Standard: IS 3156
4. Rated maximum system voltage: 7.2 kV, 12.5 kV, 72.5 kV, 145 kV, etc.
5. Rated frequency: 50 Hz
6. Overvoltage factor: 110% of rated voltage (continuous)
7. Rated secondary voltage: 110 V
8. Rated transformation ratio: 11 kV/110 V, 66 kV/110 V, 400 kV/110 V, etc.
9. Rated burden: In terms of VA

10. Accuracy: As per the requirement
11. Insulation level: As per the rated voltage

Recapitulation

- The current transformer and potential transformer are instrument transformers required to insulate the secondary circuit of metering and protective relays from high voltage primary circuit.
- These instrument transformers produce secondary quantities, which are proportional to the primary quantities.
- Secondary quantities are of standard value, that is, 1 A or 5 A for CT secondary and 110 V or 220 V for PT secondary.
- The errors in the CT results are due to the primary component of current (exciting current) that is required to excite the core.
- The ratio error of the CT is proportional to I_r/NI_s, and the phase angle error of the CT is proportional to I_q/NI_s.
- The accuracy of CT is defined as the difference between the actual ratio and the true ratio. Hence,

$$\% \text{ error} = \frac{(N \times I_s) - I_p}{I_p} \times 100$$

- The excitation curves and saturation characteristics of CTs should be considered while selecting them for a particular protection function.
- The errors in the PT results are due to the voltage drop in both primary and secondary windings as a result of burden current and the exciting current.
- The % ratio error of PT is defined as $\dfrac{(N \times V_t) - V_p}{V_p} \times 100$, where N is the turns ratio.
- A capacitive voltage transformer can perform better than a PT for a system with voltage higher than 132 kV.
- One should know the detailed specifications of instrument transformers as they will be helpful at the time of procurement of CTs and PTs.

Multiple Choice Questions

1. The material considered for the core used in protective CTs is
   (a) hot rolled non-oriented silicon steel
   (b) cold rolled oriented silicon steel
   (c) nickel iron
   (d) none of the above

2. The material considered for the core used in metering CTs is
   (a) hot rolled non-oriented silicon steel
   (b) cold rolled oriented silicon steel
   (c) nickel iron
   (d) none of the above

3. The error introduced in CTs is mainly due to

   (a) primary current (I_p)
   (b) exciting current (I_e)
   (c) primary system voltage (V_p)
   (d) secondary induced emf of CT (E_s)

4. The excitation curve of CTs depends on
   (a) the cross section and magnetic path of the core
   (b) the number of turns in the windings
   (c) characteristics of the core material
   (d) all of the above

5. The CT saturation characteristic is drawn between
   (a) secondary voltage and burden current
   (b) primary voltage and primary current

(c) open circuit voltage and exciting current

(d) primary voltage and secondary current

6. The KPV of the protective CT should be ____ the KPV of the metering CT.

   (a) lower than

   (b) half of

   (c) higher than

   (d) double

7. Saturation of CT can be reduced by

   (a) increasing the cross-sectional area of the iron core

   (b) increasing the burden

   (c) reducing the burden

   (d) options (a) and (c)

8. In the event of the open circuit of CT secondary

(a) maximum voltage is induced on the secondary side

(b) maximum ATs are required on the secondary side

(c) minimum flux is produced in the core

(d) none of the above

9. Ferroresonance test is required for

   (a) CT

   (b) electromagnetic PT

   (c) CVT

   (d) relay

10. CVTs are preferred for voltages

    (a) below 11 kV

    (b) between 11 kV and 66 kV

    (c) beyond 132 kV

    (d) none of the above

Review Questions

1. Draw the equivalent circuit and vector diagram of CT and explain the same with respect to the performance of a CT.

2. Explain the excitation and saturation curves of a CT.

3. How are errors introduced in a CT?

4. Why can the secondary of a protective CT not be open-circuited?

5. Explain the performance of electromagnetic PT and CVT with relevant vector diagrams.

6. Explain the terms 'burden' and 'accuracy' with respect to CTs.

Protection Against Transients and Surges

10

Learning Objectives

After going through this chapter, the students will be able to:

- Discuss the sources of transients or surges in EHV lines
- Explain the concept of travelling wave and its propagation on transmission line
- List the various ways of neutral grounding and their advantages
- Explain the different ways of protecting against transients and surges
- Explain the selection procedure and rating of an arrester
- Discuss the insulation coordination of electrical components

10.1 Introduction

Today's world of industries and modern technology has become extremely dependent on the qualitative and uninterrupted supply of electrical power. In India and most developed countries, commercial power is made available to loads through national grids and by interconnecting several generating stations. Loads such as residential lighting, heating, refrigeration, air conditioning, and transportation, as well as bulk supply to governmental, industrial, financial, commercial, and communications communities are fed from national grids operating at high voltages. Many power problems originate in national grids, which are subject to weather conditions such as lightning storms with equipment failure, short circuit accidents, and major switching operations. Power surges, lightning strokes, and equipment failure are inevitable. There is no way to prevent power surges and transients, but there is a way to offer protection for the equipment used in power system networks from transient overvoltages.

This chapter describes the causes of surges and transients in a power system, their travel phenomenon, attenuation, and adoption of effective protective devices against transient overvoltage such as surge arresters and lightning arresters (LAs).

10.2 Sources of Transients or Surges in Extra High Voltage Lines

Switching of high voltage transmission lines is required for ordinary functioning of an electrical network. The switching of an unloaded line, opening and closing of isolating switches, and interruption of an inductive current (reactor) and capacitive circuit (capacitor banks) produce transient overvoltages. This abrupt increase in the voltage above the rated value for a short duration in the power system is known as a *voltage surge* or *transient voltage*. Though transients or surges are temporary, they cause overvoltage in the power system.

Transient disturbances in power systems may damage main equipment, thus affecting system reliability. The switching surges produced in an electrical system depend on how repeatedly the routine switching operations are performed in a particular system. Moreover, the existence of transients and surges in the system depends on the number of faults and their clearance as well as on the changes in the system parameters and alterations of system configuration.

Lightning strikes are the main source of overvoltage. Although they are few in number, they drastically affect the performance of the power system components, including complete failure of insulation in case of severe lightning discharge. The probable causes of transients and surges in a power system network are discussed in the following subsections.

10.2.1 Switching of Transmission Line

Whenever a high voltage overhead transmission line is energized by closing the line circuit breaker, switching surges are generated in the line and supply network. The magnitude and shape of a switching transient depend on the transmission line, system parameters, and the arrangement of the network. Even with the same system parameters and network configuration, the switching overvoltages are highly dependent on the distinctiveness of the circuit breaker operation and the amount of trapped charges in transmission lines during circuit energizing. Moreover, the severity depends on the difference between the supply voltage and the line voltage at the instant of energization. Large travelling waves are injected in the transmission lines during the closing of the circuit breaker at an instant when this voltage difference is high. When these waves reach the open far end of the line, they are reflected and a high transient overvoltage is experienced. Such switching transients are more dangerous for extra high voltage (EHV) transmission networks, especially for long transmission lines. Hence, the transient voltages generated from the switching of transmission lines have an impact on equipment design and protection.

10.2.2 Switching of Capacitor Bank

Shunt capacitor banks are commonly installed at the distribution level as well as at transmission levels because of their effective performance. Capacitor bank switching is one of the most frequent utility operations depending on the need for system voltage or VAR (volt-ampere reactive) requirement from the banks. However, switching of such utility capacitor bank can have a negative impact on the power quality, especially for supply network and consumer power systems. Larger capacitor banks have more stored energy, and hence systems with high X/R ratio are of more concern in supply networks. High magnitude and high frequency transients occur during the switching of shunt capacitor banks. When a capacitor bank is energized, it draws transient inrush current from the bus or transformer. The transient inrush is distinguished by a surge of current with a high magnitude and a frequency of several hundred hertz. This transient manifests itself as a voltage increase due to the exciting system resonances or dynamic overvoltages when a capacitor bank is energized. This leads to a transient overvoltage on the bus and the neighbouring power system, which is caused by the surge of inrush current from the system source and bank outrush current due to faults in the vicinity of the capacitor banks.

Successive switching of the capacitor bank adjoining the previously energized bank results in a much higher magnitude of current with high frequency compared to the energization of only one capacitor bank. As the series connection of capacitor banks provides lower inductance, the efficiency of the capacitor is reduced.

In case of a series-compensated transmission line using a series capacitor bank, switching surge overvoltages are produced because of the trapped charges present on the bank at the instant of the line reclosing. This phenomenon increases the potential of phase-to-ground and phase-to-phase voltage of transmission lines. Moreover, the transient recovery voltage (TRV) experienced by the first circuit breaker to clear the faults in a series-compensated line is of very high magnitude.

10.2.3 Switching of Coupling Capacitor Voltage Transformer

Coupling capacitor voltage transformers (CCVTs) are used in the power system network over many years for protective relaying and as measuring instruments. They have capacitance voltage dividing networks. Under steady-state operating conditions, the performance of the CCVT is not troublesome. However, the normal switching operation of the CCVT can create unexpected overvoltages, which can affect the reliability of the power system and even lead to system failure in some CCVT units. The ferroresonance and non-linear behaviour of the potential transformer (PT) magnetic core, and effect of stray capacitances in some CCVT elements are some of the causes of transient overvoltages generated during switching operation in several 230 kV and 400 kV CCVTs.

10.2.4 Switching of Reactor

To compensate the leading reactive power and to depress the voltage rise at the end terminal of a long EHV transmission line, shunt reactors are used in the power system. Such a high voltage reactor is frequently switched on during periods of low load and is switched off with the rise of load. However, unavoidable high frequency TRV is caused at the terminal of the circuit breaker during the switching operation. Each switching operation involves a complex interaction between the circuit breaker, the supply network, and the load side shunt reactor. This interaction results in overvoltages dependent on system parameters and characteristics of the load. Energizing the reactors rarely generates high overvoltages, but excessive over-voltages are generated during de-energization. Current chopping (current interruption before its natural zero crossing) by the switching device (circuit breaker) when it interrupts the inductive currents and discharge of energy stored in the reactor inductance will cause electromagnetic transients, which lead to the switching overvoltages.

10.2.5 Arcing Ground

In case of an ungrounded three-phase system, during line-to-ground fault, the magnitude of current is sufficient to maintain an arc. The phenomenon of intermittent arc during such faults, which results in the production of transients, is known as *arcing ground*. During arcing ground, the inductance of the generator winding or line and the system capacitance establish a cycle of charging and discharging, which results in high frequency oscillations. This oscillation builds up a very high voltage, and the arc may reignite, which results in supplementary transient disturbances. The transients produced by arcing ground are cumulative and may cause serious damage to the equipment in the power system by causing breakdown of insulation. More details of arcing ground are given in Section 10.4.1.

10.2.6 Lightning Strokes

Lightning is a huge spark caused by the electrical discharge occurring between clouds, within the same cloud, and between clouds and the earth. During the movement of hot moist air from the earth towards the clouds, the friction between the air and water particles causes build-up of charges. Hence, updraft of hot air is responsible for charge separation within the cloud, such as in a huge electrostatic generator, which leads to the creation of electric fields within and around the cloud and ultimately to the electric breakdown (discharge) called *lightning*. The air between the cloud and earth cannot withstand the electrical field created in the cloud, and hence, the air gets ionized; finally, a streamer of very high current strikes between the clouds and earth resulting in the discharge of the clouds.

There are two types of lightning strokes, direct lightning stroke and indirect lightning stroke. In the former, lightning discharge (i.e., current path) takes place directly from the cloud to the electrical equipment, as in an overhead line. In the latter, the discharge results from the electrostatically induced charges on the conductors of the overhead line because of the presence of charged clouds, that is, a positively charged cloud is above the line and induces a negative charge on the line through electrostatic induction. Most surges in the transmission lines are caused by indirect lightning strokes. However, the effect of a direct stroke in terms of overvoltage due to the lightning discharge on the transmission line may be large enough to flashover the line component if not protected.

The lightning stroke current rises to a peak value very fast and then starts decaying at a low rate as shown in Fig. 10.1. The generalized wave shape can be described as a *crest* or *peak value*, and the maximum value of this current is 400 kA; the wave front time varies from 1 to 10 μs, and the time at which the stroke current reduces to 50% of that peak value varies from 10 to 100 μs. Moreover, the waveform of the surge voltages imposed on the insulation of a system due to the lightning stroke may vary extensively because of reflection in the line and the position of the stroke on transmission lines.

A lightning surge may occur at any point and at any time. Lightning is one of the most serious causes of overvoltage. If the power equipment, especially at the outdoor substation, is not protected, the overvoltage will cause complete failure of insulation. Thus, it results in complete shutdown of the power and loss of revenue and equipment.

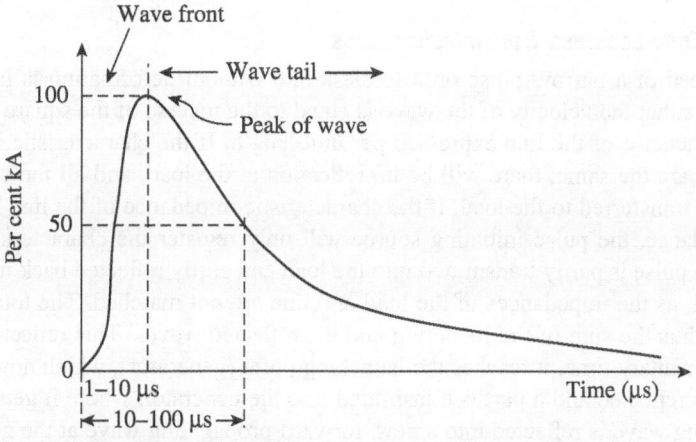

Fig. 10.1 Generalized wave shape of lightning stroke

10.3 Overvoltage Due to Lightning and Switching

Transient overvoltage is a temporary and irregular damped phenomenon observed in electrical systems. The linear impedance of a power system comprises resistance, inductance, and capacitance and may be considered as distributed or lumped. Whenever the circuit changes its condition because of short circuit, opening or closing of circuit breakers, switches, or any other disturbances in the network, new currents and voltages of very high magnitude (surges and transients) appear in the network. These are surges and transients, which appear as a travelling wave with different propagation speeds. During the travel time, the wave reflection may produce voltage considerably higher than the surge voltage generated at its source.

10.3.1 Surges and Travelling Waves

When a switch is closed to energize a transmission line with a generator or a step source, transients come into view at certain level in the transmission line. Similarly, when the line is disconnected, a transient is projected. In addition, transients or surges also occur because of lightning strokes. When a power line, which may normally operate under steady-state conditions, is subjected to a fault or when the load suddenly changes, a transient is generated. These are equivalent to initiation of a single wave of reasonable voltage or current magnitude, which travels along the length of the line with a velocity almost equal to that of light. The travelling incidence of these two waves greatly depends on the impedance of the line and load. When these travelling waves reach the other end of the line within a finite duration of time, they may either cease to exist or be partially reflected. Thus, for a finite period, the voltage and current at any point are the result of superimposed incident and reflected waves.

10.3.2 Wave Propagation on Transmission Line

The parameters of a transmission line, that is, resistance, inductance, and capacitance, are either lumped or distributed along the length of the line. The unique feature of such a line is its ability to carry the travelling waves of voltage and current. These transmission lines with distributed parameters have a finite velocity of electromagnetic field propagation. The changes in magnitude and frequency of voltage and current waves cannot happen at the same time in all parts of the network because of lightning and switching, but are spread out in the form of travelling waves or surges. For example, when a transmission line is energized from one end (source or sending end), the whole of the line is not energized suddenly, that is, the voltage does not appear immediately at the other end (receiving end). This is due to the presence of distributed parameters.

Narrow Pulses on Finite Lossless Transmission Lines

The propagation speed of a narrow pulse on a lossless line without deformation is independent of its frequency. The reason is that the velocity of the wave is equal to the inverse of the square root of the product of capacitance and inductance of the line expressed per unit length. If the characteristic impedance of the line and load impedance are the same, there will be no reflection at the load, and all the energy in the forward-propagating pulse is transferred to the load. If the characteristic impedance of the line has components other than the load impedance, the pulse initiating source will only register the characteristic impedance of the line. In this case, the pulse is partly transmitted into the load and partly reflected back into the line with some reflection coefficient, as the impedances of the load and line are not matched. The total voltage at any time and at any point is thus the sum of the incoming and the reflected waves. This reflected wave travels back, and after the supplementary time, it reaches the launching point (generator), which now acts as a load. Here, a part of the wave is reflected and a part is transmitted into the generator, where it gets dissipated. Thus, the backward-propagating wave is reflected into a new, forward-propagating wave at the generator. This process repeats itself infinitely with each reflection, with each end of the line being considered as a new wave propagating towards the other end. Instead of a single pulse, if the generator produces a train of pulses, each pulse is reflected as described earlier.

Narrow Pulses on Finite Distortionless Transmission Lines

If a distortionless transmission line is considered, then the pulses of the travelling wave do not distort. However, some losses result due to line parameters and characteristic impedance. In such a situation, during the propagation of pulses, the pulse magnitude is attenuated exponentially as it propagates from the generator to the load or from the load to the generator. This attenuation depends on the total distance travelled by the wave, regardless of how many reflections the pulse has undergone.

Transients on Transmission Lines

The condition here is that of very long pulse, the length of pulse being related to the length of the line and the speed of propagation. Consider a generator is connected to a transmission line at any instant. The generator considers the load at impedance equal to the characteristic impedance of the line since no wave has propagated to the load yet. The closing of the switch creates a disturbance on the line in the form of initiation of a transient. The forward wave now propagates towards the load at the speed of propagation on the line. For a lossless or distortionless line, this speed is independent of the frequencies in the transient pulse. After the propagation time, the forward-propagating wave reaches the load end. There are three possible conditions that may occur:

Load impedance equal to characteristic impedance In this case, the reflection coefficient at the load is zero. There will be no reflection at the load.

Load impedance greater than characteristic impedance In this case, the reflected voltage wave is in the same direction as the forward-propagating wave. The reflected current at the load is in the direction opposite to the forward current.

Load impedance less than characteristic impedance In this case, the reflected voltage wave is opposite in polarity to the forward wave, and the current is of the same polarity as the forward current wave.

After the forward wave reaches the load, it gets reflected, and this is called the *first reflection*. These two waves propagate back towards the generator, but unlike the narrow pulse situation, the forward-propagating wave still survives on the line. Thus, the voltage or current anywhere on the line is the sum of the forward-propagating wave and the backward-propagating wave. After some more time, the reflected wave reaches the generator. Although the generator has its own voltage, the generator for the reflected wave behaves as a load. Thus, the forward- and backward-propagating waves have changed roles. These waves are reflected at the generator to produce new forward-propagating waves towards the load, and theoretically, this process continues forever.

10.3.3 Reflection and Attenuation

The reflection of a travelling wave is due to different characteristic impedances offered at the ends of line where it is joined to other lines or cables. Different line parameters result in attenuation of the transient wave generated by either surges or switching.

Reflection

When a travelling wave meets any discontinuity on the line such as the end of the line, sudden change of surge impedance, junction, tap line, or lumped impedance in series or shunt, is partially or fully reflected as discussed in Section 10.3.2.

For the short-circuited condition, the reflected voltage wave is the negative of the incident voltage wave with magnitude the same as the incident wave, and the resultant being zero. The reflected current wave is equal to the incident current wave with positive reflection, with the resultant being twice the incident wave.

For an open circuit, the reflected voltage wave is equal to the incident wave, and hence, the total voltage is twice the incident voltage, and the reflected current wave is the negative of the incident current wave with the total current being zero.

The condition for the line terminated to its characteristic impedance as well as with some other impedance is described earlier. If such impedance is connected to the terminals of a transmission line, it acts as a short circuit at the moment the incident wave acts on the impedance. At that instant, the reflected voltage wave is negative, the terminal voltage is momentarily zero, and the total current is doubled momentarily.

The terminal voltage exponentially increases from zero in case of capacitive load at the end of the line as it takes some charging time. When the capacitor is fully charged, it then behaves as an open circuit for the terminal, where the current is zero and the voltage of the incident wave is doubled. Similar action is noticed in a line with an inductive load. It initially behaves as an open circuit momentarily, and the total voltage is doubled; then as the current in the inductor increases, it behaves as a short circuit, where the incident and reflected voltages become opposite.

In practice, most of the lines are terminated at the end with the power transformer and secondary circuits connected to the load. With the travelling of lightning wave in the case of an end-connected transformer, it behaves as a capacitance rather than as an inductance. Lightning voltage reaches a transformer as a travelling wave, and the front of the wave is very steep and rises so fast to the maximum that there is practically zero time for the current to start to flow through the large inductance of the transformer winding. However, there is a little capacitance in the transformer winding from turn to turn, from winding to core, and from winding to the earth, and it is for the most part this capacitance, rather than the inductance, that determines the transformer's reaction to the lightning wave.

Attenuation of Transients

The current and voltage waves get attenuated exponentially as they travel over the length of the line, and the reduction in magnitude depends greatly on the parameters of the line. Whenever the travelling waves travel along the line, they experience the following changes:

1. The peak or crest of the wave is reduced in magnitude or is attenuated.
2. The wave changes its profile, that is, gets elongated; its irregularities are smoothened out and its steepness is reduced.
3. The voltage and current waves die away in a similar manner.

These alterations occur simultaneously; however, the last two changes are together known as *distortion*. The surges are unaffected by corona till their potential exceeds the corona threshold. The initial peak is removed by the time the wave has travelled 2–3 km. After that, attenuation continues at a slightly reduced rate. The centre and tail of the waves are built at the expense of the front. Moreover, the attenuation of waves due to corona is more pronounced for positive waves than for the negative waves because of greater loss due to corona for positive waves. The effect on short waves is more than that on long waves.

Attenuation of transients using filter The installation of a filter in series with the equipment seems to be an obvious solution to the overvoltage conditions. The impedance of a low pass filter, a capacitor, for instance, forms a voltage divider with the source impedance. As the frequency components of a transient are several orders of magnitude above the power frequency of the AC circuit, the inclusion of the filter results in attenuation of the transient at high frequencies. Unfortunately, this simple approach may have some undesirable side effects such as unwanted resonances with inductive components located elsewhere in the system, leading to high voltage peaks, high inrush currents during switching, and excess reactive load on the power system voltage. However, these undesirable effects can be reduced by adding a series resistor, hence the popular use of RC snubber networks, which are circuits that consist of resistance (R) in series with capacitance (C) to suppress the rapid rise in voltage.

Attenuation of transients using isolation transformer Isolation transformers generally consist of primary and secondary windings with an electrostatic shield between the windings. The isolation transformer is placed between the source and the equipment requiring protection. As its name suggests, there is no conduction path between the primary and secondary windings. Two widely held beliefs are that isolation transformers attenuate voltage spikes and that transients do not pass through the windings of the transformer. When properly applied, isolation transformers are useful to break the ground loops, that is, block the common mode voltages. Unfortunately, a simple isolation transformer provides no differential mode attenuation. Thus, a differential mode transient is transmitted through the windings of the device. However, such an isolation transformer will not regulate voltage.

10.4 Neutral Grounding

Neutral grounding has been adopted since the beginning of power system development to maintain the phase voltages at a constant level with respect to ground. The grounding of many systems is based on past experience and the invention of the grounding methods in existing installations. The grounding of neutral has no effect on the operation of the three-phase system if it is under a balanced condition. On the other hand, the performance of the system during fault condition is directly affected by the methods of neutral grounding. Moreover, considering the stability of the system and its protection, neutral grounding provides several benefits to the power system. Hence, in a high voltage system, the neutral is solidly grounded to the earth to limit the overvoltage phenomenon. The ungrounded neutral system, its effect on system performance, and the different methods of neutral grounding are discussed in Sections 10.4.1–10.4.3.

10.4.1 Effects of Ungrounded Neutral on System Performance

To understand the effect of fault on an isolated neutral system, consider a simple three-phase system as shown in Fig. 10.2(a). Under balanced conditions, assuming a perfectly transposed line, the shunt capacitance to ground for each conductor is the same. The charging currents from the shunt capacitance in all the three phases are displaced by 120° from each other, leading by 90° with the corresponding voltages. Thus, the net sum of these three charging currents is zero as the system is under a balanced condition and the neutral is at approximately ground potential. Figure 10.2(b) shows the balanced set of voltage and current, in which phasors V_a, V_b, and V_c denote the phase value of the voltage (V_{ph}), and the currents through the individual phases I_{ca}, I_{cb}, and I_{cc} denote the charging current (I_c) through the shunt capacitor.

Consider a line-to-ground fault occurring at the point F on phase c; the voltage across the shunt capacitor of that phase reduces to zero. The voltage of the other two healthy phases raise to line-to-line value, and the shunt capacitor charging currents of that phase are displaced by 60° as against its previous value of 120°. Thus, the net charging current is the vector sum of the charging current through the healthy phase, and it is three times the phase current under balanced condition. Figure 10.2(c) shows the vector representation of voltages and currents during line-to-ground fault in an ungrounded system. The net charging current flows through the fault path into the winding of the alternator and is capable of maintaining the arc in the faulty path; this condition is known as *arcing ground*. During arcing ground, the inductance of the generator winding and the system capacitance establish a cycle of charging and discharging, which results in high frequency oscillations. This oscillation builds up a very high voltage and the arc reignites, which results in supplementary transient disturbances. The consequence of such disturbance is the breakdown of insulators. Hence, the ungrounded system is not adopted in the power system because the failure of insulation causes serious phase-to-phase fault conditions.

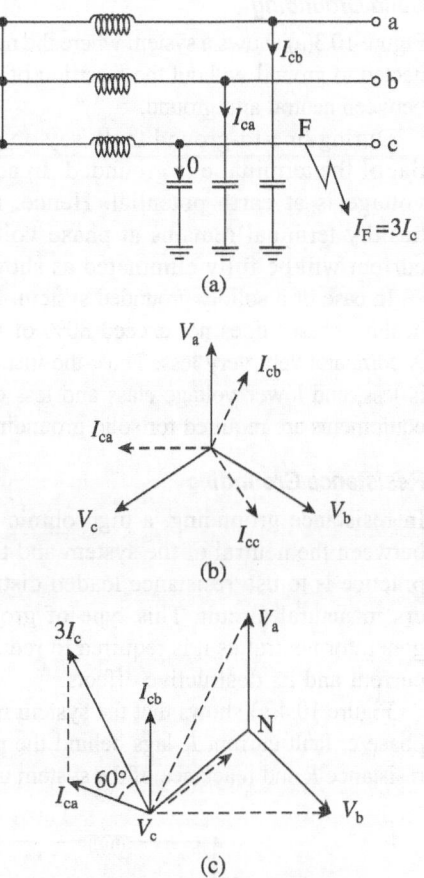

Fig. 10.2 Effect of ungrounded neutral on system performance (a) Simple three-phase system (b) Balanced voltages and currents (c) Voltages and currents during line-to-ground fault

10.4.2 Methods of Neutral Grounding

Grounding of neutral has many advantages in terms of stability and protection during short circuit:

1. Line-to-line voltage is reduced to line-to-ground voltage.
2. High voltage and transients produced during line-to-ground fault due to arcing ground can be restricted.

3. Sensitive protection against ground fault is possible.
4. The grounded neutral discharges the overvoltage produced during lightning stroke.

There are many effective and non-effective methods of neutral grounding, such as solid grounding, resistance grounding, reactance grounding, and resonant grounding. These are explained in the subsequent sections.

Solid Grounding

Figure 10.3(a) shows a system where the neutral is directly connected to ground without the insertion of any passive element between neutral and ground.

During line-to-ground fault, say on phase c, the potential of the terminal c is grounded. In addition, the neutral voltage is at earth potential. Hence, the voltage of the healthy terminal remains at phase voltage, and charging current will be fully eliminated as shown in Fig. 10.3(b).

In case of a solidly grounded system, the phase voltage of healthy phases does not exceed 80% of the line voltage and is comparatively very less. Thus, the insulation level required is less, and lower voltage class and less expensive protective equipments are required for solid grounding systems.

Resistance Grounding

In resistance grounding, a high ohmic resistor is inserted between the neutral of the system and the ground. Current practice is to use resistance loaded distribution transformers in neutral circuit. This type of grounding is used for generator neutral as it is required to reduce the ground fault current and its destructive effects.

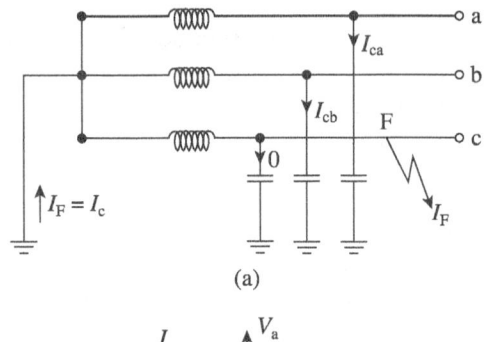

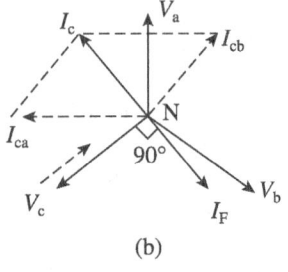

Fig. 10.3 Solid grounding (a) Neutral connected directly to ground (b) Voltages and currents during line-to-ground fault

Figure 10.4(a) shows that the system neutral is grounded through resistor R. During line-to-ground fault on phase c, fault current I_F lags behind the phase voltage of the faulted phase by an angle θ, which depends on resistance R and reactance of the system upto the fault point. Figure 10.4(b) shows the vector diagram in which

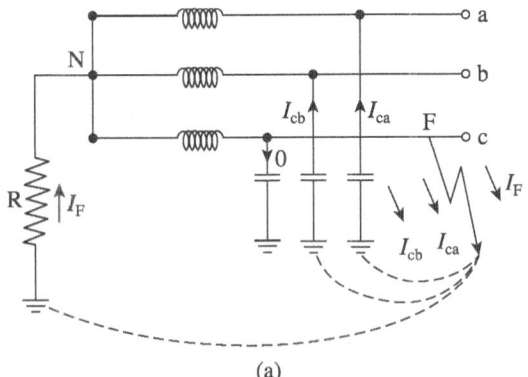

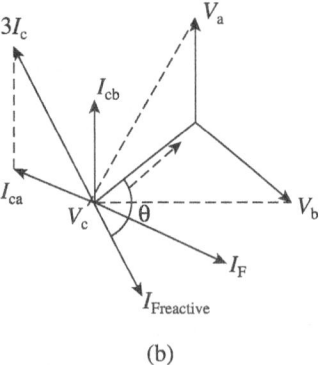

Fig. 10.4 Resistance grounding (a) Neutral connected to ground through resistor (b) Voltages and currents during line-to-ground fault

currents I_{ca} and I_{cb} lead the voltages, V_{ca} and V_{cb}, respectively, by 90°. Their resultant current is three times the charging current (I_c) through the shunt capacitor of each phase. By selecting the proper value of R, it is possible to compensate the effect of the charging current ($3I_c$) with fault current I_F so that the transient oscillation due to the arcing ground is eliminated. Thus, a resistance grounding system limits the ground fault current compared to the solidly grounded system and also permits use of selective protective scheme.

Reactance Grounding

In reactance grounding, a reactance is inserted in the ground connection. This grounding depends on the value of the ratio of X_0/X_1. If X_0/X_1 is higher than 3.0, then the system is said to be reactance grounding. The value of reactance required is to limit the fault current within the safe value as well as to ensure satisfactory relaying operation. This method is used for a circuit with large charging current such as neutral of synchronous motor and capacitors bank.

Resonant Grounding

The resonant grounding system includes an arc suppression coil (iron cored reactor) inserted between the neutral and ground. The main function of this arc suppression coil is to limit the earth fault current as also to extinguish the arc sustained during the ground fault by balancing the fault current exactly with the charging current. The arc suppression coil is also known as a *Petersen coil* or *ground fault neutralizer*.

Figure 10.5(a) shows the neutral of the system grounded through the arc suppression coil. Figure 10.5(b) shows the corresponding vector diagram of the earth fault on phase c at point F.

During the fault, the resultant capacitive current is three times the charging current (I_c) of one phase. If the desired value of inductance (L) is inserted in the neutral circuit such that $I_F = 3I_c$, then

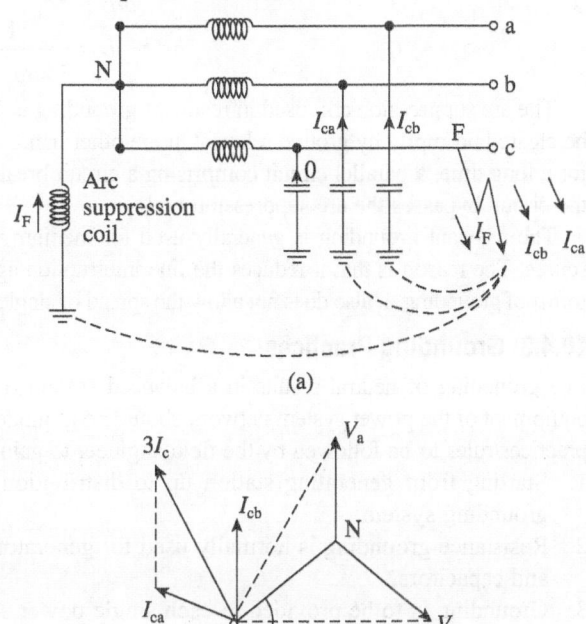

(a)

(b)

Fig. 10.5 Resonant grounding (a) Neutral grounded through arc suppression coil (b) Voltages and currents during line-to-ground fault

theoretically there is no current in the fault to sustain the arc. However, the voltage of healthy phases will be $\sqrt{3}\,V_{ph}$ during line-to-ground fault, where V_{ph} is the line-to-ground voltage of the system. If the inductive reactance of coil is X_L, then the fault current through it is given by

$$I_F = \frac{V_{ph}}{X_L}$$

If the capacitance of each phase is considered to be C, then the charging current during the fault is given by

$$3I_c = 3 \times V_{ph} \times \omega \times C$$

where $\omega = 2 \times \pi \times f$ = angular frequency
f = system frequency
Hence, the inductance of the coil can be found by

$$I_F = 3I_c$$

$$\frac{V_{ph}}{X_L} = 3 \times V_{ph} \times \omega \times C$$

$$\frac{1}{\omega L} = 3 \times \omega \times C$$

$$L = \frac{1}{3 \times \omega^2 \times C}$$

The arc suppression coil used in resonant grounding is 10 minutes time-rated for a system where the fault can be cleared promptly by ground relays. On the other hand, if the fault current magnitude is very high and persists for a long time, a parallel circuit comprising a circuit breaker, time delay relay, and resistor is arranged such that the circuit bypasses the arc suppression coil.

The resonant grounding is generally used for medium voltage transmission lines connected to the generating source. The reason is that it reduces the line interruption against transient line-to-ground faults compared to other forms of grounding. It also does not allow the spread of single-line-to-ground fault into double-line or triple-line faults.

10.4.3 Grounding Practices

The grounding of neutral results in a balanced system under normal working conditions of the system. Each equipment of the power system network should be grounded for the safe operation of the system. There are some practices/rules to be followed by the field engineer to gain the benefit of grounding.
1. Starting from generating station up to distribution, each voltage level should be provided with one grounding system.
2. Resistance grounding is normally used for generators and reactance grounding for synchronous motors and capacitors.
3. Grounding is to be provided to each single power source or multiple power sources as well as to each transformer substation.
4. If multiple power sources are working in parallel in a plant, they all are connected to a common neutral bus connected to the ground through a common resistor. However, neutral switching equipment can be used to connect a particular generator neutral to the neutral bus.
5. If a number of generators are operating in parallel, at least one generator should be grounded properly.
6. For low tension supply, that is, 220 V and 440 V, solid grounding is used, whereas for medium voltage, that is, 3.3 kV, 6.6 kV, and 11 kV, resistance or reactance grounding is used. However, for voltage rating of 33 kV and above, solidly grounded neutral system is the right choice.

Example 10.1 A 220 kV, three-phase, 50 Hz, 60 km long overhead transmission line has a capacitance of 1.2 mF/km. Determine the inductive reactance and KVA rating of the arc suppression coil suitable for this system to eliminate arcing ground effect.

Solution:

$$L = \frac{1}{3 \times \omega^2 \times C}$$

where $\omega = 2 \times \pi \times f$
$\quad\quad = 2 \times 3.14 \times 50 = 314$
and f is the system frequency in Hz

Thus,

$$L = \frac{1}{3 \times (314)^2 \times 1.2}$$
$$= 2.81 \text{ H}$$

Inductive reactance

$$X_L = 2 \times \pi \times f \times L$$
$$= 2 \times \pi \times 50 \times 2.81$$
$$= 882.78 \ \Omega$$

MVA rating of the arc suppression coil $= \dfrac{V_L^2}{3 \times \omega \times L} = \dfrac{220 \times 220}{3 \times 314 \times 2.81} = 18.28 \ \text{MVA}$

Hence, KVA rating of the arc suppression coil = 18.28 × 1000 KVA = 18280 KVA

Fault current $I_F = \dfrac{V_L}{\sqrt{3} \times X_L} \times 1000 = \dfrac{220}{\sqrt{3} \times 884.6} \times 1000 = 143.58 \ \text{A}$

10.5 Protection Against Transients and Surges

As discussed earlier, surges and transients in power systems result from the energy being released into the system. The major causes are lightning or switching events. When lightning strikes on a high voltage primary transmission system, it produces a transient overvoltage, which is transmitted through a transformer to the secondary transmission system. Lightning may strike directly on or close to the secondary transmission system, which results in an even higher energy surge to the equipment connected in the system. The second cause, switching surges, originates through breaker operation, capacitor bank switching, or a fault somewhere in the system.

Irrespective of the causes of the surges, if electric equipment in an electric network are not protected against transient overvoltages, they may go out of action or even get damaged. Protection against these surges is carried out by preventing the surges at their origin in the power system, diverting them to the ground before they spread to the extended network, and suppressing them using a surge protective device in the equipment.

Lightning arrester and surge absorber are the devices that offer protection to the electrical equipment from high magnitude transient overvoltages generated in the power system.

10.5.1 Protection Against Lightning

Electrical equipment can be damaged because of overvoltages such as switching surge overvoltage, lightning surge overvoltage, TRV, and power frequency temporary overvoltage (TOV) in the transmission line and at the receiving end of the substation. It is important to protect power equipment against them wherever possible in the system through reliable and efficient approaches. Transients or surges in a power system may originate from switching and from other causes, but the most important surges are those caused by lightning. The lightning surges may cause serious damage to the expensive equipment in the power system (e.g., generators, transformers, etc.) either by direct stroke on the equipment or by strokes on the transmission lines that reach the equipment as travelling waves. It is necessary to protect the equipment against both kinds of surges. The most commonly used devices for protection against lightning surges are earthing screen, overhead ground wires, and LAs or surge diverters.

Earthing screens protect the power station and substation against direct strokes, whereas overhead ground wires protect the transmission lines against direct lightning strokes. LAs or surge diverters protect the station apparatus against both direct strokes and the strokes that come into the apparatus as travelling waves.

Earthing Screen (Overhead Shielding)

There are many sensitive and expensive electrical equipment such as transformers, breakers, current transformers (CTs), and PTs located in generating stations and outdoor substations. The effects of direct lightning stroke on any of these electrical equipment with an unprotected substation can be destructive. Hence, some form of direct lightning stroke protection should be provided. These stations can be protected against direct lightning strikes through an earthing screen. It consists of a network of copper conductors, generally called *shield* or *screen*. The screen wires require extending over the area of substation or power station and are located above substation buses and equipment. The earthing screen is appropriately connected to the earth on at least two points through low impedance. When a direct stroke on the station occurs, the screen provides a low resistance path by which the lightning surges are discharged to the earth. In this way, station equipment is protected against damage. The limitation of this method is that it does not provide protection against the travelling waves that may reach the equipment in the station.

Overhead Ground Wires

Overhead ground wires provide the most effective method of protection to the transmission lines against direct lightning strokes. Figure 10.6 shows one ground wire and one line conductor connected to the tower, where the ground wires are located above the line conductors such that all lightning strokes are interrupted by them (i.e., ground wires). The height of the ground wire above the highest line conductor can be easily determined by the geometry of the protection zone. The effective protection given by the ground wire depends on the height of the ground wire above the ground and the protection or shielding angle. As shown in Fig. 10.6, a positively charged cloud is assumed to be above the line; it induces a negative charge on the transmission line portion below it. With the ground wire present at each tower, both the ground wire and the line conductor get the induced charge. However, the ground wire is earthed at regular intervals, and the induced charge is drained to the earth. The potential difference between the ground wire and the cloud, and that between the ground wire and the transmission line wire will be in the inverse ratio of their respective capacitances. Hence, the ground wires will take up all the lightning strokes instead of allowing them to reach upto the line conductors. However, the degree of protection provided by the ground wires depends on the earth resistance (R_e) and footing resistance of the tower.

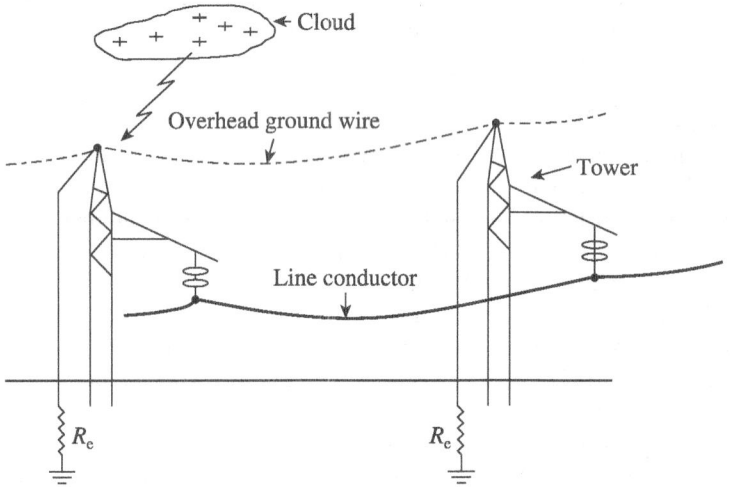

Fig. 10.6 Overhead ground wire

Surge Modifier or Absorber

Substation equipment such as breakers, transformers, and measuring CTs and PTs are very expensive and must therefore be provided with almost 100% protection against surges. Wherever lightning strokes are severe and are likely to occur frequently, overhead lines with these zones are fixed with shunt protected devices. Line terminations, junctions of lines, and substations are usually fitted with surge arresters (discussed in the following subsection). However, it is sometimes required to modify the shape of surges to reduce the slope of their wave front before they reach the substation. On reaching the substation, surges above a certain peak value are diverted into a shunt path to discharge their energies. As a result, the surges that finally reach the apparatus are so modified and reduced in strength that they are completely harmless. The shunt discharge path for the surge must be auto-clearing so as not to constitute a fault on the line. Surge modifiers comprise a surge capacitor, surge reactor, and surge absorber. A *surge modifier* is a small shunt capacitor connected between the line and earth, or a series air-cored inductor.

A capacitor connected in parallel with the equipment offers protection against surges to some extent; before the surge can impress a high voltage on the equipment, it must charge the capacitor. Temporary energy storage in the modifiers reduces the steepness of the surge wave front, which otherwise can damage the apparatus since the voltage across the equipment can rise only as quickly as the capacitor can be charged through the line. Capacitors are particularly effective against short duration surges. The energy of these surges is such that it can be absorbed by the capacitors. However, for high energy surges, the size and cost of the capacitors become prohibitive. A combination of a capacitor and an arrester can be very useful because they balance each other. The capacitor reduces the steepness of the wave.

A reactor connected in series with the equipment also serves an identical purpose. When a surge reaches the reactor, it initially appears only across the reactor; the reactor offers high impedance to the high frequency currents.

The surge energy is initially absorbed in the magnetic field of current in the reactor. Its ability to store energy is limited, but it can address low energy surges.

A surge absorber is shown in Fig. 10.7, where an inductor is enclosed in an earthed metal shield.

Damping is due to the eddy currents induced in the shield. The charging of the capacitance between the coil and earthed shield also reduces the energy of the wave. The absorber is effective not only in reducing the steepness of the wave but has some effects on the amplitude also as some energy is dissipated by the resistance of the metallic shield.

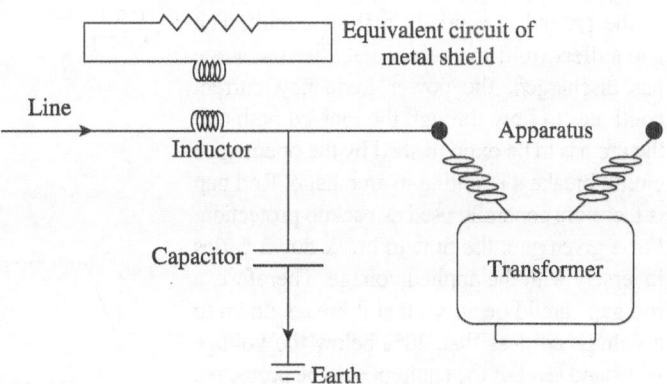

Fig. 10.7 Surge absorber equivalent circuit

Lightning Arrester (Surge Diverter)

The earthing screen and ground wires can well protect the electrical system against direct lightning strokes, but they fail to protect against travelling waves, which may reach the terminal apparatus. The LAs or surge diverters protect the system against such surges. *Lightning arrester* is the device that protects electrical equipment from damages caused by lightning overvoltage. So, it is required to connect the LA at the terminal end of the transmission line, substation, high voltage transformers, and low voltage transformers. The induction of electromagnetic transient in a power system mainly depends on the operating voltage, lengths of the lines, characteristic of circuit breakers, and substation and line configurations. Hence, the types of lighting arrester and its specification can be chosen correctly on the basis of the above-mentioned factors.

An LA can act as an open circuit during normal operation of the system. It limits the transient voltages to a safe level and brings the system back to its normal operational mode as soon as the transient voltages are suppressed. Therefore, a lightning or surge arrester must have an extremely high resistance during normal system operation and a relatively low resistance during transient overvoltages, which is due to its non-linear voltage–current (*V–I*) characteristic.

10.6 Types of Lightning Arresters

Different types of LAs are generally used for surge protection. The difference between them lies only in the structural mechanism; otherwise, the operating principle remains the same, that is, they provide a low resistance path for the surges to the ground. The different types of LAs are discussed in Sections 10.6.1–10.6.5.

10.6.1 Rod Gap Arrester

This consists of a plain air gap between two square rods (1 cm^2) bent at right angles and connected between the line and earth, as shown for the case of a transformer bushing in Fig. 10.8. The distance between the gap and the insulator (i.e., distance X) must not be less than one-third of the gap length so that the arc may not reach the insulator and damage it. When the surge voltage reaches the design value of the gap, an arc appears in the gap providing an ionized path to the ground, essentially a short circuit. The gap suffers from the defect that after the surge has discharged, the power frequency current continues to flow through the ionized path and the arc has to be extinguished by the opening of circuit breakers resulting in an outage. Rod gap is therefore generally used as backup protection. For a given gap, the time to break down varies inversely with the applied voltage. Therefore, a rod gap should be so set that it breaks down to a voltage not less than 30% below the voltage withstand level of the equipment to be protected.

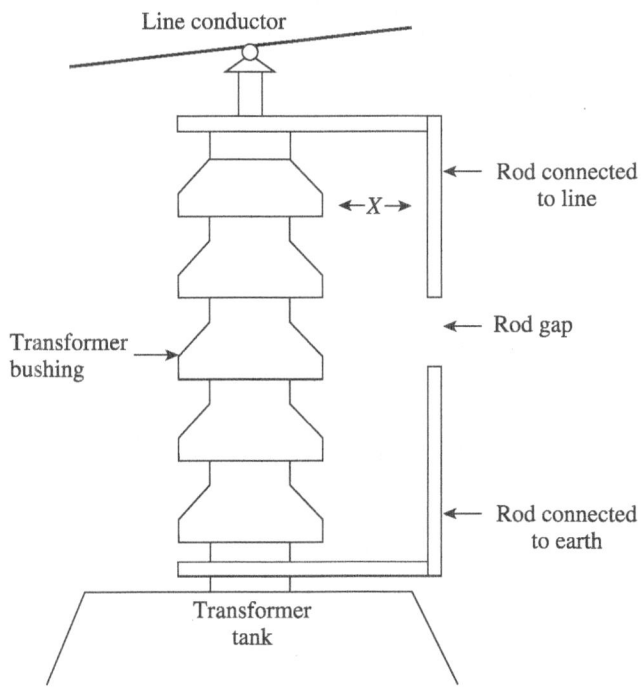

Fig. 10.8 Rod gap arrester on bushing insulator

10.6.2 Horn Gap Arrester

Figure 10.9 shows the circuit connection of a horn gap arrester. It consists of horn-shaped metal rods X and Y separated by a small air gap. The horns are so constructed that the distance between them gradually increases towards the top. The horns are mounted on porcelain insulators. One end of the horn is connected to the line through a resistance R and choke coil L while the other end is effectively grounded. Resistance R helps in limiting the current flow to a small value. The choke coil is so designed that it offers small reactance at the normal power frequency but a very high reactance at transient frequency. Thus, the choke does not allow the transients to enter the apparatus to be protected. The gap between the horns is so adjusted that the normal supply voltage is not enough to cause a spark across the gap.

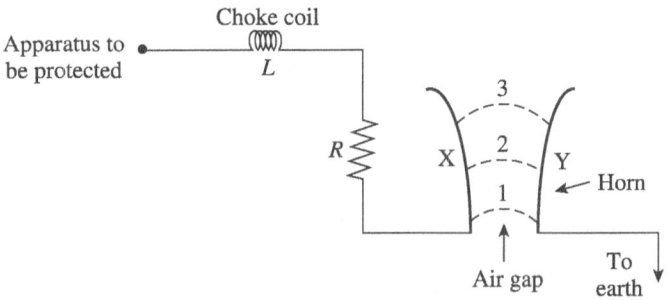

Fig. 10.9 Horn gap arrester

Under normal conditions, the gap is non-conducting, that is, the normal supply voltage is insufficient to initiate the arc between the gaps. On the occurrence of an overvoltage, spark-over takes place across the small air gap. The ionized air around the arc and the magnetic effect of the arc cause it to travel towards the upward wall of the gap. The arc moves progressively into positions 1, 2, and 3. At the top position of the horn gap, the distance may be too large, which results in a high resistance to the arc, and consequently, the arc is extinguished. The excess charge on the line is thus conducted through the arrester to the ground.

10.6.3 Multigap Arrester

In a multigap arrester, metallic cylinders (usually alloy of zinc) insulated from each other and separated by small intervals of air gaps are connected in series. Figure 10.10 shows a multigap arrester. The first cylinder (X) in the series is connected to the line and the others to the ground through a series resistance. The series resistance limits the power loss to the arc. By the addition of a series resistance, the degree of protection against the travelling waves is reduced. Therefore, some of the gaps (Y to Z) are shunted by resistance.

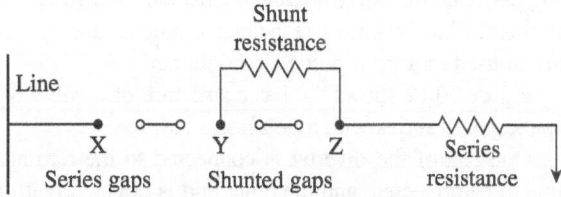

Fig. 10.10 Multigap arrester

Under normal conditions, the cylinder Y is at earth potential and the normal supply voltage is unable to break down the series gaps. During an overvoltage, the breakdown of the series gap X to Y occurs first. At this moment, high magnitude current is diverted into earth through the path of the shunted gaps by cylinders Y and Z, instead of the optional path through the shunt resistance. When the surge is over, the arcs through Y to Z extinguish and the current following the surge is limited by the two resistances (shunt resistance and series resistance) that are now in series. The current is too small to maintain the arcs in the gaps X to Y, and hence normal conditions are restored. Such arresters can be employed upto system voltage of 33 kV.

10.6.4 Expulsion Type Arrester

This type of arrester is also called *protector tube* and is commonly used on systems operating at voltages upto 33 kV. Figure 10.11(a) shows the essential parts of an expulsion type LA. It essentially consists of a rod gap X–Y in series with a second gap enclosed within the fibre tube. The gap in the fibre tube is formed by two electrodes. The upper electrode is connected to the rod gap and the lower electrode to the earth. One expulsion arrester is placed under each line conductor. Figure 10.11(b) shows the installation of an expulsion arrester on an overhead line. The series gap is set to arc over at a specified voltage lower than the withstand voltage of the equipment to be protected. The follow-on current is confined to the space inside the relatively small fibre tube. A part of the tube material vaporizes, and the high pressure gases so formed are expelled through the vent at the lower end of the tube, causing the

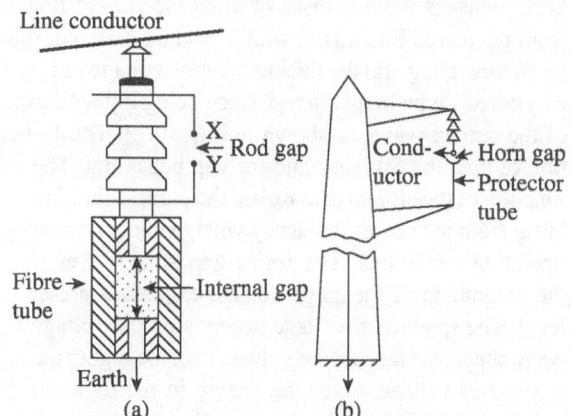

Fig. 10.11 Expulsion type arrester (a) Essential parts (b) Installation of expulsion arrester

power follow-in arc to extinguish. The device, therefore, has the desired self-clearing property. Because of vaporization of the tube material and weathering effect, the protector tube requires frequent replacement, and lack of proper maintenance may lead to occasional outage. It is, therefore, practically out of use now.

10.6.5 Valve Type Arrester

Electrode gaps arresters were used earlier to limit the overvoltages on equipment in substations at a lower voltage level. However, the characteristic of gap spark overvoltage versus surge front time does not match well with the strength versus front characteristics of most insulation. Thus, it is difficult to coordinate the protective device with the system voltage for which it is used. To resolve this, a resistive element in series is added with the gap to limit the power-follow current after an arrester discharge operation. The valve type arresters integrate non-linear resistors and are extensively used in systems operating at high voltages. The name itself suggests that the valve arresters consist of non-linear resistors, which act like valves when voltages are applied to them. The non-linear resistance elements are connected in series with the spark gaps. Both the assemblies are housed in a tight porcelain container.

Figure 10.12 shows the basic structure of a valve type LA. It consists of a spark gap in series with a non-linear resistor.

One end of the diverter is connected to the terminal of the equipment or line to be protected, and the other end is effectively grounded. The gap length of the arrester is so adjusted that the normal voltage is not enough to produce a spark. On the occurrence of an overvoltage, the air insulation across the gap breaks down and an arc is formed, providing a low resistance path for the surge to the ground. In this way, the excess charge on the line or equipment due to the surge is safely conducted through the arrester to the ground instead of being sent back over the line. There are two types of valve arresters: silicon carbide (SiC) type and metal oxide (MO) type.

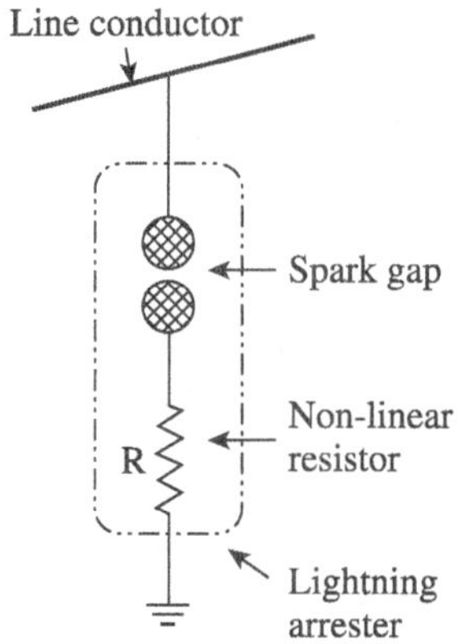

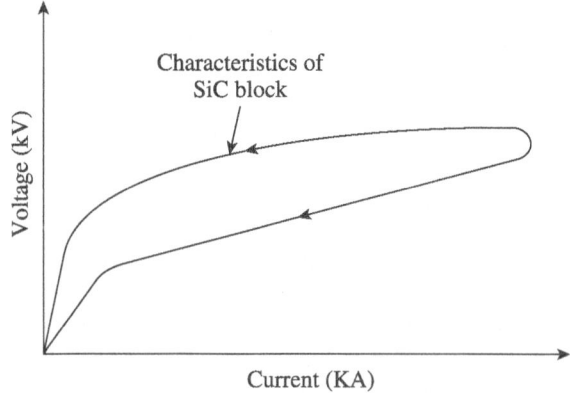

Fig. 10.12 General structure of lightning arrester

Silicon Carbide (SiC) Lightning Arrester

In this type of LA, the non-linear SiC material is connected in series with the spark gap. The spark gaps provide the high impedance during normal conditions, whereas the SiC disks obstruct the current flow through the spark over. The non-linear resistors are built from powdered SiC mixed with a binding material, moulded into a circular disk. The disk diameter depends on its energy rating and the thickness on its voltage rating. The *V–I* characteristic of a SiC non-linear valve block has a hysteresis-type loop, the resistance being higher during the rising part of the impulse current than during the tail of the current wave, as shown in Fig. 10.13. The *V–I* characteristics of SiC-type surge arresters are a combination of both the SiC disk and the gap behaviour. The function of the air gap is to isolate the current-limiting block from the power frequency voltage under normal operating conditions. The series gap sparks over if the magnitude of the surge voltage exceeds a preset level. The spark-over voltage depends on the voltage wave shape: the steeper the voltage rise, the higher the spark-over voltage. Once the energy in the transient voltage is dissipated, the power-follow current will flow through the valve elements. A relatively high current will continue to flow through the arrester after the spark over, and after the overvoltage has stopped, if the power voltage across the arrester is high enough, since the gap arc has a low resistance and the current is primarily determined by the *V–I* characteristic of the SiC block.

Fig. 10.13 *V–I* characteristics of SiC block

SiC arresters are successfully applied on transmission systems upto 220 kV, but there are limitations with regard to switching surge protection, energy discharge capability, and pressure relief capability. Due to the presence of the gaps and SiC blocks, the arrester height increases to the point where it is difficult to expel the pressure built up during a fault, which limits the arrester's pressure relief rating. Due to their discharge characteristics versus frequency or front time, the SiC surge arresters are limited for lightning protection. Some investigators have noted high silicon carbide arrester failure rates, due to moisture ingress, after several years of service on medium voltage distribution systems. They are less effective for steep-fronted surges and slow-fronted switching surges. However, many silicon carbide arresters are still in service.

Metal Oxide Lightning Arrester

The metal oxide (MO) lightning arrester utilizes metal oxide varistor material for its manufacturing. Varistors are made of zinc oxide (ZnO) powder and traces of oxides of other metals bound in a ceramic mould. Their characteristic avoids the need for series spark gaps. Therefore, the electrical behaviour is determined solely by the properties of the MO blocks.

The *V–I* characteristics of both SiC and ZnO valve elements are shown in Fig. 10.14. Here, the noticeable point is the difference in the leakage currents in both valve types for a given applied voltage. The series gap is essential for an SiC valve arrester to prevent thermal runaway during normal operation, whereas gapless operation is possible with ZnO arresters because of the low leakage current during normal operation.

The MO blocks are much more non-linear than silicon carbide blocks so that they conduct only a few milliamperes at a nominal AC voltage. Figure 10.15 shows the general use of gapped and gapless MO LAs. Figure 10.15(a) shows the typical shunt gapped arrester used in early MO arresters. At a steady state, both the non-linear elements would support the nominal voltage, reducing the current to an extent. During a surge expulsion, that is, when the surge current magnitude reaches 250–500 A, the shunt gap would spark over to bypass the smaller section of MO, thereby reducing the discharge voltage and providing a better protection. Some distribution class arresters use series gaps that are shunted by a linear component impedance network. These types of arresters adopt a series gap with a capacitive grading in early MO arresters as shown in Fig. 10.15(b). At a steady state, this decreases the voltage on the MO. During a surge discharge, the gap sparks over immediately owing to the capacitive grading. The latest generations of MO do not need these gaps

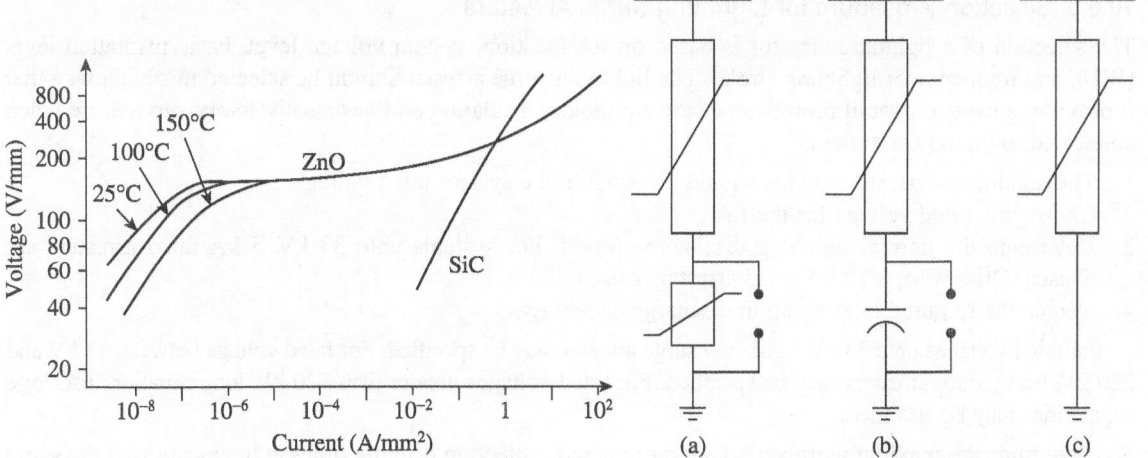

Fig. 10.14 *V–I* characteristic of ZnO and SiC elements

Fig. 10.15 Metal oxide arrester (a) Shunt gapped arrester (b) Series and shunt gapped arrester (c) Gapless arrester

as shown in Fig. 10.15(c). Thus in a gapless arrester, the discharge voltage for a given current magnitude is directly proportional to the height of the valve element stack and is more or less proportional to the arrester rated voltage. The operation of an arrester is sensitive to the rate of rise of the incoming surge current: the higher the rate of rise of the current, greater is the rise in the arrester limiting voltage. It has been suggested that the higher discharge voltage of the MO block for higher rates of current rise is caused by the negative temperature coefficient of the valve resistance as shown in Fig. 10.15.

Figure 10.16 shows the sectional view of one unit of a porcelain type LA. The lightning or surge arresters are basically composed of the ZnO varistors, spacers, and support (locating) rings. These parts are housed in a porcelain insulator cylinder, and hermetic sealing of the housing is done using a weatherproof synthetic rubber. The pressure relief diaphragm is prepared with a special metal plate, which is ruptured when a sudden pressure rise occurs during internal failure. The air void is evacuated and replaced with dry air and then sealed. A compression spring is used to hold the ZnO blocks tightly.

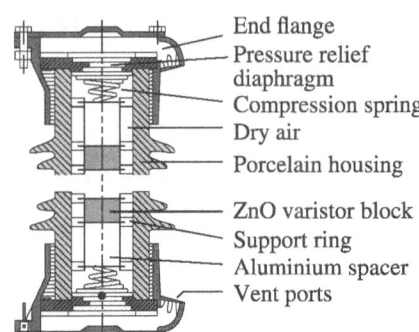

Fig. 10.16 Sectional view of a lightning arrester

Metal oxide surge arresters have several major advantages over silicon carbide arresters; these are as follows:

1. Active gaps are not necessary, which leads to improved reliability.
2. It can discharge much more energy per unit volume than silicon carbide.
3. It provides better protection across the range of surge wave fronts than silicon carbide, and in fact, protects the equipment effectively against switching surges.
4. The decrease in arrester height, caused by eliminating spark-over gaps, leads to higher pressure relief ratings.
5. It has enabled new applications, such as series capacitor protection and overhead line switching surge control, which were not possible with silicon carbide.

Virtually all new applications use MO surge arresters for high voltage and EHV systems upto 500 kV.

10.6.6 Selection Procedure for Lightning/Surge Arresters

The selection of a lightning arrestor is based on the location, system voltage level, basic insulation level (BIL), and frequency of lightning stroke. The lightning/surge arrester should be selected in such a way that it provides adequate overall protection of the equipment insulation and has a satisfactory service life when connected to the power system.

1. The continuous arrester voltage should be equal to the system rated voltage.
2. Choose the rated voltage for the LA.
3. Determine the normal lightning discharge current. For systems upto 33 kV, 5 kA rated arresters are chosen. Otherwise, a 10 kA rated arrester is used.
4. Decide the required long duration discharge capability.

For rated voltages upto 33 kV, light duty surge arrester may be specified. For rated voltage between 33 kV and 220 kV, heavy duty arresters may be specified. For rated voltages greater than 220 kV, long duration discharge capabilities may be specified.

5. Determine the maximum probable fault current and protection tripping times at the location of the surge arrester and match with the surge arrester duty.
6. Select the surge arrester with porcelain creepage distance in accordance with the environmental conditions.
7. Determine the surge arrester protection level and match with the standard IEC 99 recommendations.

10.6.7 Common Ratings of Lightning/Surge Arresters

Surge arresters have both voltage rating and class. For all types of LAs, the important voltage rating is the maximum continuous operating voltage (MCOV), which is the steady-state voltage the arrester could support for an indefinite time. This rating is most important for MO, because most of these arresters are now gapless and carry continuously a few milliamperes of current. The MCOV should be at least 1.05 times the system's nominal line-to-ground voltage. Short-term TOVs also play an important role in selecting the arrester rating, but still the basic rating is MCOV. A few more ratings of the arresters are discussed.

Rated voltage   The power frequency voltage across the arrester must never exceed its rated voltage. If it happens for a long duration, the arrester may not reseal and may disastrously fail after absorbing the energy of the surge. Hence, for an effectively earthed system, the maximum phase-to-earth voltage should not exceed 80% of the maximum line voltage in any circumstance.

Rated current   Arresters are tested with 8/20 microsecond discharge current waves of varying peak magnitudes of 1.5 kA, 3 kA, 5 kA, 10 kA, and 20 kA, with the resulting peak discharge voltages.

Normal voltage   Normal voltage is the nominal continuous voltage that the arrester can withstand before failing or flashover.

Basic impulse insulation level   Basic impulse insulation level is defined by the values of the test voltages that the insulation of the arrester under test must be able to withstand. The test voltage is the maximum impulse for a 1.2×50 μs waveform.

Discharge voltage   When the magnitude of overvoltage impulse reaches the discharge voltage value, the arrester begins to conduct (discharge) energy to earth.

10.7 Concept of Basic Insulation Level

The causes of switching and power frequency overvoltages are not common. Hence, it is important to consider the combined effect of the power frequency overvoltages as well as the transient overvoltages together while designing the insulation. There is a low chance for lightning and switching overvoltages to coincide and hence its effect can be neglected. The magnitude of overvoltage induced on the transmission line due to lightning does not depend on the line design as it is an external phenomenon. Thus, at a given system voltage, lightning inclines to improve the expanding insulation level of the system. On the other hand, local switching overvoltages are proportional to the operating voltage of the system. Hence, the criterion for the arrester design changes from lightning to switching surge for systems above 500 kV. In the range of 220 kV to 765 kV, both switching overvoltages and lightning overvoltages have to be considered, whereas for ultra high voltages (UHVs) (>700 kV), switching surges are considered the chief condition for design considerations.

Basic insulation level is defined by the values of test voltages which the insulation of equipment under test must be able to withstand. BIL is the reference insulation level expressed as an impulse crest (or peak) voltage with a standard wave not longer than a 1.2×50 μs wave. For each system voltage, BIL has been fixed by most of the national and international standards. The major substation equipments are manufactured for the same insulation level, except for transformers.

10.7.1 Selection of Basic Insulation Level

Proper selection of insulation focuses not only on the insulating the protective equipment but also on ensuring that it is not damaged. Standard insulation levels are recommended for the coordination of insulation. Basic insulation levels are the reference levels expressed in impulse crest voltage with a standard wave.

The insulation level decided for substation and power plant equipments should be equal to or greater than the BIL. The number of arresters at different locations on line and in substation depends on the severity of the steep-fronted lightning waves at substations and at different points on lines. Moreover, the voltages at substations may exceed the protective level, depending on the distances involved and the arrester locations.

The LA is located on the terminal side of the PT and the associated breaker. The closeness of the arrester location to the transformer provides the greatest protection as it reduces the effect of overvoltage created because of current chopping. On the other hand, by locating the arrester right on the terminal of the transformer, some other substation equipment may fall outside the protective zone, which is decided from the voltage withstand level of the equipment, discharge voltage of the LAs, and the distance between the equipment and the LA. Thus, the BIL is often determined by giving a margin of 30% to the protective level of the surge arrester and selecting the next nearest standard BIL. When a surge arrester is used to provide switching surge protection, the margin given is only 15%.

The insulation levels for lines and other equipment are to be selected separately. The atmospheric conditions, lightning activities, insulation pollution, numbers of line outage, and failure rates of the line play a vital role in deciding the insulation level for lines. The protective level of the substation depends on the location of the station, the arrester protective level, and the line shielding used. The line insulation in the end, which spans close to the substation, is normally reduced to limit the lightning overvoltages reaching the substation. In a substation, the busbar insulation level is the highest to ensure continuity of supply. The circuit breakers, isolators, and instrument transformers are given the next lower level. Since the power transformer is an expensive and sensitive device, its insulation level is the lowest.

Thus, in general, the insulation level of substation equipment, such as circuit breakers, switches, busbars, and instrument transformers, is assumed to be 10% higher than the transformer BIL. The insulation level across the open poles of isolator switches may be kept 10–15% higher than that provided between the poles and earth.

10.7.2 Impulse Ratio

The breakdown of insulation in any equipment depends on the peak magnitude of voltage, shape of the waveform, and the time for which it is applied to the equipment. The larger will be the magnitude of the voltage required to cause the breakdown the sharper the shape of the voltage wave. Thus, an appreciable time and energy expenditure are involved in the rupture of any dielectric material. The impulse ratio is related to the energy expenditure and the voltage withstand level. The *impulse ratio* is defined as the ratio of the breakdown voltage due to an impulse of a specified shape to the breakdown voltage at the power frequency. The insulators should have a high impulse ratio for an economic design, whereas the LAs should have a low impulse ratio so that a surge incident on the LA may be passed to the ground instead of to the apparatus.

10.7.3 Standard Impulse Test Voltage

Transient overvoltages due to lightning and switching surges cause steep build-up of voltage on transmission lines and other electrical apparatus. Experimental investigations showed that the peak rise time of these waves varies from 0.5 to 10 μs and decay time to 50% of the peak value, of the order of 30–200 μs. The wave shapes are random, but mostly unidirectional. It is to be observed that the wave shape of lightning overvoltage can be represented as a double exponential wave, which usually has a rapid rise to the peak value and slowly falls to zero value. The general impulse wave shape of 160 kV is given in Fig. 10.17. Impulse waves are specified by defining their rise time (or front time) and fall time (or tail time) to 50% peak value, and the value of the peak voltage. Thus the 1.2/50 μs, 750 kV wave represents an impulse voltage wave with a front time of 1.2 μs, fall time to 50% of the peak value of 50 μs, and a peak value of 750 kV. Indian standard specifications define the 1.2/50 μs wave to be the standard lightning impulse. The tolerance allowed in the peak value is ±3%, and the tolerances that can be allowed in the front time and tail time are ±30% and ±20%, respectively.

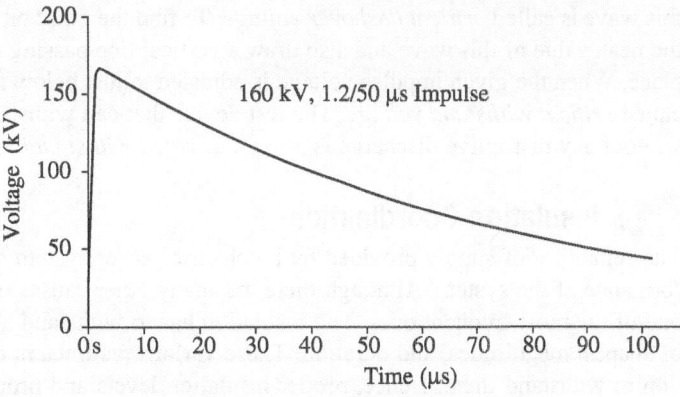

Fig. 10.17 Standard impulse wave shapes

10.7.4 Volt–Time Characteristic

The breakdown voltage of a particular insulation or flashover voltage is a function of both magnitude of voltage and time of application of the voltage. The volt–time curve is a graph showing the relation between the crest flashover voltages and the time to flashover for a series of impulse applications of a given wave shape. The construction of volt–time curve and the related terms are illustrated in Fig. 10.18.

The test device whose volt–time curve is required has been subjected to waves of the same shape but of different peak values until the flashovers occur on the front of the wave; this is known as *front flashover*. If the insulation flashover occurs just at the peak value of the wave, this gives another point on the volt–time (*V–T*) curve and is known as *crest flashover*. The third possibility is that the flashover occurs on the tail side of the wave which is known as *tail flashover*. The portion between front flashover and crest flashover is known as wave front flashover voltage range whereas the range between crest flashover voltage and the critical withstand voltage of the equipment is known as wave tail flashover voltage range. If an impulse voltage of the same wave shape is applied so that the test device flashover occurs on the tail of the wave at 50% of the application and tails to flashover occurs on the other 50% of the application, the crest value of

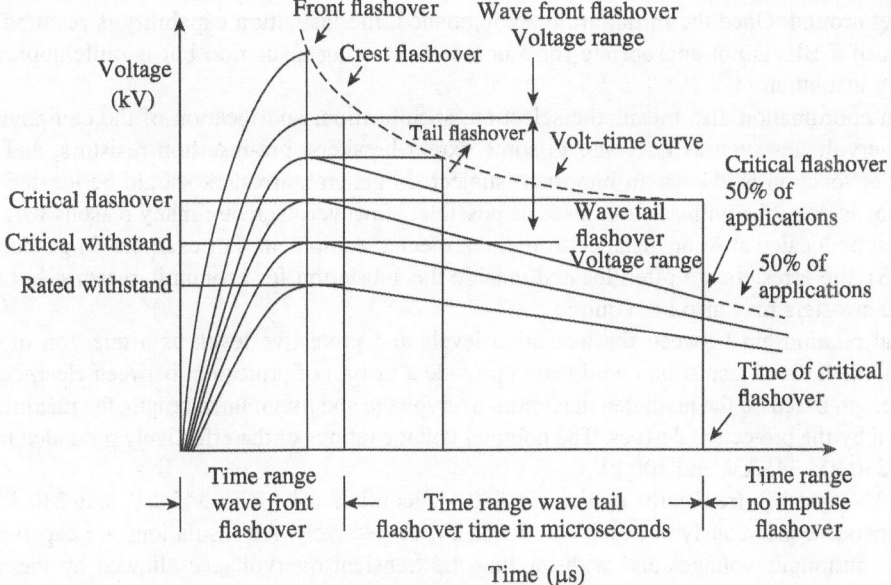

Fig. 10.18 Volt–time curves

this wave is called *critical flashover voltage*. To find the point on the *V–T* curve, draw a horizontal line from the peak value of this wave and also draw a vertical line passing through the point where the flashover takes place. When the given impulse voltage is adjusted to just below the flashover voltage of the test device, it is called *critical withstand voltage*. The test device that can withstand the crest of the applied impulse voltage without any disruptive discharge is known as *rated withstanding voltage*.

10.8 Insulation Coordination

The reliability of supply provided by an electric power system depends to a great extent on the surge performance of the system. Although there are many other causes of interruptions, breakdown of insulation is one of the most frequent ones. The insulation has to withstand a variety of overvoltages with a large range of shapes, magnitudes, and duration. These various parameters of overvoltages affect the ability of insulation to withstand them. Hence, proper insulation levels and proper overvoltage protection are required for the equipment. The insulation level of substation and line equipment is examined with reference to BIL as well as the selection, specification, and location of arresters.

Insulation coordination refers to the art of selecting appropriate insulation levels for the equipment and the corresponding overvoltage protection system (arresters) such that the insulation is protected from excessive overvoltages. In a substation, for example, the insulation of transformers, circuit breakers, bus supports, etc., should have insulation strength in excess of the voltage provided by protective devices. The overvoltage in the system can be tackled by choosing proper insulation levels for components so that it withstands all kinds of overvoltages or by selecting protective devices that could be installed at sensitive points in the system to limit the overvoltages at that point. The first option is undesirable, especially for EHV and UHV operating levels because of the requirement of excessive insulation. Hence, the second option motivates to expand and utilize protective devices.

A fundamental distinction must be made in the design process between *self-restoring* and *non-self-restoring insulations*. Solid insulation, such as in many cables systems and transformers, must be replaced once it fails; this is non-self-restoring insulation. Open air insulator strings, on the other hand, usually fail by arcing around. Once the initial arc is extinguished, the insulation capability is restored. Statistical specification of a BIL is not appropriate for a non-self-restoring insulation but is quite appropriate for a self-restoring insulation.

Insulation coordination also means the selection, specification, and location of the equipment intended to reduce overvoltages such as LAs and to some extent breakers, pre-insertion resistors, and other such items. Arrester location is always an important subject. In general, arresters should be located as close to the equipment as possible, with as short leads as possible. However, there are many reasons why sometimes arresters must be located at some distance from the protected equipment. For example, in gas insulated substations (GIS), the arresters are often located outside the substation for economic reasons, but sometimes, encapsulated arresters may also be required.

The actual relationship between the insulation levels and protective levels is a question of economics. Conventional methods of insulation coordination provide a margin of protection between electrical stress and electrical strength based on the predicted maximum overvoltage and minimum strength, the maximum strength being allowed by the protective devices. The nominal voltage ratings of the effectively grounded transmission systems are 230 kV, 345 kV, and 500 kV.

The 500 kV systems frequently operate at 550 kV continuously. The 345 kV and 230 kV systems, frequently operate continuously at 362 kV and 242 kV, respectively. All insulations are capable of operating at these continuous voltages and withstanding the transient overvoltages allowed by the overvoltage protection. Tables 10.1 and 10.2 show the standardized test voltages for ≤ 245kV and ≥ 300 kV, respectively, as suggested by the IEC for testing equipment. These tables are based on a 1992 draft of the IEC document on insulation coordination (IEC document 28 CO 58, 1992, Insulation Coordination Part I: Definitions, Principles, and Rules).

Table 10.1   Standard insulation levels for Range I (1 kV < U_m ≤ 245 kV)

Highest voltage for equipment U_m kV (RMS value)	Standard power frequency short-duration withstand voltage kV (RMS value)	Standard lightning impulse withstand voltage kV (peak value)
3.6	10	20
		40
7.2	20	40
		60
12	28	60
		75
		95
17.5	38	75
		95
24	50	95
		125
		145
36	70	145
		170
52	95	250
72.5	140	325
123	(185)	450
	230	550
145	(185)	(450)
	230	550
	275	650
170	(230)	(550)
	257	650
	325	750
245	(275)	(650)
	(325)	(750)
	360	850
	395	950
	460	1050

U_m: Highest voltage for the equipment in kV

Table 10.2 Standard insulation levels for Range II (U_m > 245 kV)

Highest voltage for equipment U_m kV (RMS value)	Longitudinal insulation (+) kV (peak value)	Standard lightning impulse withstand voltage phase-to-earth kV (peak value)	Phase-to-phase (ratio to the phase-to-earth peak value)	Standard lightning impulse withstand voltage kV (peak value)
300	750	750	1.50	850
				950
	750	850	1.50	950
				1050
362	850	850	1.50	950
				1050
	850	950	1.50	1050
				1175
420	850	850	1.60	1050
				1175
	950	950	1.50	1175
				1300
	950	1050	1.50	1300
				1425
525	950	950	1.70	1175
				1300
	950	1050	1.60	1300
				1425
	950	1175	1.50	1425
				1550
765	1175	1300	1.70	1675
				1800
	1175	1425	1.70	1800
				1950
	1175	1550	1.60	1950
				2100

U_m: Highest voltage for the equipment in kV   *(Contd)*

Recapitulation

- Whenever a circuit changes its condition because of short circuits, opening or closing of circuit breakers and isolators, and any other disturbances, it sets up a new current and voltage of very high magnitude in the network, which appear as a travelling wave with different propagation speeds.
- The travelling wave can reflect back from the other end of the line during its travel time; the wave reflection may produce voltage considerably higher than the surge voltage generated at its source.

- The transient waves get attenuated exponentially as they travel over the length of the line and the reduction in magnitude depends greatly on the parameters of the line.
- From the stability of the system and protection point of view, neutral grounding provides a number of benefits to the power system.
- Solid grounding, resistance grounding, reactance grounding, resonant grounding, and transformer groundings are the few methods used in practice for the grounding of system neutral.
- The switching of unloaded lines, opening and closing of isolating switches, arcing ground, interrupting an inductive current (reactor), and capacitive circuits (capacitor banks) produce transient overvoltages.
- Lightning strikes are the main source of overvoltage and drastically affect the performance of the power system component leading to complete failure of insulation in the case of severe lightning discharge.
- If equipment in an electric network are not protected against transient overvoltages, they may be put out of service or even damaged; therefore, surge protective devices are installed close to the equipment.
- Lightning arresters and surge absorbers are the devices that protect electrical equipment from high magnitude transient overvoltages generated in power systems.
- Rod gap arrester, horn gap arrester, multigap arrester, expulsion type arrester, and valve type arrester are the widely used arresters for the protection of apparatus against surges and transients in power systems.
- Basic insulation level (BIL) is the value of test voltage that the insulation of an electric device or specimen under test must be able to withstand. The test voltage is the maximum impulse for a 1.2×50 μs waveform.
- Insulation coordination refers to the art of selecting appropriate insulation levels for the equipment and the corresponding overvoltage protection systems (arresters) such that the insulation is protected from excessive overvoltages.

Multiple Choice Questions

1. The maximum current value of lightning stroke is
   (a) 10 A
   (b) 100 A
   (c) 10 kA to 100 kA
   (d) 1000 kA

2. Switching surges are
   (a) high voltage DCs
   (b) high voltage ACs
   (c) power frequency voltages
   (d) short duration transient voltages

3. The cause of surge voltage in power systems is
   (a) lightning
   (b) switching operations
   (c) faults
   (d) any of the above

4. Surge diverters are
   (a) non-linear resistors in series with spark gaps acting as fast switches
   (b) arc quenching devices
   (c) shunt reactors to limit the voltage rise due to Ferranti effect
   (d) overvoltages of power frequency harmonics

5. The function of lightning arresters is to protect the electric equipment against
   (a) power frequency voltage
   (b) direct strokes of lightning
   (c) fault current
   (d) overcurrent due to power frequency harmonics

6. Neutral grounding is required to
   (a) reduce the ground fault current
   (b) reduce the interference to the communication line

(c) improve the system stability and protection

(d) limit the ground relay operation

7. The arcing ground is

(a) an overvoltage phenomenon due to ungrounded neutral

(b) a resonance due to system parameter

(c) an undervoltage incident

(d) an overcurrent due to fault

8. Insulators for high voltage applications are tested for

(a) power frequency

(b) impulse

(c) both (a) and (b)

(d) none of the above

9. Basic insulation level of power system equipments is

(a) the maximum power frequency voltage of the system

(b) the minimum power frequency voltage of the system

(c) the nominal power frequency voltage that the equipment can withstand

(d) the peak impulse voltage of standard wave of 1.2 × 50 μs

10. During the insulation coordination in a substation, the lowest insulation strength is provided to the

(a) CT and PT

(b) circuit breaker

(c) transformer

(d) busbar

Review Questions

1. Explain the different sources of transients and surges in an EHV system.

2. Write a note on travelling wave and its propagation on transmission line.

3. Explain the reflection and attenuation of travelling wave.

4. Why is an ungrounded neutral system unsafe for a power system?

5. Explain different methods of neutral grounding.

6. Write the steps for grounding practices.

7. Explain the different schemes used for the protection of lightning strikes.

8. Explain the different types of LAs along with their construction.

9. Explain SiC and ZnO arresters along with *V–I* characteristic.

10. Mention the selection procedure and rating of an arrester.

11. What is a BIL? Write a note on the selection of BIL.

12. Explain the terms 'impulse ratio,' 'standard impulse test voltage,' and 'volt–time characteristic' in relation to BIL.

13. Discuss the insulation coordination of electric components.

Answers to Multiple Choice Questions

1. (c) 2. (d) 3. (d) 4. (a) 5. (b) 6. (c) 7. (a) 8. (b) 9. (d) 10. (c)

Auto-reclosing and Synchronizing

11

Learning Objectives

After going through this chapter, the students will be able to:

- Explain the concept of auto-reclosing and list its advantages
- Discuss the classification of auto-reclosing relays
- List the various factors to be considered while installing the reclosing and synchronizing scheme
- Discuss the function of synchronism check relay
- Explain the application of the auto-reclosing relays (79) to line along with tripping scheme

11.1 Introduction

Several research papers and surveys have analysed that 80% of faults on overhead transmission lines are transient in nature. In fact, several utility reports (including those by the IEEE Power Systems Relaying Committee) show that hardly 20% of the faults are permanent or semi-permanent in nature. Temporary faults, also known as *transient faults*, can be cleared by momentarily de-energizing the line. They occur because of lightning (insulator flashover), swinging wires, and temporary contact with foreign objects. Hence, there is a need to use an automatic reclosing system to improve service continuity, particularly for distribution systems. Use of high-speed auto-reclosing in interconnecting the lines greatly improves the stability of the system. On the other hand, faults on underground cables are predominantly permanent in nature. Hence, these faults should be cleared without the use of the auto-reclosing system. At the same time, auto-reclosing operation in permanent faults, particularly at transmission levels, results in a fault that has not been cleared, which may have adverse effects on system stability. Hence, knowledge of proper selection and application of auto-reclosing system is extremely necessary.

This chapter describes the applications of auto-reclosing and synchronizing relays in transmission and distribution systems.

11.2 History of Auto-reclosing

American Electric Power introduced the concept of high-speed reclosing in 1935 for radial lines by rapidly reclosing a single line, rather than providing a second, redundant path for power to flow. Based on the IEEE Power System Relaying Committee 1984 report, automatic reclosing was first applied on radial distribution feeders, ring networks, and sub-transmission networks protected by instantaneous relays. In this scheme, a multishot reclosing relay is used, which recloses the circuit in two or three attempts before final locking. The success rate of this scheme was found to be 73–88%.

Modern power systems contain a number of single-circuit radial lines, ring networks, double-circuit transmission lines, and series-compensated lines. Hence, a high-speed reclosing system is necessary for maintaining stability rather than continuity of supply at every point. At present, the single-phase high-speed auto-reclosing system is widely used for 220/400 kV transmission lines. Conversely, the three-phase auto-reclosing system is used for transmission lines emanating from power stations.

11.3 Advantages of Auto-reclosing

The use of the auto-reclosing system in transmission and distribution lines is quite common. It has been accepted worldwide from various surveys and reports (including those by the IEEE Power Systems Relaying Committee) that only about 5–10% of the faults out of the total are permanent in nature. The auto-reclosing system therefore provides significant advantages:

1. Use of the auto-reclosing system minimizes the interruption of supply at the consumer side.
2. Employing a high-speed auto-reclosing system on long extra high voltage (EHV) and ultra high voltage (UHV) lines increases system stability as well as synchronism.
3. Use of the auto-reclosing system restores system capacity and reliability. These are achieved considering the minimum number of outages and the least expenditure of manpower.
4. In modern power systems, some of the loads are very critical, and continuity of power supply is extremely important. The use of auto-reclosing system restores continuity of power supply to critical loads.
5. The modern power system network is very complex in nature. Therefore, interconnections as well as continuity of service for certain lines are very important. Tripping of these interconnecting lines is not allowed because of temporary faults. Use of auto-reclosing system restores critical system interconnections.
6. If the auto-reclosing system is employed, then there is a higher probability of recovery of continuity of power supply in the case of multiple contingency outages.
7. Use of a high-speed auto-reclosing system reduces the duration of fault. It also eliminates most temporary faults. Hence, damage due to persistent faults for a long duration reduces.
8. The maximum benefit of an auto-reclosing system is achieved particularly in distribution systems in which permanent outages in case of transient faults beyond tap fuses can be reduced.
9. Employing delayed auto-reclosing schemes increases the chances of clearance of semi-permanent faults in distribution systems where fuse is used.
10. The maximum benefit of auto-reclosing systems can be achieved for unattended substations. The use of auto-reclosing system reduces the requirement of staff and thereby the monetary resources required.
11. The use of auto-reclosing system gives relief to system operations in power system restoration during system outages.

11.4 Classification of Auto-reclosing Relays

Auto-reclosing relays are classified on the basis of the number of poles, speed, and number of attempts.

11.4.1 Auto-reclosing Based on Number of Phases

Auto-reclosing relays can be classified into two types:

1. Single-phase auto-reclosing
2. Three-phase auto-reclosing

In single-phase auto-reclosing, every pole of the breaker must be equipped with its own closing and tripping mechanisms. Hence, except in distance protection schemes, additional phase selection logic is required by tripping and reclosing mechanisms. Whenever a single line-to-ground fault occurs on any phase of the transmission line, the single-phase auto-reclosing scheme trips and recloses only the faulted pole of the breaker. Therefore, the single-phase auto-reclosing relay has three separate elements, one for each phase. Operation of any element energizes the corresponding dead timer, which in turn initiates a closing pulse for the appropriate pole of the circuit breaker.

If the fault is transient, then it vanishes after a successful reclosing attempt of the auto-reclosing relay. At the end of the reclaim time, the auto-reclosing relay resets and is ready for the next attempt. On the other hand, in case of a permanent fault, the reclosing attempt is not successful. Hence, all three poles of the breaker trip and lockout.

Advantages

1. It maintains the system's integrity in a better manner.
2. It has negligible interference with the transmission of load, particularly for multiple earth systems. This is because the fault current in the faulted phase can flow to the earth through various earthing points. This is true until the fault is cleared and the faulty phase is restored.

Disadvantages

1. Due to capacitive coupling between the faulty and healthy lines, the single-phase auto-reclosing scheme requires longer deionization time, which further results in a longer dead time.
2. Due to zero-sequence mutual coupling between the faulty and healthy lines, there is a possibility of maloperation of ground distance relays installed on double-circuit lines.

11.4.2 Auto-reclosing Based on Number of Attempts

The auto-reclosing relay is classified into two types on the basis of the number of attempts:

1. Single-shot auto-reclosing relay
2. Multishot auto-reclosing relay

A single-shot reclosing relay executes only one reclosing attempt. Thereafter, it remains in the lockout stage irrespective of the type of fault, that is, transient or permanent. In order to avoid transient faults, long EHV and UHV lines, particularly those located in the area of high probability of lightning incidence, use single-shot auto-reclosing relay. On the other hand, multishot reclosing relays pursue two or three reclosing sequences within a specified time interval. They are used in distribution systems to improve service continuity.

11.4.3 Auto-reclosing Based on Speed

On the basis of the speed, the auto-reclosing relay can be classified into two types:

1. High-speed auto-reclosing
2. Low-speed/delayed auto-reclosing

Knowledge of the system disturbance time that can be tolerated without loss of system stability is the first and foremost requirement of the high-speed auto-reclosing scheme. Therefore, for a defined set of power system configurations and fault conditions, a transient stability study is required. With knowledge of protection schemes, circuit breaker operating characteristics, and fault arc deionization time, the feasibility of high-speed auto-reclosing can be assessed. The factors to be considered for the application of high-speed auto-reclosing are protection characteristics, circuit breaker characteristics, number of shots, deionization of fault arc, choice of dead time, and choice of reclaim time.

On the other hand, delayed auto-reclosing scheme is used in interconnected transmission systems, particularly in the network where the loss of a single line can create large system disturbances and can affect synchronism. Dead time of the order of 5–60 s is commonly used. In this situation, there would not be any problem of system decay or loss of synchronism because of the fault arc deionization time, operating characteristic of the breaker, and power swing before reclosing. The three-phase auto-reclosing scheme simplifies the complex control circuit with reference to the control circuit of the single-phase auto-reclosing scheme.

Moreover, in a system where delayed auto-reclosing is permissible, the chances of success with reclosing attempts are greater with delayed reclosing than with high-speed reclosing.

11.5 Sequence of Events in Single-shot Auto-reclosing Scheme

Figure 11.1 shows the operation of a single-shot auto-reclosing relay during transient fault. Here, after the first reclosing attempt, the circuit breaker remains in closed condition. Therefore, the auto-reclosing relay can improve the stability of the system. On the other hand, Fig. 11.2 (given on page 297) shows the behaviour of a single-shot auto-reclosing relay during permanent fault. The auto-reclosing relay remains in the lockout condition after the first unsuccessful reclosing attempt, which is an indication of a permanent fault.

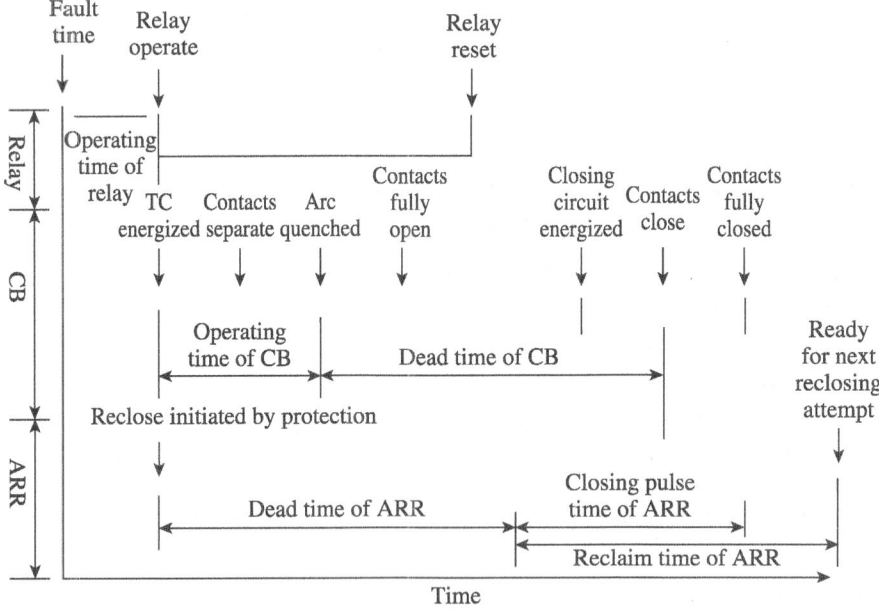

ARR, Auto-reclosing relay; CB, Circuit breaker; TC, Trip coil of CB

Fig. 11.1   Operation of single-shot reclosing relay during transient fault

11.6 Factors to be Considered during Reclosing

There are many parameters to be considered during reclosing and synchronizing. The following are the factors to be considered at the time of installation of the reclosing and synchronizing scheme in the actual field.

11.6.1 Choice of Zone in Case of Distance Relay

If distance protection is used for the protection of EHV/UHV transmission lines, then the reclosing relay should be normally kept in zone 1. Hence, the reclosing relay is activated for faults in zone 1 only, whereas it is blocked for faults in zone 2 and zone 3.

In case of distance protection, zone 1 is set to cover 80–85% of the line length instantaneously to avoid overreach, whereas the remaining section (15–20%) of the line is covered by the time-delayed zone 2 protection. It is quite possible that the fault occurring at the end of zone 1 will be cleared instantaneously. However, at the same time, the circuit breaker at the other end opens in 0.3–0.4 s, which is the operating time of the distance relay in zone 2. In this case, the application of high-speed auto-reclosing to the circuit breakers at each end of the line may result in no dead time or in a dead time that is insufficient to allow deionization of the fault arc. Hence, a transient fault can be considered as a permanent fault, which results in locking out of both breakers of the line.

The solution to this problem is to use either the transfer-tripping or blocking scheme, which involve the use of an inter-tripping signal between the two ends of the line. Alternatively, zone 1 extension scheme is another remedy that gives instantaneous tripping over the whole line length.

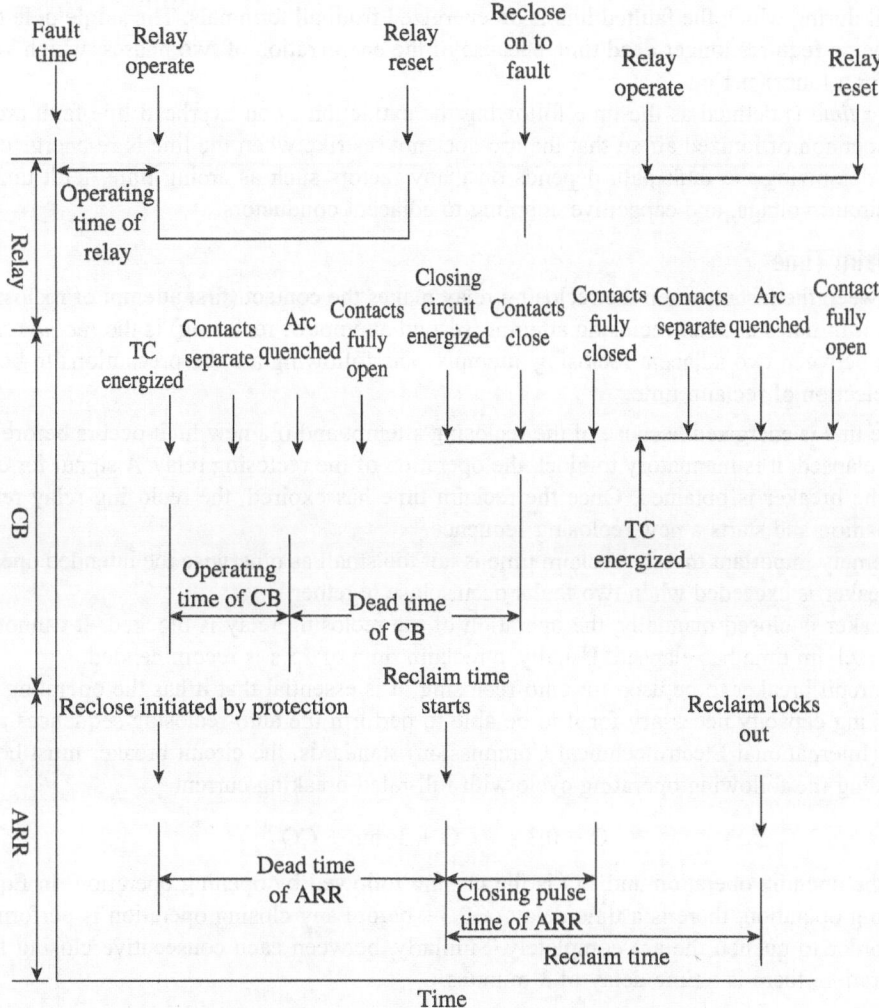

ARR, Auto-reclosing relay; CB, Circuit breaker; TC, Trip coil of CB

Fig. 11.2   Operation of single-shot reclosing relay during permanent fault

11.6.2 Dead Time/Deionizing Time

During high-speed reclosing of circuit breakers, the dead time must be considered. The minimum dead time can be represented by the following equation:

$$t = 10.5 + \frac{kV}{34.5} \text{ cycle} \qquad (11.1)$$

where kV is the rated line–line voltage of the line. Using this equation, a dead time of 0.3–0.4 s is obtained for 220 kV and 400 kV systems.

Dead time is defined as the time between the energization of auto-reclosing scheme and the operation of the contacts that energize the breaker closing circuit. The value of dead time for auto-reclosing must be greater than the deionization time. The dead time of an arcing fault on a reclosing operation is not necessarily the same as the dead time of the circuit breakers involved. This is because the dead time of the fault is the interval during which the faulted line is de-energized from all terminals. The single pole tripping and reclosing scheme requires longer dead time because of the energization of two phases, which keeps the arc conducting for a longer period.

Deionizing time is defined as the time following the extinction of an overhead line fault arc, necessary to ensure dispersion of ionized air so that the arc does not restrike when the line is re-energized. The time required for deionizing the fault path depends on many factors such as arcing time, fault duration, wind conditions, circuit voltage, and capacitive coupling to adjacent conductors.

11.6.3 Reclaim Time

The time between the instant when the reclosing relay makes the contact (first attempt of reclosing) and the instant when it initiates another reclosing attempt (second attempt of reclosing) is the *reclaim time*. Hence, it is the time between two adjacent reclosing attempts. The following are the precautions to be considered during the selection of reclaim time:

1. When the line is energized because of the reclosing attempt and if a new fault occurs before the reclaim time has elapsed, it is mandatory to block the operation of the reclosing relay. A signal for definite tripping of the breaker is obtained. Once the reclaim time has expired, the reclosing relay returns to the initial position and starts a new reclosing sequence.
2. It is extremely important that the reclaim time is not too small as otherwise the intended operating cycle of the breaker is exceeded when two faults occur close together.
3. If the breaker is closed manually, the operation of the reclosing relay is blocked. It cannot start again until the reclaim time has elapsed. Usually, a reclaim time of 25 s is recommended.
4. For the circuit breaker to be used for auto-reclosing, it is essential that it has the operating mechanism and breaking capacity necessary for it to be able to perform the auto-reclosing sequences required. As per IEC (International Electrotechnical Commission) standards, the circuit breaker must be capable of withstanding the following operating cycle with full-rated breaking current.

$$O + 0.3 \text{ s} + CO + 3 \text{ min} + CO \tag{11.2}$$

where O is the opening operation and CO is the closing followed by opening operation. In Eq. (11.2), after the opening operation, there is a time delay of 0.3 s before any closing operation is performed. This is required in order to quench the arc completely. Similarly, between each consecutive closing followed by opening operation, there is a time delay of 3 minutes.

11.6.4 Instantaneous Trip Lockout

As explained earlier, about 80–90% of the faults taking place on the distribution systems are temporary and disappear in a short time. Therefore, reclosers, in coordination with fuses, are used in the distribution system in such a manner that fuses operate only for permanent faults, thus improving the reliability of the power supply. This is explained with the help of Fig. 11.3, which shows a single line diagram of a portion of the distribution system.

For all faults on the tapped line, the fuse should not operate first; instead, the recloser should operate first to clear the transient faults, if any. Thereafter, the fuse is allowed to blow if the fault is permanent.

After the first attempt, the recloser remains in lockout condition. This is known as *instantaneous trip lockout*. These two devices (recloser and fuse) must coordinate with each other for all possible faults on the feeder.

11.6.5 Intermediate Lockout

Tappings from the transmission line are often chosen to give in-between connections to the load. This is usually carried out through a transformer and is known as *line-in-line-out* (LILO). This type of configuration

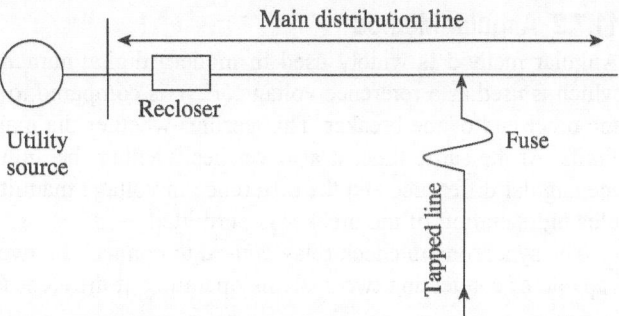

Fig. 11.3 Single line diagram of a portion of a distribution system

is widely known as *tapped transmission line* or *multiterminal line*. These types of lines are usually connected to attended substations. Whenever a fault occurs on the tapped lines, most of the attended substations are disconnected as there is no provision for intermediate lockout in the reclosing relay. Hence, manual reclosing is attempted by the operator at the remote end after the lockout, subject to the condition that the fault no longer exists. In this condition, the service at the local end is restored after successful operation of the reclosing relay in conjunction with the synchronism check relay. Conversely, for unattended substations, intermediate lockout features are available in the reclosing relay. This feature is activated on a permanent fault and bypasses all the upcoming reclosing operations of the relay.

11.6.6 Breaker Supervision Function

In order to maintain the stability of the power system, the breaker supervision function, which decides the breaker maintenance schedule, is very essential in a reclosing relay. Regular maintenance of the breaker depends on the wear withstanding capacity of the breaker. In modern digital relays, a separate function is provided, which issues an alarm/warning whenever there is a need for breaker maintenance. These are specified in terms of the maximum number of allowable reclosing operations and the time span for which the reclosing operations are permitted.

11.7 Synchronism Check

Supervision of the closing operation of a circuit breaker is carried out by a synchronism check relay. In order to reduce the impact on the breaker as well as on the power system, the synchronism check relay verifies the phasor values of voltage on both sides of the circuit breaker. Hence, this relay controls the phasor values of the voltage appearing across the contacts of the breaker by several methods such as phasing voltage method and angular method. These methods are explained in Sections 11.7.1 and 11.7.2.

11.7.1 Phasing Voltage Method

In phasing voltage method, bus voltage is used as a reference voltage, whereas line voltage supplied to the relay is used as a controlling voltage. These two voltages are compared with each other. The difference between these two voltages is compared to the predetermined value of the angular threshold. The value of threshold varies from 20° to 60°. If the difference between the two voltages is within the predetermined value of the threshold, then the closing operation of the breaker is allowed.

The normal practice in most synchronism check relays is to set long time delays of the order of 5–10 s. This long time delay confirms that voltages across the contact of the breaker are almost equal. This situation avoids excessive impact on the breaker contact, which otherwise appears on it because of the closing operation in a non-synchronous condition. However, this long time delay is reduced if high-speed reclosing is required.

11.7.2 Angular Method

Angular method is widely used in modern digital/numerical relays. In this method too, the bus voltage, which is used as a reference voltage (V_{ref}), is compared to the input voltage (V_{in}) supplied to the relay from the other end of the breaker. This verifies whether the magnitudes of V_{ref} and V_{in} are within the prescribed limits. At the same time, it also verifies whether the angular difference is within the preset limit. If both the angular difference and the difference in voltage magnitude are within the predetermined limits, then the closing operation of the breaker is permitted.

The synchronism check relay is used to connect the two ends of the line with the bus. However, it is not capable of connecting two systems operating at different frequencies.

11.8 Automatic Synchronization

In order to monitor the manual synchronizing process or to achieve automatic synchronization, the automatic synchronizing system is used in the field. The automatic synchronizing system compares the reference voltage of the bus on one side of the breaker to the input voltage supplied to the relay from the other side of the breaker. The following are the criteria to be fulfilled to permit the closing operation of the breaker.

1. The angular difference is within the prescribed limit.
2. The difference in frequency is within the preset limit.
3. The difference between the reference voltage and the initial voltage is within the predetermined threshold value.

Initially, the operator activates the process, but later on all the parameters are controlled and monitored automatically to synchronize the system without contribution of operator. Nowadays, state of the art synchronizers are developed using electronics and digital devices which facilitates precise measurement of system parameters and smooth synchronization. The electronics/digital synchronizer senses voltage magnitude, phase angle, and frequency from both sides of circuit breaker and takes control of the synchronizing process. Auto synchronizer wait till the difference of voltage (Δv), phase angle ($\Delta \theta$), and frequency (Δf) fall within the limit set in the respective comparison unit. As shown in Fig. 11.4, when the desired condition is satisfied altogether, the auto synchronizer issues closing command to the breaker.

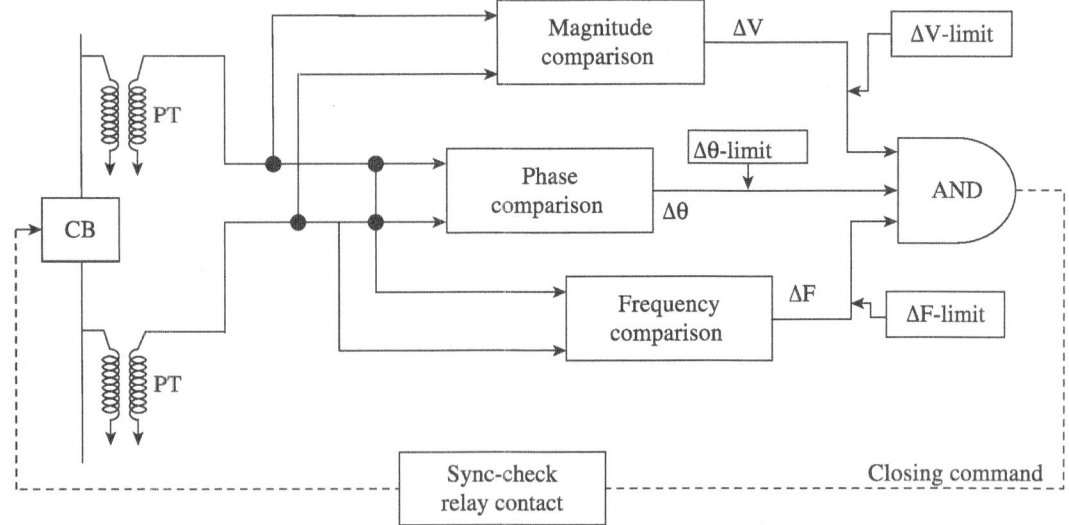

Fig. 11.4 Block diagram of automatic synchronization

11.9 Auto-reclosing Scheme (79) for Transmission Line Protection

An attempt has been made by the authors of this book to demonstrate the concept of auto-reclosing scheme for the protection of EHV system. Figure 11.5 shows the flowchart of the protection scheme for auto reclosing during temporary and permanent fault. Initially, samples of current signals of feeder CTs, voltage signals of line CVT, and bus PT are acquired by data acquisition system. The fault detection algorithm discriminates between the fault condition and the normal condition. Whenever fault is detected by the fault detection algorithm, the relay issues trip signal and as a result the breaker located in the line to be protected will open (isolate the fault).

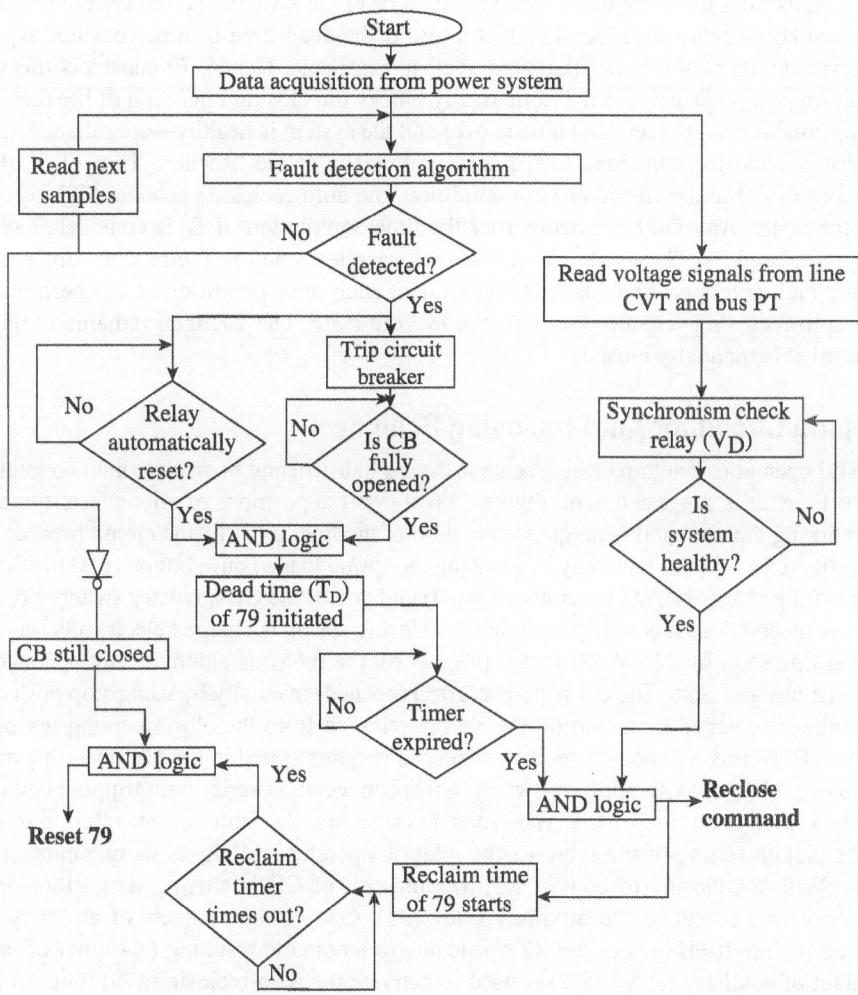

Fig. 11.5 Flowchart of the protection scheme for auto-reclosing

The status of breaker auxiliary switches (normal open to close) is simultaneously changed and are used for auto-reclosing purpose. The relay is an auto-reset relay which resets within a few milliseconds (less than 40 ms) afterwards it has issued a tripping signal. Hence, the contacts of CB to be fully opened and automatic resetting of relay conditions are checked using AND logic to initiate 'Dead Time (T_D)' of auto-reclosing scheme (79). If both the above-mentioned conditions are not satisfied, the dead-time of timer will not start.

In order to monitor the dead time, a timer continuously counts upto the set time (T_D). Dead time can be adjusted independently as per requirement.

A voltage check relay (synchronism check) is used to check the system healthy condition. The voltage check relay takes the voltage samples of line CVT and bus PT. The logic of this relay is so developed that it checks all the conditions described in Sections 11.7 and 11.8. If system healthy condition is detected by synchronism check relay, this indicates that fault no more persists and also gives an indication that the fault was transient in nature. If the fault still persists then the potential at the contacts of CB differ widely (synchronism check fails). Thus it indicates unhealthy condition of the system.

The output signals from the timer block (used to count dead time) of 79 and the system healthy indication from synchronism check relay are given to AND logic. If the dead-time of timer has not expired or if the fault is still persisting, then 79 will not issue any reclose command. Hence, 79 considers this situation as a permanent fault on power system and it automatically blocks the closing operation of breaker. On the other hand, if both the conditions—the set dead time is over and the system is healthy—are satisfied, auto-reclosing scheme (79) issues a closing command to CB and simultaneously the 'Reclaim Time (T_R)' of timer starts.

After reclosing of CB under transient fault condition, the auto-reclosing scheme (79) is reset when the reclaim timer times out. Any fault occurring after the time completion of T_R is considered as a new fault. Once the CB is reclosed by 79 and after that if a fault reappears before the reclaim timer times out, the relay detects the fault and opens the CB. However, under such open condition of CB before reclaim timer times out, the auto-reclosing scheme goes into the lockout state. The CB then remains in open condition permanently until it is manually closed.

11.10 Tripping Circuit for Auto-reclosing Scheme

For the successful operation of auto-reclosing scheme during transient and permanent fault on transmission line, tripping (control) circuit is suggested here. Figure 11.6 shows the positions of all contacts symbolized in the tripping circuit during the normal de-energized condition of the line. Initially, the circuit breaker of the line to be protected is energized (closed) manually by pressing the spring loaded push button (PB1) with the condition that the spring is fully charged (SW1 previously closed) and one of the CB auxiliary switches (CB1) is closed under breaker open condition. It is to be noted that the closing spring is charged electrically, immediately following a closing operation by 220 V AC motor (Fig. 11.6), the motor is automatically operated to keep the closing spring in a charged state. The CB is de-energized (opened) manually by using stop push button (PB2). The opening spring is charged by acquiring the energy released from the closing spring during the closing operation of the CB. Hence, an open-close-open sequence remains stored in the CB operating mechanism.

As shown in Fig. 11.6, one CB auxiliary switch (CB2) connected in series with tripping coil (TC) remains closed under the closed condition of CB. Whenever fault occurs, the fault current referred to secondary of CT exceeds the pickup setting of the relay R; the relay R operates and closes its two contacts R1 and R2 simultaneously. With the closing of contact R1, tripping coil of CB is energized and thus opens the line breaker. Closing of R2 energizes the auxiliary relay (33). One of the contacts of auxiliary relay (33-1) provides hold on path to itself (33) as the R2 opens due to automatic resetting (<40 ms) of main relay R. The other contact of auxiliary relay (33-2) is used to activate the auto-reclosing (79) unit. To issue closing command to CB by 79, three conditions are to be monitored. These are: (1) to check whether closing spring is fully charged (SW2), (2) to check whether CB is fully opened (CB3) and (3) to check whether main relay has automatically reset (R3). If all the above-mentioned three conditions are satisfied, that is, the contacts SW2, CB3, and R3 are closed then the dead time (T_D) of 79 is initiated. As and when dead time elapses, one of its contacts (79-1) provides hold-on path to itself (79) for keeping the reclaim timer in the on state. Simultaneously closing of the other contact (79-2) which is connected in closing circuit of breaker energizes the closing coil (CC). Hence, line breaker (B1) is closed successfully by the auto-reclosing scheme.

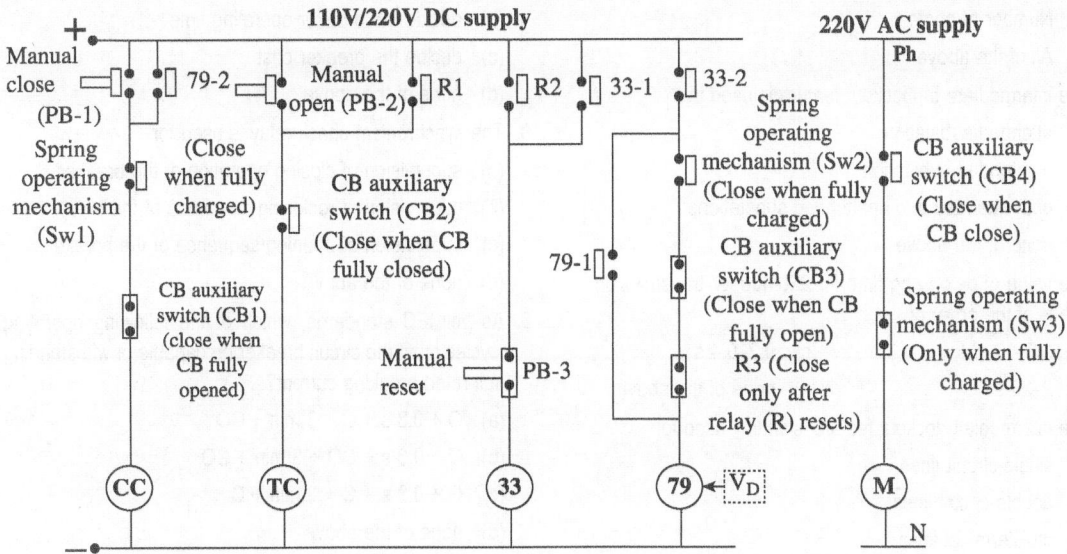

Fig. 11.6 Tripping circuit for auto-reclosing scheme

CB = Circuit breaker, **CC** = Closing coil of CB, **TC** = Tripping coil of CB, **33** = Auxiliary relay, **R** = Main relay contact, **79** = Auto-reclosing scheme, **M** = Spring charging motor, **PB** = Push button

Recapitulation

- Temporary faults created because of switching or lightning can be cleared by momentarily de-energizing the line.
- Automatic reclosing system can be used to improve service continuity and stability of a power system.
- Long EHV transmission lines utilize various reclosing schemes such as single three-phase auto-reclosing schemes and single and multishot auto-reclosing relays based on high-speed or low-speed operations.
- During the selection of a reclosing relay, various factors such as choice of zone in case of distance relay, dead time, and reclaim time should be considered.
- Supervision of closing operation of a circuit breaker is carried out by synchronism check relay.
- The synchronism check relay controls the phasor values of the voltage appearing across the contacts of the breaker by several methods such as phasing voltage method and angular method.

Multiple Choice Questions

1. Single-shot auto-reclosing relay performs
   (a) one reclosing sequence
   (b) two reclosing sequences
   (c) more than two reclosing sequences
   (d) none of the above

2. Reclosing relay is kept in
   (a) zone 1 of the distance relay

   (b) zone 2 of the distance relay
   (c) zone 3 of the distance relay
   (d) all the three zones

3. Which are the factors considered for the application of high-speed auto-reclosing?
   (a) Circuit breaker characteristic
   (b) Relay characteristic

(c) Number of shots

(d) All of the above

4. The intermediate trip lockout feature is used in

(a) attended substation

(b) unattended substation

(c) both attended and unattended substations

(d) none of the above

5. The value of deionizing time for 220/400 kV transmission line is of the order of

(a) 1.5–15 s   (c) 0.3–0.4 s

(b) 2–20 s   (d) none of the above

6. The intermediate lockout feature is widely used for

(a) single-circuit lines

(b) double-circuit lines

(c) multiterminal lines

(d) none of the above

7. Breaker supervision function is required to

(a) decide the breaker maintenance schedule

(b) decide the breaker operating time

(c) decide the breaker cost

(d) none of the above

8. The synchronism check relay is used for

(a) supervision of closing operation of the breaker

(b) supervision of reclosing sequence of the breaker

(c) supervision of opening sequence of the breaker

(d) none of the above

9. As per IEC standards, which of the following operating cycles must the circuit breaker be capable of withstanding full-rated breaking current?

(a) O + 0.3 s + C + 3 min + CO

(b) O + 0.3 s + CO + 3 min + CO

(c) O + 0.3 s + C + 3 min + C

(d) none of the above

10. For a delayed reclosing scheme, the delay time is of the order of

(a) 1–10 s   (c) 5–60 s

(b) 0.1–1 s   (d) none of the above

Review Questions

1. Discuss the various advantages of auto-reclosing with respect to power systems.

2. What is the difference between high-speed and low-speed auto-reclosing schemes?

3. Discuss the relative merits and demerits of single-phase auto-reclosing relays.

4. How does the choice of various zones (zone 1, zone 2, and zone 3) affect the reclosing scheme used for a transmission line?

5. Enlist the relative advantages and disadvantages of multiphase auto-reclosing relays.

6. What is the difference between single-shot reclosing relay and multishot reclosing relay?

7. What is the impact of deionizing time on reclosing?

8. Define reclaim time. What are the precautions to be considered during the selection of reclaim time?

9. How is the stability of the distribution system improved using the instantaneous trip lockout feature available in a reclosure?

10. Explain the importance of the intermediate lockout feature available in a reclosing relay.

11. Why is breaker supervision function essential?

12. What is the function of synchronizing check relay?

13. Explain the different methods of synchronizing.

14. Explain the application of auto-reclosing scheme (79) to transmission line with tripping circuit.

Answers to Multiple Choice Questions

1. (a)   2. (a)   3. (d)   4. (b)   5. (c)   6. (c)   7. (a)   8. (a)   9. (b)   10. (c)

System Response to Severe Upsets

12

Learning Objectives

After going through this chapter, the students will be able to:

- Explain electrical islanding and severe upsets in power system
- List the factors to be considered for load-shedding scheme
- Discuss the applications of frequency relays
- Explain the different methods of detecting islanding

12.1 Introduction

Minimization of system outages, load interruptions, and equipment damage is the main objective of power system planning and operation in case of emergency conditions. Currently these goals are becoming more challenging because of competition, privatization, and restriction of power systems. Therefore, major emphasis is laid on reducing operating and maintenance costs and maximizing profit. In addition, life of equipment is also extremely important. This requires better understanding of coordination as well as interaction of various equipment under emergency conditions. Hence, the objective is to minimize the probability of outages and to achieve faster recovery of the system during outages.

The behaviour of a power system under emergency conditions, as well as during the restoration process, depends on its characteristics. This is related to the balance between active and reactive power and efficiency of the installed control and protective systems. The process of power system restoration is also a function of predisturbance conditions, post-disturbance status, and target systems. Therefore, to maintain proper operation of modern power systems, it is extremely important to understand the factors involved in power system collapse and the restoration process.

The normal practice is to design the interconnected power systems in such a manner that no uncontrolled widespread interruptions occur. However, faults are inevitable, and hence, it is not possible to design a 100% reliable system. Blackouts are bound to occur. A *blackout* is defined as a condition where a major part or most of an electrical network is de-energized, with much of the system tied together through closed breakers. Power system disturbances are responsible for blackouts, which lead to disconnection of electric power supply to all loads within an area. Though disturbances are bound to occur, the system should be restored to normal condition as fast as possible.

12.2 Nature of System Response to Severe Upsets

Usually, power systems can withstand either a single contingency or multiple contingencies. The propagation of the disturbance to other healthy parts of the network is prevented by protection and automatic control systems. In certain situations, there is a possibility of separation of one portion of the interconnected system. This creates a situation commonly known as *electrical islanding*. These contingencies are more severe and may be taken into account at the time of design of the system. This may lead to severe stress on the system, which finally causes uncontrolled cascading outages. Owing to the wide variations in frequency (47–53 Hz) and voltage (60–120% of normal voltage), there is a possibility of deterioration of the system, which may lead to the loss of significant portions of the system load. However, the protection and control systems can control the system response in such conditions. The main task of the protection and control systems is to save as much of the system as possible from total collapse.

12.2.1 System Response to Islanding Conditions

A sustained frequency transient phenomenon is the system response during an islanding condition. In this condition, the speed control of prime mover and generator and their response are the main parameters that play a vital role in determining the nature of the dynamic response of the system. This situation is particularly endemic during undervoltage or overvoltage conditions.

12.2.2 Undergenerated Islands

If the total generation is less than the required load demand and if sufficient spinning reserve is available, then the system frequency returns to the normal value within a few seconds after an islanding condition. However, if the generation is not sufficient with respect to load, then the frequency reduces drastically, which leads to the tripping of generators through underfrequency relays. Therefore, underfrequency load-shedding schemes are usually used, which reduces the connected load that can be satisfactorily supplied by the available generators. Consequently, in an undergenerated islanding condition, the initial transient depends on the load-shedding scheme, whereas the response of the system frequency depends on the characteristics of the prime mover.

12.2.3 Overgenerated Islands

In an overgenerated islanding condition, the system frequency increases and the mechanical power generated by the turbines reduces. Therefore, the ability of power plants to sustain a partial load rejection will decide the performance of the system.

12.2.4 Grid Integration of Renewable Energy Sources

The power generation through Renewable Energy Sources (RES) has increased since last few years in India and worldwide. Motivating activities are also arranged to boost up the total installed capacity of RES in India. The government extends support to install local power generation and also provides subsidiary to establish large-scale non-conventional power generation. The wind and solar power generation contribute the majority of the RES power generation and together share around 14% of the total installed capacity as on date. However, there are some constraints of grid interconnections due to staggered location and variable nature of power generation through RES. Moreover, the accurate weather forecasting techniques are still in development stage for predicting wind velocity and sunshine. Thus variability of major power generation through RES is unpredictable and leads to system unbalance. Therefore, it is important to exercise smooth interconnection of bulky variable power generation in view of grid security. While integrating the RES with the grid, state distribution and regional power transmission facilities play a key role. Hence, while integrating the RES with the grid, it is essential to design and extend new electric network to handle the large contribution of renewable energy. Moreover, there are various factors affecting the interconnection to the grid such as geographical distribution of RES, penetration of small-scale Distributed Generation (DG), and intermittent power generation which depend on the variability of nature.

Figure 12.1 shows the interconnection of various types of DGs at different voltage levels. IEEE 1547, 2003 has given guidelines for interconnection and requirements of DGs in to the grid.

These requirements are divided into three categories as shown below.

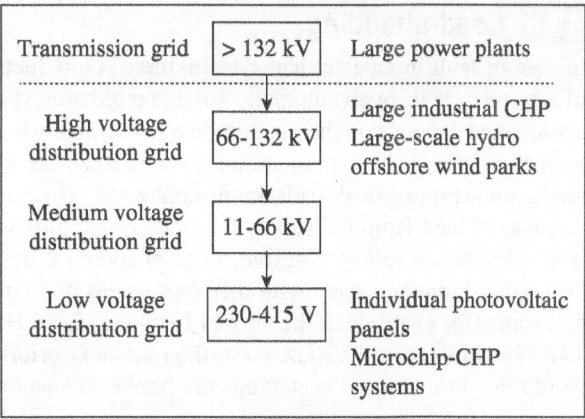

Fig. 12.1 Interconnection of various types of DGs at different voltage levels

General requirements

The general requirements are related to voltage regulations of the distribution system/network, synchronization requirements, integrity between two systems, grounding requirements, and isolation requirements of the device connected in the system.

Response of electrical power system (EPS) network during abnormal conditions

It deals with faults in EPS network, synchronizing and reclosing requirements, voltage and frequency requirements, and reconnection with the system/network.

Power quality requirements

These are related to the limitation of DC injections, voltage flicker produced by DG, and harmonics introduced by inverter fed DGs.

12.2.5 Reactive Power Balance

Reactive power balance also affects the system performance. A significant difference between the total reactive power generated and that absorbed may lead to overvoltage or undervoltage conditions. In the worst case, the protective relay can trip generators. This situation arises when islanding occurs on lightly loaded extra high voltage (EHV)/ultra high voltage (UHV) lines, which cause the generator to absorb a huge amount of reactive power. This can lead to tripping of generators due to the operation of loss of excitation protection.

12.2.6 Power Plant Auxiliaries

The performance of power plant auxiliaries can deteriorate because of the reduction in power supply voltage and frequency. This situation is very common for thermal power plants, which use auxiliaries such as circulating water pump, feed water pump, and drain pump, which are driven by induction motors. The auxiliaries used in nuclear power plants are also affected by this. Moreover, voltage dips over long periods may affect the performance of contactors and coils of relays, and in the worst case, they may disconnect the motors. Hence to resolve this, undervoltage and underfrequency relays that trip the plant at low voltages and at low frequencies are used.

12.2.7 Power System Restoration

At the time of restoration of the whole interconnected power system, several steps are required to be taken by the operators, such as restoration of generating units and loads, balance between generation and load, and resynchronization of islands and other equipment. During system restoration, it may be possible that full power may not be available for a certain period. This is due to problems in starting and reloading of thermal power plants. The restoration process can be speeded up by adopting various methods such as drawing power from captive power plants, diesel generator sets, or gas based generator sets.

12.3 Load-shedding

In case of fault in an electrical system, there is a reduction in voltage and frequency. The generating station tries to recover through automatic voltage regulation system and speed governor control system. The system disturbance depends on the magnitude of fault, area where the fault occurs, and the time taken to trip through protection relays. In the meantime, to normalize the system quickly and also to safeguard the remaining loads, some non-critical loads are disconnected. This process is known as *load-shedding*. It is defined as the removal of load from the supply over certain periods when demand becomes greater than power supply. It is also known as *rolling blackout*. Load-shedding is achieved using an efficient load-shedding arrangement through a frequency relay with different stages of frequency setting. The non-critical loads connect to the first stage (for example, if the normal frequency is 50 Hz, the first stage setting can be 49.8 Hz, second stage 49.5 Hz, third stage 49.2 Hz, etc.), then the next priority of loads to the second stage, and the subsequent one to the third stage. The settings are based on a study of the electrical system.

12.4 Factors to be Considered for Load-shedding Scheme

The following are the factors to be considered when designing load-shedding schemes.

12.4.1 Maximum Anticipated Overload

There is no limit of percentage of load to be shed by any load-shedding scheme. However, it is recommended that the percentage of shed loads, which is equal to the maximum anticipated overload, be defined. Moreover, it is also important to evaluate the cost of the load-shedding scheme with respect to the probability of occurrence of severe overload. Hence, a study is also required with reference to this severe overload, which results in failure of generators, big transmission lines, and buses. The area or part of a system or network that encounters severe generation deficiency needs more comprehensive load-shedding schemes. Further, it is also necessary to consider the load reduction factor, which reduces overload in case of reduction in frequency. Therefore, to design an effective load-shedding scheme, it is safe to consider a load reduction factor equal to zero. This is because the load reduction factor is usually not known and varies with time. In addition, it should be remembered that an isolated system, which does not have spinning reserve, cannot be resynchronized.

12.4.2 Number of Load-shedding Steps

Although there is no ideal limit to the number of load-shedding steps, most utilities use two or three of them. In a load-shedding scheme with two steps, one step is for low-set relay, whereas the other is for high-set relay. The high-set unit trips initially as long as the overload is less than the preset value. In case of more severe overloads, the frequency reduces at a slower rate until the low-set relay operates to shed the other half of the load.

12.4.3 Size of Load-shed at each Step

The size of each load-shedding step depends on the expected percentage overload. This should be decided in such a manner that there is minimum possibility of loss of certain generators or big transmission lines. Moreover, each step is chosen so that it sheds only enough loads to handle the next more serious contingency. Each load-shedding step should be evenly spread over the system by dropping loads at diverse locations.

12.4.4 Frequency Setting

The frequency at which each step will shed load depends on the following parameters.

1. Normal operating frequency range of the system
2. Operating speed
3. Accuracy of the frequency relays
4. Number of load-shedding steps

In order to allow possible variations in the tripping frequency of the relay, the first frequency of the first step is usually set just below the normal operating frequency band of the system. The usual range of the modern digital relay is from 46 Hz to 49.9 Hz with 0.01 Hz as the minimum expected normal frequency reduction step. Therefore, the frequency should be selected in such a manner that it avoids load-shedding during minor disturbances. Hence, the system can satisfactorily recover on its own.

12.4.5 Time Delay

Considering the security of the relay, it is a usual practice to use the minimum possible time delay. As the time is less, the system can easily cope with severe overloads. Moreover, all unnecessary interposing auxiliary devices should be avoided.

12.5 Rate of Frequency Decline

Frequency is the main criteria of system quality and security due to the following reasons:

1. It creates a balance between supply and demand.
2. It is the only variable quantity that remains constant in all parts of the network.
3. It is extremely important for all users.
4. Reduction in frequency is responsible for total blackout of all or part of the network of an interconnected system because of the failure of power station or transmission line.

Therefore, it is necessary to estimate the variations in system frequency in case of disturbances.

If the total load along with losses is the same as total generation, then the system frequency remains constant. Conversely, in case of loss of a big transmission line due to permanent fault, the rate of change of frequency is given by

$$\frac{df}{dt} = -\frac{\Delta P}{2H} \tag{12.1}$$

where ΔP is the decelerating power in kVA and H is the inertia constant in (MW-sec)/MVA.

The reduction in frequency is slower for a given overload for a large value of inertia constant. As the frequency reduces, the load power also decreases. A 1% reduction in frequency, that is, 0.5 Hz for a 50 Hz fundamental frequency, corresponds to a 2% reduction in load. For a small change in load, the governor is capable of correcting the deficiency in the system frequency. In case of large reduction in frequency, there is a possibility of collapse of the power station auxiliaries. In this situation, it is necessary to trip circuit breakers to disconnect minor loads.

12.6 Frequency Relays

Frequency relays (81) are used to detect deviations with reference to nominal system frequency. Deviation in frequency can be harmful to connected objects, such as generators and motors. These relays are also used in case of system abnormalities, such as faults or overload, during which the frequency can be increased or decreased beyond the prescribed limits. Various types of frequency relays have been used by the utilities over the years. Currently, digital frequency relays are used. This relay is capable of measuring both the absolute frequency and its rate of change with great accuracy. It is suitable for various applications such as the following:

1. It is used for graded load-shedding in overloaded systems.
2. It is used as a part of relaying schemes, which open an overloaded power system at the predetermined points to prevent complete system blackout.
3. It is also suitable for isolation of small island networks in the event of a fault in the supply authority system.

Figure 12.2 shows the block diagram of a static frequency relay. The input voltage, acquired from a potential transformer (PT), is further stepped down (preferably to 6 V or 12 V) and passed through a low-pass filter. This signal is passed to a multimegahertz up/down counter, which contains zero crossing detectors (ZCDs). The ZCD detects a zero crossing of the voltage and begins a counter that continues counting until the next voltage zero or the next positive zero crossing is reached. The frequency is predicted on the basis of the accumulated count. An accuracy of around 0.1% of the fundamental frequency (say, 50 Hz) is desired for any type of frequency relay. The application principle of the digital relay is the same as that of the static relay. The only difference between the digital frequency relay and the static relay is the inclusion of certain additional features such as self-checking provisions, alarms, lockouts, records, and multiple set points. Multiple set point is used for overfrequency or underfrequency applications.

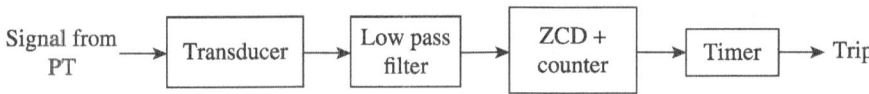

Fig. 12.2 Block diagram of static frequency relay

12.7 Islanding

Islanding is defined as the situation in which a system (especially distribution) is electrically isolated from other parts of the system but still continues to be supplied by other sources of generation (especially small-scale generation). Figure 12.3 shows the situation of islanding in which no active power source is connected to the distribution system. Further, the distribution line does not receive any power in case of a fault upstream of the transmission line. However, by using small-scale generation, there is a possibility of providing supply to the distribution line.

Islanding can be intentional or non-intentional. In case of a scheduled maintenance on the utility grid,

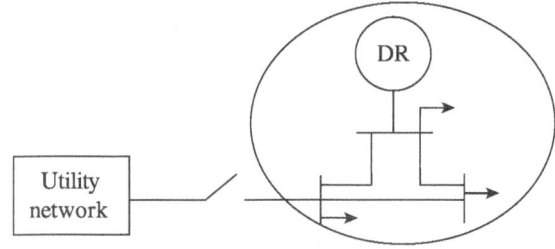

DR, Distributed resource

Fig. 12.3 Example of islanding

shutdown of the utility grid may cause islanding of generators. IEEE 1547-2003 standards recommend a maximum delay of 2 s for the detection of an unintentional islanding. The condition of islanding is usually considered undesirable because of the potential damage to the existing equipment, liability of utility, and reduction of system reliability and power quality.

12.7.1 Hazards and Risk of Islanding

The following are the hazards and risks of islanding operation.

Unregulated power system

The island is an unregulated power system. Its behaviour is unpredictable due to the power mismatch between the load and the generation and lack of 'voltage' and 'frequency' control.

Deterioration of equipment life

The voltage and frequency provided to the customers can vary significantly, and may be out of the statutory limits. It leads to high risk to the customer's equipment; yet the utility has no control over them.

Personal safety

After islanding, a section of network which is assumed to be dead can remain energized by DG units. Utility personnel sent out for maintenance work may get in contact with the live part of equipment.

Out of phase reclosing

An auto-recloser is commonly used in a distribution network to restore service after fault. The necessity of auto-recloser is also important in order to achieve the benefit of fuse saving concept. More details have been given in Chapter 11. However, DG in the island could be damaged when the island is reconnected to the utility supply system. In addition, due to out-of-phase reclosing of small-scale sources, large mechanical torques and currents are produced, which can damage the generators or prime movers or loads (for example, induction motors). This is explained as under.

Figure 12.4 shows the single line diagram of a distribution network with DG for out-of-phase reclosing. The recloser (R) has a reclosing time (T_{off}) of 2.5 s. Now, before islanding, the frequency on grid side and the frequency on DG side are same (say, 50 Hz). However, after islanding, the frequency of grid remains 50 Hz. But the frequency on DG side may change. Let us assume that the frequency on DG side after islanding is 49.8 Hz assuming minimum variation with reference to base frequency (50 Hz). Therefore, the difference in frequency (Δf) is 0.2 Hz (50–49.8). Hence, the phase shift between grid side and DG side is calculated using Eq. (12.1).

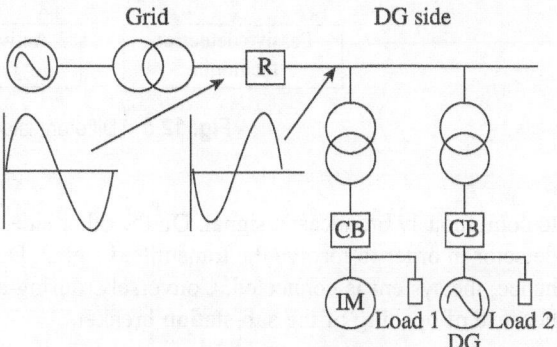

Fig. 12.4 Single line diagram of distribution network with DG for out-of-phase reclosing

$$\begin{aligned} Phase\ shift &= \Delta f \times T_{off} \times 360 \\ &= 0.2 \times 2.5 \times 360 \\ &= 180° \end{aligned} \tag{12.1}$$

Therefore, due to 180 degree phase shift between grid and DG side, large currents will flow through DG and there are chances of damage of DGs. Current in load such as induction motor (IM) is twice the normal value due to 180 phase shift (flux opposition). Hence, severe torque transients may occur. Therefore, it is mandatory to detect the islanding situation as quickly as possible.

12.7.2 Methods of Islanding Detection

The main methodology adopted to detect an islanding situation is to monitor the parameters of utility and small-scale sources, and decide whether an islanding situation has occurred from a change in these parameters. Figure 12.5 shows the classification of different islanding detection techniques.

Remote Technique

Remote technique is based on the communication between utilities and other small-scale sources. Although this technique has a better reliability than the local technique, it is more expensive because of communication equipment. Some of the remote techniques are as follows:

Power line signalling scheme This is the most widely used scheme by the various state electricity boards. In this scheme, carrier signal is used along with the power frequency signal to transmit islanded or non-islanded information. Usually, in the sub-station, a signal generator is used, which is coupled to the network in order

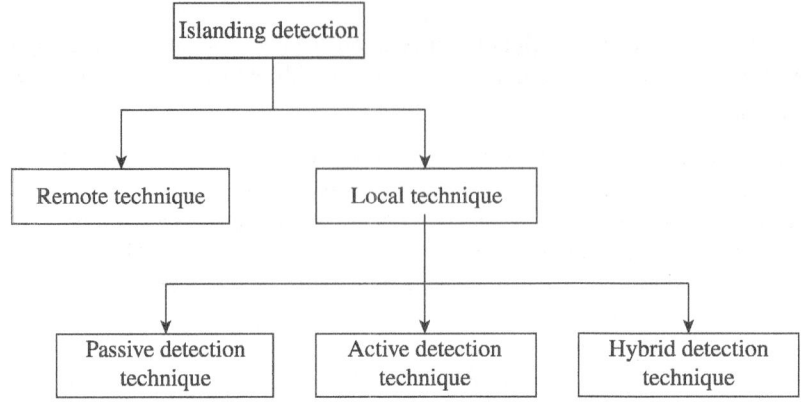

Fig. 12.5   Different islanding detection techniques

to continuously broadcast a signal. On the other side, distributed resources (DRs) are equipped with a signal detector in order to receive the transmitted signal. During normal condition, the DR receives the signal and hence, the system is connected. Conversely, during an islanding situation, no signal is received by the DR because of opening of the sub-station breaker.

Advantages

1. It is simple in control.
2. It has higher reliability.

Disadvantages

1. In order to connect a device to a sub-station, additional coupling transformer is required.
2. This technique is not economically viable if applied to a non-radial system. In this case, there is a requirement of multiple signal generators which increases the cost.

Transfer trip scheme In this scheme, the carrier signal from the remote end is accessed at the local end with the help of any communication media such as radio frequency, micro-wave frequency, or global positioning system. This scheme monitors the status of all the circuit breakers, isolators, and reclosers (if any) with the help of Supervisory Control and Data Acquisition (SCADA) Systems. Whenever the operation of breakers is detected in a sub-station, this scheme sends a signal to the DRs.

Advantages

1. It is the most common scheme used for islanding detection.
2. Implementation of this scheme is easy and it can be applied to radial system with few DRs.

Disadvantages

1. It is difficult to use this scheme for larger systems due to its complexity.
2. The control of this scheme is difficult and needs special arrangement.

Local Technique

This technique is based on the measurement of system parameters of the small-scale sources, such as voltage and frequency. It is classified as follows:

Passive detection technique Passive detection technique works on the measurement of system parameters such as variations in voltage, frequency, and harmonic distortion. These parameters are changed to a great extent during islanding. The difference between an islanding and non-islanding situation is based on the thresholds set for these parameters. Special care should be taken while setting the threshold value so as to differentiate islanding from other disturbances in the system. This technique is very fast and does not introduce disturbances in the system. However, it has a large *non-detectable zone*, which is defined as a zone in which the islanding condition is not detected. There are various passive detection techniques such as rate of change of output power, rate of change of frequency, rate of change of frequency overpower, voltage unbalance, and harmonic distortion.

Advantages

1. It has a short detection time.
2. It does not perturb the system.
3. It is very accurate when there is a large mismatch between generation and load.

Disadvantages

1. The threshold setting is difficult, and hence, there is a possibility of nuisance tripping.
2. Using this technique, it is difficult to detect islanding when load and generation in the islanded system closely match each other.

Active detection technique In this method, a small perturbation results in a significant change in system parameters when the small-scale sources are islanded, whereas the change is negligible when the small-scale sources are connected to the grid. There are various active detection techniques such as reactive power export error detection, and phase or frequency shift method.

Advantages

1. It is capable of detecting islanding in case of a perfect match between the generation and load demand in the islanded system.
2. This technique has a small non-detection zone.

Disadvantages

1. This technique can introduce perturbation in the system.
2. It has a slow detection time owing to the extra time needed to sense the system response for perturbation.
3. Due to perturbation, there is a possibility of degradation of power quality and system stability.

Hybrid detection technique Hybrid detection technique uses the concepts of both active and passive detection methods. Some of the hybrid detection techniques are as follows:

1. Technique based on positive feedback and voltage imbalance
2. Technique based on voltage and reactive power shift

Advantages

1. Perturbation is introduced only when islanding is suspected.
2. This technique has a small non-detection zone.

Disadvantage

It has long islanding detection time owing to the implementation of both active and passive techniques.

Recapitulation

- Wide variations in frequency and voltage sometimes result in the decline of the system, which may lead to the loss of significant parts of the system.
- It is important to identify the causes of frequency and voltage variations (severe upsets) and the nature of system responses during such severe upsets.
- During severe upsets, the system may run in islanding mode with underload and overload conditions, so there may be a failure of system due to power unbalance condition.
- Load-shedding using various techniques is the corrective step against severe upsets.
- Frequency relays are used for load-shedding in an overloaded system to detect the deviation in frequency with reference to nominal system frequency.
- The rate of change of frequency is given by $\dfrac{df}{dt} = -\dfrac{\Delta P}{2H}$ where ΔP is the decelerating power in kilovolt-ampere and H is the inertia constant in megawatt second per megavoltage ampere.
- Various techniques such as the remote technique and the local technique, which includes active, passive, and hybrid techniques, can be useful for islanding detection.

Multiple Choice Questions

1. As the load on the line increases, the system frequency
   (a) decreases
   (b) increases
   (c) remains unchanged
   (d) none of the above

2. The main objective of power system planning and operation in case of emergency conditions is
   (a) enhancement of load demand
   (b) minimization of system outages and load interruptions
   (c) reduction in generated power
   (d) none of the above

3. Blackouts are due to
   (a) power system disturbances
   (b) allowable overload condition
   (c) light load condition
   (d) none of the above

4. A mismatch in reactive power balance may lead to the tripping of generators due to the operation of
   (a) negative sequence protection
   (b) unbalanced protection
   (c) loss of excitation protection
   (d) none of the above

5. Load-shedding is carried out to
   (a) normalize the system quickly and safeguard the remaining loads
   (b) reduce the burden on the system
   (c) both (a) and (b)
   (d) none of the above

6. Reduction in frequency is slower for a given overload for
   (a) a small value of the inertia constant
   (b) a moderate value of the inertia constant
   (c) both (a) and (b)
   (d) a large value of the inertia constant

7. The standard time to detect an unintentional island is of the order of
   (a) 1–2 s
   (b) 10–20 s
   (c) 40–50 s
   (d) none of the above

8. Islanding condition is preferable because it
   (a) minimizes cost
   (b) improves stability
   (c) enhances system reliability and power quality
   (d) all of the above

9. Which technique is not suitable for detecting islanding situation when the load and generation closely match each other?

   (a) Passive detection technique

   (b) Active detection technique

   (c) Both (a) and (b)

   (d) None of the above

10. Which technique is capable of detecting islanding in case of a perfect match between the generation and load demand in an islanded system?

    (a) Active detection technique

    (b) Passive detection technique

    (c) Hybrid technique

    (d) None of the above

Review Questions

1. What do you mean by power system restoration?

2. What is the response of a system during severe upset?

3. Explain the role of frequency decline for an interconnected power system.

4. Why is load-shedding required?

5. Discuss the various load-shedding techniques used in an interconnected power system.

6. Explain the importance of frequency relay for a power system.

7. Draw the block diagram of a digital frequency relay.

8. What does islanding mean?

9. Explain the disadvantages of islanding.

10. Discuss the various islanding detection techniques with their relative advantages and disadvantages.

Answers to Multiple Choice Questions

1. (a) 2. (b) 3. (a) 4. (c) 5. (a) 6. (d) 7. (a) 8. (c) 9. (a) 10. (a)

Theory of Arc Interruption in Circuit Breaker

13

Learning Objectives

After going through this chapter, the students will be able to:

- Explain the fundamentals of circuit breaker
- Analyse the factors responsible for the formation of arc
- Discuss the characteristics of arc
- Explain arc quenching in AC circuit
- Discuss the various arc interruption theories
- Explain the concept of resistance switching
- Explain arc interruption in DC circuits

13.1 Introduction

A circuit breaker (CB) is a device that can open or close a high voltage circuit in a fraction of a second. The opening and closing of a circuit is achieved by the separable contacts of the CB. During normal and abnormal conditions in a circuit, the closing of contacts ascertains the flow of current, and the opening of contacts interrupts the circulation of current. The function of a CB, under the control of protective gears, is to open or close the circuit as per the requirements. When the movable contacts begin to separate, the CB begins interrupting the current. As a result, the contact area decreases. This results in high current density, which finally vapourizes the metal, and an arc is generated between the switching contacts. In spite of the physical separation of the switching contacts, current flows continuously because of the sustained arc. This arc plasma must be cooled and extinguished in a systematic way so that the gap between the contacts can again withstand the voltage in the circuit. The study of this phenomenon is very composite but has enormous importance while selecting the operating characteristic of the CB. This chapter describes the arc phenomenon and its interruption methods under different system conditions.

13.2 Fundamentals of Circuit Breaking

The separation of the switching contacts of any CB leads to the formation of gas and metal vapour between them in the current carrying condition. Any kind of gas or vapour always contains positive and negative charge carriers that are directed to the discharge process in a CB. Even when no potential is applied across the electrode, the gas conducts and sets up a small current due to natural ionization. This current is known as *leakage current*. When an electrical potential is applied across the two electrodes, the charge carriers gain mobility, and their motion depends on the applied electrical field intensity. When the moving charges collide with the electrode (ions move towards the cathode and electrons move towards the anode), they disperse their charges, and a current flows between the electrodes. This process of current conduction in a gaseous medium is due to an ionization process such as photoelectric or thermionic emission and remains continuous till a potential is applied. Initially, when a low electrical potential is applied, low current is set up because of a small number of charge carriers. Figure 13.1 shows the voltage and current relationship during electric discharge for different values of applied voltage. The part OP in Fig. 13.1 shows the linear relationship between the current and the voltage in the initial stage with a small value of applied voltage. The discharge current through the medium is proportional to the applied potential until an equilibrium is reached, where the production of charge carriers is equal to the charge carriers received by

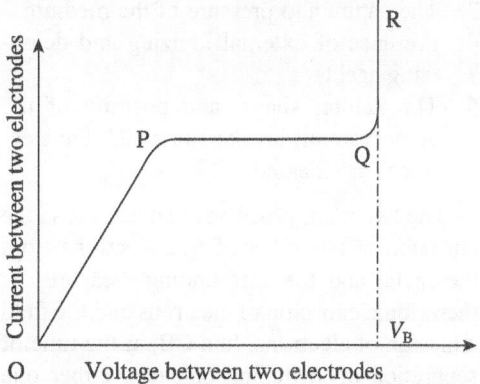

Fig. 13.1   Voltage and current relationship during discharge process

the electrodes. Afterwards, any increase in the applied voltage does not give rise to a significant current. This phenomenon depends on the intensity of ionization, the quantity of gas between the electrodes, and the gas pressure. The part PQ represents the saturation current limit. The current in the saturation limit remains constant in spite of the increase in the applied voltage because the saturation current is entirely dependent on the presence of charge carriers that are to be supplied by external ionization means.

If the electrical potential across the electrode increases to a high level, ionization occurs freely and free positive charges gain a high velocity. In this situation, they strike the cathodes with enough force to knock out a number of free electrons that maintain the discharge. Such discharge remains self-sustained because it does not require any external excitation. This process rises exponentially, and the current continues to increase between two electrodes even when the applied voltage remains constant. The voltage that forces such a high current density (million charges) through a gas medium is known as *breakdown voltage*. The part QR shown in Fig. 13.1 gives an indication of such a high current at the breakdown voltage (V_B). The gases between the electrodes no more remain insulators but provide a current conducting path. Owing to this, a continuous arc is formed between the electrodes, which are surrounded by hot ionized gases. This phenomenon is applicable for both AC and DC voltages across the electrodes. The quenching or extinction of high current is done externally. Thus, it is very important to decide the breakdown voltage and insulating medium to quench the arc while designing the CB.

13.3 Arc Phenomenon

As described in Section 13.2, the discharge occurs in AC and DC circuits because of the collision of molecules with high velocity under the voltage applied across the electrode upto the level of breakdown voltage as does the detachment of a number of free electrons from the cathodes. This is due to the increase in voltage due to the self-inductance of the circuit at the time of contact separation. Another way of arc formation is due to the transient recovery voltage, which appears across the separated contacts. It reaches a sufficiently high value, which is equal to the breakdown voltage at the time of every current zero crossing in the AC CB. The conductor of the arc uses gases and vapours initiated partly from the electrodes and partly from the surrounding medium. The electrical arc has a low voltage drop across it, and hence, it induces a large value of current.

The following are factors responsible for the formation of an arc.

1. The voltage across the electrode and its variation with time
2. The nature, shape, and separation of electrodes

3. The nature and pressure of the medium
4. Presence of external ionizing and deionizing agents
5. The nature, shape, and position of the vessel (circuit breaker) in which the electrodes are located

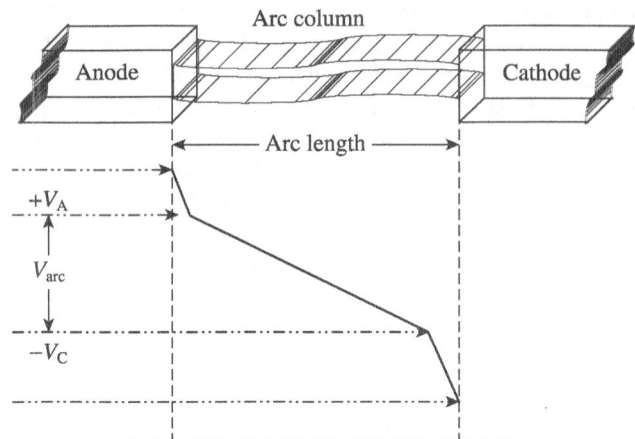

Fig. 13.2   Voltage distributions across an arc column

The two main processes that can cause the emission of a number of free electrons from the metal and the surrounding area are the thermionic emission of electrons and the field emission of electrons. In a CB, at the time of separation of switching contacts, either one or both of these processes take place. In this condition, the area and the pressure between the separating contacts decrease quickly, which creates a momentary increase in the electrical resistance. This will lead to localized heating between the contacts. This localized heating marks the initiation of thermionic emission. During this instant, the current may be of several thousands of amperes, which leads to a voltage drop of a few volts as the increase in resistance is small (a few ohms). This results in a high voltage gradient, which is sufficient to cause the emission of electrons from the cathode. This process is known as *field emission*. Both thermionic and field emission processes are responsible for the formation of an arc. However, the voltage gradient between the contacts is of the order of 10^3 kV/cm, and hence, it is more accountable for the field emission and initiation of arc at the time of contact separation. During this process of emission, a number of electrons are released from the cathode and they move towards the anode with sufficient force to ionize the medium. The ionization of the medium between the contacts creates an avalanche of electrons, which serve to maintain the arc. Even after field emission stops completely, the generation of electrons in multiple numbers continues by receiving energy from the field. The density of electrons constitutes a very high value of current because of which the discharge appears in the form of an arc, with a temperature high enough to maintain thermal ionization. This thermal ionization leads to continuous electrical conductivity (arc current) through the medium at a low voltage gradient (arc voltage decreases at very high current).

13.4 Characteristic of Arc

Arc characteristic is the curve between the instantaneous voltage across the electrode and the corresponding current through the arc. As mentioned earlier, the increase in arc current leads to dynamic ionization and reduction in the arc voltage. The rate at which the current changes its value in the upper range has no such effect on the decrease of the arc voltage. This is due to the non-linear resistance characteristic of the arc, and the sluggish rate at which the heat content of the nearby area of the arc establishes the new heat content. On the other hand, when the current falls from a very high value to a low value, the change in voltage drop is entirely due to the temperature effect. The voltage drop across the electrode depends on the collection of positive and negative charges in front of the cathode and the anode, respectively, and it also depends on the electrode material.

The voltage gradient produced by the accumulation of these charges at the electrode is higher than the voltage drop across the main arc column. The voltage drop across the main arc column mainly depends on the types of gases surrounding it, the gas pressure and magnitude of the arc current, and the length of the column. The voltage gradient across the main arc column is uniformly distributed across the length of the arc. Figure 13.2 shows the voltage distribution of an arc column, where $+V_A$, $-V_C$, and V_{arc} represent the voltage drop across the anode, cathode, and main arc column, respectively. The current density at the cathode is not largely dependent on the arc current magnitude, but it depends on the material used for the cathode. Therefore, carbon, tungsten, and molybdenum materials are preferred because of their high boiling points. If materials such as copper and mercury, which have low boiling points, are used for the cathode there may be chances of melting of the material because of high current density of the order of 10^6 A/cm^2. The temperature at the middle of the arc column is of the order of 5000–8000°C, and the same at the periphery of the arc rod is of the order of 2000–3000°C. Once again, it depends on the amount of current, types of gases surrounding the arc, and the material used for the electrodes.

13.5 Theory of Arc Quenching in AC Circuit

The quenching of an arc in an AC circuit requires a mechanism that converts the conducting path across the contacts of a CB into a non-conducting path. Processes such as reduction of velocity of charged particles, cooling of arc rod, and recombination lead to rapid deionization of the gaseous path. The AC current passes through zero twice during each cycle (100 times for 50 Hz). Thus, with variation in the current value, the ionization process also changes. As mentioned in Section 13.4, the voltage drop across the contacts depends on the temperature alteration and the rate at which the current changes. Figure 13.3 shows the wave shape of an arc current and arc voltage during every current zero with respect to time.

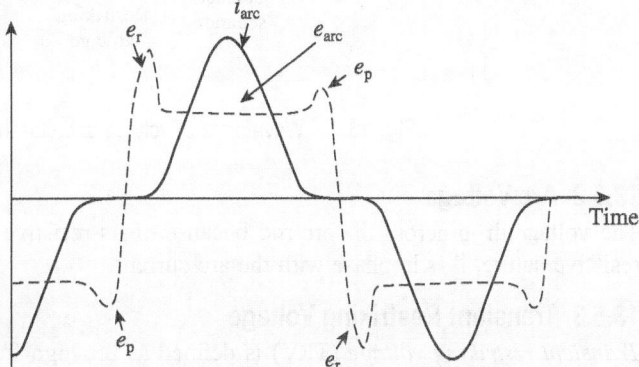

Fig. 13.3 Peak voltage (e_p) and restriking voltage (e_r) during zero crossing of arc current

There are two major peak voltage alterations that occur at the time of high current zero crossing. When an AC current crosses its zero, the effect of the deionization and cooling of the arc is major, which causes the arc voltage to rise. The arc voltage rises suddenly upto the value of the instantaneous peak voltage (e_p) known as *extinguishing voltage for the previous current loop*. For the next current loop at the same zero crossing, that is, reversal of arc current (i_{arc}), if deionization does not occur at a sufficient level, then the arc voltage reignites and reaches the peak value known as *restriking voltage* (e_r). Immediately after the restriking, the arc voltage becomes relatively constant with a magnitude lower than the restriking voltage, known as *arc voltage* (e_{arc}). This process may be repeated for several cycles, and finally, the arc interruption is to be assumed after the restriking voltage appears as a high frequency transient oscillation. Later on, this voltage rapidly decays and reaches the normal system frequency voltage. Figure 13.4 shows the waveform of the voltage and current at the time of arc extinction. Different terms related to arc quenching are discussed in Sections 13.5.1–13.5.5 with reference to Fig. 13.4.

13.5.1 Restriking Voltage

At the time of zero crossing of the arc current, the arc tries to get quenched. If the deionization process does not achieve enough dielectric strength, the arc restrikes, and the voltage at this instant is known as the *restriking voltage*.

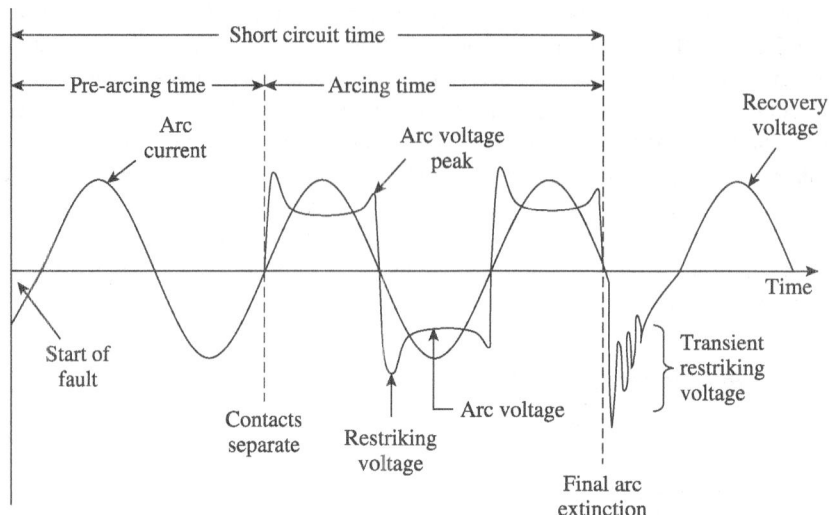

Fig. 13.4 Waveforms of voltage and current at the time of arc extinction

13.5.2 Arc Voltage

The voltage drop across the arc rod because of its resistive nature is known as *arc voltage*. Owing to its resistive nature, it is in phase with the arc current.

13.5.3 Transient Restriking Voltage

Transient restriking voltage (TRV) is defined as the high frequency restriking voltage that appears across the contact of CB just after the arc extinction. The restriking voltage (V_C) developed across the open contact of CB is given by

$$V_C = E_{max} \times (1 - \cos \omega t) \tag{13.1}$$

where $\omega = 1/\sqrt{LC}$, in which L and C are the system parameters and E_{max} is the peak value of the recovery voltage. The maximum value of the restriking voltage is $2 \times E_{max}$, and it occurs at $t = \pi/\omega$, which is 2 times the peak value of the normal voltage appearing across the contacts of the CB. The oscillatory transient voltage has a natural frequency $f_n = \dfrac{1}{2\pi \times \sqrt{LC}}$ Hz. This natural frequency is of the order of 1–10 kHz depending on the system condition and parameters (L and C). There may be an incidence of breaking of the line with L and C on both sides of the CB; both circuits oscillate at their own natural frequency. Hence, on contact of the CB, a combined double frequency transient oscillation occurs.

13.5.4 Rate of Rise of Restriking Voltage

The higher the frequency of the transient voltage, the sharper will be the slope of its first voltage rise from zero to peak. The slope of this steepest tangent to the restriking voltage curve is defined as the *rate of rise of restriking voltage* (RRRV), and it is expressed in kilovolts per microsecond. For a single frequency transient oscillation, RRRV is measured by dividing the maximum amplitude of the first peak to the time required to reach the first peak. The expression for RRRV is given by

$$RRRV = \frac{dV_C}{dt}$$

Using the value of V_C from Eq. (13.1), we get $\text{RRRV} = \dfrac{d(E_{max} \times (1 - \cos \omega t))}{dt}$

$$\text{RRRV} = E_{max} \times \omega \times \sin \omega t \tag{13.2}$$

The maximum value of RRRV can occur at $\omega t = \dfrac{\pi}{2}$, and hence, $t = \dfrac{\pi}{2\omega}$.

Thus, at time $t = \dfrac{\pi}{2}\sqrt{LC}$, the value of RRRV is maximum.
Therefore,

$$\text{RRRV} = E_{max} \times \omega \tag{13.3}$$

13.5.5 Recovery Voltage

The power frequency steady-state voltage appearing across the contacts of the CB after the final arc extinction is known as *recovery voltage*.

13.6 Arc Interruption Theories

The multifaceted behaviour of the arc in a CB at the time of short circuit has led to the invention of different arc interruption methods. Various theories have been suggested by many researchers for the fast and reliable treatment of arc interruption, such as the high resistance and the low resistance interruption theories. In all the theories, the concept of AC arc interruption near zero current is utilized because it is the most significant region where the AC arc restrikes or extinguishes. These arc interruption methods are discussed in more detail in Sections 13.6.1 and 13.6.2.

13.6.1 High Resistance Interruption

In this type of interruption, the arc is restricted by increasing its effective resistance with respect to time. With the increase of arc resistance, the current is reduced to a value inadequate to sustain the arc across the contacts of the CB. The increase of arc resistance can be achieved by several methods such as lengthening the arc column, cooling, and splitting the arc in many subsegments. Such a high resistance interruption method is not appropriate for a high power AC CB because of high energy losses at the time of arc interruption. Thus, the use of this method is limited to low power AC and DC CBs.

13.6.2 Low Resistance Interruption

This method is mainly used for the interruption of an arc in an AC circuit as the arc current passes through zero twice in a cycle (100 times per second for a 50 Hz system). The arc tries to either die out at every current zero or reignite with rising current. Low resistance interruption is explained by two theories, namely, Slepian's Theory and Cassie's Theory.

Slepian's Theory (Race Theory)

During the separation of contacts of the CB, the medium is ionized because of high field intensity, and it sets up a very hot gaseous path. To stop this ionization process, it is necessary to remove the ionized gases by prohibiting electron generation and advancing the recombination process. The rate of ionization is very low in the zero current region. Hence, it is easier to increase the resistance of the arc in this region and build up high dielectric strength across the contacts of the CB. Joseph Slepian discovered the theory of arc interruption in 1928, which is widely known as *race theory*. According to Slepian's theory, at each current zero, there is a race between the RRRV and the rate at which the insulating medium recovers its dielectric strength. If the rate at which the dielectric strength progresses is faster than the rate at which the voltage rises, the arc will be quenched; otherwise, the arc restrikes and is not interrupted. However, this theory assumes that the build-up of the restriking voltage and dielectric strength during the interruption are totally different processes. This assumption is not valid as it does not calculate the rate at which the dielectric strength recovers. Moreover, it does not consider the energy relation at the time of the interruption. Figure 13.5 shows the graphical representation of Slepian's theory.

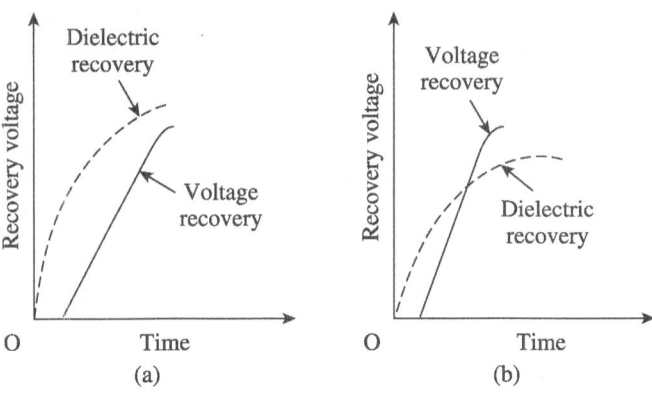

Fig. 13.5 Representation of Slepian's theory (a) Arc interruption (b) Dielectric failure

Cassie's Theory (Energy Balance Theory)

In 1939, Cassie developed the theory of arc interruption based on the assumption that the conductivity of arc due to high current is directed by the energy losses in the form of convection. The temperature of the arc is assumed to be constant during this high current period. However, the change in current creates the necessary change in the diameter of the arc, maintaining almost constant temperature at the centre of the arc column. Moreover, when current decays towards the zero current regions, the cross section of the arc is a small fraction of a millimetre, still maintaining high temperature. This high temperature can reignite the arc with a bigger cross-section if the electric field intensity reappears across the contacts of CB. Cassie declared that the *interruption of arc* is a process of energy balance. At current zero, if the rate at which the energy input to the arc column is higher than the rate at which maximum energy is lost from the arc column, the arc restrikes; if not, the arc can be interrupted. This idea of energy balance shows more potential than Slepian's theory.

At current zero, the hot arc column between the contacts of the CB needs to be cooled down to such a low temperature that it no longer conducts. Owing to the stored thermal energy, the arc has certain inertia, and when the current approaches zero, there is still some electrical conductivity left in the arc path. This gives rise to a *post arc current*. Figure 13.6 shows the phenomenon of the post arc current. The race between the energy removed from the arc by cooling and the energy input to the arc path by the post arc current, which is determined by the recovery voltage, determines whether the interruption will be successful. The time taken by this process is very short (microseconds). The input of power to the arc, and therefore, the difficulty

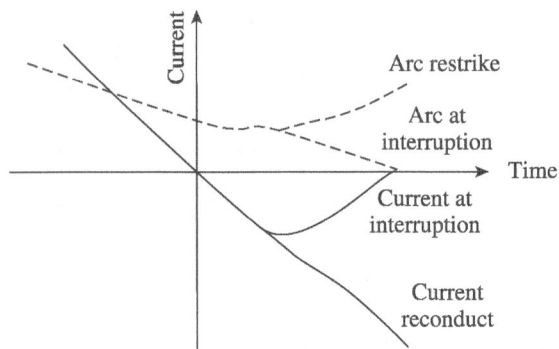

Fig. 13.6 Establishment of thermal interruption with post current zero

to interrupt the arc is related to the rate of reduction of current towards zero and to the rate of rise of recovery voltage after current zero.

Finally, in an AC CB the arc interruption process is achieved at current zero, by promptly removing the arc energy in the form of heat and rapidly increasing the dielectric strength between the contacts of the CB. However, the arc quenching process depends on the magnitude of arc current, material of contacts, speed of contact separation, medium used to recover the dielectric strength, rate at which the transient voltage restrikes, and the system condition.

13.7 Factors Affecting RRRV, Recovery Voltage, and TRV

The various factors affecting the RRRV, recovery voltage, and TRV are discussed here.

13.7.1 Power Factor of the Circuit

The instantaneous value of recovery voltage depends on the power factor of the circuit. Higher the value of the power factor, lower will be the value of the voltage stresses across the contacts of the CB at the time of current zero interruption. During normal condition of a circuit, the interruption of load current with high power factor (0.8) produces low voltage stress, and hence, the value of restriking voltage and RRRV is less. However, during the clearance of fault, which is reactive in nature with a low power factor (0.2) and of high magnitude, the instantaneous value of the recovery voltage is very high. Even the interruption of a small value of a reactive short circuit current is more difficult compared to the resistive short circuit current. Figure 13.7 shows the effect of high and low power factors on the recovery voltage. In Fig. 13.7(a), the instantaneous value of recovery voltage is very high at current zero because of low power factor, whereas its value is low for resistive current interruption (Fig. 13.7(b)).

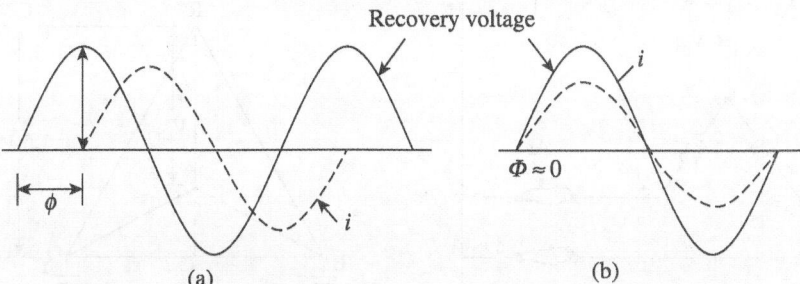

Fig. 13.7 Effect of power factor on recovery voltage (a) Low power factor
(b) High power factor

13.7.2 Circuit Condition and Types of Fault

The circuit conditions such as whether the system neutral is connected and types of fault can affect the performance of current interruption and voltage appearing across the contact of CB. If a fault on an earthed system involves the ground (L–g, L–L–g, and L–L–L–g), the voltage appearing across the contact of the CB in which the arc extinguished first is only the phase voltage (line-to-earth). On the other hand, in a situation in which a fault does not involve the ground (L–L and L–L–L) or the system itself is not earthed, the voltage across the contact of the CB where the arc extinguished first is 1.5 times the phase voltage. Hence, in the latter case, the duty of the CB is heavy as the voltage stress that appears across the contact of the CB is much more than in the former case. Figure 13.8 demonstrates the recovery voltage for different circuit conditions and fault types.

13.7.3 Asymmetry of Short Circuit Current

Most fault cases have a DC component in their short circuit current, and hence, at least one or more phases can be asymmetrical in nature by some degree. With the presence of an asymmetry in the short circuit current to be interrupted, it is possible to set up varying values of recovery voltage at current zero. As the degree of asymmetry increases, the recovery voltage at current zero reduces, and it further depends on the interruption, which follows either the major current loop or the minor current loop. Figure 13.9 shows the waveform of the recovery voltage appearing across the contacts of the CB at current symmetry and asymmetry with considerable low power factor.

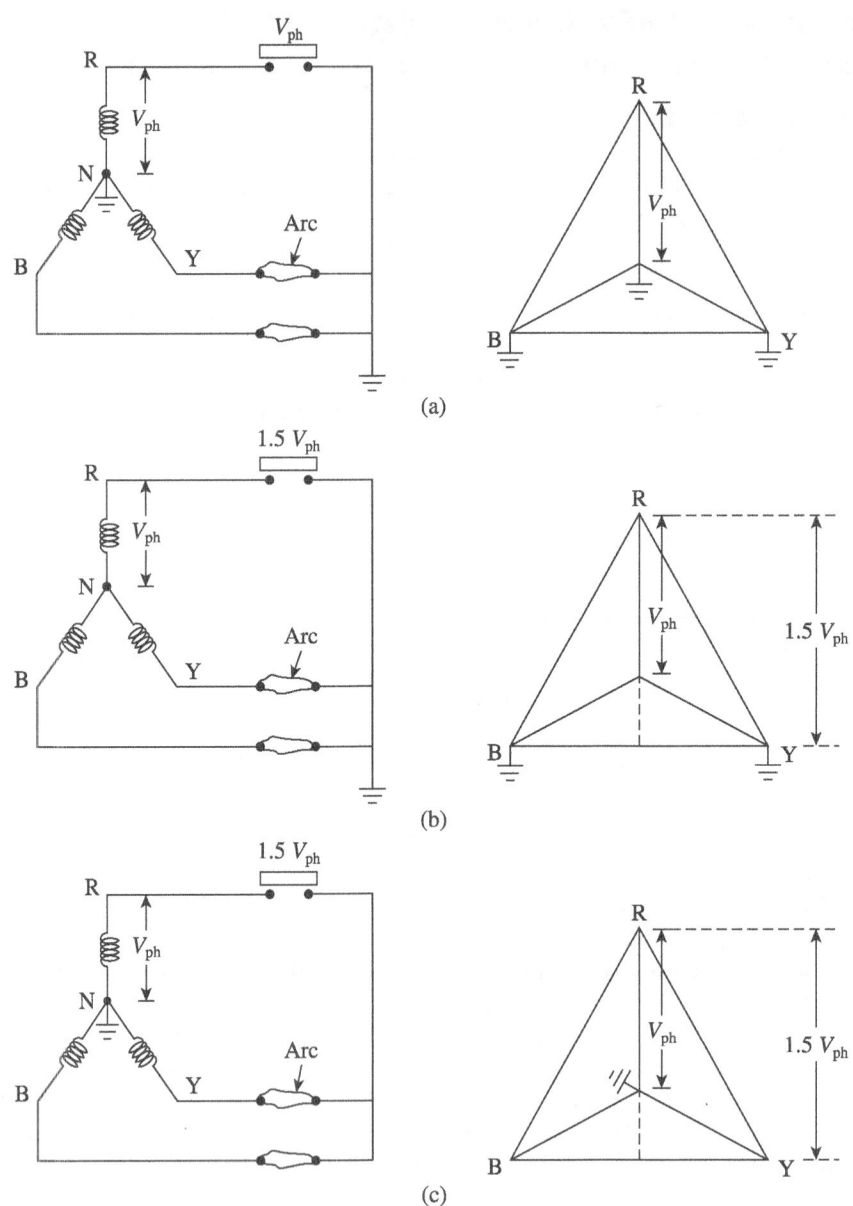

Fig. 13.8 Recovery voltages for different circuit conditions and fault types (L–L–L–g)
(a) Neutral earthed and fault grounded (b) Neutral not earthed and fault grounded
(c) Neutral earthed and fault ungrounded

13.7.4 Short Line Fault

A short line fault or a fault very close to the line side terminal of a CB enforces a severe duty on the CB. The interruption of a short line fault creates a very high saw tooth shaped TRV on the line side terminal of the CB owing to line side components. On the other hand, at the supply side, the restriking voltage is normal and gradually increases. The higher frequency components in the recovery voltage, which can contain

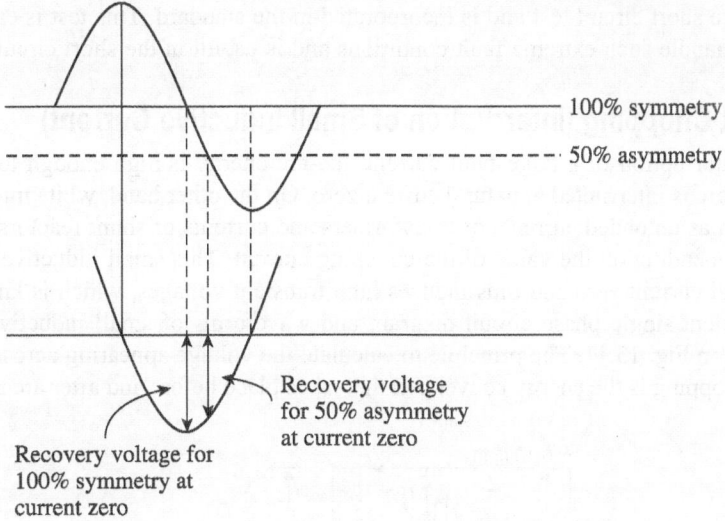

Fig. 13.9   Effect of current asymmetry on recovery voltage

more than one frequency, lead to a dangerously rapid RRRV. If this steep RRRV, which can eventually reach several numbers, occurs at a faster rate than the rate at which the breakdown strength of the breaker gap increases, an arc reignition will take place. The magnitude of transient voltage is proportional to the mag-

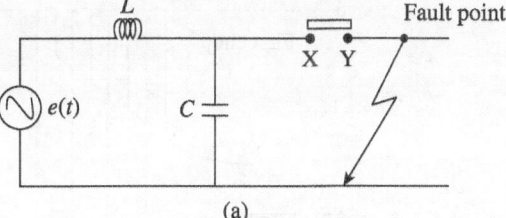

(a)

nitude of the short circuit current. Further, its frequency is inversely proportional to the distance of the line upto the fault point. After the interruption of fault current, the charges pass by the line side component because the high magnitude of current decays in the form of a transient wave, which has a high oscillating frequency (100 kHz). Thus, the TRV is of multiple frequencies with very steep RRRV for its initial peak. Moreover, the shape of the recovery transient departs from the usually estimated high frequency sinusoidal waveforms and exhibits several superimposed high frequency ramp functions of both positive and negative slopes. The value of RRRV also depends on the surge impedance of the line upto the fault point, and its value reaches several kV/μs in case of kilometric fault (fault within 1–2 km).

$$RRRV = Z_S \frac{di}{dt}$$

$$RRRV = Z_S \omega \sqrt{2} I_F \qquad (13.4)$$

where,

I_F = fault current

ω = angular frequency

Z_S = surge impedance of line for the faulted section

The explanation of short line fault is illustrated in Fig. 13.10. The short line fault test for a CB is considered

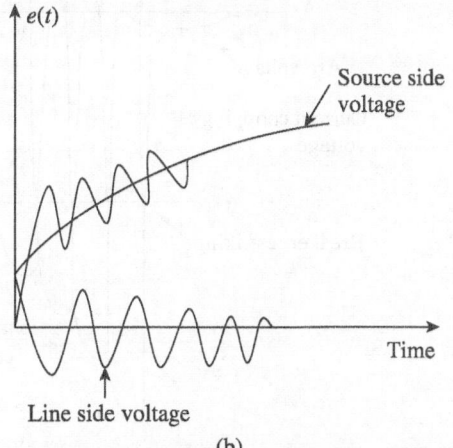

Fig. 13.10   Short line fault interruption (a) Single line diagram of circuit (b) Waveform for the short line fault

as the most severe short circuit test and is incorporated in the standard. This test is carried out to prove the ability of CB to handle such extreme fault conditions and is useful in the short circuit rating of CBs.

13.8 Current Chopping (Interruption of Small Inductive Current)

At the time of interruption of a large fault current, the arc energy is high enough to keep the arc column ionized until the arc is interrupted at natural current zero. On the other hand, while interrupting small inductive currents such as unloaded currents of transformers and currents of shunt reactors, there is a possibility of overvoltage depending on the value of the chopping current. This small inductive current is interrupted just before natural current zero and thus induces high transient voltages, which is known as *current chopping*. The equivalent single phase circuit diagram and waveforms of small inductive current interruption are demonstrated in Fig. 13.11. The principle to calculate the voltage appearing across the contacts of a CB during current chopping is the energy conversion that takes place before and after arc interruption. When the

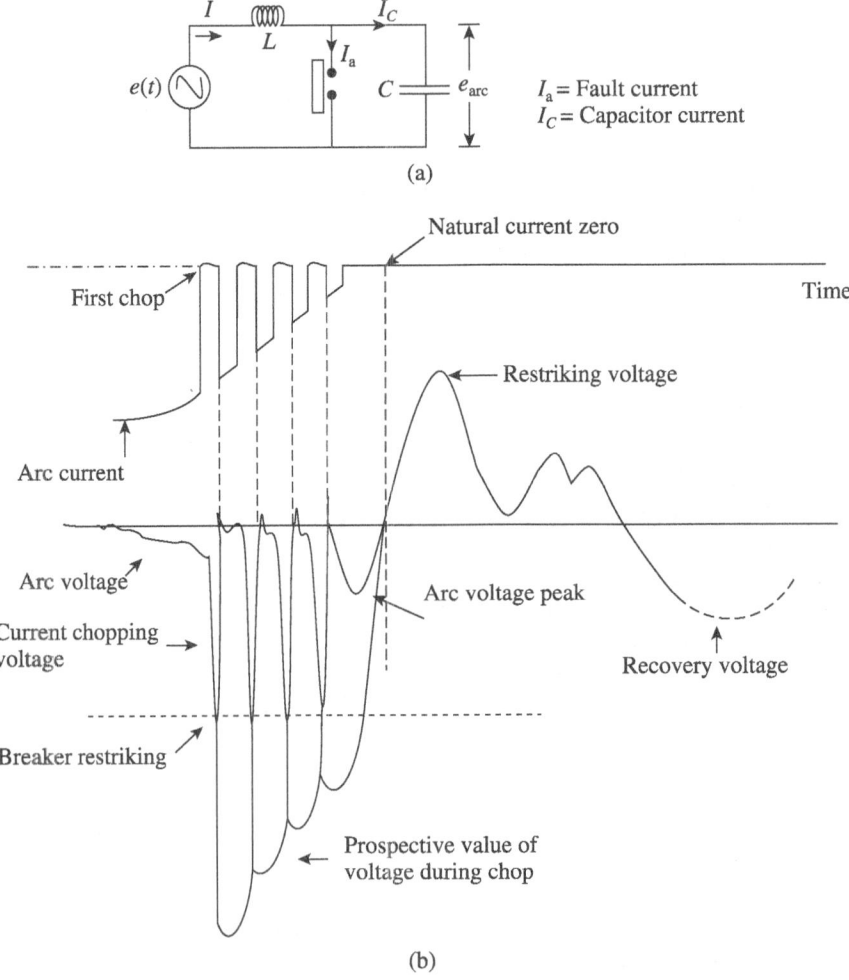

Fig. 13.11 Effect of current chopping while interrupting small inductive current
(a) Circuit diagram (b) Waveform of voltages and currents

CB interrupts an arc current I, the electromagnetic energy stored in the inductance L $(\frac{1}{2}LI^2)$ is transferred (discharged) into electrical energy in the stray capacitance C $(\frac{1}{2}CV^2)$. Thus, the energy balance equation for the transient voltage across the contacts of the CB is given by

$$\frac{1}{2}LI^2 = \frac{1}{2}CV^2$$

$$V = I\sqrt{LC} \tag{13.5}$$

The prospective value of this voltage may reach a dangerous level, even higher than the dielectric strength, and leads to the next current conduction by restriking. Finally, when enough dielectric strength recovers after selective current chop, the current is concealed without restrike.

The total amount of ionization is less at the contact gap at current zero, but the suppression of current occurs before current zero at a higher instantaneous value, which induces a very high voltage across the contacts of the CB. The overvoltage caused because of this sudden interruption of current before natural zero can be 2–3 times higher than the normal recovery voltage as shown in Fig. 13.11. The interruption of the small inductive current is made possible by providing enough thermal clearance across the contact gap at the time of rising voltage. Vacuum CBs are preferred for the interruption of the small inductive current as the number of chops is low.

13.9 Interruption of Capacitive Current

There is an increasing demand for utilities to improve the power factor. Utilities struggle to maximize the power transfer efficiency of lines by maintaining a power factor close to unity. The application of shunt capacitor banks has become a special tool for improving the power factor. It is a common practice for utilities to switch *on* and *off* the shunt capacitors as per daily load variations. Switching *on* and *off* of unloaded transmission line and connecting or disconnecting shunt capacitor banks from the system create a few unique challenges since the voltage across the capacitor cannot change instantaneously. This switching causes unwanted high frequency voltage and current transients across the contacts of CB, which may damage the equipment.

Figure 13.12 represents the waveform of the interruption of capacitive current. When the capacitive current reaches its zero value, the line voltage is at its peak value. Hence, when interruption occurs at current zero, the line remains in charged condition at this peak voltage of utilities. In Fig. 13.12, after point P, the voltage difference appearing across the contacts of the CB is due to two voltages, namely, utilities voltage (V_u) and line side capacitor voltage (V_C). After an interval of half a cycle, at instant Q, V_u reverses and hence, the voltage across the breaker is twice the peak value of V_u ($2 \times V_m$) where V_m denotes the maximum voltage. Such a high voltage at the breaker contact leads to restriking of an arc, and the circuit will reclose in an oscillating manner by developing the voltage of $-3 \times V_m$. Thus, the line is charged at a voltage of $-3 \times V_m$ after the interruption of a restriking current. At this time, the voltage across the contacts of the breaker is twice V_m, as the value of V_u is at its maximum negative. Afterwards, the voltage across the contacts of CB continues to increase, and at point S, the value becomes $4 \times V_m$. If the breaker restrikes at this instant, high frequency oscillation of V_C will occur at a voltage of $5 \times V_m$. Theoretically, this process is repeated indefinitely by increasing the voltage to its extreme value across the capacitance at a rate of twice the peak value at an interval of half a cycle. The severity of voltage is practically limited only by leakage or breakdown of insulators.

Restrike-free or forced blast CBs are generally designed with high power frequency voltage withstanding capacity to interrupt the capacitive load. Interruption of capacitive currents is easily handled by modern SF_6

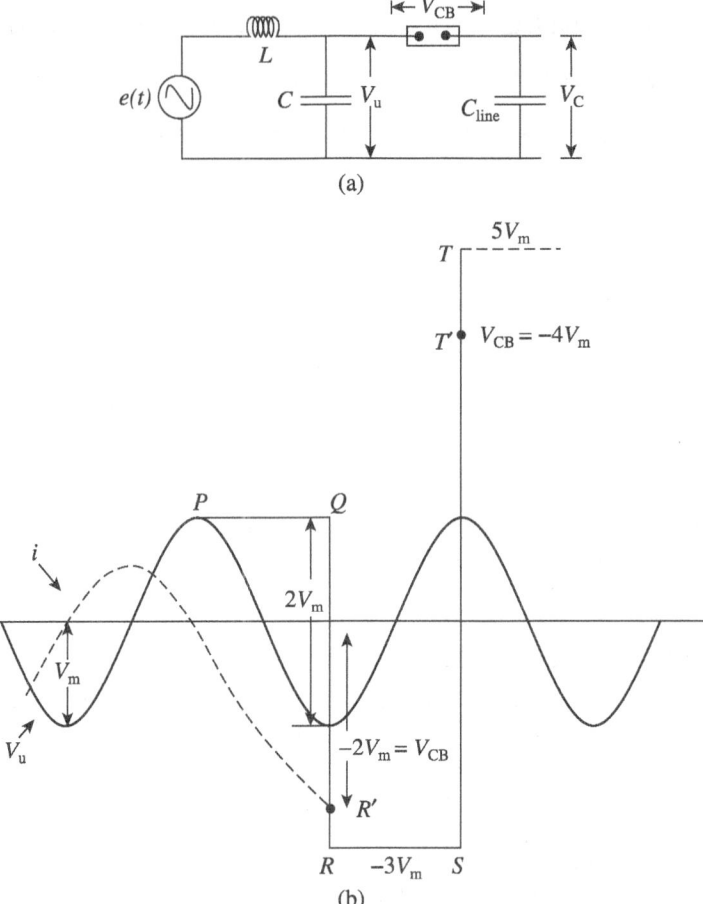

Fig. 13.12 Interruption of capacitive current (a) Circuit (b) Waveform representation

interrupting devices because of the interruption of low magnitudes of current with a very slow recovery voltage. This is also true for vacuum interrupters at medium voltage levels.

13.10 Resistance Switching

The value of RRRV and TRV depends on many circuit parameters and different circuit conditions at the time of interruption after each current zero. Moreover, it also depends on the relative position of resistance connected with the capacitance offered by circuit components. Intentional insertion of resistance across the contacts of CB after its separation is known as *resistance switching*. As discussed in Section 13.9, the voltage appearing across the contacts of a CB in case of inductive and capacitive current interruptions is very high. Utilization of shunt resistance across the contacts of a CB leads to reduction of the restriking voltage and RRRV. By incorporating artificial resistance, it turns away part of the arc current and hence, reduces the arc intensity across the contacts of the CB.

In case of air blast and SF_6 CBs, the pressure of the arc quenching medium is independent of normal current or fault current. Hence, sometimes, low current interruption is achieved with high pressure of insulating medium. In this situation, it chops the current before natural zeros. This creates a very severe transient voltage across the contact gap of the CB. Consequently, these transient voltages can cause flashover on the insulation. Therefore, resistance switching is used to damp out such extreme high voltages. Furthermore, if the value of the added resistance is higher than twice the surge impedance of the line, the natural frequency oscillation of the circuit can be easily suppressed. Moreover, RRRV is directly proportional to the natural frequency of the circuit, and it mainly depends on the value of the inserted shunt resistance. Thus, insertion of deliberate shunt resistance across the contact of the CB increases the rupturing capacity of the breaker.

13.11 Theory of Arc Interruption in DC Circuits

The most obvious difficulty in interrupting DC current is that there are no natural current zeros as in AC supplies. In order to force the current to zero, the breaker must generate an arc voltage, greater than the system voltage, across the contacts of the CB. Further, increase of arc voltage also requires removal of energy from the arc during the flow of arc current. As the behaviour of the load and fault current is highly inductive, the CB interrupting device must be capable of dissipating all of the energy in the circuit until the arc extinguishes. Resistance switching and efficient cooling are used to dissipate the arc energy. A momentary natural extinction of the DC arc due to current zero conditions is not present in most CBs. Typically, the CBs that are being used for DC circuit switching have limited features. The arc voltage can be increased by lengthening the arc column by creating the necessary force surrounding the arc so as to decrease the arc diameter. Until the early 1970s, the main method of DC arc extinction was to stretch the arc and cool it in an arcing chamber under the influence of the magnetic field produced by a series-connected *magnetic blow-out coil*. It has been observed that the voltage drop in DC arcs is around 1 V/mm. This value obviously depends on arcing conditions, but for system voltages of above a few hundreds of volts, the arc needs to be fairly long.

13.11.1 Circuit Description

Figure 13.13 shows a simple equivalent circuit with DC voltage V, circuit resistance R, and arc resistance connected in series during the closing condition of switch S. The arc current flowing through the circuit is i and depends on the circuit parameters and arc characteristic. It is to be noted that the arc voltage drop, e_a, is smaller than the supply voltage V by a value equal to $i \times R$; thus, $e_a = V - i \times R$.

Figure 13.14 represents the relationship between DC arc characteristic and voltage variation during direct current interruption. As shown in Fig. 13.14, the straight line represents the voltage drop e_a, which cuts off

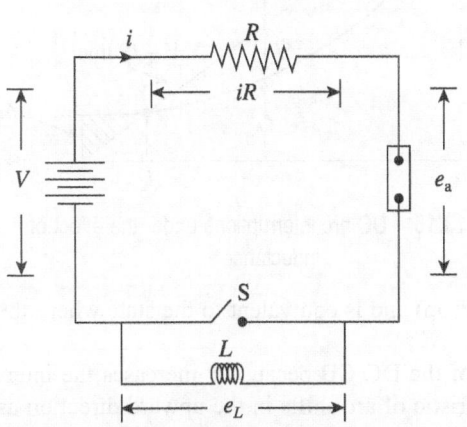

Fig. 13.13 Equivalent circuit of DC arc interruption

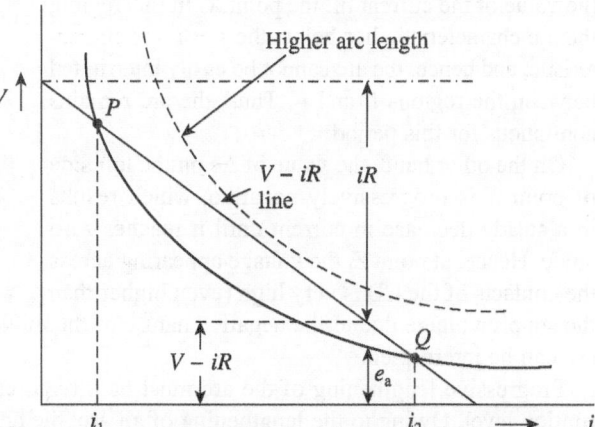

Fig. 13.14 Arc characteristic during interruption of DC current

the curve of the arc characteristic at points P and Q. At these intersection points, the arc remains stable with respect to their respective currents (i_1 and i_2).

On examining the current variation around point Q, it is seen that if the current increases corresponding to the point Q, then the respective arc voltage drop decreases and $i \times R$ drop increases. Conversely, if the current decreases, then the voltage across the arc increases at a rate faster than $i \times R$ voltage drop. In the latter case, the current slips back and reaches its stable state (point P) or close to its zero value so that the arc is interrupted. By lengthening the arc, the net resistance increases, which further reduces the net current through the circuit, thus maintaining constant supply voltage. If the arc length further increases, then the arc characteristic shifts upwards compared to the relative position of e_a. It has been observed that there is no intersection between them, and hence, the available voltage is not enough to maintain the arc. Therefore, it can be easily quenched.

13.11.2 Effect of Circuit Inductance

In Fig. 13.13, if the switch S is opened out, the inductance of the circuit comes into play. With the addition of inductance in a circuit, the voltage balance equation of the circuit is given by

$$L\frac{di}{dt} + Ri + e_a = V \tag{13.6}$$

Thus, the voltage drop across the inductance is given by

$$L\frac{di}{dt} = (V - Ri) - e_a = \Delta e \tag{13.7}$$

This voltage drop (Δe) is equal to the deduction of the voltage drop across the total resistance, including circuit and arc, from the source voltage V. This is shown in Fig. 13.15.

As shown in Fig. 13.15, the value of Δe, right to the point Y, is negative. Hence, the current decreases and reaches the point Y. On the left side of point Y, the positive value of Δe increases with the decrease in the value of the current till the point X. In this region, the arc characteristic lies below the $V - i \times R$ characteristic, and hence, the arc cannot be easily interrupted between the regions i_1 and i_2. Thus, the arc remains continuous for this period.

On the other hand, the value of Δe on the left side of point X is progressively negative, which results in a steady decrease in current until it reaches zero value. Hence, at point Z, the voltage appearing across the contacts of the CB is very high (even higher than

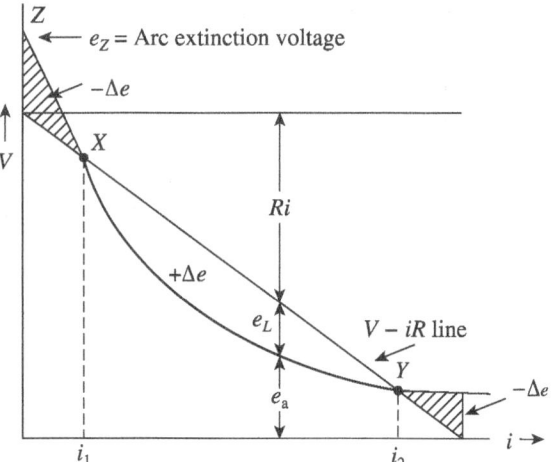

Fig. 13.15 DC arc interruptions under the effect of inductance

the supply voltage due to the negative nature of the inductive drop) and is equivalent to the state where the arc can be interrupted.

Progressive lengthening of the arc must be a requirement of the DC CB because it increases the interruption level. Owing to the lengthening of an arc, the characteristic of arc shifts in the upward direction as shown by the dotted line in Fig. 13.14.

13.12 Examples

Example 13.1 A CB is meant for the discontinuation of a 220/132 kV, 250 MVA, 50 Hz power transformer during no-load condition. Find out the worst condition of overvoltage induced across the contacts of the CB. The no-load current of transformer is 1% of the full load current of the transformer. The capacitance of the system is 10,000 pF/phase.

Solution:

The full load current of transformer is given by

$$I_{FL} = \frac{250 \times 10^6}{\sqrt{3} \times 220 \times 10^3} = 656.079 \text{ A}$$

The no-load current of transformer is given by

$$I_{NL} = 0.01 \times 656.079 = 6.56 \text{ A}$$

Now, the reactance offered because of this no-load current is

$$X_0 = \frac{220 \times 10^3}{\sqrt{3} \times 6.56} = 19.36 \times 10^3 \ \Omega$$

Thus, the inductance of the system is

$$L_0 = \frac{X_0}{2\pi f} = \frac{19.36 \times 10^3}{2 \times \pi \times 50} = 61.656 \text{ H}$$

Now, assuming that the interruption occurs at peak value of the current, the peak value of the current is given by

$$I_p = \sqrt{2} \times 6.56 = 9.277 \text{ A}$$

With the system capacitance of 10,000 pF/phase, the value of overvoltage induced at the contacts of CB is given by

$$V = \sqrt{\frac{L}{C}} \times I_p = \sqrt{\frac{61.656}{10,000 \times 10^{-12}}} \times 9.277 = 728.443 \text{ kV}$$

Example 13.2 A 50 Hz, 13.8 kV, three-phase generator with grounded neutral has an inductance of 15 mH/phase and is connected to a busbar through a CB. The capacitance to earth between the generator and the CB is 0.05 μF/phase.

Determine the following:

(a) Maximum restriking voltage
(b) Time for maximum restriking voltage
(c) Average RRRV upto the first peak
(d) Frequency of oscillations

Neglect the resistance of generator winding.

Solution:

(a) The maximum value of recovery voltage is given by

$$E_{max} = \sqrt{2} \times \frac{13.8}{\sqrt{3}} = 11.27 \text{ kV}$$

Now, $V_C = E_{max}(1 - \cos\omega t)$, and maximum restriking voltage occurs at $t = \pi/\omega$.

Hence, maximum restriking voltage is given by

$$2E_{max} = 2 \times 11.27 = 22.54 \text{ kV}$$

(b) The first peak of restriking voltage occurs at time

$$t = \frac{\pi}{\omega} = \pi \times \sqrt{LC} = \pi \times \sqrt{15 \times 10^{-3} \times 0.05 \times 10^{-6}} = 8.6 \times 10^{-5} = 86 \mu s$$

(c) Now, the average RRRV for $t = 86$ μs is given by

$$\text{RRRV} = \frac{\text{Maximum restriking voltage}}{\text{Time upto the first peak}}$$

$$= \frac{22.54}{86 \times 10^{-6}}$$

$$= 262.09 \times 10^3 \text{ kV/s}$$

(d) The frequency of oscillation is given by

$$f_n = \frac{1}{2\pi\sqrt{LC}} = \frac{1}{2 \times t}$$

$$= \frac{1}{2 \times 86 \times 10^{-6}} = 5.814 \text{ kHz}$$

Recapitulation

- The formation of an arc is due to the increase in voltage because of the presence of self-inductance of the circuit at the time of contact separation.
- It also depends on the transient recovery voltage, which appears across the separated contact.
- The properties and characteristics of an arc are required to be studied at the time of design of a CB.
- In a high resistance interruption process, it is required to increase the arc resistance by several methods such as lengthening the arc column, cooling, and splitting the arc in many subsegments.
- In low resistance arc interruption process,
  - at each current zero, there is a race between the RRRV and the rate at which the dielectric strength is recovered by the insulating medium. If the rate at which the dielectric strength's progress is faster than the rate at which the voltage rises, the arc will be quenched; otherwise, the arc restrikes and is not interrupted.
  - at each current zero, if the rate at which the energy input to the arc column is higher than the rate at which the maximum energy is lost from the arc column, the arc restrikes; if not, the arc can be interrupted.
- The effect of system parameters and circuit conditions such as power factor, asymmetry in fault current, types of fault, and short line fault can also affect the performance of arc interruption.
- The interruption of a small value of reactive short circuit current is more difficult compared to the resistive short circuit current.
- The duty of CB is heavy in case of a fault on an earthed system, which does not involve ground, or if the system itself is not earthed, as the voltage across the contact of CB where the arc extinguished first is 1.5 times the phase voltage.
- The interruption of short line fault creates a very high TRV of sawtooth shape on the line side contacts of CB because of the line side component.

- Switching *on* and *off* of unloaded transmission lines and shunt capacitor banks from the system creates unwanted high frequency voltage and current transients across the contacts of CB, which may damage the equipment. Restrike-free or forced blast CBs are generally required for such interruption.
- Shunt resistance across the contacts of CB leads to reduction of restriking voltage and RRRV. By incorporating artificial resistance, it turns away part of the arc current and hence reduces the arc intensity across the contacts of CB.
- In DC circuit breaking, to force the current to zero, the breaker must generate an arc voltage across the contact of CB, which should be greater than the system voltage.
- The CBs that are being used for DC circuit switching have limited features. The arc voltage can be increased by lengthening the arc column by creating the necessary forces surrounding the arc so as to decrease the arc diameter.

Multiple Choice Questions

1. The basic function of a CB is to
   (a) produce the arc
   (b) extinguish the arc
   (c) transmit voltage by arcing
   (d) ionize the surrounding air

2. Ionization process during arc is generally accompanied by the emission of
   (a) light
   (b) heat
   (c) sound
   (d) all of the above

3. The normal frequency root mean square (RMS) voltage that appears across the contacts of a CB after the final arc extinction is known as
   (a) recovery voltage
   (b) restriking voltage
   (c) supply voltage
   (d) peak voltage

4. The transient voltage that appears across the contacts at the instant of arc interruption is known as
   (a) recovery voltage
   (b) restriking voltage
   (c) supply voltage
   (d) peak voltage

5. In a CB, the active recovery voltage depends on
   (a) power factor
   (b) armature reaction
   (c) circuit conditions
   (d) all of the above

6. Ionization in a CB is independent of the
   (a) high temperature of the surrounding medium
   (b) material of contacts
   (c) increase of field strength
   (d) increase of mean free path

7. The voltage drop across the arc in an AC CB is
   (a) leading the arc current by 90°
   (b) lagging behind the arc current by 90°
   (c) in phase with the arc current
   (d) in phase opposition to the arc current

8. The RRRV appearing across the contacts at the time of interruption can be reduced by
   (a) inserting a resistance in the line and in parallel with the contacts
   (b) inserting a capacitor in series with the contacts
   (c) inserting a capacitor in parallel with the contacts
   (d) inserting an inductor in parallel with the contacts

9. Current chopping phenomenon is related with
   (a) resistance switching
   (b) capacitance switching
   (c) small inductive current
   (d) asymmetrical current

10. The voltage appearing across the contact of CB, which first extinguished the arc, in case of L–L–L–g fault in an insulated neutral fault system is equal to
    (a) 1/2 times the peak voltage of line-to-neutral
    (b) the peak voltage of line-to-neutral
    (c) 2 times the peak voltage of line-to-neutral
    (d) 3/2 times the peak voltage of line-to-neutral

Review Questions

1. Explain the fundamental theory of circuit breaking.
2. Explain the arc phenomenon and discuss the factors on which arc formation depends.
3. Explain the characteristic of arc and cause of initiation of arc at the time of contact separation.
4. Define the following terms:
   (a) Restriking voltage
   (b) Arc voltage
   (c) Transient recovery voltage (TRV)
   (d) Rate of rise of restriking voltage (RRRV)
   (e) Steady-state recovery voltage
5. Explain the various arc interruption methods used in a CB.
6. Why is it difficult to interrupt a fault current with low lagging power factor compared to load current with high power factor?
7. 'Interruption of asymmetrical fault current is easier for CB'. Justify the statement.
8. Explain the adverse effects of short line fault on the performance of CB.
9. Discuss the interruption of small inductive current (current chopping) with a relevant diagram.
10. Why are restrike-free or forced blast CBs generally used to interrupt the capacitive load?
11. What is meant by suppression of RRRV? Explain resistance switching in connection with this.
12. Explain the phenomenon of arc interruption in DC circuit along with arc characteristic.

Numerical Exercises

1. A CB is designed to disconnect a transformer with peak magnetizing current of 8 A. The system capacitance and inductance are 5 H/phase and 1 nF/phase, respectively. At the time of interruption, the inductive energy is discharged into the capacitance. Find the overvoltage appearing across the CB.

   $[V = 565.68 \text{ kV}]$

2. A 50 Hz, 11 kV, three-phase alternator with earthed neutral has a reactance of 6 Ω/phase and is connected to a busbar through a CB. The distributed capacitance upto the CB between phase and neutral is 0.01 μF.

   Determine the following:
   (a) Peak restriking voltage across the CB
   (b) Frequency of oscillations
   (c) The average RRRV upto the first peak

   $[V_P = 17.92 \text{ kV}, F_n = 11.54 \text{ kHz}, RRRV = 413.86 \times 10^3 \text{ kV/s}]$

Answers to Multiple Choice Questions

1. (b), 2. (d), 3. (a), 4. (b), 5. (d), 6. (b), 7. (c), 8. (a), 9. (c), 10. (d)

Types of Circuit Breakers and their Testing

14

Learning Objectives

After going through this chapter, the students will be able to:

- Explain the ratings of circuit breaker
- Discuss the function of high rating CBs
- Explain the various types of low voltage and high voltage CBs
- Compare the merits and demerits of conventional CBs
- Discuss the different tests adopted to test CBs
- Explain the process of selection of CBs

14.1 Introduction

A circuit breaker (CB) is an electrical device designed to operate automatically during short circuit or overload. This task is performed in conjunction with pilot devices such as relays and transducers. As discussed in Chapter 13, the magnitude of fault current is very high compared to the normal load current. Therefore, it is necessary to interrupt such a large current as early as possible. In this situation, the contacts of the CB separate, and there is a continuous path of current due to an arc across the contacts of the CB. Thus, a CB is designed so that it can easily extinguish the arc.

In small and medium voltage circuits, generally, fuses and miniature circuit breakers (MCBs) are employed to interrupt the short circuit current, whereas in high voltage and extra high voltage circuits, high power CBs are used to interrupt an extremely high fault current. Small rating CBs are generally installed directly in the circuit (fuses and MCBs), whereas large rating CBs are usually equipped with pilot devices, which provide a signal to operate the tripping mechanism of the CB.

This chapter deals with the design and function of different types of small rating protective devices such as switches, fuses, MCBs, and earth leakage circuit breakers (ELCBs). Large power rating CBs are also discussed, along with their merits, demerits, maintenance, and testing.

14.2 Ratings of Circuit Breakers

The performance of a CB depends on the circuit condition and the circuit parameters. The main function of a CB is to continuously carry the full load current, and at the same time, interrupt the fault current without any damage. Thus, it is necessary to define the characteristic values of different parameters of a CB at the time of design. These values refer to the rating of a CB, and it is also useful while selecting any CB for a particular application. The standard ratings of a CB are discussed in Sections 14.2.1–14.2.5.

14.2.1 Rated Current and Rated Voltage

A CB is capable of carrying continuous full load current without any excessive rise in temperature. The temperature rise inside the CB should not exceed the limit prescribed by national and international standards. Thus, the *rated current* is defined as the highest root mean square (RMS) current-carrying capability of a CB without exceeding the limit of temperature rise.

Owing to continuous fluctuation of load, the system voltage never remains constant. Therefore, satisfactory operation is achieved by designing the CB with a voltage rating higher than the nominal system voltage. The maximum RMS voltage of a CB above the nominal system voltage for which the CB is designed is known as the *rated voltage* of the CB.

14.2.2 Rated Breaking Capacity

During heavy short circuit, the total fault current involves the AC and DC components of current. As the DC component dies out rapidly, the value of fault current also decreases with time. Moreover, the protective device also takes some time to issue the trip signal to the CB. Thus, the CB starts to open its switching contacts after some time (*t*), starting from the inception of the fault. Figure 14.1 shows the waveform of a short circuit current and the value of current at the time of contact separation. The real value of fault current to be interrupted by a CB is quite less than the initial value at the time of fault inception. Therefore, the highest value of

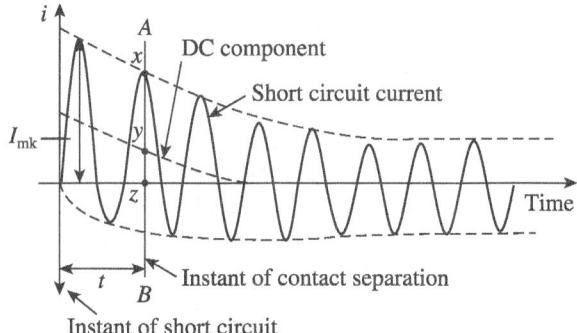

Fig. 14.1 Waveform of current during short circuit

fault current that flows through the switching contacts at the instant of contact separation is known as the *breaking current* of the CB.

There are two types of breaking current.

1. *Symmetrical breaking current* is the RMS value of the AC component of current flowing through the CB at the instant of contact separation. With reference to Fig. 14.1, it is given by $I_{\text{symmetrical}} = \dfrac{xy}{\sqrt{2}}$.

2. *Asymmetrical breaking current* is the RMS value of current (including both AC and DC components of the current) flowing through the CB at the instant of contact separation. From Fig. 14.1, it is given by

$$I_{\text{asymmetrical}} = \sqrt{\left[\left(\frac{xy}{\sqrt{2}}\right)^2 + (yz)^2\right]}$$

The breaking capacity of a CB is based on the symmetrical and asymmetrical breaking capacities. Breaking capacity is expressed in MVA by taking into account the rated breaking current and the rated system voltage. Thus, if I is the rated breaking current in kA and V is the rated system voltage in kV, then for a three-phase circuit, the breaking capacity $= \sqrt{3} \times V \times I$ (MVA).

14.2.3 Rated Making Capacity

The rated making capacity of a CB is related to its ability to withstand maximum current, when it is closed under the fault condition. The electromagnetic forces produced during this time are proportional to the square of peak instantaneous current. Thus, the peak RMS value of short circuit current measured for the first cycle of the current wave after the closure of the CB under fault condition is known as the *making capacity*. In

Fig. 14.1, I_{mk} represents the making current. The peak of making capacity is given by $\sqrt{2} \times \kappa \times$ symmetrical breaking capacity, where κ is the maximum asymmetry, whose value is 1.8. Thus, I_{mk} equals 2.55 times the symmetrical breaking current.

14.2.4 Short Time Rating

Short time rating is the duration for which the CB can carry maximum fault current under its fully closed condition without any damage. The CB should not trip during temporary faults as they persist for a short period. Therefore, the CBs must be capable of holding a high value of current for some specified period of time during closed condition. They should have short time rating, which depends on the electromagnetic force withstanding capacity of the CBs and thermal limitations.

14.2.5 Rated Standard Duty Cycle

The standard duty cycle for medium and high voltage CBs is specified in the ANSI/IEEE standards. The specification of the standard duty cycle changed with the approval of the 1999 edition of the standards. The standard duty cycle has always been estimated as the frequent operation of a CB for a particular application. The CBs are usually able to follow an open-close-open cycle with an energized spring charged mechanism without any further manual or electrical charging.

The standard operating duty of a CB is as follows:

$$O - t - CO - t' - CO$$

where, O = Open

CO = Close–Open

$t' = 3$ min

$t = 15$ s for CBs not rated for rapid reclosing

$t = 0.3$ s for CBs rated for rapid reclosing

The standard operating duty of a breaker meant for auto-reclosing is as follows:

$$O - t_D - O,$$

where t_D is the dead time of the CB in cycles.

14.3 Function of High Rating Circuit Breakers

During the normal operation of a power system, the CB provides isolation between circuits and power supplies. Moreover, it is also used to enhance system availability by allowing more than one source to feed a load. The most basic function of a CB is to interrupt the circuit during short circuit and overload conditions.

With the fast development in the field of power systems, with the voltage level of high and extra high voltage levels, the CB has to be versatile under abnormal conditions. A CB is called upon to perform interruption under a widely varying system condition from no-load to full load upto fault current. The function of a CB is to open its contacts within a predetermined time period as fast as possible to limit the amount of energy diverted into any unnecessary path. This function must be performed from no-load current to the current interrupting capacity of CB. The CB has to perform its opening and closing operations from a small no-load transformer current upto the high short circuit current.

Further, it has to handle different types of currents such as the resistive load current, inductive short circuit current and capacitive current in unloaded lines. Moreover, the CB has to close under faulty conditions, during which severe arc with peak instantaneous fault current appears across the contacts before they physically touch. As a result, the material of contact melts before its closure. In addition, the CB has to perform the most important duties of circuit interruption in the following situations:

1. Terminal faults
2. Short line faults
3. Transformer magnetizing current and reactor currents
4. Energization of long transmission lines
5. Switching of unloaded transmission lines and capacitor banks
6. Out-of-phase switching

In each of these switching duties, the characteristics of the transient recovery voltage and the rate of rise of restriking voltage (RRRV) are mainly subjected to a number of system parameters. The influence of these parameters is described in Chapter 13 with reference to each duty.

14.4 Low Voltage Circuit Breakers

A CB is a connecting device that closes and breaks an electric circuit upto its ultimate breaking capacity. Although its main function is to break the short circuit and overload currents by self-energized action, it also breaks normal currents and overload currents by voluntary action from external sources. Moreover, after opening, it provides voltage insulation to the broken circuit. The purpose of this section is to analyse the functions, technologies, and performance of low voltage CBs such as switches, fuses, and MCBs.

14.4.1 Switches

A switch is an electrical device that can make or break an electrical circuit. It allows current to flow from one conductor to another or interrupts the current from the circuit. It is an electromechanical manually operated device with one or more sets of electrical contacts. Each set of contacts can be in one or two states: either *closed*, which means that electricity flows between them because of the touching of contacts, or *open*, meaning the contacts are separated and electricity does not flow. A manual switch can be operated by a human to supervise the electrical power flow in a circuit. However, automatically operated switches can be used to control various system parameters such as the motion of electrical machines.

Switches are normally composed of many contact elements, known as *poles*. These poles are operated simultaneously or in a prearranged sequence by a switching mechanism. Switches are often categorized by the number of poles and are referred to as *single pole, double pole, triple pole*, and so on. They can also be categorized by the number of possible switch positions per pole such as a single-throw or double-throw switch. Knife switches, rotary switches, contactors, and magnetic starters are often considered as *domestic* and *industrial switches*. The most useful switches are load-break switches (LBSs) and isolating switches in power systems at the highest voltages (several hundred thousand volts).

Load-break Switch

While supplying electrical power to the domestic applications and industrial equipment, switches are required to carry load current continuously without overheating. Further, it also provides enough insulation to the equipment from the supply.

Load-break switches are vital components of an electrical circuit as they allow the desired current flow required to the circuit continuously. Moreover, they are also capable of interrupting the load current during normal/full load conditions. Although this load interruption is easily carried out in low voltage and low current electrical circuits or equipment, specially designed arc interrupters are required for high voltage and high current circuits. The load-break switch is designed to carry a large amount of load current without overheating and provides sufficient insulation in its fully open position. In a circuit with more than a few thousands of volts, the LBS is equipped with arc interrupters to interrupt the load current. In medium and high voltage applications, the most popular interrupter is the air magnetic type, where the arc is driven into an arc chute by the magnetic field produced by the load current in a blowout coil. Sometimes, LBSs are used along with CBs because of their capability to hold the contacts in closed position during short circuit condition. During such combinations, the CB in the system interrupts the short circuit current, and the LBS remains in closed position against the electromagnetic forces produced by the blowout coil. The LBS, in conjunction with the high rupturing capacity (HRC) fuse, can tackle high fault current and offers very good protection against dead short circuit current upto 40 kA. The fault clearing time and isolation for this is of the order of a few milliseconds if proper selection of LBS and fuse is done.

Isolating Switch

Some switches are used to isolate electric power from a high voltage electrical network to ensure complete separation. The use of isolating switches (either alone or in combination with high-speed short circuit inter-rupting devices) makes it possible to simplify the arrangement of switching points and eliminates expensive CBs. In case of damage to certain sections of a network, the isolating switches are manually operated to disconnect the damaged section. An isolating switch must provide reliable disconnection in case of random occurrence of a short circuit in the network. Thus, a disconnector or isolating switch is used to make sure that an electrical circuit is completely de-energized either for service or for maintenance. Isolating switches are commonly used in domestic and industrial wiring to isolate the circuit at the time of maintenance and repair of home appliances or electrical equipment.

Such switches are often found in electrical distribution and industrial applications where the machinery must have its source of driving power removed from the supply for adjustment or repair. High voltage isolation switches are used in electrical substations to isolate ap paratus such as CBs, transformers, and transmission lines for maintenance. The major difference between an isolating switch and contactor or a CB is that an isolating switch is an off-load device intended to be opened only after the current has been interrupted by some other control device. Thus, the isolation switch is not intended for the normal control of the circuit and is only used for isolation.

14.4.2 Fuses

A short metal device that can melt when excessive current flows through it and isolates the circuit within sufficient time is known as a *fuse*. It was invented by Edison in 1890. This device finds major protection application in low and medium voltage circuits. Moreover, it is also used where frequent isolation of circuits is not desired and the installation of CB is not cost-effective. A fuse is inserted in series with the circuit to be protected. It carries normal load current continuously without generating excessive temperature. When the magnitude of current exceeds the level that results in the melting of fuse material due to temperature effect, the circuit breaks and disconnects from the mains. The fuse material is of low melting point and is highly conductive in nature, such as silver and copper. The melting time of the fuse element depends on the severity of fault current through it and the material used. The higher the value of current the smaller is the time required to melt the fuse link.

Fuse is a complete device that consists of a fuse casing (base) and a fuse link, where the fuse link is connected to the terminals. In case of short circuit, the fuse link blows out and protects the circuit.

Advantages
1. Unlike a CB, a fuse itself detects short circuit and operates automatically to interrupt the circuit.
2. It can operate without noise while breaking a large current. Further, it does not produce any smoke.
3. The inverse time–current characteristic (TCC) of the fuse provides faster operation than CBs of the same rating. The minimum operating time is below 5 ms.
4. It is the cheapest form of protection applied to low and moderate voltage circuits.
5. It requires no maintenance.
6. The size of fuse element is small and hence does not require large space.

Disadvantages

1. After every operation, the fuse element should be replaced, and this takes a long time.
2. It is difficult to achieve proper discrimination between the ratings of fuses connected in series.
3. The protection against small overcurrents (around 5% of overcurrent) cannot be achieved by fuses. Thus, a fuse gives comparatively poor protection against small overcurrents.

Materials used in Fuses

In general, lead, tin, copper, and silver are used as fuse elements. For a small value of current, say below 20 A, tin or an alloy of lead and tin is used as a fuse element. For a large value of current, copper and silver are preferred. Silver is the best choice for a fuse element as it offers high conductivity and low rate of deterioration due to oxidation. The following are the reasons for selecting silver as a fuse element even though it is expensive.

1. As the coefficient of expansion of silver is small, the fuse element can carry normal rated current continuously for a longer period.
2. As the conductivity of silver is very high, the mass of silver material required is less than that of other materials for a given rating of fuse element. This reduces the problems of clearing the mass of vapourized material at the time of its operation and allows faster operation.
3. Owing to a comparatively low specific heat, silver fusible elements can be raised from normal temperature to vapourization quicker than other fusible elements. Hence, the operation becomes much faster at a higher value of short circuit currents.
4. Silver vapourizes at a lower temperature than the temperature at which its vapour will rapidly ionize. Thus, the resistance of the arc path increases, and hence, short circuit current is quickly interrupted.

Properties of Fuse Elements

Fuse elements allow normal current without overheating, and disconnect the circuit when the current exceeds its normal value. During a high value of current, the fuse element heats up and blows out by melting itself. The fuse element is capable of performing this function satisfactorily. Further, it should have the following properties:

1. Low melting point; for example, tin, lead
2. High conductivity; for example, silver, copper
3. Free from deterioration due to oxidation; for example, silver
4. Low cost; for example, lead, tin, and copper

It has been realized that no material possesses all these properties. For example, lead has low melting point but it has high specific resistance, and hence, it is liable to oxidation. On the other hand, copper has high conductivity but it has high melting point compared to lead, and hence, it oxidizes rapidly. Since all these properties are contradictory to one another, it is necessary to maintain a balance amongst them when choosing a fuse element for a particular application.

Characteristics of Fuse

Fuse characteristics are categorized as *thermal* and *interrupting* characteristics. Thermal characteristics are comparatively more responsive and are related to the following factors:

1. Current rating
2. Melting characteristics

Interrupting characteristics depend on the following factors:

1. Voltage rating
2. Interrupting rating

Thermal characteristics As the magnitude of the current increases, melting time reduces. It is clear that larger magnitude of current leads to a higher power dissipation (I^2R) in the fuse and hence, a faster rise of temperature in the element. This would imply that the melting time of the fuse should be inversely proportional to the magnitude of the square of the current. The relationship between the magnitude of the current that causes melting and the time required for melting is given by the melting TCC of the fuse. The breaking of fuse element is primarily due to the thermal effect. It does not depend on mechanical forces and inertia. Thus, there is no limit on how short the melting time can be. This extremely small melting (fast operation) of a fuse at very high currents tends to distinguish it from other protective devices.

Very inverse melting characteristic Fuse melting time characteristic is usually described as very inverse melting characteristic. When the magnitude of current is small, the rate at which heat is generated in the element is low. As a result, the temperature of the element increases gradually. As the current increases, the melting time reduces at a rate which is more than the expected and inversely proportional to the rate of heat generation (I^2R). This is because the heat generated in the reduced cross section of the element cannot be removed as fast as it is produced. This gives the fuse a very inverse characteristic. At very short melting times, no heat is lost from the smaller cross section of the element.

Interrupting characteristics It is important to realize that power apparatus and systems contain inductive elements. Hence, melting of a fusing element is not sufficient to interrupt the current. Consequently, there is always some period of arcing before the current is completely interrupted. During this period, the fuse must withstand any immediate transient voltage condition and subsequent steady-state recovery voltage. Addition of melting time and arcing time gives the total clearing time. For lower values of current, the melting time is large and the arcing time is small because of the lower amount of energy stored in the inductive element $\left(\frac{1}{2}LI^2\right)$. In contrast, for a large value of current, the melting time is small and the arcing time is large. Hence, the TCC for the melting time and the total clearing time depart as $|I|$ increases.

Both these characteristics are required to coordinate with another fuse or overcurrent relay, or any other protective devices. The backup protective device must provide sufficient opportunity (time) to the primary fuse for clearing the fault. This ensures selectivity as it minimizes loss of service.

The fuse possesses inverse TCC. Thus, higher the magnitude of current through it lesser is the time required to blowout the fuse. The electrical equipment is designed to withstand some percentage of overloads (10–20% of normal rated current). There is no damage to the equipment as well as insulation during such overload condition (normal + overload = 120% of rated current). If the magnitude of the current exceeds this value, then heat generation increases, which results in a temperature upto the withstanding temperature of the insulation. The heat generated is proportional to $I^2 \times R \times t$. Thus, for a given value of current, certain time is required to reach upto the withstanding temperature for a given value of current. If the magnitude of current is high, then this time will be less. Therefore, the fuse used to protect the equipment must operate before the temperature reaches the withstanding limit of the protected equipment. Thus, the characteristic of the fuse must be below the thermal withstanding characteristic of the equipment to be protected.

Figure 14.2 shows the characteristic of fuse and thermal withstand characteristic of the equipment to be protected. It is the curve of operating time versus current. The selection of fuse depends on the overload

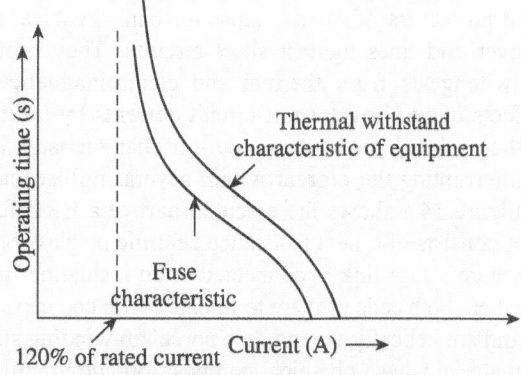

Fig. 14.2 Fuse characteristic

capability of the equipment, and it does not operate under normal as well as for allowable overload conditions. The speed at which a fuse blows depends on how much current flows through it and the material from which the fuse is made. In general, a standard fuse requires twice its rated current to open in 1s. On the other hand, a fast-blow fuse requires twice its rated current to blow in 0.1s, whereas a slow-blow fuse requires twice its rated current for tens of seconds to blow. During a heavy short circuit, the fast-blow fuse must operate within the duration of the order of 5 ms (prescribed time as per its characteristic).

Kit-Kat Fuse

In a low voltage circuit, the magnitude of current drawn by the domestic appliances as well as by the various industrial equipment is only a few tens of amperes. The protection of such low current circuit can be achieved by the kit-kat fuse, which is also known as a *semi-enclosed rewirable fuse*. The design of the kit-kat fuse includes a base and a fuse carrier. Figure 14.3 shows the pictorial representation of a kit-kat fuse. The base is made of porcelain and accommodates fixed contacts at which the

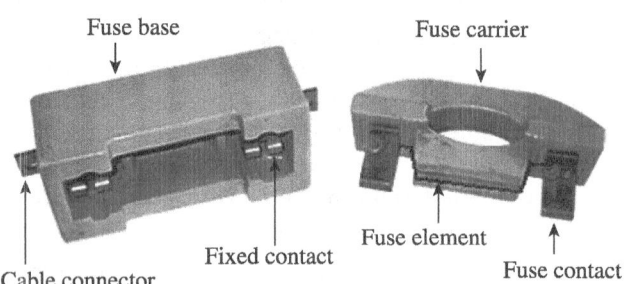

Fig. 14.3 Kit-Kat fuse

incoming wire and the outgoing wire are connected. The fuse carrier is also made of porcelain and holds the fuse element between its terminals. The fuse carrier can be inserted into the base and taken out from the base as per the requirement. During a short circuit, the fuse element in the carrier is blown out and the faulted circuit is disconnected. The blown out fuse element is replaced by a new one in the fuse carrier and again inserted in the base to restore the supply. The fuse element is wired on the inner side of the fuse carrier. Hence, at the time of replacement of the blown out fuse, it permits complete isolation of the live wire through the porcelain material. Moreover, the cost of replacement is very small. On the other hand, the renewal of the fuse element takes some time, which leads to the possibility of improper size of the fuse element.

The fuse element is of tinned copper wire with low breaking capacity. It cannot be used for a circuit with high fault level. Owing to its low breaking capacity, the applications of kit-kat fuses are limited to domestic and lighting loads.

High Rupturing Capacity Fuse

The low and uncertain breaking capacity of a kit-kat fuse is overcome by the HRC fuse. They are used to protect transformers, capacitor banks, cables, and overhead lines against short circuits. They protect switchgears from thermal and electromagnetic effects due to heavy short circuit currents by limiting the peak value of current (cut-off characteristic) and interrupting the current within several milliseconds. Figure 14.4 shows the essential part of a HRC fuse. It consists of a heat resistance ceramic or glass body where a fuse link is connected to an insulation tube,

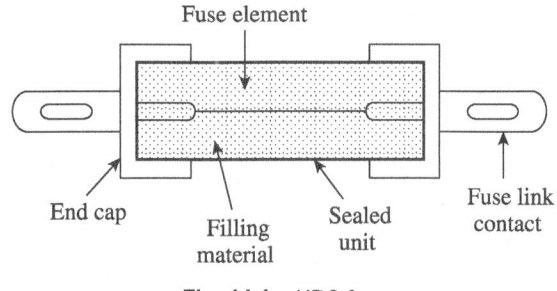

Fig. 14.4 HRC fuse

where both ends of it are terminated with end caps. Fuse elements are made from specially profiled silver wire and are helically wound on a porcelain winding stick. The interior of the fuse is filled with an arc-quenching material whose chemical composition and granularity are appropriately chosen. The filling material can be chalk, plaster of Paris, or quartz, which can also act as a cooling medium.

Under normal operating conditions, the temperature of the fuse element is within the tolerable limit, that is, below its melting point. Thus, it allows normal load current continuously without overheating. On the other hand, the operation of fuse depends on the automatic interruption of the fault current by melting the fuse element and quenching an arc produced in the fuse link interior. Similar to the kit-kat fuse, the HRC fuse has to be replaced after each operation. The following are the main features of the HRC fuse.

1. It has low to high short circuit current-limiting capability.
2. It requires very less maintenance.
3. It has low switching overvoltages.
4. It can be used with switch disconnectors.

Expulsion Type Fuse

The expulsion type fuse is used where expulsion gases cause no problem as in overhead circuits and equipment. These fuses can be termed as *current awaiting*, and the function of the interrupting medium is similar to that of an AC circuit breaker. They are used for high voltage systems upto 33 kV. Thus, considering

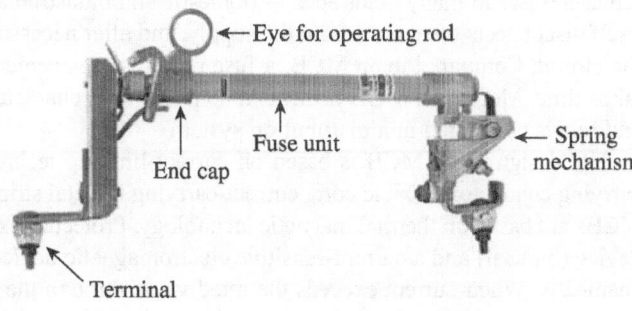

Fig. 14.5 Expulsion type fuse

its operation in a high voltage network, the design of expulsion fuse is either pole mounted or of the outdoor type. Figure 14.5 shows the pictorial view of the expulsion fuse. They comprise tin or copper fuse elements in series with a flexible braid in a tube. The tube forms one side of a triangle, with a latched connection at the top and a hinge at the bottom. The braid emerges from one end of the fuse link and is held in tension by a spring. When the element melts, the braid is no longer under tension; the latch is released and the fuse swings downwards under gravity, breaking the circuit. As the fuse swings downwards, the arc is lengthened, extinguished, and prevented from restriking. The temperature of an arc is of the order of 4000–5000 K. At this temperature, special materials located close to the fuse element rapidly create gases. Preferred gas generating materials are fibre, melamine, and boric acid. These gases help to create a high pressure turbulent medium surrounding the arc. When the current reaches a natural zero, the arc channel reduces to a minimum. Hence, the generated gases rapidly mix with the remaining ionized gas, thereby deionizing them as well as removing them from the 'arc area'. In sequence, this leads to a rapid buildup of dielectric strength that can withstand the transient recovery voltage and steady-state power system voltage.

Liquid Quenched Fuse

Liquid quenched fuses (LQFs) are based on early non-current limiting fuse links that use a liquid to quench the arc. The LQF is used for a high voltage system upto 132 kV and particularly, for the circuit with a rated current of 100 A and breaking capacity of the order of 6 kA. Figure 14.6 shows the various parts of the LQF. The LQF consists of glass tubes filled with liquid and sealed with an end cap. The fuse element is anchored to the top ferrule of a glass tube filled with a quenching liquid—usually a hydrocarbon or carbon tetrachloride. The rest of the tube is filled with a spring that holds the element in tension. Sometimes, a strain wire may be used in very high

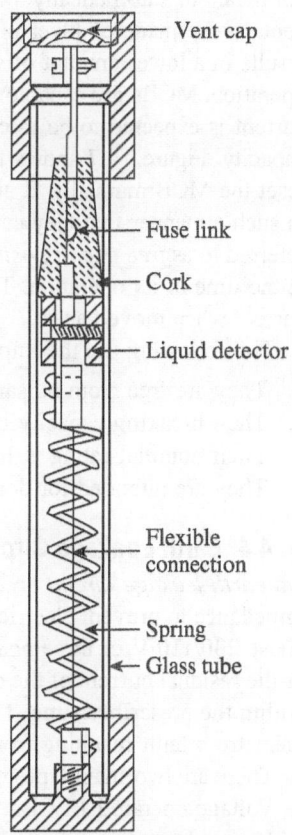

Fig. 14.6 Liquid quenched fuse

voltage fuse links. The fuse elements blow out when the current exceeds the prescribed limit. When the element melts, the spring pulls the two parts of the element apart, extending the arc and quenching it in the liquid.

14.4.3 Miniature Circuit Breakers

Miniature circuit breakers are the latest development in CB technology for low voltage systems. They have replaced fuses in many fields such as domestic and industrial applications. In an abnormal situation, the MCB itself disconnects the circuit from the supply, and after necessary repairs, one has to just switch it *on* to restore the circuit. Compared to an MCB, a fuse requires replacement of the fuse element, and restoring the circuit takes time. Moreover, MCB presents a better distinct characteristic that can be easily coordinated with fuses and relays connected in a distribution system.

The design of an MCB is based on current-limiting technology. The design of MCB includes a current-carrying conductor, flexible cord, current-carrying bimetal strip, contacts, arc chute, and a tripping mechanism. MCBs are based on thermal magnetic technology. Protection is provided by combining a temperature-sensitive device (bimetal) and a current-sensitive electromagnetic device. Both components trigger the mechanism mechanically. When current exceeds the rated value through the bimetal, because of the temperature generation by heating, it bends and triggers the tripping mechanism to trip the MCB. At the time of tripping, the arc chute splits the arc. Thus, the arc is quenched by the lengthening and cooling phenomena. The tripping mechanism is not meant to automatically switch on the MCB. It requires manual operation while supplying the circuit. The contacts are made of silver inlaid copper, which ensures longer life. Further, they have low resistance, which results in a lower amount of wattage loss. The contacts are designed to have zero bounce during the closing operation. MCBs are capable of handling fault current upto 10 kA. However, for installations where the fault current is expected to be more than 10 kA, a backup can be achieved by using HRC fuse links of suitable capacity. Figure 14.7 shows the schematic view of an MCB. Here, an operating handle is used to trip and reset the MCB manually. It also indicates the status of the CB (on or off/tripped). Most MCBs are designed in such a manner that they can trip even if the handle is held or locked in the *on* position. This is sometimes referred to as *free trip* or *positive trip* operation. The actuator mechanism forces the contacts together or apart at the time of its operation. The main contact of the MCB allows the current while touching and breaks the current when moved apart.

The following are the important features of the MCB.

1. They are free from nuisance tripping caused by vibrations.
2. Their breaking capacity is as per the requirements (below 10 kA on average).
3. Their nominal rating is 1.5–125 A, according to the loads to be supplied.
4. They are intended for domestic applications.

14.4.4 Earth Leakage Circuit Breakers

An *earth leakage circuit breaker* (ELCB) is a safety device used in electrical installations with high earth impedance to prevent electric shocks. It is used for protection against electrical leakage in circuits of single phase 230/110 V or three phase 415/440 V rated current upto 60 A. When somebody gets an electric shock or the residual current of the circuit exceeds the prescribed limit, the ELCB automatically cuts off the power within the prescribed time. Thus, it protects the human body against electric shock and prevents the equipment from fault resulting from the residual current.

There are two main types of ELCB, namely voltage-operated ELCB and current-operated ELCB.

Voltage-operated ELCBs were introduced in the early twentieth century. These ELCBs have been in widespread use since then, and many are still in operation. Current-operated ELCBs are generally known as residual current devices (RCDs). Though the details and methods of their operation are different, they provide protection against earth leakage. When the term ELCB is used, it usually means a voltage-operated device. Similar devices that are current operated are called RCDs.

The connection of earth circuit is modified when an ELCB is used. The connection to the earth rod is passed through the ELCB by connecting to its two earth terminals. One terminal goes to the installation earth circuit protective conductor (CPC) and the other to the earth rod (or sometimes other type of earth connection). Thus, the earth circuit passes through the sensing coil of ELCB. It is a specialized type of latching relay. The incoming mains power of a building is connected through its switching contacts so that the ELCB disconnects the power during an earth leakage (unsafe) condition. The ELCB detects fault currents from live to the earth (ground) wire within the installation. If sufficient voltage appears across the sensing coil of ELCB, it will switch off the power, and remains *off* until the resetting operation is carried out manually.

On the other hand, a current-operated ELCB measures differential (residual) current using the core balance transformer and trips a switching device through an electromagnetic tripping relay. Figure 14.8 shows the essential connection of such an ELCB for single-phase and three-phase supplies.

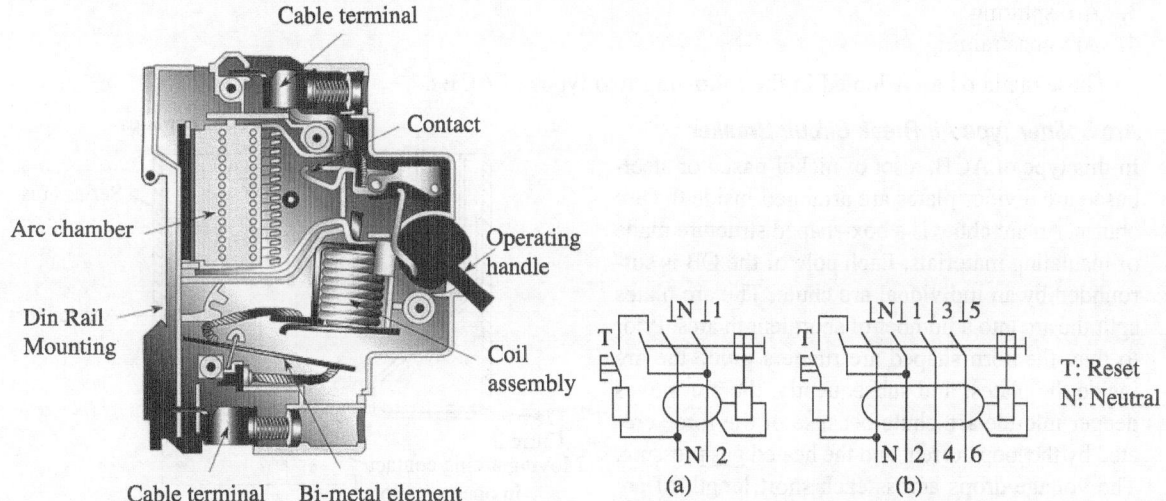

Fig. 14.7   Cross-sectional view of MCB

Fig. 14.8   Connection of current-operated ELCB
(a) Single-phase supply (b) Three-phase supply

14.5 High Voltage Circuit Breakers

Since the beginning of electric energy distribution, low voltage CBs have been used. Later on, the development of high voltage heavy-duty power circuit has necessitated the invention and implementation of high voltage CB. The first CBs were very simple. It was open air and hand-operated, but soon after, the arc-quenching mechanism was developed. Early plain break type CBs used oil to quench the arc. In as early as 1902, a CB design using compressed air was introduced because it had comparable quenching characteristics. In addition, as early as the 1920s, vacuum was investigated as an arc-quenching medium. However, in those days, industrial processes were incapable of producing a bottle that could maintain a vacuum over an extended period of time and for a number of switching operations. In 1950, sulphur hexafluoride (SF_6) was tested for its quenching properties as this gas had been known for its excellent dielectric properties. It was found to be superior over oil and air. Therefore, SF_6 has now virtually taken over the entire high voltage range of applications and for the extra high voltage and ultra high voltage range. This section deals with the currently installed high voltage CBs such as the air break circuit breaker (ACB), oil circuit breaker, air blast circuit breaker, SF_6 circuit breaker, and vacuum circuit breaker.

14.5.1 Air Break Circuit Breaker

The ACB is used for low voltage to medium voltage distribution systems. The process of arc interruption in ACB is achieved by natural deionization of the arc or by cooling action. The elongation of the arc and its cooling increases the resistance of the arc at the time of short circuit interruption, which reduces the angle between the voltage and the short circuit current. Thus, in an ACB, the current is interrupted at zero value with low recovery voltage. On the other hand, the energy dissipated in the arc is very high, and this prohibits the use of ACB in high voltage application above 12 kV. It is used only for low and medium power AC and DC systems.

In order to increase the effective resistance of an arc in an ACB, the following methods are used.

1. Arc lengthening
2. Arc cooling
3. Arc splitting
4. Arc constraining

These methods are adopted in the following two types of ACBs.

Arc Splitter Type Air Break Circuit Breaker

In this type of ACB, a set of nickel-based or steel-based arc divider plates are arranged inside the arc chutes. An arc chute is a box-shaped structure made of insulating materials. Each pole of the CB is surrounded by an individual arc chute. The arc plates split the arc into a number of short length arcs. Prior to that, the horn-shaped arc runners guide the arc inside the plates, and subsequently, the arc moves deeper into the arc chute because of the force created by the loop current and the heated gas pressure. The voltage drops across each short length of arc depend on the spacing between the arc plates. The sum of anode and cathode drops of all short arcs in series is more than the circuit voltage to be interrupted. Moreover, when the arc comes in contact with the cooled surface of the arc chute, it gets rapidly and effectively cooled. Thus, a condition for quick extinction of the arc is automatically established. Figure 14.9 shows the arc splitting arrangement in an ACB. The arc splitter type ACB is generally used for low voltage applications.

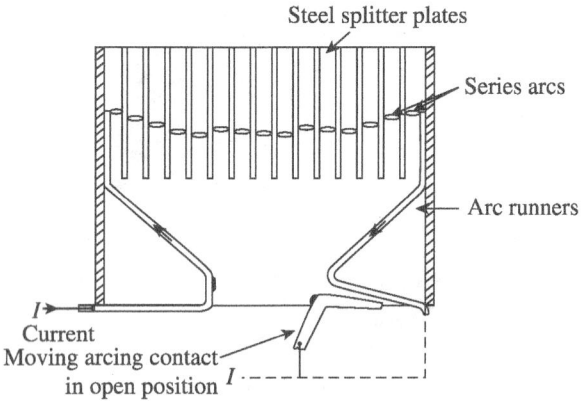

Fig. 14.9 Arc splitter type ACB

Magnetic Blowout Type Air Break Circuit Breaker

This type of ACB has similar construction as the arc splitter type. The arc chute used in a magnetic blowout type circuit is made of ceramic material. In this type of ACB, arc extinction is carried out through a magnetic blast. A magnetic blowout coil is connected in series with the circuit to be interrupted, which sets up a magnetic field. Owing to the magnetic field set-up, the arc is magnetically blown out. Figure 14.10 shows an outline of the magnetic blowout type ACB. The arc is

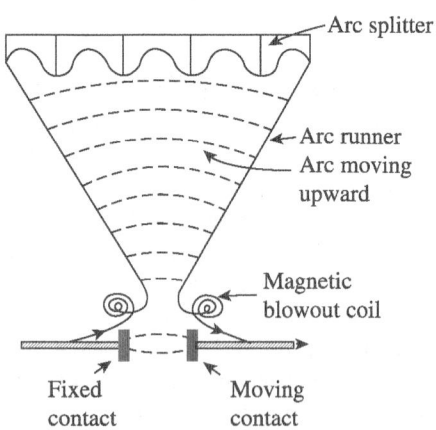

Fig. 14.10 Magnetic blowout type ACB

initially moved towards the arc chute by the arc runner and magnetically blown deeply into the arc chute, where it is lengthened, cooled, and finally extinguished. In this type of ACB, the arc is constricted as it travels through the slots in the arc plates, and it fills the narrow space between the plates. Thus, the arc breaking phenomenon becomes more efficient with heavy current. These types of magnetic blowout ACBs are intended for medium voltage application upto 16 kV and fault current of the order of 50 kA.

General construction of ACB In an ACB, contact separation and arc extinction take place at atmospheric pressure. When the contacts are opened, an arc is drawn between them. The arc resistance is increased to such an extent that the system cannot maintain the arc, and thus, the arc is finally extinguished.

An ACB consists of two contacts, namely main contact and arcing contact. The *main contact* conducts current in the closed position. It has low contact resistance and is silver plated. The main contact consists of a moving contact and a fixed contact assembly. The arcing contact is hard, heat resistant, and made of copper alloy. While opening, the main contact is dislodged first and the current is shifted to arcing contacts, and then they are dislodged. Thus, the arc is drawn between the arcing contacts. The arc is now forced by electromagnetic forces and thermal action. The ends of the arc move along the arc runners, and they are divided by the arc splitter plates in the arc chute. The dimension of the arc chute depends on the number of arcing contacts. Hence, the arc is extinguished by lengthening, cooling, and splitting. The operating mechanism is usually designed for manual operation. Further, it is a trip-free operation.

14.5.2 Oil Circuit Breaker

Oil CBs are widely used for a medium voltage system. This CB utilizes the insulating properties of oil for arc extinction as the oil has dielectric strength approximately seven times that of air at atmospheric conditions. When the contacts of the CB open, the energy of the arc cracks the oil molecules. This phenomenon generates hydrogen gas at a very high pressure. The high volume and pressure of hydrogen gas sweeps, cools, and compresses the arc column, and thus, the arc is de-ionized. Hence, in the oil CB, arc quenching is achieved in two ways:

1. Hydrogen gas reduces the ionization process by cooling the arc.
2. The gas sets up turbulence in the oil, which forces the oil into the space between the contacts; hence, it drives the arc outside and away from the generating source.

As a result of these two processes, the arc is extinguished and the large short circuit current is interrupted.

Advantages of oil The following are the advantages of oil as an arc-quenching device:
1. It provides insulation between the live parts and the earthed portion of the containers.
2. It produces hydrogen gas during arcing, which helps in extinguishing the arc.
3. It provides insulation between contacts after the arc has been finally extinguished.

Disadvantages of oil The following are the disadvantages of oil as an arc-quenching device:
1. The decomposed products of dielectric oil are inflammable and explosive.
2. Oil absorbs moisture. Thus, its dielectric strength reduces by carbonization, which occurs during arcing. Therefore, oil CB requires replacement and maintenance.
3. It is not suitable for repeated operations owing to the deterioration of oil.
4. Oil leakage, loss, replacement, and purification are troublesome.

Oil CBs can be classified into the following types:
1. Bulk oil circuit breakers (BOCB): Such CBs are classified into two categories.
   (a) Plain break oil CB
   (b) Arc control CBs

Arc control CBs are further classified into two categories:

(i)   Self-blast oil CBs

(ii) Forced blast oil CBs

2. Minimum oil circuit breaker (MOCB)

Bulk Oil Circuit Breakers

Bulk oil circuit breakers are enclosed in the metal-grounded weather proof tanks, which are referred to as *dead tanks*. The original design of BOCB was very simple and inexpensive. They are suitable only for voltages upto 33 kV; as for higher voltages, the size becomes too large. Thus, this type of breaker is rapidly being replaced by other types of breakers. It has three separate tanks for capacities of 72.5 kV and above. For capacities of 33 kV and below, single tank construction is used. It uses oil (transformer oil) as a dielectric medium for the purpose of arc extinction. In order to assist in arc extinction process, the arc control devices are fitted with the contact assembly. Construction and venting of arc control devices is such that the gases flow axially or radially with respect to the arc.

Plain break oil circuit breaker In plain break oil CBs, a huge quantity of oil is used. In this CB, oil quenches the arc column and provides insulation between the live current conducting parts and the earthed container of the CB. The contacts of this CB are separated under the whole of the oil in the closed chamber. This type of CB is not provided with any kind of arc control devices as the aim of this breaker is to increase the length of the arc column by separating the contacts at a larger distance. The process of arc destruction occurs only when a significant gap is reached between the contacts. Figure 14.11 shows the double break plain oil CB.

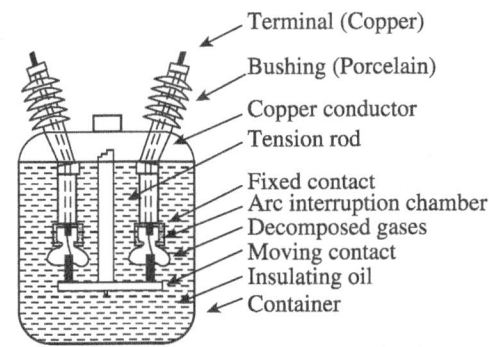

Fig. 14.11   Double break plain break oil circuit breaker

In a conventional dead tank, an extended contact passes through an arc interrupter chamber, and its upper end is engaged with a stationary contact (fixed contact) at the top of the chamber. The interrupter chamber is totally immersed in oil, which is carried in a grounded tank. The lower end of the elongated contact is moved between an engaged and a disengaged position within the interrupter chamber by means of an external insulating tension rod connected to the operating mechanism of the breaker. Under normal operating conditions, the fixed and moving contacts remain in a closed condition, and the CB carries the normal circuit current. During interruption, the conductive crossbar moves down, and the top of the movable, elongated, or bayonet contact leaves the stationary contact within the interrupter, and an arc is drawn between the separating contacts. The arc extinguishing process causes the creation of hot exhaust gases that vent through ports in the interrupter chamber and into the large bulk of oil within the grounded tank. Moreover, the rapidly expanding gas bubble displaces large quantities of oil, resulting in substantial reaction forces causing strain on many breaker parts, such as the tank and bushings, which support the interrupters. However, the hydrogen bubble generated around the arc cools the arc and helps in deionizing. Moreover, these gases set up turbulence in the oil, which helps in

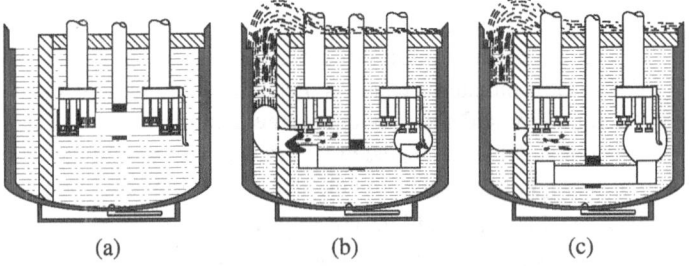

Fig. 14.12   Bubble of hydrogen surrounding an arc drawn under oil (a) Breaker closed (b) Breaker opening (c) Interruption

eliminating the arcing products from the arc path. This action can be observed from Fig. 14.12. Initially, the breaker is in a closed condition during a normal load as shown in Fig. 14.12(a). When a fault occurs, the moving contact departs from the stationary contact as shown in Fig. 14.12(b). At this time, an arc is drawn between the two contacts, which creates high volume of hydrogen gas bubble. This bubble thrusts the oil to sweep away from the arc-producing devices and sets up turbulence forces. Therefore, the oil surrounding the bubble conducts the heat away from the arc and thus contributes to the deionization of the arc. Finally, the dielectric strength of the medium increases because of the lengthening of the arc, and it extinguishes because of the formation of hydrogen gases as shown in Fig. 14.12(c).

Disadvantages

1. Interruption of arc requires a larger separation of switching contacts because of the absence of arc control devices.
2. The arc persists for a long period of time, and thus, it is not suitable for high-speed interruption.

Owing to these disadvantages, plain break oil CB is used in low capacity installations and in low voltages upto 11 kV systems.

Arc control circuit breakers As the external arc control devices are not used in plain oil CBs, interruption requires elongation of the arc. With the aid of an artificial arc control device, the final arc extinction can occur with a shorter gap between contacts during arc quenching. These breakers are called *arc control circuit breakers*. There are two types of such breakers.

Self-blast oil circuit breakers In self-blast oil CBs, arc control is provided by internal means, that is, the arc itself facilitates its own extinction efficiently. Arc-controlled oil breakers have an arc control device surrounding the breaker contacts. The purpose of the arc control device is to improve the operating capacity, speed up the extinction of the arc, and decrease the pressure on the tank. The arc control devices can be classified into two groups: cross blast and axial blast interrupters.

In cross blast interrupters, the arc is drawn in front of several lateral vents as shown in Fig. 14.13. The gas formed by the arc causes high pressure inside the arc control device. The arc is forced to blow into the lateral vents in the pot, which increases the length of the arc and shortens the interruption time. When the moving contact uncovers the arc splitter ducts, fresh oil is forced across the arc path. The arc is therefore driven sideways into the arc splitter, which increases the arc length and finally, the arc extinguishes.

Axial blast interrupters use a similar principle as that of cross blast interrupters as shown in Fig. 14.14. However, the axial design has a better dispersion of the gas from the interrupter. In an axial blast interrupter,

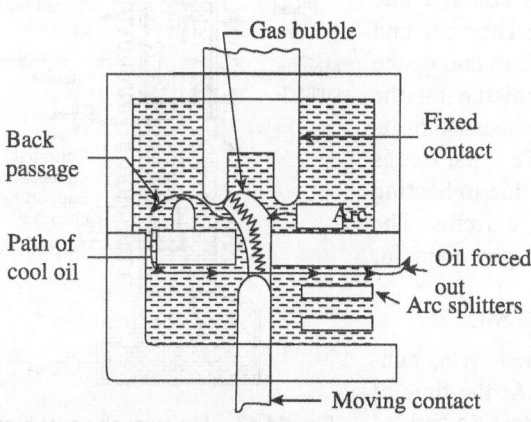

Fig. 14.13 Cross blast circuit breaker

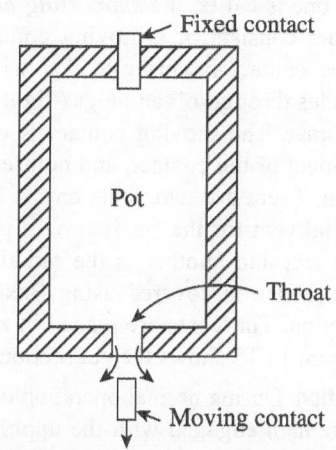

Fig. 14.14 Axial blast circuit breaker

the moving contact is a cylindrical rod passing through a restricted opening called *throat* at the bottom. When a fault occurs, the contacts get separated and an arc is struck between them. The heat of the arc decomposes the oil into a gas at a very high pressure in the pot. This high pressure forces the oil and gas through and around the arc to extinguish it. A limitation of this type of pot is that it cannot be used for very low or very high fault currents. With low fault currents, the pressure developed is small, which increases the arcing time. In the case of high fault currents, the gas is produced so rapidly that the pot may burst because of high pressure. So, this pot is used for moderate short circuit currents only where the rate of gas evolution is moderate.

Forced blast oil circuit breakers In this CB, an arc control device is provided to the CB externally by mechanical means. There is a piston attached to a moving contact. When a fault occurs, the moving contact moves, and hence, the piston associated with it also moves, producing pressure inside the oil chamber. So, the oil gains mobility or tabulates and quenches the arc.

Disadvantages of bulk oil circuit breakers
1. A large quantity of oil is necessary, although only small quantity of oil is required for arc extinction.
2. The entire oil deteriorates because of sludge formation close to the arc thus needing replacement.
3. The tank sizes for capacities of 33 kV and above become too large, expensive, and unmanageable.

Minimum Oil Circuit Breaker

A minimum oil circuit breaker uses a minimum amount of oil. In such CBs, oil is mainly used for arc extinction. The current conducting parts are insulated by air or porcelain or organic insulating material. For voltage upto 33 kV, MOCBs are generally enclosed in the draw out type metal clad switch gear. For 33–145 kV, MOCBs are outdoor type with one interrupter per pole and a single operating mechanism for the three poles. For 245 kV and above, modular construction is used. In this, twin interrupter units are connected in series in 'T' or 'Y' formation. Current interruption takes place inside the *interrupter*. The enclosure of the interrupter is made of insulating material such as porcelain. Hence, the clearance between the live parts and the enclosure can be reduced, and lesser quantity of oil is required for internal insulation.

In an MOCB, the current interruption devices are conventionally mounted in live tank housing where the housing is at high potential above the ground at the top of a porcelain column. Hence, there are two chambers in a low oil CB. The oil in each chamber is separated from each other by an insulating partition. The main advantage of this breaker is that it requires less oil, and the oil in the second chamber does not get polluted. The upper chamber is called the *circuit-breaking chamber* and the lower one is called the *supporting chamber*. The circuit-breaking chamber consists of a moving contact and a fixed contact. The moving contact is spring-loaded within the arcing chamber and protrudes through oil and a gas-tight seal at the bottom end of the interrupter. The moving contact is connected to a piston for the movement of the contact, and no pressure builds up because of its motion. There are two vents on the fixed contact. The first one is the axial vent for the small current produced in oil due to heating of the arc, and another is the radial vent for large currents. The whole device is covered using Bakelite paper and porcelain for protection. The vents are placed in a turbulator.

Figure 14.15 shows the cut section of a single pole MOCB.

Operation During normal operation of the MOCB, the moving contacts remain engaged with the upper fixed contact. At the time of fault, the tripping mechanism pulls the moving contact downward

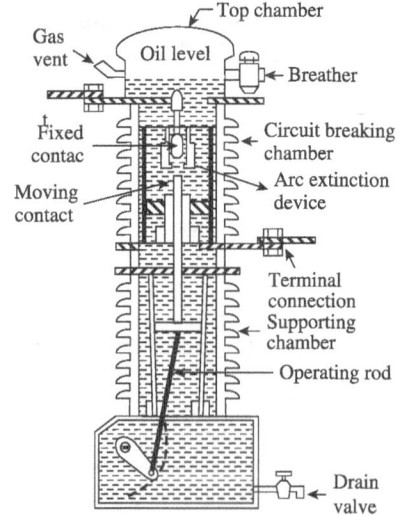

Fig. 14.15 Minimum oil circuit breaker

and an arc is struck between the switching contacts. High pressure is generated by the expanding gases created from the oil because of the arcing in the interrupting chamber. This gas pressure is relieved by allowing the gas to vent axially, across the arc, and to the external regions. These interrupters do not encounter high forces perpendicular to the axis of the turbulator because of the jet reaction force, which accompanies the operation of the bulk oil immersed interrupter described earlier. Figure 14.16 shows the operation of MOCB for the axial flow of oil and cross flow of oil during arc interruption under small and heavy fault conditions. The process of turbulation is an orderly one, where the sections of the arc are successively quenched by the effect of separate streams of oil moving across each section in turn and bearing away its gases.

Advantages

1. It requires less quantity of oil.
2. It requires smaller space.
3. There is reduced risk of fire.
4. Maintenance problems are reduced.

Disadvantages

1. Owing to the use of smaller quantities of oil, the degree of carbonization is increased.
2. It is difficult to remove the gases from the contact space in time.
3. The dielectric strength of oil deteriorates rapidly because of a high degree of carbonization.

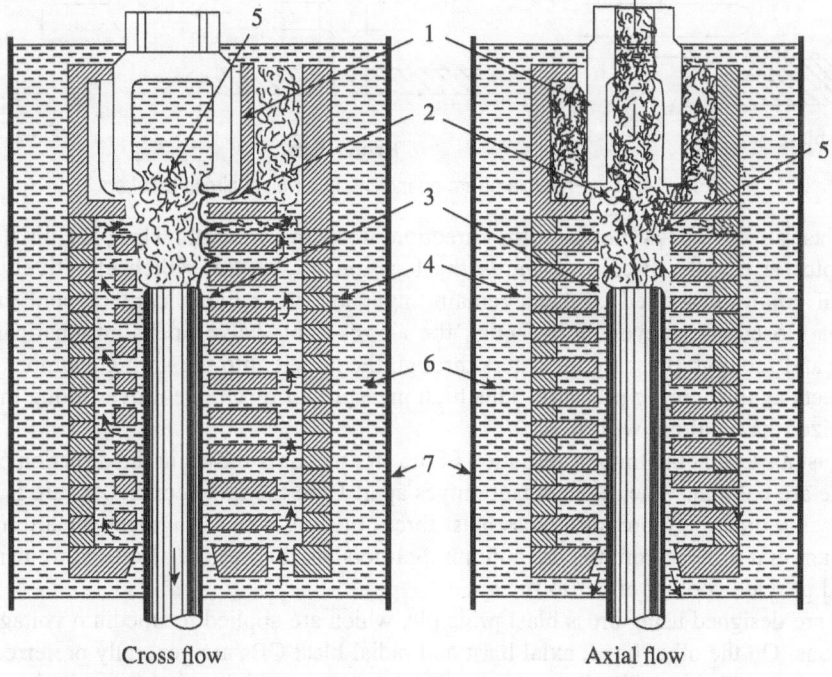

Cross flow   Axial flow

1. Fixed contact assembly   2. Arc
3. Moving contact with tungsten-copper tip   4. Fiber reinforced tube
5. Gases evolved by decomposition of oil   6. Dielectric oil
7. Outer enclosure (Porcelain)

Fig. 14.16 Axial flow of oil and cross flow of oil during arc interruption

14.5.3 Air Blast Circuit Breaker

Air blast CBs have been in use since 1940 and are still in operation throughout the complete range of high voltage. Air blast CB employs high pressure air for arc extinction. In its design, the interruption process is started by the establishment of arc across the contacts of the breaker, and at the same time, a flow of air blast is established by the opening of the pneumatic valve. This high pressure air cools the arc and sweeps away the arcing products to the atmosphere. This process rapidly increases the dielectric strength of the medium between the contacts and prevents re-establishment of the arc. Hence, the arc is extinguished and the flow of current is interrupted. Depending on the direction of the air blast in relation to quenching of the arc, air blast CBs are classified into the following types:

1. Axial blast type
2. Cross blast type
3. Radial blast type

Figure 14.17 shows the principle of arc quenching in axial, radial, and cross blast type air blast CBs.

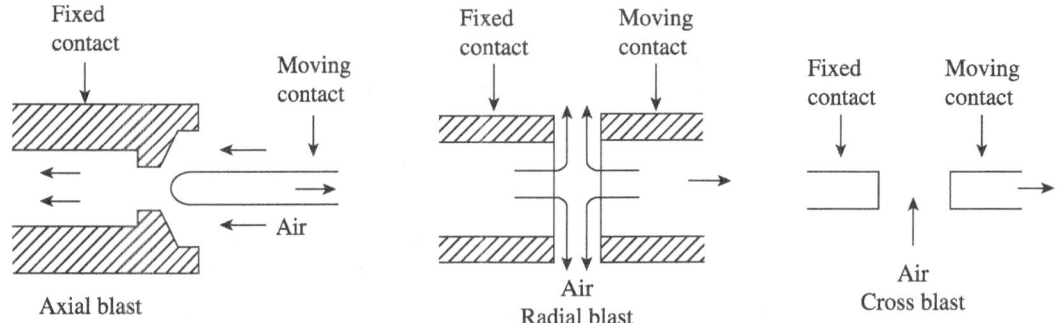

Fig. 14.17 Principle of arc quenching in air blast circuit breakers

In axial blast air CB, air is blown in the direction along the arc path. Thus, the air flow in the axial blast interrupter is exactly directed towards the location of the arc, which effectively cools the arc. Under normal conditions, fixed contacts remain engaged with moving contacts and the air reservoir valve remains closed. Whenever fault occurs, the air reservoir valve opens automatically by the tripping impulse and the high pressure air enters through the nozzle into the arcing chamber and pushes the moving contacts against spring pressure. This high pressure air discharge moves along the arc and takes away the ionized gases along with it.

In cross blast air CB, at the time of initiation of the arc, the air is blown in the direction perpendicular to the axis of the arc column. When the contact moves apart from the fixed contact, an arc is struck between them. At the same time, high pressure cross blast forces the arc, which lengthens the arc rod and removes considerable amount of heat from the arc column. Splitters and baffles are also used to increase the length of the arc and provide cooling effect to the arc.

Most CBs are designed using cross blast principle, which are applied for medium voltage and high current applications. On the other hand, axial blast and radial blast CBs are generally preferred for extra high voltage applications. Air blast CBs find wide applications in very high voltage installations. Most CBs are designed for 132 kV and 220 kV voltage systems. Figures 14.18 and 14.19 show the operating principle of axial and cross blast CBs, respectively.

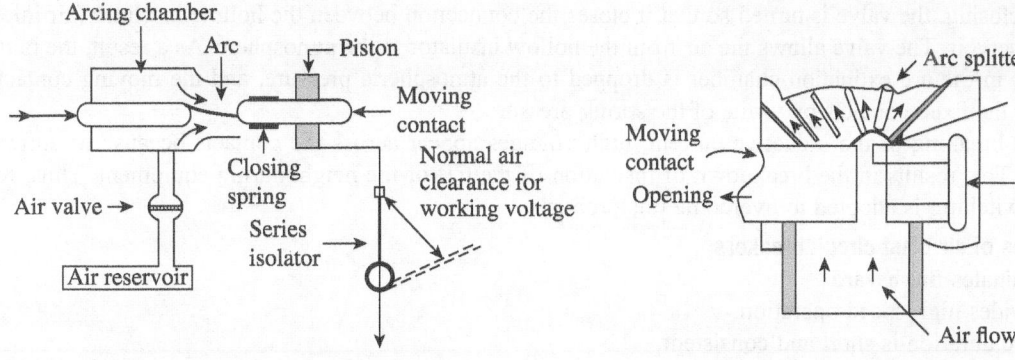

Fig. 14.18 Axial blast circuit breaker   **Fig. 14.19** Cross blast circuit breaker

Construction

High pressure air at a pressure of 20–30 kg/cm^2 is stored in the air reservoir. Three hollow insulator columns are mounted on the reservoir with valves at their bases. The double arc extinguishing chambers are mounted on top of the hollow insulator assembly. This serves as a support to the main arcing chamber and also allows the air to flow from the reservoir to the arcing chamber. Figure 14.20 shows the schematic diagram of one pole of air blast CB. The current-carrying parts connect the three arc extinction chambers to each other in series and the pole to the adjacent equipment. There are two breaks per pole. Each arc extinction chamber consists of one twin fixed contact and two moving contacts.

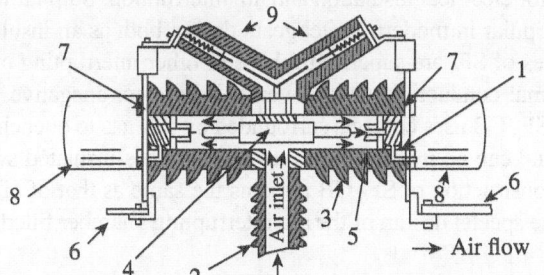

1. Port
2. Hollow insulator
3. Double arc extinction chamber
4. Fixed contact
5. Moving contact
6. Connection for current
7. Compressor spring
8. Air outlet opening
9. Resistance switching unit

Fig. 14.20 Schematic diagram of air blast circuit breaker

Operating Principle

The operating mechanism operates the rod when it gets a pneumatic or electric signal. The valves open to send high pressure air into the hollow insulator. The high pressure air rapidly enters the double arc extinction chamber. As the air enters into the arc extinction chamber, the pressure on the moving contacts becomes more than the spring pressure, and the contacts start opening. Hence, the contacts are separated and the air blast takes away the ionized gases along with it and assists in extinction. At the end of the contact travel, the port for outgoing air is closed by the moving contact. After a few cycles, the arc is extinguished by the air blast, and the arc extinction chamber is filled with high pressure air (30 kg/cm^2). The high pressure air has a higher dielectric strength than the atmospheric pressure. Hence, a small contact gap of a few centimetres is enough to quench the arc column.

While closing, the valve is turned so that it closes the connection between the hollow insulator (air inlet) and the reservoir. The valve allows the air from the hollow insulator to the atmosphere. As a result, the pressure of air in the arc extinction chamber is dropped to the atmospheric pressure, and the moving contacts close over the fixed contacts by virtue of the spring pressure.

For the breaking of the inductive current, high voltages appear across the contacts because of current chopping. This results in the breakdown of insulation of the CB or the neighbouring equipment. Thus, resistance switching is adopted to overcome this problem.

Advantages of air blast circuit breakers
1. It eliminates fire hazard.
2. It provides high-speed operation.
3. The arc duration is short and consistent.
4. Maintenance of the interrupter is less.
5. The size of the device is comparatively lower than that of oil and air CBs.
6. The arcing time is very small because of the rapid buildup of dielectric strength between the contacts.

Disadvantages of air blast circuit breakers
1. Regular maintenance of compressor plant is required.
2. Air leakages can occur at the pipeline fittings.
3. It is sensitive to variations in the rate of rise of restriking voltage.
4. Its operation is noisy.

14.5.4 Sulphur Hexafluoride Circuit Breaker

Most modern high voltage CBs use SF_6 gas for electrical insulation and arc interruption. Sulphur hexafluoride is an inert insulating gas that is increasingly popular in modern switchgears design both as an insulating medium and an arc-quenching medium. The properties of SF_6 are superior to those of other interrupting mediums, such as its high dielectric strength and high thermal conductivity. Moreover, it is an electronegative gas and has a strong tendency to absorb free electrons. As SF_6 CB uses contacts surrounded by SF_6 gas to quench the arc. They are most often used at transmission levels and can be incorporated into compact gas-insulated switchgear. For medium and low voltage applications, the construction of SF_6 CB remains the same as that of oil and air CBs, as mentioned in Section 14.5.3, except for the special design of the arc interrupting chamber filled with SF_6 gas.

Physical properties of SF₆ gas
1. SF_6 gas is 5 times heavier than air.
2. It does not have any colour or smell.
3. It is non-toxic, but decomposes into SF_2 and SF_4, which are toxic.
4. The gas starts liquefying at 100°C at a gas pressure of 15 kg/cm^2 and remains stable upto 5000°C.
5. There is no risk of fire as SF_6 gas is non-inflammable.
6. The heat transfer rate of SF_6 is about 2.5 times that of air at atmospheric pressure.

Dielectric properties of SF₆ gas
1. The dielectric strength of SF_6 gas is about 2.5–3 times that of air at atmospheric pressure.
2. The dielectric strength of SF_6 gas is equal to oil at a pressure of 0.650 kg/cm^2, and at a pressure of 1.25 kg/cm^2, it is 15% higher.
3. SF_6 gas is strongly electronegative, because of which it attracts free electrons and recovers the dielectric strength at a very faster rate.
4. The arc time constant of SF_6 CB is very low because of its ability to capture free electrons. This reduces the diameter of the arc rod, which results in the reduction of the time constant. Hence, arc quenching is done quickly.

Non-puffer Type SF₆ Circuit Breaker

Figure 14.21 shows a part of a typical non-puffer type SF_6 CB. The CB houses two contacts, namely fixed contact and moving contact, which are enclosed in an arc interruption chamber. The interruption chamber is connected to the reservoir filled with high pressure SF_6 gas. The fixed contact has hollow cylindrical contacts and contains a set of current-carrying contacts with an arcing horn. The moving contact is also a hollow cylinder with rectangular holes at the sides to let the SF_6 gas out through these holes after flowing along and across the arc. The tips of fixed contact,

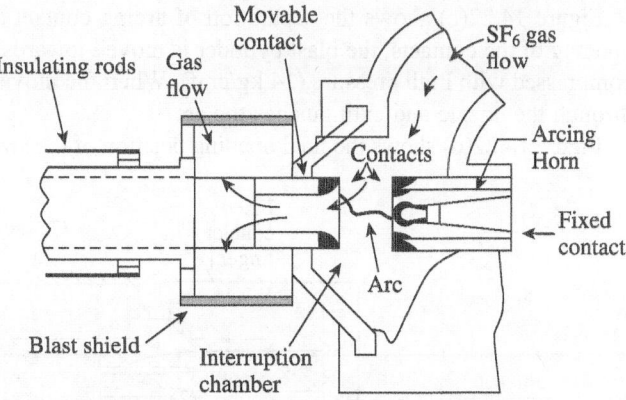

Fig. 14.21 Non-puffer type SF_6 circuit breaker

moving contact, and arcing horn are coated with copper–tungsten arc-resistant material. The contacts are surrounded by interrupting nozzles and a blast shield, which control the arc displacement and the movement of hot gases.

During normal operation of a CB, the switching contacts remain in a closed position with SF_6 gas surrounded at a pressure of about 2.8 kg/cm². Under short circuit conditions, the switching contacts of the breaker are opened, and an arc is struck between them. At the same time, the valve mechanism permits a high pressure SF_6 gas at 15 kg/cm² from the reservoir, which flows towards the arc interruption chamber. The arc is extended by separation of contacts as well as the flow of gas from high pressure to low pressure system. The high pressure flow of SF_6 rapidly absorbs the free electrons from the arc column within the interruption chamber. This leads to the building up of a high dielectric strength between the contacts which causes the extinction of the arc. After the breaker operation, the valve is closed by the action of a set of springs. As SF_6 gas is expensive, it has to be reconditioned and reclaimed by a suitable auxiliary system after each operation of the breaker. Such types of CBs are designed for the voltage level up to 132 kV.

Puffer Type SF₆ Circuit Breaker

In a non-puffer type SF_6 CB, there is a closed gas circuit for arc quenching. Further, it also contains two pressure systems (low and high pressure). The puffer type CB is designed in its simplest form with a single pressure system. In *puffer* designs, the blast of gas necessary for cooling of the arc is achieved through mechanical compression during the opening operation of the breaker. The energy, which is required to move the puffer, comes from the opening mechanism. Thus, the compressor essential in a non-puffer type system is not required in this type of CB, which further reduces the installation time and maintenance of the CB.

The puffer type CB consists of two main parts, namely the poles and the mechanism. The poles consist of contact and arc extinguishing devices. The mechanism is the part used to open or close the contacts in the poles. Figure 14.22 illustrates the technique of operation of puffer type CB.

Figure 14.22(a) shows the contacts in its closed position. The current flows through the fixed contact to the moving contact through the contact finger of the fixed contact to the cylinder to the nozzle and arcing contact of the moving contact. Here, the contact fingers and supporting parts are connected to the terminal of the pole.

Figure 14.22(b) shows the opening of the contacts with gas compressed at the back side of cylinder. During the movement of the cylinder away from the finger contact, current flows through the arcing rod of the fixed contact to the arcing contact and nozzle of the moving contact. The supporting part behaves as a fixed piston, which gradually compresses the gas inside the cylinder. On the other hand, the arcing rod of fixed contact blocks the opening of the nozzle.

Figure 14.22(c) shows the separation of arcing contact and nozzle from the arcing rod. At the time of opening of the contacts, the blast cylinder is moved towards the fixed piston, and the trapped SF_6 is thereby compressed with high pressure (14 kg/cm^2). When the moving contact opens, the compressed SF_6 gas flows through the nozzle and extinguishes the arc.

Figure 14.22(d) shows the final opening position of the switching contact after the complete arc extinction.

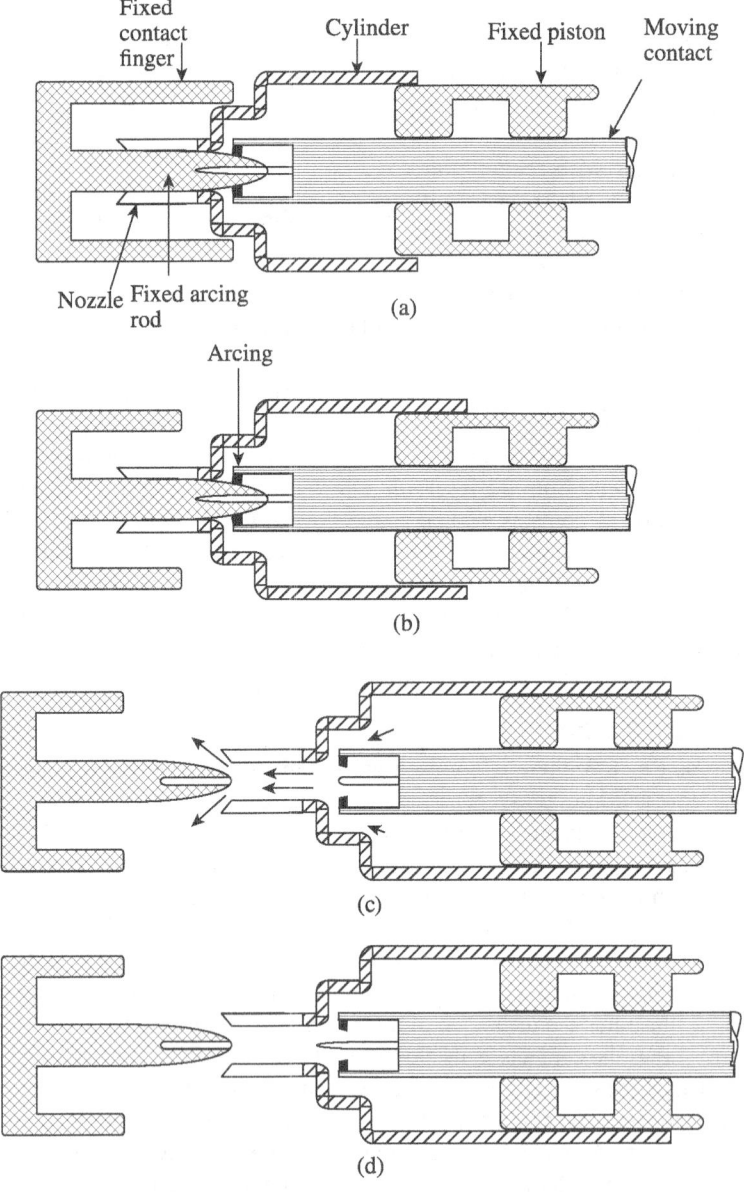

Fig. 14.22 Operation of puffer type circuit breaker (a) Breaker in closed position
(b) Breaker in opening stage (c) Breaker in opening stage
(d) Breaker in open position

In a puffer type CB, a relatively small contact clearance is required as the pressurized gas offers high dielectric strength within a short duration. This results in a small arc energy and high extinguishing ability with a short interrupting time.

Self-blast Type SF₆ Circuit Breaker

Due to its special gas properties, SF_6 allows the use of the energy stored in the arc during the high current and helps extinguish the arc at a natural current zero. This phenomenon created the third generation of SF_6 breakers, the so-called *self-pressurizing* (auto-puffer) *breaker* or *self-blast breaker*. The idea of self-blast is derived from the explosion pot used in oil CBs. In auto-puffer or self-blast design, some of the energy used for compression of the gas is also taken from the thermal energy released by the electric arc itself. In self-blast CBs, heated high pressure gases are released together with the arc after the arcing contact departs from the moving contact during the arcing period.

Another development is that the superior dielectric properties of SF_6 make it quite easy to build all three phases together in one enclosure. This is known as *gas-insulated switchgear* (GIS). Nowadays, equipment such as buses, line conductors, and isolators/disconnector are used in completely closed substations of the GIS type for voltages upto and above 400 kV.

Advantages of SF₆ circuit breakers

1. It provides restrike-free interruption of capacitive currents due to the high inherent dielectric strength of SF_6 gas and optimized contact movement.
2. It generates low overvoltages when switching inductive currents due to the optimum quenching of arc at current zero.
3. As the dielectric strength of SF_6 gas is 2–3 times that of air, such breakers are capable of interrupting much larger currents.
4. There is no risk of fire as SF_6 is non-inflammable.
5. It has low operating energy, which reduces the mechanical stress on the breaker and gives low reaction forces on the foundation.
6. It has low noise level and is hence suitable for installation in sensitive areas.
7. It has high making capacity even in the case of parallel connected capacitor banks.
8. It has low maintenance cost. Further, the foundation requirements are also very less.
9. The SF_6 breaker is totally enclosed and sealed from atmosphere. Hence, it is more suitable where explosion hazard exists.
10. It can be used along with auto-reclosers.

Disadvantages of SF₆ circuit breakers

1. SF_6 breakers are expensive due to the high cost of SF_6.
2. SF_6 gas has to be reconditioned after every operation of the breaker and hence, additional equipment is required for this purpose.
3. The arc products are toxic if they come in contact with air.

Applications Non-puffer type SF_6 breakers have been used for the voltage level of 66–230 kV. Conversely, puffer type and self-blast type SF_6 CBs are used upto the voltage level of 765 kV. Their interrupting time is less than three cycles.

14.5.5 Vacuum Circuit Breaker

The vacuum CB uses the rapid dielectric recovery and high dielectric strength of vacuum. A pair of contacts is hermetically sealed in a vacuum envelope. The actuating motion is transmitted through bellows to the movable contact. In a vacuum CB, the vacuum interrupters are used for breaking and making the load and fault currents. When the contacts in a vacuum interrupter separate, the current to be interrupted initiates a metal vapour arc discharge and flows through the plasma until the next current zero. The arc is then extinguished and the conductive metal vapour condenses on the metal surfaces within a period of microseconds. As a result, the dielectric strength in the breaker builds up very rapidly.

The properties of a vacuum interrupter depend on the material and form of the contacts. Over the period of their development, various types of contact materials have been used. It is accepted that an oxygen-free copper chromium alloy is the best material for high voltage CBs. This material combines good arc extinguishing characteristic with a reduced tendency of contact welding and low inductive current chopping. The use of this special material is that the current chopping is limited to 4–5 A.

Operating Principle

When current-carrying contacts are separated in a vacuum interrupter, cathode spots are formed depending on the severity of the current. These spots constitute the main source of vapour in the arc. Thus, because of the high electric field, heat is produced at the time of contact separation, which leads to the formation of an arc. The vacuum arc results from the neutral atoms, ions, metallic vapours, and electrons emitted from the electrodes. These metallic vapours, electrons, and ions produced during the arc condense quickly on the surfaces of the contacts of the CB because of the negative pressure (10^{-7} to 10^{-5} torr) inside the interrupter. This results in a quick recovery of dielectric strength between the contact gaps. Hence, the arc is quickly extinguished because of the fast rate of recovery of dielectric strength in the vacuum medium.

Construction

Figure 14.23 shows the schematic diagram of a typical vacuum interrupter. It consists of a fixed contact, a moving contact, and an arc shield mounted inside a vacuum chamber. The outer envelope is generally a glass vessel or a ceramic vessel, which works as an insulating body. The moving contact is connected to the operating mechanism through stainless steel bellows. The design of the bellow is very important as the number of repeated operations of the vacuum interrupter depends on it. The metal facilitates everlasting

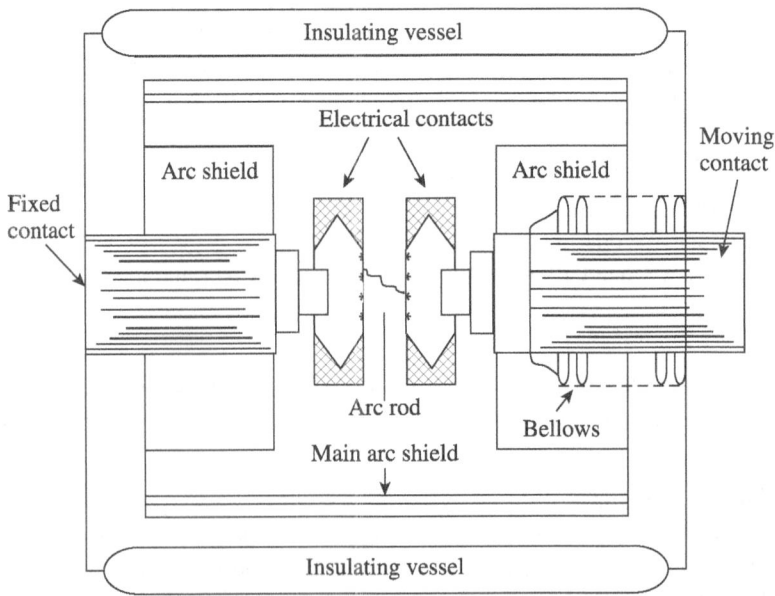

Fig. 14.23   Schematic diagram of a vacuum interrupter

sealing of the vacuum chamber to remove the possibility of vacuum leakage. The arc shield is provided between the contact and the outer insulating body, which prevents deterioration of the internal dielectric strength by preventing the metallic vapours reaching the internal surface of the outer insulating cover. One end of the fixed as well as moving contacts is brought out of the chamber for terminal connection.

Figure 14.24 shows the vacuum interrupter connection along with the operating mechanism. The operating mechanism can provide sufficient pressure so that the switching contact does not bounce at the time of the closing operation. Moreover, it should provide good connection between contacts during normal operation of the vacuum CB. The lower end of the breaker is connected with a spring-loaded mechanism so that it allows upward and downward movements of the metallic bellow during closing and opening operations, respectively.

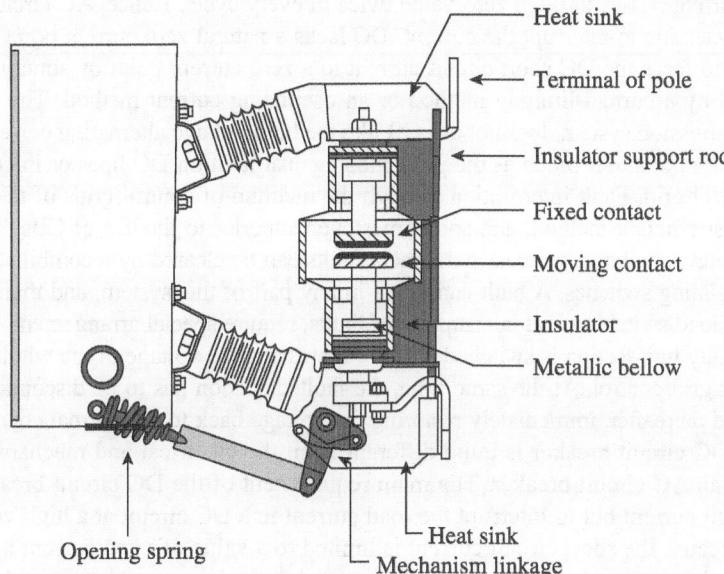

Fig. 14.24 Vacuum interrupter connections with operating mechanism

Working

During a short circuit condition, when the operating mechanism separates two current-carrying contacts in a vacuum module, an arc is drawn between them. An intense hot spot or sparks are created at the instant of contact separation from which metal vapour shoots off, constituting arc plasma. The amount of vapour in the arc plasma is proportional to the rate of vapour emission from the electrodes (arc current). With an AC arc, the current decreases during a portion of the wave and tends to zero. Thus, the rate of vapour emission tends to zero, and hence, the amount of plasma tends to zero. Soon after the natural current zero, the remaining metal vapour condenses and the dielectric strength builds up rapidly, and the restriking of arc is prevented. Thus, the vacuum CB develops a very fast rate of recovery of dielectric strength within a short contact distance of 5–10 mm depending on the operating voltage.

Advantages

1. They are entirely self-contained, need no supplies of gases or liquids, and emit no flame or gas.
2. They require no maintenance.
3. They are used in any orientation. The main feature of a vacuum CB is that it can break any heavy fault current perfectly just before the contacts reach the final open position.
4. They are not flammable.
5. They have a very high commutating ability and need no capacitors or resistors to interrupt short line faults.

6. They require relatively small mechanical energy for operation.

7. They are silent in operation.

Applications Due to the short gap and its excellent recovery characteristic, vacuum CBs can be used where the switching operation is very frequent. They can be used for capacitor switching and reactor switching too. The outdoor applications of vacuum CBs range from the voltage level of 22–66 kV. As the maintenance required is very less, vacuum CBs are most suitable for complex rural electrification. They are preferable for distribution systems in which the voltage level is 11–33 kV.

14.5.6 High Voltage DC Circuit Breakers

In case of AC, the current passes through zero value twice in every cycle. Hence, AC circuit breakers take advantage of this characteristic to interrupt the current. DC lacks a natural zero current point of time unlike AC. Hence, it is required to force the DC short circuit current to a zero current point by some means. Such means is generally provided by a current-limiting method or an oscillating current method. The high voltage direct current (HVDC) transmission system does not use CB like the high voltage alternating current (HVAC) system. The HVDC circuit does not suffer much as the protection against fault on DC lines or in converters is cleared by means of the control grid. Fault interruption through the medium of control grids of the converters is considered to be the most efficient method, and sometimes even superior to the use of CBs. Transient faults can be cleared by grid control methods, whereas permanent faults can be cleared by a combination of grid control, fault locators, and isolating switches. A fault can occur in any part of this system, and therefore, the DC CBs, which are capable of load switching and interruption of faults, require special arrangement. At the time of fault in any section, the faulty line section is switched out by suppressing the voltage of the whole system to zero by means of a converter grid control. At the same time, the faulted section has to be disconnected by fast-acting isolating switches and thereafter, immediately restoring the voltage back to the normal condition.

The design of a DC circuit breaker is quite different from the electrical and mechanical points of view compared to that of an AC circuit breaker. The main requirement of the DC circuit breaker is to not break the actual short circuit current but to interrupt the load current in a DC circuit at a high voltage with respect to ground. This is because the short circuit current is limited to a value of rated current by grid control. The design skill requires development of such DC switches so that the lines could be switched into or out of a healthy DC network without suppressing the voltage to a lower value.

Such switches have been suggested in which an artificial current zero is created by means of oscillatory discharge of a capacitor and operation of contact switches. The peak value of the oscillatory current should be higher than the direct current to be interrupted. The apparatus of high voltage DC circuit breaker comprises a commutation switch, a series circuit of a capacitor, a reactor, a switch connected in parallel with the commutation switch, and a resistor connected to charge the capacitor by the line voltage. The rise time constant of the line voltage is being made smaller than the charging time constant of the capacitor, which is determined by the values of the capacitor and the resistor. Conversely, the rise time constant of the line current is being made larger than the charging time constant of the capacitor. Figure 14.26 shows the schematic diagram of such a DC circuit breaker.

In Fig. 14.25 (a), A is normally an open contact, whereas CB and B are normally closed contacts. Thus, the capacitor C will charge upto the magnitude of line voltage through the resistor R. Since it is necessary to interrupt the normal DC current I_{DC}, the operating mechanism opens contact B and closes contact A. This initiates the oscillation of the circuit consisting of CB, A, C, and L. Thereafter, the contact CB opens immediately, which interrupts the current at a zero point (point T in Fig. 14.25 (b)). After this, contact A is opened and contact B closed.

The high voltage DC circuit breaker is capable of controlling and protecting the high voltage DC transmission system considering that the relation between the time constants is determined appropriately. The HVDC circuit breaker should be capable of clearing a direct current of 3 kA at the rated voltages of 400–1200 kV within a few seconds.

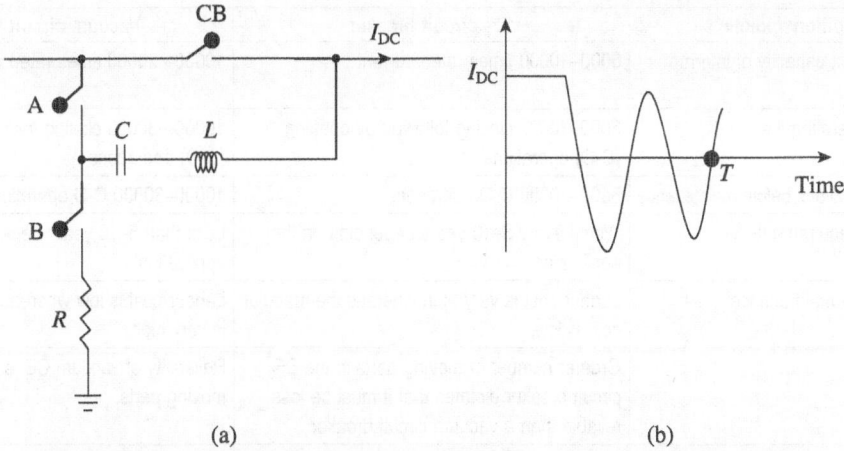

Fig. 14.25 DC circuit breaker (a) Schematic diagram (b) Vector representation

14.6 Comparison of SF$_6$ and Vacuum Circuit Breaker (Merits and Demerits of Conventional Circuit Breakers)

The minimum oil circuit breakers were mostly used for medium voltage system. However, the high maintenance cost and combustible properties are a few demerits of using oil as an arc extinguishing medium. Thereafter, in search of a better arc quenching medium, air blast and magnetic air CBs were developed. They did not survive much in medium voltage application due to heavy weight (burdensome) and have other disadvantages too. Later, exploration of SF$_6$ and vacuum as an arc quenching medium were done for medium/high voltage applications. Due to the arc interrupting properties and other advantages of these two types of CBs, these have totally replaced the previous generation CBs within a short period of time. Table 14.1 shows the comparative evaluation of SF$_6$ circuit breaker with vacuum circuit breaker.

Table 14.1: Comparison of SF$_6$ circuit breaker with vacuum circuit breaker

Conditions/points	SF$_6$ circuit breaker	Vacuum circuit breaker
Dielectric strength	The dielectric strength of SF$_6$ gas is very high compared to oil and pressurized air. It is widely used for voltages above 66 kV.	The dielectric strength of vacuum is not uniform between contact gaps. It vigorously decreases in case of increase of gap length. Thus, it is used for voltages below 33 kV.
Environmental aspects	SF$_6$ gas decomposes in to SF$_2$ and SF$_4$ which are highly toxic.	These are not affected by the production of contaminated gas. They are rather more environment-friendly than SF$_6$ CBs.
Operating energy requirements	SF$_6$ circuit breakers require more energy than vacuum CBs. This is due to the fact that a typical SF$_6$ circuit breaker has twice the number of parts in the high voltage circuit as the equivalent vacuum circuit breaker.	Operating energy requirements of vacuum CBs are lower than that of SF$_6$ CBs.
Arc energy requirements	Arc energy requirement of SF$_6$ circuit breakers are low as the arc voltage is between 150 and 200V.	Arc energy requirement of vacuum CBs are lower than SF$_6$ CBs as the arc voltage is between 50 and 100V.

Conditions/points	SF$_6$ circuit breaker	Vacuum circuit breaker
Breaking current capacity of interrupter	5000–10000 times rated current	10000– 20000 times rated current
Mechanical operating life	5000–20000 closing followed by opening (C-O) operations	10000– 30000 closing followed by opening (C-O) operations
Number of operations before maintenance	5000–20000 C-O operations	10000–30000 C-O operations
Maintenance requirement	Within every 5–10 years depending on the application	Less than 5–10 years depending on the application
Expenditure for maintenance	Labour cost is very high whereas the material cost is low.	Labour cost is low whereas the material cost is very high.
Reliability	Greater number of moving parts in the SF$_6$ circuit breaker dictates that it must be less reliable than a vacuum circuit breaker.	Reliability of vacuum CB is high due to less moving parts.
Number of short-circuit operations	Ranges from 10 to 50	Ranges from 30 to 100
Number of full load operations	Ranges from 5000 to 10000	Ranges from 10000 to 20000
Switching of short circuit current with high DC component	Suitable to switch current with high DC components	Appropriate to switch current with high DC components
Switching of short circuit current with high RRRV	Well suited under certain conditions (RRRV>1-2 kV per ms)	Best suited for high RRRV
Switching of transformers	Very well-matched	Moderately matched
Switching of reactors	Appropriate to reactor switching	Well suited. Steps to be taken when current <600A. to avoid overvoltage due to current chopping
Switching of capacitors	No restrike	Well suited. Low probability of restrike
Switching of capacitors back to back	Most suitable during switching of capacitor back to back. However, they require current limiting reactors to limit inrush current.	They are suitable during switching of capacitor back to back but they require current limiting reactors to limit inrush current.
Switching of arc furnace	Suitable for limited operation	Well suited. Steps to be taken to limit overvoltage.
Cost	Costlier than vacuum CBs.	Cheaper than SF$_6$ CBs.

14.7 Maintenance of Circuit Breakers

Circuit breakers are just meant to only open and close the contacts. However, if a breaker does not operate when it is supposed to, the power system can go awry. Hence, maintenance of circuit breakers requires a particular consideration as it provides continuous protection of equipment and their switching. The preventive maintenance of circuit breaker saves the breakup and destruction of equipment.

14.7.1 Maintenance of Medium Voltage Circuit Breaker

Periodic maintenance of CBs is recommended to obtain the necessary performance from the breakers. Medium voltage CBs are used for indoor application upto a voltage level of 33 kV and for outdoor application upto the voltage level of 66 kV. The CBs installed for indoor applications are assembled into metal enclosed switchgear, whereas they are exposed as individual components for outdoor applications. Nowadays, for most of the medium voltage indoor applications, vacuum CBs are used in place of ACBs and oil-filled CBs. Medium voltage CBs, which operate in such ranges, must be checked and maintained half-yearly or annually depending on the number of operations.

Maintenance Procedure for Medium Voltage Oil Circuit Breaker

The following steps are carried out during the maintenance of medium voltage oil CBs:

1. Drain the oil from the chamber and carefully clean the tank and other parts that have been in contact with the oil.
2. Examine the proper position and condition of the switching contacts.
3. Perform the dielectric test for drained oil. If the breakdown voltage (BDV) is found to be below the specified value, then replace the oil.
4. Inspect the breaker parts and the operating mechanisms for unfastened hardware.
5. Adjust the pole assembly of the breaker as per the manufacturer's instruction book.
6. Clean and lubricate the operating mechanism.
7. Check the electrical connection to the closing and tripping coil and test the breaker operation manually and electrically.
8. During retrofitting the tank and filling it with oil, ensure that the gaskets are not damaged and all nut-bolts and valves are tightened properly to avoid oil leakage.

Maintenance Procedure for Medium Voltage Vacuum Circuit Breaker

As vacuum interrupter is a closed container, visual inspection for switching contacts is not possible. The operating mechanisms are similar to those of oil CBs and may be maintained in the same manner. Further, the following are the maintenance steps carried out for vacuum CBs:

1. After prolonged use of a vacuum breaker, its contacts can wear down to some degree. Thus, it is required to set the shaft position of the fixed contact as per the manufacturer's instruction book.
2. The level of the vacuum pressure can be checked by a Hipot test according to the manufacturer's instruction book.

14.7.2 Maintenance of High Voltage Circuit Breaker

Maintenance of high voltage CBs requires complete external and internal inspections at half yearly to annual intervals. Proper external inspection reduces the expense required, time delay, and work effort of internal inspections.
The following are the external inspections required to be carried out for high voltage CBs.

1. Directly inspect the parts and operating mechanism of the CB for its proper adjustment and connection to each other.
2. Check the lubrication of the bearing surfaces of the movable hardware parts and operating mechanism for smooth operation of the breaker.
3. Monitor the closing and opening operations of the breaker under load conditions.
4. Check the electrical connection of tripping and closing coils.
5. Perform the operating test manually and electrically. In addition, rip the breaker through protective relays and observe for any malfunction.
6. Measure the contact resistance using Kelvin's double bridge during off-line condition. In addition, measure the insulation resistance (IR) value of the breaker using a megger to check the insulation level.
7. Measure the contact travel speed using. motion analysers, which provide graphical records of close or open initiation signals, closing or opening time of the contact with respect to initiation signals, movement and velocity of the contact, and contact's bounce or rebounce.

The following are the internal inspections required to be carried out for high voltage CBs.

1. Check the surface of the contacts and the interrupting parts by opening the breaker tanks or contact heads.
2. Check whether the inclination of keys, bolts, cotter pins, etc., appear loose.
3. Check the tendency of the loose clamps or mountings of the operating rods, supports, or guide channels.
4. Check whether carbon or sludge is formed and accumulated in the interrupter or on the bushings. If they are found, remove them.
5. Check whether the parts of the interrupter or barriers are damaged, worn, or burnt.
6. Check the pressure of the compressor and its leakage, if any, in case of failure of bushing gaskets for air blast CB.
7. Check the pressure of SF_6 gas and its leakage, if any, in case of failure of interrupters for SF_6 CBs.

These problems are most likely to appear in the older design of high voltage CBs. However, the recent design of high voltage CBs such as the SF_6 CB provides maintenance-free operation to some extent and reduces the frequency of internal inspections.

14.8 Testing of Circuit Breakers

The effectiveness of a protective device is experienced during its proper operation in domestic appliances, industries, and high power transmission lines and substations. During overload or a short circuit condition, the high voltage CB trips to limit the undesirable current. The high voltage CBs used in the industry and power system network is reliable and operates safely. Proper performance of such high voltage electric equipment during its normal working as well as during abnormal conditions depends on its testing at the time of fabrication and manufacturing.

Knowledge of the theory of CBs or the creation of an electric arc alone cannot provide a complete understanding of the interrupting process. Testing procedure is important to ensure that a particular breaker can bear certain operating conditions. Thus, to verify the ability of CBs to do their role of interruption of normal load current, fault current, small inductive current, and capacitive current in practice, different tests are to be carried out. These are *type tests* and *routine tests*. These tests are carried out in high power sophisticated laboratories with different test circuits.

The different tests carried out on a CB are classified as follows:

1. Short circuit tests
   - (a) Making test
   - (b) Breaking test
   - (c) Short time current test
   - (d) Operating duty test
2. Dielectric tests
   - (a) One-minute power frequency overvoltage withstand dry test
   - (b) One-minute power frequency overvoltage withstand wet test
   - (c) Impulse voltage withstand dry test
3. Thermal test (temperature rise test)
4. Mechanical test
5. Routine test
   - (a) Operation test
   - (b) Millivolt drop test
   - (c) Power frequency voltage withstand test (before and after the installation)
6. Synthetic test

Since the development of CBs, there is continuous improvement in the modification and realization of the interrupting process. Various innovative techniques have been developed to test the CBs. Basically, there are two types of test circuits used in the laboratory to test CBs. One is the *direct test circuit*, where the circuit with only one power source to supply the necessary current before interruption and the necessary voltage after the interruption is used. The other circuit is the *synthetic test circuit*, where the necessary current before the interruption and the voltage after the interruption are put together by means of different power sources or through different paths of the test circuit.

Short circuit tests These are the standard tests used for the verification of the short circuit performance of breakers. It consists of sequences of making and breaking operations. They are carried out to assess the fitness of CB during reduced and severe short circuit conditions.

The main parts of a basic short circuit plant are assembled as shown in Fig. 14.26. The short circuit power to the plant is fed from a special-design high capacity generator (2000 MVA) with low stator winding reactance and is driven by a motor. According to the voltage requirements, the generator winding is designed in two parts, which may be connected either in series or parallel and also arranged for either star or delta connection. One master CB is provided to interrupt the short circuit current in case of failure of the test CB. The breaking capacity of this master breaker should be higher than the short circuit MVA of the plant, Variable reactors and resistors are connected in series with the master breaker to control the magnitude of test current and power factor, respectively. The next component is a making switch. The function of this switch is to establish the short circuit test by closing it just after the closing operation of the master breaker. It is operated at a very high speed and is capable of withstanding heavy fault current without pre-arcing. Immediately after the making switch, there are step-up and step-down transformers used for testing at higher or lower voltage, respectively. The leakage reactance of these transformers is very low. Further, they are specially designed to withstand repeated short circuit tests. If the available voltage of the main generator terminal is enough to test the CB, then the outgoing terminals of the making switch are directly taken to the test bay bypassing the step-up/down transformers.

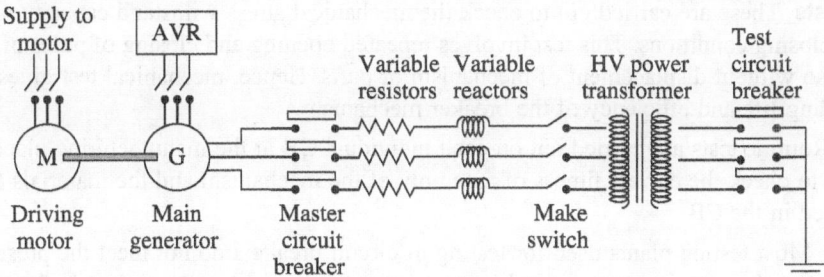

Fig. 14.26 Basic short circuit plant

Before performing the breaking test, all the apparatuses are adjusted to obtain the desired value of voltage, short circuit current, and power. The test components are connected to measuring equipment and an oscillograph to record the various parameters. Initially, the motor is operated to its full speed to provide kinetic energy to the generator. Further, it is switched off just after the closure of the making switch to avoid disturbances to the system during the short circuit test. Subsequently, the master breaker is closed, the making switch is closed, and finally the CB under test clears the short circuit. The decrement in voltage at the time of test (due to reduction in speed of the generator) is compensated by improving the generator field excitation

During the making test of a CB, the master breaker and making switch are initially in the closed condition. Later, the breaker under test is closed to create a short circuit. Thereafter, the same breaker under test clears the short circuit. The test parameters such as breaking current, making current, rate of rise of restriking voltage, power factor, and AC and DC components of fault current are measured with the oscillograph.

Dielectric tests The dielectric strength of a particular CB depends mainly on the insulating material used for quenching the arc. However, other factors such as clearance between two contacts, insulating material used, and assembly procedure also play a vital role in describing the dielectric characteristic of a CB. The purpose of this test is to check the ability of the CB to withstand the overvoltages at power frequency as well as the undesired value of voltage during lightning and other discharge phenomenon.

The power frequency dielectric withstand test is to be carried out only for indoor CBs. This is because the indoor CBs are not exposed to the impulse voltage as in the case of a high voltage system. Moreover, these CBs offer safe operation in case of high frequency switching surges as the effect of such phenomenon in indoor systems is very small.

As the operation of these CBs is related to overvoltage phenomenon caused by lightning and transient surges, both impulse voltage withstand test and power frequency withstand test are carried out for outdoor CBs.

The power frequency tests are carried out for the duration of 1 minute. Test voltages are applied between phases with the breaker in closed position and between phases and earth with the breaker in open position. The dry and wet tests are carried out to check the ability of CB to withstand the power frequency overvoltage during varying atmospheric conditions.

In impulse voltage withstand test, the test voltages of the standard wave shape of 1/50 μs according to the peak value of rated voltage of CB are applied between each phase and earth with other phases earthed. Also, the test voltage of the same wave shape is applied between phase-to-phase on one side terminals of the breaker and all other side terminals are earthed.

During these types of dielectric tests, the CB has to withstand the applied voltage for specific conditions and must not flashover or puncture.

Thermal tests These are carried out to check the temperature rise and thermal behaviour of CBs. During this test, normal rated current is allowed to pass through all poles of the CB, and the temperature rise is recorded using a thermocouple. In addition, the resistance between switching contacts is also measured.

Mechanical tests These are carried out to check the mechanical stress withstand capacity of the CB during opening and closing conditions. This test involves repeated opening and closing of poles of the CB without failure and also without displacement of mechanism or parts. Hence, mechanical test gives further consideration regarding life and efficiency of the breaker mechanism.

Routine test Routine tests are carried out on each individual CB at the manufacturing place. This test is to be performed to check the correct fitness of assembly of the mechanism and the materials (conducting and insulating) used in the CB.

Synthetic test Most testing plants used for testing of circuit breakers do not meet the prescribed standards and hence do not give satisfactory results. Moreover, due to ever-increasing demand of loads, both voltage and current ratings of CBs are increasing significantly. At the same time, the requirement for short-circuit power in testing plants is also increasing, which further increases the cost of CBs. To rectify the said problems, the synthetic testing method is used. In this method, arc current and restriking voltage are supplied through separate sources.

Figure 14.27 shows the test layout of synthetic testing of CBs. In synthetic testing, initially, short-circuit current is supplied by a high current injection generator. In order to avoid the influence of arc voltage on the short-circuit current, voltage of sufficient magnitude is required. At the time of interrupting the current, the correct magnitude of voltage pulse will be provided by a high voltage injection circuit. A realistic magnitude of voltage impulse must be provided for valid synthetic testing.

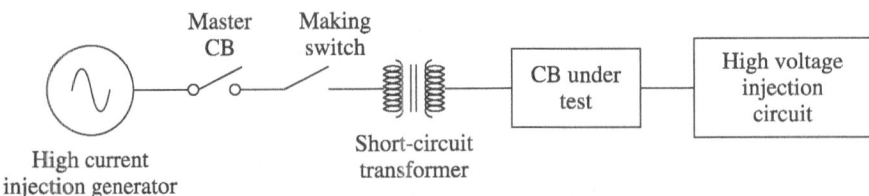

Fig. 14.27 Test layout of synthetic testing of CB

14.9 Selection of Circuit Breaker

The selection of CB depends on the application for which it is utilized, that is, for repeated switching application, capacitor switching, or inductive element switching. Moreover, the most useful purpose of CB is the sudden interruption of short circuit current. Thus, the withstanding (interrupting) capability of CB is calculated based on symmetrical rms AC amperes during a three phase bolted short circuit. A factor of 1.6 is generally multiplied to sub-transient current of first half to full cycle for calculating the withstanding current of the CB. Generally, the required symmetrical interrupting capability of CB is calculated by multiplying the ratio of rated maximum voltage to the nominal operating voltage by rated short circuit current.

The protective device takes definite time to sense the fault after its occurrence. This time is called *relay operating time*. Once the relay issues trip signal to the tripping coil of CB, the contact of CB starts to depart from each other. During this action, very high voltage gradient forms on the tip of departing contacts resulting in the formation of electric arc. In spite of the physical separation of switching contacts, current flows continuously because of the sustained arc. This arc plasma must be cooled and extinguished in a systematic way so that the gap between the contacts can again withstand the voltage in the circuit. Hence, the circuit breaker is selected in such a way that it can withstand varying magnitudes of fault current starting from the inception of the fault to the time by which the arc extinguishes successfully. Thus, the selection of CB depends on the highest magnitude of current that it can withstand during normal operation as well as during contact separation. Moreover, the selection of CB also depends on the following factors: (1) Nominal voltage, (2) Maximum rated voltage, (3) Nominal current rating of CB, (4) Rated short circuit current at maximum voltage, (5) Ratio of rated maximum voltage to the lowest limit of range of operating voltage (6) rated operating time, and (7) specified duty cycle.

14.10 Examples

Example 14.1 The rating of an oil CB is 2000 A, 2500 MVA, 66 kV, 3 s, three-phase. Determine the rated breaking current, rated making current, and short time rating.

Solution:
The rated normal current is 2000 A. The breaking current can be found from the given MVA rating.

$$\text{Breaking current} = \frac{2500}{\sqrt{3} \times 66} = 21.87 \text{ kA}$$

Making current = 2.55 × breaking current = 2.55 × 21.87 = 55.77 kA
Short time rating of CB = 21.87 kA for 3 s.

Example 14.2 A CB is tested for its make–break test. The values obtained during the testing are as follows:

(a) Under a faulty condition, the CB is closed and the peak of the first envelop of current is recorded as 50 kA.
(b) The peak to peak system voltage is 36 kV.
(c) During the break test, the AC component of the breaking current is 25 kA, and the DC component of the breaking current is 10 kA.

Determine the following:

(i) Rated line voltage for which the breaker is to be installed
(ii) Peak making current
(iii) Symmetrical breaking current
(iv) Asymmetrical breaking current

Solution:

(i) Peak to peak system voltage is 36 kV.

Phase voltage of the system (RMS) $= \dfrac{36}{2\sqrt{2}} = 12.72$ kV

Rated line voltage $= \sqrt{3} \times 12.72 = 22$ kV

(ii) Peak value of making current = 50 kA

(iii) Symmetrical value of breaking current $= \dfrac{25}{\sqrt{2}} = 17.68$ kA

(iv) Asymmetrical value of breaking current $= \sqrt{\left(\dfrac{25}{\sqrt{2}}\right)^2 + (10)^2} = 20.31$ kA

Recapitulation

- A circuit breaker (CB) is an electrical device designed to operate automatically in conjunction with pilot devices in the event of short circuit or overload.
- Rated current, rated voltage, rated breaking capacity, rated making capacity, short time rating, and rated standard duty cycle are the ratings of a CB, and they help in selecting the appropriate CB for a particular application.
- Symmetrical breaking current is given by $I_{\text{symmetrical}} = \dfrac{xy}{\sqrt{2}}$, and asymmetrical breaking current is given by

$$I_{\text{asymmetrical}} = \sqrt{\left[\left(\dfrac{xy}{\sqrt{2}}\right)^2 + (yz)^2\right]}; \text{ making capacity is given by } \sqrt{2} \times \kappa \times \text{symmetrical breaking capacity,}$$

where κ is the maximum asymmetry, whose value is 1.8.
- The function of a CB is to open its contacts within a predetermined time period as fast as possible to limit the amount of energy diverted in an unnecessary path.
- It is important to know the operating characteristic, design, and functions of low voltage power controlling devices such as switches, fuses, MCBs, and ELCBs.
- The high voltage CB plays a vital role in the protection of high voltage and extra high voltage networks.
- From the protection viewpoint, it is important to learn different types of high voltage CBs such as air break, oil break, SF_6, vacuum, and high pressure gas CBs along with their constructional features, merits, and demerits.
- The HVDC transmission system has deficiency of DC circuit breakers, but the fault or abnormality on a DC circuit can be cleared by grid control.
- The operating person should be aware of the periodic maintenance procedure of a high voltage CB as maintenance is recommended to ensure proper working of the CB during abnormal conditions.
- Proper performance of high voltage CB during its normal working as well as during abnormal conditions depends on testing it at the time of fabrication and manufacturing as well as field testing.
- Short circuit test, dielectric test, thermal test, mechanical test, and routine test are a few standard tests to be carried out on high voltage CBs.

Multiple Choice Questions

1. A fuse wire should have
   - (a) low specific resistance and high melting point
   - (b) low specific resistance and low melting point
   - (c) high specific resistance and high melting point
   - (d) high specific resistance and low melting point

2. The material best suited for manufacturing of fuse wire is
   - (a) aluminium
   - (b) silver
   - (c) lead
   - (d) copper

3. The breaking capacity of a CB is usually expressed in terms of
   - (a) amperes
   - (b) volts
   - (c) MW
   - (d) MVA

4. Which of the following CBs is preferred for extra high voltage (EHT) applications?
   - (a) Air blast CBs
   - (b) MOCBs
   - (c) BOCBs
   - (d) SF_6 gas CBs

5. The medium employed for extinction of arc in air break circuit breaker is
   - (a) SF_6
   - (b) oil
   - (c) air
   - (d) water

6. In air blast CBs, the pressure of air is of the order of
   - (a) 100 mmHg
   - (b) 1 kg/cm^2
   - (c) 20–30 kg/cm^2
   - (d) 200–300 kg/cm^2

7. SF_6 gas
   - (a) is yellow in colour
   - (b) has pungent odour
   - (c) is highly toxic
   - (d) is non-inflammable

8. The pressure of SF_6 gas in CBs is of the order of
   - (a) 100 mmHg
   - (b) 1 kg/cm^2
   - (c) 3–5 kg/cm^2
   - (d) 30–50 kg/cm^2

9. In a vacuum CB, the vacuum is of the order of
   - (a) 10^{-5}–10^{-7} kg/cm^2
   - (b) 10^{-1}–0^{-2} kg/cm^2
   - (c) 1–5 kg/cm^2
   - (d) 10^5–10^7 kg/cm^2

10. While selecting the gas for a CB, the property of gas that should be considered is its
    - (a) dielectric strength
    - (b) non-inflammability
    - (c) non-toxicity
    - (d) all of the above

Review Questions

1. Explain the different ratings and functions of CBs.
2. Explain the isolating and load-break switches.
3. Enumerate the different types of fuses for low voltage applications and write a detailed note on HRC fuse.
4. Explain the construction and operating principle of MCB.
5. Why is the ELCB used in domestic supply?
6. Write a note on ACBs.
7. Discuss the advantages of oil as an insulating medium in a CB.
8. Draw a well-labelled diagram of an MOCB and explain each part and how it works.
9. Write a brief note on vacuum CB and compare its advantages over oil CB.

10. State the merits of SF_6 gas compared to other arc-quenching mediums.
11. Explain the working of the following SF_6 CB:
    - (a) Non-puffer type SF_6 breaker
    - (b) Puffer type SF_6 breaker
12. Explain the maintenance procedures for medium voltage CBs.
13. Give the classification of tests to be carried out on CB.
14. Explain the short circuit test plant and procedures of testing for high voltage CBs.

Answers to Multiple Choice Questions

1. (d) 2. (b) 3. (d) 4. (d) 5. (c) 6. (c) 7. (d) 8. (c) 9. (a) 10. (d).

Testing, Commissioning, and Maintenance of Relays

15

Learning Objectives

After going through this chapter, the students will be able to:

- Explain different tests to be performed at the time of relay manufacturing
- Explain various commissioning and maintenance tests of relay
- Discuss test set-up of different relays
- Explain computer-based testing procedure of relay

15.1 Introduction

Regular and systematic operation of protective devices is extremely important as these devices need frequent checking and verification. This is because the actual relay operation during its whole life is considerably small. Therefore, as the protective gear is solely concerned with fault conditions and cannot readily be tested under normal system operating conditions, the testing of protective schemes has always been a problem. This situation has been aggravated in recent years by the complexity of protective schemes and relays. In order to test any protective device, a good set of testing equipment and relay tools are required. Several manufacturers are now producing portable relay test sets. However, the use of the portable relay test kit requires a skilled person who is familiar with the relays and the circuits involved. In recent installations, particularly where test blocks are used, proper precautions need to be taken at the time of removing or inserting plugs as otherwise there is a possibility of building up of a high voltage that may be dangerous to human beings. As far as the frequency of the test is concerned, it is usually recommended that the protective device must be given a complete calibration test and inspection test at least once a year. There are three types of tests to be performed on a protective device during its lifetime. These are type tests, acceptance tests, and routine maintenance tests (also known as *periodic tests*).

15.2 Type Tests

Type tests are usually meant for manufacturers of relays. All manufacturers must perform the following tests on each relay.

15.2.1 Operating Value Test

Operating value test is related to the operating quantity used in a particular relay. In this test, operating quantities such as current, voltage, power, and frequency change with reference to their preset value, and the operating value at which the relay operates is noted. Before performing this test, special care should be taken that the relay is completely reset before taking the next reading of a particular quantity. Normally, for voltage, the allowable value is within ±10% of the normal value. Similarly, for current, the permissible value is within 90–110% of the normal value as indicated on the name plate of the relay.

15.2.2 Operating Time Test

Operating time test is related to the operating time of a particular relay. It is defined as the time from the instant when the relay coil gets energized and its contact operates. For inverse time relays, operating times are obtained at different values of plug setting multipliers (PSMs) at different time dial settings (TDSs). According to the standards, one must get the operating time within the permissible limit with reference to its declared value for different ranges of PSMs. For example, for a PSM between 2 and 4 and from 4 to 20, the permissible deviation from the declared value must be within 12.5% and 7.5%, respectively, at the highest value of TDS. In case of definite time delay relays, the allowable error in the operating time is ±5% upto 0.1 s. In certain cases, several readings are taken and the average of those readings is carried out.

15.2.3 Reset Value Test

Reset value is defined as a percentage of the nominal set value of the relay. In this test, the operating quantity is set in such a manner that the relay remains in operating condition. Thereafter, by changing the actuating quantity, the value at which the relay just fails to operate is noted down.

15.2.4 Reset Time Test

In the reset time test, initially the operating quantity is set in such a manner that the relay remains in operating condition. Thereafter, the energizing quantity is removed suddenly, and the time taken by the relay contacts to return to their unoperating state from the fully operating state is noted.

15.2.5 Temperature Rise Test

The temperature rise test is carried out for all relays to check the withstand capability of insulation used in relays against a temperature rise according to its class. In this test, the rated value of a particular quantity is passed through the relay. For example, in current-based relays only the rated current is passed, whereas for voltage-operated relays 10% more voltage than the rated voltage is given. For the two input relays, that is, distance or directional, both the quantities are given to the coil of the relay. These quantities are passed through the relay until the relay circuit attains an ambient temperature.

15.2.6 Contact Capacity Test

As the contacts of a relay have to actuate the tripping coil of the circuit breaker, they must possess high volt–ampere capacity. This is also important to carry out making and breaking operations. These operations are achieved using appropriate loading such as resistive loading and inductive loading. For example, as per international standards, breaking capacity is decided considering inductive loading at a very low power factor (0.4 lagging) and at a very small time constant (two to three cycles). The whole making and breaking process is repeated within a short time span, usually 25–30 s, between two successive operations.

15.2.7 Overload Test

In the overload test, for inverse time relays, 20 times the plug setting current or the pickup value of current is injected into the relay coil at maximum TDS, and the continuous current-carrying capacity is checked. Usually, several operations are performed and averaging is carried out. For a thermal relay, the same procedure is repeated but at 8–10 times the plug setting current. In case of the two input relays, that is, distance and directional, 20 times the plug setting current is injected into the current coil of the relay, and at the same time, rated voltage is also given to the voltage coil of the relay.

15.2.8 Mechanical Test

Mechanical test is usually meant for electromechanical relays that contain moving parts. In this test, double the plug setting current is injected into the coil of current-based relays at maximum TDS, whereas rated voltage is given to the voltage coil of voltage-based relays. Several hundreds of operations are performed, and after these many operations, the moving parts should be in proper condition without any damage.

15.2.9 Stability Test

Stability test is carried out on relay used for unit protection (differential scheme) and has restraining feature for external fault condition. The stability of relay is tested for external ground and phase fault. To check whether the relay will operate or not, an anticipated load variation must be performed on either side of unit protection. If the load variation is inadequate, primary injection will be required to check the stability of the differential scheme. Hence, stability test determines the effective relay setting for internal fault and it remains inoperative for any kind of through fault condition.

15.2.10 Overshoot or Overtravel Test

The overshoot test is done on time delayed electromechanical relays. The moment of inertia in electrometrical relay keeps the disc rotating even after the fault current is shut off. The overshoot or overtravel of relay is checked by injecting very high current (approximately 20 times the relay pick-up setting) with maximum time dial setting (TDS = 1) for specified time period slightly lower than actual operating time. During this test, it is observed that the fixed and moving contact of relay should not be bridged.

15.2.11 Voltage Withstand Test

The voltage withstand test is carried out on each and every relay as a routine test to ensure that the relay is manufactured as per design. A high voltage rated frequency supply for one minute is given between terminals and earth to check the dielectric strength so that the relay is safe to operate in normal conditions. For voltage operated relay, high voltage impulse for very short duration is given and the parts of relay (coil) must survive it to pass this test. High impulse voltage withstand test is carried out to make sure that the relay can withstand overvoltage during transient and lightning conditions.

15.3 Commissioning and Acceptance Tests

Commissioning tests are carried out to ensure the following.

1. No damage has been done to the relay during transit.
2. The relay has been installed correctly.
3. The protection system works correctly as per the design and purchase order.
4. A set of test data for future reference is achieved.

On the other hand, acceptance tests are carried out at the manufacturer's premises. These tests are very detailed in nature and are done to verify the relay performance as per the declared/agreed customer's specifications along with the relevant standards.

15.3.1 Insulation Resistance Test

In insulation resistance test, initially all the earth connections are removed from the wiring. Thereafter, a 500 V megger is used to measure the insulation resistance to earth of the current transformer (CT) circuits. It is connected across the appropriate earth links before they are opened. They should be removed after the reclosing of the earth link. The ideal value of an insulation resistance is of the order of 5 MΩ. The measured value of insulation resistance may be used in future to determine the deterioration of insulation. The measured value of insulation resistance depends on the amount of wiring circuits involved, grade of insulation, and humidity at the site.

15.3.2 Secondary Injection Test

Secondary injection test is done to ascertain that the relay calibration is correct during an injection of current in the coil. This test is carried out using the secondary injection kit. In early stage, it is made up of test blocks or test sockets in the relay circuits so that connections can be made to the test equipment without disturbing the wiring. At present, all manufacturers provide a separate secondary injection kit that contains modular type case through which the entire testing is performed.

15.3.3 Primary Injection Test

Primary injection test ensures the correct installation and operation of the whole protection scheme. Hence, primary injection tests are always carried out after the secondary injection tests. The primary injection test involves CT secondary winding, relay coils, trip and alarm circuits, and all intervening wiring circuits. This test is carried out usually with a portable injection transformer. It uses local mains supply and possesses several low voltage heavy current windings. These windings can be connected in series or parallel according to the current requirement. For example, considering that 10 kVA is the rating of the injection transformer with a ratio of 250/10 + 10 + 10 +10 + 10, one can achieve up to 1000 A when four windings are connected in parallel and 250 A when they are connected in series. To control the current of the injection transformer, either a tapped reactor or a heavy current variable autotransformer is used. The use of a resistor for current control is not good as it causes a great deal of power loss through dissipation.

15.3.4 Tripping Test

The complete sequence of tripping should be checked from the protective relays to the tripping of the circuit breaker. This test is carried out just after the primary and secondary injection tests. During these tests, the trip and alarm circuits are not operative. Hence, after the completion of primary and secondary injection tests, it is necessary to check the tripping and alarm circuits. This is done by manually closing the circuit breaker contact. Moreover, during this test, it is important to check that the relay contacts are clean and secure. Further, all flags must be checked for positive operation.

15.3.5 Impulse Test

Impulse test is specifically carried out for static relays. If any transient waves are present in the incoming circuits, then there is a possibility of maloperation of static relays. As recommended by the International Electrotechnical Commission (IEC), an exponential wave with a rise time of 1.2 ms and a fall time of 60 µs is applied and the components of the relay must withstand it to pass this test. The peak value of the voltage wave applied to the relay varies between 1 kV and 5 kV. This value depends on the nature and location of the external wiring connected to the relay.

15.4 Maintenance Test

Maintenance test is required to ensure that the protective gear senses the fault and operates at the expected speed. At the same time, the same protective gear remains inoperative in case of external or out-of-zone faults. Therefore, it is mandatory to carry out routine maintenance tests on protective gear to ensure that the equipment is always ready to perform its duty in a fully discriminative manner.

15.4.1 Requirement of Routine Maintenance Test

Usually, a protective gear does not deteriorate with reference to its age of installation. However, there is a possibility of damage to such protective gear due to many adverse conditions, such as the following:

1. Continuous vibration may affect the pivots or bearing of the electromechanical relay.
2. Insulation resistance of the multicore cable and wiring may be reduced because of dampness in the junction boxes and circuit breaker kiosks.

3. There is a possibility of deterioration of relay ligaments, relay contacts, and auxiliary switches due to polluted atmosphere.
4. Insulation may deteriorate because of continuous energization of the coil, which produces heat.
5. There is a possibility of open circuiting of coils or contacts due to electrolysis, which causes a green spot.

15.4.2 Frequency of Routine Maintenance

The frequency of routine maintenance test varies with the type of equipment. However, it depends on the fault history and fault liability of the equipment. Certain types of equipment are continuously checked, whereas others may be checked/tested annually. A typical schedule is as follows:

1. Continuous checking
   (a) Pilot supervision
   (b) Trip circuit supervision
   (c) Relay voltage supervision
   (d) Battery earth fault supervision
   (e) CT supervision of busbar
2. Daily
   All relay flags and semaphore indicators must be inspected daily or on every shift. If a carrier protection is used, then test on carrier protection must be performed daily.
3. Monthly
   Water level of liquid earthing resistances must be checked every month.
4. Bimonthly
   Channel tests must be performed bimonthly.
5. Half-yearly
   All the tripping tests must be performed every 6 months.
6. Yearly
   The following operations must be performed annually:
   (a) Operating level check, sensitivity check, and tripling angle check
   (b) Secondary injection test on more complex and important relays
   (c) Insulation resistance test
   (d) Inspection of gas and oil-actuated relays
   (e) Inspection of battery biasing equipment

15.4.3 Records of Commissioning and Maintenance

Good records are essential for proper management of protective gear. This helps in assessing the performance of protective gear for long and short durations. Typical records to be maintained are described as follows:

Commissioning log   It is essential to keep a running record of progress and all test results.

Test record schedules   Schedules of test records are essential for recording the results of secondary injection test, tripping test, insulation resistance test, and other tests. These schedules may vary depending on the type of test. For example, if it is a commissioning test, then it is performed in detail but only once, whereas routine maintenance test is carried out many times but not in too much detail.

15.5 Test Set-up of Different Types of Relays

The laboratory set-up of different types of relays is described in the following sections.

15.5.1 Overcurrent Relay Test Bench

All types of overcurrent and ground relays are available with a rated current of either 1 A or 5 A. The plug setting of electromechanical or static phase and ground relays is in the range 5–200%. Therefore, there is a need for a current source with large coverage of current starting from a very low value (0.01 A) to a very high value (250 A). Moreover, if the test voltage is given to the relay coil through an autotransformer, then the waveform of the test current is distorted because of the non-linear impedance of most of the relay coil. This is not permitted as per standards as it recommends third and fifth harmonics within 5%. This problem can be rectified by using a tapped non-saturating reactor, which effectively suppresses the harmonics and yields a non-distorted current waveform. This technique is used by most manufacturers who supply portable relay test kits. Portable overcurrent relay test equipment has been designed primarily for the testing of inverse time overcurrent relays, at 50 Hz particularly, onsite, where portability and steady current output are essential. The salient features of this unit are undistorted output waveform and continuously adjustable output from 50 mA to 200 A.

In the portable overcurrent test kit, the current is controlled by a series reactor. A tapped non-saturating reactor suppresses the harmonics so that a good waveform can be obtained with minimum power dissipation because the resistance component is relatively small. This, in turn, means that the whole equipment can be made much smaller and lighter, for ease of transport. The test equipment contains a primary supply circuit to which the relay is connected by a current injection transformer. The primary circuit current is varied by means of coarse, medium, and fine controls between 1 A and 40 A. This current is matched to the relay setting by the current injection transformer. In other words, the relay appears to the primary circuit as having the same impedance no matter what its setting may be. This simplifies the testing procedure. Figure 15.1 shows the connection diagram for any type of overcurrent relay testing using a conventional standard test kit.

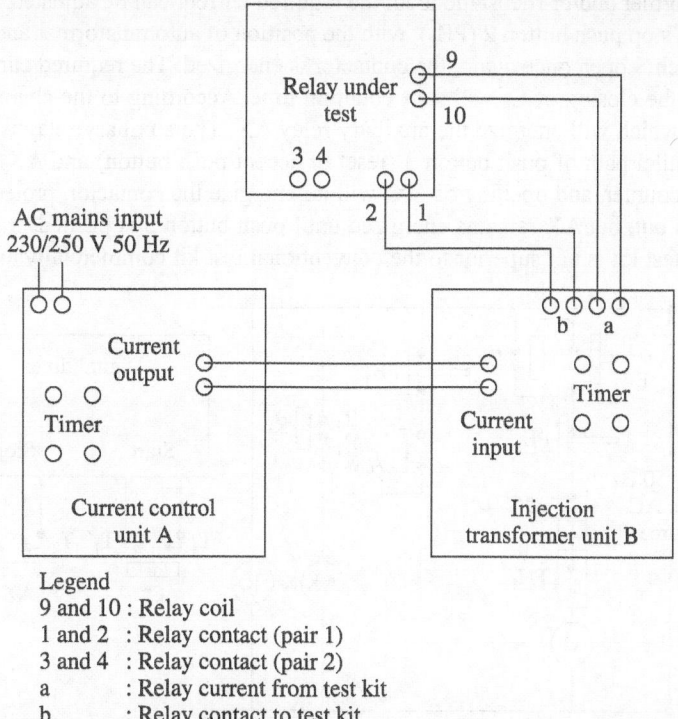

Legend
9 and 10 : Relay coil
1 and 2 : Relay contact (pair 1)
3 and 4 : Relay contact (pair 2)
a : Relay current from test kit
b : Relay contact to test kit

Fig. 15.1 Connection diagram of relay testing using conventional standard test kit

This relay test kit is, however, very expensive. Hence, the authors have designed a new, but economical, relay test set to test the overcurrent and undercurrent relays for engineering colleges and industries. The circuit diagram of this economical test set-up for obtaining different characteristics such as overcurrent, undercurrent, thermal, and differential relay is shown in Fig. 15.2.

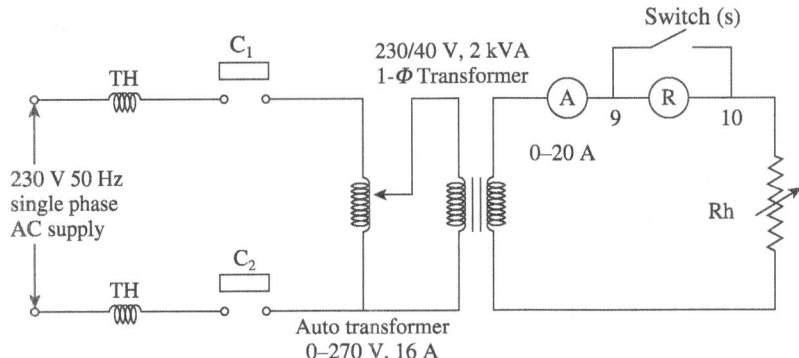

Fig. 15.2 Circuit diagram of economical test set-up designed in laboratory

Figure 15.3 shows the control circuit and timer counter circuit, respectively, for the same. The operation of the test set-up can be easily understood. Observing Figs 15.2 and 15.3 together, by pressing the spring loaded pushbutton 1(PB$_1$), contactor (C) energizes. The hold on path is being provided by the auxiliary contact (C$_3$) of the contactor.

Hence, the main contacts C$_1$ and C$_2$ in Fig. 15.2 close. By keeping the relay shorting switch S closed and varying the autotransformer and/or rheostatic load, the required current can be adjusted. Then, the contactor is de-energized using the stop push button 2 (PB$_2$), with the position of autotransformer and/or rheostat unvaried. After keeping the switch S open once again, the contactor is energized. The required current will pass through the relay coil. Hence, the closing of C$_4$ will start counting time. According to the characteristics of the relay, contact R$_1$ will close, which will energize the auxiliary relay AX. The auxiliary relay is made independent of R$_1$ by providing a parallel path of push button 3 (reset or accept push button) and AX$_2$. Closing of AX$_3$ will stop the time interval counter, and opening of AX$_1$ will de-energize the contactor, protecting the relay against high current. R$_1$ opens out, but AX remains energized until push button 3 is pushed. It is to be noted that the proposed economical test kit is not superior to the conventional test kit commercially available in the market.

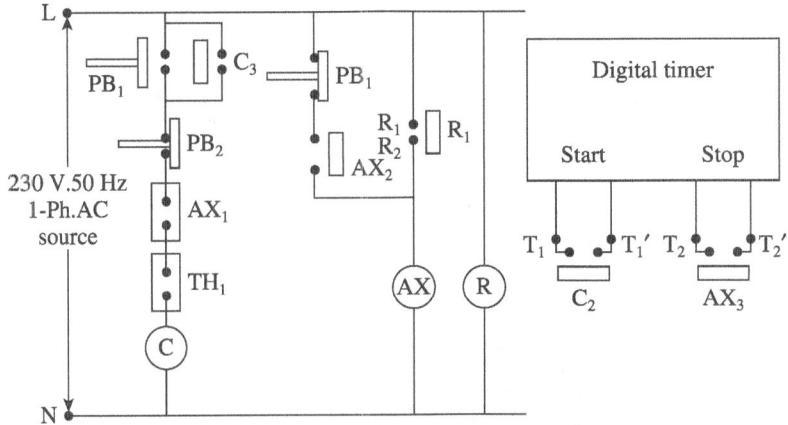

Fig. 15.3 Control and timer circuits of economical test set-up

Furthermore, for testing of the overcurrent relay, another circuit, as shown in Fig. 15.4, can be used.

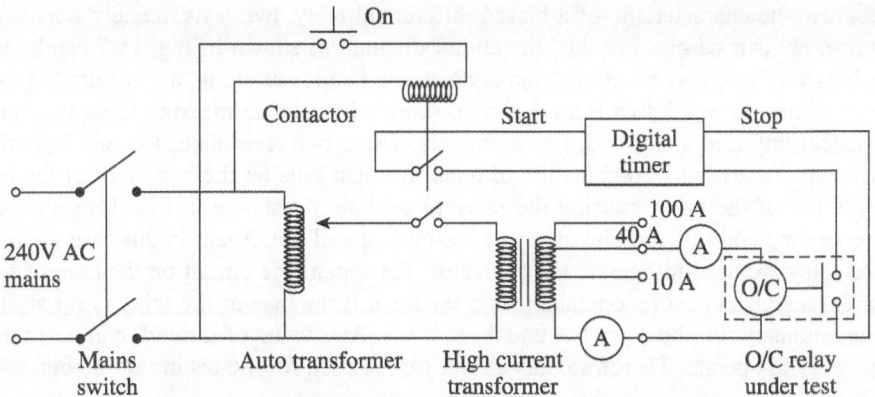

Fig. 15.4 Secondary injection testing of overcurrent relay

The normally open contacts are used to stop the timer. A contactor is used to close the circuit of the high CT. The same contactor simultaneously also closes the contacts to start the high precision timer. Once the circuit is closed, current is injected into the relay and the timer starts to measure the time duration. When the relay operates, the normally open contacts of the relay are closed causing the timer to stop, and the high CT is also de-energized. The timer is designed in such a manner that its display is latched; thus, it gives the operating time for the relay corresponding to the current being fed to the relay.

15.5.2 Overvoltage Relay Test Bench

Overvoltage relays with different relay characteristics such as instantaneous, definite time, and inverse time are manufactured by different manufacturers. For testing these relays, the circuit shown in Fig. 15.5 can be used. The normally open contacts are used to stop the timer. A contactor is used to close the circuit of the voltage injection transformer. The same contactor simultaneously also closes the contacts to start the high precision timer. Once the circuit is closed, the voltage is injected into the relay, and the timer starts measuring the time duration. When the relay operates, the normally open contacts of the relay are closed causing the timer to stop, and the voltage injection transformer is also de-energized. The timer is designed in such a manner that its display is latched; thus, it gives the operating time for the relay corresponding to the voltage being fed to the relay.

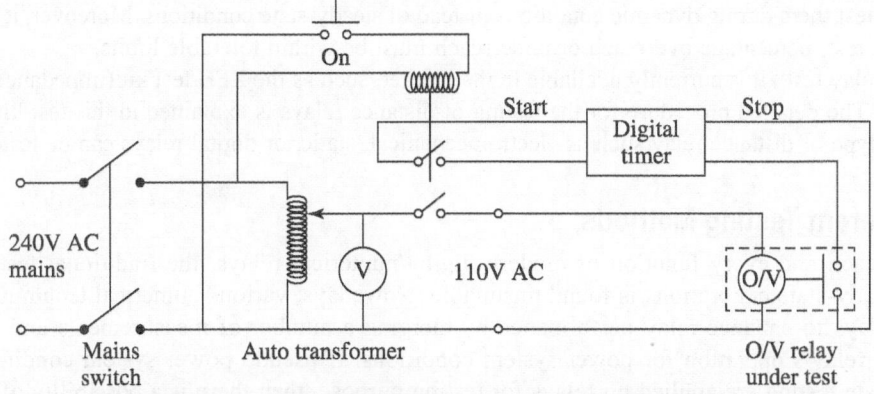

Fig. 15.5 Secondary injection testing of overvoltage relay

15.5.3 Biased Differential Relay Test Bench

In order to confirm the characteristic of a biased differential relay, two tests, namely *sensitivity threshold test* and *bias test*, are carried out. For this, the circuit diagram as shown in Fig. 15.7 can be used. For the testing of the biased relays, two quantities, namely operating and restraining, are required. As shown in the circuit, these two currents are drawn from the same source. In order to measure these two quantities, two variable resistances are used along with two ammeters. These two resistances (R_1 and R_2) are adjusted in such a manner that the resulting combination of quantities just falls on the boundary of the operating and non-operating zones of the relay, causing the relay to operate. As shown in Fig. 15.6, the relay has two coils namely operating coil (O) and biased coil or restraining coil (B). Again in this case the normally open contacts of the relay are used to operate the contactor. This opens the circuit on the current source so that the relay gets de-energized post its operation. For example, if the bias of the relay is set at 30%, then the operating zone boundary will be $A_1 = 1$ A and $A_2 = 0.3$ A. Any value of through current more than 1.0 A will cause the relay to operate. Therefore, this circuit can be used for the testing of the biased relays.

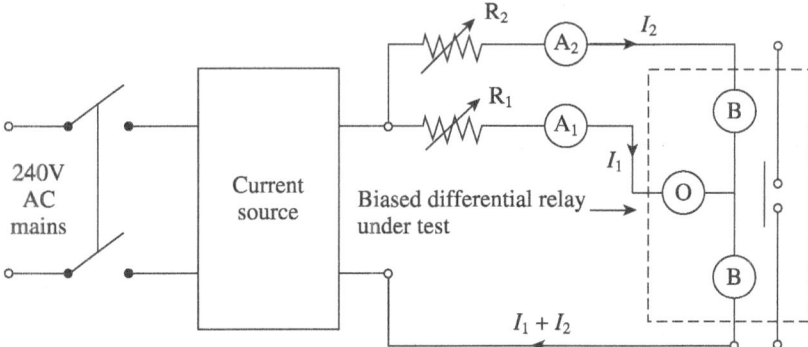

Fig. 15.6   Bias test on biased differential relay

15.5.4 Distance Relay Test Bench

Distance relays are widely used for the protection of ultra high voltage (UHV) and extra high voltage (EHV) lines. They have different characteristics such as mho, impedance, reactance, quadrilateral, quadra-mho, elliptical, and lenticular. These relays with different characteristics are manufactured by various manufacturers. Their operation needs to be checked for all the two/three zones. While testing these relays, it is extremely necessary to test them during dynamic conditions instead of steady-state conditions. Moreover, it is important that after the test, percentage overreach or underreach must be within tolerable limits.

Portable relay test kit is currently available in the market, such as the ZFB test kit (impedance testing kit) of ABB Ltd. The detailed procedure for the testing of distance relays is explained in this test kit. Using this test kit, any type of distance relay such as electromechanical, static, or digital relays can be tested.

15.6 Different Testing Methods

For testing each and every function of modern digital/numerical relays, the traditional testing method that uses steady-state calibrations is found unsuitable. Nowadays, various numerical techniques are used in digital relays to enhance relay performance by merging a number of measurements and also by optimizing the relay's operation for power system conditions. If pseudo power system conditions created by steady-state testing are applied on relays for testing purpose, then there is a possibility of occurrence of problems in testing and understanding the relay's operation. Moreover, the time involved to test the individual functions of the relay would be excessive. This is because the time required to reconfigure each individual element is very high.

On the basis of a report from IEEE entitled *Relay Performance Testing*, the following relay testing methods are used in the field:

1. Steady-state testing
2. Dynamic-state testing
3. Transient-state testing

15.6.1 Steady-state Testing

In steady-state test, to determine the relay settings, phasors are applied by slowly varying the relay input. For example, the injected quantity (current, voltage, or frequency) is maintained at a predetermined value for a duration longer than the planned time for the relay. Thereafter, it is changed gradually at a rate much smaller than the resolution of the relay. This can be achieved either manually or by an automated system. However, the main disadvantage of this method is that it is not useful as it lacks the relation with the actual power system faults.

15.6.2 Dynamic-state Testing

Modern digital/numerical relays are designed to provide complete protection to a power system component, which includes many settings. These require extensive configuration and relay setting procedures. The conventional steady-state method of testing, which suggests individual calibrations one at a time, is not a viable method because of the excessive time requirement to reconfigure each individual testing of the element. Moreover, steady-state testing methods were developed using basic test equipment components such as auto-transformers, phase shifters, and load boxes. However, power system conditions can be simulated very easily using different types of modern test equipment. This is achieved by dynamic-state testing. In this test, the fundamental frequency components of voltage and current are applied simultaneously, which represent the power system states of pre-fault, fault, and post fault. Later, the time of operation of the relay is measured. In certain cases, DC offset can be included.

Advantages of dynamic-state testing

1. The performance of the complete relaying scheme can be tested for each simulated power system event.
2. As the relay elements do not need to be disabled for testing the function of a relay, it results in faster relay testing.
3. As this type of test method considers the effect of the system impedance ratio (SIR), the realistic relay operating time test can be easily carried out.
4. This test method provides reliable test results with a reduced test time. Hence, a database can be created, and it will also be of use for future relay operations.

15.6.3 Transient-state Testing

Transients are inevitable in real power systems because of which the behaviour of relays is random in nature. When input signals are applied with transients, there is a possibility that the relay behaviour may not match the one observed in the case of phasor-based test waveforms. In particular, the response of the relay and its operating time continuously change and depend on the content of the fault transients. Measuring the direct-trip time and determining if the relay should have tripped constitute the assessment approach in this case. This type of testing is called *application* or *transient testing*. In this test, fundamental and non-fundamental frequency components of voltage and current are applied simultaneously, which represents the power system conditions obtained from digital fault recorders (DFR), electromagnetic transient programs (EMTP), or power system computer-aided design (PSCAD).

15.7 Computer-based Relay Testing

Manual testing of relay functions is obsolete now. All the modern digital relays are tested using computer programs because of the following reasons.

1. Slow speed of amplifier injection is not capable of testing modern digital relays.
2. More stresses are developed on relays because of heat production due to lack of speed of relays.
3. No skilled person is required for relay testing as all the parameters such as type of tests, test procedure, and frequency are decided by the computer.
4. Specific test report format is available.
5. Uniformity is maintained.

Figure 15.7 shows the test set-up of computer-based relay testing. Three-phase voltages and currents are given to the relay under test through injection amplifier. Communication between injection amplifiers and the PC is governed by the IEEE 488 interface bus.

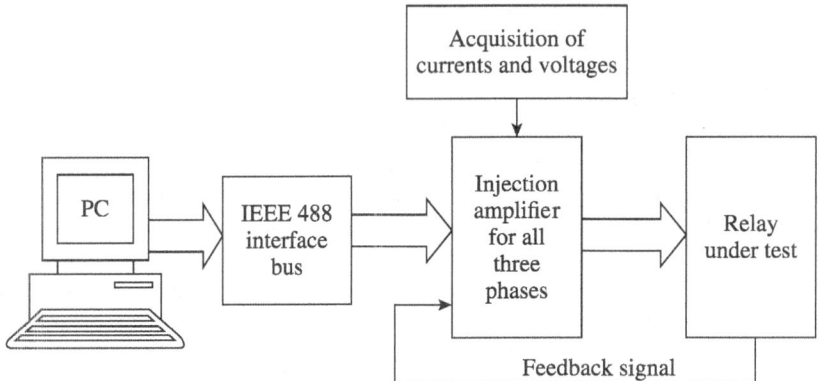

Fig. 15.7 Hardware configuration for computer-based relay testing

Computer-based relay testing can be achieved using two different ways.

Model-specific In model-specific program, checking of a particular type of relay is carried out. It checks all the functions of the particular relay produced by a certain manufacturer. This governs the uniformity, and the test results can be stored to find the trends.

Application-specific An application-specific program checks the primary side of a CT, that is, the actual circuit relay protects the system around it. No focus is given to the relay and its functions.

The main drawbacks of automation of testing are lack of flexibility and high cost.

Recapitulation

- Reliable and efficient operation of protective schemes depends on proper commissioning, frequent testing, and maintenance of the protective devices.
- Testing of protective relays can be done by various types of test kits available in the market as well as the economical test kit developed by the authors.
- Every protective device should be tested for complete calibration and inspection at least once in a year as per recommendation.
- Different tests to be performed on a protective device during its lifetime are type tests, acceptance tests, and routine maintenance tests (also known as periodic tests).

- To enhance the knowledge of relay testing, some laboratory test setup for different types of relays such as overcurrent, overvoltage, differential relay, and distance relay are explained in this chapter.
- The relay testing methods used in field are steady-state testing, dynamic-state testing, and transient testing.
- Computer-based relay testing is necessary for modern digital/numerical relays as it offers complete protection to a power system component that includes many settings.

Multiple Choice Questions

1. Commissioning tests are carried out to ensure
   (a) correct relay characteristic
   (b) that no damage has been done to the relay during transit
   (c) capacity of the relay
   (d) none of the above

2. Acceptance tests are tests that are
   (a) performed to ascertain no damage during transit
   (b) not mandatory to be performed
   (c) performed at the laboratory in the presence of customers
   (d) none of the above

3. Temperature rise test is carried out for all relays
   (a) to check the withstand capability of insulation used in relays
   (b) to ensure integrity of the relay
   (c) to ascertain correct relay characteristic
   (d) none of the above

4. The typical value of insulation resistance is of the order of
   (a) 5–100 kΩ
   (b) 500–1000 Ω
   (c) 5–50 kΩ
   (d) 5 MΩ

5. The main disadvantage of economical overcurrent relay test kit is its susceptibility to
   (a) third and fifth harmonics
   (b) high power loss in resistors
   (c) both (a) and (b)
   (d) none of the above

6. Which of the following tests is performed in detail but only once?
   (a) Commissioning test
   (b) Acceptance test
   (c) Both (a) and (b)
   (d) None of the above

7. All the tripping tests must be performed
   (a) every year
   (b) twice in a year
   (c) everyday
   (d) once in six months

8. Which test is carried out many times but not in too much detail?
   (a) Acceptance test
   (b) Routine maintenance test
   (c) Commissioning test
   (d) All of the above

9. All relay flags and semaphore indicators must be inspected
   (a) daily or on every shift
   (b) monthly
   (c) bimonthly
   (d) none of the above

10. According to IS, in operating time test of overcurrent relays for maximum TDS and PSM between 2 and 4, the permissible deviation must be within
   (a) 12.5%
   (b) 7.5%
   (c) 5.0%
   (d) none of the above

Review Questions

1. Why is testing of each relay required to be carried out?
2. Discuss the various types of tests to be performed on a protective device.
3. Why are commissioning tests performed on relays?
4. What is the frequency of routine maintenance tests?
5. Discuss the factors affecting the frequency of routine maintenance tests.
6. How are the records of commissioning and maintenance tests maintained?
7. Draw the connection diagram of relay testing using the conventional overcurrent relay test kit.
8. Draw and explain the circuit diagram of overvoltage relay test kit.
9. Draw and explain the circuit diagram of biased differential relay test kit.
10. Discuss the various steady-state and dynamic testing methods used for relays.

Answers to Multiple Choice Questions

1. (b), 2. (c), 3. (a), 4. (d), 5. (a), 6. (a), 7. (d), 8. (b), 9. (a), 10. (a)

Recent Developments in Protective Relays 16

Learning Objectives

After going through this chapter, the students will be able to:

- Explain the detection, classification, and location of faults
- Discuss the architectures of wide-area protection
- Explain the concept of synchronized sampling
- Discuss the application of AI and wavelet transform in protective relays
- Explain phasor measurement unit (PMU)

16.1 Introduction

Power stations and numerous substations are connected by transmission lines into a vast network. In order to obtain the direct benefit of this large network through the transfer of electric power from the producer to the consumer, the modern power system should function properly. This is heavily dependent on the healthy operation of the transmission lines within it. Long extra high voltage (EHV) and ultra high voltage (UHV) transmission lines are likely to be subjected to faults due to treacherous weather. If faults are not cleared promptly, the power system can even be driven into instability, resulting in shutdown of either a large portion of the network or the entire network. Since modern networks are increasingly operated closer to their stability limits, fault clearing time becomes more and more important. Therefore, there is a need to detect fault as fast as possible. Furthermore, it is equally important to identify the types of faults and location of fault on the line and other parts of network.

This chapter describes the recent advances in protective relays. The concepts of fault detection, classification, and location technique are discussed. Thereafter, different relay algorithms based on artificial intelligence (AI) techniques are also discussed, along with their relative merits and demerits. The chapter also introduces Phasor Measurement Unit (PMU).

16.2 Fault Detection, Classification, and Location Scheme

One of the major threats to uninterrupted electric power supply is the system faults. Faults on electric power systems are unavoidable. Hence, a well-coordinated protection system must be provided to detect and isolate faults rapidly so that the damage and disruption caused to the power system is minimized. The clearing of faults is usually accomplished by the devices that can sense the fault and quickly isolate the faulty section from the healthy ones.

To protect electrical power systems, faults must be detected accurately and then isolated. The operators in the control centres have to deal with a large amount of data to get the required information about the faults. This indeed takes a long time; however, time is the crucial factor for satisfactory performance of the protection systems. In addition, the protection system may itself fail because of disruption in the communication networks and corruption of transferred data. All these situations put forth challenges to the protection of electric power systems. Important among them are the detection, classification, and location of the faults as fast as possible when they occur.

Fast detection of transmission line faults enables quick isolation of the faulty line from the service and, hence, protecting it from the harmful effects of the fault. Classification of faults means identification of the type of fault which is required for fault location and assessment of the extent of repair work to be carried out. Figure 16.1 shows a generalized block diagram of fault detection, classification, and location scheme. Depending on the algorithm used, data on current and/or voltage of all phases are acquired by the data acquisition system through the current transformer (CT) and the potential transformer (PT). This data is given to the fault detection algorithm. The fault detection algorithm is either current-based or voltage-based or both current and voltage based, and uses the data window

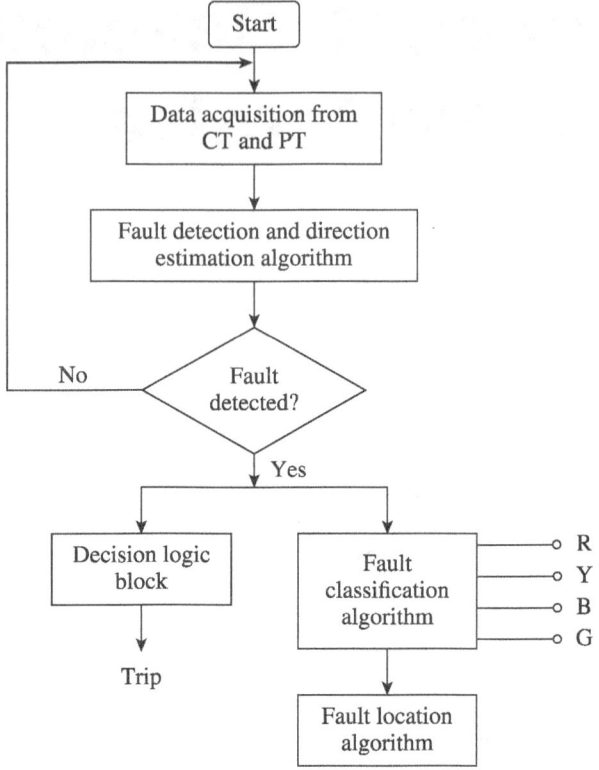

Fig. 16.1 Basic block diagram of fault detection, classification, and location scheme

concept to identify faults. If the value of a particular quantity derived on the basis of the algorithm is less than a certain threshold, a healthy state is assumed and the output is taken as zero; otherwise, a fault is detected.

The value of the threshold chosen is based on extensive simulations of the transmission line, considering different types of faults, with different fault resistances, different fault inception angles, and variable loading conditions. In order to address the noise in the input signal and to enhance the reliability and security of the system, a trip signal is provided after considering three consecutive samples. Fault detection is an online process. After the detection of a fault, the next task is its classification and location. These are offline processes and are carried out by different algorithms based on the data recorded during the fault. Depending on the logical value of R (1/0), Y (1/0), B (1/0), and G (1/0), fault classification is carried out.

16.3 Wide-area Protection and Measurement

In a conventional protection system, decisions are made on the basis of local measurements only, that is, by measuring local currents and voltages. Hence, as only the local system could be protected, it is very difficult to maintain the stability and security of the whole system. This is due to the fact that only local measurements are employed in the protection system. In order to protect the entire system, wide-area protection based on wide-area measurements is developed, and is deemed very promising.

16.3.1 Definition of Wide-area Protection

Wide-area protection is a new type of protection system based on wide-area measurements. It coordinates with conventional protection to isolate faulty electrical components rapidly, reliably, and accurately. This is required to perform online security analysis for the post-fault or post-disturbance system, and to take some proper measures whenever necessary to prevent the power system from outage or, in the worst case, blackout.

16.3.2 Architectures of Wide-area Protection

The wide-area protection system contains three major design architectures. These are described as follows:

Enhancement in SCADA/EMS

In case of a fast-evolving disturbance in the system, it is impossible for the supervisory control and data acquisition (SCADA) system and energy management system (EMS) to extract all important signals. Hence, it is necessary to locate the phasor measurement unit (PMU) at a suitable location all over the grid to improve the accuracy of the state estimation algorithm. As the possibility of building new functions is limited in SCADA/EMS, it is important to provide stand-alone solutions.

Flat Architecture with System Protection Terminals

All modern protective devices that include an intelligent electronic device (IED) have a facility of control and communication. Using the communication facility between different IEDs, synchronized measurements from PMUs can be obtained at different locations in grids. Thus, a decentralized structure of protection terminals can be obtained and robust wide-area protection systems can then be designed.

Multilayered Architecture

Multilayered architecture integrates the functions of protective device and EMS. It consists of three layers, namely the bottom layer, the middle layer, and the top layer. The bottom layer consists of many PMUs, which are used to collect synchro-phasors. The middle layer consists of several local protection centres, which involve data concentrating, protection, and control functions. The top layer includes a system protection centre, which coordinates with the local protection centres.

16.4 Concept of Synchronized Sampling

Phasors are basic tools of AC circuit analysis that are used when the waveforms are in steady-state condition, when the power system is undergoing oscillations during power swings, or when the waveforms are changing rapidly during transient conditions. Hence, in relaying applications, phasors of voltage or current over a half cycle or full cycle are measured. It is possible to measure phasor and phase angle difference in real time using time-synchronizing techniques along with computer-based measurement techniques.

Phasors are measured for each of the three phases, and the positive sequence phasor is computed using the following expression:

$$X_1 = -(X_a + aX_b + a\ X_c), \text{ where } a = e^{j2\pi/3}$$

These voltage and current phasors are sampled precisely at the same instant. This is very easy to achieve for small distance substations by distributing the common sampling clock pulses to all the measuring systems. To measure phasors at a common reference point, the task of synchronizing the sampling clocks is very important. Previously developed techniques have severe service limitations. This can be rectified through a synchronized phasor measurement technique using global positioning satellite (GPS) communication media. This technique is devised using wide-area phasor measurement technology, which is described in Section 16.5.

16.5 Wide-area Phasor Measurement Technology

The technology of synchronized phasor measurements is well established and widely used in state estimation, adaptive relaying, fault and disturbance recordings, and instability prediction. The essential feature of the technique is that it measures positive sequence (and negative and zero sequence quantities, if needed) voltages and currents of a power system in real time with precise time synchronization. This allows accurate comparison of measurements over widely separated locations as well as potential real-time measurement-based control actions. Very fast recursive discrete Fourier transform (DFT) calculations are normally used in phasor calculations. Synchronization is achieved through a GPS system, which provides continuous precise timing of more than 1 μs. Other synchronization signals such as microwave or fibre optics can be used, with sufficient accuracy of synchronization.

Phasor measurement unit (PMU) is the device used to measure synchronized phasors. It is situated at different locations in a power system to measure positive sequence voltages and currents of a power system with precise time synchronization. Afterwards, this data is transferred to a centre for comparison, evaluation, and further processing.

Figure 16.2 shows a typical configuration of synchronized PMU. The GPS transmission is received by the receiver section. Thereafter, it delivers a phase-locked sampling clock pulse to the analog-to-digital (A/D) converter system. The sampled data is converted into a complex number, which represents the phasor of the sampled waveform. The phasors of the three phases are combined to produce a positive sequence measurement.

Modern digital relays are capable of developing positive sequence measurement of voltages and currents. By using a synchronizing pulse derived from the GPS receiver, the measured values are placed on a common time reference frame. To achieve this task, an A/D converter with high resolution is required so that it is capable of providing good accuracy of representation of data even during light load conditions as well as fault conditions.

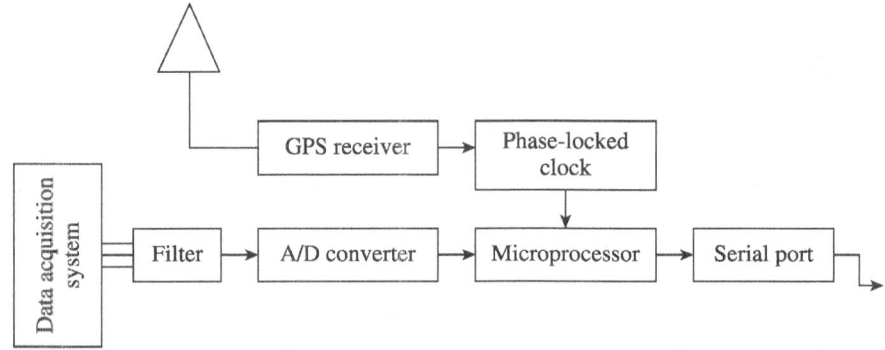

Fig. 16.2 Block diagram of the synchronized phasor measurement unit

For the most effective use of phasor measurements, some kind of data concentrator is required. The simplest is a system that retrieves files recorded at the measurement site and then correlates with files from different sites by the recording time stamps. This allows system and event analysis to utilize the precise phasor measurement of the quantity.

16.6 Introduction to Phasor Measurement Unit

Phasor Measurement Unit (PMU) is a device widely used in modern smart grid environment. Its prototype was developed at Virginia Tech in early 1980s in which a clock signal is given externally through Global

Positioning System (GPS). The first commercial manufacturing of PMUs with Virginia Tech collaboration was started by Macrodyne in 1991.

It is universally accepted that six fault loop equations, during each sample time, have to be solved by the protective device (or more widely Intelligent Electronic Devices (IEDs)) in order to detect any type of fault out of ten types of fault in a power system network. It is possible to perform all fault calculations by utilizing a single equation with the help of symmetrical components and a few quantities derived from them. The state vector of a power system constituted by positive sequence voltages of a network is extremely important for power system fault analysis.

In order to maintain accuracy in the absence of visible satellites, the clock must be equipped with a precision internal oscillator. The format of data files generated and transmitted by the PMU was published by Institute of Electrical and Electronics Engineers (IEEE) in terms of a standard in 1991. Later on, a revised version of the standard was also published in 2005. Some of the important features of PMU are as follows:

(i)   The measurements acquired by the PMU are time-stamped with high precision at the source. This is due to the fact that the data transmission speed is no longer a critical parameter in making use of this data.

(ii) All PMU measurements with the same time-stamp are used to understand the state of the power system at the instant defined by the time-stamp.

(iii) Depending upon the propagation delays of the communication channel in use, PMU data is arrived at a central location at different times.

(iv) In order to create a coherent picture of the power system out of such data, the time tags associated with the phasor data provide an indexing tool.

16.6.1 Generic PMU

Figure 16.3 shows the basic block diagram of a generic PMU. A generic PMU will capture the essence of its principal components of PMUs developed by different manufacturers. The development of this type of PMU is based on symmetrical component based digital relay. In contrast to a relay, a PMU may have currents in several feeders originating in the substation and voltages belonging to various buses in the substation.

As shown in Fig. 16.3, currents and voltages of all the three phases are acquired from the secondary of CTs and PTs/CVTs. These analog signals are passed through isolation/instrument transformer and surge protection circuit. Here, current and voltage signals are stepped down to a lower scale (typically within the range of ±5). Then, signals are given to anti-aliasing filter (AAF) which removes the lower order harmonics. These scaled down analog values are converted into digital form by analog to digital convertor (ADC) and given to microprocessor for further processing.

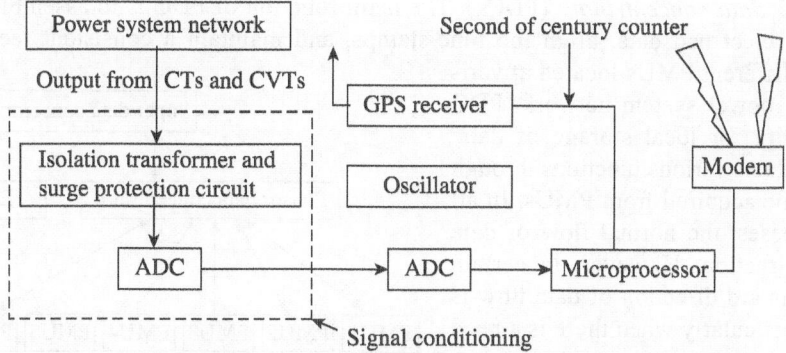

Fig. 16.3 Block diagram of generic PMU

Now, for any data acquisition system, selection of sampling rate is very crucial and directly indicates the frequency response of the anti-aliasing filters. Generally, anti-aliasing filters are analog-type filters with a cut-off frequency less than half the sampling frequency in order to satisfy the Nyquist criterion. However, in modern digital/numerical relays, the normal practice is to use a high sampling rate which is known as *oversampling* with corresponding high cut-off frequency of the analog anti-aliasing filters. If high sampling rate is used then it is followed by a digital 'decimation filter'. The function of this filter is to convert the sampled data into a lower sampling rate. In this way, it provides a digital anti-aliasing filter concatenated with the analog anti-aliasing filters. The main advantage of oversampling of the filter is that it remains immune to ageing and temperature variations. It is to be noted at this juncture that phase angle differences and relative magnitudes of signals remain unaltered due to the same phase shift and attenuation of all the analog signals. The clock pulse of the GPS receiver is phase-locked with the sampling clock pulse. It has been observed from the literature that the sampling frequency used in real field varies between 12 samples/ cycle and 128 samples/cycle. It is understood that higher sampling rates offer better accuracy but at the same time require more computation.

Microprocessor is the main processing unit in which all the calculations are performed. Here, positive sequence value of all currents and voltages are computed (by means of algorithm constructed using set of equations) along with some other values such as frequency and rate of change of frequency. All these values are received as an output of PMU. Based on the signals obtained from the GPS, the time-stamp is generated which recognizes the uniqueness of the universal time coordinated second (UTC). At the same time it also identifies the boundary of one of the power frequency periods. Finally, the output data provided by PMU is transmitted at a next level substation through communication medium with proper modems.

16.6.2 Hierarchy for Phasor Measurement Systems

PMUs installed in substations constitute an important component of power system networks. The substations in which they are installed depend on the type and use of measurements recorded by the substations. Generally, phasor data is used at remote locations and hence, in order to achieve the complete benefit of PMUs, communication links and data concentrators are also available with PMUs. This type of architecture is shown in Fig. 16.4. Figure 16.4 indicates that time-stamped measurements of positive sequence voltages and currents along with frequency and rate of change of frequency of various buses and bays are measured by PMUs and stored in local data storage devices. Such stored data can be accessed by PMUs located at remote locations for diagnostic purposes. As the capacity of local storage devices is limited, some of the important data is permanently stored in it with proper flag marking.

The data stored by different PMUs is then transferred to the next level of the hierarchy which is known as *phasor data concentrators* (PDCs). The main function of PDC is to assemble the data from various PMUs, reject bad data, align the time-stamps, and maintain a consistent record of all data available from different PMUs located at various parts of the power system network. PDC also has a facility for local storage of data. Moreover, it contains various functions through which data can be acquired from PMUs. In all the above processes, the normal flow of data is in upward direction. However, in certain situations, downward direction of data flow is also possible, particularly when there is a need to configure downstream components or when data is required in a particular format.

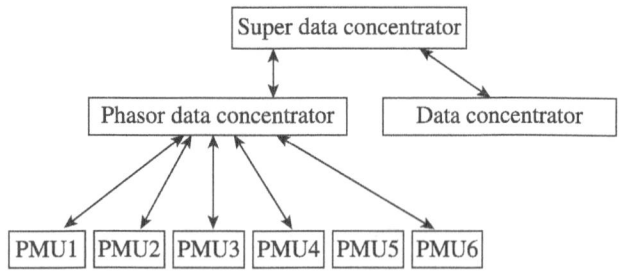

Fig. 16.4 Hierarchy for phasor measurement systems

16.6.3 Communication Options for PMUs

There are many communication mediums available in the field such as powerline carrier, microsatellite, fiber optic, and GPS. Transfer of data, for any communication line, depends on the capacity of channel which is the measure of the data rate usually in kilobits per second or megabits per second. Moreover, it also depends on latency which is defined as the time lag between the time at which the data is created and when it is available for the desired application. Out of the above communication mediums, fiber-optic is widely used due to its high rate of data transfer, unsurpassed channel capacity, and immunity to electromagnetic interference.

It is recommended by IEEE standard for PMU that the communication system may apply any protocol, encryption, or change the ordering of the data, as long as it is restored to its original format at the receiver end. The whole data stream from the PMU must be mapped in proper sequence on the serial communication port particularly when serial communication over an RS-232 is used. PMU messages may also be mapped in their entity into transmission control protocol (TCP) or user datagram protocol (UDP). They can be accessed by using standard Internet Protocol (IP) functions. The IP may be carried over Ethernet or other available transport means.

16.6.4 Functional Requirements of PMUs and PDCs

In actual practice, there is a possibility that PMUs installed in the substations are from different manufacturers. Hence, in order to achieve interoperability among PMUs made by different manufacturers, it is essential that all PMUs perform to a common standard.

IEEE Power System Relaying Committee has given their guidelines regarding interoperability of PMUs in IEEE standard for 'Synchrophasors for Power Systems', C37.118-2005. A working group of the Power System Relaying Committee of IEEE undertook the revision of the standard, and the result is the current standard which clarified the requirements for PMU response to off-nominal frequency inputs.

16.6.5 PMU Performance for Input Signals of any Frequency

The output given by the PMU must be correct at all frequencies irrespective of a balanced or unbalanced input. Moreover, the output phasor of PMU has a magnitude equal to the rms value of the input signal and its phase angle θ must be the angle between the reporting instant and the peak of the sinusoid in case when the input signals are pure sinusoids of any frequency with time-tag in phasor estimate. This is depicted in Fig. 16.5.

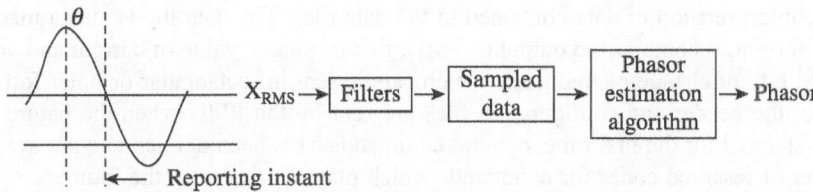

Fig. 16.5 Output of PMU during difference in phase time-tag

Furthermore, input signals given to PMU are passed through a number of filters. Hence, delays are inevitable and these should be compensated before reporting the phasor output of PMU.

In actual practice, the PMU standard calls for this specification to hold over a frequency deviation of $\pm$ 5 Hz from the nominal frequency.

16.6.6 File Structure of Synchrophasors Standard

As per IEEE Standard C37.111-1991, Common Format for Transient Data Exchange (COMTRADE) file format is used by all digital/numerical relays, fault locators, synchrophasors, and various transient data power producers. This is the only file format recommended by International Electrotechnical Commission (IEC) and widely used for transient data collection and distribution.

PMUs generate the following four types of files for transmission of data.

(i) Header files
(ii) Configuration files
(iii) Data files
(iv) Command file

Out of the above four files, the command file is used by PDC for communication with PMU. All the said four files have a common structure. Figure 16.6 shows the file structure of synchrophasor standard used by digital relays

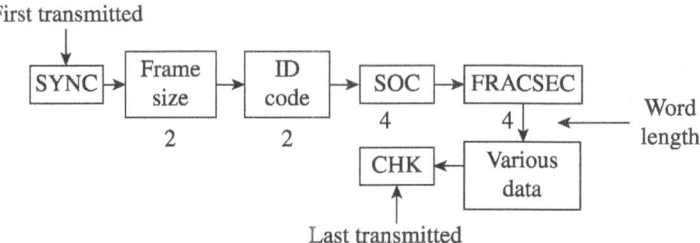

Fig. 16.6 File structure of synchrophasors standard

Synchronization of data transfer is carried out by the first word which is of 2 bytes in size. The second word defines the size of the total record whereas the third word indicates the data originator uniquely.

The next two words provide the 'second of century' (SOC) and the 'fraction of a second' (FRACSEC) at which the data is being reported. The specifications given in the configuration file decide the length of the data words 'fraction of a second'. Any type of error in the data transmission is determined by the last word, that is, checksum. The first file, that is, header file is a file that can be read by humans. Sharing of any kind of data by the data producer with the user data is done by this file.

Configuration file is a machine readable file with specific format and used for providing information regarding the interpretation of data contained in the data files. The data file is also a machine readable file with specific format. It contains the output of PMU, that is, phasor value of current and voltages along with frequency and rate of change of frequency which can be sent in rectangular or polar form.

In practice, the header and configuration files are sent by the PMU when the nature of the data being transferred is defined for the first time. Several commands have been defined and are available at this time, with a number of reserved codes for commands which may be needed in the future.

16.7 Travelling Wave-based Algorithm

Whenever a fault occurs, it propagates as a travelling wave on a transmission line. These travelling waves, when travelling along a transmission line, encounter discontinuities such as buses and transformers. When they reach a discontinuity, a part of the wave is reflected back and the remaining part of the wave passes through the line. The magnitude of the reflected and refracted waves depends on the characteristic impedance (Z_c) of the transmission line as well as the impedance beyond the discontinuity. The amplitude of the reflected and refracted waves is such that the proportionality of the voltage and current is preserved. The phenomenon of reflection and refraction of travelling waves is usually shown by the Bewley's Lattice diagram.

Many protection schemes based on the travelling wave have been proposed by researchers. Figure 16.7 shows the voltage and current during a fault on the transmission line, which is connected between bus A and bus B. During a fault on the transmission line at point F, the post-fault voltage (V) and current (i) can be split into four components. Two components are the pre-fault voltage and current and the other two components are the changes in the voltage and current due to the fault. Pre-fault voltage and current are shown in Fig. 16.8. Figure 16.9 shows a model of a transmission line in which a fault is replaced by a fictitious source.

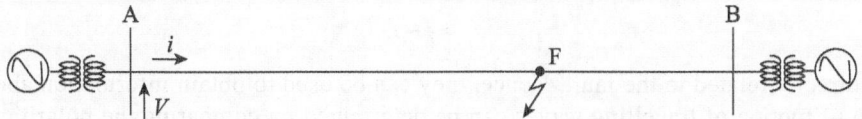

Fig. 16.7   Voltage and current during a fault on the transmission line

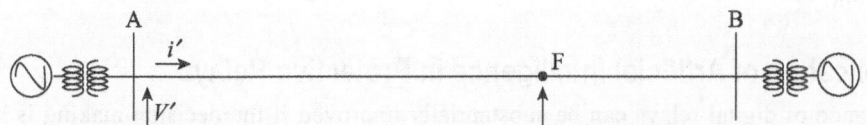

Fig. 16.8   Voltage and current before fault on the transmission line

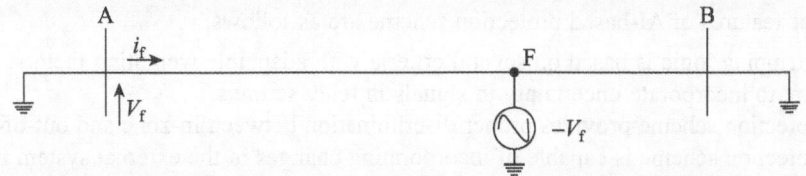

Fig. 16.9   Components of voltage and current injected by the fictitious source at the fault point

Fault-injected components of voltage (V_f) and current (i_f) can be expressed in terms of forward and backward travelling waves.

$$V_f(x, t) = f^+\left(t - \frac{x}{v}\right) + f^-\left(t + \frac{x}{v}\right) \tag{16.1}$$

$$i_f(x, t) = \frac{1}{Z_0}\left[f^+\left(t - \frac{x}{v}\right) + f^-\left(t + \frac{x}{v}\right)\right] \tag{16.2}$$

where f^+ and f^- are functions representing the forward and backward travelling waves,

v is the velocity of propagation of travelling waves,

Z_0 is the surge impedance of the transmission line, and

x is the distance travelled by the travelling waves.

Rearranging Eqs (16.1) and (16.2) provides

$$2f^+\left(t - \frac{x}{v}\right) = V_f(x, t) + Z_0 i_f(x, t) \tag{16.3}$$

$$2f^-\left(t + \frac{x}{v}\right) = V_f(x, t) - Z_0 i_f(x, t) \tag{16.4}$$

If the pre-fault voltage and current are V' and i', and the fault-injected voltage and current are V_f and i_f, then

$$V_f = V - V' \tag{16.5}$$

$$i_f = i - i' \tag{16.6}$$

V_f and i_f are directly related to the fault. Hence, they can be used to obtain information about the fault. The direction of motion of travelling waves can be determined by comparing the polarities of the pre-fault voltage and current with the polarities of the fault-injected voltage and current. Discrimination between internal and external faults is obtained by comparing the polarities of the fault-injected voltage and current.

16.8 Application of Artificial Intelligence in Protective Relays

The performance of digital relays can be substantially improved if the decision-making is based on AI. Various AI techniques used for the design of the intelligent relay are artificial neural network (ANN), fuzzy logic, and expert systems. This section deals with the application of these three techniques as applied to protection.

The important features of AI-based protection scheme are as follows:

1. Decision of tripping logic is based on several criteria with adaptable weighting factors.
2. It is very easy to incorporate uncertainty in signals in relay settings.
3. AI-based protection scheme provides proper discrimination between in-zone and out-of-zone faults.
4. AI-based protection scheme is capable of incorporating changes in the external system in relay settings and is also capable of adding an adaptive feature in the relay settings.

16.8.1 Neural Network-based Scheme

During the last decade, digital protective relaying of transmission lines has greatly benefitted from the development of AI and signal processing techniques. The use of ANN for solving problems in power systems is rapidly increasing. The generalization and fault tolerance capabilities of a neural network make it a reliable tool to be used to handle unseen fault patterns. The neural network applications in transmission line protection are mainly concerned with improvements in effective and efficient fault detection, classification, and location schemes.

The block diagram of ANN-based fault detection, classification, and location scheme for transmission line is shown in Fig. 16.10. The proposed scheme provides trip signal only in the case of forward faults that occur in the transmission line when the power flows in the forward direction, that is, from bus to line. The forward direction needs to be identified so that the relay does not operate for reverse faults, that is, the faults that occur behind the relay, either at the bus or on the line. To estimate the direction of the fault, an approach based on the status of the normal power direction along with directional discrimination functions is used.

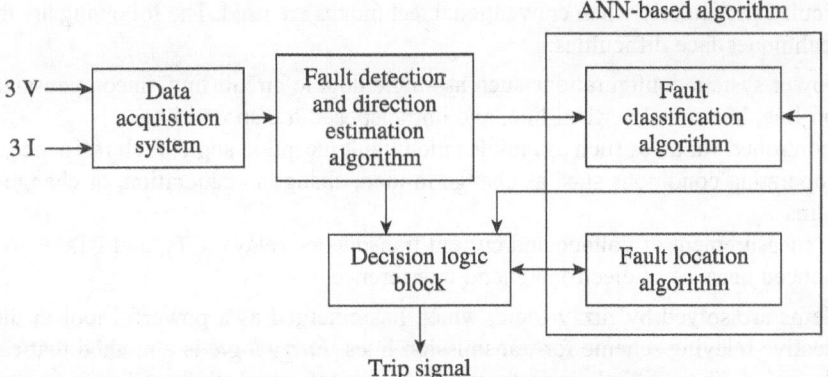

ANN-based algorithm

Trip signal

Fig. 16.10 Block diagram of ANN-based scheme

The ANN-based algorithm is initiated by collecting a one-cycle sampled data window for each signal. In general, a sampling frequency of 2–4 kHz can be used for a 50 Hz system. For each new sample to enter the window, the oldest one is rejected. The current signals are decomposed through wavelet transform (WT) (explained in Section 16.9) for fault detection in the first stage of the proposed scheme.

Once the fault is detected in the forward direction, fault classification and location follows. For fault classification and location, three voltages and three current signals are fed into the ANN algorithm. The fault classification unit classifies the fault type first and then sends a signal to the decision logic. The decision logic sends a signal for fault location and fires the appropriate neural network (different for various types of fault). The decision logic block derives the trip or not-to-trip decision from the output signals of the classifier and locator. Once the fault detection unit detects a fault in the forward direction, it triggers the fault classifier unit. The fault classifier unit classifies the type of fault using the ANN.

Figure 16.11 shows the block diagram of the fault classification unit using the ANN. It consists of 11 outputs, numbered from 1 to 11, representing different types of faults with pre-fault condition. During training of ANN, these outputs are assigned 1 or 0 considering whether or not a particular fault is involved. This classification approach considers a particular phase to be involved with fault if its corresponding value is greater than 0.5; else, it categorizes the phase to be undisturbed.

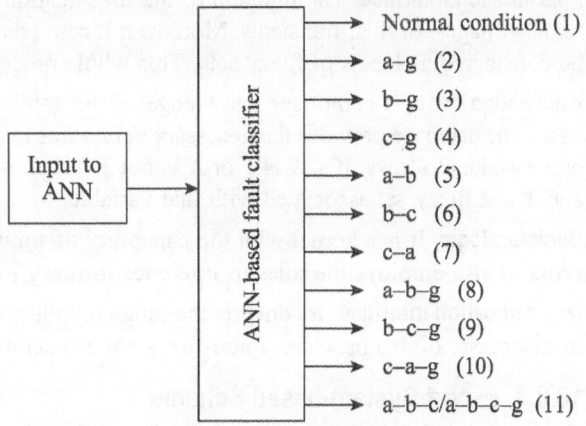

Fig. 16.11 Block diagram of ANN-based fault classifier

16.8.2 Fuzzy Logic-based Scheme

Most of the conventional protection techniques define states of the equipment by identifying the patterns of the associated voltage and/or current waveforms. However, there is a possibility of a large element of uncertainty and vagueness due to the complex relationships between the vast numbers of system variables. Moreover, because of the regulatory reforms and environmental concerns, many utilities worldwide are investigating novel ways of better utilization and control of existing transmission systems. This imposes a significant difficulty, particularly when conventional techniques are used. The following are the reasons why conventional techniques face difficulties:

1. Different power system configurations such as single/double circuit line, uncompensated/compensated transmission line, horizontal/vertical line, and untransposed/transposed line
2. Various uncontrolled variables such as fault location, fault inception angle, fault resistance, and fault types
3. Change in operating conditions such as change in load, change in generation, or change in topology of power systems
4. Error in the measurement of voltage and current transducers, relays, CTs, and PTs
5. Noise introduced because of electromagnetic interference

These problems are solved by fuzzy logic, which has emerged as a powerful tool in the development of a novel protective relaying scheme for transmission lines. *Fuzzy logic* is a method that easily represents *human expert knowledge* on a digital computer where mathematical or rule-based expert systems experience difficulty. Figure 16.12 shows the generalized structure of a fuzzy logic-based relaying scheme.

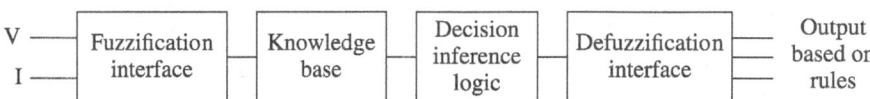

Fig. 16.12 Generalized structure of a fuzzy-logic-based relaying scheme

It contains four major components, namely fuzzification interface, knowledge base, decision logic, and defuzzification interface.

Fuzzification interface The function of the fuzzification interface is to measure input quantities such as currents, voltages, or their transients. Moreover, it converts input data into suitable linguistic values, which may be considered as labels of fuzzy sets. This whole process is known as *fuzzification*.

Knowledge base It comprises knowledge of the application domain and consists of a database and a rule base. The database provides the necessary definitions to define linguistic control rules. Fuzzy rules are usually expressed as follows: if $< X$ is Y or X is not $Y >$ then $< X$ is Y or X is not $Y >$, where X is a scalar variable and Y is a fuzzy set associated with that variable.

Decision logic It is a kernel with the capability of simulating human decision-making based on fuzzy concepts. It also employs the rules of inference in fuzzy logic.

Defuzzification interface It converts the range of values of the output variables into corresponding universes of discourse. It also provides a non-fuzzy control action from an inferred fuzzy control action.

16.8.3 Expert System-based Scheme

The setting of the operating parameters of relays requires expertise in relays and experienced relay engineers. This problem is compounded when various types of relays from different manufacturers are involved. Further, changes in the external system conditions also impose an additional burden on the relay setting process. As the relay settings are available in a rule style, an expert system is capable of optimizing relay settings. Figure 16.13 shows the generalized structure of an expert system-based relay.

It contains four main components.

Knowledge base It consists of frames and production rules used in representing a system's knowledge. This is provided by the expert to enhance the reasoning process.

Inference engine Once the expert gives effective reasoning, knowledge process is done by the inference engine. It interacts with the user interface to accept the description of the network from the user. It also supplies the relay setting results. With the available information on a given problem, the inference engine takes the help of the knowledge stored in the knowledge base and draws conclusions, which will be further used for recommendations. It also takes the help of short circuit analysis algorithm to perform short circuit analysis.

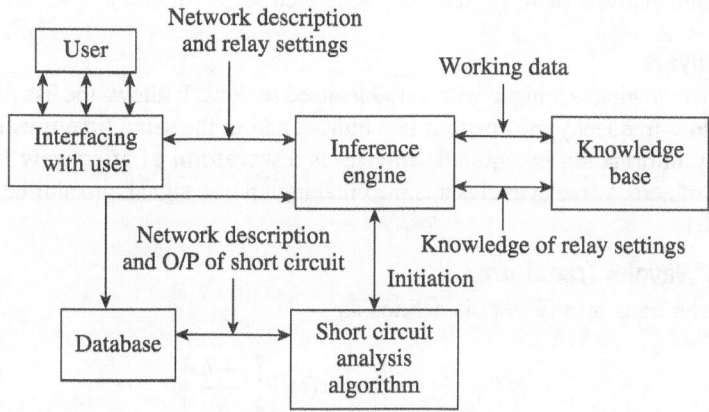

Fig. 16.13   Generalized structure of an expert system-based relay

Database It helps users by providing the required information to solve their problem.

User interfacing It helps users to solve their problems by consulting the expert system. It also assists users by explaining the concepts of various functions.

16.9 Application of Wavelet Transform in Protective Relaying

Wavelet transform (WT) was introduced in the beginning of the 1980s, and is still continuing to grow independently in the field of mathematics, quantum physics, electrical engineering, and seismic geology. As WT is better suited for the analysis of certain types of transient waveforms, it has received great attention in the power community. It can be effectively used for fault detection, fault discrimination, fault classification, and phasors calculations of the measured signals. Literature review reveals that the use of WT in power system protection is increasing by the day.

16.9.1 Introduction

For the last several years, Fourier transform has been extensively used by many researchers in the field of power system protection. However, when a signal is transformed to the frequency domain, the time domain information is lost, which is a serious drawback with Fourier transform. In the Fourier transform of a signal, it is impossible to predict when a particular event has taken place. If the signal properties do not change much over time (in case of a stationary signal), this drawback is not significant. However, fault signals contain numerous non-stationary or transitory characteristics. These characteristics are often very significant in the signal, and Fourier analysis is not suited for their detection. Wavelet transforms are capable of revealing those aspects of data that are usually missed by other signal analysis techniques (such as Fourier). These aspects include trends, breakdown points, discontinuities in higher derivatives, and self-similarity. Furthermore, as wavelet analysis provides information in both frequency and time domains, it can compress or de-noise a signal without appreciable degradation. Indeed, within their short span of application in the signal processing field, wavelets have proved themselves to be an indispensable addition to the signal analysts' collection of tools, and today they enjoy burgeoning popularity.

Wavelets are mathematical functions that decompose data into different frequency components, and then study each component with a resolution matched to its scale. There are two main approaches which are used to present the wavelet theory:

1. The integral transform approach (continuous time)
2. The multi-resolution analysis (MRA)/filter bank approach (discrete time)

16.9.2 Wavelet Analysis

Wavelet analysis is a windowing technique with variable-sized regions. It allows the use of long time intervals, where more precise low-frequency information is required, and at the same time uses short time intervals, where high-frequency information is required. Wavelet is a waveform of effectively limited duration that has an average value of zero. Wavelet analysis is the breaking up of a signal into shifted and scaled versions of the original wavelet.

16.9.3 Continuous Wavelet Transform

The continuous wavelet transform (CWT) is defined as

$$\text{CWT}(a,b) = \frac{1}{\sqrt{a}} \int_{-\infty}^{\infty} x(t)\, \Psi\!\left(\frac{t-b}{a}\right) dt \tag{16.7}$$

where the signal $x(t)$ is transformed by an analysing function $\Psi\!\left(\dfrac{t-b}{a}\right)$.

The analysing function $\psi(t)$ is known as mother wavelet. It is not limited to the complex exponential, but it is short and oscillatory. Figure 16.14 shows some examples of the mother wavelet.

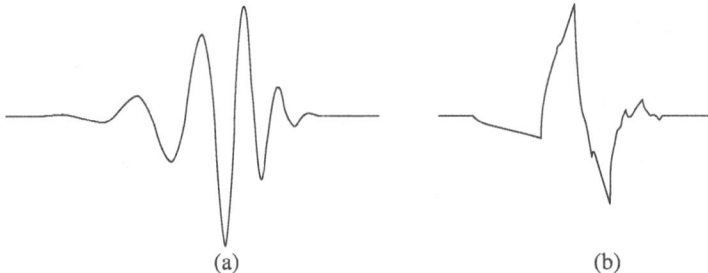

(a) (b)

Fig. 16.14 Example of mother wavelets with (a) low scale factor and (b) high scale factor

Equation (16.7) can be rewritten as the inner product of the original signal with scaled, shifted versions of the basic mother wavelet function $\psi(t)$ and is given by

$$\text{CWT}(a, b) = \int x(t) \cdot \Psi(t)\, dt = \langle x, \Psi_{a,b} \rangle$$

where $\emptyset_{a,b} = \dfrac{1}{\sqrt{a}} \emptyset\!\left(\dfrac{t-b}{a}\right)$; a indicates the scale that determines the oscillating behaviour of the particular daughter wavelet, and b represents the shifting of the mother wavelet, which provides time localization information of the original signal.

It is difficult to use CWT in practice. Hence, for easy computer implementation, the discrete wavelet transform is used.

16.9.4 Discrete Wavelet Transform

Discrete wavelet transform (DWT) involves two fundamental equations: scaling function and wavelet function, as given in Eqs (16.8) and (16.9), respectively.

$$\varphi(t) = \sqrt{2}\sum_n h_n \varphi(2t - n) \tag{16.8}$$

$$\psi(t) = \sqrt{2}\sum_n g_n \varphi(2t - n) \tag{16.9}$$

where h_n is a low band filter and g_n is a high band filter. $\varphi(t)$ and $\psi(t)$ are scaling function and wavelet function, respectively. For the orthogonal wavelets, the relation between the filter coefficients of the low band filter and the high band filter is

$$g(k) = (-1)^k \bar{h}(1 - k), \, k \in z \tag{16.10}$$

A discrete signal can be decomposed to obtain information in different scales by DWT. With the given coefficients, if the sampling data is $\{C_{j+1,k}\}$, then the decomposition equation is given in Eqs (16.11) and (16.12).

$$C_{j,k} = \sum_l h_{l-2k} \, C_{j+1,l} \tag{16.11}$$

$$D_{j,k} = \sum_l g_{l-2k} \, C_{j+1,l} \tag{16.12}$$

where $0 \le k \le N - 1$, $k \in z$; $C_{j,k}$ is the kth calculation value on scale j of low band frequency; $D_{j,k}$ is the kth calculation value on scale j of high band frequency, and l is the location index.

Wavelet transform can be implemented with a specially designed pair of finite impulse response (FIR) filters called a *quadrature mirror filters* (QMFs) pair. QMFs are distinctive because the frequency responses of the two FIR filters separate high-frequency and low-frequency components of the input signal. The dividing point is usually halfway between 0 Hz and half the data-sampling rate (the Nyquist frequency). The tree or pyramid algorithm can be applied to the WT by using the wavelet coefficients as the filter coefficients of the QMF filter pairs.

Recapitulation

- The modern protective relaying technology has many advanced features such as fault detection, classification, and location in the network.
- The operation of recently developed relays based on synchronized sampling and wide-area measurement technique along with application of PMU is noteworthy.
- Different algorithms used by modern digital relays such as the travelling wave, artificial neural network (ANN), fuzzy logic, and expert system are extremely important.
- A new pre-processing technique known as wavelet transform (WT), which is superior to the discrete Fourier transform (DFT), is widely used in modern digital/numerical relays.

Multiple Choice Questions

1. Fault detection is an
   - (a) online process
   - (b) offline process
   - (c) both (a) and (b)
   - (d) none of the above

2. A fault detector unit is
   - (a) a distance protection unit
   - (b) an overcurrent unit
   - (c) both (a) and (b)
   - (d) none of the above

3. Fault classification and location are

   (a) online processes   (c) both (a) and (b)

   (b) offline processes   (d) none of the above

4. The wide-area protection scheme measures

   (a) local voltages and currents

   (b) partial local voltages and currents

   (c) voltages and currents at each substation

   (d) none of the above

5. The phase measurement unit is a device used to measure

   (a) synchronized phasors   (c) impedance of the line

   (b) voltage and frequency   (d) none of the above

6. A wavelet gives

   (a) more precise low-frequency information only

   (b) more precise high-frequency information only

   (c) more precise low-frequency and high-frequency information only

   (d) none of the above

7. Wavelet transform measures information in

   (a) time domain only

   (b) frequency domain only

   (c) both time and frequency domains

   (d) none of the above

8. For easy computer implementation,

   (a) DWT is more suitable

   (b) CWT is more suitable

   (c) both (a) and (b)

   (d) none of the above

9. Fourier analysis is not suitable to detect

   (a) trends, breakdown points, discontinuities in higher derivatives, and self-similarity

   (b) information in frequency domain

   (c) both (a) and (b)

   (d) none of the above

10. The accuracy of synchronization achieved through a GPS system is

   (a) more than 1 s   (c) more than 10 ms

   (b) more than 1 ms   (d) more than 1 μs

Review Questions

1. Why is fault detection carried out separately along with protective device?

2. What is the importance of fault classification and fault location techniques?

3. Explain the basic flowchart of fault detection and classification schemes used in a power system.

4. How is the wide-area protection scheme different from the traditional protection scheme?

5. Define the wide-area protection scheme.

6. What is the function of PMU?

7. Explain the basic architecture of the wide-area protection scheme.

8. Explain the concept of the travelling-wave-based protection scheme.

9. Explain the various protective relaying schemes based on neural network, fuzzy logic, and expert system.

10. Discuss the various similarities and differences between Fourier transform and WT.

Answers to Multiple Choice Questions

1. (a)   2. (b)   3. (b)   4. (c)   5. (a)   6. (c)   7. (c)   8. (a) 9. (a)   10. (d)

PSCAD and Its Application in Power System

17

Learning Objectives

After going through this chapter, the students will be able to:

- List the key applications of PSCAD
- Discuss the various library models of PSCAD
- Explain the various components of continuous system model functions
- Explain stepwise the procedure for the construction of a power system model
- List and explain the various relay characteristics available in the PSCAD library
- Explain special converter and inverter model of power electronics

17.1 Introduction

Power systems CAD (PSCAD) is a powerful and flexible graphical user interface to the world-renowned EMTDC (electromagnetic transients including DC) solution engine. It enables the user to schematically construct a circuit, run a simulation, analyse the results, and manage the data in a completely integrated, graphical environment. Online plotting functions, controls, and meters are also included so that the user can alter system parameters during a simulation run and view the results directly. PSCAD comes complete with a library of pre-programmed and tested models, ranging from simple passive elements and control functions to more complex models, such as electric machines, FACTS (flexible AC transmission systems) devices, transmission lines, and cables. If a particular model does not exist, PSCAD provides the flexibility of building custom models, either by assembling them graphically using existing models, or by utilizing an intuitively designed *design editor*.

The following are some common models found in systems studied using PSCAD:

1. Resistors, inductors, and capacitors
2. Mutually coupled windings, such as transformers
3. Frequency-dependent transmission lines and cables (including the most accurate time domain line model in the world)
4. Current and voltage sources
5. Switches and breakers
6. Protection and relaying
7. Diodes, thyristors, and gate turn-off thyristors (GTOs)
8. Analog and digital control functions
9. AC and DC machines, exciters, governors, stabilizers, etc.

10. Meters and measuring functions
11. Generic DC and AC controls
12. High voltage direct current (HVDC), static var compensator (SVC), and other FACTS controllers
13. Wind source, turbines, and governors

PSCAD and its simulation engine EMTDC have enjoyed close to 30 years of development, inspired by ideas and suggestions by its ever-strengthening, worldwide user base. This development philosophy has helped establish PSCAD as one of the most powerful and intuitive CAD software packages available.

PSCAD can find applications in the simulation of the following cases:

1. Insulation coordination of AC and DC equipment
2. Contingency studies of AC networks consisting of rotating machines, exciters, governors, turbines, transformers, transmission lines, cables, and loads
3. Relay coordination
4. Transformer saturation effects
5. Optimal design of controller parameters
6. Investigation of new circuit and control concepts
7. Traditional power system studies, including transient overvoltage (TOV), transient recovery voltage (TRV), faults, reclosure, and ferroresonance
8. Relay testing (waveforms) and detailed analysis of the current transformer (CT)/voltage transformer (VT)/coupling capacitor voltage transformers (CCVT) responses and their impact on operation
9. Waveforms generated by PSCAD can be saved using the PSCAD real time playback (RTP)/comtrade recorder. Then, by using the RTP playback system, these waveforms can be used to test physical protection and control equipment.
10. Design of power electronic systems and controls, including FACTS devices, active filters, low voltage series, and shunt compensation devices
11. Incorporation of the capabilities of MATLAB/Simulink directly into PSCAD/EMTDC
12. Subsynchronous oscillations, their damping, and resonance
13. Effects of DC currents and geomagnetically induced currents on power systems, inrush effects, and ferroresonance
14. Distribution system design, including transient overvoltages, with custom power controllers and distributed generation
15. Power quality analysis and improvement, including harmonic impedance scans, motor starting sags and swells, non-linear loads, such as arc furnaces, and associated flicker measurement
16. Design of modern transportation systems (ships, rail, automotive) using power electronics
17. Design, control coordination, and system integration of wind farms, diesel systems, and energy storage
18. Variable speed drives, their design, and control
19. Industrial systems
20. Intelligent multiple-run optimization techniques can be applied to both control systems and electrical parameters.

The workspace can be divided into three sections. The project section lists all the loaded cases and libraries. The output section shows the status of each project and any errors or warnings that the project might have. The design editor section consists of the actual circuit diagram that is under investigation. The 'circuit element toolbar' provides an easy drag-and-drop feature to insert circuit elements although a larger selection is available via the master library. The 'general toolbar' consists of standard Windows features such as cut, copy, paste, and zoom. All these features provide for a comfortable user-oriented interface that is easy and fun to work with. Figure 17.1 shows the user interface of PSCAD v4.2.1.

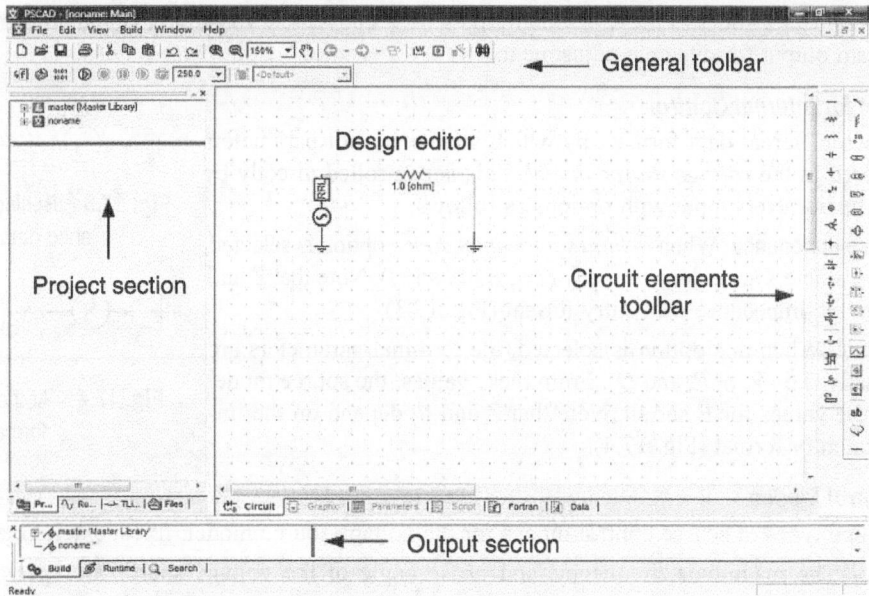

Fig. 17.1 Graphical user interface of PSCAD® v4.2.1

17.2 Different Library Models

The library section contains various system components and a quick description of each model. For more details on these components, one can refer to the PSCAD online help system. The following are a few models available in PSCAD.

17.2.1 Voltage Source Model

As shown in Fig. 17.2, the voltage source component models a three-phase AC voltage source, with specified source and/or zero-sequence impedance. A zero-sequence impedance branch may be added directly within the component. The RL and RRL blocks indicate the type of internal impedance used in the source. In addition, this component allows the regulation of the bus voltage on a remote location on the network, or the internal phase angle can be regulated to control the source output power.

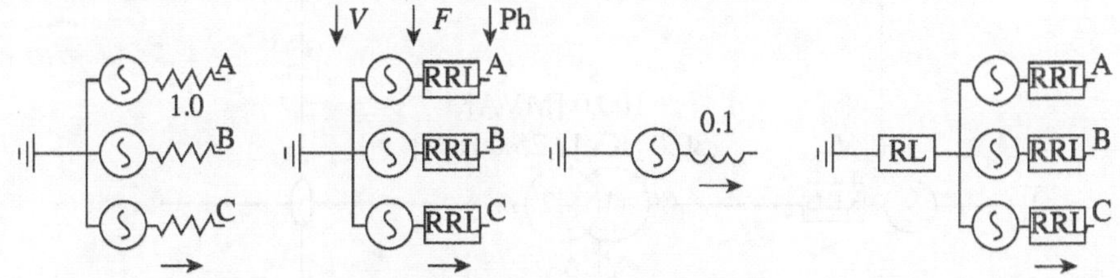

Fig. 17.2 Voltage source model

This source may be controlled through either fixed internal parameters or variable external signals. The external inputs are described as follows:

1. *V*—Line-to-line, RMS voltage magnitude (kV)
2. *F*—Frequency (Hz)
3. Ph—Phase angle (° or rad)

One can connect a slider to these external inputs for a convenient run-time manual adjustment, or use a control system output for dynamic adjustment.

Data Format for Internal Control

There are two different data formats by which source control parameters may be entered in the *voltage source model*. This is controlled directly by the specified parameters input with options as follows:

Fig. 17.3 Behind source impedance data format

Behind source impedance When *behind source impedance* option is selected, the source parameters are entered directly (i.e., E, Φ, and f). Note that Z and Φ depend on the impedance data entry format (Fig. 17.3).

At the terminal When this option is selected, the terminal parameters are entered directly (i.e., V, δ, P, and Q). From these values, the source model determines the values for E and Φ. Note that Z and Φ depend on the impedance data entry format (Fig. 17.4).

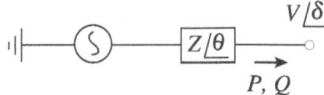

Fig. 17.4 At the terminal data format

Source Control Modes

There are three types of source control modes for the voltage source model: fixed, external, and auto.

Fixed control The magnitude, frequency, and phase angle of the voltage source are specified internally through the *source values for fixed control* dialog. The base voltage and base frequency specified in the *configuration* dialog are not meant for controlling.

External control *External control* option provides external input connections for specifying magnitude, frequency, and phase values. You can connect a slider to this input for a convenient run-time manual adjustment, or use a control system output for dynamic adjustment. The external input signals must be in the following units:

1. *Magnitude*—kV, line-to-line, RMS
2. *Frequency*—Hz
3. *Phase angle*—Degrees or radians, depending on the *external phase input unit* setting in the configuration dialog.

Auto control (three-phase only) In auto control mode, the voltage magnitude can be adjusted automatically to regulate the voltage at a selected bus and/or adjust the source phase angle internally to regulate the real power leaving the source. Figure 17.5 shows how the source is connected to allow automatic voltage control.

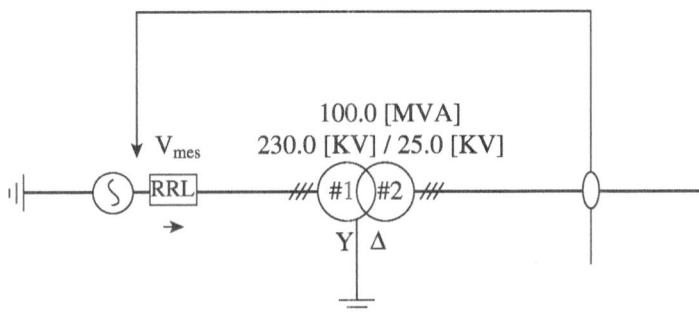

Fig. 17.5 Automatic voltage control

17.2.2 Transmission Lines

There are two basic ways to construct an overhead line in PSCAD. The first is the original method referred to as the *remote ends* method, which involves a transmission line configuration component with two overhead line interface components, representing the sending and receiving ends of the line. The purpose of the *interface* components is to connect the transmission line to the greater electric network (Fig. 17.6).

As shown in Fig. 17.7, the second and more recently introduced method is to use the *direct connection* method, where the interfaces and the corridor properties are housed within a single component. This method, however, can only be used for one-phase, three-phase, or six-phase, single-line systems, where the maximum number of conductors is six.

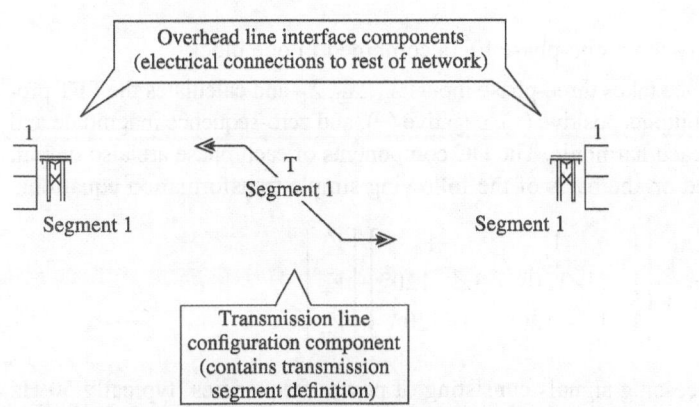

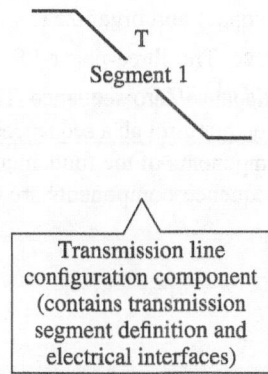

Fig. 17.6 An overhead transmission line (remote ends method)

Fig. 17.7 An overhead transmission line (direct connection method)

17.2.3 Machines and Transformers

The library section has synchronous machines (single-line diagram view and three-phase view) such as induction machine model (single-line diagram view and three-phase view), DC machine model, permanent magnet machine model, AC and DC exciter models, static exciter models, version 2 compatible governor and exciter models, internal combustion engine, wind turbine, source, governor model, turbine control models, power system stabilizers, steam governor models, and hydro governor models.

Moreover, single-phase transformer and three-phase transformer models (two winding, three winding, and four winding), unified magnetic equivalent circuit (UMEC) transformer models, and autotransformer models are also available.

17.2.4 Online Frequency Scanner (Fast Fourier Transform)

Figure 17.8 shows an online fast Fourier transform (FFT) that can determine the harmonic magnitude and phase of the input signal as a function of time. The input signals are first sampled before they are decomposed into harmonic constituents.

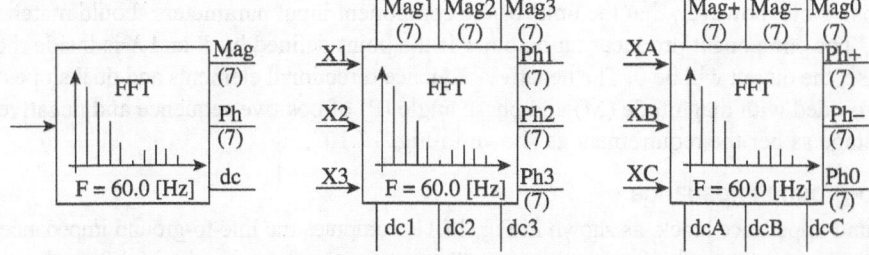

Fig. 17.8 Online frequency scanner

Options are provided to use one, two, or three inputs. In the case of three inputs, the component can provide output in the form of sequence components.

The user may select one of the following three FFT block types:

One-phase This is a standard one-phase FFT. The input is processed to provide the magnitude 'Mag' and phase angle 'Ph' of the fundamental frequency and its harmonics (including the DC component, dc).

Two-phase The two-phase FFT is equivalent to two one-phase FFTs combined in a single block to keep things compact and organized.

Three-phase The three-phase FFT is merely three one-phase FFTs combined in one block.

Positive/Negative/Zero-sequence The sequence takes three-phase inputs X_A, X_B, X_C and calculates the FFT preliminary output through a sequencer, which outputs positive (+), negative (−), and zero-sequence magnitude and phase components of the fundamental and each harmonic. The DC components of each phase are also output.

The sequence components are computed on the basis of the following simple transformation equation:

$$\begin{bmatrix} V_a \\ V_b \\ V_c \end{bmatrix} = \frac{1}{3} \cdot \begin{bmatrix} 1 & 1 & 1 \\ 1 & 1\angle 120° & 1\angle -120° \\ 1 & 1\angle -120° & 1\angle 120° \end{bmatrix} \cdot \begin{bmatrix} V_0 \\ V_+ \\ V_- \end{bmatrix}$$

Note: This component is meant for processing signals consisting of power frequencies (typically 50 Hz and 60 Hz) and its harmonics and therefore, is not well tested for higher frequencies.

17.2.5 Sequence Filter

The sequence filter calculates the magnitudes and phase angles of sequence components when those of the phase quantities are given (Fig. 17.9).

17.2.6 Protection Components

Protection components contain elements related to protection such as impedance zone elements, out-of-step elements, negative-sequence directional elements, dual slope differential elements, inverse time overcurrent elements, CT models, two CTs in differential mode, potential transformer (PT), capacitive voltage transformers, overcurrent detection element block, average phase comparator, phase quantities to sequence component filter, L–L impedance calculation element, and L–G impedance calculation elements. Figure 17.10 shows some of the main components used for each type of protection scheme. Out of these, for distance protection, the widely used relay characteristic is *mho circle*. The mho circle component is classified as an 'impedance zone element' that checks whether a point described by the inputs R and X lies inside a specified region on the impedance plane. R and X represent the resistive and reactive parts of the monitored impedance, and may be input in per unit or ohms. Note, however, that the units of the component input parameters should match that of the R and X inputs. The component produces an output 1 if the point defined by R and X is inside the specified region; otherwise, the output will be 0. The negative-sequence directional elements and dual slope differential elements are provided with magnitude (M) and phase angle (P) of positive sequence and negative sequence current and voltage as per the requirement as shown in Fig. 17.10.

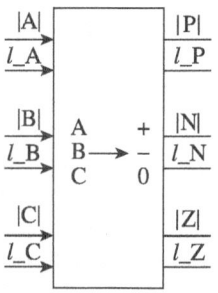

Fig. 17.9 Sequence filter

17.2.7 Line-to-ground Impedance

The line-to-ground impedance block, as shown in Fig. 17.11, computes the line-to-ground impedance as per the given input sequence quantities of voltage and current. The output impedance as obtained from the said block is in rectangular format (R and X). These values are utilized for plotting various distance relay characteristics such

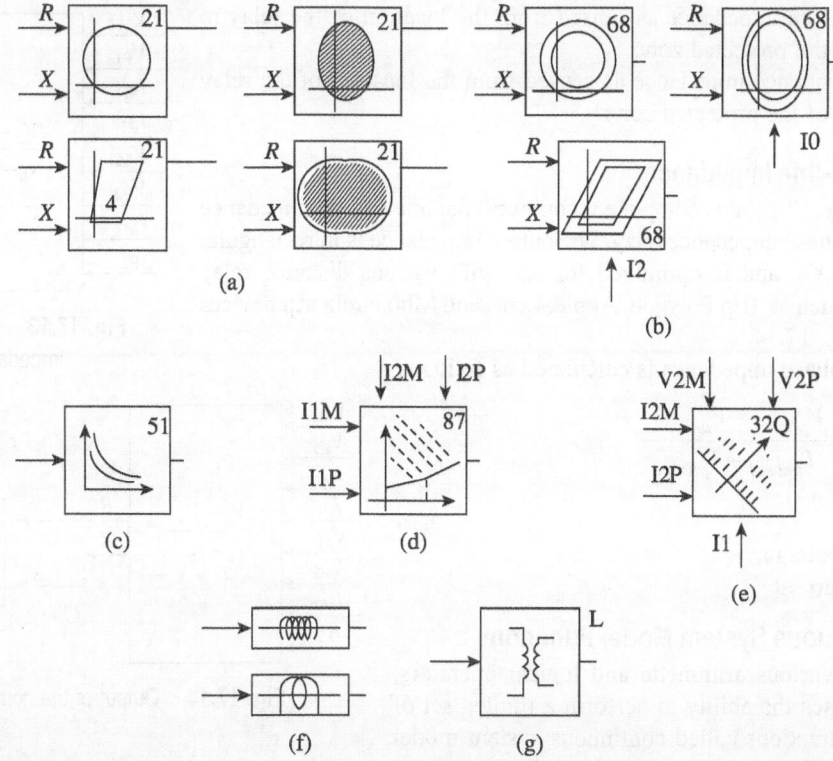

Fig. 17.10 Protection elements (a) Impedance zone (21) elements (b) Out-of-step (68) elements (c) Inverse time overcurrent (51) element (d) Dual slope differential (87) element (e) Negative sequence directional (32Q) element (f) Current transformer models (g) Potential transformer

as mho characteristics (polygon), quadrilateral characteristics, apple characteristics, and lens characteristics. Among the aforementioned characteristics, the block for achieving the mho characteristics is connected to the output of line-to-ground impedance block. This is shown in Fig. 17.12.

The online ground impedance is calculated as follows:

$$Z_{L0} = \frac{V_{phase}}{I_{phase} + k \cdot I_0}$$

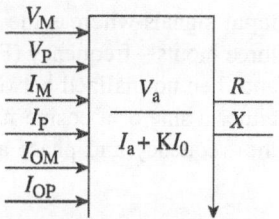

Fig. 17.11 Line-to-ground impedance

where,

V_{phase} = phase voltage

I_{phase} = phase current

$$I_0 = \frac{1}{3}(I_A + I_B + I_C)$$

$$k = \frac{Z_0 - Z_1}{Z_1}$$

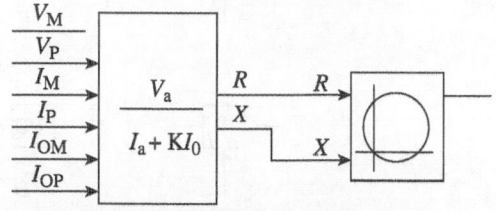

Fig. 17.12 Output of the component

Z_0 = Zero-sequence impedance as sensed from the location of the relay to the end of the protected zone

Z_1 = Positive sequence impedance as sensed from the location of the relay to the end of the protected zone

17.2.8 Line-to-line Impedance

As shown in Fig. 17.13, this component computes the line-to-line impedance as sensed by a phase impedance relay. The output impedance is in rectangular format (R and X), and is optimized for use with various distance relay characteristics such as Trip Polygon, Apple, Lens, and Mho circle trip devices as shown in Fig. 17.14.

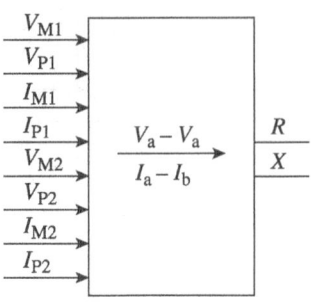

Fig. 17.13 Line-to-line impedance

The online phase impedance is calculated as follows:

$$Z_{LL} = \frac{V_{phase1} - V_{phase2}}{I_{phase1} - I_{phase2}}$$

where,

V_{phase} = phase voltage

I_{phase} = phase current

17.2.9 Continuous System Model Functions

In addition to various arithmetic and logical operators, PSCAD possesses the ability to perform a limited set of mathematical functions called continuous system model functions (CSMF). This is shown in Fig. 17.15.

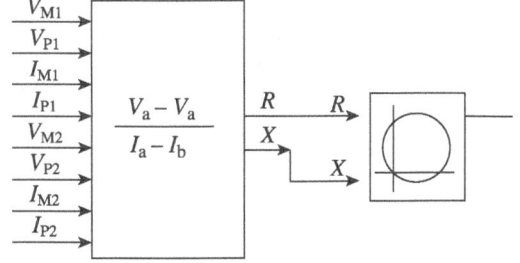

Fig. 17.14 Output of the component

Mathematical expressions are used mostly in the computation segment, whereas arithmetic and logical operators are used throughout for component definition. The summing/difference junction allows the circuit designer to linearly combine several signals together. A maximum of seven inputs can enter into the junction. Each input can be added to or subtracted from the sum. The multiplier component is used to multiply two input signals whereas the divider component is used to divide two signals. The AM/FM/PM function has three inputs—frequency (Freq), phase (Phase), and magnitude (Mag). Freq is integrated with respect to time, and then normalized between -2π and $+2\pi$. Phase is added to this, and the sum is used as an argument of either a sine or a cosine function. This result is then multiplied by Mag, and appears on the output line. If the frequency and phase are kept constant then the output is an amplitude modulation (AM) of the input

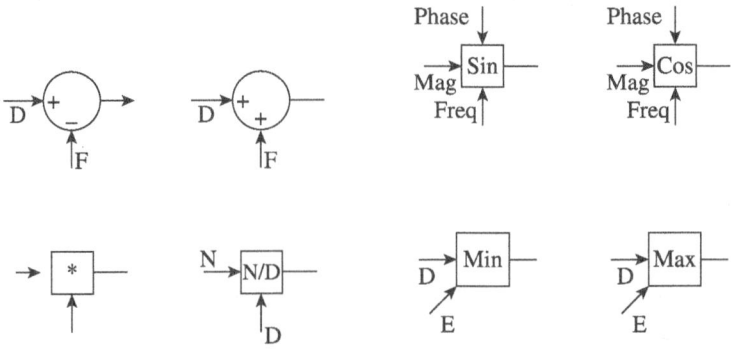

Fig. 17.15 CSMF component

Mag. On the other hand, if phase and magnitude are kept constant then the output is a frequency modulation (FM) of the input Freq. Conversely, if frequency and magnitude are kept constant then the output is a phase modulation (PM) of the input Phase. The maximum/minimum component allows the circuit designer to select the maximum or minimum of several signals and up to seven inputs may enter into the block.

17.2.10 Breakers

The component shown in Figs 17.16(a)–(d) simulates the three-phase circuit breaker operation. The *on* (closed) and *off* (open) resistance of the breaker must be specified along with its initial state. This component is controlled through an input signal named as BRK, which stands for breaker. The breaker logic is as follows:

1. 0 = *on* (closed)
2. 1 = *off* (open)

Three-phase breaker operation is virtually identical to that described for the single-phase breaker. The breaker control can be configured automatically by using the timed breaker logic component or the sequencer components. The breaker may also be controlled manually through the use of online controls or through a more elaborate control scheme.

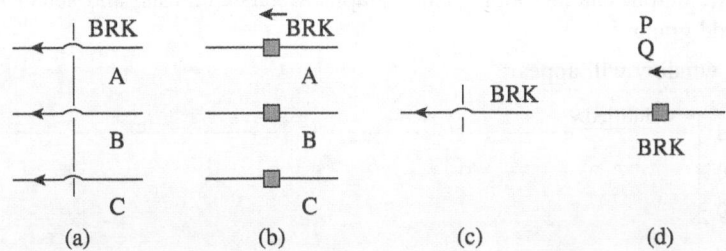

Fig. 17.16 Circuit breaker representation (a) Three-phase view of low voltage breaker (b) Three-phase view of high voltage breaker (c) Single-line view of low voltage breaker (d) Single-line view of high voltage breaker

17.2.11 Control Panels

A *control panel* is a special component used for accommodating 'control' or 'meter' interfaces and can be placed anywhere in a project page. Once a control panel has been added, you may then proceed to add as many control or meter interfaces to it as you wish.

Each type of control component will have a different control interface when added to a control panel. Figure 17.17 shows the available control components and their corresponding control interfaces.

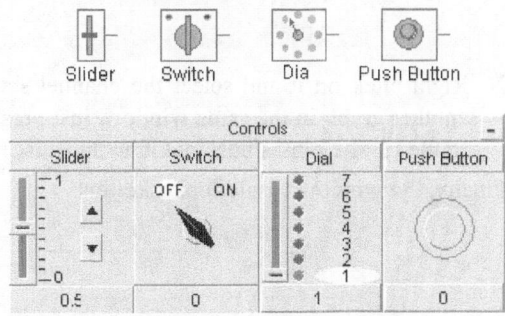

Fig. 17.17 Control components

17.3 Making a Simple Circuit in PSCAD

The following are the steps for constructing a circuit in PSCAD:

1. Go to 'File' in the menu bar and select 'Case' from new option.
2. Now, 'noname' would appear on the workspace window. Click on it.
3. Then, select 'Master library' and then select the appropriate elements from each module as per the requirement of the circuit.

Consider the circuit of a simple voltage divider that requires an AC source, two resistive elements, a current meter, a volt meter, and a graph.

4. To select any object from the master library, right click on that element, select 'Copy', and then paste on the design editor. The value can be edited by double clicking on it or by right clicking and selecting 'Edit parameters'.

The circuit looks like this:

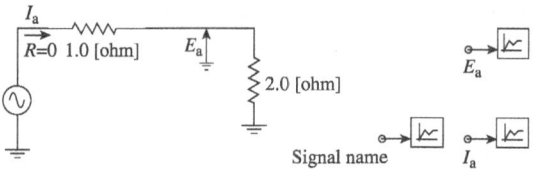

5. Use the output channel and signal input/output node for plotting the graph. Change the name of the signal node to E_a by double clicking on it.
6. Right click on the output channel and select 'Graphs/Meters/Controls' and select 'add overlay graph with signal' to add graph.

Then, the following window will appear:

7. Right click on it and select the channel setting from the drop down menu and edit the title. To add another graph in the same window, just press the control key and drag the output channel on the graph frame till the pink sinusoidal icon appears; release the mouse button, and then release the control key.

Finally, the window would look like this:

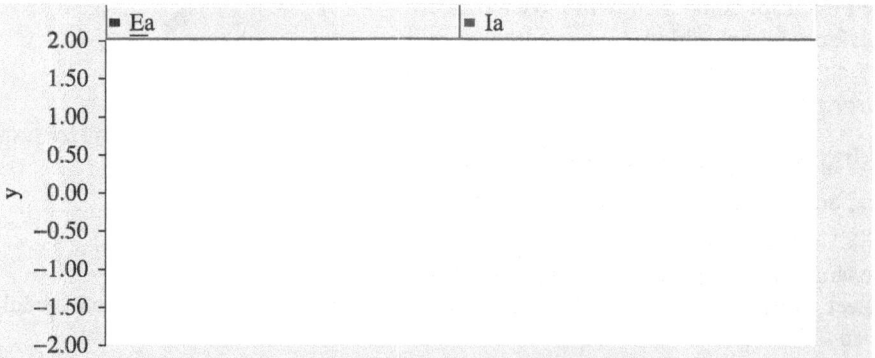

8. Click on 'run' button from the compile menu or from the tool bar.

The output will appear something like this:

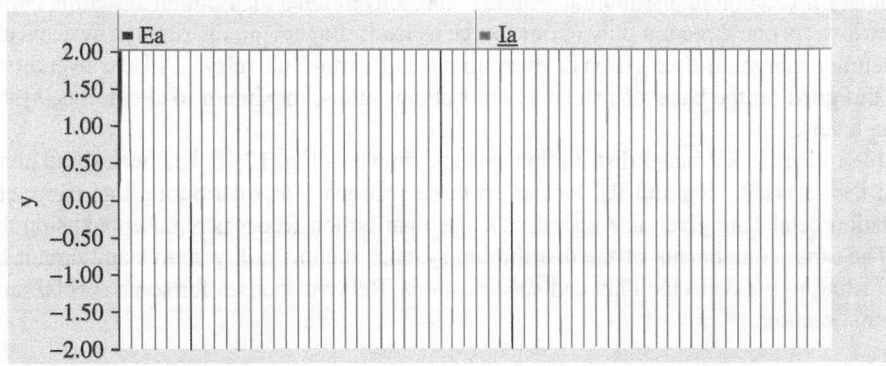

9. Now, press 'R' from the keyboard on the graph; the output will then look like this:

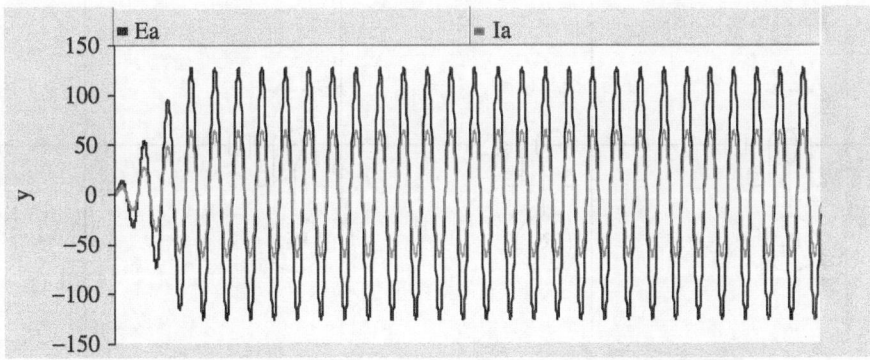

10. Now, drag the area you require.

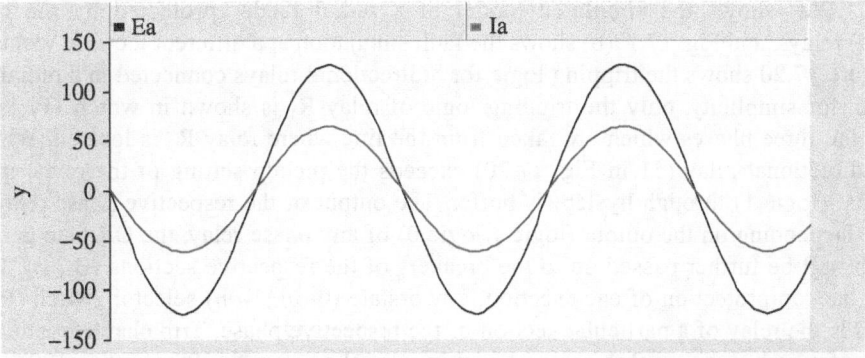

11. Save the file (Case) in the hard disc drive. The case project should be saved with the file extension '.psc'.

17.4 Case Study for Overcurrent Relay Coordination

Overcurrent relaying, which is simple, economic, and relatively easy in coordination, is commonly used for providing primary protection in distribution systems. The conventional distribution system is radial in nature, that is, power flows in one direction only (from source to load). In general, distribution systems are protected by inverse definite minimum time (IDMT) overcurrent and earth-fault relays. All the overcurrent relaying schemes are designed on the basis of available short circuit ratios, maximum load currents, system voltage, and insulation levels.

A part of the Indian 11 kV radial distribution system, shown in Fig. 17.18, has been simulated in PSCAD to coordinate the relays R_1, R_2, and R_3, located in three sections. The distribution line parameters and the generating station details are given in Appendix D. The distribution feeder is represented using the Bergeron line model. The other components of the distribution system, such as utility source and circuit breakers are designed according to the collected data and specifications. Relay responses for some special fault cases are described in this section.

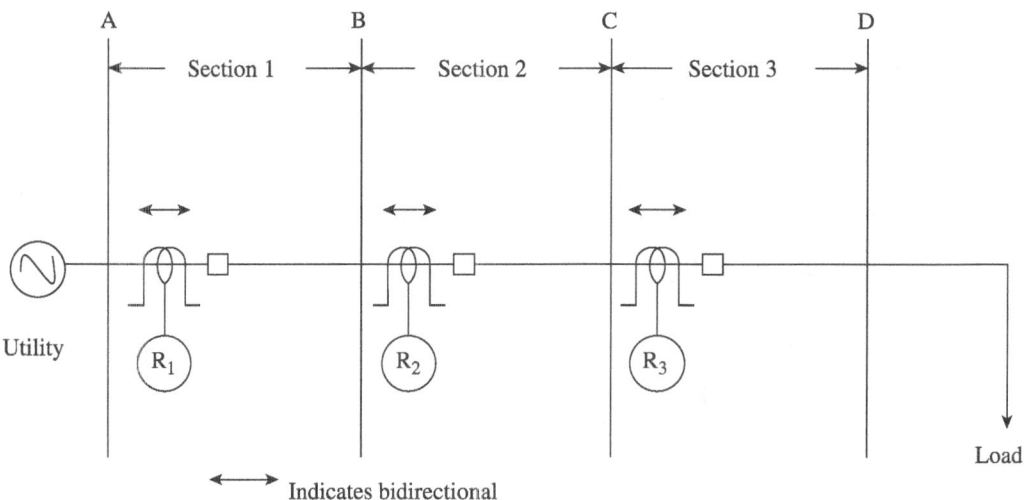

Fig. 17.18 Relay coordination of distributed feeder

Figure 17.19(a) shows the simulated model of a radial feeder protected by the conventional (bidirectional) relays, and Fig. 17.19(b) shows the fault simulation at a different location within these three sections. Figure 17.20 shows the tripping logic for bidirectional relays connected in a radial distribution system. Here, for simplicity, only the tripping logic of relay R_1 is shown in which IT1 is the current signal of all the three phases which are taken from the line where relay R_1 is located. When a current through a bidirectional relay (51 in Fig. 17.20) exceeds the pickup setting of the relay, it generates a constant signal (logic 1) through hysteresis-buffer. The output of the respective phase relay is given to the OR gate. Depending on the output (logic 1/logic 0) of any phase relay, the OR gate generates a trip signal, which will be further passed on to the breakers of the respective sections (B_1, B_2, B_3). In order to check the backup protection of each section, a two-state (0-*off*, 1-*on*) selector switch (R_1A) is used, which bypasses the relay of a particular section of the respective phase. Trip characteristic constants of each relay are set such that the relay gives normal inverse characteristic. Plug setting (PS) and time dial setting (TDS) of each relay are set such that relay R_1 provides backup to relay R_2, and relay R_2 provides backup to relay R_3 for downstream faults.

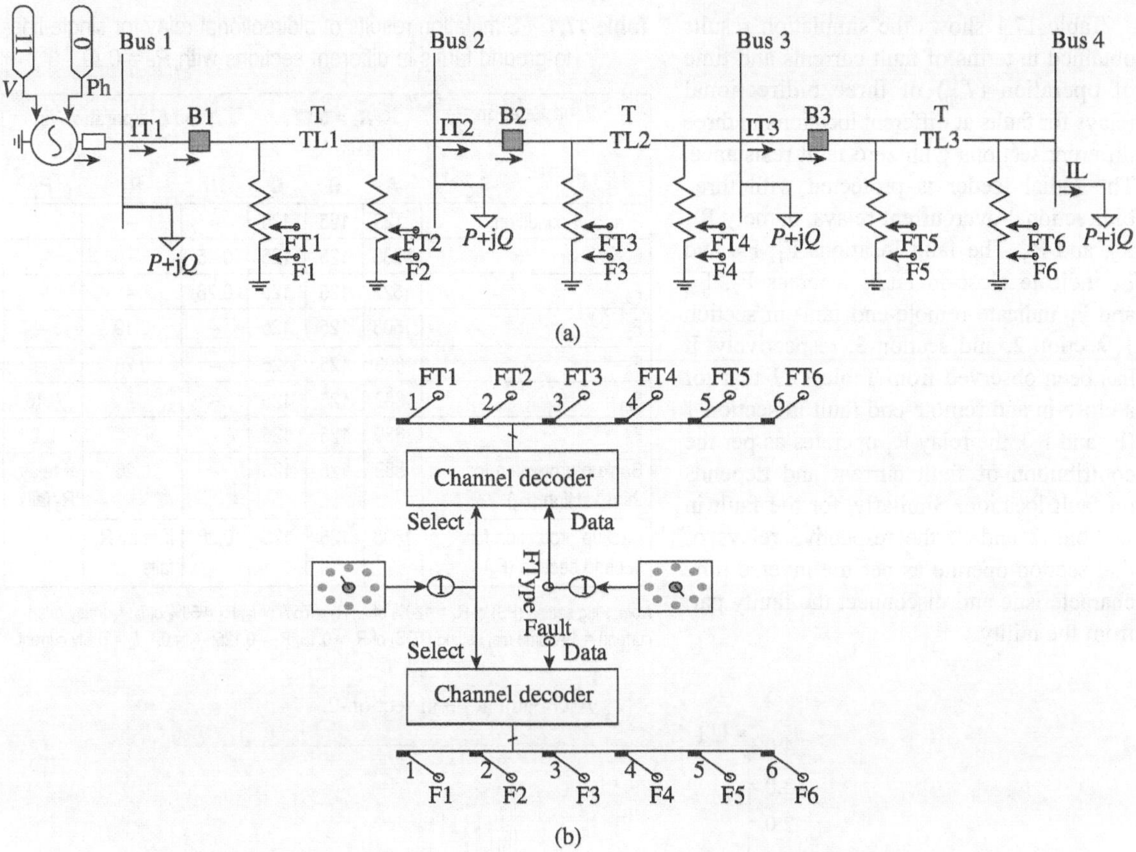

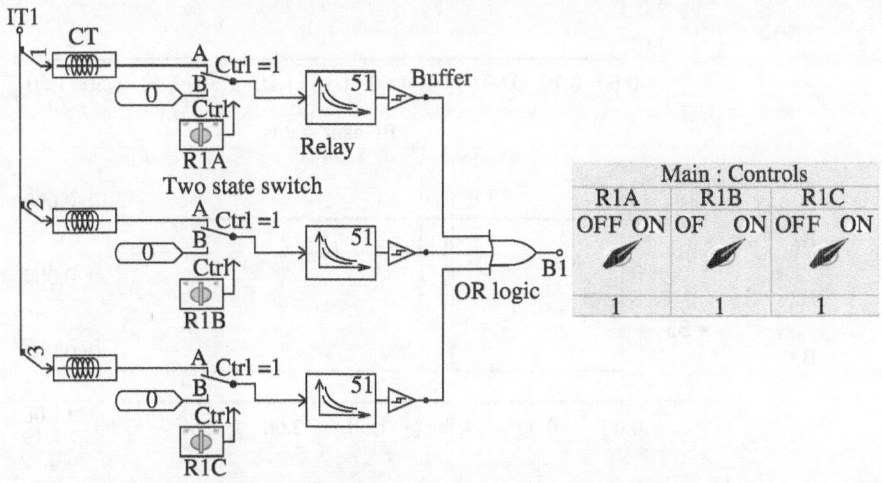

Fig. 17.19 Overcurrent relay coordination (a) Simulation of radial feeder protected by conventional (bidirectional) relays (b) Fault simulation with location

Fig. 17.20 Tripping logic of bidirectional relays

Table 17.1 shows the simulation results obtained in terms of fault currents and time of operation (T_{op}) of three bidirectional relays for faults at different locations in three different sections with zero fault resistance. The radial feeder is protected with three bidirectional overcurrent relays, namely R_1, R_2, and R_3. The fault locations F_1, F_3, and F_5 indicate close-in fault, whereas F_2, F_4, and F_6 indicate remote-end fault in section 1, section 2, and section 3, respectively. It has been observed from Table 17.1 that for a close-in and remote-end fault in section 1 (F_1 and F_2), the relay R_1 operates as per the contribution of fault current and depends on fault location. Similarly, for the fault in sections 2 and 3, the respective relays of that section operate as per the inverse time characteristic and disconnect the faulty part from the utility.

Table 17.1 Simulation results of bidirectional relay for single-line-to-ground faults in different sections with $R_F = 0\ \Omega$

Fault location	$R_F = 0\ \Omega$ I_U (A)			Time of operation (s)		
	A	B	C	R_1	R_2	R_3
Pre-fault condition	125	125	125	–	–	–
F_1	5236	125	125	0.15	–	–
F_2	1522	125	125	0.26	–	–
F_3	1508	125	125	–	0.19	–
F_4	893	125	125	–	0.26	–
F_5	889	125	125	–	–	0.16
F_6	639	125	125	–	–	0.2
Backup protection for third section (F_5)	889	125	125	–	0.26	If relay R_3 fails
Backup protection for second section (F_3)	1508	125	125	0.26	If relay R_2 fails	–

Note: Plug setting (PS) of R_1 = 100% of I_R; R_2 = 75% of I_R; R_3 = 50% of I_R; I_R (relay rated current) = 1 A; time dial setting (TDS) of R_1 = 0.15; R_2 = 0.125; R_3 = 0.1; I_U = Utility current.

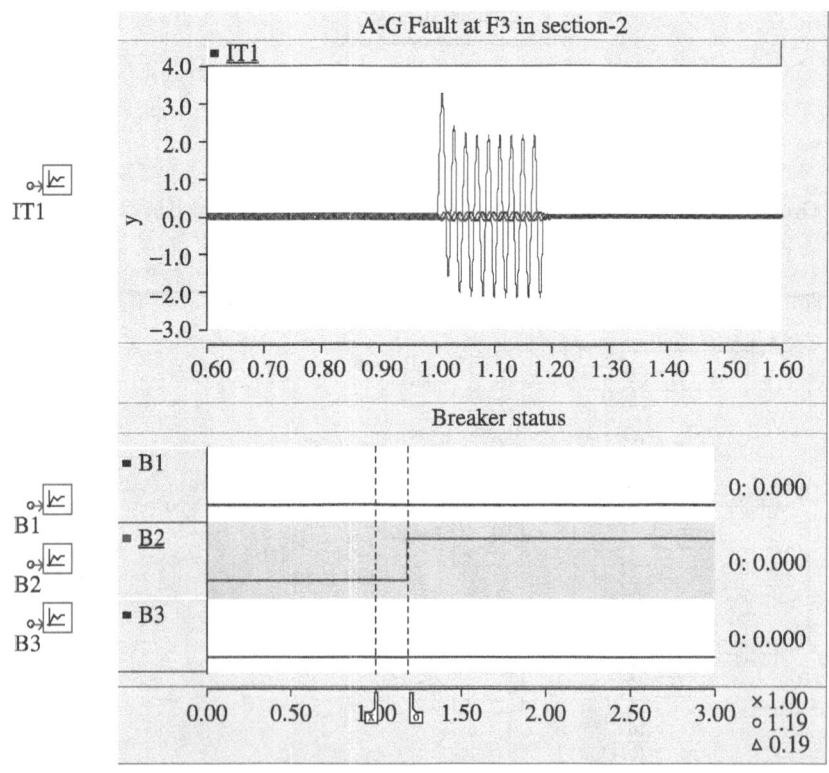

Fig. 17.21 Response of the relay R_2 for a single-line-to-ground fault (F_3) in section 2

Figure 17.21 shows the response of the bidirectional relay R_2 for a single-line-to-ground fault (F_3) in section 2, which has been obtained from PSCAD simulation. It has been observed from Fig. 17.21 that the relay operates when the fault current (first window) exceeds the predetermined threshold value depending on the types of faults. Accordingly, the respective relay generates a tripping signal, which will be passed to the circuit breaker connected in that section where the fault occurs. It is to be noted from the second window that the breaker status will change from low to high for breaker B_2, and this clears A–G fault within 0.19 s in section 2.

17.5 Case Study for Implementation of Distance Relaying Scheme

This case study demonstrates the implementation of distance relaying scheme for a transmission line using the various components of PSCAD.

17.5.1 EMTP/PSCAD Models

MODELS is a symbolic language interpreter for the EMTP/PSCAD that has recently gained popularity for the electromagnetic transient's phenomenon modelling. MODELS enables monitoring and controllability of power systems as well as some other algebraic and relational operations for programming. It models the power system by describing the physical constants and/or the subsystems functionally for the target systems. Even though it has such strong features for programming in simulation tasks, it has a drawback of limited memory allocation for data arrays. At the pre-processing stages, the anti-aliasing low-pass filter and the DC-offset removal filter are implemented to produce the voltage and current values for the extraction of the fundamental frequency component, which in turn is used for impedance calculations. For the fundamental frequency signals, the FFT method is used in the simulation.

17.5.2 Study of Simulation Case

Here, the theory and structure of the interactive relay test systems are described, beginning with an example of a power transmission line of fault simulation to test relay operation. Figure 17.22 depicts the one-line diagram of the 345 kV, 60 Hz

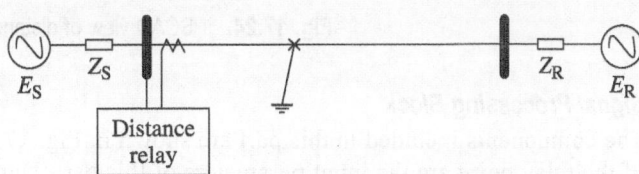

Fig. 17.22 One-line diagram of simulation system

simulated system. Figure 17.23 shows the simulated system model by PSCAD. The other related parameters of the simulated system are shown in Appendix D. Zone 1 is set to 85% of the total line length. This example uses mho type distance characteristics to explain the relay operation performance. Mimic filters with a time constant of two cycles are used. The phase difference between E_S and E_R is 15° and the sampling frequency is 1920 Hz. The transmission line length is 100 km. The phasors are estimated by full-cycle discrete Fourier transform (DFT).

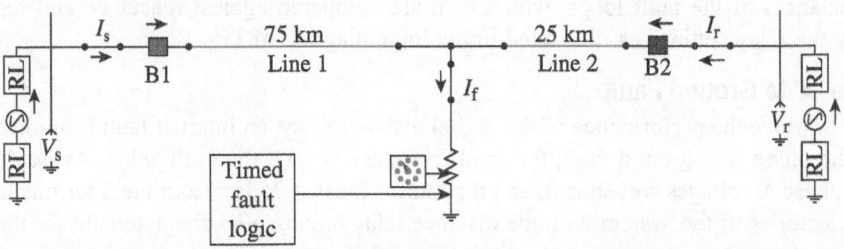

Fig. 17.23 Simulated system models by PSCAD

17.5.3 PSCAD View of Distance Relay

In this section, the PSCAD view of the distance relay is reviewed, which consists of two parts (blocks) as shown in Fig. 17.24. Here, both parts are summarized as follows:

1. In the signal processing block, the measured voltage (V_s) and current (I_s) of the output of the PT and CT at the relay point enter as inputs as shown in Fig. 17.24.
2. In the protection scheme block, the outputs of the first block enter as input parameters (Fig. 17.24).

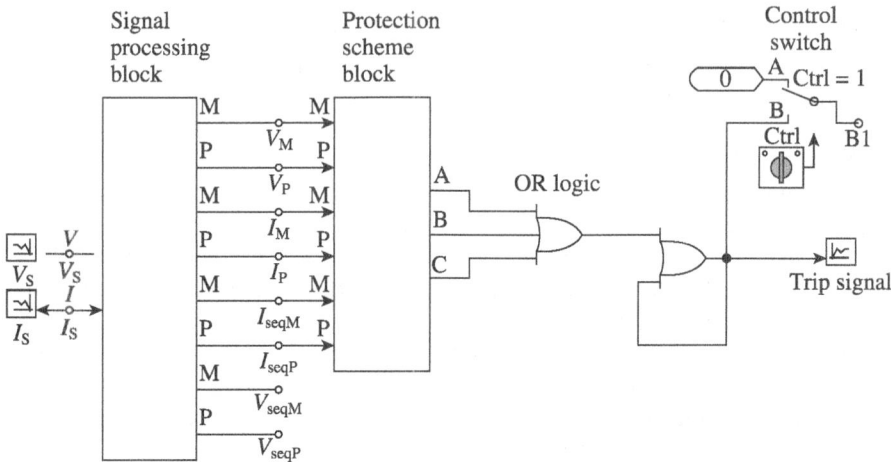

Fig. 17.24   PSCAD view of distance relay procedure

Signal Processing Block

The components included in this part are shown in Fig. 17.25, where the measured voltage and current of the relay point are the input parameters of this part. During this process, first, the input is processed to provide the magnitude 'Mag' and phase angle 'Ph' of the fundamental frequency and its harmonics (including the DC component, dc). Then, the sequence filter calculates the magnitudes and phase angles of the sequence components when the magnitudes and phase angles of the phase quantities are given. So, the final output of this block is transferred to the protection scheme block as input parameters (Fig. 17.24).

Protection Scheme Block

In this part, six mho distance elements with a positive sequence voltage polarization, three elements for phase–phase loops, and three elements for the phase–ground loops are modelled. The relay calculates the apparent impedances of the fault loops, which then are compared against reactance and resistance limits determined by the relay settings as illustrated in the logic diagram of Fig. 17.26.

17.5.4 Phase 'A' to Ground Fault

First, in order to prove the performance of the digital distance relay, an internal fault is applied to the power system with the phase A to ground fault; the fault resistance is 1 Ω, the fault angle is zero degree (refer to S terminal of phase A voltages waveform), and the fault is located 75 km from the S terminal. The apparent impedance trajectories of the system with the distance relay having *mho* characteristic for the fault located in the region of the protected zone are shown in Fig. 17.27. Moreover, the corresponding waveforms of voltages, currents, fault current, and the trip signal of the relay are shown in Fig. 17.28.

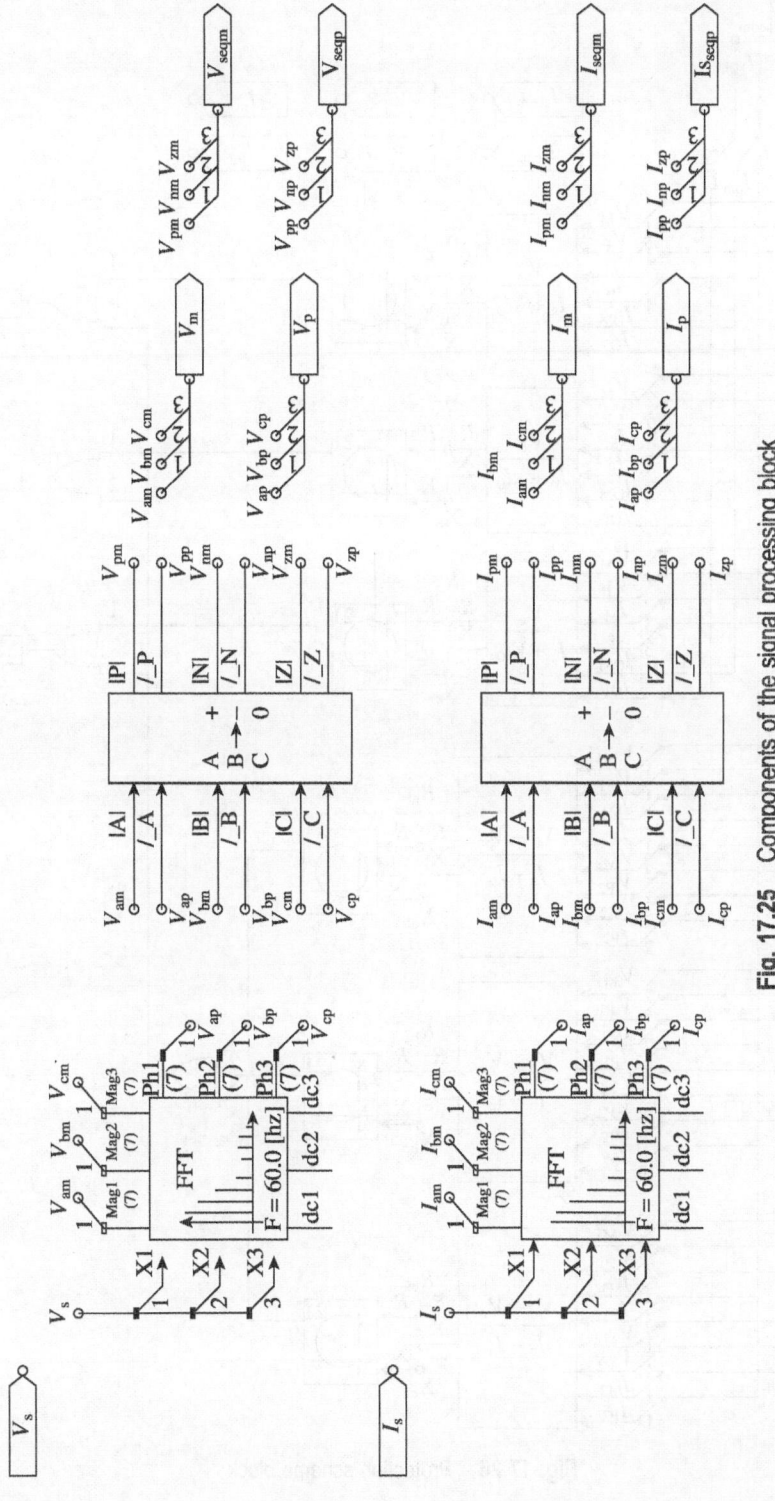

Fig. 17.25 Components of the signal processing block

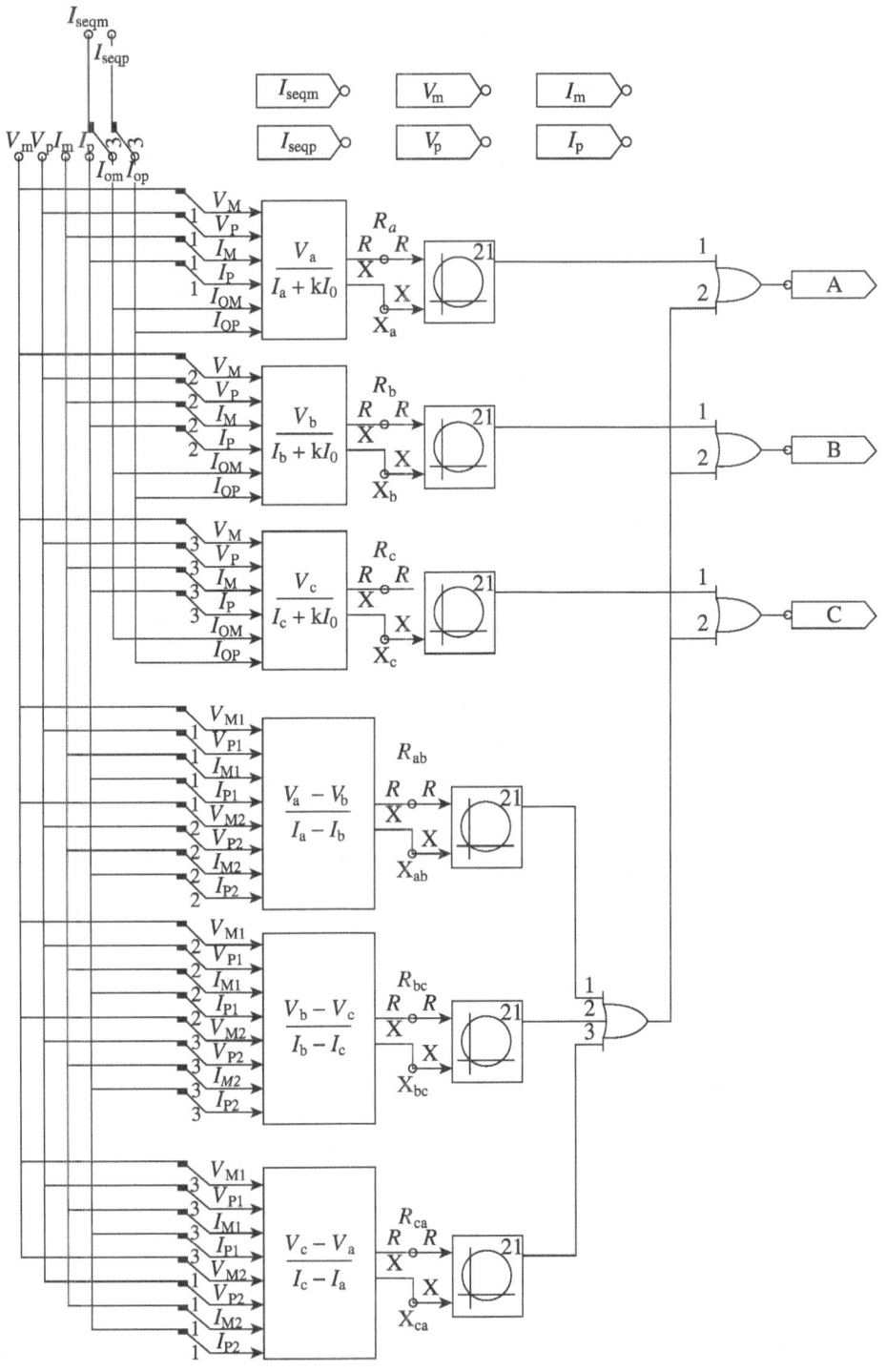

Fig. 17.26 Protection scheme block

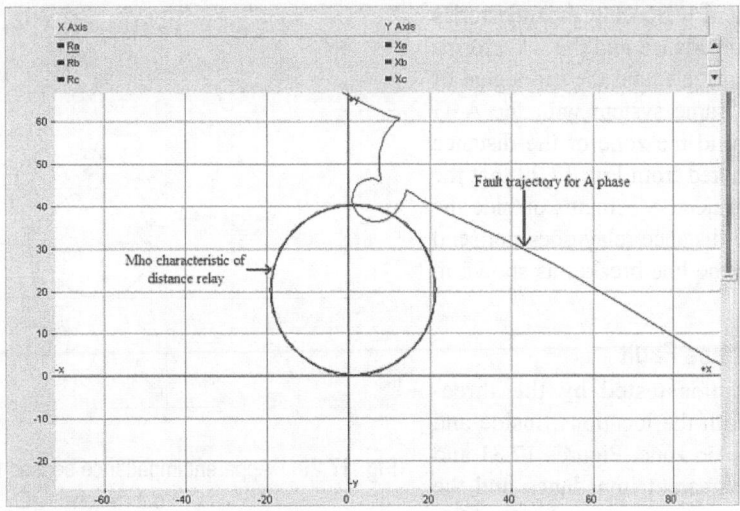

Fig. 17.27 Apparent impedance sensed by the distance relay during in-zone A–G

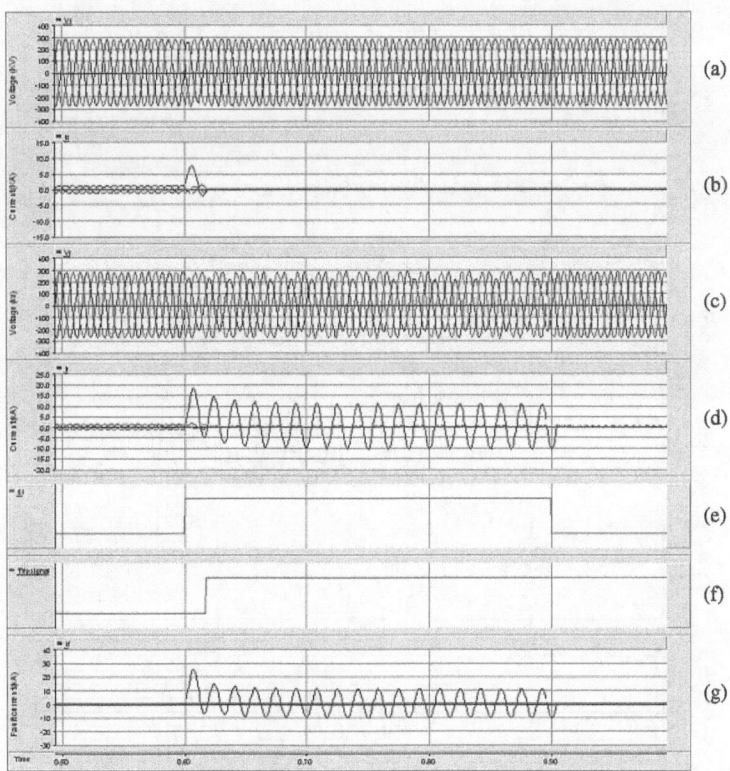

Fig. 17.28 A–G fault inside the protected zone (a) Sending-end voltage
(b) Sending-end current (c) Receiving-end voltage (d) Receiving-end current
(e) Duration of the fault (f) Trip signal of the relay (g) Fault current

Figures 17.29 and 17.30 show the trajectories of the apparent impedance and the waveforms of the voltages, currents, and the trip signal of the relay for the same system with the A–G fault applied beyond the zone of the distance relay. It is to be noted from Fig. 17.29 that the fault impedance trajectory remains outside the circle. Hence, the distance relay does not send any trip signal to the line breaker as shown in Fig. 17.30(f).

17.5.5 Three-phase Fault

The same system is tested by the three-phase fault for both the locations: inside and outside the protected zone. Figures 17.31 and 17.32 show the apparent impedance and the waveforms of the power system during ABC

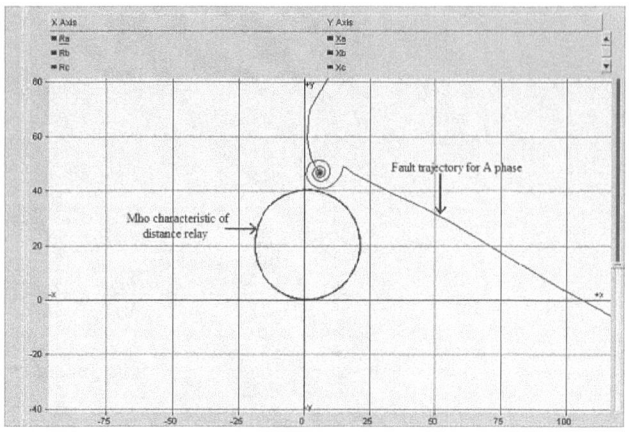

Fig. 17.29 Apparent impedance sensed by distance relay during out-of-zone A–G

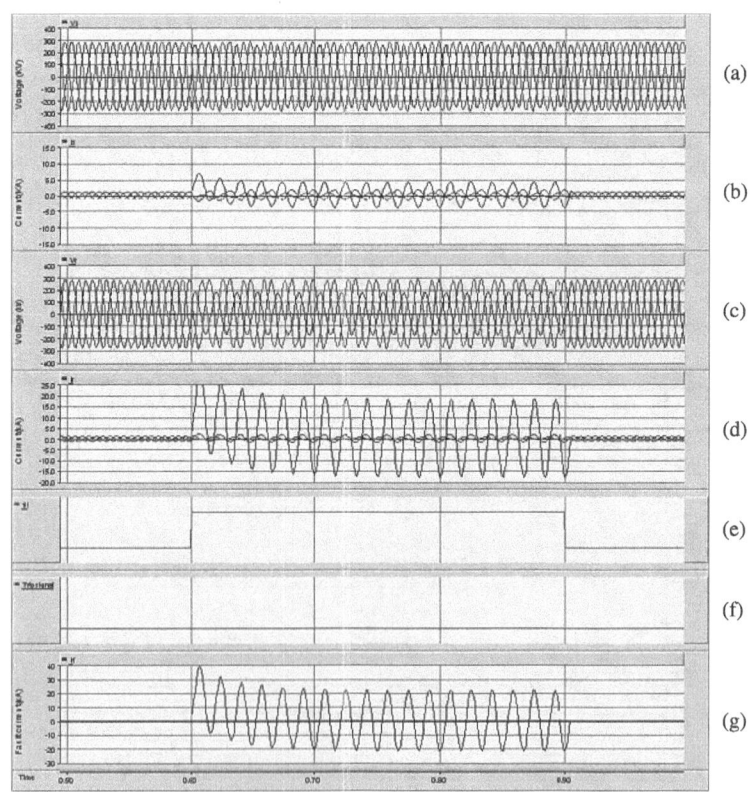

Fig. 17.30 A–G fault out of the protected zone (a) Sending-end voltage (b) Sending-end current (c) Receiving-end voltage (d) Receiving-end current (e) Duration of the fault (f) Trip signal of the relay (g) Fault current

fault applied inside the protected zone of the distance relay, from which is clear that the impedance trajectories of all the phases settle down within the circle of the mho characteristic.

Figures 17.33 and 17.34 show the trajectories of the apparent impedance and the waveforms of the voltages, currents, and the trip signal of the relay for the same system with the ABC fault applied out of the protected zone. It is clear that for such a position of the fault, the distance relay should not operate and remain stable, and the impedance sensed by the relay will not take place in the circle.

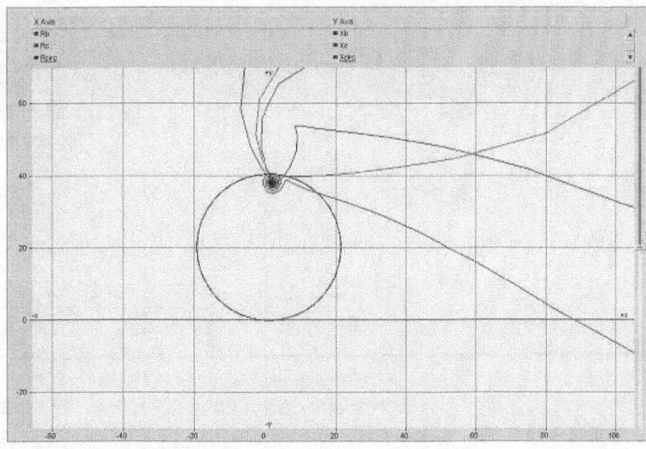

Fig. 17.31 Apparent impedance during in-zone ABC fault

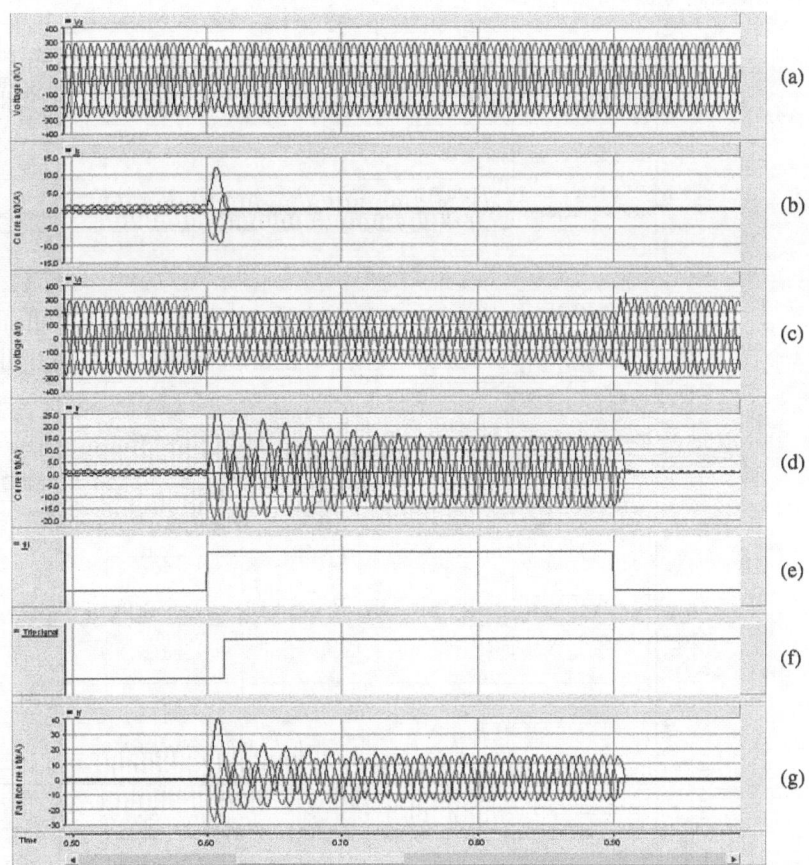

Fig. 17.32 ABC fault inside the protected zone (a) Sending-end voltage (b) Sending-end current (c) Receiving-end voltage (d) Receiving-end current (e) Duration of the fault (f) Trip signal of the relay (g) Fault current

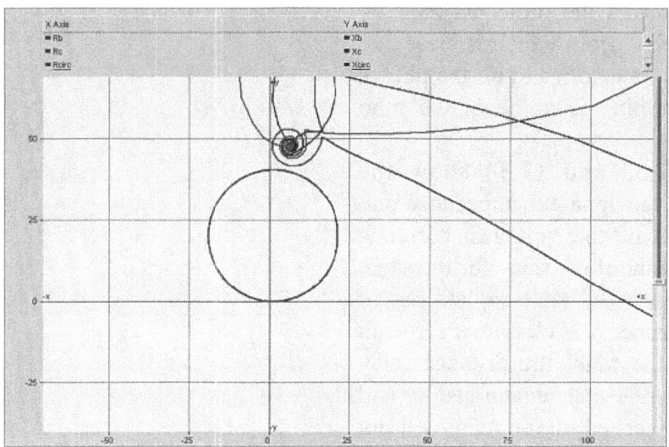

Fig. 17.33 Apparent impedance seen during out-of-zone ABC fault

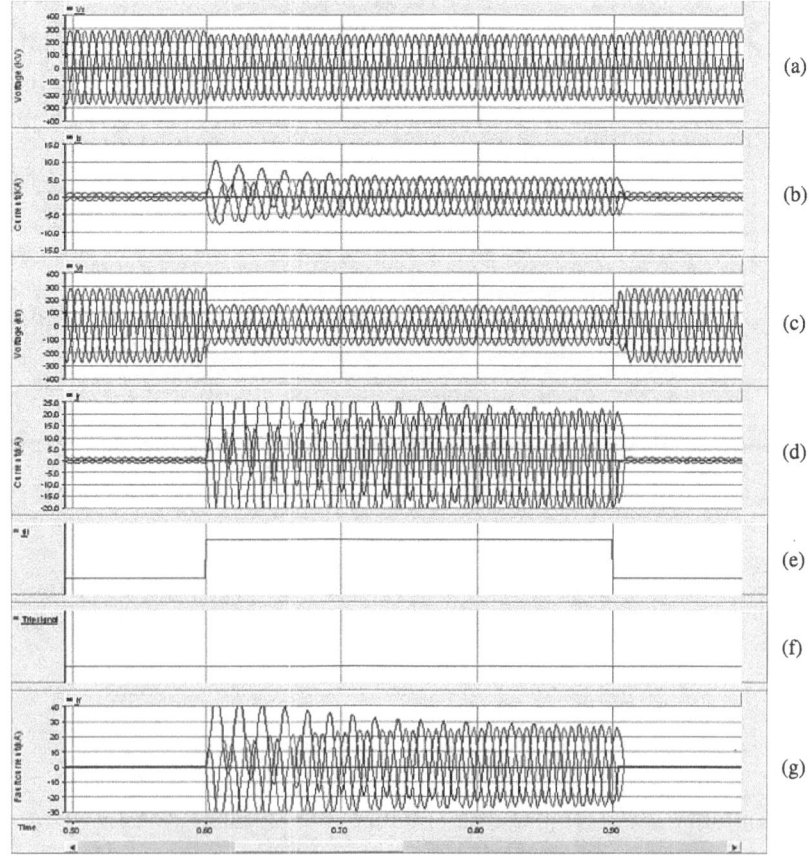

Fig. 17.34 ABC fault out of the protected zone (a) Sending-end voltage (b) Sending-end current (c) Receiving-end voltage (d) Receiving-end current (e) Duration of the fault (f) Trip signal of the relay (g) Fault current

17.6 Case Study on Buck and Boost Converter

17.6.1 Buck Converter

The buck converter is a dc–dc power converter which reduces the voltage level. A simple buck converter needs a diode, IGBT/transistor, and an energy storage device in combination of capacitor and inductor to reduce the input voltage at its output load. The aim of this case study is to examine the working of buck converter and to inspect the level of output and input voltages. Also, one can simulate the boundary between continuous and discontinuous conduction modes of the buck converter. The following circuit parameters are considered to simulate the buck convertor in PSCAD:

Input voltage (V_s) = 20 kV
Output voltage (V_o) = 10 kV (at 50% duty cycle)
Switching frequency (f_s) = 400 kHz
Inductance (L) = 21 µH
Capacitance (C) = 23 µF
Load resistance (R_L) = 10 Ω

Figure 17.35 shows the simulated model of a buck converter in PSCAD environment. The switching of IGBT is done using the control signal generated by comparing a triangle waveform with a constant set value (0.5). The duty cycle of the PWM signal is manually adjusted. Figure 17.36 shows the waveform of input voltage (V_s), output voltage (V_o), and inductor current (I_L) when the duty cycle is 50%. It is observed from the inductor current that the buck converter operates in the continuous conduction mode.

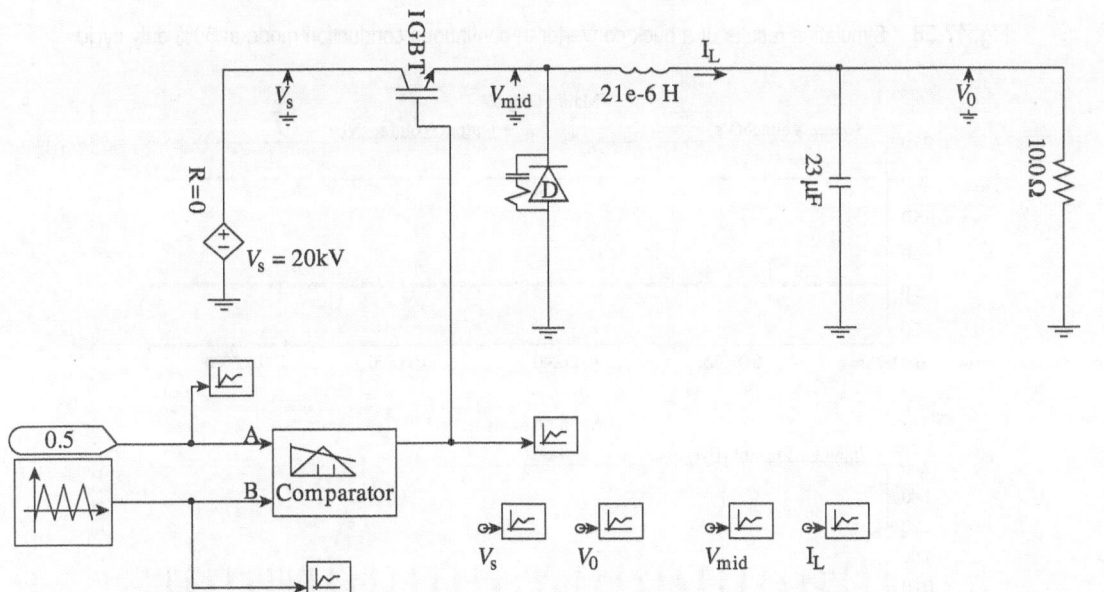

Fig. 17.35 Block diagram of buck converter simulated in PSCAD

Now, by changing the duty cycle to 30%, it is observed from Fig. 17.37 that the DC output voltage is reduced to 6 kV. Therefore, a proportional change in output voltage is observed during the change in duty cycle.

In order to simulate discontinuous conduction mode of buck converter, the inductor value has been reduced from 21µH to 3µH keeping the duty cycle at 50%. It can be observed from Fig. 17.38 that when

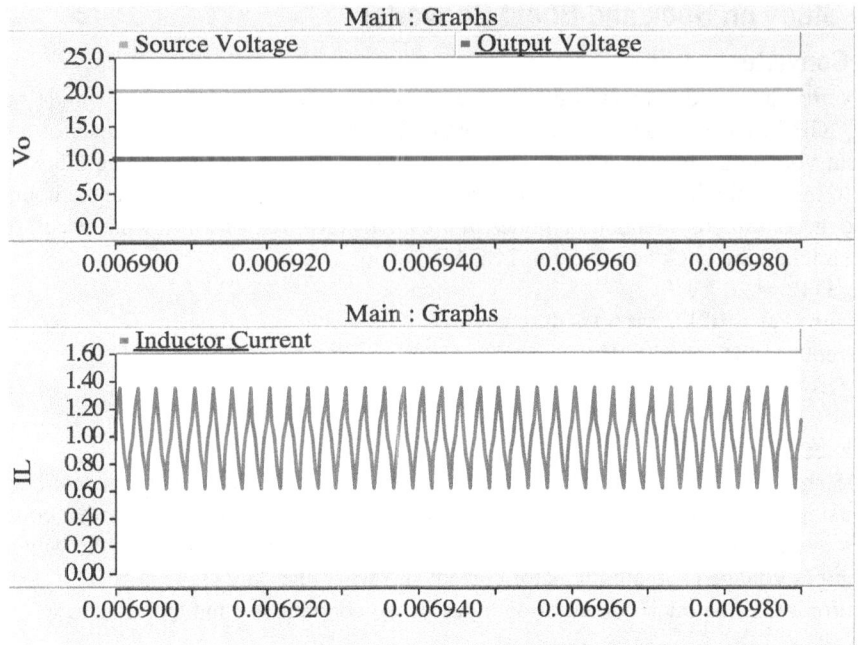

Fig. 17.36 Simulation results of a buck converter in continuous conduction mode at 50% duty cycle

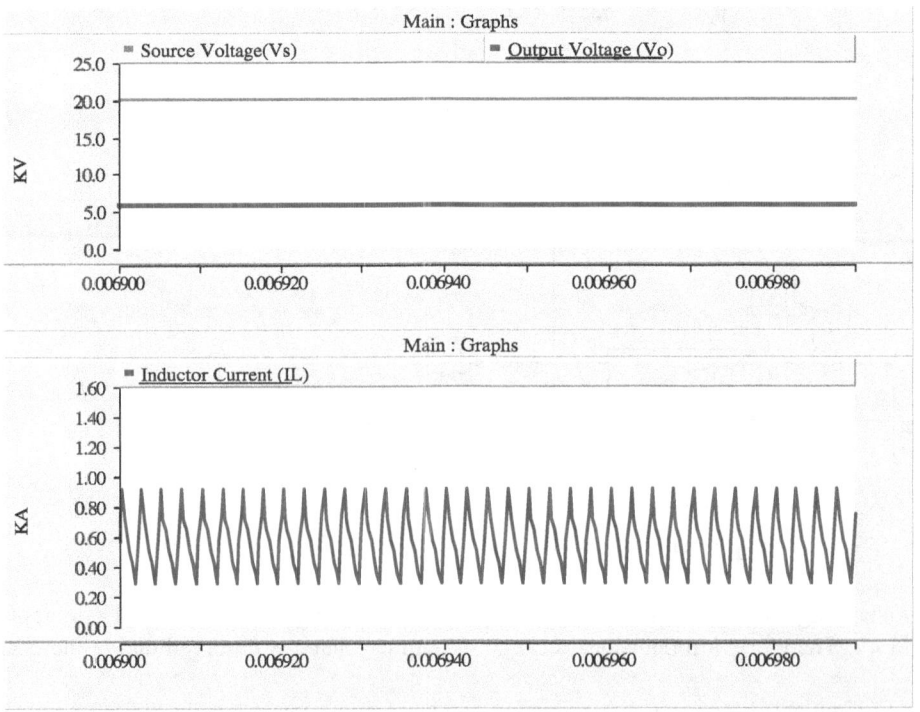

Fig. 17.37 Simulation results of a buck converter in continuous conduction mode at 30% duty cycle

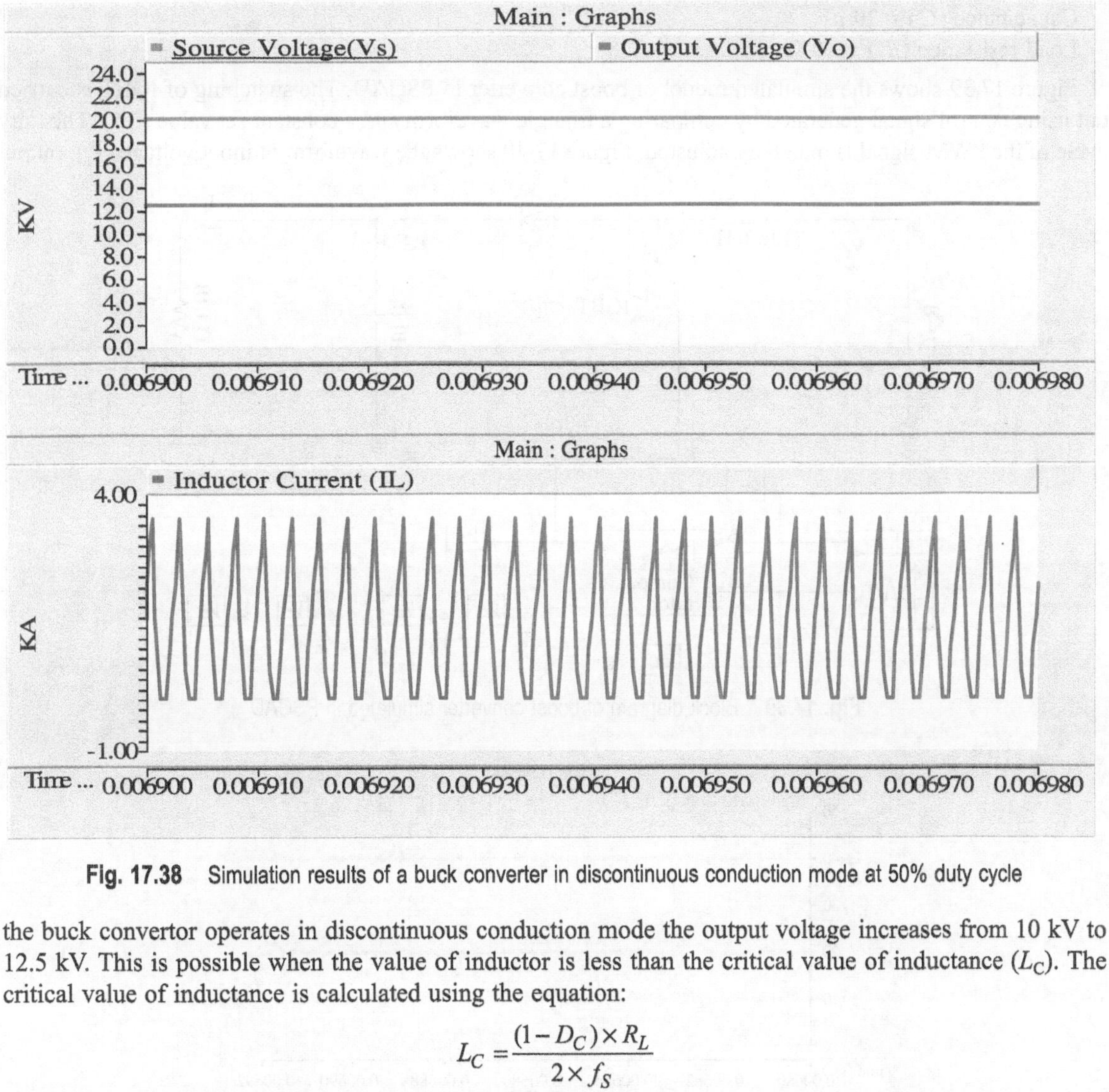

Main : Graphs

■ Source Voltage(Vs)   ■ Output Voltage (Vo)

Fig. 17.38   Simulation results of a buck converter in discontinuous conduction mode at 50% duty cycle

the buck convertor operates in discontinuous conduction mode the output voltage increases from 10 kV to 12.5 kV. This is possible when the value of inductor is less than the critical value of inductance (L_C). The critical value of inductance is calculated using the equation:

$$L_C = \frac{(1 - D_C) \times R_L}{2 \times f_S}$$

where D_C = duty cycle, R_L = load resistance, and f_S = sampling frequency

For the given case, the minimum inductance required is $L_C = \dfrac{(1-0.5)\times 10}{2 \times 400 \times 10^3} = 6.25\,\mu H$.

17.6.2 Boost Converter

The boost converter steps up the voltage level. The following circuit parameters are considered to simulate the boost convertor in PSCAD environment.

Input voltage (V_s) = 20 kV
Output voltage (V_o) = 40 kV (at 50% duty cycle)
Switching frequency (f_s) = 400 kHz
Inductance (L) = 312 μH

Capacitance $(C) = 20\ \mu\text{F}$

Load resistance $(R_L) = 10\ \Omega$

Figure 17.39 shows the simulated model of boost converter in PSCAD. The switching of IGBT is carried out using control signal generated by comparing a triangle waveform and a constant set value (0.5). The duty cycle of the PWM signal is manually adjusted. Figure 17.40 shows the waveform of input voltage (V_s), output

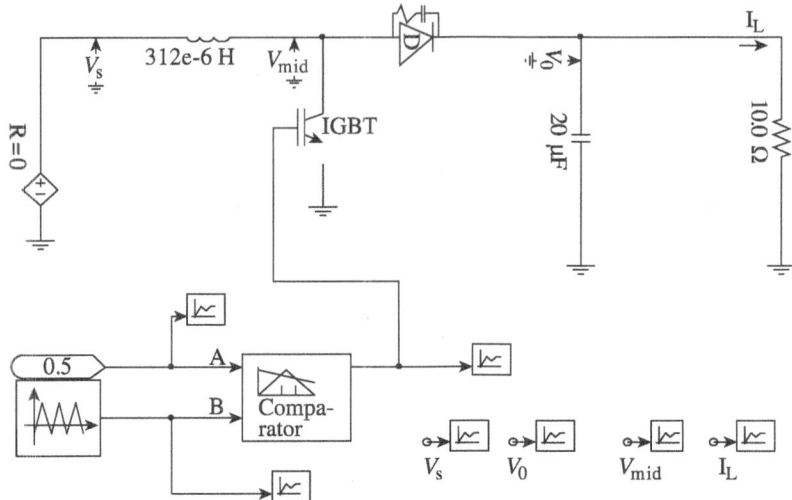

Fig. 17.39 Block diagram of boost converter simulated in PSCAD

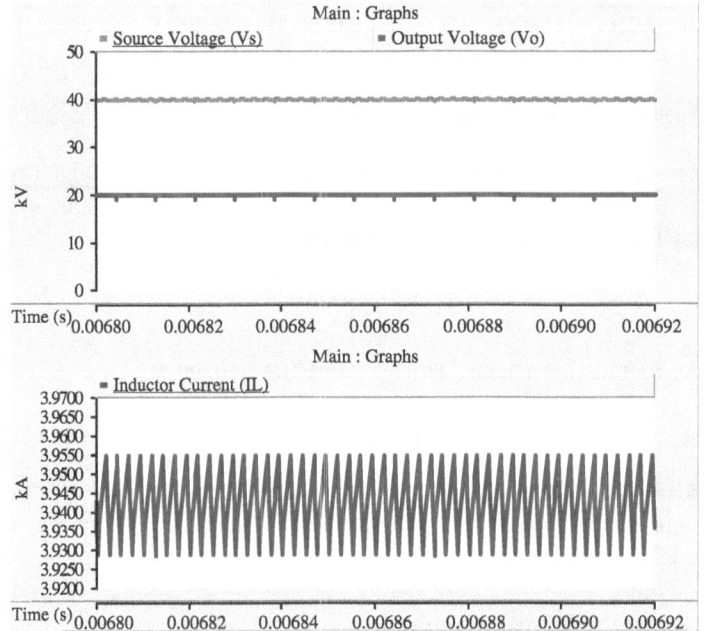

Fig. 17.40 Simulation results of a boost converter in continuous conduction mode at 50% duty cycle

voltage (V_o), and inductor current (I_L) when the duty cycle is 50%. It is observed from the inductor current that the boost converter increases the output voltage by 40 kV when it operates in the continuous conduction mode.

Now by changing the duty cycle to 30%, it is observed from Fig. 17.41 that the DC output voltage is reduced to 28 kV from 40 kV. Therefore, the proportional change in output voltage is observed during the change in the duty cycle.

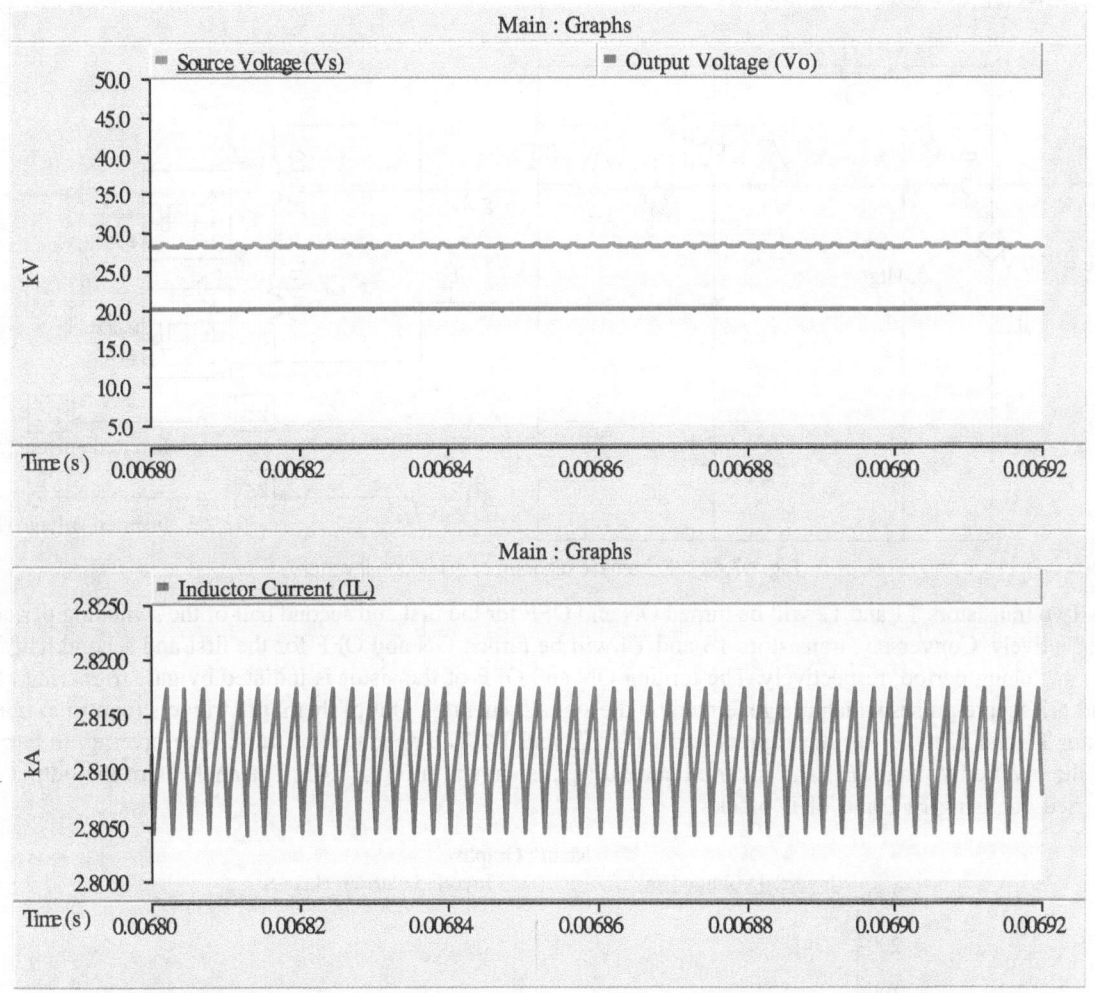

Fig. 17.41 Simulation results of a boost converter in continuous conduction mode at 30% duty cycle

Due to change in the value of inductor and capacitor, the ripple has been observed in the inductor current and output voltage.

17.7 Case Study of Full-bridge Inverters

The full-bridge inverter is a simple circuit which converts DC supply to AC supply. Here, the bridge inverter is controlled by a set of square waves generated by the signal generator. The main purpose of this case study is to understand the controlling of a full-bridge inverter using square wave. Also, fast fourier transform (FFT) and total harmonic distortion (THD) of the output voltage and current are performed to observe the magnitude of fundamental and other harmonic components. The simulation circuit for full-bridge inverter is shown in Fig. 17.42. The following parameters are considered for the case study:

Input DC voltage (*V*s): =230 V
Load (*R*) = 10 Ω, L=10 mH
Switching frequency (*f*s) = 50 Hz

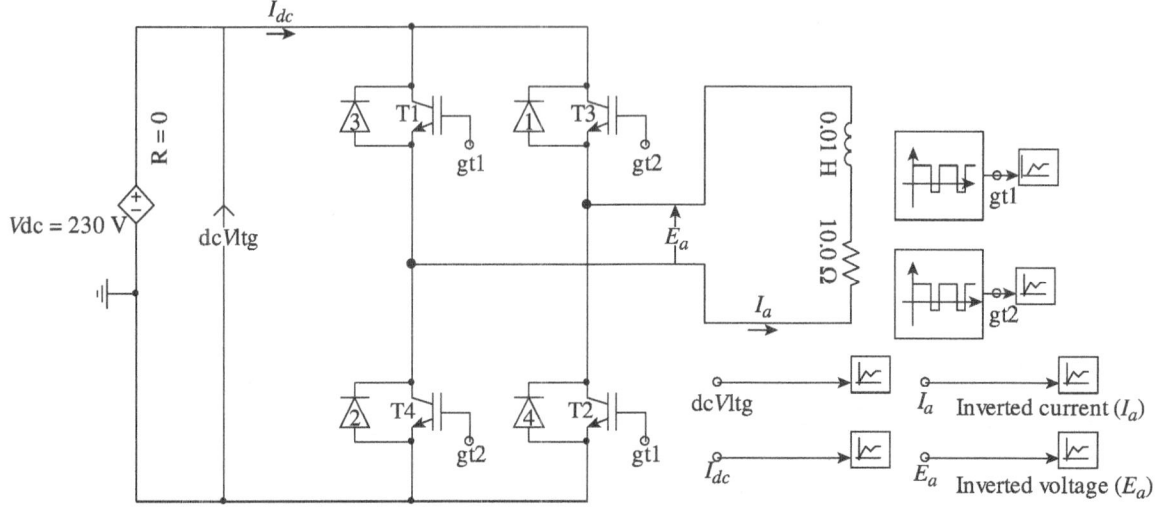

Fig. 17.42   Schematic diagram of full-bridge inverter

Two transistors T1 and T2 will be turned ON and OFF for the first and second half of the switching period, respectively. Conversely, transistors T3 and T4 will be turned ON and OFF for the first and second half of the switching period, respectively. The turning ON and OFF of transistor is initiated by gate triggering gt1 and gt2 square pulses which are generated by the signal generator. Out of them, gt1 triggers (controls) transistor T1 and T2 whereas gt2 triggers transistors T3 and T4. The operation of full-bridge inverter, in terms of the inverted voltage (E_a) and inverted current (I_a), is shown in Fig. 17.43. The analysis of the output is carried out using FFT and THD block.

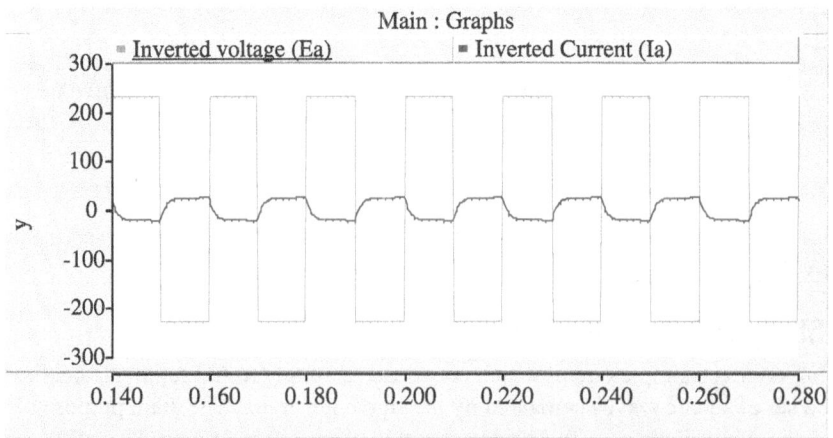

Fig. 17.43   Output voltage (E_a) and current (I_a) of inverter

The FFT block, as shown in Fig. 17.44, determines the harmonic magnitude and phase angle of the given input voltage (E_a) and current (I_a) signals as a function of time. The frequency spectras of the output voltage and current up to seventh harmonic including fundamental is shown in Fig. 17.45. The fundamental voltage and current components have the magnitudes of 206.942 V and 19.74 A, respectively. To determine the THD level in output inverted voltage and load current, the output of the FFT block is connected to the THD block. As shown in Fig. 17.46, 41.76% THD in output voltage and 28.31 % THD in output current have been observed.

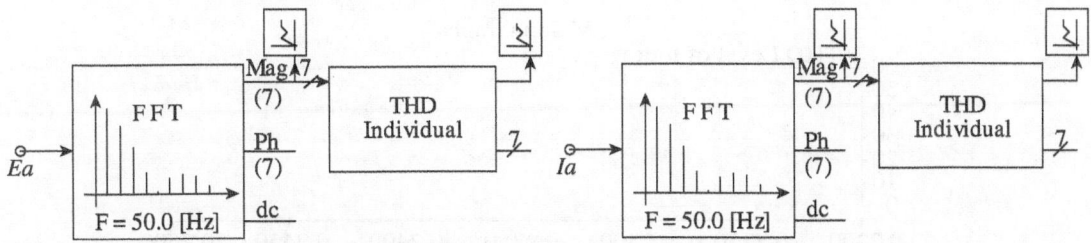

Fig. 17.44   FFT and THD analysis of inverted voltage and load current

Output voltages FFT (n=7)

300.0

0.0

[1] 206.942

Output Current FFT (n=7)

30.0

0.0

[1] 19.7415

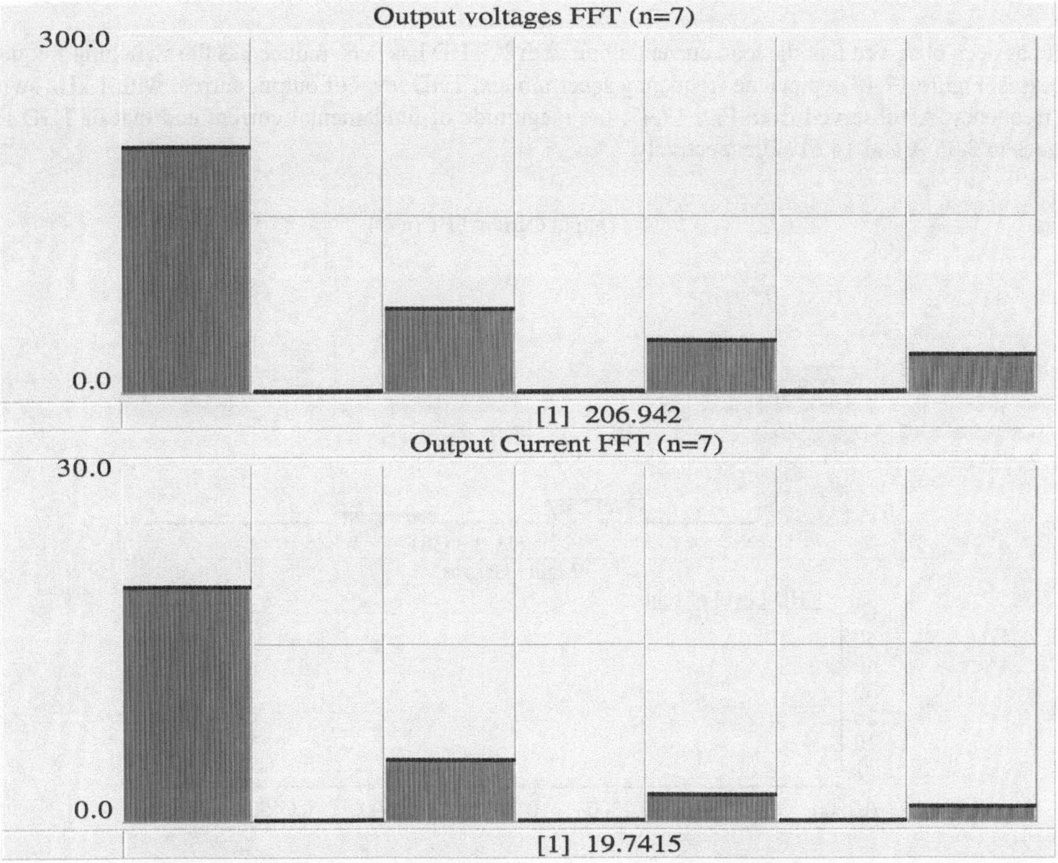

Fig. 17.45   Frequency spectrum of voltage (E_a) and current (I_a)

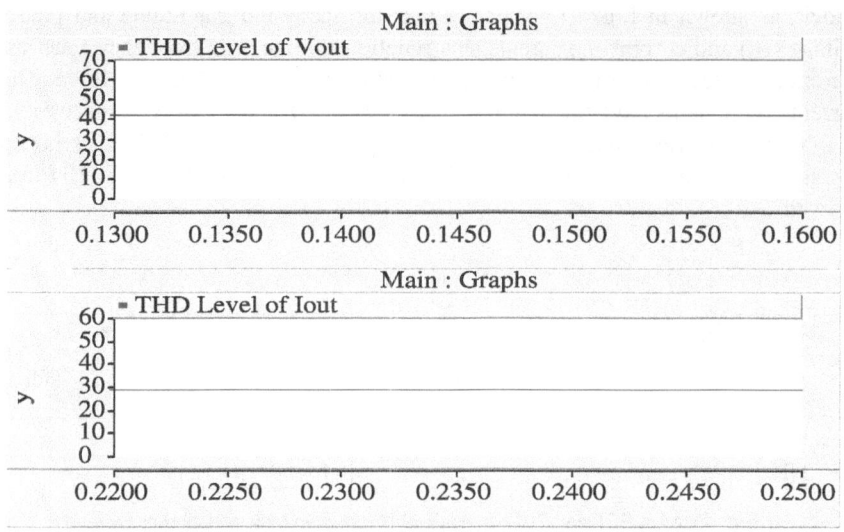

Fig. 17.46 Magnitude of THD in percentage (%) for output voltage and current

It has been observed that the load current along with % THD has been reduced as the switching frequency increases. Figure 17.47 depicts the frequency spectrum and THD level of output current with 1 kHz switching frequency. As observed from Fig. 17.47, the magnitude of fundamental current and that of THD level reduces to 3.35 A and 14.61%, respectively.

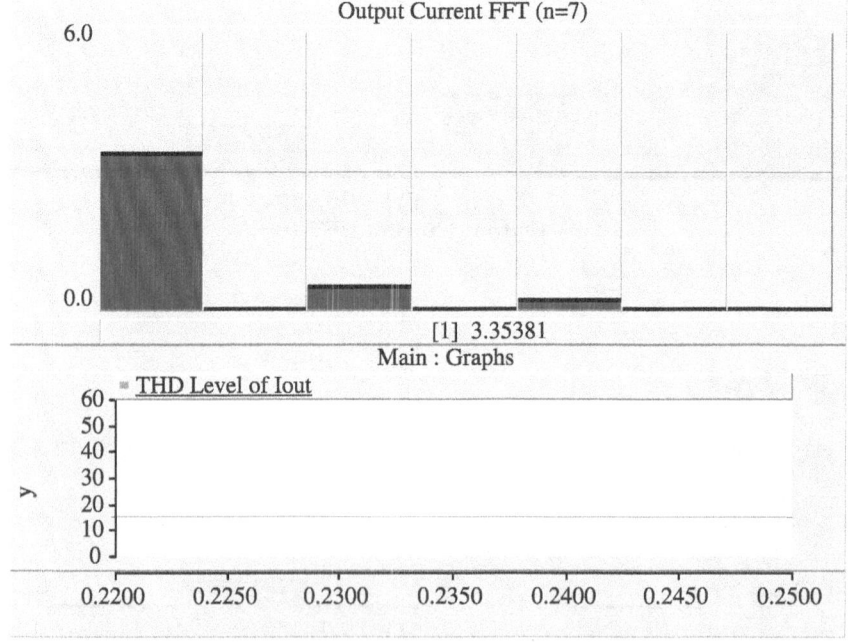

Fig.17.47 Frequency spectrum and THD level of output current with 1 kHz switching frequency

Recapitulation

- An introduction to PSCAD as a new tool is presented in this chapter.
- The applications of PSCAD are explained in brief.
- Various library models of PSCAD such as source, transformer, transmission line, frequency scanner, and machine are explained in detail.
- The procedure for construction of any case or model used in power system is discussed step-wise.
- Four case studies, namely overcurrent protection, distance protection of lines, buck/boost converter, and fullbridge inverter are explained with simulation results.

Multiple Choice Questions

1. EMTDC stands for
   (a) electromagnetic transients including DC
   (b) electrostatic transients for DC
   (c) electromagnetic transients without DC
   (d) none of the above

2. What is the mandatory requirement in transmission line component connection?
   (a) Segment names of tower and line component must be different
   (b) Segment names of tower and line component must be the same
   (c) Segment names of tower and line component have no relation
   (d) There is no mandatory requirement

3. Which component is used for the conversion of phase quantities into sequence quantities?

   (a) Channel
   (b) Summing/differential block
   (c) Online frequency scanner
   (d) Breaker block

4. Which component is used to plot a graph in PSCAD?
   (a) Control panel
   (b) Slider
   (c) Dial switch
   (d) Output channel

5. How do you send a current waveform to be plotted from the output channel?
   (a) Right click on the output channel, select 'Graphs', and then select 'Add overlay graph with signal' to add the waveform.
   (b) Copy the channel and paste it into the plot window.
   (c) Simply drag the channel and drop it into the plot window.
   (d) None of the above

Review Questions

1. Enumerate the various applications of PSCAD.
2. Describe the various types of sources available in the PSCAD library.
3. Explain the various components of continuous system model functions (CSMF) as used in PSCAD simulation.
4. Write the steps to construct any simple case (electrical circuit) in PSCAD.
5. Explain online frequency scanner available in PSCAD library and also discuss its application.
6. Discuss the different relay characteristics available in the PSCAD library.
7. Explain the control components used during simulation of a case in PSCAD.
8. Explain the opening and closing logic used in the operation of a circuit breaker.
9. Explain the governing equation of line-to-line impedance and line-to-ground impedance units used in distance relays.
10. Explain the tripping logic used for bidirectional relays.
11. Explain the role of duty cycle in buck/boost converter using PSCAD simulation.
12. Simulation of half and fullbridge inverter in PSCAD and calculate the THD of outputs.

Tutorial 17.1 To study current transformer models and fault simulation.

Objectives
- Understanding models related to fault simulation
- Comparing different CT models

Procedure
1. Create a new case by using either the 'Menu' or 'Toolbar'. A new case should appear in the *workspace settings* entitled *noname [psc]*. Right click on this workspace settings entry and select 'Save As' and give the case a name. Create a folder called *c:......./PscadTutorial/Faults*. Save the case as *test1.psc*.
2. Open the main page of your new case. The single-line diagram shown in Fig. 17.48 is a part of a sub-station feeding a shunt reactor. The reactor is modelled in two parts to enable a fault at point B inside the turns. The component data is as shown in Fig. 17.48 (make the transformer losses zero to limit the number of nodes if using the student version).

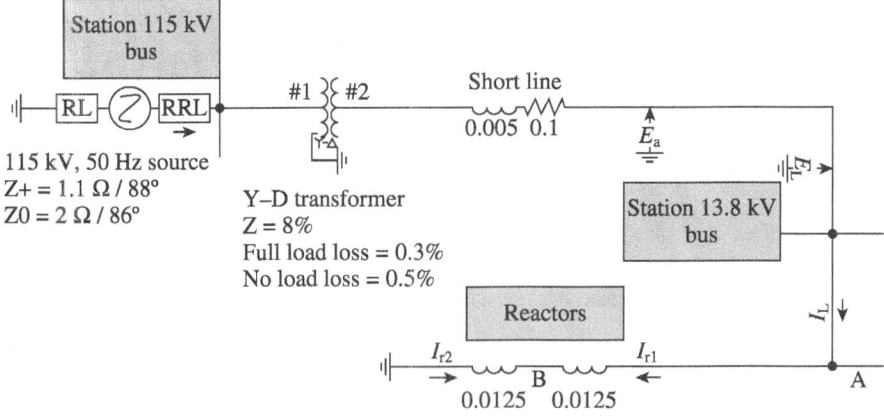

Fig. 17.48 Single-line diagram with a substation feeding a shunt reactor

You may use the wire mode to connect different components.
3. Build the case in PSCAD and enter the component data.
4. Plot the current I_L and the voltage E_L.
5. Use the fault component as shown in Fig. 17.49 to simulate a phase A to ground fault at location A at 0.1 s.
6. Observe the fault curent I_L. What is the reason for the presence of the initial DC exponential component?

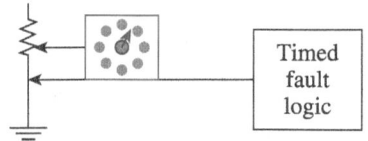

Fig. 17.49 Fault component

7. What affects the rate of decay of the DC components? Change the resistance of the short line to 1 Ω and observe the results.
8. Does the instant of the fault inception have an effect on the DC offset?
9. What negative impacts can the DC-offset have on system protection?
10. Connect the phase A line current at point A to the CT model as shown in Fig. 17.50. The CT ratio is 5:400. The CT burden is 0.15 Ω in series with 0.8 mH. Plot the secondary current and the flux density.

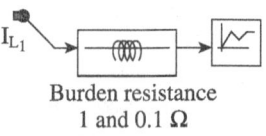

Burden resistance
1 and 0.1 Ω

Fig. 17.50 CT model

11. Increase the burden resistance to 4 Ω and observe the results. Note the half cycle saturation effects due to the DC offset in the primary current.
12. The reactor is protected by a differential relaying scheme. Use the 2-CT model in PSCAD to connect one phase of the reactor protection scheme as shown in Fig. 17.51.
13. Verify the burden current in the differential CT connection for faults at A and B.
14. Does the impedance of the connection leads have an effect on the results? How is this impedance accounted for?

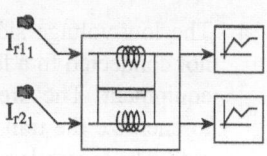

Fig. 17.51 2-CT model

Tutorial 17.2 To study the inrush phenomena when energizing a transformer.

Objectives
- Learning about the transformer and breaker operation
- Creating inrush phenomena when energizing a transformer
- Understanding plotting and control

Procedure
1. Create a new case by using either the 'Menu' or 'Toolbar'. A new case should appear in the *workspace settings* entitled *noname [psc]*. Right click on this workspace settings entry and select 'Save As' and give the case a name. Create a folder called *c:......./PscadTutorial/Inrush*. Save the case as *test2.psc*.
2. Open the main page of your new case. Build a case to study the inrush phenomena when energizing a transformer. The component data is as shown in Fig. 17.52. The transformer is rated 66/12.47 kV.
3. Plot the currents (I_a) and bus voltages (E_66) on the high voltage side of the transformer. *Note: I_a and E_a contain the three waveforms of the three phases.*

Figure 17.53 shows the basic steps to create a graph with a selected signal.

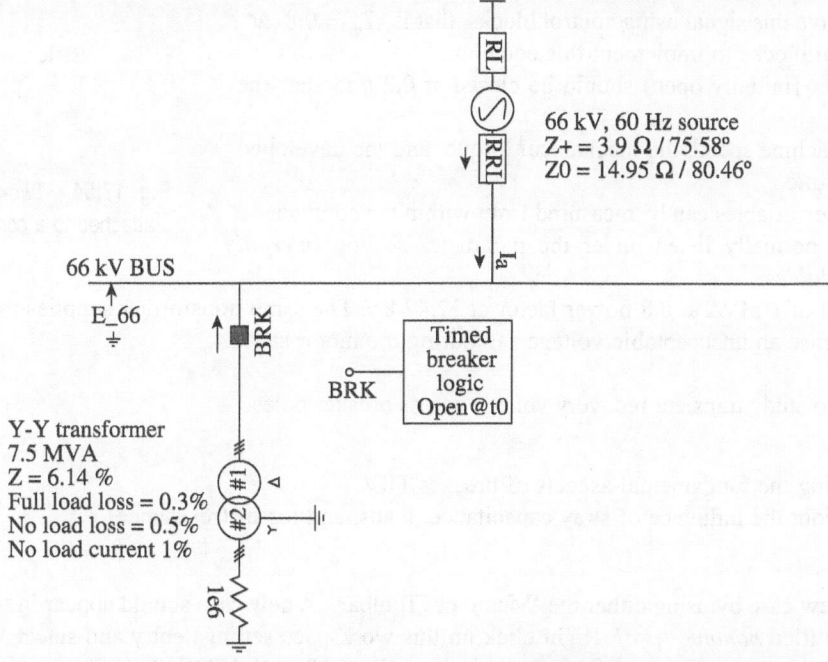

Fig. 17.52 Transformer energizing circuit

4. The low-voltage side of the transformer is not connected to a load or any other system equipment. The breaker is closed at 0.5 s to energize the transformer from the 66 kV side. The inrush is related to core saturation. Verify that saturation is included in the model used for this simulation.

Inrush current magnitude depends on the 'point on wave' switching conditions. Use a manual switch to operate the breaker. Note the point on wave dependency of the inrush peak.

Figure 17.54 shows the two-state switch attached to a control panel.

5. Modify the case to include the 12.47 kV, 0.5 MVA (wound rotor type) induction machine. This case will be used to study the process of starting an induction motor. The component data is shown in Fig. 17.55.

You may use the wire mode to connect different components.

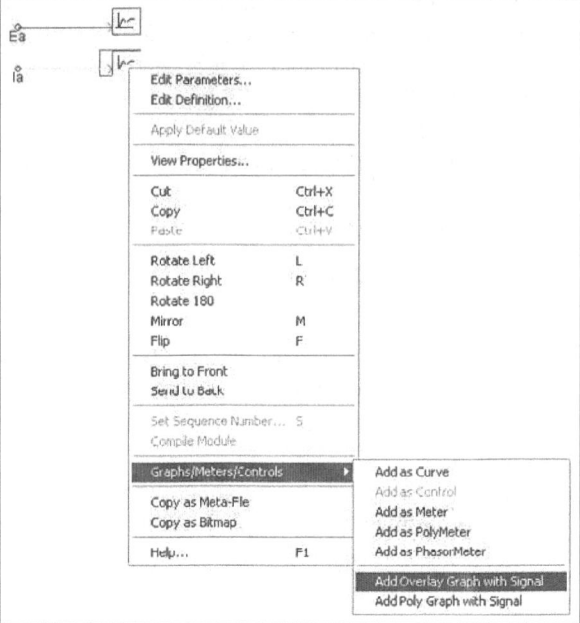

Fig. 17.53 Basic steps to create a graph with a selected signal

6. Enter the component data.
   Note: Use 'typical' data for the machine.
7. Plot the currents on either side of the transformer (i_a and i_b).
8. The input torque to the machine is equal to 80% of the square of the speed. Derive this signal using control blocks, that is, $T_m = 0.8 \cdot \omega^2$. Use control blocks to implement this equation.
9. The breaker (initially open) should be closed at 0.2 s to start the motor.
10. Plot the machine speed, the mechanical torque, and the developed electric torque.
    Note: Some variables can be measured from within the component. These are normally listed under the parameter section *internal output variables.*

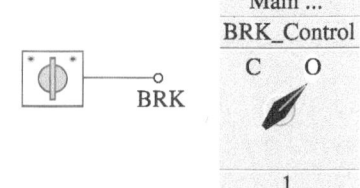

Fig. 17.54 Two-state switch attached to a control panel

11. Add a load of 1 MVA at 0.8 power factor at 12.47 kV. The same transformer supplies this load. Does the load sense an unacceptable voltage sag during the motor start?

Tutorial 17.3 To study transient recovery voltage across breaker poles.

Objectives
- Understanding the fundamental aspects of breaker TRV
- Learning about the influence of stray capacitance, loads, and losses (resistance)

Procedure
1. Create a new case by using either the 'Menu' or 'Toolbar'. A new case should appear in the workspace settings entitled *noname [psc]*. Right click on this workspace settings entry and select 'Save As' and give the case a name. Create a folder called *c:......./Pscad Tutorial/TRV*. Save the case as *test3.psc*.

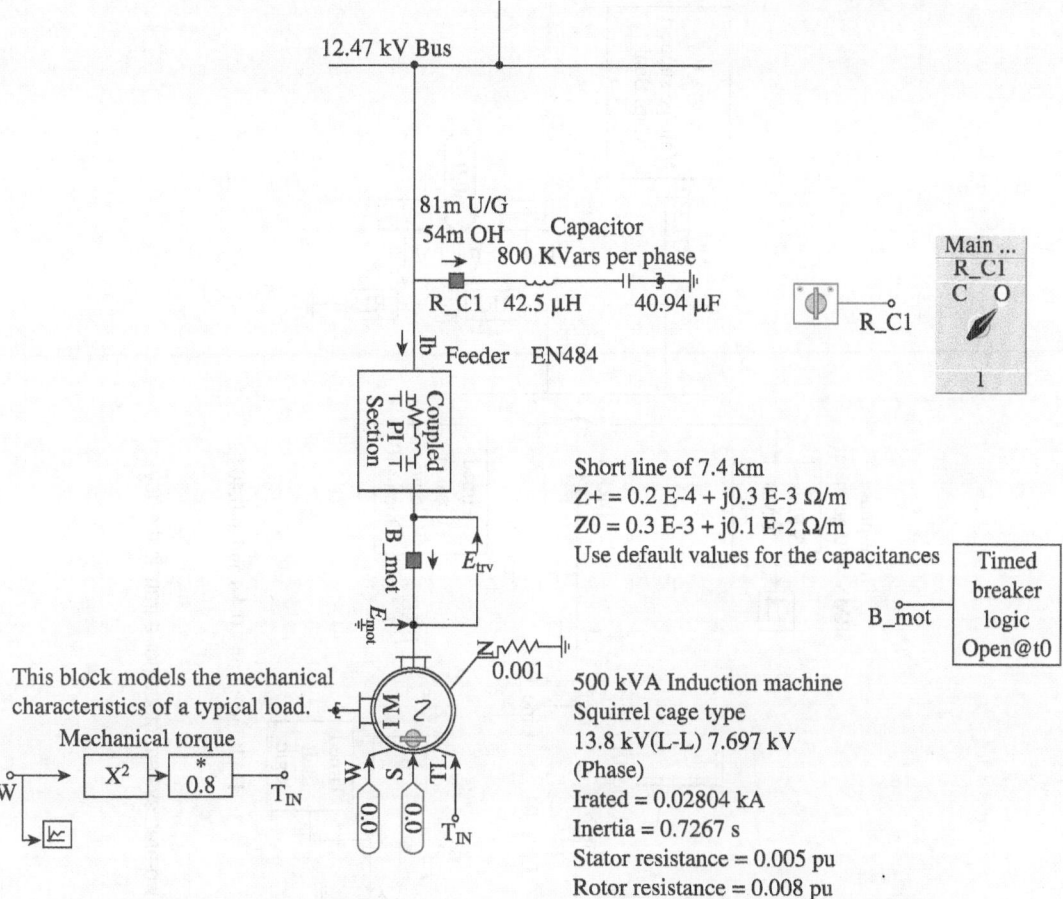

Fig. 17.55 Component data used to study the process of starting an induction motor

2. Open the main page of your new case. Build a case representing a simplified two-area power system as shown in Fig. 17.56. A 55 km transmission line connects the source station to a 100 MW wind farm. The conductor arrangement of the line is as shown in Fig.17.56. Use the frequency-dependent phase model to represent the line.

3. The wind farm is also represented by network equivalence. All other connections to the source station are represented by an equivalent 230 kV source. The equivalent source impedance is derived from a steady-state fault study at 60 Hz.

4. Use detailed models to represent the 33/230 kV transformer and the 55 km transmission line. The transformer has a Y–Y configuration and consists of three single-phase units. The no-load current is 1%. The no-load and copper losses are 0.003 pu and 0.002 pu, respectively. The positive-sequence impedance of this source at 33 kV is 1 Ω at 89°.

5. Include the breakers, breaker controls, meters, PSCAD 'fault component', and fault controls as shown in Fig. 17.56. In addition, add 300 MVars of capacitive reactive power at source station to support the 230 kV bus voltage. A transient study is required to design the equipment of this installation.

6. Initially, keep the breakers BRK1, BRK2, and BRK3 in closed position and the capacitor banks open. Run the case and make sure the power flow is as expected.

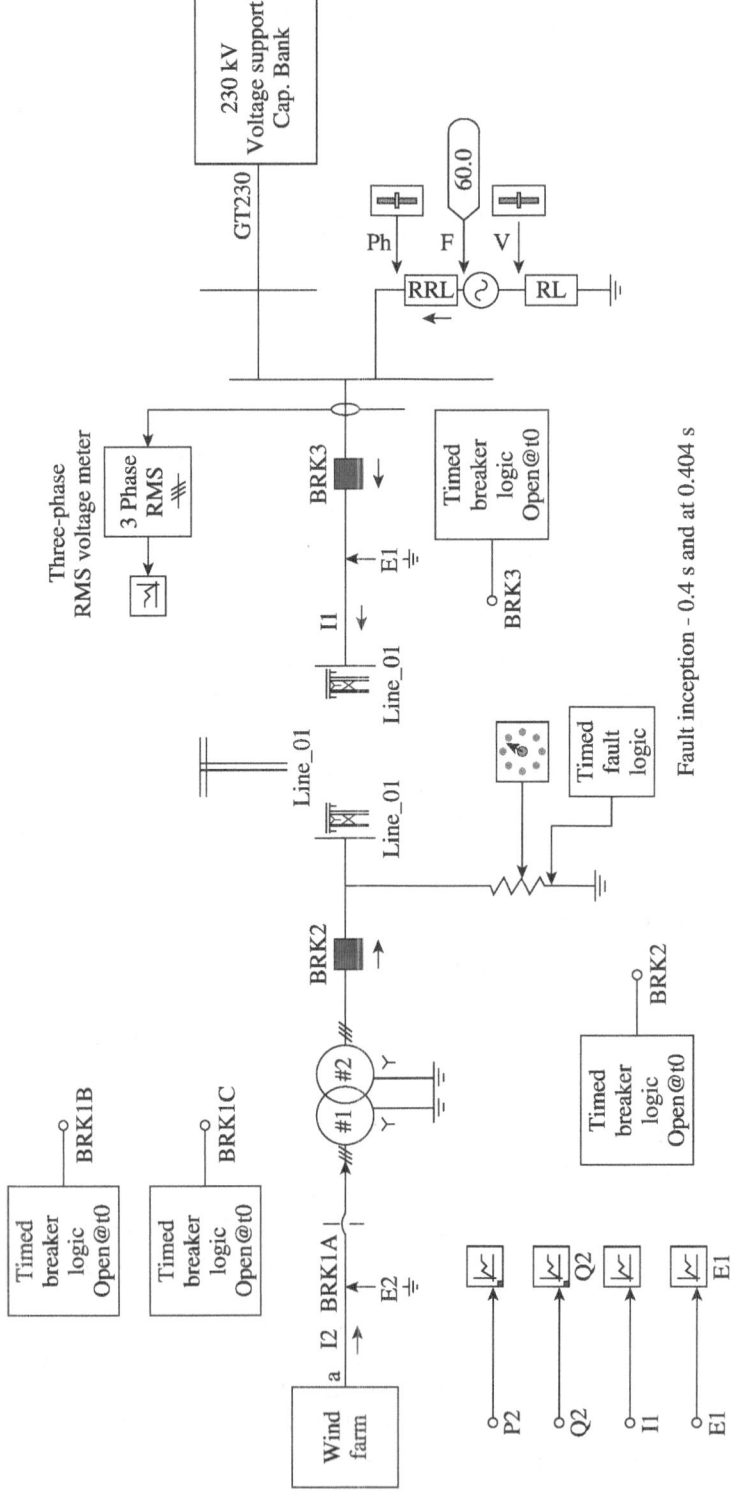

Fig. 17.56 Two-area system model for a transient study

7. Apply a three-phase fault to ground at 0.4 s. The duration is 1 s.
8. Open the breaker BRK3 at 0.44 s using timed breaker logic. Observe the voltage across the breaker poles.
9. Discuss the reason for TRV. Now, lower the time step to 2 μs and observe the results. This will make clear that for TRV studies, a small time step is necessary. Figure 17.57 shows the breaker TRV and the IEEE TRV limits.
10. In TRV studies, the stray capacitances near the breaker must be modelled adequately. How do we determine these values?
11. What are the measures available to reduce the TRV levels?

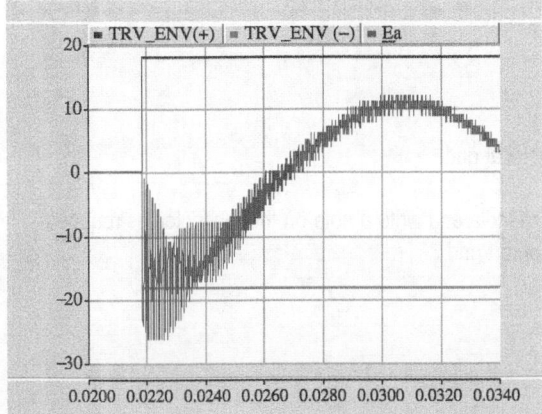 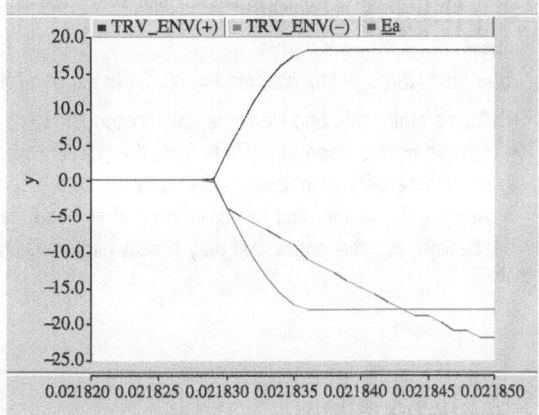

Fig. 17.57 Breaker TRV and the IEEE TRV limits

Smart Grid Technologies and Applications

18

Learning Objectives

After going through this chapter, the students will be able to:

- Define smart grid and list its various components
- Explain how a smart grid differs from the conventional electrical grid
- List the benefits of installing smart grid
- Analyse the factors that favour smart grid implementation in India and write a note on the technologies required
- Highlight the challenges that may arise while designing a smart grid

18.1 Introduction

A smart grid is an advanced version of existing electrical grid which has digital equipment integrated for energy measurement and control. The electrical smart grid is equipped with various devices such as smart meters, smart appliances, non-conventional energy resources, and conventional energy efficiency resources. The aim of this new emerging concept is to provide technical and financial advantage for both power producers and consumers. Smart grid provides a new opportunity to the customer to promptly change the status from a 'consumer' to that of 'supplier' in the electricity market. Smart grid can be defined as the 'convergence of physical infrastructure as well as information and communication technology infrastructure.' It intelligently integrates the renewable and non-renewable generators in order to distribute continuous, cheap, and reliable electricity with automation and minimum losses. The pattern of flow of electricity for the existing electrical grid and the smart grid is shown in Fig. 18.1.

18.2 Benefits of Smart Grid

The existing power grid has many constraints about energy production, transmission, and utilization. The use of automation and control in every field of nation's infrastructures is required to uplift the living standards of human beings. Hence, the implementation of smart technology will definitely modernize the power grid and transform the way we live our lives. Though, new investment is required to reform the nation's electric energy transportation, it will give benefit to power producer, utility, consumers, and the nation as a whole. The following sub-sections enumerate the various benefits of smart grid.

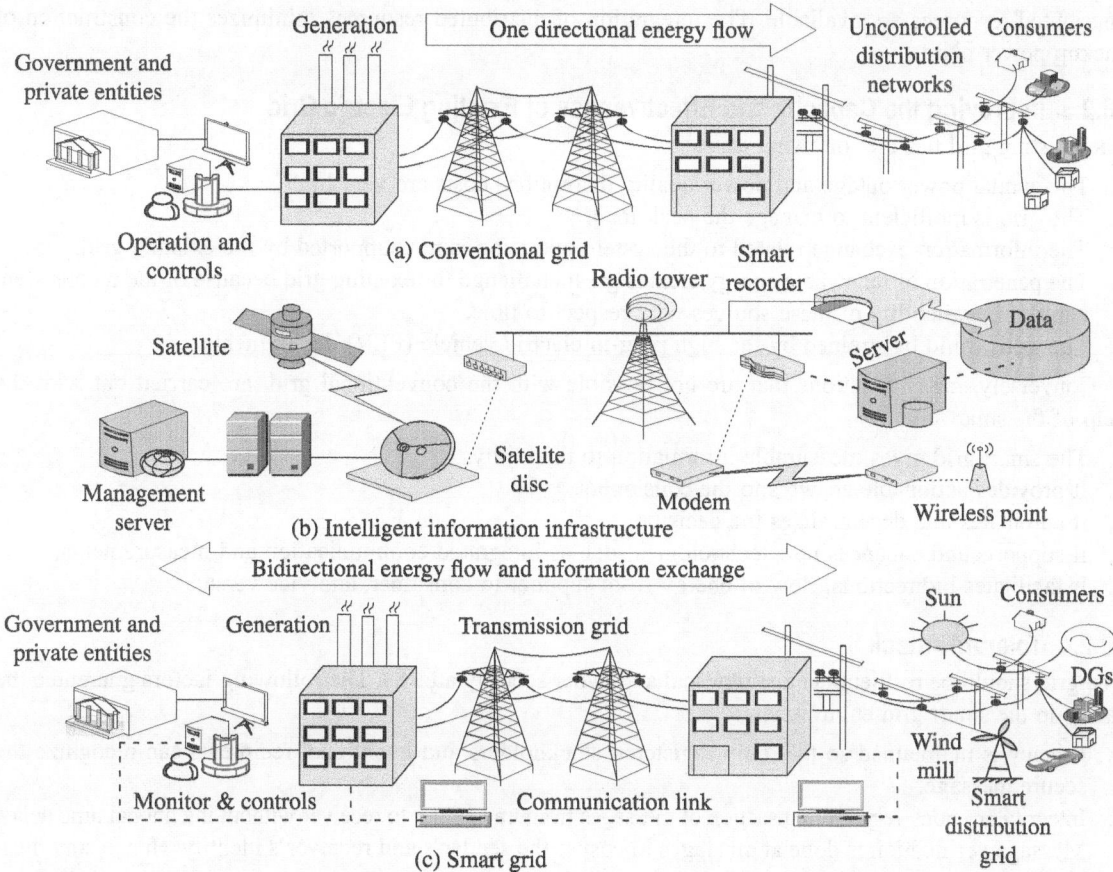

Fig. 18.1 Communication infrastructure of (a) conventional grid (b) intelligent information infrastructure (c) smart grid

18.2.1 Self-healing

The self-healing feature of smart grid takes periodic measure without human intervention to reduce power outages and problems related to power quality. Dedicated sensors/transducers, implanted at various locations in the grid, provide real-time information of system parameters which are useful to detect system problems and take automatic action for controls. Like the immune function of the human body, self-healing makes the grid capable to withstand and mitigate any internal or external hazard (fault). It also ensures the gird stability and supply quality. Self-healing occurs in two areas, namely transmission grid and distribution grid. Due to the differences in their roles in terms of network configurations and operation, the requirements of the self-healing functions of transmission and distribution grid are different. The transmission network transmits the power from the large power plant to the major load centres. It is a meshed network fed with multiple generation plants. The cutoff of one or several elements will not affect the operation of the network. Therefore, the self-healing of transmission grid is to continuously monitor the condition of the electric apparatus in the transmission grid, detect, and mitigate the apparatus' problems, and isolate the faulted apparatus by fast protection. The other functions are online security assessment, early warning, and corrective control of system stability to prevent the system from cascaded blackouts.

18.2.2 Minimizing the Need to Construct Backup Power Plants

In order to cope with the load demand during peak hours, the conventional grid uses backup power plants widely known as *spinning reserve*. The smart grid minimizes the requirement of backup power plants as distributed resources are available. The integration of distributed resources minimizes the construction of backup power plants.

18.2.3 Improving the Capacity and Effectiveness of Existing Electric Grid

The existing grid has the following issues.

1. The annual power outage and power quality disruptions costs are very high.
2. The grid is inefficient to manage the peak load.
3. The information exchange related to the system operation is not supported by the existing grid.
4. The penetration of renewable energy creates great challenge for existing grid because of the inconsistent nature of availability of these sources with respect to time.
5. The grid would be strained by the high plug-in electric vehicle (PEV) deployments.

Conversely, many functions that are not possible with the conventional grid, are carried out with the help of the smart grid.

1. The smart grid gives measurable information to the utility.
2. It provides actionable answers to the consumers.
3. It automates and decentralizes the decisions.
4. It supports and enhances new technologies such as integrated communication and measurements.
5. It facilitates bidirectional flow of energy from supplier to consumer, and vice versa.

18.2.4 Tolerant Attack

The grid should be resilient against physical and cyber security attacks. The following factors guarantee the safety in the smart grid environment:

1. Privacy is maintained so that only registered stockholders and intentional recipients can recognize the secure message.
2. Integrity provides very fast transaction of messages from transmitter to receiver without intentional time delay.
3. Message verification is done at all stages to ensure the sender's and receiver's identity. Hence, any message from unauthorized sender (hacker) is denied.
4. Non-refusal of message provides certification that the message must come from unique sender and the same sender cannot refuse to send the message.

18.2.5 Reducing Greenhouse Emission

The smart grid can make more efficient operations for utilities by introducing an opportunity to interconnect the renewable energy sources which reduces pollution level. Integration of these renewable energy sources along with deployment of hybrid vehicles reduces CO_2 level.

18.2.6 Reducing Oil Consumption

The interconnection of renewable energy sources such as wind and solar in smart grid decreases the power generation by means of oil consumption (diesel generator). The smart grid also supports the facility of electrical vehicle for transportation rather than using oil as its primary source of fuel. Thus, it reduces the oil consumption of the country.

18.2.7 Increasing Consumers' Choice

The smart grid allows consumers to select any power supplier. If the consumers are conscious about their power consumption then they can save energy. Moreover, use of smart appliances, advance metering, home automation, electric vehicles, and smart storage devices shifts the power consumption to non-peak periods and reduces the generation and transmission capacity. The smart grid also provides reliable supply by reducing outages and using renewable sources such as rooftop solar and wind.

18.3 Comparison Between Existing Grid and Smart Grid

The detailed comparison between existing grid and smart grid is shown in Table 18.1.

Table 18.1 Comparison between existing grid and smart grid

Feature	Existing grid	Smart grid
Technology used	The existing grid utilizes electromechanical devices.	It uses devices based on digital technology.
Communication of data	The path of exchange of data is in one direction only.	It utilizes two-way communication and hence, exchange of information takes place in both directions.
Type of generation	It uses the concept of centralized generation which leads to higher cost and longer construction time period.	It involves Distributed Energy Resources (DERs) which can be located at a place where actually load is situated or available. This leads to the reduction of building of big power stations and also reduces Aggregate Transmission and Commercial (AT & C) Loss.
Monitoring of the network	The existing grid dost not have self-monitoring and self-healing feature. Hence, recovery from disturbances is not as fast as in case of smart grid.	The smart grid has self-monitoring and self-healing feature. Therefore, it has better ability to withstand and recover from disturbances.
Customer choice and participation	In the existing grid, the customer has limited choice and their participation in the grid is also limited.	The smart grid provides better option of information and control of various data to the customers. Hence, customers are able to modify their consumption based on real-time pricing.
Checking and testing	The checking and testing is carried out manually in the existing grid.	The smart grid involves remote checking and testing.

18.4 Factors that Favour Smart Grid Implementation in India

The existing grid has many limitations such as real-time pricing, high aggregate technical and commercial losses, augmentation of distribution system, and islanding operation of power system network. The smart grid helps to minimize these issues and improves the effectiveness of the entire grid. Moreover, the power generation to some extent by non-conventional energy sources and its interconnection to existing grid enable smart use of electricity at a reduced price.

18.4.1 Aggregate Technical and Commercial Loss Reduction

The transmission and distribution losses in our country are of the order of around 30%. This is due to lack of transparency and non-metering. Implementation of smart grid reduces these technical and commercial losses up to a certain extent by means of better management of transmission and distribution networks.

18.4.2 Consumer Price Signal

In recent days, the cost of electricity diverges considerably, due to usage of conventional sources. The smart grid encourages consumers to use energy in an intelligent way. This factor attracts the customers to drift from static mode pricing to dynamic mode pricing.

18.4.3 Integration of Renewable Energy Resources

The majority of power block generated in India is from thermal and nuclear power plants. However, environmental concerns urge us to set up generation of power from clean energy sources such as wind and solar. Wind energy generation has been supported by India and many states like Gujarat and Tamil Nadu are encouraging wind power generation. Recently, the Government of India announced the National Solar Mission which has the goal to generate around 22,000 MW of power by the end of 2020. Thus, the Indian smart grid can be accelerated by integration of renewable energy resources. Around 67,076 square km area is projected to implement the wind energy generation and 44,105 square km area is intended to realize the solar power generation. The Ministry of New and Renewable Energy (MNRE) of India supports renewable energy to integrate and contribute the existing grid. As on date (2016), the combined renewable energy capacity of our country has reached around 45 GW, out of which nearly 63% comes from wind and 16% is contributed by solar energy. The Indian government proposes to accomplish nearly 40% to 45% of collective electric power production from non-conventional sources by 2030.

18.5 Key Areas for Smart Grid Initiatives in India

The advancement in technology and its application renovates the existing power system to an intelligent grid. Various smart meters, automatic controls, software, and the latest communication technology available can commence smart grid project in India. The following are the aspects of smart grid initiatives in India.

18.5.1 Advanced Metering Infrastructure (AMI)

AMI constitutes the whole of integrated infrastructure set up to enable transfer of real-time energy usage information and embarks on bidirectional communication linking the utility supply and its consumers in the existing network. Typically, these systems involve state-of-the-art electronic hardware and software, robust communication systems, and data reception and management systems. Worldwide, AMI has been implemented across various service providers such as electricity, water, and gas. In India, the power distribution sector has initiated actions to adopt AMI. From the utilities' perspective, getting the regulators and consumers to agree on the cost of such projects is a difficult proposition.

However, keeping cost issues aside, there are technical challenges to integrate complex AMI system with diverse IT systems in utility operations. Implementation of AMI also requires interoperability of technical standards across individual components. With the changing scenario of educated class, inclination of the people is increasing towards AMI.

18.5.2 Meter Data Management (MDM)

MDM provides a facility to analyse the data being collected and sent over the system. This analysis is useful to supplier for deciding the cost of energy consumed by the subscriber. With MDM facility, consumer can also know about the electric energy being utilized by them. The smart meter provides real-time information to supplier and consumers regarding the efficient use of energy, any accident to supply system, and restoration of supply after grid failure.

18.5.3 Geographical Information System (GIS)

Demand side management and energy security are the factors in sustainable business of electricity in smart grid. The incorporation of GIS with different features improves the efficiency of a utility intelligently. Extensive use of GIS is helpful to the Indian smart grid to tackle issues like consumption monitoring, detection of tampering, and reduction of line losses. GIS, along with India Smart Grid Knowledge Portal (ISGKP), is a platform which shares knowledge and broadcasting information among all stakeholders.

18.5.4 Enterprise Asset Management (EAM)

The proper management of available asset throughout the life of the utility from the commencement of electricity business increases the performance of grid. The management involves procurement, planning, construction, operation, maintenance, and disposal of asset. At various stages of utility operation, asset management reports need to be generated to assist further investment. The vast paperwork and manual labour make EAM process tedious, time-consuming, and lead to human errors. The design and development of a software system for EAM improves system performance and transparency in information. To manage the core business processes, India has good venture assets management software and IT solutions.

18.5.5 Distribution Automation (DA)

DA technology is implemented by some utilities as a part of pilot projects involving one or more renewable sources. Generally, electric utility uses Supervisory Control and Data Acquisition (SCADA) system for widespread control over transmission equipment, and growing control in the distribution equipment by means of distribution automation (DA). However, it is difficult to control smaller entities such as Distributed Resources (DRs), buildings, and homes. Implementation of DA provides cost-effective solution over the equipment control and communication methods. Utilities should put more effort to utilize standardized DA devices which work together appropriately.

18.5.6 Automated Call Centre (ACC)

The aim of ACC infrastructure in utility is to integrate customer information, their address, queries, and complaints. Moreover, it also provides basic information about office locations, billing information, bill payment centres, connection status, service levels, planned outages, and information on efficiency programmes, among others. ACC helps the customer to convey the day-to-day upgradation in system measure and energy efficient schemes. They also facilitate utilities to follow consumption patterns and customer payments. At present, the customer complaint solving process in India is inferior, resulting in reduced customer satisfaction. However, ACCs are striving to put up better processes and more liability within the utilities.

18.5.7 Customer Relationship Management (CRM)

A utility provides various services to enlighten the customers regarding quality, availability, and reliability of power supply. Moreover, customers should be aware of their role in the business of electricity and the utility has to regularly upgrade the customer information in the CRM process. By putting CRM and AMI data together, the benefits such as actual capacity utilization and system performance are obtained.

Customers should be well aware of about the risks involved in smart grid. The smart meter measures accurate energy consumption when compared to the electromechanical meters and hence the customers should be prepared to see higher bills. They have to revise their energy usage methods to reduce the billing amount. Hence, the participation of customer is extremely important in all of these activities.

This will also help to improve the smart grid's overall governance and utility-wide implementations.

18.5.8 Some Smart Grid Pilot Projects in India

The Indian government has approved many projects to modernize the existing power network and a step towards smart grid. Under the National Smart Grid Mission (NSGM), India has approved many smart grid pilot projects in various states of the country. The Ministry of Power (MoP) has approved smart grid pilot projects that will be implemented by various state-owned distribution utilities in India. Table 18.2 shows six major pilot projects awarded by Government of India (GOI). Table 18.3 shows some other smart grid projects to be implemented by Distribution Company of state governments.

Table 18.2 Smart grid pilot projects in India

Project name	Functionality
Smart Grid Research Laboratory, CPRI, Bengaluru, India.	The Ministry of Power, has endorsed 'Smart Grid Research Laboratory' at Central Power Research Institute (CPRI), Bengaluru with the aim to establish Smart Grid Technology Centre (SGTC) and Interoperability Laboratory. This laboratory will have various smart grid components like Intelligent Electronic Device (IED), smart meters, and testing facilities as per IEC 61850 (standard for communication networks and systems in substation) compatible devices. Moreover, the laboratory will support renewable energy integration and advance communication technology, besides promoting advanced

	research, conduction of workshops, and conferences for utility engineers and academicians. The tentative expense of this project is around INR 11.05 Crore. Also, U.S. Trade and Development Agency (USTDA) has granted $ 692000 to the Institute in June 2013 for the smooth execution of this project.
AMI Project, TATA Powers, Mumbai.	The Project includes implementation of Advanced Metering Infrastructure (AMI) in a selected area of Mumbai city by syndicate members such as L&T, Cyan and Neosilica. A good initiative by Cyan, and Neosilica to provide wireless technology for Meter Data Acquisition System (MDAS) and Tata's Meter Data Management System (MDMS) for billing and fault supervision. The project will deploy 5,000 smart meters in a selected area in its preliminary stage.
Smart Grid Pilot Project, Indiranagar, Bengaluru by BESCOM.	Bangalore Electricity Supply Company Limited (BESCOM) is planning to implement 26119 residential and commercial consumer-based smart grid pilot project for Indiranagar, a metropolitan area of Bengaluru, India. The project includes functioning of AMI, Peak Load Management (PLM), and Solar Rooftop PV Systems (RTPV). Under this project, smart meters and communication systems including MDAS, MDMS, PLM, and Demand Response (DR) software will be installed in the control centre for data collection, storage, and analysis. A number of RTPV systems are planned to be set up in Indiranagar and will be interconnected with the local distribution system using wireless and net metering facilities.
Ganjam Smart Grid Project, Odisha	The Odisha government has awarded a smart grid project for power network in cyclone prone area of Ganjam district in Odisha worth Rs. 1000 Crore. The Asian Development Bank (ADB) has agreed to provide a loan of around Rs 650 crore and the rest will be funded by the Odisha state government. This project mainly focuses on deployment of the latest technology in power networks to withstand stormy cyclones, with wind speeds more than 300 kmph. This project emphasizes on the implementation of underground distribution supply and gas insulated sub-stations with full automation and control. More than 5800 consumers would be alerted for services, schedule power cuts, and any emergency control over the distribution system under this project.
Battery Energy Storage demonstration projects	Power grid Corporation of India Limited (PGCIL) has declared a tender for 3 demonstration projects at Puducherry on grid connected battery storage devices using (1) LI-Ion, (2) Advanced Lead Acid and (3) Sodium Sulphur batteries. The aim of this project is frequency regulation and finds suitability of battery for grid interconnection of renewable energy in India. The project includes battery energy storage supply, its installation, testing and commissioning for a size of 1 MWhr.
Calcutta Electric Supply Corporation Limited (CESC)	USTDA has funded project grant to Calcutta Electric Supply Corporation Limited (CESC) for the implementation of smart grid technologies and practices across their electricity supply and distribution networks in Kolkata, India. The road map is planned by Tetra Tech and ESTA International study to develop smart grid and addresses a range of improvements, including integrating smart meters and automated meter reading into CESC's distribution system. The CESC has recommended reliability and power quality improvements, implementation of SCADA/DMS/EMS, communications infrastructure for AMI and Distribution Automation (DA) under this project.

Table 18.3 Smart grid project approved by Ministry of Power, India

Utility name	Functionality	Project area	Number of consumers benefited	Approved project cost (in Crore)
CESC, Mysore Karnataka	AMI Residential, AMI Industrial, PQM, Peak Load Management, MG/DG	VV Mohalla, Mysore	21824	Total:32.59 GOI:16.30
TSSPDCL, Telangana	AMI Residential, AMI Industrial, Outage Management, Peak Load Management, Power Quality Management	Jeedimetla Industrial Area	11904	Total:41.82 GOI:20.91
APDCL, Assam	AMI Residential, AMI Industrial, Outage Management, Peak Load Management, Power Quality Management, DG	Guwahati Distribution Region	15083	Total:29.94 GOI:14.97
UGVCL, Gujarat	AMI Residential, AMI Industrial, Outage Management	Naroda of Sabarmati circle and Deesa of Palanpur circle	22000	Total:82.70 GOI:41.35
UHBVN, Haryana	AMI Residential, AMI I, Outage Management	Panipat City Sub-division	11000	Total:20.7 GOI:10.35
TSECL, Tripura	AMI Residential, AMI I, Peak Load Management	Electrical Division No.1 of Agartala Town	42676	Total:63.43 GOI:31.72
HPSEB, Himachal Pradesh	AMI Industrial, Outage Management, Peak Load Management, Power Quality Management	KalaAmb, industrial area	1251	Total:19.45 GOI:9.73
PED, Puducherry	AMI Residential, AMI Industrial	Division 1 of Puducherry	34000	Total:46.11 GOI:23.06
PSPCL, Punjab	AMI, PLM	Industrial Division of City Circle Amritsar	2734	Total:10.11 GOI:5.06
WBSEDCL, West Bengal	AMI Residential, AMI Industrial, Peak Load Management	Siliguri Town in Darjeeling District	5275	Total:7.03 GOI:3.52

18.6 Technologies to be Used for Smart Grid

The following technologies are used for smart grid.

18.6.1 Analytical Tools

A few of the analytical tools are discussed in the following pages.

Energy Storage Computational Tool

The energy storage computational tool is useful to compute the benefit of energy storage and also to calculate the cost of it. Moreover, this tool assists to analyse how the costs and benefits vary for a given different situation and assumption of momentary energy storage. The output of the energy storage computational tool will be in the form of charts, graphs, and tables that recapitulate the benefits and costs of an energy storage deployment.

Interruption Cost Estimation Tool

The interruption cost estimation tool is basically designed to calculate service interruption cost and reliability of the system. This tool is used by the utilities, government organizations, or other entities that are interested in estimating interruption costs or the benefits associated with reliability improvements. This tool needs cause of interruption, interruption duration, amount of power not delivered/transferred, and energy rate to estimate the interruption cost and reliability improvement.

Smart Grid Computational Tools

The smart grid computational tool is required to compute the payback of smart grid projects before implementation. Moreover, this tool can also be used for evaluating deviations in expenditure and repayment in case of various circumstances of smart grid operation. Figure 18.2 shows the analytical framework of computational tool.

Weather Prediction Tool

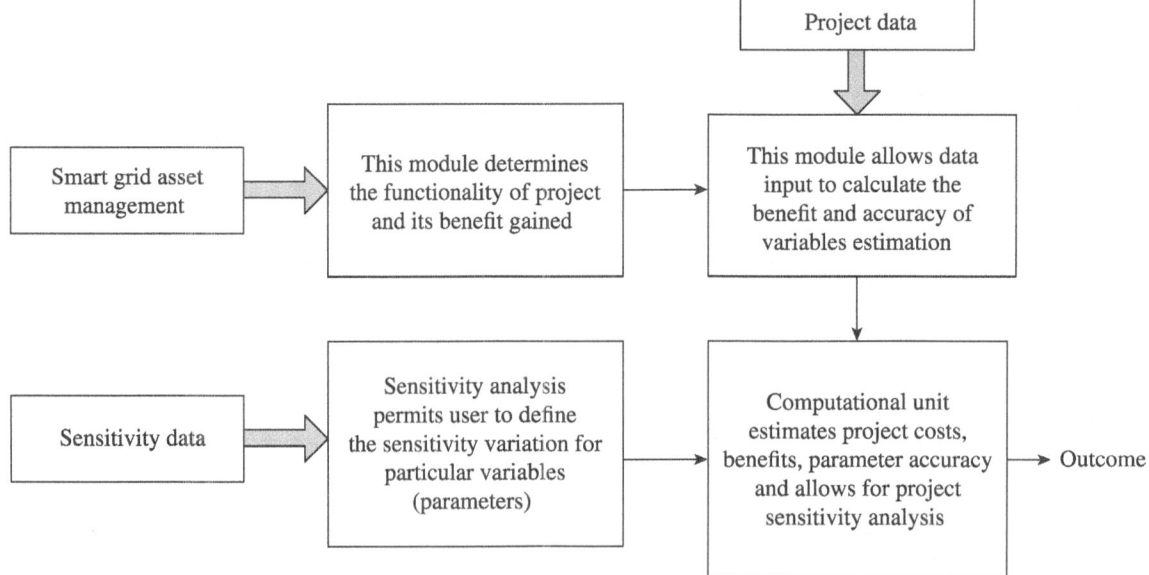

Fig. 18.2 Analytical framework of computational tool

A weather prediction tool is required to analyse the weather condition which guides against interruptions and risk assessment. Based on hourly and daily weather data collection, this tool predicts the frequency of interruptions in a particular time slot. At the time of developing the weather prediction model, various environmental conditions have to be considered such as rain, wind, temperature, lighting density, humidity, barometric pressure, snow, and ice.

Analytical Framework of Model

As discussed above, the weather condition is predicted to decide the number of interruptions in the grid. Hence, a complete structure is developed for interruption forecasting by accounting the historical weather data. Proper utilization of this framework minimizes the risk of utility and preventive maintenance programmes are scheduled, once the number of interruptions is forecasted based on historical weather data. This alertness of the power distribution corporation will absolutely help to attain an adequate level of reliability. Thus, the improvement of reliability by means of analytical framework is one of the key purposes of shifting towards the smart grid. Figure 18.3 shows the analytical framework of the model.

18.6.2 Different Communication Technology Used in Smart Grid

A communication channel is the main module of the smart grid infrastructure. The enormous data generated at various points in the smart grid has to be conveyed by this channel at appropriate place. Moreover, with the use of latest technologies and applications to form an intelligent electricity grid, the collected data is utilized for further analysis, control, and real-time pricing methods. Thus, electric utilities have to identify a

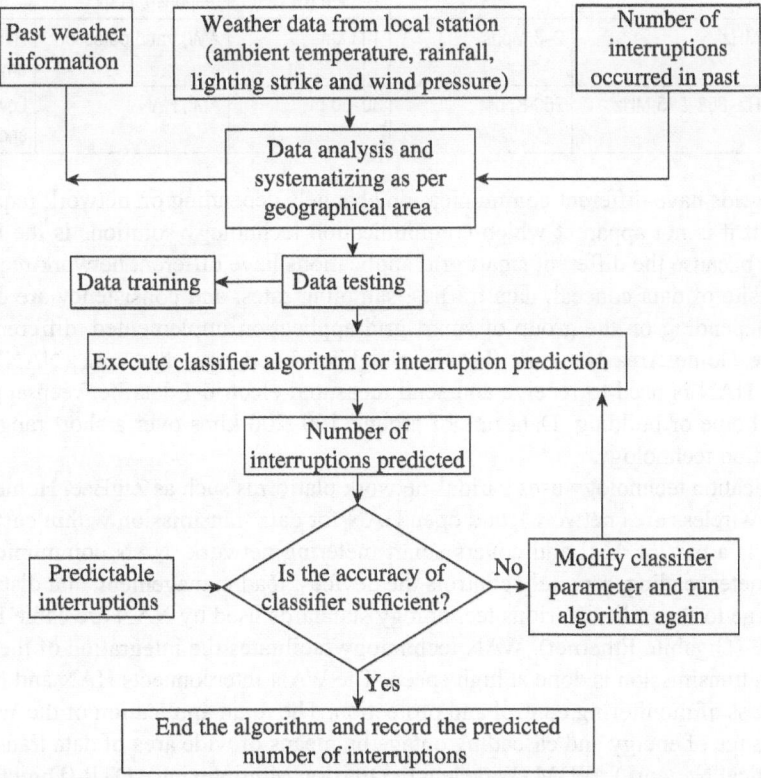

Fig. 18.3 Analytical framework of model

dedicated communication technology which is capable to handle the generated data and provide a consistent, safe, and cost-effective service throughout the entire system.

In the smart grid, for data transmission between smart meter and utilities, two communication media, namely wired and wireless have been widely used. Table 18.4 shows the different communication technologies used in smart grid.

Existing cellular networks are communicating well between smart meters and the utility. Wide area environment based cellular network gives solutions to enable smart metering positioned in scattered area. 2G, 2.5G, 3G, WiMAX, and LTE are examples of existing cellular communication technologies in utilities for smart metering deployments.

Table 18.4 Communication technologies used in smart grid

Technology	Spectrum	Data rate	Coverage range	Applications	Limitations
GSM	900–1800 MHz	Upto 14.4 Kpbs	1–10 km	AMI, demand response, HAN	Low data rates
GPRS	900–1800 MHz	Upto 170 Kpbs	1–10 km	AMI, demand response, HAN	Low data rates
3G	1.92 to 1.98 GHz 2.11 to 2.17 GHz (licensed)	384 Kpbs to 2 Mpbs	1–10 km	AMI, demand response, HAN	Costly spectrum fees
WiMAX	2.5 GHz, 3.5 GHz and 5.8 GHz	Upto 75 Mpbs	10–15 km (LOS) 1–5 km (NLOS)	AMI, demand response, HAN	Not widespread
PLC	1–30 MHz	2–3 Mpbs	1–3 km	AMI, fraud detection	Harsh, noisy channel environment
ZigBee	2.4 GHz, 868–915 MHz	250 Kpbs	30–50 m	AMI, HAN	Low data rate, short range

Various smart grids have different communication channels depending on network requirement and data broadcast. But still it is not apparent which communication technology solutions is the finest for grid applications. This is because the different smart grid applications have different network requirements. In this case, the prerequisite of data conceal, data traffics, sampling rates, and consistency are different in different technology. Depending on the group of smart grid application implemented, different communication networks used are Home Area Network (HAN), Neighbourhood Area Network (NAN), and Wide-Area Network (WAN). HAN is used to receive and send measured electrical data between appliances and controller within the home or building. Data rate of around 100–200 kbps over a short range is sufficient for HAN communication technology.

HAN communication technology uses various network platforms such as ZigBee, Home Plug, 6LowPAN (IPV6 low power wireless area network), and open HAN for data transmission within customer's premises. NAN technology is a part of AMI and covers smart metering network. NAN communication technology supports remote meter reading, remotely controls the devices, load management, and distributers' information of energy usage to the utility. Various technology standards used by NAN are cable Ethernet, wireless Ethernet, and GbE (Gigabite Ethernet). WAN technology facilitates the integration of the entire smart grid area in which data transmission is done at high speed. The WAN interconnects HAN and NAN to the utility for real-time process of monitoring control and protection. The main application of the WAN is prevention of unauthorized usage of energy and cascading outage by means of wide area of data transmission. SONET (Synchronous Optical Network), WDM (Wavelength Division Multiplexing), DTR (Digital Trunked Radio), and RoIP (Radio Over IP) are a few transmission media used in WAN technology for smart grid.

Currently, Bluetooth, and Zigbee are the most valued by the industry in smart grid automation. Zigbee supports low power consumption, low cost, and requires minimum number of nodes. Moreover, Zigbee wireless SEP2.0 specification supports multiple network interfacing features (standardized protocol) and is compatible with a variety of communication methods to connect with the smart meter.

18.6.3 Wide Area Monitoring and Control System

Wide Area Monitoring (WAM) and control system provides automatic control over the equipment of smart grid. Its function is to optimize the comprehensive operation of the grid. It is also linked indirectly to fault clearance, system protection, or equipment maintenance. The main function of WAM and control system is to ensure secure operation of the system as discussed below.

1. Automatically changes the tap position of transformer to maintain normal system voltage.
2. Reactive power and voltage of power system can be controlled by capacitor bank and inductive reactor. Hence, their switching could be automatically controlled by WAM as per the reactive power requirement in system.
3. Interlocking of switchgear equipment in substation to prevent unsafe operation. The automation and control system of smart grid assess the status of the whole grid/substation and may prevent the switching of particular equipment. For example, an operator issues a command to close the earthing switch of the transmission line connected to the live bus prior to maintenance. However, based on the data collected from line and programmed logic, WAM automation system may reject the operator's request of closing the earthing switch to prevent fault.
4. Safe and sound operation of entire grid is achieved by means of automatic sequencing controls. These facilities may eliminate the individuals of group command issued by operators to control the system. Hence, the entire system operates in sequence as pre-programmed in computer (software) by rejecting any invalid command issued by the operator.
5. WAM and automation control perform automatic load balancing in the transmission and distribution systems which eventually reduces maintenance and resistive losses. For example, one line is loaded with 80% and the nearby line is loaded with 25%. Instead of this if both may be loaded at 50%, then resistive losses reduce in the first line (losses vary with the square of the current). This arrangement improves the efficiency of the power system and reduces wear on equipment.
6. Automotive restoration services are also available with the transmission line and feeder in the event of fault. In the above two cases, the information about load unbalance in line/transformer or the inception of fault is to be notified to the substation computer by Intelligent Electronic Devices (IEDs) located at various points in the substation/grid.

18.6.4 Dynamic Line Rating Technology

The current-handling capacity of transmission line conductor confines the operation of power system. All the operators estimate the static rating of line for normal and worst operating conditions. Thus, static rating gives the maximum current withstanding capacity of the conductor under prescribed weather condition. These static ratings of line are modified frequently (hourly and daily basis) to report the maximum temperature, known as *ambient temperature rating*. On the other hand, dynamic line rating (DLR) system has communication and control system which adjusts the dynamic rating of the line during operating condition. This system includes transducers and sensors located on the line, communication channels, and computer software based energy management system (EMS). In addition to these, DLR system has weather sensors to record temperature, wind speed, and solar radiation. The communication media broadcast the collected data of line such as conductor temperature, sag, tension and line spacing to DLR software in which the dynamic rating of the line is estimated. The main advantages of DLR technology are (i) enhancement in the performance of transmission line (ii) reduction in initial investment and congestion cost by improving usage of existing assets, and (iii) augmenting the operational flexibility of transmission line.

18.6.5 Conductors in Smart Grid Technology

According to the weather conditions and varying load conditions, the selection of conductors is extremely important in smart grid. Moreover, technical and economic factors also play an important role in the selection of distribution conductors. The existing aluminum- and copper-based conductors are not sufficient for the rapidly growing smart grid technology. Hence, advanced electric conductors are needed to renovate the existing conventional power structures to distributed and interconnecting infrastructure. Superconducting wire of line and cable transmits hundred times more current than conventional copper conductor of the same size. This is due to the design of superconductor having some particle that offers no electrical resistance below assured temperature. Superconducting technology is the best choice for smart grid which provides excellent and cost-effective performance with high power transfer capacity and reduced voltage drop. Superconducting fault current limiter (SFCL) provides protection to the grid by quickly detecting fault and distributes power among efficient renewable sources.

18.6.6 Sensors in Smart Grid Technology

Generally, sensors are employed to measure the physical quantities and convert them to equivalent electrical signals. Sensors and sensor networks applications in smart grid, smart buildings, smart appliances, and industrial processes appreciably help to utilize the available resources more efficiently. With the use of smart sensors technology in power grid, they can accurately compute a variety of quantities such as temperature, wind pressure, solar intensity, chemical reaction, humidity/moisture, and variation in electrical/magnetic properties. The output of sensors is voltage, current, change of resistance/inductance/capacitance which are interfaced with electrical devices. Hence, as per the quantity sensed, sensors are categorized as electronic sensors, biosensors, and chemical sensors. Table 18.5 lists various sensors and their application at different locations in smart grids.

Table 18.5   Sensors used in smart grid and their application

Area	Component	Sensor	Application
High and low voltage substations	Substation bay	Antenna array	Identifies place and discharging components
		On line Infrared	Records thermal images of components
	Transformer	Gas sensor	To measure H_2 and C_2H_2 in transformer tank and oil
		3D acoustics	Analysis of discharge action in transformers
		Acoustics fiber optic	Measure little internal discharges in high risk regions
		Gas fiber optic	Detection of gas formation in high risk regions
		On line FRA	Continuously monitor frequency response using natural transients
	On load tap changer	OLTC gassing	Identifying overheating or damaged contacts
	Post and bushing eternal insulation	RF leakage current	Identification of high risk insulation requiring washing
	Disconnect	RF disconnect	Identifies high risk contacts wirelessly
	CTs and PTs	RF acoustic emissions	Indicates wireless mesh to identify internal discharges
	Breaker	RF SF_6 density	Demonstrating wireless mesh to trend SF_6 density
	Oil	MIS sensor	To measure H_2 and C_2H_2 gases in oil filled cable
	Underground cable system	Various sensors	Identifies thermal behaviour of cable, discharge measurement, and mechanically damaged

Underground cable system	Compression connector	RF temp and current	Measures connector temperature and current to determine risk and identify high risk components
	Conductor	RF temp and current	Measures connector temperature and current for rating
Overhead transmission lines	Line insulator	RF leakage current	Identification of leakage current and high risk insulation requiring maintenance
	Shield wire	RF fault magnitude and location	Determine fault location and severity
	Structure	RF lightning	Separation of lightning current magnitudes
		Sensor system	Integrates RF and image recognition sensors to investigate transmission line issues

18.6.7 Instruments/Transformers Used in Smart Grid

The accurate measurement of real-time data in smart grid optimizes the complexity of energy management system (EMS). A number of techniques have been developed for accurate measurement of energy flow in renewable integrated smart grid. Smart meters are connected to different sensors that help to determine the current drawn and energy consumption at various points. Current transformers, Rogowkis coils, resistive shunt, and Hall Effect sensors are widely used in conjunction with EMS in smart grid for bidirectional energy flow measurement. Privacy Enhancing Technology (PET) enables bidirectional power flow control in smart grid and also supports power factor correction, voltage sag, fault current limitation, and harmonic elimination. Intelligent distribution of power, appropriate integration of distributed resources (DRs), and suitable power utilization in transmission and distribution system can be realized by the involvement of PETs into smart grid. Service in and out of any distributed renewable energy resources (DRER), distributed energy storage devices (DESD), and load can be monitored using advanced digital instrument transformers.

18.6.8 Fault Testing Recloser

The distribution and transmission system is prone to environmental changes from time to time and precautions are required to tackle these changes. Automatic control and protection are essential functions of the smart grid. These include reclosing of circuit after being disconnected due to transient fault. The recloser automatically closes the circuit unlike conventional circuit breaker which requires manual operation. In this way, it improves stability and reliability of the distribution system. The communication technology used in the grid has facility to monitor and control these elements (recloser) from a centralized place (substation).

18.7 Challenges while Designing Smart Grid

Smart grid demands technological upgradation of existing power networks. The main constraint in this technological innovation is high installation and operating cost of large communication networks. Thus, financial issue, government regulation, profit constraints, lack of consumer awareness, skilled person to grip the system, and hi-tech developments are the main aspects while designing the smart grid.

18.7.1 Financial Resources

Smart grid demands digital technology for its efficient performance. The communication and control system required for smart grid are smart meter, intelligent sensors and equipment for data transfer, real-time measurement, and operation. These require enormous investment by the operator and government to make connection between smart grid and consumers. The initial cost to set up smart grid will be very high as they demand the latest hardware, software, and their application to smart grid. Hence, enough financial support is needed from stakeholders and government to establish the smart grid.

18.7.2 Government Support

Financial aid of government is a key aspect for initialization of any smart grid projects. Government forms policies and regulations for integrating the renewable resources. It also decides the standards of inter-operability and pricing of energy, energy exchange amongst different entities, and energy consumed. The government, along with private union, outlines the strategy for procurement and purchase of equipment from national and international markets. In general, the government prepares the generalized policies for the benefit of people's contribution to smart grid as supplier and consumers.

18.7.3 Speed of Technology Development

Technology is the backbone of smart grid which involves a wide range of communication, hardware, and software development. In certain areas, the technology is well developed whereas in others it is still in the growing phase. Many newly established companies have better solution to the communication infrastructure, protection, and control of hardware operation. As the technology progresses, it minimizes manual task, man-power, and time consumption. This reduces risk and provides speedy operation with correct decision. Hence, many stakeholders, service providers, and utilities urge well-developed hardware and software technology for the implementation of the smart grid.

18.7.4 Policy and Regulations to be Framed

The policies are formulated based on active participation of the consumer in the smart grid establishment. The regulations are framed to develop the utility with existing resources and the strategy is made to achieve smart grid facility at state/country level. Captive power plant and renewable resources interconnection to grid has to follow some rules formed by the government and different entities. Policies are also finalized for energy consumption tariff, energy efficient products, market energy pricing, and public infrastructure usage (electrical vehicle and storage devices). Some laws and regulations are also framed for investment in research, development, training programmes, consumer awareness, and efficient utilization of technology in smart grid.

18.7.5 Cooperation between Different Entities

At the planning stage, cooperation among different private sectors and individual power producers should be activated. The role of different entities is investment and profit sharing. The bidirectional information flow on open way leads to better cooperation among different entities and customer. In large-scale smart grid, better coopera-tion among entities will eventually ensure dynamic and efficient energy distribution. Hence, a well-understood interaction between different entities will aid in harmonized and competent relationship for sustained smart grid.

18.7.6 Cyber Security Challenges

Security is provided to the smart grid as its information technology is assorted in nature as a variety of devices and software are employed in the smart grid. Operation of smart grid without dedicated cyber security results in cyber-attacks and allows weakening of useful data. Insufficient security measure may lead to instability of grid communication and cause fraud, loss of information, and loss of recorded/real-time data of energy consumption. Hence, security against unauthorized changes or damage to confidential information is a challenging issue.

18.8 Basic Structure of Smart Grid

Earlier, electricity was produced, transmitted, and distributed without appropriate control over it. More-over, the number of small-scale power generation based on hydro, solar, and wind were also in growing stage. Hence, the interconnection of the renewable sources was very less. Furthermore, communication technology was not established for control of bidirectional energy flow and measurement of energy con-sumption. Even in the present era, technology development and its application for grid automation is not fully exploited to accelerate the existing utility. In future, the advancement in research, development, and technology will reform the present structure of power grid to intelligent (smart) grid. The use of digital sensors, smart metering interface, smart inverter, storage device, and automatic switching equipment leads to efficient control over the power grid. The ever-growing, renewable energy generation and its integration to the existing grid demand innovative solution to accommodate them. With the application of self-regulating technology, any fluctuating resources are controlled as per electricity demand. The use of SCADA system in transmission grid controls active and reactive power by means of automatic compensation. Similarly, AMI and EMS control power, voltage, and power factor in distribution circuit automatically. Automated fault location and restoration reduces the frequency and duration of outages with the use of sensing devices and smart algorithms. The communication infrastructure is a key component in smart grid to interface all the processes and devices working in synchronism. Various wired and wireless communication technology are used at every place in the grid for control, measurement, automation and data transfer from one place to another. Figure 18.4 shows the basic structure of a smart grid.

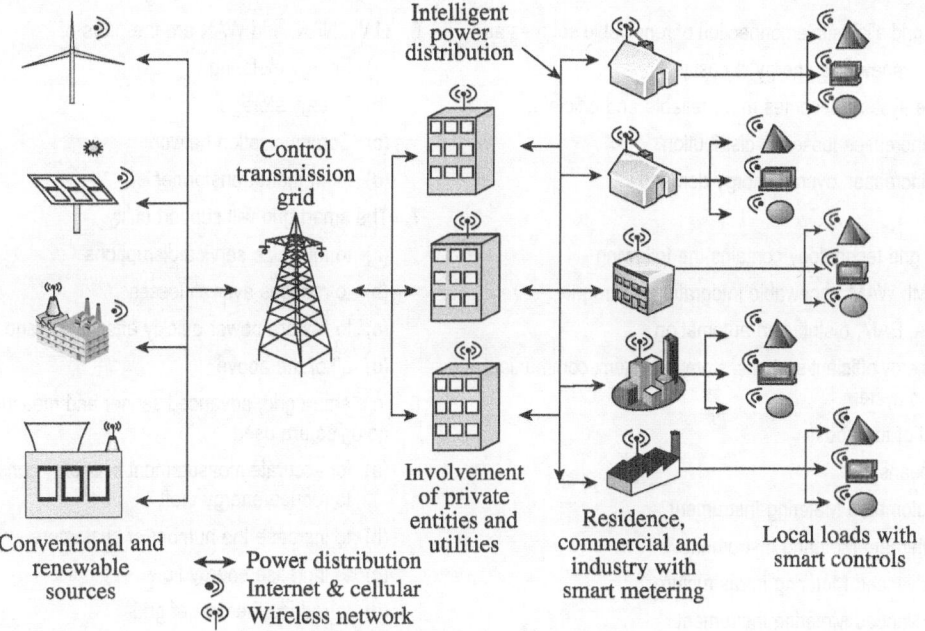

Fig. 18.4   Basic structure of smart grid

Recapitulation

- Smart grid is the reformation of an existing electrical grid which converges different operational and energy measures including communication technology, smart meters, smart appliances, renewable energy resources, and energy efficiency devices.
- Policies and regulations, financial assistance, government support, technology development, cyber security and teamwork of different entities are the major prerequisites of setting up a smart grid.
- Different technologies are required for implementation of smart grid such as smart sensor, instrument transformer, superconductors, wide area measurement, and automatic control and communication technologies.

- Advance Metering Infrastructure (AMI), Energy Management System (EMS), Demand Side Management (DSM), energy security, asset management, distribution automation, and customer relationship management (CRM) are the key aspects for smart grid development in India.
- Benefits of smart grid are self-healing, improves the effectiveness of the existing grid, makes the best use of available resources, intelligently integrates renewable power sources, reduces plant reserve capacity, minimizes loss, and trims down greenhouse emission.

Multiple Choice Questions

1. Smart grid allows interconnection of renewable sources as
   - (a) the renewable energy is costly.
   - (b) the system becomes more reliable and efficient.
   - (c) it increases losses in distribution.
   - (d) it increases overall energy demand.

2. Smart grid technology contains the following
   - (a) AMI, WAM, renewable integration, microgrid
   - (b) DA, EAM, distribution automation
   - (c) Energy efficient systems, storage system, communication system
   - (d) All of the above

3. AMI means
   - (a) Automated Metering Instrument
   - (b) Alternate Metering Instrument
   - (c) Advanced Metering Infrastructure
   - (d) Advanced Metering Instrument

4. EMS means
   - (a) Energy Management System
   - (b) Emergency Management System
   - (c) Energy Measure system
   - (d) Enterprise Management system

5. WAM means
   - (a) Whole Area Measurement
   - (b) Wide Area Monitoring
   - (c) Wide Area Maintenance
   - (d) None of the above

6. HAN, NAN, and WAN are the parts of
   - (a) Energy metering
   - (b) Energy storage
   - (c) Communication network
   - (d) Instrument transformer

7. The smart grid will support utility
   - (a) to minimize service disruptions
   - (b) to reduces aver all losses
   - (c) to restore power quickly after an outage
   - (d) all of the above

8. In a smart grid, advanced sensor and measurement technologies are used
   - (a) for accurate measurement of energy consumption and to reduce energy theft
   - (b) to increase the number of customers
   - (c) to increase energy flow
   - (d) to reduce the cost of grid

9. The function of smart meters in smart grid is
   - (a) real-time electricity measurement
   - (b) transfer of data between utility and customers
   - (c) to provide information to consumers about energy usage
   - (d) all of the above

10. Insufficient cyber security may lead to
    - (a) safe data transfer
    - (b) instability of grid communication and cause fraud
    - (c) safety of recorded and real-time data
    - (d) all of the above

Review Questions

1. What is smart grid? Why does the existing utility need to be renovated?

2. Explain the basic structure of smart grid. How does it differ from the conventional grid?

3. What are the benefits of smart grid?

4. What is the role of smart meters?

5. What is the function of GIS system in smart grid?

6. What are the different communication technologies used in smart grid?

7. Which factors need to be considered during the implementation of smart grid?

8. What are the functions of WAM and control system?

9. What are the challenges to be faced while designing the smart grid?

Symmetrical and Unsymmetrical Faults in Power Systems

19

Learning Objectives

After going through this chapter, the students will be able to:

- Explain symmetrical and unsymmetrical fault analysis
- Classify the different types of faults
- List the advantages of per unit system
- Explain symmetrical components
- Discuss the transformation of unbalanced phasors into balanced symmetrical components and vice versa
- Explain the transient phenomenon that occurs in transmission line
- Discuss the creation of sequence network for power systems

19.1 Introduction

A power system includes generating station, transmission lines, substations, and distribution system feeding electricity to various types of loads staggered over large areas. The entire power system network is very large, contains costly equipment and their functioning is complex. Power system analysis involves load flow analysis, short circuit studies, transient stability, and issues related to power network. Equipment or insulation failure, falling of tree branches on the line, bird strike, switching or lightning surges, mechanical damage, and electrical breakdown are various causes of fault or short circuits in power system. The high magnitude of current flow during such short circuit may damage the equipment and result in loss of electricity and revenue, if adequate protection is not applied. This chapter deals with the effects of symmetrical and unsymmetrical faults on power system network. The chapter also discusses symmetrical component and its application in the analysis of unsymmetrical fault. Per unit system which provides an easy way to solve all types of faults occurring in systems is also discussed in this chapter.

19.2 Nature and Causes of Faults

Transmission line faults can be classified into two categories—shunt faults and open-circuit faults.

19.2.1 Shunt Faults

Shunt faults can be divided into four categories:

Single line-to-ground faults These faults occur because of the shorting of one of the conductors of a three-phase transmission line to ground. These types of faults occur most commonly on transmission lines and their probability is approximately 85%.

Double-line faults These faults occur owing to the shorting of any of the two conductors of a three-phase transmission line (a–b/b–c/c–a). They are in the neighbourhood of 8% of the total transmission line faults.

Double-line-to-ground faults These faults occur owing to the short-circuiting of any of the two conductors of a three-phase transmission line with ground (a–b–g/b–c–g/c–a–g). The probability of these types of faults is around 5% of the total faults on the transmission line.

Triple-line faults These faults occur owing to the shorting of three conductors together of the transmission line. The probability of these types of faults is around 2% of the total faults on the transmission line.

19.2.2 Open-circuit Faults

Open-circuit faults occur because of the failure of opening or closing of one or more phases of the circuit breaker (CB) or isolator. These types of faults can be divided into two categories:

One conductor open This type of fault occurs because of the opening of a single conductor in case of a break in one of the three phases of a transmission line.

Two conductors open This type of fault occurs because of the opening of two conductors (a and b, b and c, or c and a).

19.3 Per Unit System

Normally, the power system is operated at high voltage and power in terms of kV and MW, respectively. The solution of big system network in terms of load flow or fault analysis may become complex and burdensome due to large physical values. Therefore, it is necessary to scale down the large quantities into narrow range magnitude referred to as per unit system. The per unit representation of voltage, current, impedance, and power is merely used to make calculation simpler in power system analysis.

Per unit value of any quantity is defined as the ratio of actual quantity to its base quantity. It is given by Eq. (19.1). Per unit is dimensionless as both actual and base quantities are expressed in the same unit.

$$\text{Per unit quantity} = \frac{Actual\ quantity}{Base\ quantity} \qquad (19.1)$$

In a three-phase system, the following formulas, as given from Eqs (19.2) to (19.5), are used with base voltage as line-to-line volts or kV and the base power as a three-phase power in KVA or MVA.

$$\text{Base current (A)} = \frac{Base\ KVA}{\sqrt{3} \times Base\ KV} \qquad (19.2)$$

$$\text{Base impedance } (\Omega) = \frac{Base\ volt\,(L-n)}{Base\ current} \qquad (19.3)$$

$$\text{Base impedance } (\Omega) = \frac{(Base\ KV)^2}{Base\ MVA} = \frac{(Base\ KV)^2 \times 1000}{Base\ KVA} \qquad (19.4)$$

$$\text{Per unit impedance} = \frac{Actual\ impedance\ (\Omega)}{Base\ impedance\ (\Omega)} \qquad (19.5)$$

In most calculations, the per unit impedance/reactance of a portion of power system is selected on the rating of that portion which is different from the other section. Conversely, the per unit impedance of any other section of the power system must be expressed on the new base. Thus, it is required to transform per unit impedance referred to old base to per unit impedance referred to new base with kV and MVA using Eq. (19.6).

$$Per\ unit\ impedance_{new} = \left[Per\ unit\ impedance_{old} \right] \times \left(\frac{Base\ KV_{old}}{Base\ KV_{new}} \right)^2 \times \left(\frac{Base\ MVA_{new}}{Base\ MVA_{old}} \right) \quad (19.6)$$

The following are the advantages of the per unit system.

1. Manufacturers usually give impedance in per unit of equipment rating.
2. Per unit values are expressed in small ranges thus erroneous data can be easily traced in calculations.
3. Per unit especially reduces computational burden in a large system having several voltage levels separated through transformers.
4. Per unit impedance of transformer on both side is same. It is independent of the base kV.

19.4 Single Line Diagram or One-Line Diagram

The complete power system is formed by assembling various apparatuses in a sequential manner. Under the balanced condition of a three-phase system, the analysis is done by representing any one phase out of the three phases. In this per phase analysis, the corresponding circuit is composed of one of the three phases and the neutral return. While illustrating the per phase circuit, the diagram is again simplified by removing the return path of neutral connection at appropriate place. Moreover, the rating of the circuit is not shown and the complete circuit is composed of standard symbols of component parts. This simplified diagram of power system network is called single line or one-line diagram. Such a single line diagram reveals the essential information regarding the considered system. Further, the information provided on one line-diagram differs with the issue under consideration and the purpose for which the diagram is designed. For example, the use of circuit breakers and relays is not necessary in a steady state and load flow analysis of the system. They are represented in single line diagram for transient condition and fault condition as their purpose is to protect the system. Similarly, in some circuit metering, protection is essential which includes current and potential transformers in one-line diagram of the power system. A few symbols of electrical equipment and components are shown in Fig. 19.1, which are commonly used to represent a single line diagram.

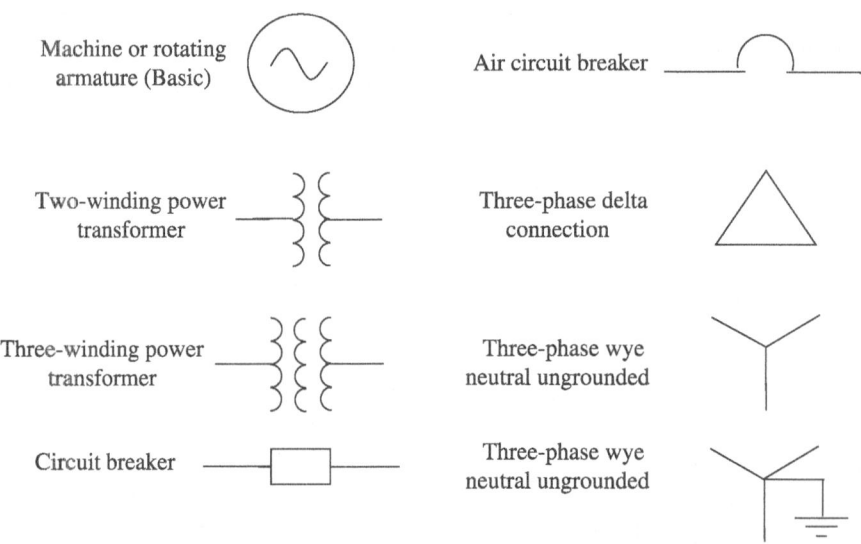

Fig. 19.1 Symbols of different equipment

Figure 19.2 shows the single line diagram of a portion of power system network. A generator is grounded through a reactor and load is connected to a bus through a step-up transformer. Two motors, one of which is grounded through reactors, are connected to a bus through step-down transformer at the remote end of the line. A load is also connected to the same bus with the motors. The information regarding the ratings of the generators, transformers, motors, and loads is usually provided on the one-line diagram.

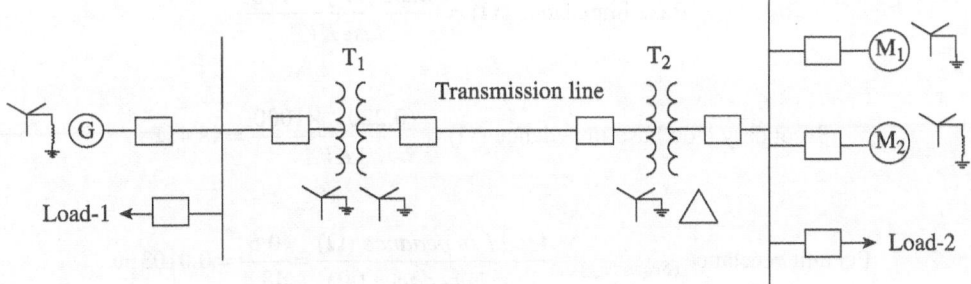

Fig. 19.2   Single line diagram of a power system

19.5 Impedance or Reactance Diagrams

To evaluate the steady-state and transient condition of any power system, the single line diagram should be constructed as shown in Fig. 19.2. The equivalent circuit of each component and its modeling is useful to develop the impedance/reactance diagram of the said power system. While constructing the reactance diagram, dynamic machines (generator and motor) are represented by constant voltage sources in series with appropriate reactance. Transformers are symbolized with three branches—two series branches representing the primary and secondary winding reactances and one shunt branch representing the magnetizing current and the effect of the no-load losses. However, the shunt branch of transformer is omitted as the effect of the magnetizing current is neglected during evaluation. Transmission line is modelled by series resistance and reactance, and a shunt capacitance at each end of the line equal to half the total capacitance of the line. However, the line reactance is much greater than the resistance of the line. Moreover, the effect of shunt capacitance is negligible for short transmission line. Consequently, only the series reactance is considered. Loads are represented by equivalent shunt impedance. Once the equivalent circuit of each component in the power system is obtained, the reactance diagram of the given system is represented in per-phase bases. Figure 19.3 shows the per-phase representation of the reactance diagram for the single line diagram shown in Fig. 19.2.

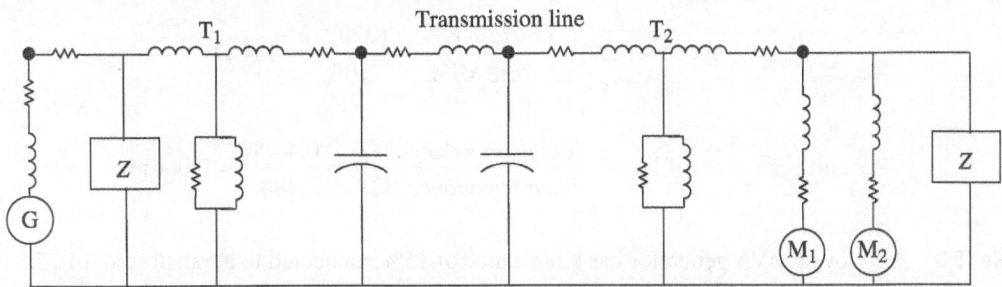

Fig. 19.3   Reactance diagram for the single line diagram of Fig. 19.2

Example 19.1 A single phase transformer of 440/220 V, 1 KVA has a leakage reactance of 0.5 Ω on the 220 V side. Determine the per unit leakage reactance of the transformer on both sides.

Solution:
Consider the base KVA = 1 KVA and base KV as 0.440 kV on the primary side and 0.220 kV on the secondary side of the transformer.

Now,
$$\text{Base impedance } (\Omega) = \frac{(Base\ KV)^2 \times 1000}{Base\ KVA}$$

$$\text{Secondary side base impedance } (\Omega) = \frac{(0.220)^2 \times 1000}{1\ KVA} = 48.4\ \Omega$$

$$\text{Per unit reactance }_{(Secondary)} = \frac{Actual\ impedance\ (\Omega)}{Base\ impedance\ (\Omega)} = \frac{0.5}{48.4} = 0.0103\ \text{pu}$$

$$\text{Leakage reactance referred to primary side} = 0.5 \times \left(\frac{440}{220}\right)^2 = 2\ \Omega$$

Now,
$$\text{Primary side base impedance } (\Omega) = \frac{(0.440)^2 \times 1000}{1\ KVA} = 193.6\ \Omega$$

$$\text{Per unit reactance }_{(primary)} = \frac{2}{193.6} = 0.0103\ \text{pu}$$

Thus, it can be observed from the above calculation that per unit reactance of the transformer remains the same on both sides of the transformer, irrespective of the actual impedance on the primary and secondary sides.

Example 19.2 An 80 km long transmission line has a reactance of 0.4 Ω/km. The line delivers a power of 100MVA at 220kV. Determine the per unit reactance of the transmission line.

Solution:
Consider the base MVA = 100 MVA and base KV as 220 kV for the given transmission line.

$$\text{Base impedance } (\Omega) = \frac{(Base\ KV)^2}{Base\ MVA} = \frac{(220)^2}{100} = 484\ \Omega$$

$$\text{Per unit reactance of line} = \frac{Actual\ impedance\ (\Omega)}{Base\ impedance\ (\Omega)} = \frac{0.4 \times 80}{484} = 0.066\ \text{pu}$$

Example 19.3 A 15 kV, 2 MVA generator has a reactance of 15% connected to a transformer of 15/3.3kV, 2 MVA, with leakage reactance of 8%. A load of 2Ω is connected to the low tension side of the transformer. Find the per unit impedance of load referred to in the circuit and draw the reactance diagram for the whole system.

Solution:

Consider the generator and high tension side as circuit 'P' and transformer low tension side and load as circuit 'Q' as shown in Fig. 19.4.

Base voltage for circuit P = 15kV

Base voltage for circuit Q = 3.3 kV

Base MVA for the whole system is 2 MVA.

Per unit reactance of generator = $j0.15$ and per unit reactance of transformer is $j0.08$.

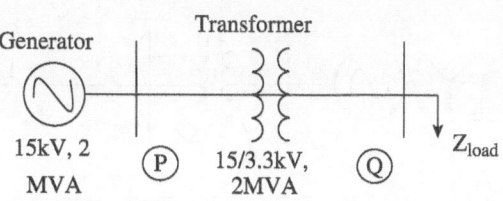

Fig. 19.4   Circuit diagram for Example 19.3

$$\text{Base impedance of circuit Q } (\Omega) = \frac{(Base\ KV)^2}{Base\ MVA} = \frac{(3.3)^2}{2} = 5.45\ \Omega$$

$$\text{Per unit impedance of load in circuit Q} = \frac{Actual\ impedance\ (\Omega)}{Base\ impedance\ (\Omega)} = \frac{2}{5.45} = 0.367\text{pu}$$

Now,

$$\text{Base impedance of circuit P} (\Omega) = \frac{(Base\ KV)^2}{Base\ MVA} = \frac{(15)^2}{2} = 112.5\ \Omega$$

$$\text{Impedance of load referred to circuit P} = 2 \times \left(\frac{15}{3.3}\right)^2 = 41.32\ \Omega$$

$$\text{Per unit impedance of load referred to circuit P} = \frac{Actual\ impedance\ (\Omega)}{Base\ impedance\ (\Omega)} = \frac{41.32}{112.5} = 0.367\text{pu}$$

The reactance diagram is shown in Fig. 19.5.

It is to be concluded that the per unit reactance of any load connected to the secondary side of transformer remains the same as referred to the primary side of transformer irrespective of the actual load impedance in ohms referred to both sides of the transformer.

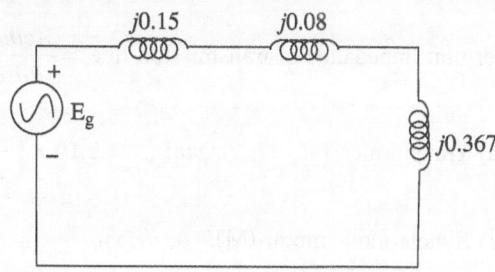

Fig. 19.5   Reactance diagram for Example 19.3

Example 19.4   A 200 MVA, 13.8 kV generator has subtransient reactance of 20%. The generator supplies a synchronous motor through a 50 km transmission line using two transformers at both ends, as shown in Fig. 19.6. The ratings of synchronous motor, transmission line, and transformer are as under:

(a) Transformer T_1 = 250 MVA, 13.8/220 kV, with $x_d'' = 10\%$

(b) Transformer T_2 = 220 MVA, 220/11 kV, with $x_d'' = 10\%$

(c) Transmission line = 0.45 Ω/km, 50km long

(d) Synchronous motor = 150 MVA, 11 kV, with $x_d'' = 15\%$

Draw the reactance diagram with all the reactances marked in per unit.

Solution:

The circuit shown in Fig. 19.6 is divided in to three sections, generator plus T_1 primary (section-1), T_1 secondary plus transmission line plus T_2 primary (section-2), and secondary of T_2 plus synchronous motor (section-3).

Base MVA = 200MVA

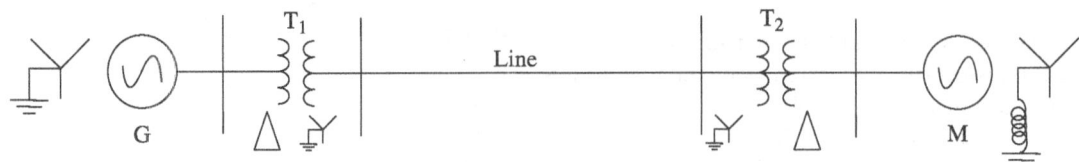

Fig. 19.6 Power system network for Example 19.4

Consider that the base kV for section-1 is 13.8kV, base kV for section-2 is 220kV, and base kV for section-3 is 11kV.

Now, the new per unit reactance for each component is found by,

$$Xpu_{(new)} = Xpu_{(old)} \times \left(\frac{base\,kV_{old}}{base\,kV_{new}}\right)^2 \times \left(\frac{base\,MVA_{new}}{base\,MVA_{old}}\right)$$

(a) Generator (G): $\qquad Xpu_{(new)} = Xpu_{(old)} = j0.2$ pu

(b) Transformer T$_1$: $\qquad Xpu_{(new)} = 0.10 \times \left(\frac{13.8}{13.8}\right)^2 \times \left(\frac{200}{250}\right) = j0.08$ pu

(c) Transmission line:

$$\text{Actual impedance} = 0.45 \times 50 = 22.5\,\Omega$$

$$\text{Base impedance of line} = \frac{(Base\,KV)^2}{Base\,MVA} = \frac{(220)^2}{200} = 242\ \Omega$$

$$\text{Per unit impedance of transmission line} = \frac{Actual\,impedance\,(\Omega)}{Base\,impedance\,(\Omega)} = \frac{22.5}{242} = j0.093\text{ pu}$$

(d) Transformer T$_2$: $\qquad Xpu_{(new)} = 0.10 \times \left(\frac{220}{220}\right)^2 \times \left(\frac{200}{220}\right) = j0.0909$ pu

(e) Synchronous motor (M): $\qquad Xpu_{(new)} = 0.15 \times \left(\frac{11}{11}\right)^2 \times \left(\frac{200}{150}\right) = j0.2$ pu

The reactance diagram in pu is shown in Fig. 19.7.

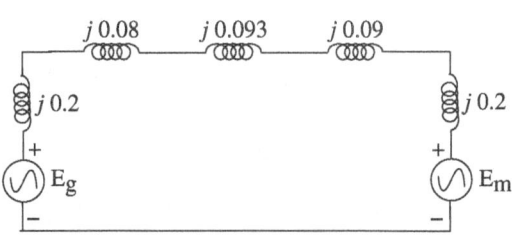

Fig. 19.7 Reactance diagram for Example 19.4

19.6 Symmetrical Short Circuit Analysis

Faults can also be categorized as symmetrical and un-symmetrical faults. Triple-line and triple-line-to-ground faults are symmetrical faults, which result in equal magnitude of fault current displaced by 120° electrical from each other. A majority of faults on power systems is unsymmetrical in nature. These are line-to-ground, double-line, and double-line-to-ground faults. When such faults occur, they result in unequal magnitudes of current in the three lines and have unequal phase displacement. At the time of calculating the fault currents and voltages during an unsymmetrical fault, a method known as *symmetrical component method* is used. Symmetrical fault analysis is explained here in the following subsection.

19.6.1 Transient Phenomenon in Transmission Line

In order to understand the transient phenomenon on transmission line, let us consider a line resistance R and inductance L connected in series and fed by an alternating voltage source (V_s). Here, the line capacitance is neglected and the unloaded line is considered for the analysis. As shown in Fig. 19.8, a constant voltage source $V_s = V_m \sin(\omega t + \alpha)$ is given to the circuit, where α is the phase angle of the applied voltage.

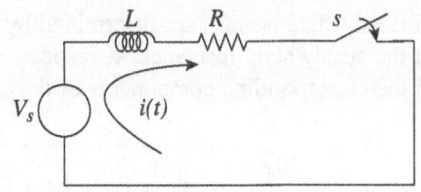

Fig. 19.8 Transient on unloaded transmission line

In order to analyse the three-phase short circuit, the switch 's' is closed at time t = 0. We will see the effect of variation of parameter α which controls the instant of short circuit on the voltage wave. It is observed that the short circuit current is composed of the two terms after fault is applied. The expression for fault current is $i(t) = i_{ac} + i_{dc}$, where i_{ac} is the steady state current and i_{dc} is the transient current or DC decaying current.

Consider,

$$V_s = V_m \sin(\omega t + \alpha) = iR + L\left(\frac{di}{dt}\right) \tag{19.7}$$

The solution to Eq. (19.7) is

$$i(t) = i_{ac}(t) + i_{dc}(t) \tag{19.8}$$

and $i_{ac} = \dfrac{V_m}{Z}\sin(\omega t + \alpha - \theta)$ and

$$i_{dc} = \frac{V_m}{Z}\sin(\alpha - \theta)e^{-\frac{Rt}{L}} \tag{19.9}$$

where

$$Z = \sqrt{R^2 + (\omega L)^2}, \theta = \tan^{-1}\left(\frac{\omega L}{R}\right)$$

The first term i_{ac} varies sinusoidal with time and the second term i_{dc} is asymmetric and decays exponentially with a time constant L/R. When the switch 's' is closed at time t = 0, the value of the steady-state current i_{ac} is not zero and i_{dc} is also present in the solution. The DC current does not appear in the solution if the switch is closed at a point on the voltage wave such that α − θ = 0 or = π. Figures 19.9(a) and (b) illustrate the waveform of current *i(t)* when α − θ = 0 and α − θ = π/2.

For the first case, that is, Fig. 19.9(a) only sinusoidal current is present whereas for the second case (Fig. 19.9(b)) the dc component has its maximum initial value. It can be seen from Fig. 19.9(b) that the maximum momentary current is twice the maximum value of symmetrical short circuit current. Thus, the component i_{dc} may have any value from 0 to V/Z, depending on the power factor angle (θ) and instantaneous value of the voltage at the time of closing the switch (α). A similar effect of short circuit has been observed when a three-phase fault occurs on the terminals of a synchronous generator. However, due to the presence of damper winding and armature winding, the flux in the air gap during fault is not only of dc field winding. Thus, if we ignore the dc decaying component and consider only the sinusoidal component of current, the resulting plot of each phase current versus time is as shown in Fig. 19.10 for a synchronous machine. This is due to the fact that in a synchronous machine, the flux across the air gap is not constant at the time of fault. The change in flux is characterized by the joint effect of the field, the armature, and the damper windings of rotor. During pre-fault condition of a synchronous machine, it offers normal reactance which is proportional to leakage reactance and reactance due to armature reactance. However, just after the fault, there is a mutual combination of stator, rotor, and damper windings reactance. Later, the effect of damper winding reactance will disappear after a few cycles due to its low time constant. Hence, after the occurrence of fault, sub-transient, transient, and steady-state periods are determined by the respective so called reactance X″, the transient reactance X′, and the steady-state reactance X, respectively. These reactances have increasing values (that is, X″<X′<X) and the corresponding components of the short circuit current have decreasing magnitudes (I″>I′> I).

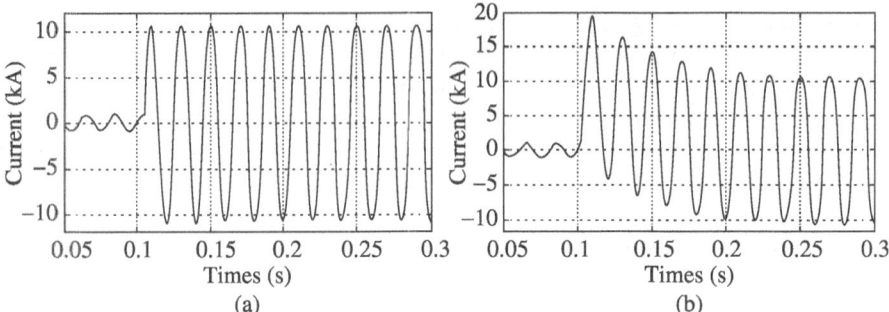

Fig. 19.9   (a) Steady state current, $\alpha - \theta = 0$ and (b) Transient current, $\alpha - \theta = \pi/2$

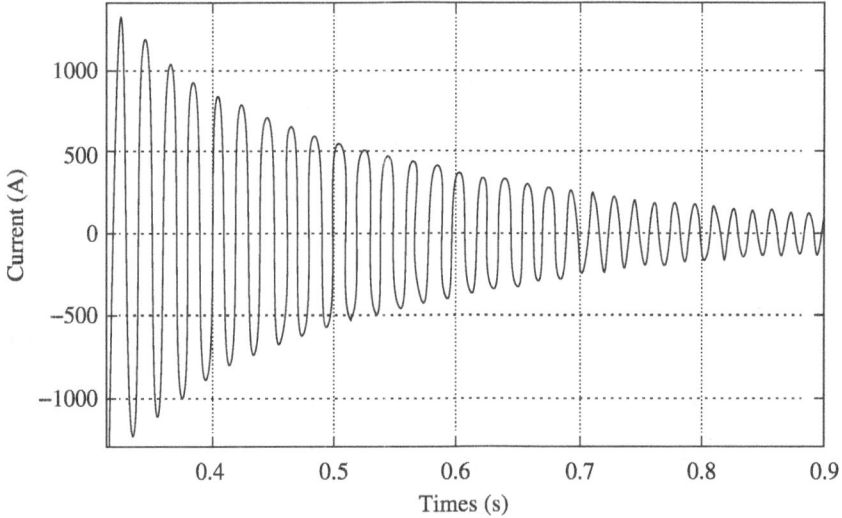

Fig. 19.10   Short circuit current of synchronous machine

19.6.2 Symmetrical Faults on Power System

Symmetrical faults are very severe faults. However, they occur rarely in power systems. It is a situation of power system wherein all three phases of transmission line are together or grounded at a common point. Figure 19.11 shows two types of three-phase symmetrical faults. The calculation of symmetrical fault is simple and usually done on per phase basis. This analysis is useful to decide the fault MVA of power system, phase setting of relays, and selection of circuit breakers.

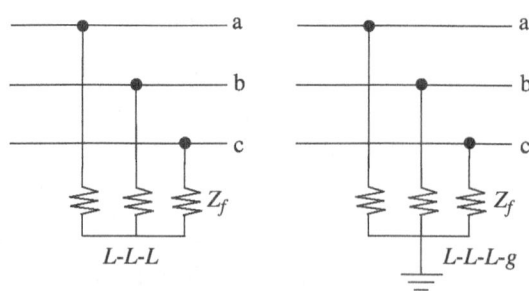

Fig. 19.11   Symmetrical fault

Due to the balance nature of fault, only positive sequence component is required for per phase calculation. Moreover, zero sequence and negative sequence voltages and related currents in the network remain absent (zero). The analysis is performed using Thevenin's equivalent approach. Figure 19.12 shows the single line diagram of a simple system consisting of the balanced three-phase load and source connected through a transmission line.

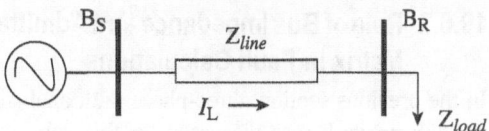

Fig. 19.12 Single line diagram of a simple system

Consider a balanced three-phase fault through identical impedance Z_f at receiving end bus (B_R) for which Thevenin's equivalent impedance Z_{th} is computed. This is illustrated by the one-line diagram in Fig. 19.13 considering sub-transient phenomenon.

In the figure, Z_S is source impedance, Z_{line} is transmission line impedance, Z_{load} is load impedance, Z_f is impedance involved in fault path, E_g is sub-transient internal generator voltage, V_t is generator terminal voltage, and I_f is the sub-transient current in faulted path.

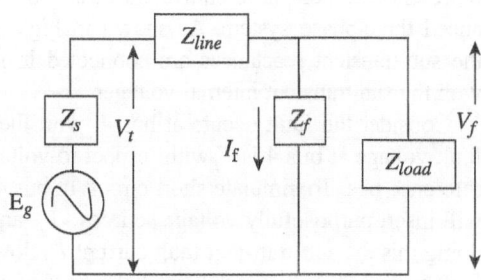

Fig. 19.13 Impedance diagram of power system

As stated above, using Thevenin theorem, the equivalent impedance (Z_{th}) is calculated by shorting all voltage sources and looking from faulted point in a given network (Fig. 19.13). Remember that the equivalent impedance (Z_{th}) is calculated excluding fault impedance (Z_f). Moreover, during bolted three phases to ground short circuit (anywhere in system), the $Z_f = 0$. Hence, as shown in Fig. 19.13, the calculated value of Z_{th} is given by:

$$Z_{th} = \left(Z_{line} + Z_S\right) // Z_{load} = \frac{\left(Z_{line} + Z_S\right)Z_{load}}{Z_{line} + Z_S + Z_{load}} \tag{19.10}$$

It is worthwhile to note that the load impedance Z_{load} is extremely high compared to line impedance (Z_{line}) and source impedance (Z_S). Hence, by neglecting the effect of the small impedance offered by $Z_{line} + Z_S$ in the denominator of Eq.19.10, we get,

$$Z_{th} = \frac{\left(Z_{line} + Z_S\right)Z_{load}}{Z_{line} + Z_S + Z_{load}} \cong \frac{\left(Z_{line} + Z_S\right)Z_{load}}{Z_{load}} \cong Z_{line} + Z_S \tag{19.11}$$

Now, the Thevenin voltage (V_f) is the voltage at the faulted point before the occurrence of fault on the network. It is given by:

$$V_f = E_g - I_L\left(Z_{line} + Z_s\right) \tag{19.12}$$

This Thevenin voltage V_f along with equivalent impedance (Z_{th}) is in series with the fault impedance Z_f during fault condition. Let us rearrange the network of Fig. 19.13 to carry out this information including Z_{th}, as shown in Fig. 19.14.

From Fig. 19.14, the three-phase sub-transient fault current (I_f) during symmetrical fault is given by

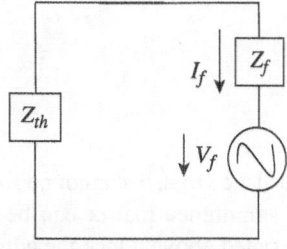

$$I_f = \frac{V_f}{Z_f + Z_{th}} \tag{19.13}$$

Fig. 19.14 Thevenin's equivalent circuit

19.6.3 Role of Bus Impedance and Admittance Matrix in Fault Calculations

In the previous section, three-phase fault analysis for a small network was discussed. In this sub-section, the analysis has been extended for large networks. A more meaningful equation is generalized to specific circuit for symmetrical fault analysis. Figure 19.15 represents the per-phase equivalent circuit of a balanced three-phase system. As shown in Fig. 19.15, the sub-transient reactances are connected in series with the sub-transient internal voltages.

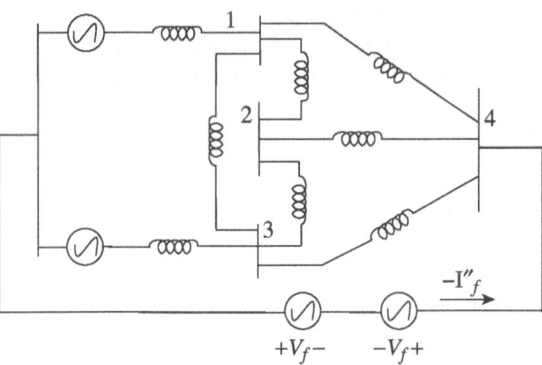

Fig. 19.15 Apart of power system representing a symmetrical fault at bus-4

Consider the fault occurs at bus 4, with the pre-fault voltage at bus 4 is V_f with respect to voltage at reference bus. To simulate short circuit at bus 4, we will insert purposefully voltage sources $-V_f$ and in series with V_f between bus 4 and the reference bus. While doing this the sub-transient fault current I''_f flows from bus 4 to the reference bus or $-I''_f$ flows into bus 4. This fault current I''_f is due to the insertion of $-V_f$ which will flow through the different lines of the considered network and will also cause a change in the voltages at various buses. If we remove available sources and V_f, then $-V_f$ is the only source present in the network which allows a current $-I''_f$ into bus 4. The changes in voltages of the various buses that are caused by the insertion of voltage $-V_f$ and the current $-I''_f$ are given by

$$I_{bus} = [Y_{bus}] \times E_{bus} \tag{19.14}$$

$$\begin{bmatrix} 0 \\ 0 \\ 0 \\ -I''_f \end{bmatrix} = Y_{bus} \begin{bmatrix} \Delta V_1 \\ \Delta V_2 \\ \Delta V_3 \\ -V_f \end{bmatrix} \tag{19.15}$$

where, Δ denotes the changes in the bus voltages due to the current $-I''_f$. Y_{bus} is the bus admittance matrix; I_{bus} and E_{bus} are the bus current and bus voltage vectors, respectively. The rule of examining the network to form the bus admittance matrix includes the diagonal elements and off-diagonal elements. A diagonal element Y_{ii}, is equal to the sum of the admittance values of all the lines connected to the bus i; Whereas, an off-diagonal element Y_{ij} is equal to the negative of the admittance value of the connecting lines present between the buses i and j, if any. Hence, the Y_{bus} matrix can be given as:

$$Y_{bus} = \begin{bmatrix} Y11 & -Y12 & -Y13 & -Y14 \\ -Y21 & Y22 & -Y23 & -Y24 \\ -Y31 & -Y32 & Y33 & -Y34 \\ -Y41 & -Y42 & -Y43 & -Y44 \end{bmatrix} \tag{19.16}$$

Bus impedance matrix cannot be formed by direct inspection of the given power system network. However, the bus admittance matrix can be determined by examining the network and by following the standard steps as stated above. Once the admittance matrix is formed, it can be inverted to obtain the bus impedance matrix (Z_{bus}).

$$\begin{bmatrix} \Delta V_1 \\ \Delta V_2 \\ \Delta V_3 \\ -V_f \end{bmatrix} = Z_{bus} \begin{bmatrix} 0 \\ 0 \\ 0 \\ -I''_f \end{bmatrix} \tag{19.17}$$

From the above equations, we can write

$$V_f = Z_{44}I''_f \ \ \Delta V_1 = -Z_{14}I''_f \ \ \Delta V_2 = -Z_{24}I''_f \ \text{ and } \ \Delta V_3 = -Z_{34}I''_f \tag{19.18}$$

Further we can generalize the equation:

$$\Delta V_i = -Z_{i4}I''_f = -\frac{Z_{i4}}{Z_{44}}V_f, \quad i = 1,2,3 \tag{19.19}$$

Here also it is assumed that the system is previously unloaded and the magnitude and phase angles of all the generator internal emfs are the same. Thus, no current circulates during pre-fault condition, and voltages at all the bus are same and equal to V_f. After the fault, the changed value of all the bus voltages can be given by superposition principle:

$$V_i = V_f + \Delta V_i = V_f \left(1 - \frac{Z_{i4}}{Z_{44}} \right), \quad i = 1,\cdots,4 \tag{19.20}$$

Example 19.5 A generator is rated at 75 MVA 13.8 kV having sub-transient reactance 20%. It is connected to the transmission line followed by a transformer as shown in Fig. 19.16. The transformer is rated at 75 MVA 13.8/66 kV having sub-transient reactance of 10%. The transmission line having reactance of 20 Ω is connected to bus 3. Find the sub-transient current in the generator if a 3-phase fault occurs on bus 3.

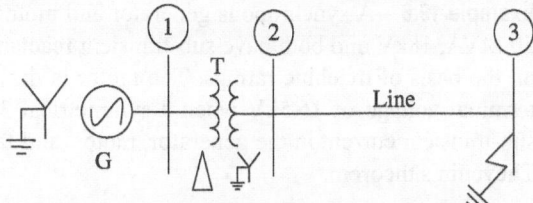

Fig. 19.16   Single line diagram for Example 19.5

Solution:
Per unit method is used to find the sub-transient current during symmetrical three-phase fault.
Let us consider the base MVA for entire system = 75 MVA and base kV = 13.8 kV in generator section and 66 kV in transmission line section.

$$\text{Base current in generator section} = \frac{Base\ KVA}{\sqrt{3} \times Base\ KV} = \frac{75000\ KVA}{\sqrt{3} \times 13.8\ KV} = 3137.8\ A$$

$$\text{Base current in line section} = \frac{Base\ KVA}{\sqrt{3} \times Base\ KV} = \frac{75000\ KVA}{\sqrt{3} \times 66\ KV} = 656\ A$$

The pu reactance referred to new base quantities are:
For generator (G):

$$Xpu_{(new)} = Xpu_{(old)} = j0.2 \text{ pu}$$

For transformer (T):

$$Xpu_{(new)} = Xpu_{(old)} = j0.1 \text{ pu}$$

For transmission line:

Actual impedance = 20 Ω

$$Base\,impedance\,of\,line\,(\Omega) = \frac{(Base\,KV)^2}{Base\,MVA} = \frac{(66)^2}{75} = 58.08\,\Omega$$

$$Per\,unit\,impedance\,of\,transmission\,line = \frac{Actual\,impedance\,(\Omega)}{Base\,impedance\,(\Omega)} = \frac{20}{58.08}$$

$$= j0.3443\,\text{pu}$$

The sequence network is shown in Fig. 19.17.
For a three-phase symmetrical fault at bus 3, the fault current (I_f'') is given by,

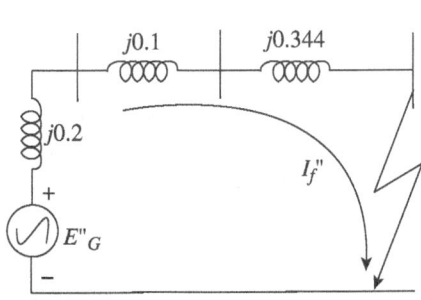

$$I_f'' = \frac{V_f}{Z_{th}} = \frac{1\angle 0°}{j0.2 + j0.1 + j0.3443} = \frac{1\angle 0°}{j0.6443} = -j1.552\,\text{pu}$$

Thus, the actual current in fault at bus-3 = 1.552 × 656

= 1018.112 A

The fault current delivered by generator = 1.552 × 3137.8

= 4869.86 A

Fig. 19.17 Representation of fault on reactance diagram for Example 19.5

Example 19.6 A synchronous generator and motor are rated at 50 MVA, 18kV and both have sub-transient reactance of 25%. The line connecting them has 10% reactance on the basis of machine ratings. The motor is drawing 30 MW power at 0.85 power factor leading with a terminal voltage of 16.5kV when a symmetrical 3-phase fault occurs on the terminals of motor. Find the sub-transient current in the generator, motor and fault by using (a) the internal voltage of machines and (b) Thevenin's theorem.

Solution:

Per unit method is used to find the sub-transient current during 3-phase fault on motor terminal.

Assume 50MVA and 18kV as base value for the calculation. Let us consider voltage V_f as reference phasor at fault point,

$$V_f = \frac{16.5}{18} = 0.9167\,\angle 0°\,\text{pu}$$

$$Base\,current = \frac{Base\,KVA}{\sqrt{3} \times Base\,KV} = \frac{50000\,KVA}{\sqrt{3} \times 18\,KV} = 1603.75\,A$$

(a) Using internal voltage of machines

During pre-fault condition the normal load current (I_L) is flowing in the circuit from generator to motor and its per unit value is given,

$$I_L = \frac{30000}{\sqrt{3} \times 16.5 \times 0.85} = 823.31\angle 31.788°A$$

$$I_L(pu) = \frac{823.31\angle 31.788°}{1603.75} = 0.513\angle 31.788° = 0.436 + j0.270 \text{ pu}$$

In generator circuit, the terminal voltage (V_t) and sub-transient emf (E''_g) are given as,

$$V_t = V_f + j0.1 \times (I_L)$$

$$E''_g = V_t + j0.25 \times (I_L) = V_f + j0.35 \times (I_L)$$

Now, the sub-transient current (I''_g) contributed by generator in the fault as shown in Fig. 19.18 and using internal voltage of generator is given by,

$$I''_g(pu) = \frac{E''_g}{j0.1 + j0.25} = \frac{V_f + j0.35 \times (I_L)}{j0.35} = \frac{0.9167 + j0 + j0.35 \times (0.436 + j0.270)}{j0.35}$$

$$I''_g(pu) = \frac{0.9167}{j0.35} + \frac{j0.35 \times (0.436 + j0.270)}{j0.35} = -j2.62 + 0.436 + j0.270$$

$$I''_g(pu) = 0.436 - j2.35 \text{ pu}$$

$$I''_g(amp) = 1603.75 \times (0.436 - j2.35) = 699.23 - j3768.81 \text{A}$$

Similarly, the sub-transient current (I''_m) contributed by motor in the fault as shown in Fig. 19.18 and using internal voltage of motor is given by,

Since, $$V_t = V_f = 0.9167\angle 0^0 \text{ pu}$$

$$E''_m = V_t - j0.25 \times (I_L) = V_f - j0.25 \times (I_L)$$

$$I''_m(pu) = \frac{E''_m}{j0.25} = \frac{V_f - j0.25 \times (I_L)}{j0.25} = \frac{0.9167 + j0 - j0.25 \times (0.436 + j0.270)}{j0.25}$$

$$I''_m(pu) = \frac{0.9167}{j0.25} - \frac{j0.25 \times (0.436 + j0.270)}{j0.25} = -j3.67 - 0.436 - j0.270$$

$$I''_m(pu) = -0.436 - j3.94 \text{ pu}$$

$$I''_m(amp) = 1603.75 \times (-0.436 - j3.94) = -699.23 - j6318.77 \text{A}$$

The sub-transient current in the fault (I''_f) as shown in Fig. 19.18 along with I''_g and I''_m are given by,

$$I''_f = I''_g + I''_m = 0.436 - j2.35 - 0.436 - j3.94 = -j6.29 \text{ pu}$$

$$I''_f = 699.23 - j3768.81 - 699.23 - j6318.77 = 10087.58 \text{A}$$

(b) using Thevenin's theorem

For a three-phase symmetrical fault at the terminal of motor, as per Thevenin's theorem all the generated voltages are short circuited and the equivalent circuit has a single new generator with voltage V_f in series with a single equivalent impedance (Z_{th}) of the circuit. The fault current (I_f'') is given by,

$$I_f'' = \frac{V_f}{Z_{th}} = \frac{0.9167\angle 0°}{Z_{th}}$$

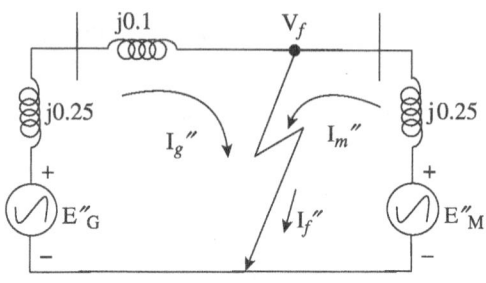

Fig. 19.18 Equivalent circuit for Example 19.6

where, Z_{th} is calculated by considering Fig. 19.18, in which the combination of series reactance of generator (x_g) and line (X_L) is in parallel with motor reactance (X_m)

Hence,
$$Z_{th} = \frac{(X_g + X_L) \times X_m}{(X_g + X_L + X_m)} = \frac{(j0.25 + j0.1) \times j0.25}{(j0.25 + j0.1 + j0.25)} = j0.146 \text{ pu}$$

Thus, the per unit current in fault $I_f'' = \dfrac{V_f}{Z_{th}} = \dfrac{0.9167\angle 0^0}{j0.146} = -j6.28 \text{ pu}$

The fault current deliveresd by generator $= I_g'' = -j6.28 \times \dfrac{j0.25}{j0.6} = -j2.616 \text{ pu}$

The fault current delivered by motor $= I_m'' = -j6.28 \times \dfrac{j0.35}{j0.6} = -j3.664 \text{ pu}$

If these currents are added to pre-fault load current (I_L) then the current supplied by individual machines is as,

$$I_g'' = 0.436 + j0.527 - j2.616 = 0.436 - j2.089 \text{ pu}$$

In case of motor, the I_L is in opposite direction to that of fault current from motor, thus per unit value of I_L is subtracted,

$$I_m'' = -0.436 - j0.527 - j3.664 = -0.436 - j4.1 \text{ pu}$$

The per unit values found for I_g'', I_m'', and I_f'' in this case are almost same as that of case shown in (a). However, it is to be revealed from Thevenin's equivalent method that the load current is neglected in each branch during the inception of fault. So, if we ignore the pre-fault load current the current from both machines towards the fault is given by,

$$I_g'' = -j2.616 \times 1603.75 = 4195.41 \text{ A}$$

$$I_m'' = -j3.664 \times 1603.75 = 5876.14 \text{ A}$$

Also the current in the fault is,

$$I_f'' = 4195.41 + 5876.14 = 10071.55 \text{ A}$$

Thus, it is to be concluded that the magnitude of current in the fault path remains the same whether load current is considered or not. However, the sharing of current from branches connected to the fault point is different.

Example 19.7 Find the sub-transient current for three-phase fault at bus 4 using bus impedance matrix for the network shown in Fig. 19.19. Consider the internal voltages of both the generators are equal to $1.0\angle 0°$ and all reactance of elements are in per units. Neglect all resistances. Also, determine the alerted voltage at remaining buses and sub-transient current supplied by both generators in per unit.

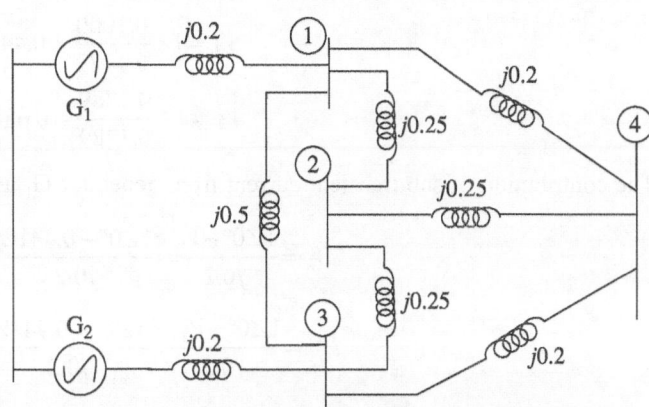

Fig. 19.19 Reactance diagram for Example 19.7

Solution:

If the fault occurs at bus 4 then the voltage at bus 4 in the given single phase circuit equivalent of a three-phase system is V_f. The network with admittance marked in per unit is shown in Fig. 19.20.

The admittance matrix is given by,

$$Y_{bus} = j\begin{bmatrix} -16 & 4 & 2 & 5 \\ 4 & -12 & 4 & 4 \\ 2 & 4 & -16 & 5 \\ 5 & 4 & 5 & -14 \end{bmatrix}$$

This admittance matrix is inverted by means of computer programming to give impedance matrix for fault calculation. It is given by,

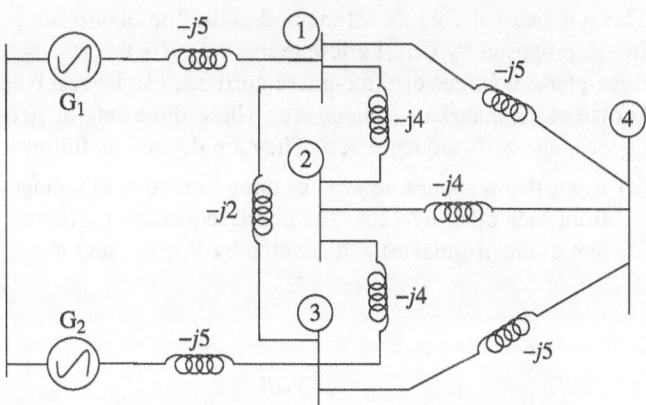

Fig. 19.20 Admittance diagram for Fig. 19.19

$$Z_{bus} = j\begin{bmatrix} -0.1278 & -0.1000 & -0.0722 & -0.1000 \\ -0.1000 & -0.1921 & -0.1000 & -0.1263 \\ -0.0722 & -0.1000 & -0.1278 & -0.1000 \\ -0.1000 & -0.1263 & -0.1000 & -0.1789 \end{bmatrix}$$

Now the sub-transient current in a three-phase fault at bus 4 is given by,

$$V_f = Z_{44} I_f''$$

$$I_f'' = \frac{V_f}{Z_{44}} = \frac{1\angle 0°}{j0.1789} = -j5.59 \text{ pu}$$

Also, the altered bus voltages for a symmetrical fault in bus 4 are given by,

$$V_1 = 1 - \frac{0.1000}{0.1789} = 0.4412 \text{ pu}$$

$$V_2 = 1 - \frac{0.1263}{0.1789} = 0.2941 \text{ pu}$$

$$V_3 = 1 - \frac{0.1000}{0.1789} = 0.4412 \text{ pu}$$

$$V_4 = 1 - \frac{0.1789}{0.1789} = 0 \text{ pu}$$

The contribution of sub-transient current from generator G_1 and G_2 is given by,

$$I''_{G_1} = \frac{1\angle 0° - V_1}{j0.2} = \frac{1\angle 0° - 0.4412}{j0.2} = -j2.794 \text{ pu}$$

$$I''_{G_2} = \frac{1\angle 0° - V_{31}}{j0.2} = \frac{1\angle 0° - 0.4412}{j0.2} = -j2.794 \text{ pu}$$

19.7 Symmetrical Components

The symmetrical component method is used to obtain the solution of an unbalanced three-phase network. It was proposed by C.L. Fortescue in 1918. His theory suggests that any set of unbalanced phasors, either three-phase voltages or three-phase currents, can be resolved into three balanced sets or networks, widely known as *symmetrical components*. These three sets or networks contain positive sequence, negative sequence, and zero sequence sets. They are defined as follows:

(a) A *positive sequence network* of three symmetrical voltages or currents is numerically equal but displaced from each other by 120°. The phase sequence or direction of a positive sequence network is the same as that of the original set and denoted by V_{a1}, V_{b1}, and V_{c1} or I_{a1}, I_{b1}, and I_{c1}. This is shown in Fig. 19.21(a).

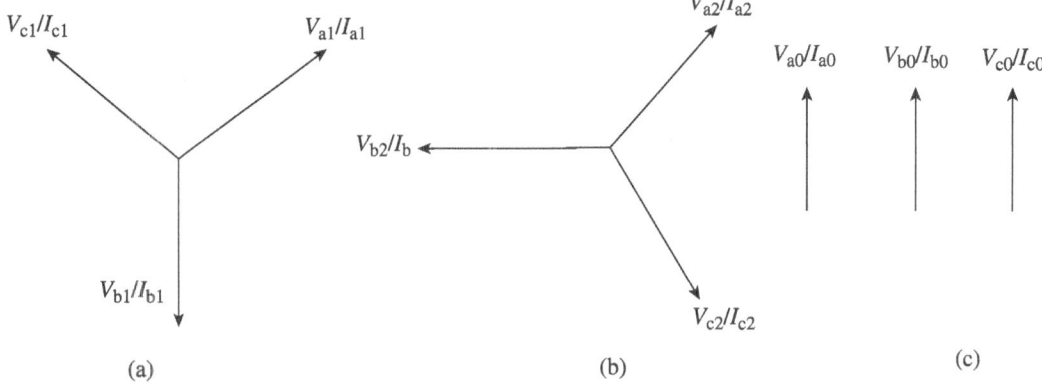

Fig. 19.21 Representation of symmetrical components (a) Positive sequence components
(b) Negative sequence components (c) Zero sequence components

(b) A *negative sequence network* of three symmetrical voltages or currents is numerically equal, but the voltages/currents are displaced from each other by 120° and have opposite phase sequence to that of the original set. It is denoted by V_{a2}, V_{b2}, and V_{c2} or I_{a2}, I_{b2}, and I_{c2}. This is shown in Fig. 19.21(b).

(c) A *zero sequence set* of three symmetrical voltages or currents is equal in magnitude and phase. The voltages and currents are denoted by V_{a0}, V_{b0}, and V_{c0} or I_{a0}, I_{b0}, and I_{c0}. They are shown in Fig. 19.21(c).

The actual meaning of the symmetrical component may be understood by applying these three sequence voltages to a rotating machine (induction motor). When only positive set of sequence voltages is applied to the terminal of the motor it rotates in one particular direction. Conversely, on the application of negative set of sequence voltages to the motor, it rotates in the reverse direction to that of the previous case. This reveals that the negative set of sequence produces rotating field which rotates in opposite direction to that produced by positive sequences. The set of zero sequence voltages does not produce any rotating field as they are in phase and motor windings are displaced by 120 degree electrical. If only positive and negative sequence components of current are present in the phasor, the sum of each will be zero as they are balanced. The zero sequence component of current flows only when the system has a return path through which the residual current will flow which is three times the zero sequence current of any one phase.

The symmetrical components are significantly useful for power system analysis under unbalance and faulty conditions. The unbalance phasors are transformed into symmetrical components; the system response is worked out using the derived components on simple circuit representation and the outcome is converted again into original phasor values. This procedure is easier and better than the direct solution of unbalanced or faulty three-phase system.

19.7.1 Transformation of Unbalanced Phasors to Symmetrical Components

The original three-phase voltages or currents are obtained by adding the symmetrical components as follows:

$$
\begin{aligned}
V_a &= V_{a1} + V_{a2} + V_{a0} \\
V_b &= V_{b1} + V_{b2} + V_{b0} \\
V_c &= V_{c1} + V_{c2} + V_{c0}
\end{aligned}
\tag{19.21}
$$

$$
\begin{aligned}
I_a &= I_{a1} + I_{a2} + I_{a0} \\
I_b &= I_{b1} + I_{b2} + I_{b0} \\
I_c &= I_{c1} + I_{c2} + I_{c0}
\end{aligned}
\tag{19.22}
$$

While working with three-phase quantities, it is convenient to have a simple phasor operator such as 'a', which will add 120° to the phase angle of the phasor without changing its magnitude. It is defined as

$$a = 1\angle 120° = e^{j2\Pi/3}$$

The function of operator 'a' is to rotate a phasor by +120° (counter clockwise/anti-clockwise direction) as shown in Fig. 19.22.

Various linear combinations of operator 'a' are as follows:

$$a^2 = 1\angle -120°,\ a^3 = 1\angle 0°,\ a^4 = a,\ a^5 = a^2,\ a^6 = 1$$

Using operator 'a', the phasors of voltages and currents can be written as follows:

$$
\begin{aligned}
V_{b1} &= a^2 \times V_{a1},\ V_{c1} = a \times V_{a1} \\
V_{b2} &= a \times V_{a2},\ V_{c2} = a^2 \times V_{a2} \\
V_{b0} &= V_{a0},\ V_{c0} = V_{a0}
\end{aligned}
$$

$$
\begin{aligned}
I_{b1} &= a^2 \times I_{a1},\ I_{c1} = a \times I_{a1} \\
I_{b2} &= a \times I_{a2},\ I_{c2} = a^2 \times I_{a2} \\
I_{b0} &= I_{a0},\ I_{c0} = I_{a0}
\end{aligned}
$$

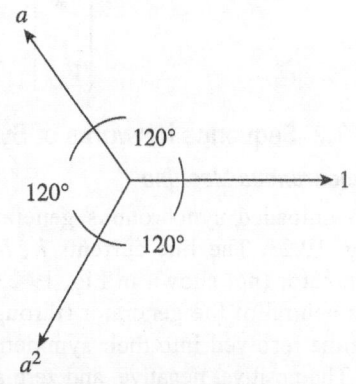

Fig. 19.22 Phasor diagram of operator *a*

Putting the said values in Eqs 19.21 and 19.22, we get,

$$\left.\begin{array}{l} V_a = V_{a1} + V_{a2} + V_{a0} \\ V_b = a^2 V_{a1} + a V_{a2} + V_{a0} \\ V_c = a V_{a1} + a^2 V_{a2} + V_{a0} \end{array}\right\} \qquad (19.23)$$

$$\left.\begin{array}{l} I_a = I_{a1} + I_{a2} + I_{a0} \\ I_b = a^2 I_{a1} + a I_{a2} + I_{a0} \\ I_c = a I_{a1} + a^2 I_{a2} + I_{a0} \end{array}\right\} \qquad (19.24)$$

In matrix form, these equations are written as follows:

$$\begin{bmatrix} V_a \\ V_b \\ V_c \end{bmatrix} = \begin{bmatrix} 1 & 1 & 1 \\ 1 & a^2 & a \\ 1 & a & a^2 \end{bmatrix} \begin{bmatrix} V_{a0} \\ V_{a1} \\ V_{a2} \end{bmatrix}, \quad \begin{bmatrix} I_a \\ I_b \\ I_c \end{bmatrix} = \begin{bmatrix} 1 & 1 & 1 \\ 1 & a^2 & a \\ 1 & a & a^2 \end{bmatrix} \begin{bmatrix} I_{a0} \\ I_{a1} \\ I_{a2} \end{bmatrix} \qquad (19.25)$$

In compact form, they are written as follows:

$$V_{abc} = [A] V_{012} \text{ and } I_{abc} = [A] I_{012}$$

where, $\quad A = \begin{bmatrix} 1 & 1 & 1 \\ 1 & a^2 & a \\ 1 & a & a^2 \end{bmatrix}, \ V_{abc} = \begin{bmatrix} V_a \\ V_b \\ V_c \end{bmatrix}, \ V_{012} = \begin{bmatrix} V_{a0} \\ V_{a1} \\ V_{a2} \end{bmatrix}, \ I_{abc} = \begin{bmatrix} I_a \\ I_b \\ I_c \end{bmatrix}, \text{ and } I_{012} = \begin{bmatrix} I_{a0} \\ I_{a1} \\ I_{a2} \end{bmatrix}$

To resolve the unbalanced phasor into their symmetrical component, taking inverse of A,

$$A^{-1} = \frac{1}{3} \begin{bmatrix} 1 & 1 & 1 \\ 1 & a & a^2 \\ 1 & a^2 & a \end{bmatrix}$$

By multiplying both sides of Eq. 19.25 by A^{-1},

$$\begin{bmatrix} V_{a0} \\ V_{a1} \\ V_{a2} \end{bmatrix} = \frac{1}{3} \begin{bmatrix} 1 & 1 & 1 \\ 1 & a & a^2 \\ 1 & a^2 & a \end{bmatrix} \begin{bmatrix} V_a \\ V_b \\ V_c \end{bmatrix} \quad \begin{bmatrix} I_{a0} \\ I_{a1} \\ I_{a2} \end{bmatrix} = \frac{1}{3} \begin{bmatrix} 1 & 1 & 1 \\ 1 & a & a^2 \\ 1 & a^2 & a \end{bmatrix} \begin{bmatrix} I_a \\ I_b \\ I_c \end{bmatrix} \qquad (19.26)$$

19.7.2 Sequence Networks of Synchronous Machine, Transmission Line, and Transformer

Synchronous Machine

An unloaded synchronous generator with neutral grounded through an impedance Z_n is shown in Fig. 19.23. The line currents I_a, I_b, and I_c flow whenever a fault takes place at the terminals of the generator (not shown in Fig. 19.23). If ground is involved in the fault, the current that flows through the neutral of the generator (through Z_n) is specified by 'I_n'. The unbalanced line currents due to fault can be resolved into their symmetrical components.

The positive, negative, and zero sequence networks of the generator can be drawn with the voltage and impedance components. The generated voltages are of positive sequence only as the generators are symmetrical and designed to produce balanced three-phase voltages. Thus, positive sequence network is composed of an emf in series with the positive sequence impedance Z_1. The negative and zero sequence networks are composed of only the respective sequence impedances as there is no corresponding sequence emf. The reference bus for the positive and negative-sequence networks is the neutral of the generator which is at ground potential. The neutral impedance Z_n does not involve in the positive or negative circuit. Hence, no positive or negative-sequence current can flow through Z_n. The current flowing through the impedance Z_n between neutral and ground is "$3I_{a0}$" as shown in Fig.19.24. Thus, the zero-sequence voltage drop from the terminal of generator 'a' to the ground, is given by $-3I_{a0}Z_n - I_{a0}Z_{g0}$, where, Z_{g0} is the zero-sequence impedance of the generator. Thus, the equivalent zero-sequence network carries only current of one phase and has impedance of $Z_0 = Z_{g0} + 3Z_n$.

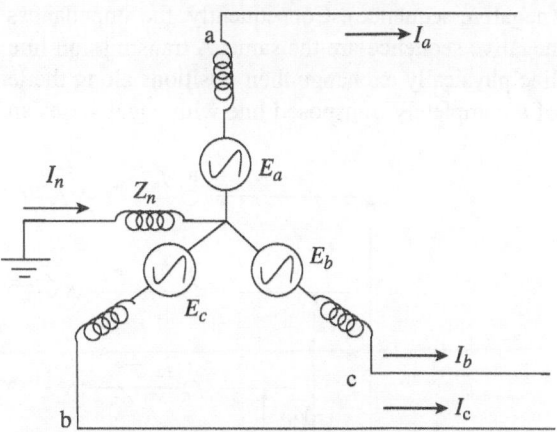

Fig. 19.23 Circuit diagram of unloaded generator grounded through Z_n

From the sequence networks, as shown in Fig. 19.24, the voltage drops from point 'a' to reference bus/ground are as follows:

$$\left. \begin{array}{l} V_{a1} = E_a - I_{a1} \times Z_1 \\ V_{a2} = -I_{a2} \times Z_2 \\ V_{a3} = -I_{a0} \times Z_0 \end{array} \right\} \qquad (19.27)$$

where, E_a is no load terminal voltage to neutral or transient/sub-transient internal voltage. Z_1, Z_2, and Z_0 are the positive, negative and zero sequence impedance of the generator, respectively. I_{a1}, I_{a2}, and I_{a0} are the respective positive, negative and zero sequence current, respectively.

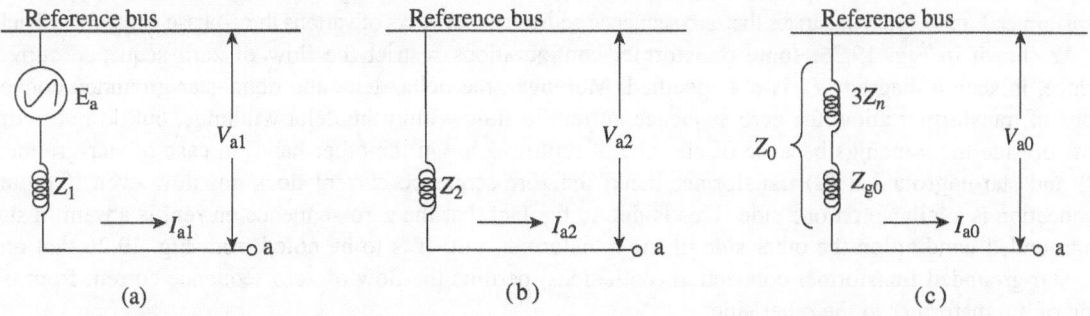

Fig. 19.24 (a) Positive sequence network (b) Negative sequence network (c) Zero sequence network

Transmission Line

A transmission line is a linear passive device which has no voltage or current source in its equivalent model. The behaviour of the line remains the same as it is a bilateral device which means line current can flow in any direction. With these features, the phase sequence of the applied voltage does not make any difference in voltage drop irrespective of the phase sequence of voltages, that is, a-b-c (positive sequence) or a-c-b (negative sequence). Consequently, the impedances of a transposed transmission line for its positive and negative sequence are the same. A transmission line is called *transposed* when the phase conductors of the line physically exchange their positions along the length of the line. Figure 19.25 shows the representation of a completely transposed line with equal series and mutual impedances.

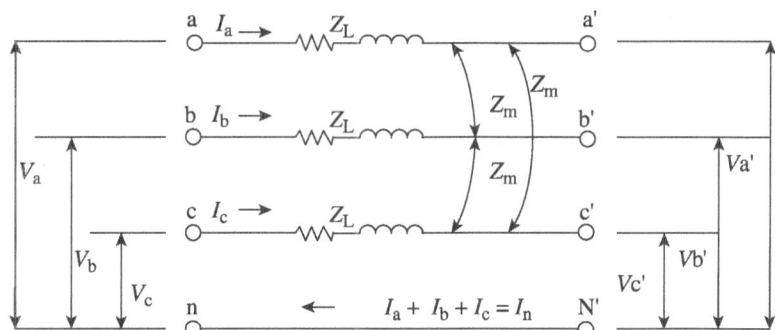

Fig. 19.25   Representation of transmission line with identical
series and mutual impedances

The zero sequence impedance of the line is dissimilar to positive and negative sequence impedance and has a higher value. This is because the magnetic field set up by the zero sequence current is incredibly different from that of the positive or negative sequence currents due to no phase difference among zero sequence currents flowing in three phases. The zero sequence impedance of the line is generally two to four times greater than the positive-sequence impedances.

Transformer

A three-phase transformer bank is composed of three equivalent single-phase transformers. The magnitudes of positive and negative sequence impedances of a three-phase power transformer are the same. However, the zero sequence impedance slightly differs from the positive- and negative sequence series impedance. For the practical purpose, all three sequence impedances are considered to be identical ($Z_1 = Z_2 = Z_0$) irrespective of the types of the transformer. Figure 19.26 illustrates the zero sequence equivalent networks of various three-phase transformer banks.

As shown in Fig. 19.26, some transformer configurations restrict the flow of zero sequence current. Hence, in such a diagram, I_{a0} is not specified. Moreover, the delta–delta and delta–star-grounded connections of transformer allow the zero sequence current to flow within the delta windings, but do not permit flow outside the windings because of absence of return path. On the other hand, in case of star-grounded (P) and star-ungrounded (S) transformer bank, the zero-sequence current does not flow even if ground connection is available on one side. This is due to the fact that the zero-sequence current is absent in star-ungrounded winding on the other side of the transformer unit. It is to be noted from Fig. 19.26 that only the star-grounded transformer connection (both sides) permits the flow of zero sequence current from one side of a transformer to the other side.

Phase shift of symmetrical components in transformer

The following are some facts about phase shift in transformer bank:

(1) For star–star and delta–delta transformer banks, the phase shift in positive and negative sequence voltages and currents are 0°.

(2) However, for star–delta and delta–star transformer banks, the phase shift is sequence-dependent, that is, +30°phase shift is observed in case of positive sequence than −30° phase shift in negative sequence.

(3) For a particular transformer connection, zero sequence voltage and current does not introduce a phase shift, but they may obstruct the circulation of zero sequence as shown in Fig. 19.26.

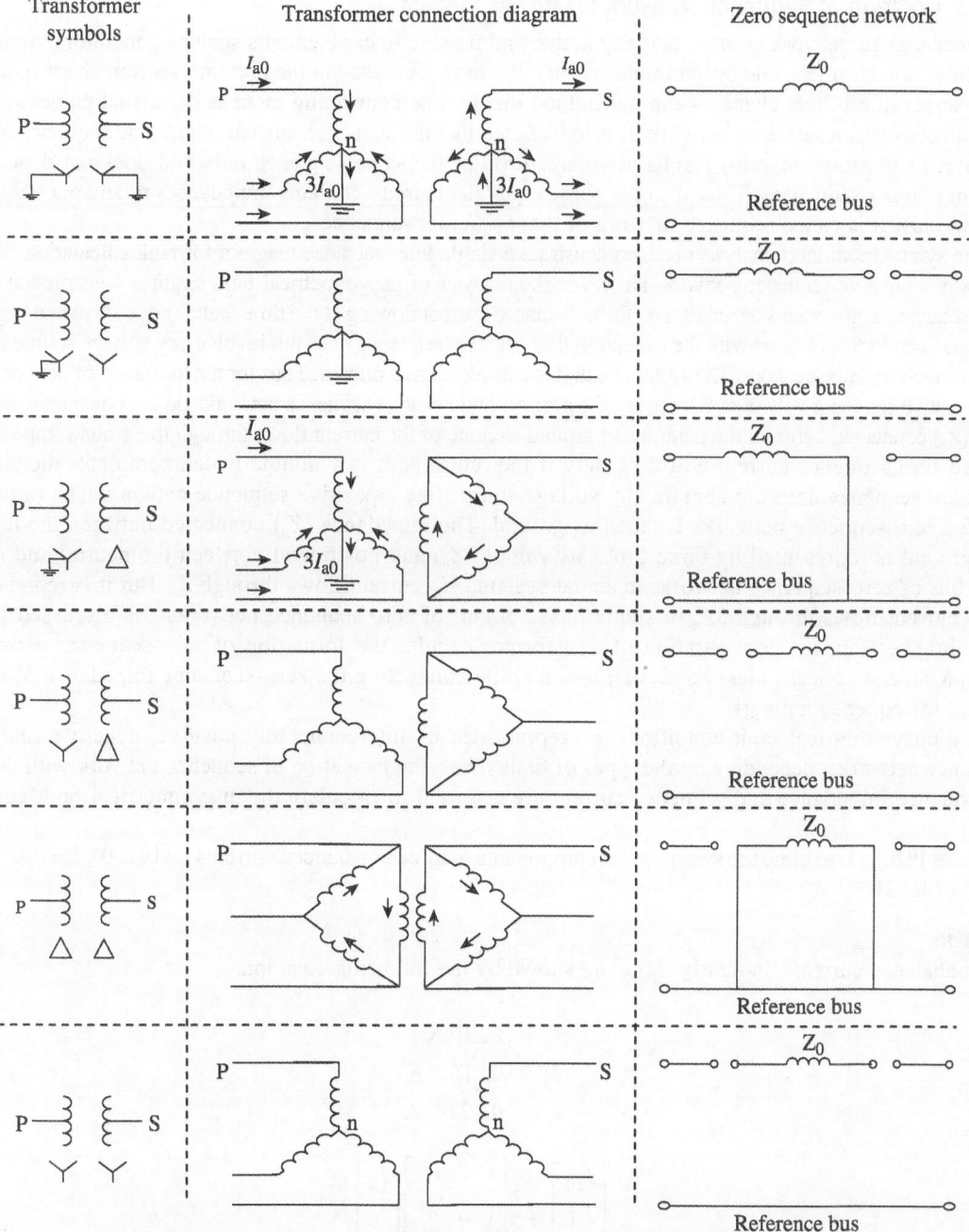

Transformer symbols	Transformer connection diagram	Zero sequence network

Fig. 19.26 Zero sequence network equivalents of three-phase transformer bank

It is to be stated that due to phase shift in sequence component of voltages/currents and the probable prevention of zero sequence current, the transformers noticeably play an important function in unbalanced as well as faulty condition of power system.

19.7.3 Creation of Sequence Network for Power System

A power system network comprises many active and passive linear elements such as generators, transmission lines, transformers, and synchronous motors. We have discussed in the previous section about sequence impedances of all these elements and understood that by interconnecting them as per actual power system, one can construct a sequence network. It is to be noted that the sequence current of any one sequence (either positive, or negative, or zero) results in voltage drop of the same sequence only and does not depend on current of other sequences. Thus, a single phase equivalent circuit contains impedances to any one sequence current only. It is called sequence network of that particular sequence.

The symmetrical fault analysis needs to construct a single line reactance diagram for fault calculation. This is basically a positive sequence network. However, the analysis of unsymmetrical fault requires construction of all three sequence components depending on the unbalanced current flowing at the time fault. The positive and negative sequence networks are same with the exception that negative sequence does not involve any voltage source unlike the positive-sequence network. The system neutral is considered as a reference bus for the formation of both positive and negative sequence networks. Moreover, the positive and negative sequence networks do not contain the impedance (Z_n) connected between the neutral and ground as none of the current flows through the ground impedance.

The zero sequence current will flow only if the return path is available which completes the circuit. The zero sequence does not contain any voltage source like a positive sequence network. The reference bus for zero sequence networks is taken as ground. The impedance (Z_n) connected between the neutral and ground is represented by three times its value ($3Z_n$) and positioned between the neutral and reference bus of zero-sequence network. In actual system, $3I_{a0}$ current flows through Z_n. But it is represented as I_{a0} current flows through $3Z_n$ in single phase circuit of zero sequence networks. As discussed in the preceding section, the configuration of transformer decides the formation of zero-sequence networks. Thus, a special concern must be accomplished while connecting the zero-sequence impedance of transformer in sequence network.

The unsymmetrical fault conditions are represented by interconnecting positive, negative, and zero sequence networks, depending on the types of fault. Thus, the formation of sequence network with the use of sequence impedance and sequence current are essential to calculate the unsymmetrical fault current.

Example 19.8 Determine the symmetrical components of three unbalanced currents $I_a=10\angle 0°$, $I_b=10\angle 150°$, and $I_c=10\angle 210°$.

Solution:
The unbalance currents, in matrix form, are shown by the following equation.

$$I_a = 10\angle 0° \, \text{A}$$
$$I_b = 10\angle 150° \, \text{A}$$
$$I_c = 10\angle 210° \, \text{A}$$

$$\begin{bmatrix} I_{a0} \\ I_{a1} \\ I_{a2} \end{bmatrix} = \frac{1}{3} \begin{bmatrix} 1 & 1 & 1 \\ 1 & a & a^2 \\ 1 & a^2 & a \end{bmatrix} \begin{bmatrix} I_a \\ I_b \\ I_c \end{bmatrix}$$

$$a = 1\angle 120° \text{ and } a^2 = 1\angle 240°$$

$$a = -0.5 + j0.866 \text{ and } a^2 = -0.5 - j0.866$$

$$I_{a0} = \frac{1}{3}(I_a + I_b + I_c)$$

$$I_{a0} = \frac{1}{3}(10 + j0 - 8.66 + j5 - 8.66 - j5) = \frac{1}{3}(-7.5) = -2.5\angle 0°$$

$$I_{a1} = \frac{1}{3}(I_a + aI_b + a^2 I_c)$$

$$I_{a1} = \frac{1}{3}(10\angle 0° + 10\angle 150° \times 1\angle 120° + 10\angle 210° \times 1\angle 240°)$$

$$= \frac{1}{3}(10\angle 0° + 10\angle 270° + 10\angle 90°)$$

$$I_{a1} = \frac{1}{3}(10 + j0 + 0 - j10 + 0 + j10) = 3.33\angle 0°$$

$$I_{a2} = \frac{1}{3}(I_a + a^2 I_b + aI_c)$$

$$I_{a2} = \frac{1}{3}(10\angle 0° + 10\angle 150° \times 1\angle 240° + 10\angle 210° \times 1\angle 120°)$$

$$= \frac{1}{3}(10\angle 0° + 10\angle 30° + 10\angle 330°)$$

$$I_{a2} = \frac{1}{3}(10 + j0 + 8.66 + j5 + 8.66 - j5) = 9.106\angle 0°$$

Hence,

$$I_{a0} = -2.5\angle 0°, I_{b0} = -2.5\angle 0° \text{ and } I_{c0} = -2.5\angle 0°$$
$$I_{a1} = 3.33\angle 0°, I_{b1} = 3.33\angle 120° \text{ and } I_{c1} = 3.33\angle 240°$$
$$I_{a2} = 9.106\angle 0°, I_{b2} = 9.106\angle 120° \text{ and } I_{c2} = 9.106\angle 240°$$

Example 19.9 The symmetrical components are $V_{a1} = 30\angle 10°$, $V_{a2} = 20\angle 100°$, and $V_{a0} = 10\angle 190°$. Determine the value of the phasor voltages V_a, V_b, and V_c with respect to the neutral.

Solution:
For the given symmetrical components of voltage, the unbalanced phasors are given by,

$$\begin{bmatrix} V_a \\ V_b \\ V_c \end{bmatrix} = \begin{bmatrix} 1 & 1 & 1 \\ 1 & a^2 & a \\ 1 & a & a^2 \end{bmatrix} \begin{bmatrix} V_{a0} \\ V_{a1} \\ V_{a2} \end{bmatrix}$$

$$V_a = V_{a1} + V_{a2} + V_{a0}$$
$$V_a = 30\angle 10° + 20\angle 100° + 10\angle 190°$$
$$= 29.54 + j5.2 - 3.47 + j19.7 - 9.84 - j1.74 = 16.23 + j23.16 = 28.28\angle 55°$$

$$V_b = a^2V_{a1} + aV_{a2} + V_{a0}$$
$$V_b = 30\angle10° \times 1\angle240° + 20\angle100° \times 1\angle120° + 10\angle190°$$
$$= -10.26 - j28.19 - 15.32 - j12.86 - 9.84 - j1.74 = -35.42 - j42.79 = 55.55\angle -129.6°$$

$$V_c = aV_{a1} + a^2V_{a2} + V_{a0}$$
$$V_c = 30\angle10° \times 1\angle120° + 20\angle100° \times 1\angle240° + 10\angle190°$$
$$= -19.28 + j22.98 + 18.79 - j6.8 - 9.84 - j1.74 = -10.33 + j14.44 = 17.72\angle125.65°$$

Hence, unbalanced phasors are obtained as under.

$$V_a = 28.28\angle55°, \ V_b = 55.55\angle -129.6° \text{ and } V_c = 17.72\angle125.65°$$

Example: 19.10   For the system shown in Fig. 19.27, draw zero sequence networks.

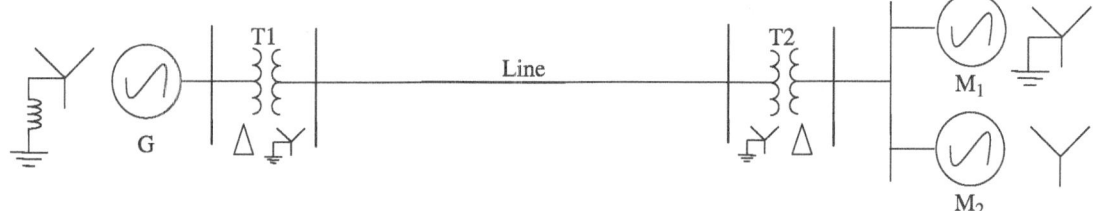

Fig. 19.27   One-line diagram for Example 19.10

Solution:
The zero sequence network is shown in Fig. 19.28.

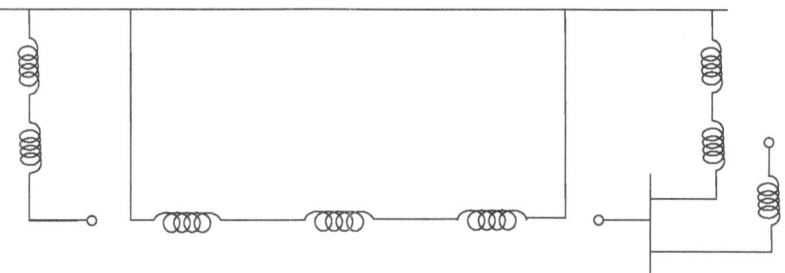

Fig. 19.28   Zero sequence network for the system of Example 19.10

Example 19.11   Figure 19.29 shows the single line diagram of a portion of the power system network. Draw zero sequence networks for the power system network shown in Fig. 19.29.

Solution:
The zero sequence network is shown in Fig. 19.30.

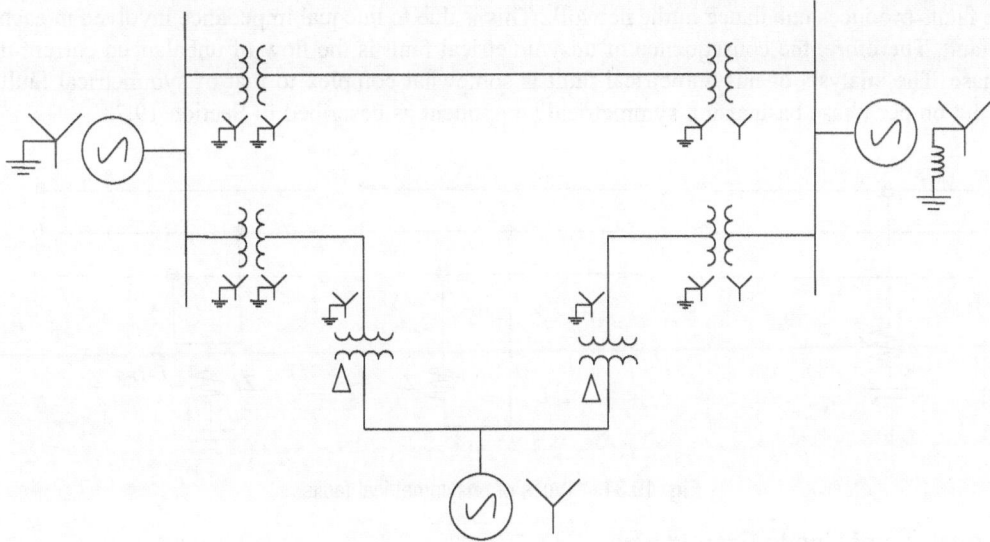

Fig. 19.29 One-line diagram for Example 19.11

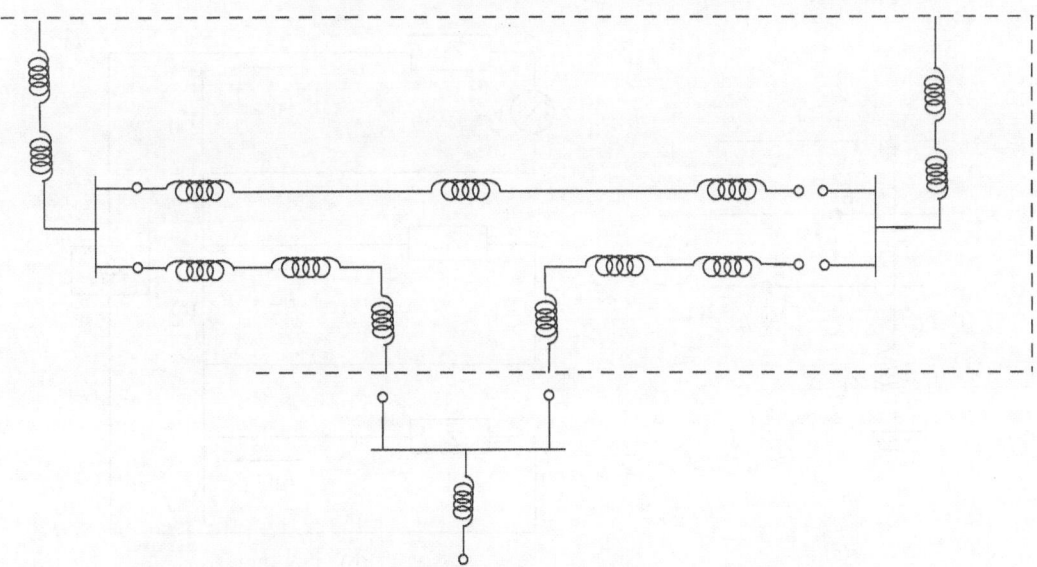

Fig. 19.30 Zero sequence network for the system of Example 19.11

19.8 Unsymmetrical Faults Analysis

Unsymmetrical faults are normally less severe than symmetrical faults. They are mainly categorized as: line-to-ground (*L-g*), line-to-line (*L-L*), and double line-to-ground (*L-L-g*) faults as shown in Fig. 19.31.

Out of these faults, L-g fault is the most frequent fault that occurs on the transmission line/feeder. This type of fault occurs due to shorting of live conductor with earth or ground may be through low or high resistance path (Z_f). Swinging of lines due to winds or birds flying in between two lines generate *L-L* and *L-L-g* fault which makes two conductors in contact with each other or together with ground. The occurrence of these faults produces unbalance in the network. This is due to unequal impedance involved in each phase during fault. Therefore, the consequence of unsymmetrical fault is the flow of unbalanced current through each phase. The analysis of unsymmetrical fault is somewhat complex to that of symmetrical faults. It is carried out on per phase basis using symmetrical component as described in Section 19.7.

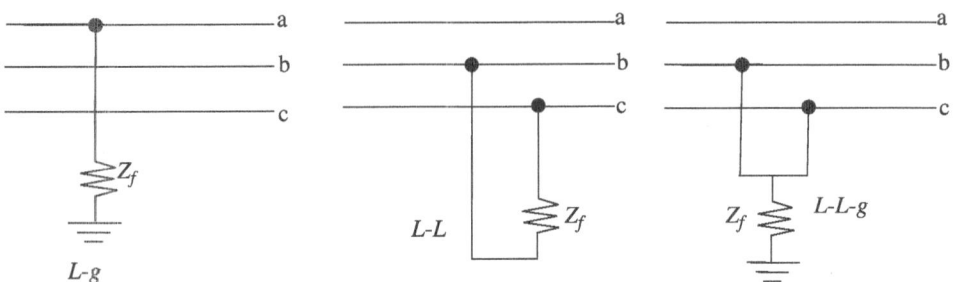

Fig. 19.31 Types of unsymmetrical faults

19.8.1 Analysis of Line-to-Ground Fault

Consider a single line-to-ground fault occurring on phase-a on a network as shown in Fig. 19.32.

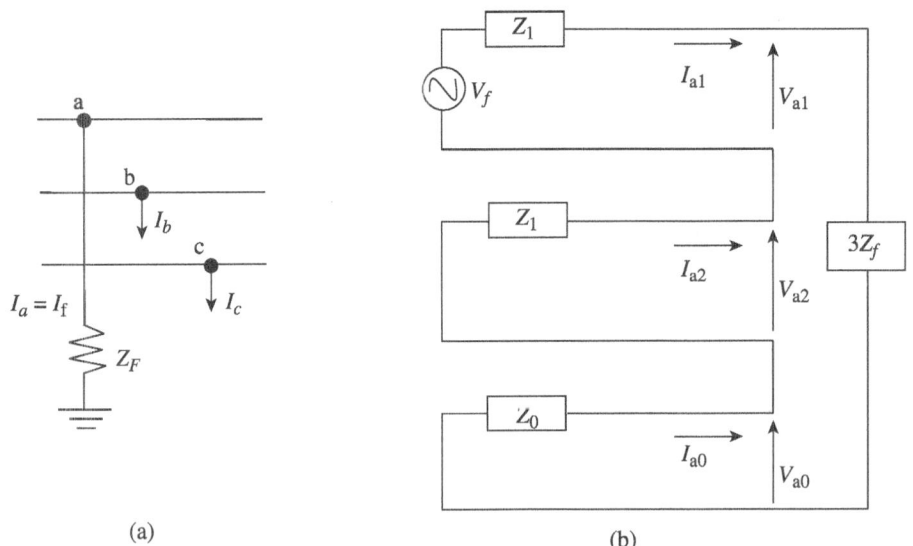

(a)   (b)

Fig. 19.32 Representation of (a) single line-to-ground fault and (b) Thevenin's equivalent network

Assume that the system is not loaded prior to the occurrence of fault. Also, the fault path has an impedance of Z_f.

$$I_b = I_c = 0 \ and \ I_a = I_f \tag{19.28}$$

Also, the voltage of phase a at the fault point is given by

$$V_a = Z_f I_a = Z_f I_f \tag{19.29}$$

The sequence components of current are given by Eq. 19.30. Utilizing Eq. 19.28 in this equation we obtain the equation as given here.

$$\begin{bmatrix} I_{a0} \\ I_{a1} \\ I_{a2} \end{bmatrix} = \frac{1}{3} \begin{bmatrix} 1 & 1 & 1 \\ 1 & a & a^2 \\ 1 & a^2 & a \end{bmatrix} \begin{bmatrix} I_a \\ 0 \\ 0 \end{bmatrix} \tag{19.30}$$

Thus,
$$I_{a0} = I_{a1} = I_{a2} = \frac{I_a}{3} \tag{19.31}$$

Equation 19.31 specifies that the three sequence currents are in series for L-g fault. By taking Z_0, Z_1, and Z_2 as the zero, positive, and negative sequence Thevenin's impedance, respectively, at the fault point. Also, the Thevenin's voltage of phase a at the faulted point is V_f. Hence, we get three sequence networks from which the following set of equations is obtained.

$$\begin{aligned} V_{a0} &= -Z_0 I_{a0} \\ V_{a1} &= V_f - Z_1 I_{a1} \\ V_{a2} &= -Z_2 I_{a2} \end{aligned} \tag{19.32}$$

Now, from Eqs 19.31 and 19.32, we derive the following equations.

$$\begin{aligned} V_a &= V_{a0} + V_{a1} + V_{a2} \\ V_a &= V_f - (Z_0 + Z_1 + Z_2) I_{a0} \end{aligned} \tag{19.33}$$

Again, from Eqs 19.29 and 19.30,

$$V_a = Z_f I_a = Z_f \left(I_{a0} + I_{a1} + I_{a2} \right) = 3 Z_f I_{a0} \tag{19.34}$$

Finally, substitution of the value of Eqs 19.34 into 19.33, we get

$$I_{a0} = \frac{V_f}{Z_0 + Z_1 + Z_2 + 3 Z_f} \tag{19.35}$$

19.8.2 Analysis of Line-to-Line Fault

Consider line-to-line fault between phases b and c through fault impedance Z_f as shown in Fig. 19.33.

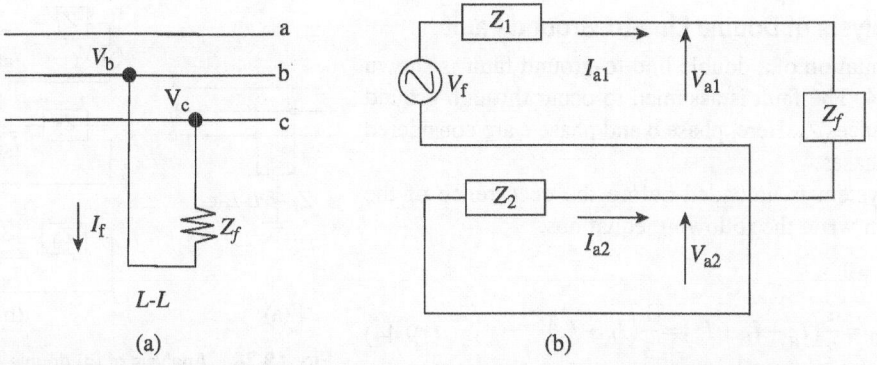

Fig.19.33 Representation of (a) line-to-line fault and (b) Thevenin's equivalent network (L-L)

By shorting phase b with phase c, the conditions of current are given by,

$$I_a = 0, \; and \; I_b = -I_c \qquad (19.36)$$

Since $I_c = -I_b$, the positive, negative-, and zero-sequence currents are calculated and given by,

$$\begin{bmatrix} I_{a0} \\ I_{a1} \\ I_{a2} \end{bmatrix} = \frac{1}{3} \begin{bmatrix} 1 & 1 & 1 \\ 1 & a & a^2 \\ 1 & a^2 & a \end{bmatrix} \begin{bmatrix} 0 \\ I_b \\ -I_b \end{bmatrix} \qquad (19.37)$$

Hence, from Eq. 19.37, we can write the following equations,

$$I_{a0} = 0$$
$$I_{a1} = -I_{a2} \qquad (19.38)$$

It is to be noted from Eq. 19.38 that the zero sequence circuit is absent (remains a dead network) for L-L fault. Moreover, the positive and negative sequence currents are equal in magnitude but opposite in direction. From Fig. 19.33, the following expression is written for voltage at the faulted point.

$$V_b - V_c = Z_f I_b \qquad (19.39)$$

Further, by inserting the symmetrical component of the real quantity in Eq 19.39,

$$V_{a0} + a^2 V_{a1} + a V_{a2} - V_{a0} - a V_{a1} - a^2 V_{a2} = Z_f (I_{a0} + a^2 I_{a1} + a I_{a2}) \qquad (19.40)$$

From Eqs 2.15 and 2.17, we can write

$$\left(a^2 - a\right) V_{a1} - \left(a^2 - a\right) V_{a2} = Z_f \left(a^2 - a\right)(I_{a1}) \qquad (19.41)$$

Therefore,

$$V_{a1} - V_{a2} = Z_f I_{a1} \qquad (19.42)$$

Hence, from Eqs 19.38 and 19.40, we can say that the positive and negative sequence circuits are connected in parallel as shown in Fig. 19.33. From this network, we obtain the following equation.

$$I_{a1} = -I_{a2} = \frac{V_f}{Z_1 + Z_2 + Z_f} \qquad (19.43)$$

19.8.3 Analysis of Double Line-to-Ground Fault

The representation of a double line-to-ground fault is shown in Fig. 19.34. The fault is assumed to occur through ground fault impedances Z_f. Here, phase b and phase c are considered as faulted phases.

Since the system is unloaded before the occurrence of the fault, we can write the following equations.

$$I_a = 0$$

$$I_{a0} = \frac{1}{3}\left(I_a + I_b + I_c\right) = \frac{1}{3}\left(I_b + I_c\right) \qquad (19.44)$$

$$I_b + I_c = 3I_{a0}$$

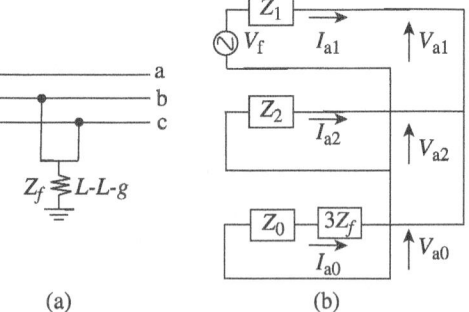

(a)   (b)

Fig. 19.34   Analysis of (a) double line-to-ground fault and (b) Thevenin's equivalent network (L-L-g)

Also, the voltages of phase b and phase c are given by,

$$V_b = V_c = Z_f \left(I_b + I_c \right) = 3 Z_f I_{a0} \tag{19.45}$$

Therefore, from Eq. 19.45 we can write the following equations.

$$
\begin{aligned}
V_b &= V_c \\
V_{a0} + a^2 V_{a1} + a V_{a2} &= V_{a0} + a V_{a1} + a^2 V_{a2} \\
V_{a1} &= V_{a2}
\end{aligned} \tag{19.46}
$$

Moreover, from Eqs 19.45 and 19.46, we can derive the following equations.

$$
\begin{aligned}
V_b &= 3 Z_f I_{a0} \\
V_{a0} + a^2 V_{a1} + a V_{a2} &= 3 Z_f I_{a0} \\
V_{a0} - V_{a1} &= 3 Z_f I_{a0}
\end{aligned} \tag{19.47}
$$

Thus, from Eqs 19.44 to 19.47 we can write the following equations.

$$
\begin{aligned}
V_{a1} = V_{a2} &= V_{a0} - 3 Z_f I_{a0} \\
I_a = I_{a0} + I_{a1} + I_{a2} &= 0
\end{aligned} \tag{19.48}
$$

With the use of Eq.19.48 we can construct the sequence network for double line-to ground fault. The Thevenin's equivalent sequence network diagram for double line-to-ground fault is shown in Fig. 19.34. From this figure, we finally obtain the following equation.

$$I_{a1} = \frac{V_f}{Z_1 + \left[Z_2 \| \left(Z_0 + 3 Z_f \right) \right]} = \frac{V_f}{Z_1 + \dfrac{Z_2 \left(Z_0 + 3 Z_f \right)}{Z_2 + Z_0 + 3 Z_f}} \tag{19.49}$$

19.8.4 Analysis of Unsymmetrical Fault on Power System

The analysis of any unsymmetrical fault on a general power system network involves the derivation of symmetrical components of current and voltage. The pre-fault voltage of the fault point is considered as V_f which is positive sequence voltage as the system is previously assumed to be balanced.

Consider a single line diagram of power system network, as shown in Fig. 19.35, for which an unsymmetrical fault is applied at point 'X'. The single phase representation of all three sequence networks is drawn with fault point X marked on each sequence network. Each sequence network can be converted to its equivalent Thevenin's circuit between the point of fault applied and reference bus. The equivalent positive sequence impedance (Z_1) of all the components is connected in series with the internal voltage of generator (V_f). Similarly the negative and zero sequence impedances Z_2 and Z_0, respectively, are connected between point X and reference bus without any emf in series.

Assuming that I_a is the current flowing into fault and its symmetrical component I_{a1}, I_{a2}, and I_{a0} will flow out of the respective sequence network from point X. Depending on the types of unsymmetrical fault, Thevenin's equivalent circuit of particular sequence networks can be connected in series, parallel, or in combination as explained in the previous section.

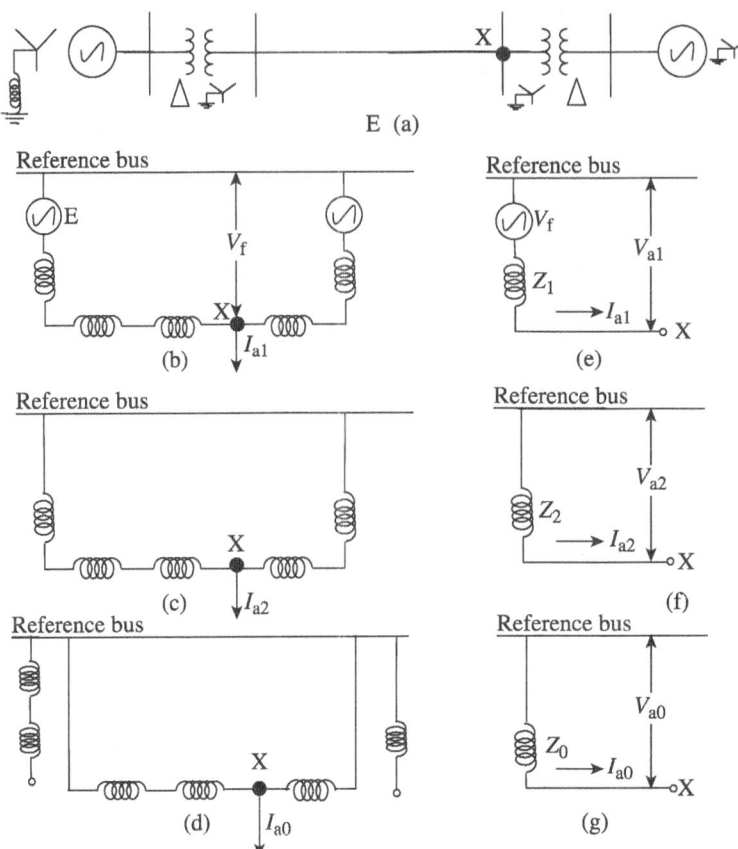

Fig. 19.35 (a) Single line diagram of power system, (b), (c), (d) are three sequence networks of power system and (e), (f), (g) are the Thevenin's equivalent circuit of sequence networks

Example 19.12 A generator of 50 MVA, 17.5 kV has direct axis sub-transient reactance of 0.20 pu. The negative and zero sequence reactance is 0.25 pu and 0.15 pu, respectively. The neutral of generator is solidly grounded. Determine the sub-transient current in the generator when single line-to-ground fault occurs on the terminal of generator. Also, determine the line-to-neutral voltage during sub-transient condition.

Solution:

Considering 50MVA and 17.5 kV as base value for per unit calculation and $V_f = 1$, the base current is given by,

$$\text{Base current, } I_B = \frac{50000}{\sqrt{3} \times 17.5} = 1649.57 \text{ A}$$

Without fault impedance ($Z_f = 0$), $Z_1 = j0.2$, $Z_2 = j0.25$, and $Z_0 = j0.15$

For line-to-ground fault $I_{a1} = I_{a2} = I_{a0}$

$$I_{a0} = \frac{V_f}{Z_0 + Z_1 + Z_2} = \frac{1\angle 0°}{j0.15 + j0.2 + j0.25} = \frac{1\angle 0°}{j0.6} = -j1.67 \text{ pu}$$

$$I_a = 3 \times I_{a0} = 3 \times (-j1.67) = -j5 \text{ pu}$$

Thus, sub-transient fault current is $I_f = I_a \times I_B = 5 \times 1649.57 = 8247.86$ A

The symmetrical components of voltage from terminal 'a' to ground are,

$$V_{a1} = E_a - I_{a1} \times Z_1 = 1 - (-j1.67) \times (j0.2) = 0.666 \text{ pu}$$

$$V_{a2} = -I_{a2} \times Z_2 = -(-j1.67) \times (j0.25) = -0.41 \text{ pu}$$

$$V_{a0} = -I_{a0} \times Z_0 = -(-j1.67) \times (j0.15) = -0.25 \text{ pu}$$

Now, line-to-neutral voltages are given by,

$$V_a = V_{a1} + V_{a2} + V_{a0}$$
$$V_a = 0.66 - 0.41 - 0.25 = 0 \text{ pu}$$

$$V_b = a^2 V_{a1} + a V_{a2} + V_{a0}$$
$$V_b = 0.66 \times (-0.5 - j0.866) - 0.41 \times (-0.5 + j0.866) - 0.25$$
$$= -0.33 - j0.57 + 0.205 - j0.355 - 0.25$$
$$= -0.375 - j0.925 = 0.998 \angle -112°$$

$$V_c = a V_{a1} + a^2 V_{a2} + V_{a0}$$
$$V_c = 0.66 \times (-0.5 + j0.866) - 0.41 \times (-0.5 - j0.866) - 0.25$$
$$= -0.33 + j0.57 + 0.205 + j0.355 - 0.25$$
$$= -0.375 + j0.925 = 0.998 \angle 112°$$

The post-fault lines to neutral voltages are,

$$V_a = 0 \text{ kV}$$

$$V_b = (0.998 \angle -112°) \times \frac{17.5}{\sqrt{3}} = 10.08 \angle -112° \text{ kV}$$

$$V_c = (0.998 \angle 112°) \times \frac{17.5}{\sqrt{3}} = 10.08 \angle 112° \text{ kV}$$

The post fault line-to-line voltages are,

$$V_{ab} = V_a - V_b = (0 + 0.375 + j0.925) \times \frac{17.5}{\sqrt{3}} = 10.08 \angle 70° \text{ kV}$$

$$V_{bc} = V_b - V_c = (-0.375 - j0.9250 + 0.375 - j0.925) \times \frac{17.5}{\sqrt{3}} = 18.7 \angle -90° \text{ kV}$$

$$V_{ca} = V_c - V_a = (-0.375 + j0.925 - 0) \times \frac{17.5}{\sqrt{3}} = 10.08 \angle 112° \text{ kV}$$

Example 19.13 Find the sub-transient current when double-line fault occurs on the terminal of unloaded generator as mentioned in Example 19.12. Also, find the change in line-to-line voltage during fault.

Solution:
The value of base current and the values of positive, negative, and zero sequence impedances remain the same (as given in Example 19.12).

For line-to-line fault $I_{a1} = -I_{a2}$ and $I_{a1} = 0$

$$I_{a1} = \frac{V_f}{Z_1 + Z_2} = \frac{1\angle 0°}{j0.2 + j0.25} = \frac{1\angle 0°}{j0.45} = -j2.23 \text{ pu}$$

$$I_a = I_{a1} + I_{a2} + I_{a0} = 0 \text{ pu}$$

$$\begin{aligned} I_b &= a^2 I_{a1} + a I_{a2} + I_{a0} \\ &= -j2.23 \times (-0.5 - j0.866) + j2.23 \times (-0.5 + j0.866) + 0 \\ &= j1.115 - 1.93 - j1.115 - 1.93 \\ &= -3.86 \text{ pu} \end{aligned}$$

$$I_c = -I_b = 3.86 \text{ pu}$$

Thus, sub-transient fault current is $I_f = 3.86 \times 1649.57 = 6367.34$ A

The symmetrical components of voltage from terminal 'a' to ground are

$$V_{a1} = E_a - I_{a1} \times Z_1 = 1 - (-j2.23) \times (j0.2) = 0.446 \text{ pu}$$

$$V_{a2} = V_{a1} = 0.446 \text{ pu}$$

$V_{a0} = 0$ pu as ground is not involved in the fault.

Now, line-to-neutral voltages are given by,

$$V_a = V_{a1} + V_{a2} + V_{a0}$$
$$V_a = 0.446 + 0.446 + 0 = 0.892 \text{ pu}$$

$$V_b = a^2 V_{a1} + a V_{a2} + V_{a0}$$
$$V_b = 0.446 \times (-0.5 - j0.866) + 0.446 \times (-0.5 + j0.866) + 0$$
$$= -0.446 \text{ pu}$$
$$V_c = V_b = -0.446 \text{ pu}$$

The post fault line-to-line voltages are,

$$V_{ab} = V_a - V_b = (0.892 + 0.446) \times \frac{17.5}{\sqrt{3}} = 13.52\angle 0° \text{ kV}$$

$$V_{bc} = V_b - V_c = 0 \text{ kV}$$

$$V_{ca} = V_c - V_a = (-0.892 - 0.446) \times \frac{17.5}{\sqrt{3}} = 13.52\angle 180° \text{ kV}$$

Example 19.14 Find the line-to-line voltages and sub-transient current when double line-to-ground fault occurs on the terminal of unloaded generator as mentioned in Example 19.12.

Solution:

The value of base current and the values of positive, negative, and zero sequence impedances remain the same (as given in Example 19.12).

For double line-to-ground fault, $V_{a1} = V_{a2} = V_{a0}$ and symmetrical component of current is given by,

$$I_{a1} = \frac{V_f}{Z_1 + \dfrac{(Z_2 \times Z_0)}{Z_2 + Z_0}} = \frac{1\angle 0°}{j0.2 + \dfrac{(j0.25 \times j0.15)}{j0.25 + j0.15}} = \frac{1\angle 0°}{j0.293} = -j3.4 \text{ pu}$$

Now, the symmetrical components of voltages from terminal 'a' to ground are given by,

$$V_{a1} = V_{a2} = V_{a0} = E_a - I_{a1} \times Z_1 = 1 - (-j3.4) \times (j0.2) = 0.32 \text{ pu}$$

$$V_{a2} = -I_{a2} \times Z_2$$

$$I_{a2} = -\frac{0.32}{j0.25} = j1.28 \text{ pu}$$

$$V_{a0} = -I_{a0} \times Z_0$$

$$I_{a0} = -\frac{0.32}{j0.15} = j2.12 \text{ pu}$$

Now, line currents are given by,

$$I_a = I_{a1} + I_{a2} + I_{a0}$$
$$I_a = -j3.4 + j1.28 + j2.12 = 0 \text{ pu}$$

$$I_b = a^2 I_{a1} + a I_{a2} + I_{a0}$$
$$= (-j3.4) \times (-0.5 - j0.866) + (j1.28) \times (-0.5 + j0.866) + j2.12$$
$$= -4.045 + j3.18 = 5.145\angle 141.8^0 \text{ pu}$$

$$I_c = a I_{a1} + a^2 I_{a2} + I_{a0}$$
$$= (-j3.4) \times (-0.5 + j0.866) + (j1.28) \times (-0.5 - j0.866) + j2.12$$
$$= 4.045 + j3.18 = 5.145\angle 384.12° \text{ pu}$$

The current flowing in ground $I_n = 3 \times I_{a0} = 3 \times j2.12 = j6.36 \text{ pu}$, which is equal to the addition of currents in two lines, i.e., $I_b + I_c = (-4.045 + j3.18) + (4.045 + j3.18) = j6.36 \text{ pu}$.

Now, the actual currents are as follows

$$I_a = 0 \text{ A}$$
$$I_b = 5.145\angle 141.8° \times 1649.57 = 84847\angle 141.8° \text{ A}$$

$$I_c = 5.145\angle 38.2° \times 1649.57 = 84847\angle 38.2° \text{ A}$$

$$\text{Current in ground} = I_n = 6.36\angle 90° \times 1649.57 = 10491\angle 90° \text{ A}$$

The post-fault lines to neutral voltages are given by,

$$V_a = V_{a1} + V_{a2} + V_{a0}$$

$$V_a = 3 \times V_{a1} = 3 \times 0.32 = 0.96 \text{ pu}$$

$$V_b = V_c = 0 \text{ pu}$$

The post fault line-to-line voltages are given by,

$$V_{ab} = V_a - V_b = (0.96 - 0) \times \frac{17.5}{\sqrt{3}} = 9.69\angle 0° \text{ kV}$$

$$V_{bc} = V_b - V_c = 0 \text{ kV}$$

$$V_{ca} = V_c - V_a = (0 - 0.96) \times \frac{17.5}{\sqrt{3}} = 9.69\angle 180° \text{ kV}$$

Example 19.15 Two synchronous motors are connected through a transformer supplied by synchronous generator as shown in Fig. 19.36. The ratings of generator, transformer, and motor are as follows. Synchronous generator: 5000 kVA, 6.6 kV with $X_1 = 0.2$ pu, $X_2 = 0.15$ pu, $X_0 = 0.05$ pu and neutral is solidly grounded; Transformer: 5000 kVA, 6.6kV/440V, star-grounded/star-grounded, with leakage reactance of 0.1 pu;

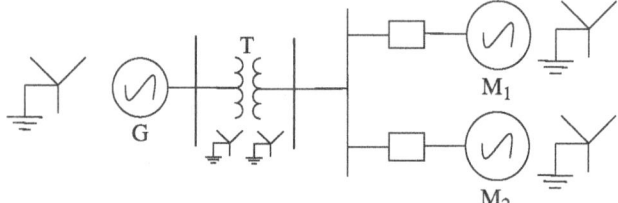

Fig. 19.36 One-line diagram for Example 19.15

Two synchronous motors: each rated as 2000 kVA, 440V, $X_1 = 0.2$ pu, $X_2 = 0.2$ pu, $X_0 = 0.05$ pu, star connection solidly grounded.

Determine the sub-transient current and draw the sequence networks showing all values of reactance in per unit when single line-to-ground fault occur on the motor bus.

Solution:

Consider 5000 kVA and 6.6 kV as base for the generator section and 0.440 kV as base for the synchronous motor section.

Now, the reactance for generator:

$$X_1 = j0.2\text{pu}, \ X_2 = j0.15\text{pu}, \ X_0 = j0.05 \text{ pu}$$

and, the reactance for transformer:

$$X_1 = X_2 = X_0 = j0.1 \text{ pu}$$

The reactance for both synchronous motors:

$$X_1 = 0.2 \times \left(\frac{0.440}{0.440}\right)^2 \times \left(\frac{5000}{2000}\right) = j0.5 \text{ pu}$$

$$X_2 = 0.2 \times \left(\frac{0.440}{0.440}\right)^2 \times \left(\frac{5000}{2000}\right) = j0.5 \text{ pu}$$

$$X_0 = 0.05 \times \left(\frac{0.440}{0.440}\right)^2 \times \left(\frac{5000}{2000}\right) = j0.125 \text{ pu}$$

The base current in motor section $= \dfrac{5000}{\sqrt{3} \times 0.440} = 6560.7 \text{ A}$

The base current in the generator section $= \dfrac{5000}{\sqrt{3} \times 6.6} = 437.38 \text{ A}$

Figure 19.37 shows the connection of all sequence networks for line-to-ground fault on motor bus. The fault point is marked as 'F'.

If pre-fault current is neglected and the sequence network shown in Fig. 19.37 is replaced by Thevenin's equivalent circuit, we get

$$Z_1 = \frac{(j0.3 \times j0.25)}{j0.3 + j0.25} = j0.137 \text{ pu}$$

$$Z_2 = \frac{(j0.3 \times j0.25)}{j0.3 + j0.25} = j0.137 \text{ pu}$$

$$Z_0 = \frac{(j0.15 \times j0.0625)}{j0.15 + j0.0625} = j0.044 \text{ pu}$$

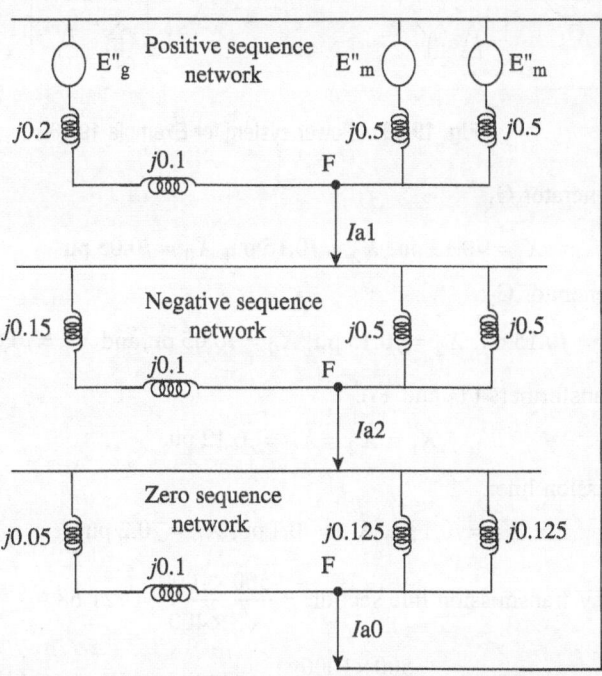

Fig. 19.37 Connection of sequence-network diagram for Example 19.15

Considering pre-fault voltage $V_f = 1\angle 0°$ pu,

$$I_{a1} = I_{a2} = I_{a0} = \frac{V_f}{Z_0 + Z_1 + Z_2} = \frac{1\angle 0°}{j0.044 + j0.137 + j0.137} = \frac{1\angle 0°}{j0.318} = -j3.145 \text{ pu}$$

Sub-transient current in fault $I_a = 3 \times I_{a0} = 3 \times (-j3.145) = -j9.435$ pu

Thus, the actual value of sub-transient fault current is $I_f = I_a \times 6560.7 = 9.435 \times 6560.7 = 61900$ A

Example 19.16 For the power system network shown in Fig. 19.38, find the sub-transient fault current, for (a) L-g (b) L-L (c) L-L-g fault occurs at bus 3. The system parameters are given as below:

G_1 and $G_2 = 500$ MVA, 25 kV, $X_1 = X_2 = 15\%$, $X_0 = 5\%$, for G_2 $X_n = 2\%$,
T_1 and $T_2 = 500$ MVA, 25 kV/400 kV, $X = 12\%$,
For line, $X_1 = X_2 = 10\%$, $X_0 = 20\%$ on the base of 500 MVA, 400 kV.

Solution:
Consider 500 MVA and 25 kV as base for both generator section and 400 kV as base for transmission line section.

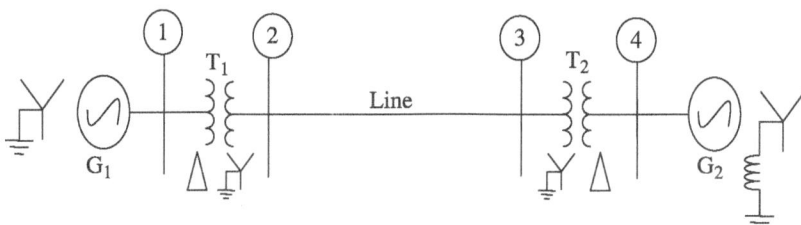

Fig. 19.38 Power system for Example 19.16

Now, the reactance for generator G_1:

$$X_1 = j0.15 \text{ pu}, X_2 = j0.15 \text{ pu}, X_0 = j0.05 \text{ pu}$$

Now, the reactance for generator G_2:

$$X_1 = j0.15 \text{ pu}, X_2 = j0.15 \text{ pu}, X_0 = j0.05 \text{ pu}, \text{and } X_n = j0.02$$

The reactance for both transformers (T_1 and T_2):

$$X_1 = X_2 = X_0 = j0.12 \text{ pu}$$

The reactance for transmission line:

$$X_1 = j0.1 \text{ pu}, X_2 = j0.1 \text{ pu}, X_0 = j0.2 \text{ pu}$$

The base current in 400 kV transmission line section $= \dfrac{500 \times 1000}{\sqrt{3} \times 400} = 721.68$ A

The base current in generator section is $= \dfrac{500 \times 1000}{\sqrt{3} \times 25} = 11547$ A

Figure 19.39 shows the formation of all sequence networks for fault on bus 3 as indicated by point 'A'. It also illustrates the Thevenin's circuit and the equivalent impedance for all three sequence networks.

The corresponding equivalent impedance is given by,

$$Z_1 = \frac{(j0.37 \times j0.27)}{j0.37 + j0.27} = j0.156 \, \text{pu}$$

$$Z_2 = \frac{(j0.37 \times j0.27)}{j0.37 + j0.27} = j0.156 \, \text{pu}$$

$$Z_0 = \frac{(j0.32 \times j0.12)}{j0.32 + j0.12} = j0.087 \, \text{pu}$$

(a) Line-to-ground fault:

Consider pre-fault voltage, $V_f = 1\angle 0° \, \text{pu}$

$$I_{a1} = I_{a2} = I_{a0} = \frac{1\angle 0°}{Z_0 + Z_1 + Z_2} = \frac{1\angle 0°}{j0.087 + j0.156 + j0.156} = \frac{1\angle 0°}{j0.4} = -j2.5 \, \text{pu}$$

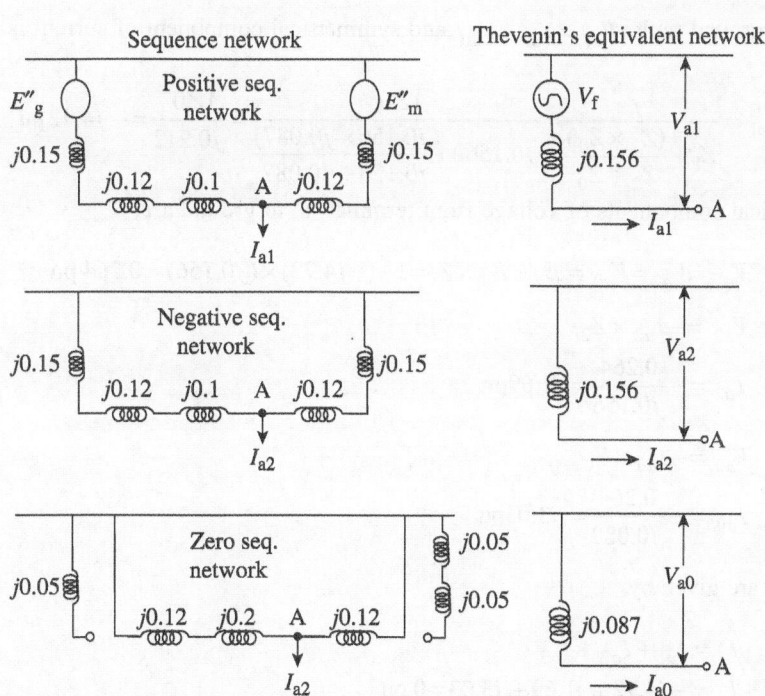

Fig. 19.39 Sequence networks and Thevenin's equivalent circuit for Example 19.16

Sub-transient current in fault $I_a = 3 \times I_{a_1} = 3 \times (-j2.5) = -j7.5 \, \text{pu}$

Thus, sub-transient fault current in ampere is $I_f = I_a \times 721.68 = 7.5 \times 721.68 = 5412.66 \, \text{A}$

(b) Line-to-line fault:

For line-to-line fault $I_{a1} = -I_{a2}$ and $I_{a1} = 0$

$$I_{a1} = \frac{V_f}{Z_1 + Z_2} = \frac{1\angle 0°}{j0.156 + j0.156} = \frac{1\angle 0°}{j0.312} = -j3.2 \text{ pu}$$

$$I_a = I_{a1} + I_{a2} + I_{a0} = 0 \text{ pu}$$

$$
\begin{aligned}
I_b &= a^2 I_{a1} + a I_{a2} + I_{a0} \\
&= -j3.2 \times (-0.5 - j0.866) + j3.2 \times (-0.5 + j0.866) + 0 \\
&= j1.6 - 2.77 - j1.6 - 2.77 \\
&= -5.44 \text{ pu}
\end{aligned}
$$

$$I_c = -I_b = 5.44 \text{ pu}$$

Thus, the actual value of sub-transient fault current is $I_f = 5.44 \times 721.68 = 3926$ A

(c) Double line-to-ground fault:

For double line-to-ground fault, $V_{a1} = V_{a2} = V_{a0}$ and symmetrical component of current is given by,

$$I_{a1} = \frac{V_f}{Z_1 + \dfrac{(Z_2 \times Z_0)}{Z_2 + Z_0}} = \frac{1\angle 0°}{j0.156 + \dfrac{(j0.156 \times j0.087)}{j0.156 + j0.087}} = \frac{1\angle 0°}{j0.212} = -j4.72 \text{ pu}$$

Now, the symmetrical components of voltage from terminal 'a' to ground are,

$$V_{a1} = V_{a2} = V_{a0} = E_a - I_{a1} \times Z_1 = 1 - (-j4.72) \times (j0.156) = 0.264 \text{ pu}$$

$$V_{a2} = -I_{a2} \times Z_2$$

$$I_{a2} = -\frac{0.264}{j0.156} = j1.69 \text{pu}$$

$$V_{a0} = -I_{a0} \times Z_0$$

$$I_{a0} = -\frac{0.264}{j0.087} = j3.03 \text{pu}$$

Now, line currents are given by,

$$I_a = I_{a1} + I_{a2} + I_{a0}$$
$$I_a = -j4.72 + j1.69 + j3.03 = 0 \text{ pu}$$

$$
\begin{aligned}
I_b &= a^2 I_{a1} + a I_{a2} + I_{a0} \\
&= (-j4.72) \times (-0.5 - j0.866) + (j1.69) \times (-0.5 + j0.866) + j3.03 \\
&= -4.687 + j4.545 = 6.53 \angle 135.88°
\end{aligned}
$$

$$I_c = aI_{a1} + a^2 I_{a2} + I_{a0}$$
$$= (-j4.72) \times (-0.5 + j0.866) + (j1.69) \times (-0.5 - j0.866) + j3.03$$
$$= 4.687 + j4.545 = 6.53\angle 44.11°$$

The current flowing in ground $I_n = 3 \times I_{a0} = 3 \times j3.03 = j9.09\,\text{pu}$, which is equal to the addition of currents in two lines, i.e., $I_b + I_c = (-4.687 + j4.545) + (4.687 + j4.545) = j9.09\,\text{pu}$.

Now, the actual values of line currents are given by,

$$I_a = 0\,\text{A}$$
$$I_b = 6.53\angle 135.88° \times 721.68 = 4712.57\angle 135.88°\,\text{A}$$
$$I_c = 6.53\angle 44.11° \times 721.68 = 4712.57\angle 44.11°\,\text{A}$$
$$\text{Current in Ground} = I_n = 9.09\angle 90° \times 721.68 = 6560\angle 90°\,\text{A}$$

Recapitulation

- Faults in power system can be classified into symmetrical faults and unsymmetrical faults.
- Per unit method is used to reduce the calculation burden in the fault analysis of small to large power system.
- One-line diagram and impedance diagram are used in symmetrical and unsymmetrical fault calculations.
- Thevenin's equivalent impedance calculated from impedance diagram is used to find out sub-transient current in fault.
- Bus admittance and impedance matrix are used to solve large system with computer programming.
- Symmetrical components play an important role in solving unsymmetrical faults.
- Any unbalanced phasors can be transformed into a set of three balance components, i.e., positive sequence, negative sequence, and zero sequence.
- Positive and negative sequence networks are almost same, but zero sequence network is exaggerated by the way natural grounding is carried out.
- Formation of sequence network with the use of sequence impedance is essential for unsymmetrical fault calculation.
- The effect of impedance in the fault can be taken into account because most of the line-to-ground fault involves impedance in the fault path due to insulation and tower footing resistance.

Multiple Choice Questions

1. In per unit method, base current is given by

   (a) $\dfrac{Base\ KVA}{Base\ KV}$

   (b) $\dfrac{(Base\ KV)^2}{\sqrt{3} \times Base\ KVA}$

   (c) $\dfrac{Base\ KVA}{\sqrt{3} \times Base\ KV}$

   (d) $\dfrac{Base\ MVA}{Base\ KV}$

2. During fault, zero-sequence impedance at fault point is given by

   (a) $Z_1 + Z_2$

   (b) $Z_0 + Z_2$

   (c) $3Z_n - Z_{g0}$

   (d) $3Z_n + Z_{g0}$

3. Zero sequence current exists in the line when
   (a) the system is delta connected.
   (b) the system is star connected and star is grounded.
   (c) the system is star connected and star is ungrounded.
   (d) the system is perfectly balanced.

4. For Δ-Y transformer, if Y-side system voltage lags by 30^0 to Δ-side, then Y-side negative sequence voltage
   (a) lags by 30^0 to Δ-side
   (b) leads by 30^0 to Δ-side
   (c) lags by 0^0 to Δ-side
   (d) leads by 0^0 to Δ-side

5. The zero-sequence current is absent in
   (a) L-g fault
   (b) L-L fault
   (c) L-L-g fault
   (d) None of the above

6. The positive sequence current alone is present in
   (a) L-g fault
   (b) L-L fault
   (c) L-L-g fault
   (d) L-L-L

7. A 5kVA 440/220V transformer has 12% reactance on primary side. The reactance referred to secondary side is
   (a) 12%
   (b) 24%
   (c) 6%
   (d) infinite

8. During line-to-ground fault, the zero sequence current is _____ that of the fault current in ground.
   (a) three times
   (b) same as
   (c) one-third times
   (d) zero times

9. The negative sequence current is not present on the occurrence of
   (a) L-g fault
   (b) L-L fault
   (c) L-L-g fault
   (d) L-L-L-g

10. The most severe fault that occurs on the terminal of unloaded solidly grounded generator is
    (a) L-g fault
    (b) L-L fault
    (c) L-L-g fault
    (d) L-L-L

11. The reference bus for zero sequence network is
    (a) terminal of generator
    (b) neutral of generator
    (c) ground
    (d) none of the above

12. Decaying DC component of fault depends on
    (a) types of fault
    (b) fault duration
    (c) fault inception time
    (d) AC component of fault

13. Phasor displacement among zero sequence voltage or current component is
    (a) 0°
    (b) 90°
    (c) 120°
    (d) 180°

14. The zero-sequence impedance of the transmission line is _____ the positive sequence impedance.
    (a) four times greater than
    (b) four times lower than
    (c) 20 times lower than
    (d) same as

15. During open circuit fault in any one phase of a 3-phase power system, the sequence components present are
    (a) all three
    (b) positive and zero
    (c) negative and zero
    (d) positive and negative

Review Questions

1. What are symmetrical and unsymmetrical faults?
2. Classify the natural faults that occur in power systems.
3. What do you mean by per unit system? State the advantages of per unit system.
4. Discuss the theory of symmetrical components.
5. How are the unbalanced phasor quantities determined into balanced symmetrical components?
6. How are the symmetrical components transformed into unbalanced phasors?

7. Explain the transient phenomenon that occurs in transmission line.
8. Explain the representation of reactance diagram for any power system network.
9. Derive an equation for three-phase fault through identical impedances.
10. What is the function of bus admittance and impedance matrix in fault calculation?
11. Draw the zero sequence network for various connections of a two-winding transformer.

12. Derive the essential equation to calculate sub-transient current while line-to-ground fault occurs in the power system.

13. Derive an equation for sequence current while line-to-line fault occurs through impedance Z.

14. Draw a figure demonstrating the interconnection of sequence

15. Compare three-phase fault, line-to-ground fault, line-to-line fault, and double line-to-ground fault with reference to their occurrence on the terminal of generator.

network for double line-to-ground fault. Also, obtain the necessary equation for sub-transient current for such a fault.

Numerical Exercises

1. A 100 MVA, 11 kV, generator with $x_d'' = 25$ % is connected to a transformer rated 125 MVA, 13.8/220 kV with leakage reactance of 10%. If the base of 150 MVA and 230 kV is used on HV side of transformer, determine the per unit value to be used for the generator and transformer.

2. The single line diagram of an unloaded power system is shown in Fig. 19.40. The rating of each component is given as below.

Generator1 (G1): 30 MVA, 18 kV, $x_d'' = 0.2$pu, Generator2 (G2): 30 MVA, 15 kV, $x_d'' = 0.15$pu, Transformer (T1) is composed of three single phase unit, each rated 10 MVA, 127/18 kV, X = 10%, Transformer (T2): 35 MVA, 230/15 kV, X = 10%, Transmission line-1 has total reactance of 50Ω and line-2 has total reactance of 70Ω. Compute the per unit reactance of all components and draw reactance diagram marking all reactances in per unit.

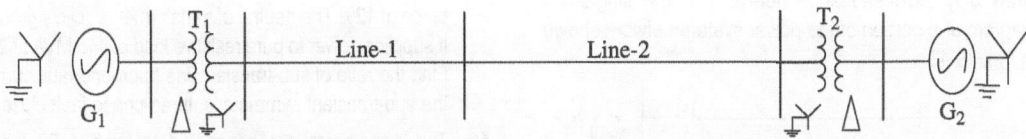

Fig. 19.40 One-line diagram of power system for Problem-2

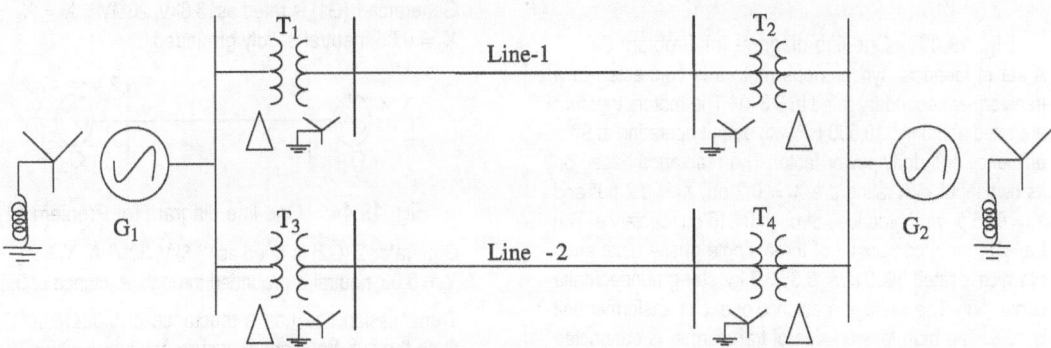

Fig. 19.41 Single line diagram of network for Problem 3

3. For the single line diagram given in Fig. 19.41, draw a reactance diagram with all reactances marked in per unit. Use a base of 132 kV and 40MVA for transmission line. The parameters of each component are as under:

Generator G_1 and G_2 = 35 MVA, 20 kV, x_d''= 20%

Transformer T_1 and T_3 = 15 MVA, 20 kV/132 kV x_d'' = 10%

Transformer T_2 and T_4 = 15 MVA 132 kV/20 kV, x_d'' = 10%

Line L1 and L2 = 30Ω

4. A 30,000 kVA, 11kV generator with $x_d'' = 20\%$ is connected to a synchronous motor through transformer. The transformer is rated at 35,000 kVA, 11 kV/6.6kV with leakage reactance of 10%. The motor is rated at 30,000 kVA, 6.6 kV with sub-transient reactance of 25%. Find the sub-transient current when a symmetrical fault occurs at the terminals of motor using (a) Thevenin's impedance method and (b) Bus impedance matrix method.

5. A 1 MVA, 5kV generator with x_d" = 25% is connected to a bus through a circuit breaker as shown in Fig. 19.42. Three synchronous motors rated at 750 kVA, 5kV having x_d" = 20% are connected through circuit breakers to the same bus. Find the symmetrical short circuit current for a 3-phase fault at point P as indicated in Fig. 19.42. Simplify the calculation by neglecting pre-fault current.

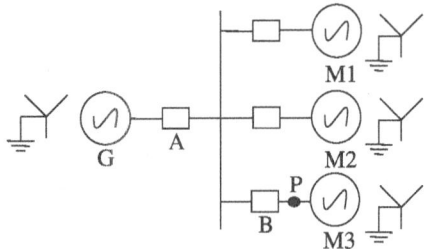

Fig.19. 42 One-line diagram for Problem 5

6. Draw only zero sequence network for the single-line diagram of a portion of the power system network shown in Fig. 19.43.

Fig. 19.43 One-line diagram for Problem 6

7. A set of identical synchronous motors is connected to a transformer secondary rated at 3.3 kV. The motors together are rated at 3.3 kV, 10,000 HP with output operating at 90% efficiency and unity power factor. The reactance based on its own input kVA rating are $X_1 = 0.2$ pu, $X_2 = 0.2$ pu, and $X_0 = 0.05$ pu and grounded through 0.015 pu reactance. The transformer is composed of three single phase units each of which is rated 3000 kVA, 6.35/3.3 kV star-grounded/delta connection. The leakage reactance of each transformer unit is 10%. The high tension side of transformer is connected to a generator rated 9000 kVA, 11kV with $X_1 = 0.2$pu, $X_2 = 0.15$pu, $X_0 = 0.05$pu, and reactance from neutral to ground is equal to 0.02pu. Considering the set of motor as a single equivalent motor, draw the reactance diagram of the sequence network and find the sub-transient line currents in all parts of the system when single line-to-ground fault occurs on the secondary of transformer.

8. A generator rated 210 MW, 20kV, 0.95 power factor has $X_1 = X_2 = 0.25$ pu and $X_0 = 0.05$ pu. Its neutral is grounded through reactance of 0.3 Ω. The generator is operating at rated voltage without load when single line-to-ground fault occurs at its terminal. Find the sub-transient current in the fault path. Also, find the sub-transient current when three-phase fault occurs on its terminal and compare the two faults.

9. In Problem 8 find the sub-transient current if double-line fault and double line-to-ground fault occurs, respectively at the same point.

10. A generator is rated as 500 MVA, 25kV has $X_1 = X_2 = 15\%$ and $X_0 = 6\%$. It is connected to delta/star-grounded transformer rated at 500 MVA, 25kV/225 kV with leakage reactance of 12%. The neutral of transformer is solidly grounded. It supplies power to pure resistive load of 445 MVA at 225kV. Find the ratio of sub-transient line to ground fault current to the sub-transient symmetrical three-phase fault current.

11. Two generators are connected through a 80 km long transmission line as shown in Fig. 19.44. The parameters of other equipment are as under.

Generator-1 (G1) is rated as13.8kV, 20MVA, $X_1 = X_2 = 0.2$, $X_0 = 0.05$, neutral solidly grounded.

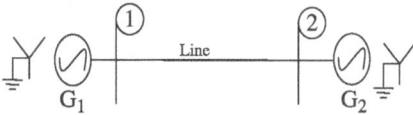

Fig. 19.44 One-line diagram for Problem 11

Generator-2 (G2) is rated as13.8kV, 30MVA, $X_1 = X_2 = 0.2$, $X_0 = 0.05$, neutral is grounded through reactance of 0.02pu.

Transmission line has a reactance of 0.02Ω/km. Determine the sub-transient fault current when a single line-to-ground fault occurs on bus 2. Consider 30 MVA, 13.8kV base for this example.

Answers to Multiple Choice Questions

1. (c) 2. (d) 3. (b) 4. (b) 5. (b) 6. (d) 7. (a) 8. (c) 9. (d) 10. (c) 11. (c) 12. (c) 13. (a) 14. (a) 15. (d)

Basic Concept and Application of Controlled Switching

20

Learning Objectives

After going through this chapter, the students will be able to:

- Explain the basic concept of controlled switching
- Understand the philosophy of controlled switching
- Explain energization targets for various power system equipment

20.1 General Background

Controlled switching is becoming more and more popular in recent days for improving the thermal and dielectric performance of various power system components due to reduction in stresses imposed on them. It is already employed for improving the performance of all major power system components such as transformer/reactor energization (magnetic inrush current reduction) and de-energization (reduction in overvoltages due to current chopping), capacitor bank energization (capacitive inrush current reduction), and de-energization (capacitive overvoltage reduction) and uncompensated and shunt compensated transmission line energization (reduction in switching overvoltages) and energization of partially discharged and fully discharged transmission lines. The aforesaid applications of controlled switching also lead to improvement in power quality of power utility. Intelligent devices facilitating controlled switching for EHV circuit breakers (CBs) are already available in the market from various manufacturers. Application of controlled switching to assist retrofitting of old circuit breakers is also becoming popular as the pressure built up in old circuit breakers is reduced due to ageing effect and will result in the reduction of committed operating duties of the circuit breaker. Application of controlled switching in such cases will reduce the stresses imposed on the circuit breaker and hence, will help in restoring the committed operating duties of the circuit breaker.

20.2 Introduction to Controlled Switching

Controlled switching is the term used to describe the application of electronic control devices to control the mechanical closing or opening of circuit breaker contacts. It has been a desirable method for stress reduction and in particular for reduction of switching overvoltages during de-energization of shunt reactors, and shunt capacitor banks, and energization of transmission lines. In addition, it assists in reducing the magnitude of inrush current during energization of transformers and shunt capacitor banks. On the other hand, with controlled switching application, the decaying DC component can be eliminated during energization of shunt reactor. Hence, it is becoming an issue of widespread interest to the utilities and manufacturers, especially for voltage levels of 220 kV above. Its benefits and feasibility were presented by CIGRE Task Force 13.00.1, with emphasis on mitigation of switching surges and related economical features due to the reduction of insulation levels of large capacitor banks, compaction of transmission lines, and reduction of arresters rating, and thereby reduction in dielectric and thermal stresses on CBs. Regarding CBs, the controlled switching may provide an increase in the lifetime of power apparatus and improvement in power quality of power utility. In addition, it can eliminate the need for pre-insertion resistors, and may limit switching overvoltages to acceptable values, especially when used in conjunction with surge arresters for transmission lines. The optimal making instant for controlled switching of an unloaded line is the instant at which the voltage across the circuit breaker contacts for each phase is zero and the predicted time span between the closing instant of the first and the last poles is as small as possible.

Apart from the technical challenges, the application of controlled switching becomes tricky due to other practical issues. The controlled switching will be performed through EHV CBs and hence, it will have consistent behaviour for a large number of operations. This controls the crucial parameters such as Rate of Rise of Dielectric Strength (RRDS) and Rate of Decay of Dielectric Strength (RDDS). These parameters are defined as follows

(i) RRDS: It is defined as the ratio of peak value of restriking voltage to the time to reach to that peak.
(ii) RDDS: In case of closing of circuit breaker, the rate at which the dielectric strength deteriorates across the breaker contacts is defined as the rate of decay of dielectric strength.

Furthermore, accurate switching target will necessitate close co-ordination between breaker characteristics and behaviour of equipment to be switched. In order to meet the controlled switching requirements of a large variety of equipment having different design philosophies and connection configurations, it should be possible to achieve electrical breaking and making at any point on the non-sinusoidal gap voltage for individual poles of the CB. This in turn, requires a certain RRDS and RDDS of the CB. Moreover, the choice of energization targets will also depend upon the voltage level, behaviour of equipment to be switched, gap voltage across individual poles of CB, and the target point on the gap voltage wave. Usually, CBs are designed to have higher operating speed, small operating time, and comparatively lower statistical scatter during opening than the closing operation. Hence, they typically offer suitability in terms of RRDS and operating time along with their statistical scatter during controlled de-energization operation. On the other hand, it has been observed that the breakers have comparatively slow closing speed and hence, large operating time and low RDDS with a wide range of their statistical scatter. Therefore, it is difficult to achieve a large range of energization targets on the gap voltage wave with smaller RDDS and large operating time deviation.

20.3 Concept of Controlled Switching

Figure 20.1 shows the basic block diagram of controlled switching philosophy. As shown in Fig. 20.1, the controller collects the system inputs from the instrument transformers and accordingly, the intentional time delay is determined to achieve the optimal target. Afterwards, the controller will raise closing/opening command to CB.

Figure 20.2 shows time sequence of controlled opening. As shown in Fig. 20.2, the opening command is initiated haphazardly with respect to the reference signal at 't_R' instant. The controller postpones the opening command by 'T_D' seconds

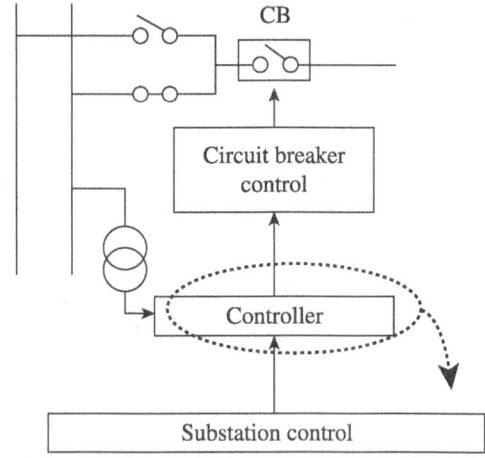

Fig. 20.1 Block diagram of controlled switching philosophy

which is the addition of deliberate lingering time ('T_C') and a definite waiting time interval ('T_W'). The controller utilizes the waiting time interval ('T_W') to judge the next current zero instant from the real field. The time 'T_C' is a function of the opening time and it determines the appropriate time of zero crossing instant ('t_{SEPT}'). 'T_P' is the time defined by the mechanical opening time duration from activation of the trip coil to the starting instant for contact separation of the CB. However, the arcing time ('T_A') is defined as the time of contact separation till current interruption occurs at natural current zero. NT_{cycles} indicates the number of half cycles used to attain the optimal target.

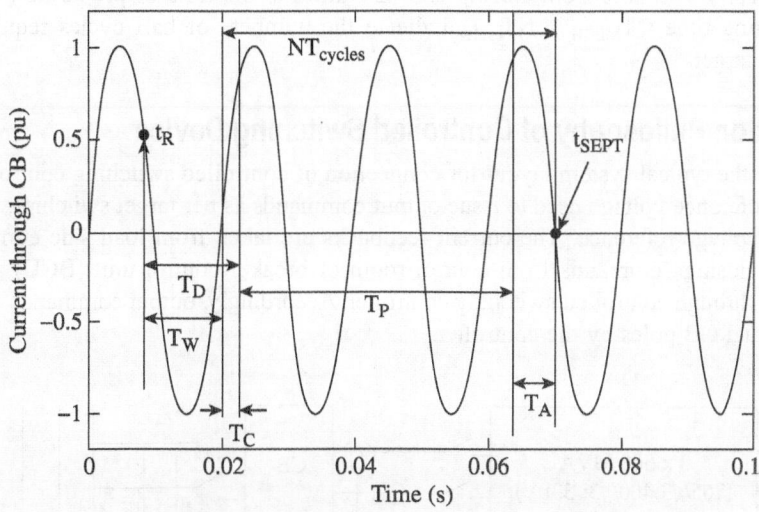

Fig. 20.2 Time sequence of controlled opening

Figure 20.3 shows the time sequence of controlled closing. This strategy controls the closing instant for each pole of the CB with respect to the phase angle of the voltage. In this regard, the controller has to monitor the voltages at the source side and compare it with the reference voltage signal. As shown in

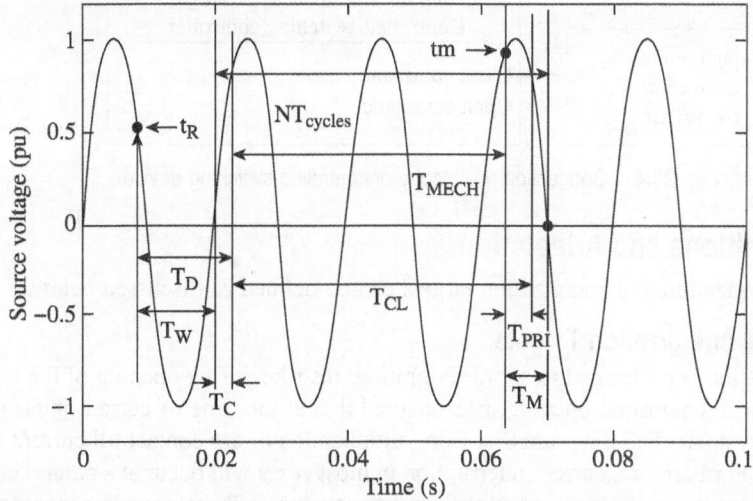

Fig. 20.3 Time sequence of controlled closing

Fig. 20.3, the closing command is raised arbitrarily at an instant 't_R'. The controller introduces delay ('T_D') at the arbitrarily acquired closing command, which is the sum of a deliberate time delay ('T_C') and a certain waiting-time interval ('T_W'). The time 'T_C' is an intentional delay introduced by controller, whereas T_W is the time taken by the controller to detect the natural voltage zero of the bus voltage.

The time for pre-striking ('T_{PRI}') is the time duration from the instant at which the flow of current started in the main circuit until contact touches physically. The 't_m' is the target-making instant (electrical closing). The time from the activation of the closing coil to the instant at which contact physically touches is defined as closing time ('T_{CL}'). The time from closing coil activation to the time of pre-strike ('T_m') is known as mechanical operating time ('T_{MECH}'). NT_{cycles} indicate the numbers of half cycles required to attain the controlled closing target.

20.4 Connection Philosophy of Controlled Switching Device

Figure 20.4 shows the typical system layout for connection of controlled switching controller with the auto transformer. The reference voltage used to issue output commands as per target switching instances is taken from busbar side voltage reference. The current feedbacks are taken from load side current transformers. The opening and closing commands from control room or breaker control unit (BCU) are executed in a controlled manner through controlled switching controller. Accordingly, output commands at target instances are sent to individual CB poles by the controller.

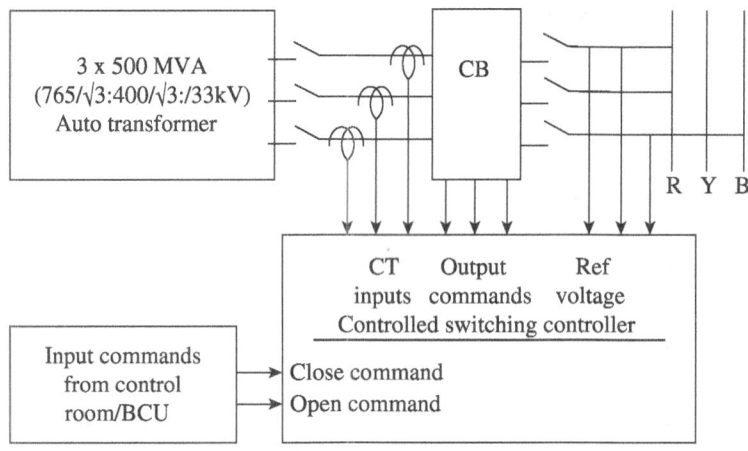

Fig. 20.4 Connection philosophy of controlled switching device

20.5 Target Definitions and Adaptation

The controlled de-energization and energization targets can be defined as discussed below.

20.5.1 Controlled De-energization Targets

Controlled de-energization is performed to avoid re-ignition/ restrike during opening of the connected load. To achieve the same with controlled opening, it is ensured that at the time of current interruption, the CB contacts are sufficiently apart. The time duration from initialization of arc contact till current interruption is termed as *arcing time*. Furthermore, current interruption in most cases will occur at a natural current zero for SF_6 circuit breakers, except for very low current interruption (example: Transformer no load de-energization) due to design of arcing chamber and arc extension mechanism of the breakers. In case breakdown occurs, the arcing time will be extended to avoid repetitive breakdowns. The parameters for controlled opening target at peak of reference wave is demonstrated in Fig. 20.5. In Fig. 20.5, T_{ARC} denotes the arcing time to achieve successful controlled de-energization. Also, the arcing time can be further extended ($T_{ARC-EXTENDED}$) in the event of breakdown detected by controlled switching device during a controlled de-energization operation known as 'adaptation during controlled de-energization'.

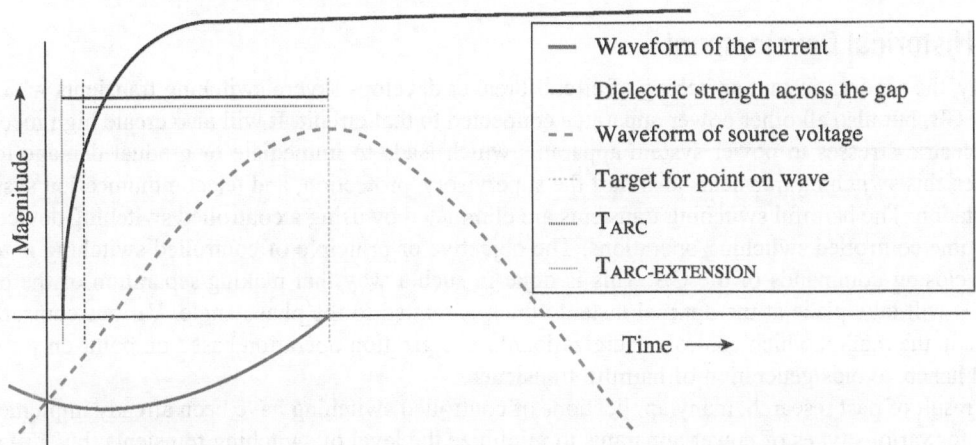

Fig. 20.5 Switching target definition during controlled opening operation

20.5.2 Controlled Energization Targets

The energization target, also known as *making instant*, for controlled switching is shown in Fig. 20.6. As shown in Fig. 20.6, the electrical closing happens when RDDS crosses over the gap voltage across the circuit breaker. Here, 'Ts$_1$' denotes the electrical target point with respect to positive going gap voltage reference and 'Ts$_2$' denotes the time from the start of electrical conduction through arc till the breaker contacts physically touch each other.

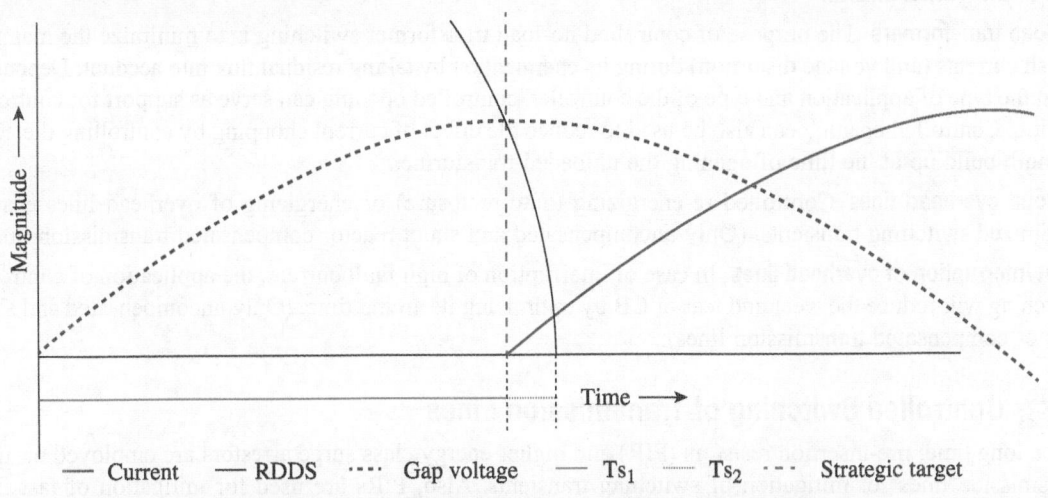

Fig. 20.6 Controlled energization on peak gap voltage target

Therefore, to achieve the desired energization target, the RDDS gap voltage, and pre-arcing time shall be considered. The RDDS depends upon the speed of the breaker during closing operation and the breakdown strength of the contact gap. Like controlled de-energization, feedback of actual closing time of circuit breaker (CB auxiliary contacts) is fed to the controlled switching device. The discrepancy in actual closing time of circuit breaker estimated with the feedback is corrected in subsequent controlled energization operations, known as 'adaptation during controlled energization'.

20.6 Historical Development

Normally, the random instant switching of circuit breaker develops severe switching transients which hurts not only CB, but also all other power apparatus connected to that circuit. It will also create high mechanical and di-electric stresses to power system apparatus which leads to immediate or gradual damage to them. Moreover, this switching transients also hurt the supervisory, protection, and telecommunication systems of the substation. The harmful switching transients are eliminated by using a controlled switching device which utilizes time controlled switching operations. The objective or principle of controlled switching is to delay opening/closing commands of the CB. This is done in such a way that making/separation of the contacts of the CB will take place at the optimal instant which is related to the phase angle. Various controllers are available in the market which controls energization/de-energization operation based on point-on wave position and hence, avoids generation of harmful transients.

As a result of past research, many applications of controlled switching have been already implemented in practice for various types of power apparatus to minimize the level of switching transients; but still there is a wide scope of work to be explored in this area. Hence, it will be challenging for the researcher to identify the optimal instant for various power system apparatus.

The application of controlled switching for various power system apparatus are as follows:

Shunt capacitor banks The aim is to control closing to minimize the energizing transients (voltage transients as well as inrush currents). To improve interrupting performance with respect to overvoltage stresses, controlled opening can also be employed.

Shunt reactors The basic aim is to control de-energizing, to ensure re-ignition free behaviour. In addition, controlled closing also serves as a useful method for minimizing decaying DC asymmetry in charging current of individual phases.

No-load transformers The purpose of controlled no-load transformer switching is to minimize the magnetic inrush currents (and voltage distortion) during its energization by taking residual flux into account. Depending upon the type of application and type of the controller, controlled opening can serve as support for controlled closing. Controlled opening can also be used to reduce the effect of current chopping by controlling dielectric strength build up at the time of opening the unloaded transformer.

No-load overhead lines Controlled re-energizing (auto-reclosure) or energizing of overhead lines ensures minimized switching transients. (Only uncompensated and shunt reactor compensated transmission lines).

Fault interruption of overhead lines In case of interruption of high fault current, the application of controlled switching will reduce the wear and tear of CB by optimizing its arcing time. (Only uncompensated and shunt reactor compensated transmission lines).

20.7 Controlled Switching of Transmission Lines

Since long time, pre-insertion resistors (PIR) and higher energy class surge arrestors are employed on EHV transmission lines for mitigation of switching transients. Also, PIRs are used for mitigation of fast front overvoltages and temporary overvoltages (TOV) generated due to high inrush currents during energization of transformers. However, PIRs can turn out to be the weakest link in the performance of circuit breakers due to involvement of additional mechanical linkage. Also, in the case of transmission lines, multiple restrikes can damage surge arresters if severe switching surges appear across the arresters.

Due to wide range of design and connection configurations for various power system equipment, the application of controlled switching is comparatively challenging. Further, in case of transmission lines, inclusion of auto-reclosure for enhancement of system stability makes the application more challenging as the effects of trapped charge, inclusion of multiple frequency components due to faults, and type of compensation needs to be carefully investigated while deciding the targets for switching of transmission lines.

20.8 Coordination of Circuit Breaker Characteristics

Apart from technical challenges, the implementation of controlled switching becomes tricky due to other practical issues. The controlled switching shall be performed through EHV circuit breakers, and accurate switching target will necessitate consistent behaviour of the CB for a large number of operations. These will demand mechanical operating time and arcing/pre-arcing time to vary in a narrow band. Moreover, permissible limits are decided by equipment electrostatic and electromagnetic behaviour and connected power system parameters. Also effects of variations in external parameters such as temperature, auxiliary supply voltage, idle time, and others on CB operating time must be taken in to account while deciding controlled switching strategy. Furthermore some type tests need to be included or modified as per latest report released by International Electrotechnical Commission (IEC). In addition, limited availability of necessary measuring instruments (also having accuracy and consistency in measurements) along with a large number of diversified substation configurations makes the implementation of controlled switching more challenging.

20.9 Diversified Targets for Various Load Configurations

The application of controlled switching to various power system equipment (capacitor and reactor banks, transformers and transmission lines) and the electrical and magnetic coupling along with the type of connection configuration (isolated neutral, solidly grounded and grounded with impedance) demands a wide range of controlled de-energization and energization targets. The targets may vary from circuit breaker gap voltage zero to peak. In certain applications, the polarity of gap voltage also needs to be considered while deciding the target points. Therefore, usage of same circuit breaker having specific electrical and mechanical characteristics for controlled switching of different power system equipment with diversified design and connection configurations will be quite challenging.

20.10 Controlled Fault Interruption

Controlled fault interruption is used to optimize the arcing time of self-blast circuit breakers as it plays a key role for dielectric built up in EHV circuit breakers. This is achieved by separating CB contacts at a predefined instant prior to prospective current zero on which arc is supposed to be quenched. This will result into adequate dielectric built up across CB contacts/inside arcing chamber, which ensures restrike free/re-ignition-free operation of CB. Insufficient arcing time may result into poor dielectric build up inside the interrupting chamber of CB. This can instigate re-strike across contacts of CB, and hence, arc will be continued till the next prospective current zero. On other hand, larger arcing time will cause fault current to flow through arcing contacts for comparatively higher duration. Both higher and lower arcing times may result in increase in contact erosion and wear and tear of CB components and hence, will lead to reduction in life cycle duration of CB.

20.11 Controlled Switching of Transformers

Power transformers are the most expensive and vital components in electrical power system networks. They are used in a variety of configurations and can be switched ON occasionally (yearly basis) or frequently (daily basis). Persistently, the energization at levels of 765 kV and even at 400 kV results into heavy inrush currents. This would impose high thermal stresses on power transformers as well as on circuit breakers used to switch ON the same. The inrush currents contain high amount of decaying DC component and harmonics; among which the second, third, and fourth harmonics components are the most predominant ones. The magnitude of inrush current can go up to as high as short circuit level and its decay may take hundreds of cycles due to high X/R ratio of the system at EHV and UHV levels. Specifically, switching at unfavourable instances, transformer energization can create inter-winding forces and heavy mechanical vibrations due to high level of asymmetric inrush currents which may last for seconds till inrush current almost goes to steady state value. Furthermore, if their energization occurs near gap voltage peak, steep front overvoltages can be generated in case of highly capacitive elements (long HV cables, STATCOM, etc.) present in the vicinity of the transformer. This may impose severe electrical stresses on the transformer inter-winding insulation which may slowly result into inter-winding short-circuit faults. Furthermore, the high amount of harmonic content and decaying DC component in the magnetic inrush currents can cause false operation of fuses and protective relays. It can also deteriorate power quality of the interconnected grid system. Nevertheless, mitigation of inrush currents for transformers has been achieved using pre-insertion resistors (PIR) in the past, similar to their application to achieve reduction of overvoltages during no-load energization for long lines. However, controlled switching is being promoted due to added advantage of higher reliability compared to PIR and limitation on number of consecutive operations due to heat dissipation issues of resistor discs for PIR. In this context, controlled switching can be used to effectively mitigate magnetic inrush of power transformers, if the effect of residual flux and interphase coupling between phases due to design and connection configuration of transformers is considered properly. Furthermore, inrush current mitigation can be effectively achieved considering residual flux effect without availability of transformer side voltage measurement which is usually employed to evaluate residual fluxes post transformer de-energization.

20.12 Controlled Switching Targets for Various Power System Equipment

As discussed in the previous section, for controlled de-energization, the targets are decided so as to ensure successful arc interruption at prospective current zero without breakdown due to possible overvoltages to be imposed by the system. Moreover, controlled closing is employed for inrush current mitigation for transformers and capacitor banks and for removal of exponentially decaying component of current for reactors during closing operation. Furthermore, the energization targets are decided based on the type of equipment and its design philosophy and connection configuration. Also, the switching transients can be mitigated by employing controlled energization on transmission lines. Table 20.1 shows commonly adapted controlled energization targets for various power system equipment having different design and connection configurations.

Table 20.1 Suggested controlled energization targets for various equipment

Load	Connection configuration	Design philosophy	Targets (Closing sequence 1-2-3)		
			Pole 1	Pole 2	Pole 3
Capacitor bank	Grounded star	-	Positive going Ph-g voltage zero of individual phase		
	Ungrounded star or Δ connected	-	Ph1-Ph2 zero crossing of voltage		1.5 Cycle after poles 1 & 2
Line/Bus reactor	Star connected with neutral grounding reactor	3/4/5 Limb core or 1-Φ bank	Positive going Ph-g voltage peak	Φ_c after Pole 1	Positive going Ph-g voltage peak

(Contd.)

Load	Connection configuration	Design philosophy	Targets (Closing sequence 1-2-3)		
			Pole 1	Pole 2	Pole 3
Line/Bus reactor or Power transformer**	Grounded star	4/5 Limb core or 1-Φ bank without any delta winding	Positive going Ph-g voltage peak of individual phase		
		3-limb with any design	Positive going Ph-g voltage peak	Quarter cycle after pole 1 target*	
		4/5 Limb core or 1-Φ bank with one delta connected winding			
	Ungrounded star or Δ connected	3/4/5 Limb core or 1-Φ bank	Peak of Ph1-Ph2 voltage		Quarter cycle after poles 1 & 2 targets
Power transformer⁺	Any connection configuration	Any design philosophy	When source side flux equal to dynamic load side flux considering residual flux for pole to be closed		
Transmission line	Uncompensated/ compensated lines***	No line side voltage measurement	Gap voltage zero of individual breaker pole		
	Uncompensated line/ compensated line	With line side voltage measurement			

For Power transformers, side of energization is column 2 of Table 20.1.
*Closing sequence for these cases are considered as 1-3-2.
**Targets are shown neglecting residual fluxes may be little modified to consider residual flux effects+.
***Targets are suggested for uncompensated line without auto re-closure or for compensated lines, having surge arresters at both ends of line in combination with controlled switching.

It can be observed from Table 20.1 that target for controlled energization of major equipment configurations are either gap voltage zero or gap voltage peak. Moreover, for certain loads for which it is possible to provide load side voltage measurement, (transformers where the residual flux is measured or for transmission lines where line side voltage is measured), the optimum energization targets may be at any point on the gap voltage wave.

Recapitulation

- Due to wide range of design and connection configurations for various power system equipment, the application of controlled switching is comparatively challenging. Apart from technical challenges, the implementation of controlled switching becomes tricky due to other practical issues.
- The controlled switching shall be performed through EHV circuit breakers, and accurate switching target will necessitate consistent behaviour of the circuit breaker for a large number of operations.
- The application of controlled switching to various power system equipment (capacitor and reactor banks, transformers and transmission lines) and the electrical and magnetic coupling along with the type of connection configuration (isolated neutral, solidly grounded and grounded with impedance) demands a wide range of controlled de-energization and energization targets.
- The targets may vary from circuit breaker gap voltage zero to peak. In certain applications, the polarity of gap voltage also needs to be considered while deciding the target points.

Multiple Choice Questions

1. In practice, at which voltage level, the application of controlled switching strategy is advisable?
   (a) LV level
   (b) HV level
   (c) EHV level
   (d) All of these

2. The conventional controlled switching strategy is applicable for
   (a) energization of only power system apparatus
   (b) both energization and de-energization of power system apparatus
   (c) energization of power transformer only
   (d) none of these

3. Based on conventional controlled switching strategy, the making targets in case of energization of three-phase uncompensated transmission line are
   (a) at peak of gap voltages across the contacts of circuit breaker
   (b) at gap voltage zero across the contacts of circuit breaker
   (c) at peak of load side voltages for each phase
   (d) at zero load side voltages for each phase

4. RDDS stands for
   (a) Rate of decay in electric strength
   (b) Rate of decay in diversified strength
   (c) Rate of decay in dielectric strength
   (d) Rate of decay in thermal strength

5. Controlled switching is precisely analogous to
   (a) point on wave switching
   (b) inductive switching
   (c) pre insertion resistance switching
   (d) capacitive switching

6. The prime limitation of controlled switching is
   (a) mechanical& time scattering
   (b) rate of rise of recovery voltage
   (c) speed of circuit breaker
   (d) type of circuit breaker

7. Controlled switching is not used for
   (a) mitigation of switching surges
   (b) mitigation of inrush current
   (c) controlled fault interruption
   (d) identification of type of fault for the transmission line

8. Pre-arcing time is defined as the time laps between
   (a) electrical making and thermal making of breaker contacts
   (b) mechanical making and thermal making of breaker contacts
   (c) electrical making and mechanical making of breaker contacts
   (d) arcing contacts and main contacts of breaker

9. All the controlled closing targets are
   (a) electrical closing targets
   (b) mechanical closing targets
   (c) arcing targets
   (d) die-electric targets

10. The random closing command is converted to controlled closing command by introducing delay of
    (a) T_w (b) T_d (c) T_{MECH} (d) none of these

Review Questions

1. Explain the concept of controlled switching.

2. Discuss the basic philosophy of controlled switching.

3. Explain the phenomenon of controlled closing.

4. Discuss the role of a circuit breaker for successful implementation of controlled switching strategy.

5. What are the different barriers for implementation of controlled switching?

6. Explain de-energization and energization targets for controlled switching.

7. Discuss the energization target used for controlled switching of power transformer.

8. Which energization target is utilized for controlled switching of transmission line?

9. List out various challenges to implement the controlled switching strategy.

10. Explain the advantages of controlled switching strategy over pre insertion resistor.

Answers to Multiple Choice Questions

1. (c)   2. (b)   3. (b)   4. (c)   5. (a)   6. (a)   7. (d)   8. (c)   9. (a)   10. (b)

Codes of Protective Devices Used in Control Circuits

All the devices used in switching apparatus are denoted by specific numbers. They are based on IEEE Standard C37.2 and are widely used in connection diagrams and control wiring. The most commonly used device numbers along with their description are as follows.

Code number	Description of device
1	Master element
2	Starting or closing time delay element
3	Interlock relay
13	Synchronous speed switch
15	Speed or frequency matching device
20	Electronically operated valve
21	Distance relay
23	Control device (say, temperature)
25	Synchronizing check device
26	Thermal device
27	Undervoltage relay
29	Isolating contactor
30	Annunciator relay
32	Directional power relay
36	Polarizing voltage device
37	Undercurrent relay
40	Field relay
41	Field circuit breaker
42	Running circuit breaker
43	Manual transfer switch
46	Phase balance or reverse phase relay
47	Phase sequence voltage relay
48	Incomplete sequence relay
49	Machine thermal relay
50	Instantaneous or rate of rise relay

Code number	Description of device
51	AC time-delay overcurrent relay
52	AC circuit breaker
53	Exciter or DC generator relay
55	Power factor relay
59	Overvoltage relay
60	Voltage or current balance relay
61	Phase unbalance relay
62	Time delay stopping or opening relay
63	Pressure switch
64	Ground protection relay
65	Governor
67	AC directional overcurrent relay
69	Permissive control device
70	Rheostat
71	Liquid or gas level relay
72	DC control breaker
74	Alarm relay
76	DC overcurrent relay
78	Phase angle measuring or out-of-step relay
81	Frequency relay
85	Carrier or pilot wire receiver relay
86	Lookout relay
87	Differential relay
90	Regulating device
91	Voltage differential relay
94	Trip-free relay
98	Pole slipping relay
99	Overfluxing relay
101	Control switch

Manuals/Data Sheets of Various Types of Relays

B.1 Different Characteristics of Overcurrent Relays

Although the characteristics of overcurrent relays tend towards infinity when the current approaches threshold value, the minimum guaranteed value of the operating current for all relays with the inverse time characteristic is 1.1 times the threshold value with a tolerance of ±0.05 of threshold value. The following are the standard characteristics of digital relays used by the utilities in the field:

1. Standard inverse characteristic
2. Very inverse characteristic
3. Extremely inverse characteristic

Sections B.1.1–B.1.3 show these characteristics, Section B.1.4 provides a comparison of these characteristics.

B.1.1 Standard Inverse Characteristic

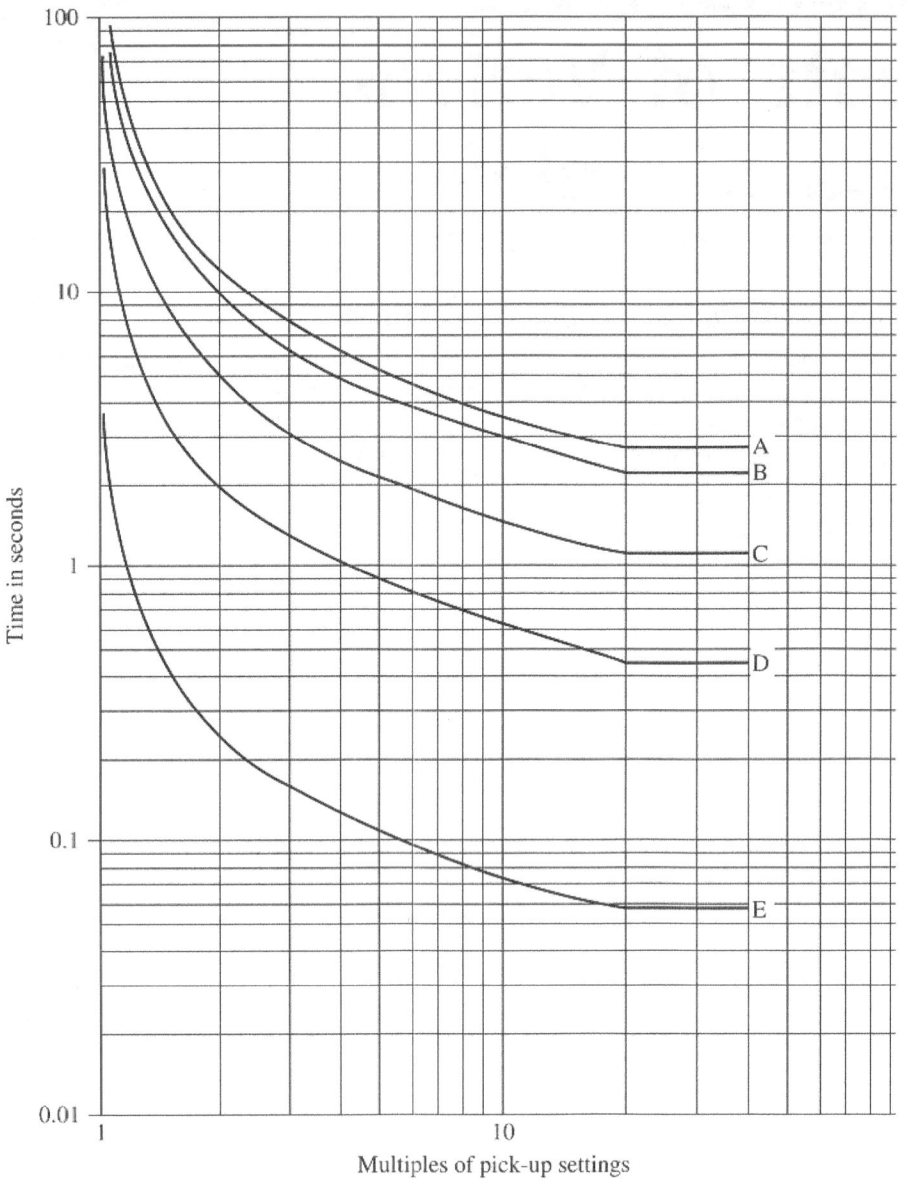

Fig. B.1 Standard inverse characteristic of overcurrent relay

Note:
A: TDS = 1.25
B: TDS = 1.0
C: TDS = 0.5
D: TDS = 0.2
E: TDS = 0.025

B.1.2 Very Inverse Characteristic

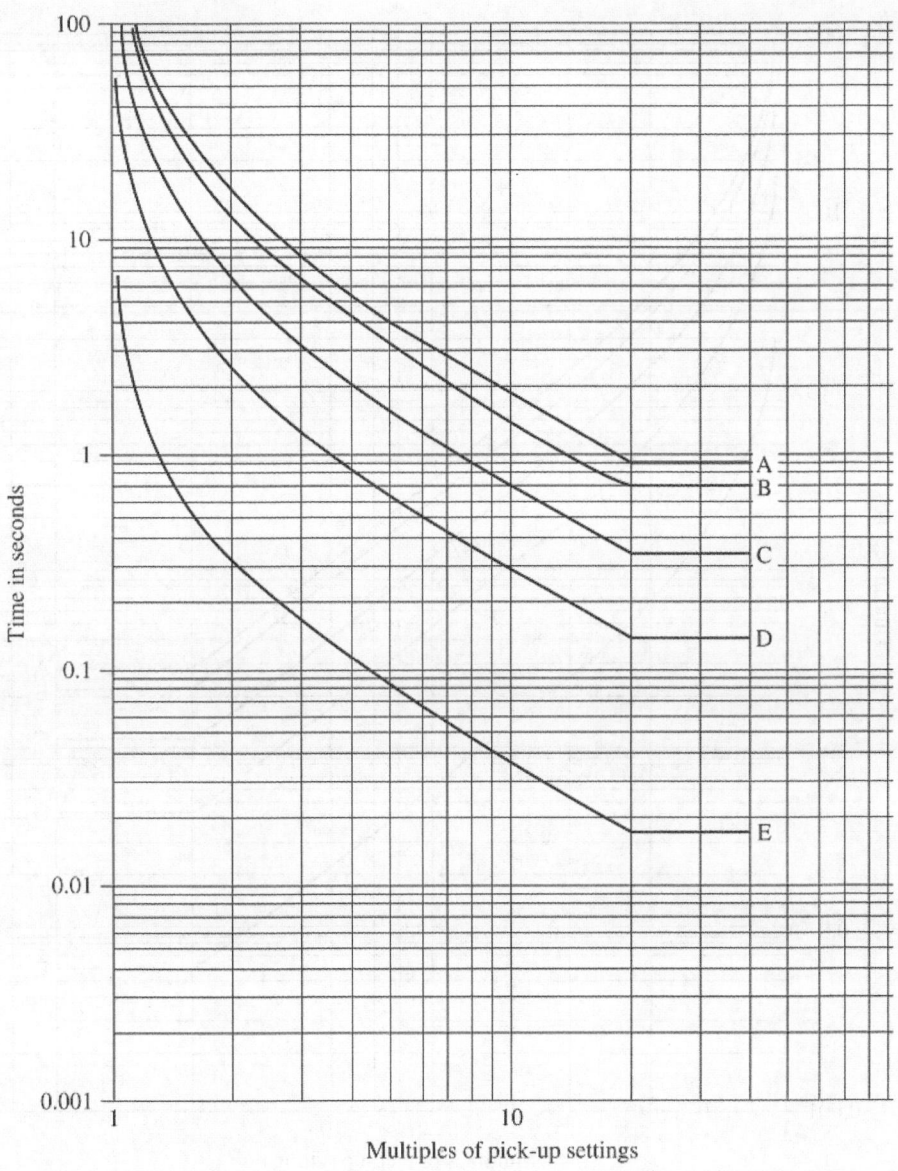

Fig. B.2 Very inverse characteristic of overcurrent relay

Note:
A: TDS = 1.25
B: TDS = 1.0
C: TDS = 0.5
D: TDS = 0.2
E: TDS = 0.025

B.1.3 Extremely Inverse Characteristic

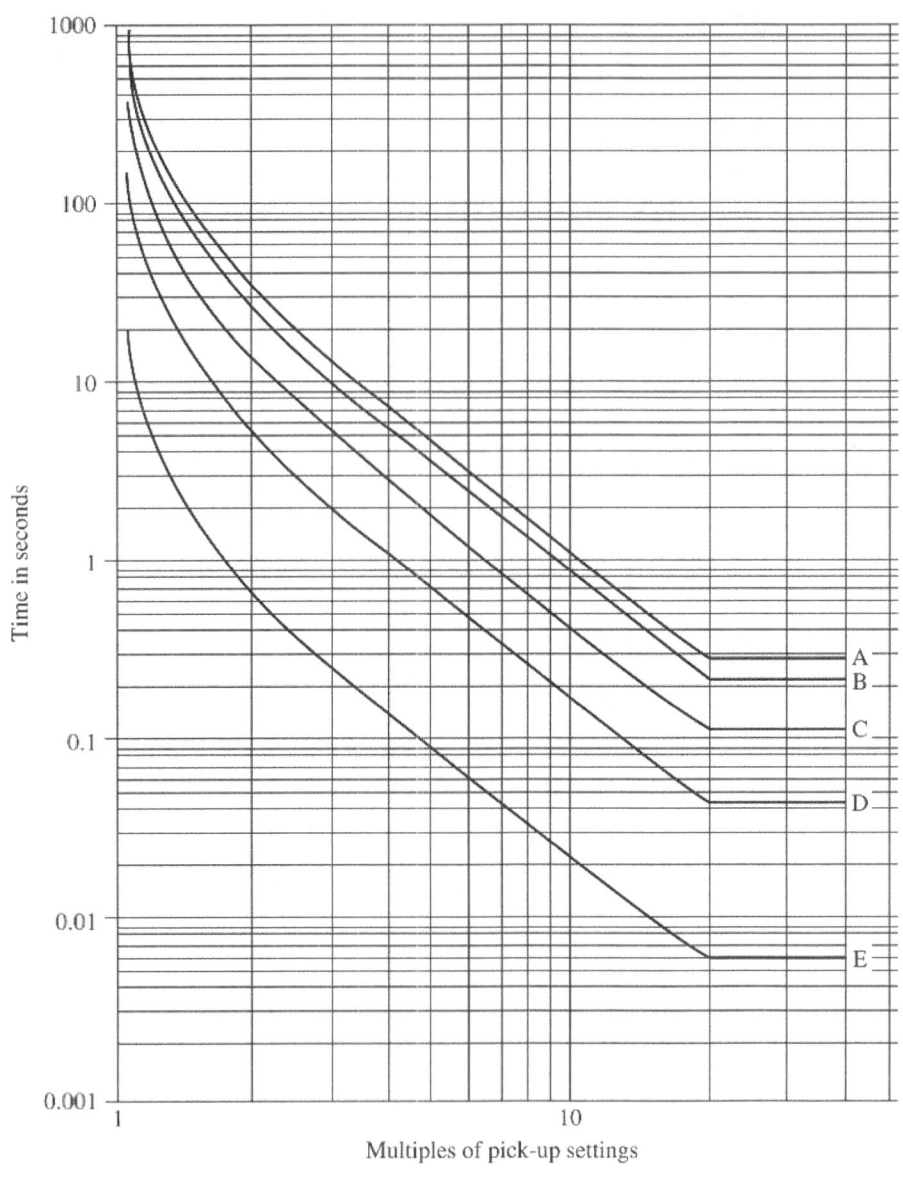

Fig. B.3 Extremely inverse characteristic of overcurrent relay

Note:
A: TDS = 1.25
B: TDS = 1.0
C: TDS = 0.5
D: TDS = 0.2
E: TDS = 0.025

B.1.4 Comparison of Various Characteristics

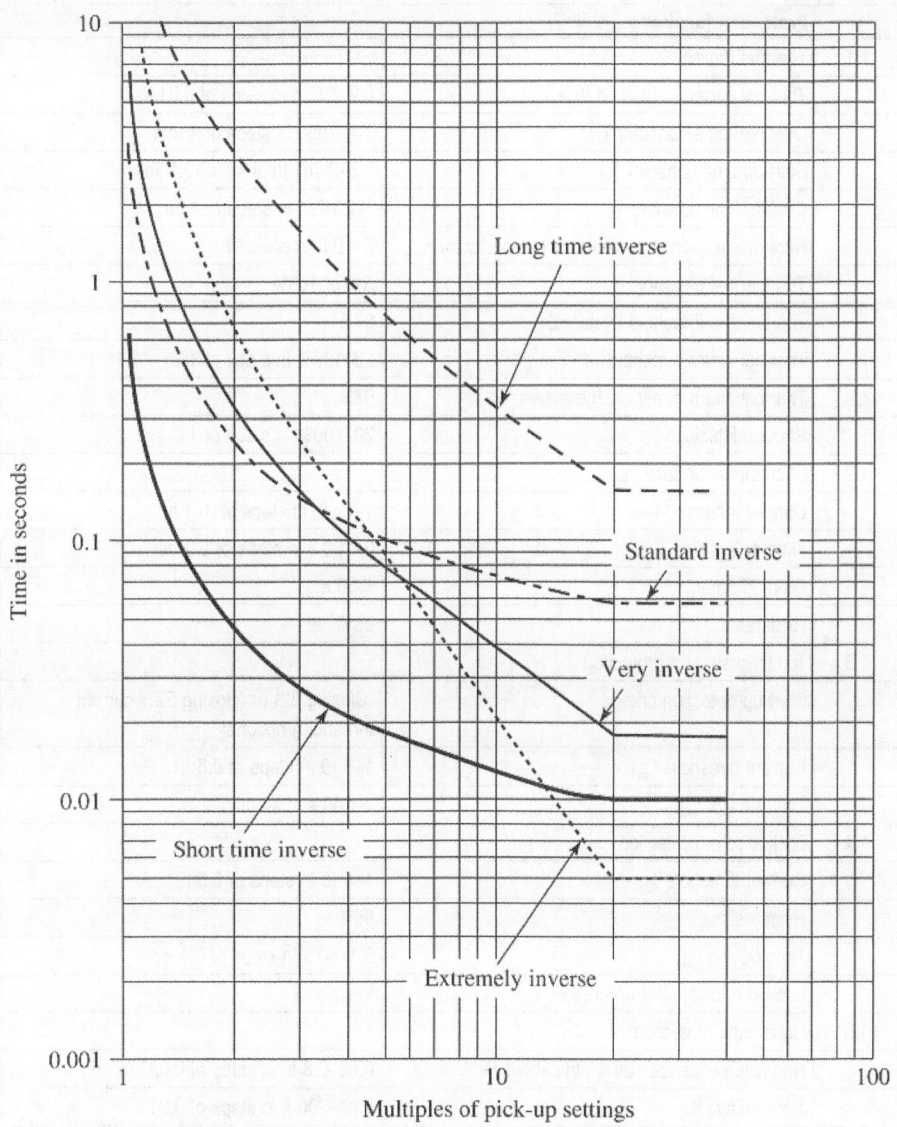

Fig. B.4   Comparison of various overcurrent characteristics

B.2 Technical Guide of Digital Induction Motor Protection Relay

1.	Protection functions	
1.1	Thermal replica	
	Thermal current threshold Iθ >	0.2–1.5 in in steps of 0.01 in
	Overload time-constant T_{e1}	1–64 min in steps of 1 min
	Start-up time constant T_{e2}	0.5–2 Te1 in steps of 0.1 Te1
	Cooling time constant T_r	1–20 Te1 in steps of 1 Te1
	Negative sequence current recognition factor K_e	0–10 in steps of 1
	Trip thermal threshold	Set of 100%
	Trip thermal threshold hysteresis	97%
	Thermal alarm threshold	20–100% in steps of 1%
	Thermal alarm threshold hysteresis	97%
	Start-up inhibition	20–100% in steps of 1%
1.2	Short-circuit protection	
	Current threshold I>>	1–12 in in steps of 0.1 in
	Time-delays tI>>	0–100 s in steps of 0.01 s
	Drop-off time	< 30 ms
	Hysteresis	95%
1.3	Too long start-up protection	
	Start-up detection criteria	(Closing 52) or (closing 52 + current threshold) optional
	Current threshold I_{start}	1–5 Iθ in steps of 0.5 Iθ
	Time delays t_{start}	1–200 s in steps of 1 s
1.4	Locked rotor protection	
	Current threshold I_{stall}	1–5 Iθ in steps of 0.5 Iθ
	Hysteresis	95%
	Time delays I_{stall}	0.1–60 s in steps of 0.1 s
	Locked rotor at start up detection	Yes/No
1.5	Unbalance protection	
	Negative sequence current threshold Ii>	0.05–0.8 in in steps of 0.025 in
	Time delays tIi>	0.04–200 s in steps of 0.01 s
	Negative sequence current threshold Ii>>	0.2–0.8 in in steps of 0.05 In
	IDMT Time delay	Operating time $t = 1.2/(I_2/I_1)$
	Hysteresis	
1.6	Current threshold Io>, Io>>	0.002–1 ion in steps of 0.001 ion
	Time delays tIo>, tIo>>	0–100 s in steps of 0.01 s
	Operating time	< 30ms
	Drop off time	< 30ms
	Hysteresis	95%

1.7	Under current protection	
	Current threshold I<	0.01–1 In in steps of 0.01 In
	Time delays tI<	0.2–100 s in steps of 0.1 s
	Inhibition time at start-up T_{inhib}	0.05–300 s in steps of 0.1 s
	Hysteresis	105%
2.	**Automation functions**	
2.1	Limitation of the number of start ups	
	Reference period $T_{reference}$	10–120 min in steps of 5 min
	Number of cold starts	0–5 in steps of 1
	Number of hot starts	0–5 in steps of 1
	Restart inhibition time $T_{interdiction}$	1–120 min in steps of 5 min
2.2	Time between two start-ups	
	Inhibition time $T_{between\ two\ starts}$	1–120 min in steps of 5 min
2.3	Reacceleration authorization	
	Voltage collapse duration $T_{reacceleration}$	0.2–10 s in steps of 0.05 s
2.4	Auxiliary timers	
	Logic inputs with alarm message on occurrence	2 external signals EXT1 and EXT2
	Logic inputs without alarm message on occurrence	2 external signals EXT3 and EXT4 (from V3.A software version)
	Timers t_{EXT1}, t_{EXT2}, t_{EXT3}, and t_{EXT4}	0–200s in steps of 0.01 s
2.5	Programmable scheme logics	
	AND logic scheme	
	Pick-up time delay	0–60 min in steps of 0.1 s
	Reset time	0–60 min in steps of 0.1 s
2.6	Latching of the output relays	
	Latching of the output relays on	Short-circuit, earth fault, unbalance, AND logical gates
2.7	Control and monitoring of the breaker device	
	Close command hold	0.1–5 s in steps of 0.1 s
	Open command hold	0.1–5 s in steps of 0.1 s
	Number of operations alarm	0–50,000 operations in steps of 1
	Summated contact breaking duty	10^6–4000.10^6 in steps of 10^6
	Adjustment of the exponent (n)	1 or 2
	Opening time alarm	0.05–1 s in steps of 0.05 s
3.	**Optional functions**	
3.1	Optional analogue output	
	Rating	0.20 mA, 4–20 mA
	Insulation	2 kV

	Maximum load with active source mode	500 Ω for ratings 0–20 mA and 4–20 mA
	Maximum voltage with passive source mode	24 V
	Accuracy	±1% full scale
3.2	Optional 6 RTD inputs	
	RTD type	PT100, Ni120, Ni100, Cu10
	Connection type	3 wires + 1 shielding
	Insulation	2 kV, active supply
	Setting of threshold	0–200°C in steps of 1°C
	Settings of timings	0–100 s in steps of 0.1 s
	Influence of thermal image	Yes/No
3.3	Optional two thermistor inputs	
	Thermistor type	PTC or NTC
	Setting of thresholds	100–30,000 Ω in steps of 100 Ω
	Time delay	Set to 2 s
4.	**Recording functions**	
4.1	Event recorder	
	Capacity	75 events
	Time-tag	To 1 ms
	Triggers	Any protection alarm and threshold Any logic input change of state Self-test events Any setting changes
4.2	Fault recorder	
	Capacity	5 records
	Time-tag	To 1 ms
	Triggers	Any trip order (relay RL 1 operation)
	Data	Fault date Active setting group Faulty phase(s) Fault type, protection threshold Magnitude of the fault current Phase and earth currents magnitudes
4.3	Oscillography	
	Capacity	5 records of 3 s each
	Sampling rate	32 samples per frequency cycle
	Pre-time setting	0.1–3 s in steps of 0.1 s
	Post-time setting	0.1–3 s in steps of 0.1 s
	Triggers	Any protection threshold overreach or any trip order (relay RL1 operation) Logic input Remote command

			4 analog channels (3 phase currents + earth current) Logic input and output states Frequency value
5.	**Communication**		
5.1	Modbus communication		
	Mode		RTU (standard)
	Transmission mode		Synchronous
	Interface		RS485
	Data rate		300 to 38,400 baud (programmable)
	Relay address		1–255
	Connection		Multi-point (32 connections)
	Cable type		Half duplex (screened twisted wire pair)
	Maximum cable length		1000 meters
	Connector		Connector screws or snap-on
	Insulation		2 kV RMS
5.2	Front communication		
	Interface		RS232
	Protocol		Modbus RTU
	Connectors		Sub-D9 pin female connector
	Cable type		Crossed
6.	**Inputs and outputs**		
6.1	Inputs		
	Phase current in		1 and 5 A
	Earth current ion		1 and 5 A
	Frequency	Range Nominal	45–65 Hz 50/60 Hz
	Burdens	Phase currents inputs Earth current inputs	< 0.3 VA (5A) < 0.025 VA (1A) < 0.01 VA at 0.1 ion (5A) < 0.04 VA at 0.1 ion (1A)
	Thermal withstand of the phase and earth current inputs		100 in for 1 s 40 in for 2 s 4 in for continuous
6.2	Logic inputs		
	Type		Independent optical isolated
	Number		5 (3 programmable, 2 fixed)
	Burden		< 140 mA for each input
	Recognition time		< 5 ms

The logic inputs shall be powered with a dc voltage.				
	Logic input operation			
Cortec code	**Relay auxiliary voltage range (V_{dc})**	**Auxiliary voltage Range for the logic inputs* (V_{dc})**	**Minimum voltage level (V)**	**Minimum current level (mA)**
A	24–60	19–60	15	3.35
F	48–150	32–150	25	3.35
M	130–250, 110–250	48–250	38	2.20

*The tolerance on the auxiliary voltage variations for the logic inputs is −20%/+20%

6.3	Output relays	
	Type	Dry contact AgCdO
	Number	6 (5 programmable + 1 watchdog)
	Communication capacity Make Carry continuously Break	30 A for 3 s 5 A 135 Vdc, 0.3 A (L/R = 30 ms) 250 Vdc, 50 W resistive 220 Vdc, 5 A (50/60Hz-cosφ = 0.6)
	Operation time	< 7 ms
	Durability	> 100,000 operations
6.4	Auxiliary voltage	
	Auxiliary voltage 3 ranges	24–60 Vdc 48–150 Vdc 130–250 Vdc/110–250 Vdc
	Variations	−20%/+20%
	Residual peak to peak triple	12%
	Power off withstand	50 ms
	Burden	< 3 W standby +0.25 for each output relay energized in Vdc < 6 VA in Vac
7.	**Accuracy**	
	Protection threshold	± 2%
	Time delays	± 2% with a minimum of 10 ms
	Measurements	Typical ± 2% at in for currents ± 2 °C for temperatures
	Pass band for measurements of true RMS values	500 Hz
8.	**CT data**	
	Phase CT primary	1–3000 in steps of 1
	Earth CT primary	1–3000 in steps of 1
	Phase CT secondary	1 or 5
	Earth CT secondary	1 or 5
	Recommended phase CT	5P10 (class of CT), 5VA (CT burden)
	Recommended earth CT	Residual connection or core balanced CT (preferred in isolated neutral systems)

9.	High voltage withstand capability		
	Dielectric withstand (50/60 Hz)	IEC 60255-5 BS 142 ANSI C37.90	2 kV in common mode 1 kV in differential mode
	Impulse voltage (1.2/50 µs)	IEC 60255-5 BS 142	5 kV in common mode 1 kV in differential mode
	Insulation resistance	IEC 60255-5	> 100 MΩ
10.	Electrical environment		
	High frequency disturbance	IEC 61000-4-1	2 kV in common mode, class 3 1 kV in differential mode, class 3
	Fast transient disturbance	IEC 61000-4-4 ANSI C37.90.1	4 kV auxiliary supply, class 4 2 kV other, class 4
	Electrostatic discharge	IEC 61000-4-2	8 kV, class 4
	Radio frequency impulse	ANSI C37.90.2 EC 61000-4-3	35 V/m 10 V/m
11.	Environment		
	Temperature	IEC 60255-6 Storing and transportation operation	−40°C to +70°C −25°C to +55°C
	Humidity	IEC 60068-2-3	56 days at 93% RH and 40°C
	Enclosure protection	IEC 60529	IP 52, IK 07
	Vibration	IEC 60255-21-1	Response and endurance, class 4
	Shock and bump	IEC 60255-21-11	Response and withstand, class 1
	Seismic withstand	IEC 60255-21-3	Class 2

System/Line Parameters— Overcurrent Relay Coordination

The system and line parameters used in overcurrent relay coordination problem are as follows:

System data	Value
System voltage	11 kV RMS (L–L)
System frequency	50 Hz
Utility impedance	$Z_1 = 0.104 + j1.195$ (Ω), $Z_0 = 0.104 + j\,1.195$ (Ω)
Impedance of each line section	$0.585 + j2.9217$ (Ω)
Length of each section	2 km
Load connected at each bus ($P + jQ$)	0.75 MW
Tripping characteristic and constants of relays	$$T_{op} = TDS \times \left(\frac{A}{(M^p - 1)} + B \right) + K$$ where: TDS is the time dial setting and usually in the range of 0–1 s; $A = 0.05$, $B = 0.1$, $K = 0.1$, and $p = 0.02$ are constants whose values depend on the characteristics of relays; M is the plug setting multiplier (PSM) and given by, $$PSM = \frac{\text{Fault current referred to CT secondary}}{PS}$$ where PS is the plug setting of the relay.

System/Line Parameters— Simulation of Transmission Line

The system and line parameters used in the simulation of transmission line system protected by distance relaying scheme are as follows:

System data	Value
Voltage rating	345 kV
System frequency	60 Hz
Equivalent voltage per unit	$E_S = 1 \angle 15°$, $E_R = 1 \angle 0°$
Equivalent source impedance	$Z_{S1} = 0.238 + j5.72\ (\Omega)$, $Z_{S0} = 2.738 + j10\ (\Omega)$ $Z_{R1} = 0.238 + j6.19\ (\Omega)$, $Z_{R0} = 0.833 + j5.12\ (\Omega)$
Length of the line	100 km
Line parameters	$R_0 = 0.275\ (\Omega)$, $L_0 = 1.345\ (mH)$, $C_0 = 6.711\ (nF)$ $R_1 = 0.0275\ (\Omega)$, $L_1 = 3.725\ (mH)$, $C_1 = 9.4831\ (nF)$

Bibliography

Books

Anderson, P.M., *Power System Protection*, IEEE Press, New York, 1999.

Blackburn, J.L., *Applied Protective Relaying*, Westinghouse Electric Corporation, New York, 1982.

Burrus, C.S., R.A. Gopinath, and H. Guo, *Introduction to Wavelets and Wavelet Transform: A Primer*, New Jersey, Prentice Hall, 1998.

Chakrabarti, A., M.L. Soni, P.V. Gupta, et al., *A Text Book on Power System Engineering*, Dhanpat Rai & Co. Pvt. Ltd, Delhi, 2010.

Chui, C.K., *An Introduction to Wavelets*, Academic Press Inc., San Diego, 1992.

Elmore, W.A. *Protective Relaying*, New York, Marcel Dekker Inc., 1994.

Garzon, R.D., *High Voltage Circuit Breakers Design and Applications*, Marcel Dekker Inc., New York, 1996.

GEC Measurement, *Network Protection and Automation Guide*, Morrison & Gibb Ltd; Edinburgh, Scotland, 1987.

Grigsby, L.L., *Electric Power Engineering Handbook*, CRC Press, Taylor & Francis Group, New York, 2007.

Gupta, B.R., *Power System Analysis and Design*, S.Chand, New Delhi, 2006.

Hewitson, L., Mark Brown, and Ramesh Balakrishnan, *Practical Power Systems Protection*, Newnes, IDC Technologies, Burlington, MA, 2004.

Holtzhausen, J.P. and W.L. Vosloo, *High Voltage Engineering Practice and Theory*, [online]. Available: http://ebookbrowse.com/ee-1402-high-voltage-engineering-pdf-d106048477.

Horowitz, S.H. and A.G. Phadke, *Power System Relaying*, John Wiley & Sons, New York, 1996.

Johns, A.T. and S.K. Salman, *Digital Protection for Power Systems*, Peter Peregrinus Ltd, UK, 1995.

Kalam, A. and D.P. Kothari, *Power System Protection and Communications*, New Age International Publishers, New Delhi, 2010.

Kuffel, E., W.S. Zaengl, and J. Kuffel, *High Voltage Engineering Fundamentals*, Reed Educational and Professional Publishing Ltd, UK, 2000.

Kundur, P., *Power System Stability and Control*, Tata McGraw-Hill, New Delhi, India, 2008.

Lythal, R.T., *J&P Switchgear Book*, Aditya Books Pvt. Ltd, New Delhi, 1994.

Martinez-Velasco, J.A., *Power System Transients Parameter Determination*, Taylor & Francis Group, New York, 2010.

Mason, C.R., *The Art and Science of Protective Relaying*, Wiley Eastern Ltd, New Delhi, 1987.

Naidu, M.S. and V. Kamaraju, *High Voltage Engineering*, Tata McGraw-Hill, New Delhi, 2005.

Oza, B.A., N.C. Nair, R.P. Mehta, et al., *Power System Protection & Switchgear*, Tata McGraw Hill, New Delhi, 2010.

Paithankar, Y.G. and S.R. Bhide, *Fundamentals of Power System Protection*, PHI Learning Pvt. Ltd, New Delhi, 2009.

Phadke, A.G. and J.S. Thorp, *Computer Relaying for Power Systems*, Research Study Press Ltd, John Wiley & Sons, Taunton, UK, 1988.

Prévé, C., *Protection of Electrical Network*, ISTE Ltd, London, UK, 2006.

Ravindranath, B. and M. Chander, *Power System Protection and Switchgear*, New Age International Publishers, New Delhi, 2009.

Singh, L.P., *Digital Protection*, Wiley Eastern Limited, New Delhi, 1994.

The Electricity Council, *Power System Protection*, Peter Peregrinus Ltd, UK, 1981.

Van C. Warrington, A.R., *Protective Relays: Their Theory and Practice*, Vol. 1, Chapman & Hall Ltd, London, 1962.

Wadhwa, C.L., *Electrical Power System*, New Age International Publishers, New Delhi, 2010.

Warwick, K., Arthur Ekwue, and Raj Aggarwal, *Artificial Intelligence Techniques in Power Systems*, The Institution of Electrical Engineers, UK, 1997.

Others

'Controlled Switching of HVAC Circuit-Breakers: Benefits & Economic Aspects', *Cigré Working Group A3.07*, January 2004.

'Controlled Switching of HVAC Circuit-Breakers: Guidance for further applications including unloaded transformer switching, load and fault interruption and circuit-breaker uprating', *Cigré Working Group A3.07*, December 2004.

'IEEE Guide for AC Generator Protection', ANSI/IEEE C37.1021995.

'IEEE Recommended Practice for Protection and Coordination of Industrial and Commercial Power Systems', IEEE Industry Applications Society, IEEE Std 242-1986.

'IEEE Standard Common Format for Transient Data Exchange (COMTRADE) for Power Systems', Sponsored by the Power System Relaying Committee of the Power Engineering Society, IEEE C37.111-1991.

'Power Swing and Out-of-step Considerations on Transmission Lines', *IEEE Power System Relaying Committee*, 2005 Report, [online]. Available: http://www.pes–psrc.org.

'Transformer Energization in Power Systems: A Study Guide', *Cigré Working Group C4.307*, February 2014.

A Report to the Line Protection Subcommittee, Power System Relay Committee IEEE Power Engineering Society, Prepared by Working Group D8, 'Justifying Pilot Protection on Transmission Lines,' Draft 1.4, 2006, pp. 1–21.

Andow, F., N. Suga, Y. Murakami, et al., 'Microprocessor-based Busbar Protection Relay', in *5th International Conference on Developments in Power System Protection*, IEE Publication No. 368, 1993, pp. 103–106.

Andrzej, W. and Kasztenny Bogdan, 'A Multi-criteria Differential Transformer Relay based on Fuzzy Logic', *IEEE Transactions on Power Delivery*, Vol. 10, No. 4, October 1995, pp. 1786–1792.

Atefi, M.A., and M. Sanaye-Pasand, 'Improving Controlled Closing to Reduce Transients in HV Transmission Lines and Circuit Breakers', *IEEE Transactions on Power Delivery*, Vol. 28, No. 3, pp. 733–741, July 2013.

Aujla, R.K., 'Generator Stator Protection, Under/Overvoltage, Under/Over Frequency and Unbalanced Loading', Universtiy of Western Ontario, Canada, 5 May 2008.

Benmouyal, G., 'The Protection of Synchronous Generators', *Schweitzer Engineering Laboratories*, Taylor & Francis Group, 2006.

Benmouyal, G., D. Hou, and D. Tziouvaras, 'Zero-setting Power Swing Blocking Protection', [online]. Available: http://www.selinc.com/techpprs/6172_Zerosetting_20050302.pdf.

Bhalja, B. and N.G. Chothani, 'Electrical Busbar Protection Philosophy: Past, Present and Future', *Electrical India Industrial Magazine*, Vol. 50, No. 1, January 2010, pp. 122–129.

Bhalja, B. and R. P. Maheshwari, 'Trends in Adaptive Distance Protection of Multi–terminal and Double–Circuit Lines', *International Journal of Electric Power Components & Systems*, Vol. 34, No. 6, June 2006, pp. 603–617.

Bhalja, B. and R.P. Maheshwari, 'A New Differential Protection Scheme for Tapped Transmission Line', *IET Generation, Transmission & Distribution*, Vol. 2, No. 2, March 2008, pp. 271–279.

Bhalja, B. and R.P. Maheshwari, 'An Adaptive Distance Relaying Scheme using Radial Basis Function Neural Network', *International Journal of Electric Power Components and Systems*, Vol. 35, No. 3, March 2007, pp. 245–259.

Bhalja, B. and R.P. Maheshwari, 'Challenges in Line Protection Philosophies', *Electrical India Industrial Magazine*, Vol. 47, No. 2, April 2008, pp. 30–38.

Bhalja, B. and R.P. Maheshwari, 'Digital Protection of Power Transformers: Issues & Trends', *Electrical India Industrial Magazine*, Vol. 56, August 2008, pp. 56–60.

Bhalja, B. and R.P. Maheshwari, 'High Resistance Faults on Two Terminal Parallel Transmission Line: Analysis, Simulation Studies and an Adaptive Distance Relaying Scheme', *IEEE Transactions on Power Delivery*, Vol. 22, No. 2, April 2007, pp. 801–812.

Bhalja, B. and R.P. Maheshwari, 'Philosophy of Protection for Multi-terminal and Double-circuit Lines', In Proceeding CERA-2005: International Conference, *Computer Application in Electrical Engineering—Recent Advances*, September/October, IIT Roorkee, 2005, pp. 582–588.

Bhalja, B. and R.P. Maheshwari, 'Protection of Transmission Line using Distance Measurements: Problems & Solutions', In Proceedings of *National Conference on Advancement of Technologies—Global Scenario* (ADTECH—GLOS), Mathura, February 2007.

Bhalja, B. and R.P. Maheshwari, 'Protection Philosophies: Past, Present and Future', *Electrical India Industrial Magazine*, Vol. 47, No. 2, February 2007, pp. 30–38.

Bhalja, B., B.A. Oza, and P.H. Shah, 'Coordination of Overcurrent relay for Cascaded Parallel Feeder', In Proceeding APSCOM–06: *International Conference on Advances in Power System Control, Operation and Managements*, Hong Kong, October/November, 2006, pp. 1–4.

Bhalja, B., B.A. Oza, and P.H. Shah, 'Testing of Overcurrent/Undercurrent Relays', In Proceedings of *National Conference on Electrical Engineering Developments*, Vishakapatnam, June 2005.

Bhalja, B., N.G. Chothani, and R.P. Maheshwari, 'A New Digital Differential Relaying Scheme for the Protection of Busbar', *4th International Conference on Computer Applications in Electrical Engineering—Recent Advances*, IIT Roorkee, February 2010.

Bhalja, B., N.G. Chothani, and R.P. Maheshwari, 'A Review on Busbar Protection Philosophy: Past, Present and Future', *Journal of Institution of Engineers*, November 2010, pp. 1–16.

Bhalja, B., R.P. Maheshwari, and B.A. Oza, 'Stator Earth Fault Protection for Generator', *Electrical India Industrial Magazine*, Vol. 47, No. 12, December 2007, pp. 116–120.

Bhalja, B., R.P. Maheshwari, and N.G. Chothani, 'A Review on Busbar Protection Philosophy: Past, Present and Future,' Journal of The Institution of Engineers, India, Vol. 90, June 2010, pp. 1–17.

Bhalja, B., R.P. Maheshwari, and V. Makwana, 'Real Time Implementation of Numerical Relay for Induction Motor', *32nd National System Conference*, NSC 2008, IIT Roorkee, December 2008.

Bhalja, B., R.P. Maheshwari, Ashesh Shah, et al., 'Experimental Test Setup for Measuring Transient Overreach of Instantaneous Overcurrent Relay', In Proceedings of *32nd National System Conference on Energy Systems: Optimization & Conservation*, December 2008, pp. 319–324.

Bhalja, B., R.P. Maheshwari, B. Das, et al., 'A New Fault Classification Technique for Protection of Series Compensated Transmission Lines', In Proceedings of *International Conference on Power System Protection*, CPRI Bangalore, India, February 2007, pp. 17–26.

Bhalja, B., R.P. Maheshwari, B.A. Oza, et al., 'Development of a New Overcurrent/Undercurrent Relay Testing Kit', *International Journal of Electric Power Components and Systems*, USA Vol. 37, No. 11, November 2009, pp. 1208–1218.

Bhalja, B., R.P. Maheshwari, D. Birla, and Manoj Tripathy, 'Advances in Line and Transformer Protection Schemes', In Proceedings of NCEC-2005, *National Conference on Emerging Computational Techniques & their Applications*, Jodhpur, October 2005, pp. 236–240.

Bhalja, B., R.P. Maheshwari, Saurav Nema, et al., 'Neuro Fuzzy Based Scheme for Stator Winding Protection of Synchronous Generator', International Journal of Electric Power Components and Systems, USA, Vol. 37, No. 5, May 2009, pp. 560–576.

Bhalja, B., R.P. Maheshwari, Urmil Parikh, et al., 'Decision Tree-based Fault Classification Scheme for Protection of Series Compensated Transmission Lines', *International Journal of Emerging Electric Power Systems*, Berkeley Electronic Press, Vol. 8, Issue 6, 2007, Article 1, pp. 1–12.

Bhalja,B. and U. B. Parikh, 'SVR Based Current Zero Estimation Technique for Controlled Fault Interruption in Series-Compensated Transmission Line,'*IEEE Transactions on Power Delivery*, Vol. 28, No. 3, July 2013, pp. 1364–1372.

Bhalja, B. and U. B. Parikh, 'Challenges in Feld Implementation of Controlled Energization for Various Equipment Loads with Circuit Breakers Considering Diversified Dielectric and Mechanical Characteristics,' *International Journal of Electrical Power and Energy Systems*, Vol. 87, No. 3, 2017, pp. 99–108.

Bhalja. B. and U. B. Parikh, 'Mitigation of Magnetic Inrush Current During Controlled Energization of Coupled Un-Loaded Power Transformers in Presence of Residual Flux Without Load Side Voltage Measurements,' *International Journal of Electrical Power and Energy Systems*, Vol. 76, March 2016, pp. 156–164. Birla, D., 'Coordination of Directional Overcurrent Relay in Power Networks', Ph D Thesis, IIT Roorkee, September 2006.

Brahma, S.M., 'Distance Relay with Out–of–Step Blocking Function using Wavelet Transform,' *IEEE Transaction on Power Delivery*, Vol. 22, No. 3, July 2007, pp. 1360–1366.

Bratton, R.E., 'Optical Fiber Link for Power System Protective Relays: Digitally Multiplexed Transfer Trip Circuits', *IEEE Transactions on Power Apparatus and Systems*, February 1984, pp. 403–406.

Brewis, K., K. Hearfield, and K. Chapman, 'Theory and Practical Performance of Interlocked Overcurrent Busbar Zone Protection in Distribution Substations', Developments in Power System Protection, Conference Publication No.479, IEE 2001, pp. 475–478.

Brunke, J.H, Frohlich, K.J, 'Elimination of transformer inrush currents by controlled switching: Part I - Theoretical considerations', *IEEE Transactions on Power Delivery*, Vol. 16, No. 2, pp. 276–280, April 2001.

Bus Differential Relaying: Methods & Application, Basler Electric Company, Model No: 618/654–2341, 2005, pp. 1–57.

Bus Protection for 400 kV and 275 kV Double Busbar Switching Stations, National Grid Technical Specification NGTS 3.6.3, Issue 3, December 1996, pp. 1–19.

Cable, B.W., L.J. Powell, and R.L. Smith, 'Application Criteria for High Speed Bus Differential Protection', *IEEE Transactions on Industry Applications*, Vol. IA–19, July/August 1983, pp. 619–624.

Carreau, D., U. Habedank, D. Kopejtkova, T. Kuntze, C. Leu, F., H.P. Schmidt, and N. Trapp, 'Controlled switching of unloaded transformers - Application with 245/15/15 kV step-up transformer', *Cigré 13-110*, Session 1998.

Chamia, M. and S. Liberman, 'Ultra High Speed Relay for EHV/UHV Transmission Lines—Development, Design and Application', IEEE Transactions on Power Apparatus and Systems, Vol. PAS–97, No. 6, November/December 1978, pp. 2104–2112.

CIGRE Report by Joint Working Group 34/35.11, 'Protection using Telecommunications', Document Reference No. 192, August 2001, pp. 13–15.

Commercial Power Systems', IEEE Industry Applications Society, IEEE Std 242-1986.

Commissioning Instructions Manual, 'Portable Overcurrent Relay Test Equipment—Type CFB', GEC Alsthom India Ltd.

Controlled switching buyer's and application guide, ABB Ltd, Ludvika, Sweden, 2009.

Controlled switching of HVAC circuit breakers, Guide for application lines, reactors, capacitors, transformers, *Part I, ÉLECTRA No. 183*, April, 1999.

Controlled switching of HVAC circuit breakers, Guide for application lines, reactors, capacitors, transformers, *Part II, ÉLECTRA No. 183*, August, 1999.

Corssley, P.A. and P.G. McLaren, 'Distance Protection Based on Traveling Waves, *IEEE Transactions on Power Apparatus and Systems*, Vol. PAS-102, No. 9, September 1983, pp. 2971–78.

Dantas, K. M. C., W. L. A. Neves, D. Fernandes Jr., G. A. Cardoso, and L. C. Fonseca, 'Real Time Implementation of Transmission Line Controlled Switching', *International Conference on Power Systems Transients (IPST)*, Delft, June 2011.

Dantas, Karcius, W.L.A. Neves, and Dajo Fernandes 'An Approach for Controlled Reclosing of Shunt-Compensated Transmission Lines', *IEEE Transactions on power delivery*, vol. 29, No. 3, pp. 1203–1211, June 2014.

Darwish, H.A., Adbel-Maxoud I. Taalab, 'Development and Implementation of an ANN-based Fault Diagnosis Scheme for Generator Winding Protection', *IEEE Transactions on Power Delivery*, Vol. 16, No. 2, April 2001, pp. 208–214.

Das, B. and J.V. Reddy, 'Fuzzy-logic-based Fault Classification Scheme for Digital Distance Protection', *IEEE Transactions on Power Delivery*, Vol. 20, No. 2, April 2005, pp. 609–616.

Dash, P.K., O.P. Malik, and G.S. Hope, 'Fast Generator Protection against Internal Asymmetrical Faults', *IEEE Transactions on Power Apparatus & Systems*, Vol. 96, No. 5, September/October 1977, pp. 1498–1506.

Developments in Power System Protection, Conference Publication No.479, IEE 2001, pp. 475–478.

Dierks, K.C.A., 'Relay Specific Computer Aided Testing for Protective Relays', Eskom, Republic of South Africa.

Eissa, M.M., 'A Novel Digital Directional Technique for Busbars Protection', *IEEE Transactions on Power Delivery*, Vol. 19, No. 4, October 2004, pp. 1636–1641.

Evans, J.W., R. Parmella, K.M. Sheahan, et al., 'Conventional and Digital Busbar Protection: A Comparative Reliability Study', *Developments in Power System Protection*, March 1997 Conference Publication No. 434, IEE, 1997 pp. 126–130.

Facilities Instructions, Standards and Techniques Volume 3–16, Maintenance of Power Circuit Breakers, December 1999.

Fernandez C. Paulo, Paulo C. V. Esmeraldo, Jorge Amon Filho, and Cesar Ribeiro Zani, 'Use of controlled switching systems in power system to mitigate switching transients. Trends and benefits — Brazilian experience', *Cigre Session 2002*: 13-20.

Fernandez, Rosa M de Castro and Horacio Nelson Diaz rojas, 'An Overview of Wavelet Transforms Application in Power Systems', In Proceedings of *14th Power System Computation Conference*, Spain, Session 1, Paper No. 6, June 2002.

Funk, H.W. and G. Ziegler, 'Numerical Busbar Protection, Design and Service Experience', *The 6th International Conference on Developments in Power System Protection*, IEE Publication No. 434, 25–27 March 1997, pp. 131–134.

Gajic, Z., 'Design Principles of High Performance Numerical Busbar Differential Protection', Relay Protection and Substation Automation of Modern Power Systems, September 2007, pp. 1–7.

Gajic, Z., 'Modern Techniques for Protecting Busbars in HV Networks', Study Committee B5 Colloquium, October 2009, Jeju Island, Korea, pp. 1–7.

GEC Measurements, Protection Relay Application Guide, Morrison & Gibb Ltd, Edinburgh, Scotland, 1975.

Gill, H.S., T.S. Sidhu, and M.S. Sachdev, 'Microprocessor-based Busbar Protection System', IEE Proceedings Generation, Transmission and Distribution, Vol. 147, No. 4, July 2000, pp. 252–260.

Girgis, A.A. and R.G. Brown, 'Application of Kalman Filtering in Computer Relaying', *IEEE Transaction on Power Apparatus and Systems*, Vol. 100, No. 7, July 1981, pp. 3387–3397.

Goldsworthy Dan, Tom Roseburg, Demetrios Tziouvaras, and Jeff Pope, 'Controlled Switching of HVAC Circuit Breakers: Application Examples and Benefits', *61st Annual Conference for Protective Relay Engineers*, pp 520–535, April 2008.

Graps, A., 'An Introduction to Wavelets', *IEEE Computational Science & Engineering*, Vol. 2, No. 2, pp. 1–18, 1995.

Grid Technical Specification NGTS 3.6.3, Issue 3, December 1996, pp. 1–19.

Guide for the Application of Autoreclosing to the Bulk Power System, Northeast Power Coordinating Council, 2009.

Gupta, R.P. and M.B. Lonkar, 'Power Transformer Protection', Proceedings of International Conference on Control, Communication and Power Engineering, 2010, pp. 225–230.

Guzman, A., B.L. Qin, and Casper Labuschagne, 'Reliable Busbar Protection with Advanced Zone Selection', *IEEE Transactions on Power Delivery*, Vol. 20, No. 2, April 2005, pp. 625–629.

High Impedance Differential Relaying, GER–3184, GE Power Management, Ontario, Canada, pp. 1–19, [online]. Available: www.geindustrial.com/pm.

High-voltage switchgear and control gear – Part 302: Alternating current circuit-breakers with intentionally non-simultaneous pole operation, *IEC/TR 62271-302 ed1.0*, 06-2010.

Hope, G.S., P.K. Dash, and O.P. Malik, 'Digital Differential Protection of a Generating Unit: Scheme and Realtime Test Results', *IEEE Transactions on Power Apparatus & Systems*, Vol. 96, No. 2, March/April 1977, pp. 502–512.

Horowitz, S.H. and A.G. Phadke, 'Third Zone Revisited', *IEEE Transactions on Power Delivery*, Vol. 21, No. 1, January 2006, pp. 23–29.

Horton, J.W., 'The use of Walsh Functions for High Speed Digital Relaying', *IEEE PES Summer Meeting*, San Francisco, California, July 1975, pp. 1–9.

Htwe, N.K., 'Analysis and Design Selection of Lightning Arrester for Distribution Substation', *World Academy of Science, Engineering and Technology*, 2008, pp. 174–178.

Hughes, R. and E. Legrand, 'Numerical Busbar Protection Benefits of Numerical Technology in Electrical Engineering', In Proceedings of *7th IEE International Conference on Developments in Power System Protection*, April 2001, pp. 133–136.

I.S. 9124 Guide for Maintenance and Field Testing of Electrical Relays, Indian Standard Institution, New Delhi, 1979.

IEEE Application Guide for AC High Voltage Breakers Rated on a Symmetrical Current Basis, ANSI/IEEE C37.010-1979.

IEEE Committee Report, 'Line protection design trends in USA and Canada', *IEEE Transactions on Power Delivery*, Vol. 3, No. 4, October 1988, pp. 1530–1535.

IEEE Committee Report, 'Pilot Relaying Performance Analysis', *IEEE Transactions Power Delivery*, Vol. 5, No. 1, January 1990, pp. 85–102.

IEEE Guide for Protective Relay Applications to Power System Buses, IEEE Standard C37.97 (Reaffirmed 12/90), 1979.

IEEE Power Line Carrier Working Group Report, 'Power Line Carrier Practices and Experience', IEEE Paper 94, SM428-3, 1994.

IEEE Power Systems Communication Committee Report, 'Power Line Carrier Application Guide', *IEEE Transactions on Power Apparatus and Systems*, November/December 1990, pp. 2334–2337.

IEEE Power Systems Relaying Committee, 'Automatic Reclosing of Transmission Lines', *IEEE Transactions on Power Apparatus and Systems*, Vol. PAS-103, No. 2, February 1984, pp. 234–245.

IEEE Power Systems Relaying Committee, 'Wide Area Protection and Emergency Controls', Technical Report 2002, [online]. Available: http://www.pes-psrc.org/.

IEEE Power Systems Relaying Committee, Guide for Automatic Reclosing for Line Circuit Breakers for AC Distribution and Transmission Lines, Draft Document, 1998.

IEEE Recommended Practice for Protection and Coordination of Industrial and Commercial Power Systems, IEEE Industry Applications Society, IEEE Standard 242–1986.

IEEE Standard 37.113-1999, 'Guide for Protective Relay Applications to Transmission Lines', 1999.

IEEE Standard Definitions for Power Switchgear, IEEE Std. C37.100-1992.

IEEE Standard for Synchrophasors for Power Systems, IEEE Standard 1334-1995 (Revised in 2001), 2001.

IEEE Working Group D10 Report, 'Application of expert systems to power system protection', *IEEE PES Summer Meeting*, Paper 93, SM 384-8-PWRD, 1993.

Insulation Coordination Part 1: Definitions, Principles and Rules, IEC Document 28 CO 58, 1992.

Jafarin, P. and Majid Sanaye-Pasand, 'A Traveling Wave-based Protection Technique using Wavelet/PCA Analysis', *IEEE Transactions on Power Delivery*, Vol. 25, No. 2, April 2010, pp. 588–599.

Jamali, M., M. Mirzaie, and S. Asghar Gholamian, 'Calculation and analysis of transformer inrush current based on parameters of transformer and operating conditions', *Electronics And Electrical Engineering, T190 Electrical Engineering*, No.3 (109), pp. 17–20, 2011.

Javed, A.M. and M.T. Javed, 'A Digital Frequency Relay for Over/Under Frequency Detection', TENCON '91 IEEE Region *10th International Conference on EC3—Energy, Computer, Communication and Control Systems*, Vol. 1, August 1991, pp. 440–442.

Jeyasurya, B. and W.J. Smolinski, 'Identification of a Best Algorithm for Digital Distance Protection of Transmission Lines', *IEEE Transactions on Power Apparatus and Systems*, Vol. 102, No. 10, October 1980, pp. 3358–3369.

Jodice, J.A., 'Relay Performance Testing', *IEEE Transactions on Power Delivery*, Vol. 12, No. 1, January 1997, pp. 169–171.

Jongepier, A.G. and L. van der Sluis, 'Adaptive Distance Protection of a Double–circuit Line', *IEEE Transactions on Power Delivery*, Vol. 9, No. 3, July 1994, pp. 1289–97.

Kang, Y.C., J.S. Yun, B.E. Lee, et al., 'Busbar Differential Protection in Conjunction with a Current Transformer Compensating Algorithm', IET Generation Transmission & Distribution, Vol. 2, No. 1, January 2008, pp. 100–109.

Kasztenny, B., 'Distance Protection of Series Compensated Lines Problems & Solutions', *28th Annual Western Protective Relay Conference (GE)*, October 2001, pp. 1–34.

Kasztenny, B., Lubomir Sevov, Gustavo Brunello, 'Digital Low-Impedance Bus Differential Protection—Review of Principles and Approaches', GER-3984.

Kezunovic, M. and I. Rikalo, 'Detect and Classify Faults using Neural Nets', *IEEE Computer Applications in Power*, Vol. 9, No. 4, October 1996, pp. 42–47.

Kezunovic, M., 'Digital Protective Relaying Algorithms and Systems—An Overview,' *Electric Power Systems Research*, No. 4, 1981, pp. 167–180.

Kezunovic, M., J. Domaszewicz, V. Skendzic, et al., 'Design, Implementation and Validation of a Real-time Digital Simulator for Protection Relay Testing', *IEEE Transactions on Power Delivery*, Vol. 11, No. 1, January 1996, pp. 158–164.

Kezunovic, M., Tomo Popovic, Donald Sevcik, et al., 'Transient Testing of Protection Relays: Results, Methodology and Tools', In Proceedings of *International Conference on Power System Transients*, Hong Kong, 2003, pp. 1–6.

Kezunovic, M., Y.Q. Xia, Y. Guo, et al., 'Distance Relay Application Testing using a Digital Simulator', *IEEE Transactions on Power Delivery*, Vol. 12, No. 1, January 1997.

Khederzadeh, M. and Tarlochan S. Sidhu, 'Impact of TCSC on the Protection of Transmission Lines', *IEEE Trans. of Power Delivery*, Vol. 21, No.1, January 2006, pp. 80–87.

Kumar, A. and P. Hansen, 'Digital Bus-zone Protection', *IEEE Computer Applications in Power*, Vol. 6, No. 4, October 1993, pp. 29–34.

Legate, M. A. C., J. H. Brunke, J. J. Ray, and E. J. Yasuda, 'Elimination of Closing Resistors on EHV Circuit Breakers', *IEEE Transactions on Power Delivery*, Vol. 3, No. 1, pp. 223–231, January 1988.

Lindenmeyer, D., H.W. Dommel, and M.M. Adibi, 'Power System Restoration—A Bibliographical Survey', *International Journal of Electrical Power & Energy Systems*, Vol. 23, Issue 3, 1 March 2001, pp. 219–227.

Ma, Z., C.A. Bliss, A.R. Penfold, et al., 'An Investigation of Transient Overvoltage Generation when Switching High Voltage Shunt Reactors by SF6 Circuit Breaker', *IEEE Transactions on Power Delivery*, Vol. 13, No. 2, April 1998.

Maheshwari, R.P., 'Developments in Adaptive Relaying for Transformer and Line Protection', Ph D Thesis, IIT Roorkee, September 1995.

Manitoba HVDC Research Center, 'PSCAD v4.2.1 Online Help System', https://pscad.com/products/pscad/free_downloads/, accessed on 21 July 2011.

Mann, B.J. and I.F. Morrison, 'Digital Calculation of Impedance for Transmission Line Protection', *IEEE Transactions on PAS–90*, No. 1, pp. 270–279, Jan/Feb 1971.

Martinez-Velasco, J.A., *Power System Transients Parameter Determination*, Taylor & Francis Group, New York, 2010.

Mason, C.R., *The Art and Science of Protective Relaying*, Wiley Eastern Ltd, New Delhi, 1987.

McLaren, P.G. and M. A. Redfern, 'Fourier Series Techniques Applied to Distance Protection', *Proceedings of IEE*, Vol. 122, No. 11, November 1975, pp. 1301–1305.

McLaren, P.G., et al., 'A Real-time Digital Simulator for Testing Relays', *IEEE Transactions on Power Delivery*, Vol. 7, No. 1, 1992.

Megahed, A.I. and O.P. Malik, 'An Artificial Neural Network Based Digital Differential Protection Scheme for Synchronous Generator Stator Winding Protection' *IEEE Transactions on Power Delivery*, Vol. 14, No. 1, January 1999, pp. 86–93.

Megahed, A.I. and O.P. Malik, 'Simulation of Internal Faults in Synchronous Generators' IEEE Transactions on Energy Conversion, Vol. 14, No. 4, December 1999, pp. 1306–1311.

Megahed, A.I. and O.P. Malik, 'Synchronous Generator Internal Fault Computation and Experimental Verification', *IEE Proceedings on Generation, Transmission, Distribution*, Vol. 145, No. 5, September 1998, pp. 604–610.

MiCOM P220 Technical Guide, MiCOM Series P220 Motor Protection, ALSTOM Limited, 2003.

Mike Dragomir Manitoba HVDC Research Centre Inc. On-line Condition Monitoring of a 230 kV Minimum Oil Circuit Breaker Finepoint Circuit Breaker Conference, Atlanta 2004.

Misiti, M., Yves Misiti, Georges Oppenheim, and Jean-Michel Poggi, 'Wavelet Toolbox Users Guide for use with MATLAB,' Ver. 2, The Math Works Inc., 2002.

Mozina, C.J., 'Advanced Applications of Multifunction Digital Generator Protection', Beckwith Electric Company, USA, [online]. Available: http://www.beckwithelectric.com.

Mozina, C.J., 'Upgrading Generator Protection using Digital Technology', Beckwith Electric Company, USA, [online]. Available: http://www.beckwithelectric.com.

Nagpal, M., 'Expert Talk on Comprehensive Digital Protection of Generators, Transformers, and Busbars', In Proceedings of TENCON 2008 *International Conference on Innovative Technologies for Societal Transformation*, November 2008, Hyderabad, [online]. Available: www.tencon2008.org/Title_Comprehensive.pdf.

Naidu, M.S. and V. Kamaraju, *High Voltage Engineering*, Tata McGraw-Hill, New Delhi, 2005.

Nunes, R.R. and Wallacedo do Couto Boaventura, 'Insulation Coordination Considering the Switching Overvoltage Waveshape—Part I: Methodology', *IEEE Transactions on Power Delivery*, Vol. 24, No. 4, October 2009, pp. 2434–2440.

Nvosel, D., Arun Phadke, M.M. Saha, and S. Lindhal, 'Problems and Solutions for Microprocessor Protection of Series Compensated Lines', *Sixth International Conference on Developments in Power System Protection,* Conference Publication No. 434, March 25–27, 1997, pp. 18–23.

Okafor, E.N.C., D.C. Idoniboyeobi, and I.O. Akwukwaegbu, 'Surge Protection Practice for Equipment in Substations', *International Journal of Academic Research*, Vol. 3. No. 1. January 2011, Part I, pp. 25–31.

Oza, B.A. and S.M. Brahma, 'Development of Power System Protection Laboratory Through Senior Design Projects', *IEEE Transactions on Power Systems*, Vol. 20, No. 2, May 2005, pp. 532–537.

Oza, B.A., N.C. Nair, R.P. Mehta, et al., *Power System Protection & Switchgear*, Tata McGraw Hill, New Delhi, 2010.

Ozgonenel, O., E. Arisoy, M.A.S.K. Khan, et al., 'A Wavelet Power-based Algorithm for Synchronous Generator Protection', *IEEE Power Engineering Society General Meeting*, June 2006.

P740 Numerical Busbar Protection, Technical Data Sheet, Publication No. P740/EN TD/G22, AREVA T&D, pp. 1–22.

Pandy, S.K. and L. Satish, 'Multiresolution Signal Decomposition: A New Tool for Fault Detection in Power Transformers during Impulse Tests,' IEEE Transactions on Power Delivery, Vol. 13, No. 4, November 1998, pp. 1194–1200.

Parikh B. Urmil and Bhavesh Bhalja, 'Mitigation of Magnetic Inrush Current during Controlled Energization of coupled un-loaded Power Transformers in presence of Residual Flux without load side voltage measurements', *Electrical Power and Energy Systems*, Vol. 76, No. 1, pp. 156–164, March 2016.

Parikh B. Urmil and Bhavesh Bhalja, 'SVR Based Current Zero Estimation Technique for Controlled Fault Interruption in Series Compensated Transmission Line*', IEEE Transactions on Power Delivery*, Vol. 28, No. 3, pp. 1364–1372, July 2013.

Park, Y.M. and G.W. Kim, 'A logic based expert system for fault diagnosis of power systems,' IEEE Transactions on Power Systems, Vol. 12, December 1994, pp. 363–369.

Peck, D.M., B. Nygaard, and K. Wadelius, 'A New Numerical Busbar Protection System with Bay-oriented Structure,' in *5th International Conference on Developments in Power System Protection*, IEE Publication No. 368, 1993, pp. 228–231.

Peelo, D.F., G. Bowden, et al., High Voltage Circuit Breaker and Disconnector Application in Extreme Cold Climates, CIGRE 2006, F-75008, Paris, [online]. Available: http: //www.cigre.org.

Penman, J. and H. Jiang, 'The Detection of Stator and Rotor Winding Short Circuits in Synchronous Machines by Analyzing Excitation Current Harmonics', Proceedings of IEE International Conference on Opportunities and Advances in International Power Generation, No. 419, 1996, pp. 137–142.

Phadke , A. G., and J. S. Thorp, 'Synchronized Phasor Measurements and Their Applications,' Springer Science + Business Media, 2008

Phadke, A.G., 'Synchronized Phasor Measurements in Power Systems', *IEEE Computer Applications in Power*, Vol. 6, Issue 2, April 1993, pp. 10–15.

Phadke, A.G., T. Hibka, and M. Ibrahim, 'A Digital Computer System for EHV Substation: Analysis and Field Tests', *IEEE Transactions on Power Apparatus and Systems*, Vol. 95, No. 1, January/February 1976, pp. 291–301.

Protection Application Handbook, ABB Transmission Systems and Substation, BU TS/Global LEC Support Programme, Vasteras, Sweden.

PSCAD/EMTDC Manual, 'Getting Started', Manitoba HVDC Research Centre Inc., January 2001.

PSRC H9, 'Special Considerations in Applying Power Line Carrier for Protective Relaying', Power Engineering Society, 2004.

Pukar Mahat, Zhe Chen, and Birgitte Bak–Jensen, 'Review of Islanding Detection Mehtods for Distributed Generation', DRPT2008, Nanjing, China, April 2008.

Qin, Bai-Lin, A. Guzman, and Edmund O. Schweitzer, 'A New Method for Protection Zone Selection in Microprocessor-based Bus Relays', *IEEE Transactions on Power Delivery*, Vol. 15, No. 3, July 2000, pp. 876–887.

Rafaat, R.M., 'Considerations in Applying Power Bus Protection Schemes to Industrial and IPP Systems', *IEEE Transactions on Industry Applications*, Vol. 40, No. 6, November/December 2004, pp. 1705–1711.

Ramamoorthy, M., 'Application of Digital Computers to Power System Protection', *Journal of Institution of Engineers*, India, Vol. 52, No. 10, June 1972, pp. 235–238.

Ranjbar, A.M. and B. J. Cory, 'An Improved Method for the Digital Protection of High Voltage Transmission Line', *IEEE Transactions on Power Apparatus and Systems*, PAS-94, No. 2, March/April 1975, pp. 544–550.

Recommended Practice for Utility Interconnected Photovoltaic Systems, IEEE Standard 929-2000, 2000.

Relay Performance Testing, IEEE Power System Relaying Committee Report, Special Publication No. 96 TP 115-0, 1996, pp. 1–25.

Relaying Scheme for the Protection of Busbar, In Proceedings of *4th International Conference on Computer Applications in Electrical Engineering—Recent Advances*, IIT Roorkee, February 2010, pp. 1–5.

Report of Working Group I16 of the Relaying Practices Subcommittee, 'Understanding Microprocessor-based Technology Applied to Relaying', Power System Relaying Committee, pp. 1–77, January 2006.

Roberts, J., A. Guzman, and E.O. Schweitzer, 'Z=V/I Does Not Make a Distance Relay', *20th Annual Western Protective Relaying Conference*, Washington, 19–21 October 1993, pp. 1–20.

Sachdev, M.S. (Coordinator), IEEE Tutorial Course Text: Computer Relaying, Publication No. 79, EH0148-7-PWR, New York, 1979.

Sachdev, M.S. (Coordinator), IEEE Tutorial Course: Advancements in Microprocessor-based Protection and Communication, *IEEE Power Engineering Society*, New Jersey, 1997.

Sachdev, M.S. and H.C. Wood, 'Introduction and General Methodology of Digital Protection', Power System Research Group, Toronto, March 1986.

Sachdev, M.S. and M. Nagpal, 'A Recursive Least Error Squares Algorithm for Power System Relaying and Measurement Applications', *IEEE Transactions on Power Delivery*, Vol. 6, No. 3, July 1991, pp. 1008–1015.

Sachdev, M.S. and M.A. Baribeau, 'A New Algorithm for Digital Impedance Relays', *IEEE Transactions on Power Apparatus and Systems*, Vol. 98, No. 6, November/December 1979, pp. 2232–2240.

Sachdev, M.S., T.S. Sidhu, and H.S. Gill, 'A Busbar Protection Technique and its Impact on Power System Protection and Control Reliability', *Schweitzer Engineering Laboratories (SEL) Inc.*, 2007–0226 TP6275–01, pp. 1–6.

Sachdev, M.S., T.S. Sidhu, and H.S. Gill, 'A Busbar Protection Technique and its Performance during CT Saturation and CT Ratio Mismatch', *IEEE Transactions on Power Delivery*, 2000, Vol. 15, No. 3, pp. 895–901.

Saied, M.M., 'The Kilometric Faults: Modeling and Normalized Relations for Line Transients and the Breaker Recovery Voltage', *IEEE Transactions on Power Delivery*, Vol. 20, No. 2, April 2005.

Shutingg, W., L. Heming, L. Yonggang, et al., 'The Diagnosis Method of Generator Rotor Winding Inter-turn Short Circuit Fault based on Exciting Current Harmonics', IEEE *5th International Conference on Power Electronics and Drive Systems,* Vol. 2, November 2003, pp. 1669–1673.

Sidhu, T.S., B. Sunga, and M.S. Sachdev, 'A Digital Technique for Stator Winding Protection of Synchronous Generators', International Journal of Electrical Power System Research, Vol. 36, No. 1, January 1996, pp. 45–55.

Sidhu, T.S., et. al., 'Protection Issues during System Restoration', *IEEE Transactions on Power Delivery*, Vol. 20, No. 1, January 2005, pp. 47–56.

Sidhu, T.S., H. Singh, and M.S. Sachdev 'Design, Implementation and Testing of an Artificial Neural Network-based Fault Direction Discriminator for Protecting Transmission Lines' *IEEE Transactions on Power Delivery*, Vol. 2, No. 10, April 1995, pp. 697–706.

Sidhu, T.S., L. Mital, and M.S. Sachdev, 'A Comprehensive Analysis of an Artificial Neural Network-based Fault Direction Discriminator', *IEEE Transactions on Power Delivery*, Vol. 19, No. 3, July 2004, pp. 1042–1048.

Singh, L.P., *Digital Protection*, Wiley Eastern Limited, New Delhi, 1994.

Skendzic, V., Ian Ender, and Greg Zweigle, 'IEC 61850-9-2 Process Bus and its Impact on Power System Protection and Control Reliability', *Schweitzer Engineering Laboratories (SEL) Inc.*, 2007–0226 TP6275–01, pp. 1–6.

Smolinski, W.J., 'An Algorithm for Digital Impedance Calculation using a Single Pi Section', *IEEE Transactions on Power Apparatus and Systems,* Vol. 98, No. 5, September/October 1979, pp. 1546–1551.

Smolinski, W.J., 'Digital Distance Protection of Transmission Lines', *Electric Power System Research*, Vol. 2, No. 4, December 1979, pp. 261–268.

Speas, Jr., T.P., 'Capacitor Switching and Capacitor Switching Devices', Presented at the *45th Annual Minnesota Power Systems Conference*, Atlanta, November 2009.

Stanek Michael, 'Experiences with improving power quality by controlled switching', CIGRE WG A3.07: *Seminar and Workshop on Controlled Switching*, St. Pete Beach, FL, USA, May 2003.

Steinhauser, F., 'Automated Testing for Electomechnical Overcurrent Relay', In Proceedings of *5th International Conference (IEE) on Developments in Power System Protection*, Pub. No. 479, 2001, pp. 62–65.

Stenström Lennart, Minoo Mobedjina, 'Limitation of switching over voltages by use of transmission line surge arresters', *SC 33, p. 30, International Conference*, Zagreb, Croatia, 1998.

Stockton, M., 'Protection of Distribution Systems using Electronic Devices', IEE Colloquium on a Practical Approach to the Protection of Industrial Power Systems Networks up to 11 kV, Ref. No. 1999/062, 24 February 1999, pp. 1–4.

Switchsync™ PWC600-User Manual, ABB Ltd, Västerås, Sweden, 2015.

Takami, J., Shigemitsu Okabe, and Eiichi Zaima, 'Study of Lightning Surge Overvoltages at Substations Due to Direct Lightning Strokes to Phase Conductors', *IEEE Transactions on Power Delivery*, Vol. 25, No. 1, January 2010, pp. 425– 433.

Tawfik, M. and M. Morcos, 'A novel approach for fault location on transmission lines,' IEEE Power Engineering Review, Vol. 18, No. 1, November 1998, pp. 58–60.

Tziouvaras, D. and D. Duo, 'Out-of-step Protection Fundamentals and Advancements', [online]. Available: http://www.selinc.com/techpprs/6163.pdf.

Vankayala, V.S. and N.D. Rao, 'Artificial Neural Networks and Their Applications to Power Systems—a Bibliographical Survey,' Internaional Journal of Electric Power System Research, Vol. 28, 1993, pp. 67–79.

Völcker, O. and H. Koch, 'Insulation Coordination for Gas-insulated Transmission Lines (Gil)', *IEEE Transactions on Power Delivery*, Vol. 16, No. 1, January 2001, pp. 122–130.

Warwick, K., Arthur Ekwue, and Raj Aggarwal, *Artificial Intelligence Techniques in Power Systems*, The Institution of Electrical Engineers, UK, 1997.

Watanabe, H., I. Shuto, K. Igarashi, et al., 'An Enhanced Decentralized Numerical Busbar Protection Relay utilizing Instantaneous Current Values from High Speed Sampling', In Proceedings of *7th IEE International Conference on Developments in Power System Protection*, April 2001, pp. 133–136.

Williams, J.R. and K. Amaratunga, 'An Introduction to Wavelets in Engineering', *International Journal of Numerical Methods in Engineering*, Vol. 37, No. 14, 1994, pp. 2365–2388.

Xiao, J., Fushuan Wen, C.Y. Chung, et al., 'Wide Area Protection and its Applications—A Bibliographical Survey', In Proceedings of APSCOM–06: *International Conference on Advances in Power System Control, Operation and Management*, Hong Kong, October/November, 2006, pp. 1–10.

Xuan, Q.Y., R.K. Aggarawal, A.T. Johns, et al., 'A Neural Network Based Protection Technique for Combined 275kV/400kV Double Circuit Transmission Lines', *International Journal of Neurocomputing*, Vol. 23, July 1998, pp. 59–70.

Yi Hu, Damir Novosel, M.M. Saha, et al., 'Improving Parallel Line Distance Protection with Adaptive Techniques', *IEEE Power Engineering Society Winter Meeting*, Vol. 3, January 2000, pp. 1973–78, 23–27.

Yonggang, L., Z. Hua, and I. Heming, 'The New Method on Rotor Winding Inter-turn Short Circuit Fault Measure of Turbine Generator', *IEEE International Conference on Electric Machines and Drives*, Vol. 3, June 2003, pp. 1483–1487.

Youssef, O.A.S., 'Combined Fuzzy Logic Wavelet-based Fault Classification Technique for Power System Relaying', *IEEE Transactions on Power Delivery*, Vol. 19, No. 2, April 2004, pp. 582–589.

Ziegler, G., 'Application Guide on Protection of Complex Transmission Network Configurations', *CIGRE Report*, Paris, 1991.

Additional Multiple Choice Questions

1. Selectivity, which is one of the requirements of protection system, is also known as
   - (a) dependability
   - (b) relay coordination
   - (c) security
   - (d) none of the above

2. Economics criteria of the protective scheme indicates to combine features of
   - (a) maximum protection with minimum cost
   - (b) maximum protection with maximum cost
   - (c) minimum protection with maximum cost
   - (d) none of the above

3. The cost of the protection system should not exceed
   - (a) 15% of the cost of the equipment to be protected
   - (b) 10% of the cost of the equipment to be protected
   - (c) 20% of the cost of the equipment to be protected
   - (d) 5% of the cost of the equipment to be protected

4. According to the literature survey, 80–90% faults occurring on overhead transmission lines are
   - (a) L-L faults
   - (b) L-L-L faults
   - (c) L-G faults
   - (d) L-L-G faults

5. Out of the total faults that occur in the whole power system network, maximum percentage of faults occur in
   - (a) cables
   - (b) overhead lines
   - (c) switchgear
   - (d) CTs & CVTs

6. The VA rating of a CT is decided by considering
   - (a) burden of a relay
   - (b) resistance of relay
   - (c) reactance of relay
   - (d) none of the above

7. Different abnormal conditions that will occur in
   - (a) both generators & transformers
   - (b) transmission lines
   - (c) cables
   - (d) all of the above

8. Electromechanical relays are still used by the utilities due to their
   - (a) ruggedness and withstanding capacity of voltage spikes
   - (b) lower cost
   - (c) simple construction
   - (d) all of the above

9. Operating torque is provided in single input relay with the help of
   - (a) coil
   - (b) moving armature
   - (c) copper shading ring
   - (d) none of the above

10. The main problem with electromechanical and static relays is that
    - (a) they are very costly.
    - (b) they are not rugged in nature.
    - (c) there is no continuous check on their operational integrity.
    - (d) none of the above.

11. The apparent power (VA) required to cause the operation of relay is known as
    - (a) stability of the relay
    - (b) sensitivity of the relay
    - (c) tripping time of relay
    - (d) none of the above

12. The plug setting of ground relays are lower than phase relays due to involvement of
    - (a) tower resistance and ground resistance in the fault path
    - (b) relay resistance
    - (c) both (a) and (b)
    - (d) none of the above

13. The plug setting and time dial setting of ground relays are affected by

(a) response of the CVT

(b) excitation current of the CT

(c) arc resistance

(d) none of the above

14. The overloading condition on a line or apparatus is similar to

(a) short circuit

(b) earth fault

(c) both (a) and (b)

(d) increasing full load current of the feeder or apparatus

15. Characteristic of an overcurrent relay is always plotted between

(a) current & time

(b) multiple of pick up current & time

(c) voltage & time

(d) none of the above

16. The transient overreach phenomenon for an overcurrent relay increases if

(a) the decay of the dc component of fault current is slow.

(b) the decay of the dc component of fault current is fast.

(c) the decay of the dc component of fault current is constant.

(d) none of the above.

17. The main drawback of definite time delay relay is

(a) it takes long time to clear the fault near the source.

(b) it is unable to clear the fault near the generator.

(c) both (a) and (b)

(d) none of the above

18. Though the magnitude of fault current is lower than the full load current, it is harmful to the system and equipment due to

(a) generation of arc

(b) negative and zero sequence currents

(c) both (a) and (b)

(d) none of the above

19. Minimum coordination tome (MCT) is decided by considering

(a) errors in the relay alone

(b) operating time of breaker

(c) errors in the CTs and CVTs

(d) operating time of relay and breaker and errors in the CTs and CVTs

20. Directional relay is required at a location/bus where

(a) load is very high.

(b) load is very low.

(c) there are chances of reversal of fault current.

(d) none of the above

21. Directional relay does not operate properly during

(a) remote end fault

(b) middle end faults

(c) close in faults

(d) none of the above

22. Cross-polarized directional relay fails during

(a) single line to ground fault

(b) double line fault

(c) double line to ground fault

(d) triple line fault/symmetrical fault

23. Which zone of distance relay is affected by extreme loading conditions (also known as load encroachment phenomena)

(a) zone-1

(b) zone-2

(c) zone-3

(d) all of the above

24. For a single line to ground fault with considerable value of fault resistance, the distance relay

(a) underreaches

(b) overreaches

(c) both (a) and (b)

(d) none of the above

25. The setting of phase distance relays are done on the basis of

(a) zero sequence impedance

(b) positive sequence impedance

(c) both (a) and (b)

(d) none of the above

26. Which characteristic of distance relay provides adequate protection in case of close in fault

(a) mho

(b) reactance

(c) impedance

(d) none of the above

27. The setting of ground distance relay are done based on

(a) positive sequence impedance

(b) negative sequence impedance

(c) zero sequence impedance

(d) none of the above

28. Due to the current infeed from remote end bus, the distance relay located at local end bus,

(a) underreaches

(b) overreaches

(c) both (a) and (b)

(d) none of the above

29. Overreaching/ underreaching of distance relay in case of parallel transmission lines depends on

    (a) relative direction of the parallel line's zero sequence current with respect to the compensated current given to the relay

    (b) magnitude of opposite end zero sequence current

    (c) magnitude of opposite end phase current

    (d) none of the above

30. The value of load at which relay is on the verge of operation is known as

    (a) time of operation of the relay

    (b) overreaching of the relay

    (c) underreaching of the relay

    (d) loadability limit of the relay

31. The current/voltage inversion phenomenon in series compensated transmission line depends on

    (a) location of series capacitor

    (b) location of fault

    (c) location of relay

    (d) none of the above

32. Metal oxide varistor (MOV) is used for the protection of series compensated transmission line due to its

    (a) best characteristic (c) both (a) and (b)

    (b) small reinsertion time (d) none of the above

33. Interposing CTs for equalizing pilot current on both sides of power transformer are not required with

    (a) electromechanical relay

    (b) static relay

    (c) digital relay

    (d) none of the above

34. The protective scheme required to utilize single pole tripping facility, is

    (a) distance protection

    (b) carrier aided distance scheme

    (c) current based scheme

    (d) none of the above

35. Simultaneous opening of both (local and remote end) breakers are achieved with distance protection scheme for a fault on

    (a) the entire line

    (b) 60% from midpoint of the line

    (c) 80% from mid-point of the line

    (d) none of the above

36. The term 'pilot' for transmission lines refers to

    (a) communication medium

    (b) current flows through the line

    (c) transmission line conductors

    (d) none of the above

37. The communication of information via digital signal than the analog signal offers

    (a) low channel reliability

    (b) low channel failure

    (c) low channel density

    (d) none of the above

38. Sensitivity of wire pilot relaying scheme is reduce due to

    (a) line length

    (b) charging current between pilot wires

    (c) line voltage

    (d) all of the above

39. Wire pilot relaying scheme is used for the protection of

    (a) short transmission line

    (b) medium transmission line

    (c) long transmission line

    (d) none of the above

40. When a carrier signal is used to initiate tripping of the relay, the scheme is known as

    (a) carrier blocking scheme

    (b) directional comparison scheme

    (c) carrier tripping scheme

    (d) none of the above

41. Additional time delay or coordination is not required in

    (a) carrier blocking scheme

    (b) carrier tripping scheme

    (c) both (a) and (b)

    (d) none of the above

42. The main disadvantage of direct underreach transfer scheme is

    (a) its higher cost

    (b) its higher time of operation

    (c) its mal operation due to noise by switching

    (d) none of the above

43. Differential protection is generally applied to generator with ratings more than
    (a) 1 MVA
    (b) 0.5 MVA
    (c) 1 KVA
    (d) none of the above

44. The unequal CT secondary current (even though the primary currents are the same) that flow through the operating coil of the differential relay is known as
    (a) CT secondary current
    (b) relay pick up current
    (c) spill current
    (d) none of the above

45. The value of spill current increases due to
    (a) increase in CT ratio
    (b) the unequal length of pilot wires
    (c) decrease in CT ratio
    (d) all of the above

46. Incorporation of stabilizing resistance in series with operating coil of the differential relay reduces
    (a) sensitivity of the relay during internal fault
    (b) sensitivity of the relay during external fault
    (c) reliability of the relay during internal fault
    (d) none of the above

47. High impedance differential relaying scheme is used for the protection of generator due to its
    (a) stability during heavy through faults and external switching events
    (b) stability against CT saturation condition
    (c) both (a) and (b)
    (d) none of the above

48. The main advantage of biased differential protection scheme is its
    (a) high sensitivity during in-zone faults
    (b) lower sensitivity in case of external fault
    (c) better stability during external fault
    (d) high sensitivity during internal faults and better stability during external faults

49. Sensitivity of the relay used for the protection against stator earth fault in generator reduces due to
    (a) undetectable value of fault current because of high impedance grounding
    (b) stabilizing resistance

50. The function of ground fault detection in generator grounded with high impedance is performed by
    (a) current based protection scheme
    (b) differential protection scheme
    (c) voltage based protective scheme
    (d) none of the above

51. The complete protection to the high impedance grounded generator against ground faults at any point on the stator winding is obtained by
    (a) low impedance protection scheme
    (b) 100% stator earth fault protection scheme
    (c) both (a) and (b)
    (d) none of the above

52. The relay used for the protection of generators against the condition of reversal of power is
    (a) low forward reverse power relay
    (b) IDMT relay
    (c) differential relay
    (d) earth fault relay

53. The relay used to protect the generator against the loss of excitation condition is
    (a) differential relay
    (b) IDMT relay
    (c) earth fault relay
    (d) offset MHO relay

54. The application of overcurrent relay monitored by an under-voltage relay is preferred in case of
    (a) varying load condition
    (b) varying fault conditions
    (c) varying fault resistance
    (d) varying generating conditions

55. Buccholz realy is used for the protection against
    (a) inter-turn fault
    (b) phase to ground fault
    (c) phase to phase fault
    (d) none of the above

56. The main drawback of Bucchoz relay is
    (a) it may mal operate in case of vibrations and shocks
    (b) its higher time of operation
    (c) both (a) and (b)
    (d) none of the above

57. HV winding of a star-delta transformer is usually star connected due to

(a) reduction in current of the order of 1/square root of 3 times the phase current

(b) reduction in phase to neutral voltage to 1/square root of 3 times the line voltage

(c) both a) and b)

(d) none of the above

58. During energization of an unloaded transformer, the maximum value of inrush current flows when the voltage wave is passing through

(a) zero

(b) positive maximum

(c) negative maximum

(d) none of the above

59. Faults occurring in auxiliary equipment of a transformer are also known as

(a) external faults

(b) internal fault

(c) incipient fault

(d) none of the above

60. Which harmonic component is maximum as a percentage of fundamental component amplitude (usually in percent)

(a) 2nd harmonic component

(b) 5th harmonic component

(c) 3rd harmonic component

(d) None of the above

61. Restraining coil is used in the differential relay to avoid

(a) mal operation during internal fault

(b) unwanted tripping which occurs due to high spill current in case of heavy external fault

(c) both (a) and (b)

(d) none of the above

62. Interposing current transformer (CTs) are used with the electromechanical/static type differential relay to match

(a) CT ratio

(b) non-identical CT saturation characteristic

(c) the required pilot current

(d) all of the above

63. Interposing CTs are not required for the digital/numerical differential realy due to

(a) its capability to perform arithmetic operations internally

(b) its capability to remove spill current

(c) both a) and b)

(d) none of the above

64. Inherent phase-shift of currents in the star-delta transformer

or the problem of matching the two pilot current in the magnitude as well as in phase can be achieved/resolved by

(a) connecting CTs on the delta side of the transformer in star and in star side of the transformer in delta

(b) connecting CTs on the delta side of the transformer in delta and in star on star side of the transformer

(c) both (a) and (b)

(d) none of the above

65. Over fluxing protection is used in the transformer to avoid

(a) large value of inrush current

(b) large value of flux

(c) large value of spill current

(d) none of above

66. Percentage bias differential relay maloperates in case of

(a) an external fault

(b) CT ratio mismatch

(c) CT ratio error

(d) an over voltage on source side of the transformer

67. A 160 MVA, 132/66 kV, DY-1, 3-phase transformer is to be protected against short-circuit. An instantaneous overcurrent relay is used to achieve the above task. The magnetizing inrush current of the transformer is 8 times the rated current. The setting range of an instantaneous overcurrent relay is 400-2000% of 1 A in step of 50%. The CT ratio is 1000/1 A. The setting of an instantaneous overcurrent relay

(a) 600% of 1 A

(b) 700% of 1 A

(c) 650% of 1 A

(d) none of the above

68. The chances of damage of the insulation of the induction motor winding is more as the

(a) degree of overloading increases

(b) degree of overloading decreases

(c) starting current of motor increases

(d) none of the above

69. The phenomena in which the motor will fail to start or it will run very slowly due to abrupt increase in load on the motor is known as

(a) single phasing

(b) over loading

(c) stalling

(d) none of the above

70. Which starter is generally used to reduce the amount of inrush current during starting of an induction motor

(a) Electro-mechanical

(b) Soft

(c) Electronic

(d) None of the above

71. The induction motor thermal limit curve should be provided for

(a) hot running condition of motor

(b) cold running condition of motor

(c) both hot and cold running conditions of motor

(d) none of the above

72. The relay used for locked rotor protection in induction motor is

(a) instantaneous overcurrent relay

(b) thermal relay

(c) NPS relay

(d) inverse time overcurrent relay

73. The setting of NPS relay is decided on the basis of

(a) the ratio of Z_2/Z_1 (c) the ratio of V_2/V_1

(b) the ratio of I_2/I_1 (d) none of the above

74. The relay used for loss of load in induction motor is

(a) instantaneous overcurrent relay

(b) definite time overcurrent relay

(c) IDMT relay

(d) none of the above

75. In modern digital induction motor protection relay, protection against phase reversal is achieved through

(a) voltage measurement

(b) negative sequence current measurement

(c) positive sequence current measurement

(d) none of the above

76. To protect against unbalanced currents, the relay used for large and small induction motor is

(a) NPS and phase unbalance relay, respectively

(b) phase unbalance relay and NPS relay, respectively

(c) NPS and thermal relay, respectively

(d) thermal and NPS relay, respectively

77. The special setting given in modern digital motor protection relay to avoid temperature rise in induction motor, is

(a) NPS (c) number of starts

(b) stalling (d) none of the above

78. Single bus bar arrangement can be used for

(a) small & medium sized substations where shut down can be permitted

(b) large sub-stations

(c) both (a) and (b)

(d) none of the above

79. Main and transfer bus bar arrangement is used when

(a) small interruptions to the load is permitted.

(b) large interruptions to the load is permitted.

(c) uninterrupted power supply is required to the load.

(d) none of the above.

80. To protect bus bar against internal fault, the scheme required is

(a) non-unit protection

(b) unit protection scheme

(c) both (a) and (b)

(d) none of the above

81. The maximum percentage of faults that occur on bus bar are

(a) single line to ground fault

(b) double line to ground fault

(c) triple line to ground fault

(d) none of the above

82. The ratio of the operating coil current to the restraining coil current is expressed in percentage and it is usually known as

(a) differential current setting

(b) primary to secondary current ratio

(c) the slope of the relay characteristic

(d) none of the above

83. CT saturation phenomenon can be reduced by

(a) reducing burden of the CT

(b) increasing cross sectional area of the CT core

(c) reducing level of remnant flux

(d) all of the above

84. The differential current caused by the CT ratio mis-match is resolved by

(a) the restraining current

(b) the operating current

(c) both (a) and (b)

(d) none of the above

85. In order to prevent mal-operation of differential relay during mis-matches in the CTs, the slope of the differential relay is kept

(a) lower (c) medium

(b) higher (d) none of the above

86. The main advantage of directional protection scheme used for bus bar protection is

(a) its sensitivity during internal fault

(b) its stability during external fault

(c) its immunity against CT saturation

(d) none of the above

87. High impedance voltage differential protection scheme is used

(a) to overcome spill current due to CT saturation during external faults

(b) to avoid errors in the CTs

(c) to avoid CT ratio mis-matches

(d) none of the above

88. Which scheme is more suitable for retrofit applications

(a) decentralized bus bar protection scheme

(b) centralized bus bar protection scheme

(c) both (a) and (b)

(d) none of the above

89. The standard value of secondary current of CT is

(a) 1 A/5 A

(b) 1 A/10 A

(c) 2 A/10 A

(d) none of the above

90. Usually, a metering CT and a protective CT may be of

(a) higher class and low class, respectively

(b) both lower class

(c) low class and higher class, respectively

(d) none of the above

91. The number of secondary Ampere Turns (ATs) required for the CT is

(a) 5%

(b) 10%

(c) 20%

(d) 1%

92. The excitation of a CT depends on

(a) the magnetic properties of the core material

(b) cross-sectional area of the CT core

(c) length of the magnetic path of the core

(d) all of the above

93. The point on the CT saturation curve at which an increase of 10% in the exciting emf produces a 50% exciting current is known as

(a) knee point

(b) intersection point

(c) saturation point

(d) none of the above

94. The load connected to the secondary winding of a CT is called

(a) lead resistance

(b) load impedance

(c) burden

(d) none of the above

95. The size of the CT is directly affected by the

(a) CT ratio

(b) knee point voltage

(c) core material used in the CT

(d) burden and current rating of the secondary winding

96. For protective relaying applications, the allowed accuracy variation of CT is

(a) ±10% to ± 15%

(b) ±5% to ±10%

(c) ±1% to ±5%

(d) none of the above

97. The ratio of maximum possible current CT can faithfully transform on secondary side to rated current of the CT is known as

(a) knee point

(b) accuracy limit factor

(c) burden

(d) none of the above

98. In digital relay, the problem of development of high voltage across CT secondary against the open circuitry of the CT secondary is taken care by

(a) connecting a conductor

(b) burden

(c) CT shorting switch

(d) none of the above

99. The transient response of CVT is _____ to that of the electromagnetic PT

(a) superior

(b) inferior

(c) similar

(d) none of the above

100. The accuracy limit factor of a CT having rated primary current of 100 A and maximum fault current of 1500 A is

(a) 10

(b) 0.066

(c) 15

(d) none of the above

101. For a CT which is required to supply a relay taking 10 VA through a lead resistance of 0.1 Ω, the total burden at 5 A is

(a) 10 VA

(b) 12.5 VA

(c) 7.5 VA

(d) none of the above

102. Faults that can be cleared by momentarily de-energizing the line is known as

(a) transient faults

(b) permanent faults

(c) semi-permanent faults

(d) none of the above

103. Utilization of auto-reclosing feature will improve

(a) stability of the system

(b) service continuity of the system

(c) sensitivity of the system

(d) none of the above

104. Transient fault occurs due to
    (a) lightning
    (b) swinging of the wires
    (c) temporary contact with foreign objects
    (d) all of the above

105. Auto reclosing system is applied to
    (a) overhead lines only
    (b) underground cables only
    (c) both (a) and (b)
    (d) none of the above

106. The maximum benefit of auto reclosing systems can be achieved for
    (a) unattended sub-stations
    (b) attended sub-stations
    (c) both (a) and (b)
    (d) none of the above

107. The reclosing relay used in long EHV/UHV transmission lines is
    (a) single shot reclosing relay
    (b) multi-shot reclosing relay
    (c) two shot reclosing relay
    (d) none of the above

108. Single shot reclosing relay requires
    (a) low dead time
    (b) longer dead time
    (c) either (a) and (b)
    (d) none of the above

109. Delayed auto reclosing scheme is used for the system
    (a) where chances of loss of synchronism is less
    (b) where chances of loss of synchronism is high
    (c) both (a) and (b)
    (d) none of the above

110. If distance protection is used for the protection of EHV/UHV transmission lines, then the reclosing relay should be normally kept in
    (a) zone-2
    (b) zone-3
    (c) zone-1
    (d) none of the above

111. For 220 kV/400 kV systems, dead time is of the order of
    (a) 0.3–0.4 s
    (b) 0.5–1.0 s
    (c) 1.0–2.0 s
    (d) none of the above

112. Breaker supervision function decides the
    (a) breaker operating time
    (b) breaker closing instant
    (c) breaker maintenance schedule
    (d) all of the above

113. Synchronism check relay is required to
    (a) synchronise breaker with line
    (b) reduce impact on the breaker
    (c) maintain system stability
    (d) none of the above

114. What is the peak value of the power frequency overvoltage across contacts of a circuit breaker when an unloaded 400 kV transmission line is switched off?
    (a) 1230.94 kV
    (b) 461.88 kV
    (c) 653.19 kV
    (d) 1131.37 kV

115. A 50 Hz, 11 kV, 3-phase alternator with earthed neutral has a reactance of 6 ohm per phase and is connected to a bus bar through a circuit breaker. The distributed capacitance up to the circuit breaker between phase and neutral is 0.01 μF. Peak restriking voltage across the circuit breaker is
    (a) 17.29 kV
    (b) 17.92 kV
    (c) 18.92 kV
    (d) 18.29 kV

116. What will be the ratio of the CT on the HV side of a 3-phase Y/Δ, 33/11 kV transformer if the protecting CT on the LV side has a ratio of 300 : 5?
    (a) $200:5/\sqrt{3}$
    (b) $100:5/\sqrt{3}$
    (c) $150:4/\sqrt{3}$
    (d) $120:6/\sqrt{3}$

117. A three phase, 33 kV oil circuit breaker is rated 1200 A, the symmetrical breaking capacity is 2000 MVA ,3s. Then the making capacity is
    (a) 5000 MVA
    (b) 5100 MVA
    (c) 5300 MVA
    (d) 4800 MVA

118. A CB is design to disconnect a transformer having peak magnetizing current of 8A. The system inductance and capacitance are 5H/phase and 1nF/phase, respectively. At the time of interruption, the inductive energy is discharged in to capacitance. The overvoltage appears across the CB is
    (a) 656.68 kV
    (b) 565.68 kV
    (c) 555.68 kV
    (d) none of the above

119. For a short circuit test of circuit breaker, time to reach the peak re-striking voltage is 60 μs & the peak re-striking voltage is 120 kV. Determine average RRRV (Rate of rise re-striking voltage)

(a) 3×10^6 KV/sec   (c) 2×10^6 KV/sec

(b) 1×10^6 KV/sec   (d) 3.5×10^6 KV/Sec

120. The accuracy limit factor of a CT having rated primary current of 100 A and maximum fault current of 1000 A is

(a) 10

(b) 0.066

(c) 15

(d) none of the above

Answers to Multiple Choice Questions

1. (b)   2. (a)   3. (d)   4. (c)   5. (b)   6. (a)   7. (a)   8. (d)   9. (c)   10. (c)   11. (b)   12. (a)   13. (b)   14. (d)

15. (b)   16. (a)   17. (a)   18. (b)   19. (d)   20. (c)   21. (c)   22. (d)   23. (c)   24. (a)   25. (b) 26. (d)   27. (c)

28. (a)   29. (a)   30. (d)   31. (a)   32. (b)   33. (c)   34. (b)   35. (b)   36. (a)   37. (c)   38. (b)   39. (a)   40. (c)

41. (b)   42. (c)   43. (a)   44. (c)   45. (b)   46. (a)   47. (c)   48. (d)   49. (a)   50. (c)   51. (b)   52. (a) 53. (d)

54. (d)   55. (a)   56. (a)   57. (b)   58. (a)   59. (c)   60. (a)   61. (b)   62. (c)   63. (a)   64. (a)   65 (b)   66. (d)

67. (a)   68. (a)   69. (c)   70. (b)   71. (c)   72. (d)   73. (a)   74. (b)   75. (b)   76. (a)   77. (c)   78. (a)   79. (c)

80. (b)   81. (a)   82. (c)   83. (d)   84. (a)   85. (b)   86. (c)   87. (a)   88. (b)   89. (a)   90. (c)   91 (d)   92. (d)

93. (a)   94. (c)   95. (d)   96. (a)   97. (b)   98. (c)   99. (b)   100. (c)   101. (b)   102. (a)   103. (b)   104. (d)

105. (a)   106. (a)   107.(a)   108. (b)   109. (a)   110. (c)   111. (a)   112. (c)   113. (b)   114. (b)   115. (b)   116. (b)

117. (b)   118. (b)   119. (c)   120. (a)

Index

About the Authors

Bhavesh R. Bhalja is Associate Professor, Department of Electrical Engineering, Indian Institute of Technology Roorkee. A post-graduate in Power System Engineering, he obtained Ph.D. from IIT Roorkee in 2007. He has over 15 years of teaching experience.

A prolific researcher Prof. Bhalja has published more than 100 papers in both national and international journals. He received 'Pandit Madan Mohan Malviya memorial Prize' awarded by the Institution of Engineers, India in 2016. He is also the recipient of numerous awards and honours including 'Merit Award' and 'Young Engineers Award' by Institution of Engineers, India; 'Hari-ohm Ashram Prerit Inter-University Smarak Trust Award' by Sardar Patel University, Gujarat,; 'Best Poster Award' at IEEE Conference on Recent Advances in Intelligent Computational Systems, 2011, Thiruvananthapuram. He is the author of *Transmission Line Protection Using Digital Technology*, published by Springer Science + Business Media Singapore Private Ltd, Singapore, January 2016. He is also Associate Editor, Canadian Journal of Electrical and Computer Engineering, IEEE Canada.

Currently, Prof. Bhalja is involved in many research and development projects of the Department of Science and Technology, Ministry of Science and Technology, Government of India. His research interests include digital protection and automation, smart grid technologies and applications, distributed generation, micro-grid, power quality improvement, wide area monitoring and protection, condition monitoring, and application of artificial intelligence in power systems. He is a Senior Member of IEEE and Member of IE and ISTE.

Rudra Prakash Maheshwari received his B.E. and M.Sc. (Engg) degrees from A.M.U. Aligarh, and Ph.D. from University of Roorkee for his work on developments in protective relays.

Dr Maheswari is Professor, Department of Electrical Engineering, IIT Roorkee. He has over 31 years of academic experience and has published more than 175 research papers in national and international journals and conferences. His areas of interest include numeric power system protection and protective relay testing. He has successfully guided eleven Ph.D. work and more than 100 M.Tech. theses at IIT Roorkee.

Dr Maheshwari has been conferred numerous awards by various organizations. He is on the editorial boards as well as panels of reviewers of various international journals. He is also an expert for National Science and Technology agencies of many countries.

Nilesh Chothani is Associate Professor, Department of Electrical Engineering at A. D. Patel Institute of Technology, Gujarat. He obtained his Ph.D. from Sardar Patel University, Gujarat in 2013. He has more than a decade of teaching experience.

Dr Chothani has published several papers in journals and conferences. His areas of interest include digital protection, power system modeling and simulation, and artificial intelligence techniques.

Related Titles

Circuits and Networks Analysis, Design, and Synthesis 2e (9780199460922)

T.K. Nagsarkar, Formerly, Professor and Head, Department of Electrical Engineering, Punjab Engineering College, Chandigarh

M. S. Sukhija, Formerly, Founder Principal, Guru Nanak Dev Engineering College, Bidar, Karnataka

This second edition serves as a textbook for undergraduate students of electrical, electronics, and instrumentation engineering. The new approach using MATLAB-based problem solving enhances the book's utility among professionals and practitioners as well as for laboratory-based learning.

Key Features

- Approximately 400 new solved examples
- MATLAB-based solved and unsolved problems
- New topics on sawtooth signals, doublet, four-wire system, terminating half section, composite filters, Bode plots

Microprocessors and Microcontrollers 2e (9780199466597)

N. Senthil Kumar is Professor, Dept of Electrical and Electronics Engg, Mepco Schlenk Engineering College, Sivakasi.

M. Saravanan is Professor, Thiagarajar College of Engineering, Madurai.

S. Jeevananthan is Professor, Pondicherry Engineering College.

Textbook for engineering students pursuing a course in electrical and electronics, electronics and communication, computer science, and information technology. This second edition of the book goes one step further in providing a comprehensive coverage of topics and an application-oriented approach.

Key Features

- Further to the already covered real-life applications such as traffic light control, washing machine control and elevator control, this edition includes new case studies on microprocessor-based temperature control system and thyristor triggering control.
- Additional timing diagrams for 8085, debugging of assembly language programs, ARM microcontrollers, and programming examples for 8051 in C language.
- Numerous additional solved programming examples for 8051 to aid in better understanding of the theory.

Basic Electrical Engineering 2e (9780198068907)

T.K. Nagsarkar, Formerly, Professor and Head, Department of Electrical Engineering, Punjab Engineering College, Chandigarh

M. S. Sukhija, Formerly, Founder Principal, Guru Nanak Dev Engineering College, Bidar, Karnataka

Basic Electrical Engineering provides a lucid exposition of the principles of electrical engineering for both electrical as well as non-electrical undergraduate students of engineering. Students pursuing diploma courses as well as those appearing for AMIE examinations would find this book extremely useful.

Key Features

- Coverage of important sections on electrostatics, Biot-Savart law, and synchronous generator connected to an infinite bus bar
- Separate chapter on single-phase induction motors and special machines
- Chapter-end exercises with answers and MCQs

Power System Analysis 2e (9780198096337)

T.K. Nagsarkar, Formerly, Professor and Head, Department of Electrical Engineering, Punjab Engineering College, Chandigarh

M.S. Sukhija, Formerly, Founder Principal, Guru Nanak Dev Engineering College, Bidar, Karnataka

A basic text for undergraduate students of electrical engineering. It provides a thorough understanding of the basic principles and techniques of power system analysis as well as their application to real-world problems.

Key Features

- Contains a large number of illustrative problems that use MATLAB in the analysis of power systems
- Includes advanced topics such as contingency analysis and state estimation
- Provides an introduction to HVDC and FACTS
- Includes numerous examples with step-by-step procedures and a variety of chapter-end problems